Waterjetting Technology

Waterjetting Technology

DAVID A. SUMMERS

Curators' Professor of Mining Engineering and
Director High Pressure Waterjet Laboratory
University of Missouri-Rolla Missouri, USA

E & FN SPON
An Imprint of Chapman & Hall
London · Glasgow · Weinheim · New York Tokyo · Melbourne ·Madras

Published by E & FN Spon, an imprint of Chapman & Hall,
2-6 Boundary Row, London SE1 8HN

Chapman & Hall, 2-6 Boundary Row, London SE1 8HN, UK

Blackie Academic & Professional, Wester Cleddens Road, Bishopbriggs, Glasgow G64 2NZ, UK

Chapman & Hall GmbH, Pappelallee 3, 69469 Weinheim, Germany

Chapman & Hall USA, One Penn Plaza, 41st Floor, New York NY 10119, USA

Chapman & Hall Japan, ITP-Japan, Kyowa Building, 3F, 2-2-1 Hirakawacho, Chiyoda-ku, Tokyo 102, Japan

Chapman & Hall Australia, Thomas Nelson Australia, 102 Dodds Street, South Melbourne, Victoria 3205, Australia

Chapman & Hall India, R. Seshadri, 32 Second Main Road, CIT East, Madras 600 035, India

First edition 1995

Printed in Great Britain by the Alden Press, Oxford

ISBN 0 419 19660 9

A catalogue record for this book is available from the British Library

∞ Printed on permanent acid-free text paper, manufactured in accordance with ANSI/NISO Z39.48-1992 and ANSI/NISO Z39.48-1984 (Permanence of Paper).

To my mother, Margaret Summers, my aunts Linda McCullough -Thew and Nora Little, and most especially to my wife Barbara, as a small token of appreciation. This book is also dedicated to my sons Daniel and Joseph with thanks.

Contents

Preface

The ways in which water under pressure can now be usefully applied are limited often, it seems, only by human imagination. The formal gathering of literature into conference and symposia proceedings has been in place for over twenty years, and some of the earlier literature is now out of print and no longer available. This is in many ways a pity since, when the field was young there was often no immediate application for the ideas we discussed. Thus discussion and the reporting of experiments and their results was therefore often more free and open than it has since become as the techniques have acquired more immediate value.

At the same time, regrettably one finds that literature searches will only go back a few years and often do not review the underlying work done at the beginnings of the field, when the underlying principles were first explored. This book thus sets out to summarise and integrate that earlier research with that which has come about as the technology has moved into increasingly widespread industrial use.

Because waterjetting is now used in several different and distinct areas where there may be no immediately obvious cross-over of interest between the tasks (house cleaning and industrial cutting come easily to mind as two examples of this) it has been possible to divide the technology into generalized but separate areas of use. However many of the rules which apply in one field also apply in the others. Further much of the progress made in moving the technology forward has come from hearing a paper on an unrelated subject and realizing that this could be applied to a new field. As an example of this the use of waterjets in treating skin cancer came about from a discussion on their use in removing deteriorated concrete. Readers might therefore gain some benefit from perusing chapters which do not immediately relate to their needs.

David A. Summers

Rolla, Missouri
November, 1994

Acknowledgements

At the time that I first began my studies into waterjet cutting as a neophyte graduate student it was my father, William Summers's advice that helped choose my research topic. As a Mining Engineer who has been closely involved with the mechanization of mining operations and the development of new technology throughout his life, his has been the guidance which I have most sought and valued.

I have been very fortunate in my career to have been mentored by a succession of dedicated individuals. Beginning with Professor Hubert King at the University of Leeds as an undergraduate, my graduate studies, during which I first cut rock with water, were guided by Dr. Norman Brook and Mr. C.C. Dell. The enthusiasm which I have had for this subject since is in no small part due to the encouragement I found while a student of theirs at Leeds. I could not have continued past that point had it not been for the interest which the late George Clark found in what I was doing. Dr. Clark's help in structuring the career of a young assistant professor was exemplary. His group at the University of Missouri-Rolla were one of the first to see the benefits of waterjet use in rock cutting in the United States and led to my migration to America. His help in establishing, in turn, our High Pressure Waterjet activity and encouraging its growth provided a strong base from which we have been able to build our Laboratory and provided the training in research which we needed. It was an honor to follow him as Director of the Research Center but that effort, while building our activities into the broader spectrum of industrial waterjet cutting could not have been made without the guidance (often daily) of the late Dean Planje. The direction of research programs requires management and administration skills and any failures of mine in this area are not due to the assiduousness with which he sought to guide my path, help sadly halted by his untimely passing.

One of the literal markers to the growth of waterjetting came with the construction of the UMR Stonehenge. Dr. Joseph Marcello was the Chancellor at UMR at that time, and it was from his idea that the megalith was waterjet cut and assembled. His excitement at our work and his willingness to help us with our research and in creating the Laboratory helped make it all possible. The activity of a research oriented professor often appears to slight the teaching that we do - Nolan Aughenbaugh, Ernest Spokes and John Wilson each have helped bring back the reality that if we don't not pass on our knowledge then gaining it has no value. And to that end I must recognize Christine Blanks (nee Bloxham) who was the first of our "lot" at Leeds to write a book and give me the initial incentive to work on this.

The study of waterjets at Rolla has always been a group effort and while mine has been the privilege of leading much of that research, almost none of it could have happened, and certainly none of the successes without the help of the group members. Vicki Snelson, Jim Blaine, John Tyler and Dr. Marian Mazurkiewicz were the first to help build the team and they have been joined by Bob Fossey - now a veteran of only fifteen years, and more recently Dr. Greg Galecki and Jo Ellen Blaine. Others have come and regretfully have had to leave, most particularly the graduate and undergraduate students who form an integral part of the program and its reason for being. It has been a joy to help them move toward a degree and a career of their own.

The waterjet community has grown much and there are many individuals who have contributed to this, and hopefully the most important of that effort is contained and recognized in the Chapters that follow. But there have been some leaders of this effort and the industry thanks has been reflected in the Pioneer award which has been created by the

Water Jet Technology Association. It is fitting to recognize and thank again those who have been recognized by this:

1981 Jacob N. Frank
1983 H.S. Stephens
1985 William C. Cooley
1987 Richard Paseman
1989 Norman C. Franz
1991 John Olsen
1993 Fun-Den Wang.

Helping to form the Association and see it grow has shown the willingness of others to give freely of time to work for a communal benefit. In particular the efforts of John Wolgamott, George Savanick, Mohan Vijay, Andrew Conn and Tom Labus have from the earliest tough times helped lead the Association to its current stature, which it could not have attained without the efforts of the folks at David Birenbaum and Associates.

This book was initially started in collaboration with the British Hydromechanics Research Group, and after the usual trials and tribulations has been put in place under the guidance of Rachael Jones and Martin Tribe at Spon. I am grateful for all their patience and guidance. It was submitted as “camera ready” copy. Until now I had no idea of the work that entailed and even now, thanks to the efforts of Jo Blaine who rendered it so, I have only a poor appreciation of the hours required to bring it to its current form. The errors that remain are mine for hers was a Herculean task.

Unit Conversion Table

There was considerable discussion at several times during the preparation of this text as to the best units to use in order to make it easiest to follow for the most readers. The final choice has been to use bar as the measure of pressure rather than MegaPascals as is now common in the scientific literature or psi as is still the most common unit in industry.

Conversion factors for the most common units used in this text are:

Metres to feet	Multiply by	3.282
Metres to inches		39.37
Degrees to radians		0.01745
Metres/sec to miles/hr		2.237
Metres/sec to furlongs/fortnight		6,013
Kilograms to pounds		2.205
Kilograms (force) to newtons		0.1019
Bar to pounds/sq. inch		14.49
Cubic meters to cubic feet		35.33
Cubic meters to gallons (US)		2.64
Kg/cubic m. to lb/cu. ft		1.686
Liters/min to gpm		0.264
Joules to kWh		0.2778×10^{-6}
kWh to horsepower		1.34

1 WATERJETTING FUNDAMENTALS

1.1 INTRODUCTION

The use of waterjets under pressure has become much more common, in recent years, for an increasing variety of tasks. As their advantage has become clear, so waterjetting equipment has been developed, used and waterjets have become, in several industries, the accepted method for solving a problem. This book sets out to discuss many of those applications, and to suggest not only ways in which waterjets can be used, but also ways in which it might be possible to improve their performance. As the text develops so certain conventions will become clear. So also will several words which have acquired a specific meaning. When these words are first used they will be indicated in a bold face, with a word of explanation as to their accepted meaning.

Waterjetting is, in its simplest form, concerned with the development, the transmission and the application of power. This power is normally created in a water medium by a pump, pushing a given volume of water into a high pressure feed line and providing it with a certain amount of energy in the process. This water flows down through the line, usually a strong metal tube over at least part of its length, to a **nozzle**. The nozzle contains one or more exit holes or **orifices** which are normally of a much smaller size than the feed line. Since a constant volume of water reaches the nozzle, it must accelerate to a higher speed in order to escape through these orifices, which also serve to focus the water into a coherent stream or jet, and to direct the streams towards the required point on the target surface or work piece.

The water jet which comes out of each orifice will generally have to travel some distance (usually referred to as the **stand-off distance**) to the target, losing energy as it moves through the air or other fluid which is in its way. When it reaches the target surface the remaining power in the jet can be applied to one of a number of goals, usually related to the removal of material. This can be as simple as the cleaning of the target, by removing a layer of unwanted material from the surface. It can be the cutting into the target to create a slot, or it can be used to break out large volumes of material from the surface, as in some form of mining application.

This book sets out to discuss these various aspects of waterjet technology. Waterjets have been developed and used for many years, but are only now becoming popular for use at a higher pressure, and for a broader range of purposes. While many of these uses are new, many of the basic features of water jet use are common both to new and older applications. Lessons from one use can be learned and applied in improving jet performance in other applications. More exciting, the ways in which some forms of water jets are used can be applied in other ways to find new uses for this tool, and allow processes which have not been possible without the development of this new form of power.

But, as with other forms of power, it is easy to either use or misuse this new tool, or to misunderstand those things which make it work. Thus the text will discuss the different aspects which make up the system, what makes these systems as effective as they can be, and what pitfalls should be avoided in them. But before reviewing any of those questions, one must first decide what a waterjet is.

1.2 TERMINOLOGY

First an additional comment about the words that are going to be used. As with any new tool, or business, this growing industry has begun to develop special meanings for some of the words that are commonly used in it (Fig. 1.l). The first change has been in the name of the technology itself. Until just a few years ago water jets were spelled as two words, but within the past five years the practice of joining them together as one word, waterjets, has become more widespread. Waterjets and waterjetting will each be used as one word in this text when the tool and its use are discussed. Where the original two-word term is used it will generally be found in quotations from earlier work, or when the fluid which makes up the jet is being discussed is water, to distinguish it from other fluids.

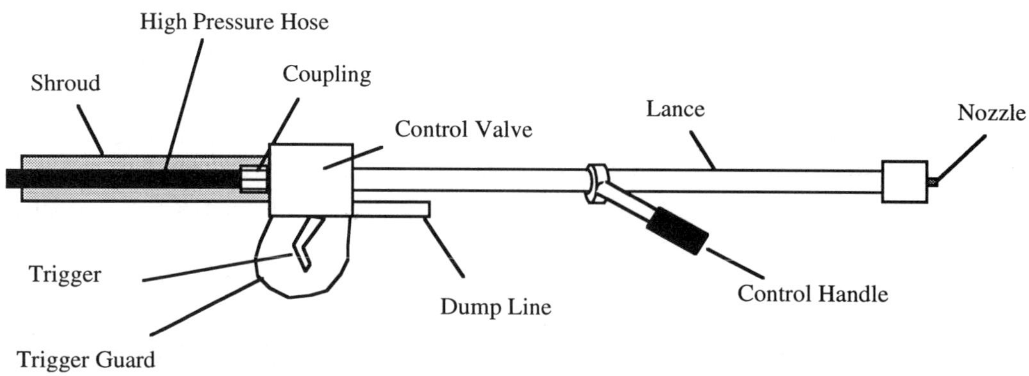

Figure 1.1 Terms to describe common components of waterjet systems.

Most new words are related to the business end of the delivery system. When the water is accelerated through the nozzle to its final speed it has to pass through a section of the line which gives some control to the resulting jet. Consider, for example, the common components used in many manual systems where the jet is to be used for cleaning. The delivery end of the line (see Fig. 1.1) will usually contain a **control valve** which controls how much water goes out of the nozzle. The valve may be operated by a **trigger** or control lever, often manually operated, which opens and closes the valve either to direct the water out along a channel with no restriction (the **dump line** which may direct the water at no pressure into open air or back to a storage tank) or down a second line to the nozzle. The passage to the nozzle will contain a straight section of tube directly behind the nozzle which will often have a **control handle** on it for the operator to use, and it may have a support built onto the line, to help the operator withstand the reaction force which the jet will apply through the line and handle to the operator. The nozzle on the end of the line may be called a **tip**. The straight section of tube, and the nozzle are sometimes known as the **lance**. When the trigger, control valve and the **guard** around the trigger are also included this may be called a **waterjet gun**. Other terms will be defined as they are needed during the development of the text. But it is important to begin by understanding the real subject and that is the waterjet itself.

1.3 WHAT IS A WATERJET ?

Water by itself has chemical and mechanical properties which make it useful in a number of ways. While these properties may be changed in ways to be discussed later for this text the water becomes of more interest and value when it begins to move. More particularly when, as a moving stream of water, it is formed into a controlled jet of a specified shape. The most common form of such a waterjet is a round stream of water which can be aimed at some object in order to get some work done. Even when the water is not moving very fast it can be quite effective. As a simple illustration of this, the water from a hose can be used to wash the dirt from a concrete floor, or to remove the soap and grime from a car.

The speed of the jet is stressed, since once it has left the nozzle orifice it will no longer be under pressure. However for the pump to push a given volume of water through the hole in the end of the nozzle within a given time it must exert a given **pressure** on that water. The pressure provided by the pump will generally be expended in two main ways, the first is in driving the water through the line from the pump to the nozzle, and the second is in sending the water through the orifice at a given velocity. The main pressure loss in the delivery line comes from the line friction as the water moves against the walls of the tube or hose, however it can also be lost in turbulence where the water flow becomes disturbed as it moves through passages of different shape.

When the waterjet reaches its target, the energy which the jet contains, as a result of its speed, is changed back into an **impact pressure** in order to get an effective amount of desired work done on that surface. Two quantities relating to the jet have, in the past, been found to be the most important in the effectiveness of this exchange. These are how much water is hitting the object, and how fast is it moving. Between them they control the power which arrives at the object, and thus how much work can be obtained from it. These values, in turn, are largely controlled by two variables in the delivery system. The volume of water delivered by the pump, and the diameter of the orifice at the end of the delivery line control both the speed of the waterjet stream, and the area over which it is applied. In a later discussion, important qualifiers to these values will be related to the stand-off distance that the jet must travel in getting from the nozzle to the work surface and the **traverse speed** with which the nozzle moves over the surface. This latter value relates to how long the jet will be aimed at any single point on the surface.

When the jet leaves the nozzle orifice, the most common shape that is assumes is cylindrical. Conventionally, however, this jet has become known as a **round jet**. This shape is able to carry the jet energy quite efficiently, but the impact area on the target surface is quite small. This makes the jet more difficult to use in cleaning large areas. The jet shape has, therefore, been modified for that purpose. The normal way of doing this is to cause the jet to spread out along a line on the target. Because of the shape the jet takes to do this, this jet is known as a **fan jet**. More recently nozzles have been introduced which direct the jet into different shapes as it issues from orifices which might be triangular, square or some other pattern. Such jets will be referred to as **shaped jets** to distinguish them from the more historically conventional cylindrical and fan forms.

The scope of what is becoming known as waterjet technology has also been broadened to include jets which are made up of other fluids. For example in the cutting of some foods which use a lot of sugar, water cannot be used since it dissolves the sugar and ruins the product. In these cases the cutting jet can be changed to a **fluid jet** which will not interact with the surface. In the case of cakes and chocolates, for example, a vegetable or other digestible oil can be used as the cutting agent, since this will not dissolve the sugar or damage the final product.

As will be discussed in more detail later, very hard materials such as glass and metal are difficult for a waterjet to penetrate on its own. For this reason abrasive particles can be added to the jet stream in order to help in the process. These particles acquire the speed of the water, and provide most of the cutting action when the jets hit the target. These combined jets are called **abrasive waterjets** and have become more popular in recent years and a small industry has developed, built around their use. The definition of an **abrasive particle** has changed from the original material, which was commonly a fine grade of sand, to include materials ranging from crushed walnut hulls, through steel shot to highly aggressive materials such as aluminum oxide. **Soluble abrasives** have also been introduced in which the dry particles are introduced into the stream just before the jet leaves the final acceleration and collimating nozzle. After striking the target these then dissolve in the water, helping with the cleanup of the site after work is completed. Two forms of injection of the particles into the jet stream have developed. In the more currently conventional method of injection the water is accelerated to its final velocity through a small orifice and then directed into a mixing chamber where it entrains abrasive particles before being formed back into a jet by a second, **collimating nozzle**. This technique will be referred to as **conventional abrasive jetting**. In contrast the abrasive can be mixed with the water before the fluid is accelerated through the orifice. This technique has been given the name of **direct injection of abrasive** or may be referred to as **slurry abrasive jetting**.

In transferring the energy in the jet from the pump to the target, through the water, it is possible to adjust the properties of the water to reduce system losses. This is normally done by including a chemical additive in the jet stream. Relatively small concentrations of **polymeric additives** have been found very useful in reducing the friction which is generated as the water flows along the walls of the delivery pipe. The additives are often long chain molecular polymers. They also help to "glue" the water together after it has left the nozzle, keeping the stream together and at its delivery speed, over a greater distance from the nozzle, and thus increasing its effectiveness. Polymeric additives have found application both in air and in underwater applications, and can improve the performance of both conventional and abrasive waterjets.

The power of a waterjet can be redistributed within its own structure. By changing the shape and the speed of consecutive segments of the jet length, the overall power in the jet can be altered into pulses which generate alternating very high, and much lower pressures, on the target surface. If this pulsation is properly controlled, and if it hits the right type of surface, it can cause a significant improvement in the effectiveness of the jet on the target. One way to create a **pulsating waterjet** stream is to break the jet into discrete segments, which, if done properly will make the jet more effective under certain circumstances. As an example of this consider the pulsating shower heads which are now found in many bathrooms. Pulsation can be caused by a special device located behind the nozzle in the delivery system, or by locating an interrupting device between the nozzle and the target.

Pulsating jets should be distinguished from **interrupted jets** such as those produced by the water cannon types of device. These latter jets are typically generated at much higher pressures. When jet pressures much above 4,000 bar are required it is not possible to obtain commercial pumps which can sustain these pressures at high volume flow rates. To resolve this problem, the **water cannon** type of delivery system was developed. The name derives from the fact that the device looks much like a cannon, and many early devices were made from old guns. The cannon is loaded with a water charge and a volume of pressurized gas is generated and released in the closed volume behind the slug of water. This drives the water out of the mouth of the cannon as a single shot or interrupted jet. In general a nozzle of fixed orifice size has been either attached to the end of the barrel or incorporated within the structure of the barrel in order to control the shape and speed of the resulting jet. However, in some equipment, the speed of the jet, over its length, is further developed by a special shaping of the acceleration section of the nozzle. Water jet pressures of over 100,000 bar can be generated in this way, when the driving gas is provided by deflagrating a small amount of energetic material behind the slug of water.

Another way of re-distributing the power within the jet is through the phenomenon of **cavitation**. It is possible, in one of several ways, to create an internal tension within a body of water. Water does not have any great strength in tension, however, and thus very small gas bubbles, formed to fill this otherwise empty space, are created in the water by this action. These very small bubbles with virtually no inside pressure collapse as soon as the water around them is pressurized. But as they collapse they do so in an irregular way, which results in a tiny jet being formed during the final stages of the bubble collapse. The jet passes across the bubble body and can create a very high impact pressure, but over a very small area, when it enters the water on the other side of the bubble. By a special choice of nozzle design it is possible to induce these cavitation bubbles within a flowing waterjet. With care the system can be designed so that the jet carries the bubbles within its stream until it hits the target, where they collapse and create a zone of damage.

If most of the bubbles can be made to collapse on the target surface, then this will improve the cutting or cleaning ability of the jet. This can be illustrated in two ways. In cleaning jewelry and intricate small shapes it is possible to buy small ultrasonic baths. The item is placed in the bath under a small amount of water. When the power is switched on, the walls of the bath vibrate ultrasonically. This causes alternate tensile and compressive waves to pass through the water, creating cavitation bubbles in the water, on the tension stroke. These are pushed against the item and collapsed by the consequent compression, and the very small jets created hit the dirt and surfaces of the jewelry. Because of this tiny scrubbing action over the whole surface even the narrow passages in very very intricate items can be cleaned by this method.

At the other extreme, when cavitation bubbles are induced in a jet stream at a pressure of 1,000 bar, the modification in the power spectrum of the jet is such that it can cut through ceramic materials which cannot otherwise be penetrated by plain waterjets at a pressure of 4,000 bar.

In this context **plain waterjets** are meant to exclude those jets whose performance has been enhanced either by pulsation, interruption or by the addition of abrasive. Waterjet technology has thus come to mean a number of different things as the technology has advanced and found a role for itself. Within the umbrella of this title, as covered by the waterjet symposia and meetings of the last eighteen years, waterjets have been described which have found a useful application at a pressure of less than 10 bar and flow rates below one liter a minute (lpm). At the other extreme waterjets have been created with a useful purpose at flow rates of over 1,000 lpm for mining applications, and there are military uses of waterjets which have been developed at impact pressures above 600,000 bar. The range of the technology is thus very broad, though they all arose from a relatively simple common beginning.

1.4 WHERE DID WATERJETS COME FROM AND HOW DID THEY DEVELOP?

A heavy rain will move the soil in a garden, and cut channels through the soil of an exposed bank. Over a longer time scale, the action of the rain will erode even the hardest rocks and wear down mountains. At the same time that this slow water action is breaking the rock, it will also move it away from the area. The great deltas which spread from the mouths of the rivers of the world show the great carrying capacity of the water, even with the relatively slow speed of a river. This means that the three aspects of waterjet use, the breakage, mobilization and removal of material, have been an integral part of the shaping of the Earth into the form it holds today.

Because this power could be seen, it could also be adapted to man's use. Early stone documents described the way in which the path of a river could be diverted in Egypt to wash the soil away from over a valuable mineral deposit [1.1]. Later, in Roman times, the power could be artificially generated, by storing the water in specially built reservoirs on hilltops. When the reservoir was tapped, the resulting stream could then be directed, as needed, down onto the ore deposits on the hill below. Here it would break out, remove and carry away the material, down into the valley where the valuable mineral could be more easily extracted [1.2]. Even recently this use of water has been popular. Hydraulic mining of coal in both the Soviet Union [1.3] and New Zealand [1.4] first involved such a collection of water from a stream, and the directing of its path, through pipes, to wash blasted coal from the solid face into flumes, through which the water carried the coal out of the working area of the mine. This technique has recently been revised and introduced into the deep gold mines of South Africa to move the blasted rock away from the mining face into collection tunnels or drifts [1.5].

In more modern times, two developments have led to the creation of the modern industry. The first was the development of hydraulic mining as the major gold production tool in the California Gold Country during the years from 1853 - 1886 [1.6]. Even though water rapidly reached a premium price the benefits from its use in excavating the soft gold bearing rock were sufficient for it to become universally adopted in those surface mines. Because of the soft nature of the rock, it was only necessary to use water at the natural head but greater productivity came with increasing pressure and flow.

One of the advantages of the technique arose because of the nature of the sites. The gold was mined from high cliffs, where manual excavation ran the risk of having the overlying cliff fall on the miner (Fig. 1.2). By using a pressurized waterjet stream it was possible to have the miner located well back from the face. Obviously the higher the pressure and jet volume, the further back the miner could stand and also the more productive the system. Success in California, led to the adoption of hydraulic mining in other states of the Union, and its adoption in other countries to mine many different minerals.

Figure 1.2 Early hydraulic mining practice in Idaho (courtesy of the Idaho State Historical Society).

By the early 20th century this development had reached the peat mining operations in Prussia, and thence it moved into Russia, where in the 1930s it was first used for mining coal. The remote aspects of this method of mining made it attractive for mining the thin steeply dipping coal seams found in parts of the Soviet Union. The ability of the water to carry the coal away with it also meant that it was possible to mine the coal out without miners needing to enter the dangerous working area around the point of excavation. Production levels were found to be higher than conventional mining, and the method became adopted as a mining system [1.3].

Problems with existing methods of mining led a number of other countries to try this method. The most successful adoptions were those in Japan, China and Canada. This success, in turn led to attempts to cut harder and more resistant rock types. Initial experiments in the Soviet Union used a water cannon type of device to generate pressures of up to 7,000 bar, in order to cut through these harder rocks [1.7]. The water cannons generated a single slug of water driven out through a nozzle threaded onto the muzzle of the device. This type of pressure generator remained the major method of ultra high pressure generation into the 1970s, where systems were developed in the United States capable of creating driving pressures of up to 40,000 bar, and their use indicated that there was some potential for rock cutting by waterjet means.

Concurrent with the development of the mining technology, waterjets had been evolving as a means for rapidly cleaning surfaces. The initial power to drive the water at high speeds was created using steam, but the cost and system losses of that mechanism led to a slow evolution toward the use of a higher pressure, cold water washing unit as an alternative device. The low reaction force exerted by a waterjet stream back on the operator meant that relatively rapid cleaning could be carried out, while the use of heat allowed a sterilization of many of the work surfaces-an advantage in operations such as meat processing plants. The use of small hand-held lances in car washes, and similar applications spread rapidly and has provided a simple demonstration of the advantages of this new tool.

With time the capabilities of the systems have increased, and pumps have been developed which can operate at a steadily increasing pressure. This has allowed the resulting jets to be able to clean or to cut through harder and harder materials. The requirements of the system have also changed. Whereas the initial applications called for increasingly large volumes of water in order to carry the resulting debris away from the working area, this is no longer as great a concern. In many operations the water which is used in the cutting or cleaning process will entrain small amounts of the cut material. This may create a waste disposal problem, particularly if the target material is classified as hazardous. Under these circumstances, given the high costs of contaminated waste disposal, the smaller the volume of water generated the better. Thus systems have been generated and found a market which operate at pressures considerably above that necessary for effective work, but with the higher pressure justified by the reduced water flow requirement generated. Waterjetting has become increasingly popular as a method of surface cleaning because of its more benign impact on the environment than many of the cleaning chemicals whose use it has displaced.

The growth of the high pressure cleaning industry provided a source of pumps and related equipment which could be exploited by those developing a market in geotechnical applications. Early work in that field, however, suggested that higher pressures would be required to provide an effective excavation tool. In the late 1960s Professor Robert Franz and his students, working in Michigan, found that very high pressure jets could be used to cut through wood products with little damage to the material on the outside of the cut surface and at relatively high cutting speeds [1.8]. Working with the McCartney Manufacturing Company (now a division of Ingersoll-Rand) a prototype system was developed for cutting material at pressures of up to 4,000 bar. The first equipment was installed in Alton Boxboard in 1972 and led to the development of a new tool for the manufacturing industry [1.9]. A new company, Flow Industries, also began to market equipment at this high pressure for industrial use. The lower reaction force from a 3.5 lpm jet, even when cutting at 3,500 bar, made it an ideal tool for use with the developing process of robotic cutting.

Yet there were limitations to the range of materials that this tool could cut. As will be discussed later, a major mechanism of material removal is through fluid penetration and exploitation of existing cracks in the material [1.10]. Where such cracks are small or virtually non existent, as in ceramic and metallic surfaces, waterjet cutting is a slow or totally impractical procedure. While some work has been carried out, aimed at exploiting cavitation as a way of cutting these products, the breakthrough came in adapting a technology used initially for cleaning surfaces.

In order to achieve a white metal finish in metal cleaning, operators had induced sand into the cleaning water stream. The sand reacted to give faster cleaning rates and better contaminant removal, particularly of harder deposits. The process could be enhanced raising waterjets at higher pressure and scientists at Flow Research adopted this technique to high pressure cutting. Although the passage of the abrasive when driven by a high pressure jet is sufficient to rapidly erode the collimating nozzle used to provide the abrasive jet stream [1.11], for many applications this cost is secondary to the benefits achieved. It was demonstrated that the use of a high pressure abrasive waterjet stream could cut through both ceramic and metallic materials with virtually no damage to the material on either side of the cutting line. As previously with the high pressure waterjet systems, abrasive waterjets could be adapted for use in robotically controlled cutting systems. Further because the jets did not damage the material on either side of the cut by imposing the thermal stresses associated with many mechanical and thermal cutting techniques which are the more conventional ways of cutting particularly metal, this technique has acquired the descriptive title of **cold cutting** and this capability has found many different applications in cutting other materials.

The development of the market from the mid 70s to the mid 80s had perhaps been dominated by the use of these very high pressure, low flow rate operations. Many of these were built around the use of an intensifier system and were most effective in a factory environment. A considerable market however still remained in the cutting and cleaning of concrete and other civil and geotechnical materials. While it has proven possible to use very high pressure equipment under these conditions, the relative delicacy of the equipment, its sensitivity to water quality, and its high cost have limited its penetration of this market. An alternate approach was needed.

In 1986 the British Hydromechanics Research Association announced such a breakthrough [1.12]. After a period of years of development they had produced a means by which an abrasive could be injected into the flow line between the high pressure pump and the initial acceleration nozzle. This development overcame a number of disadvantages of the higher pressure systems. Most immediately, it allowed use of a much lower pressure to achieve acceptable cutting rates in metallurgical and ceramic materials. Operating pressures of 350 bar were needed for this equipment, although in a tradeoff to achieve economic effectiveness, flow rates of up to 70 lpm were required. Where the abrasive was not recycled, this increased operational costs and limited its range of application, given the continued use of an abrasive concentration of approximately 125 grams per liter. Similarly, the high flow of abrasive through the nozzle gave a relatively short operational lifetime. Recent work, however, has indicated that in brittle material removal, particularly of ceramic and geotechnical materials, there are abrasives which can be virtually 100% recycled and which reduce nozzle wear by a factor of 10. Further with the use of enhanced nozzle materials significantly improved lifetimes have been achieved for the nozzles and commercially available systems can now be obtained for operation at pressures of up 700 bar. These systems use lower flow rates than those necessary at the lower pressures.

An alternative approach has been developed, over the years, which has also the potential for a major impact on that aspect of machine technology dealing with material removal. During his work for the Chamber of Mines in South Africa, Dr. Hood discovered that one of the factors which significantly reduced the working life of the drag bits used to remove rock in the mines, was that they were overheating [1.13]. In order to reduce this problem, he added a stream of water to the cutting surface, and directed this under pressure to provide a sufficient heat sink to keep the carbide cutting tool cool. He discovered that this operation significantly improved the cutting ability of the tool, much more so than could be explained by the cooling action alone. The **waterjet assisted cutting** bit that he subsequently developed, was able to cut up to five times deeper than the normal bit, even in a rock whose compressive strength was up to ten times that of the cooling jet.

The significance of this work has since been taken up by initially American and European government research investigations, by European mining concerns, and then by equipment manufacturers who have developed, and now market, mining machines which incorporate this waterjet assisted pick cutting into the designs of the excavation heads. Significant improvements in performance and reduction in equipment costs have been reported and the technique is of such benefit that is now being introduced for coal mining machines, as well as for rock cutting operations.

Studies of the behavior of this new tool suggest that it is as effective as it is because the jets wash away the crushed rock from the vicinity of the tool. This removes a plastic layer of material that deforms and distributes the cutting load. As a result the pick is more aggressive in its cutting action. However, in recent work, Dr. Mazurkiewicz has shown that a similar result occurs when the waterjet is added to the cutting action of a machine tool in cutting metal [1.14]. Significant increased performance is achieved and difficult to machine metals become easier to turn. Since metals do not, however, crush, under the cutting tool, it is apparent that more work of a theoretical nature will be required to adequately explain this phenomenon.

As Dr. Lichtarowicz has said "it is not necessary to understand the chemistry of digestion in order to enjoy a good meal". Thus, although our understanding of some waterjetting phenomena is still not fully developed, we have the experimental and developmental studies to provide data from which to move this technology into industry. This technology is evolving to produce equipment with an increasingly universal application and with the technology changing as problems are identified to resolve them and thereby to increase the market applicability of the technology.

1.5 HOW ARE WATERJETS USED ?

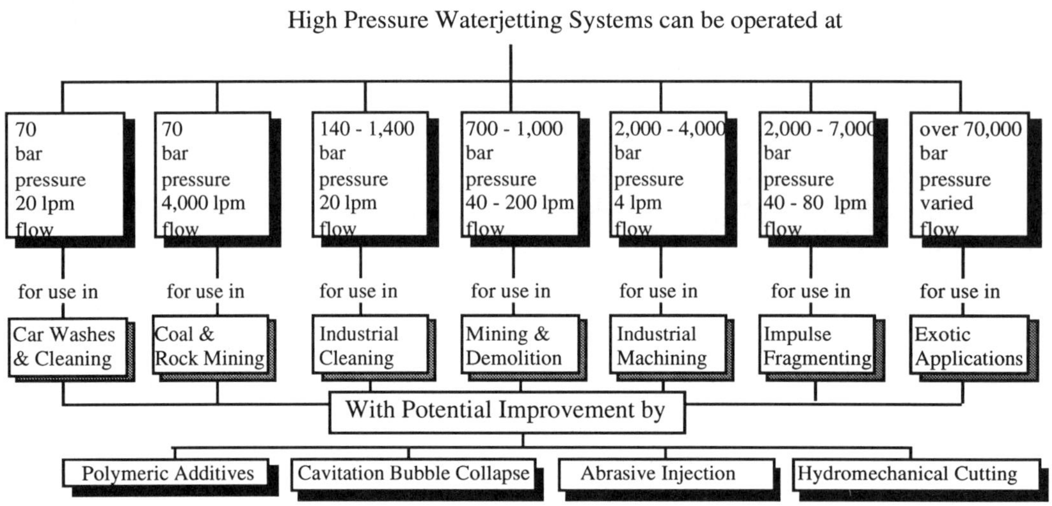

Figure 1.3 Some ranges of pressure and application of waterjets.

The current range of applications of high pressure waterjets covers a wide spectrum of use (Fig. 1.3). The most common systems are those used in industrial cleaning. Because a small 350 bar system, delivering 20 lpm system is relatively inexpensive, their use has become widespread within the last twenty years., cleaning surfaces from the outside of houses to the insides of industrial ovens. As pressures have increased, and the capabilities of units have grown with them, so the range of application of commercial cleaning units has increased to 700 bar and more recently 1,400 bar. In specialized applications small units are now available which operate at pressures as high as 3,500 bar. Given that the industrial cleaning market has been estimated as several billion dollars a year, the penetration of waterjets into this market can be anticipated to continue growing in the years ahead. This is particularly true where it has special application advantages, for example in such areas as hand-held underwater cleaning.

Although hydraulic mining was of the initial applications of waterjets, its use has not grown with the technology. While some hydraulic mines continue operation, the number has declined in recent years. Instead the growth, in mining and geotechnical applications, has been through the combination of waterjets with mechanical tools.

In this regard the use of waterjets, with a shaped drilling head has been found to have considerable advantage in drilling relatively short horizontal holes for utility installations. This technology relies on small jets, operating at pressures of up to 350 bar in order to drill and ream holes of roughly 10 cm diameter through soil and unconsolidated ground. Because it allows the hole to be drilled, and the line to be installed, without digging the intervening trench, this technique has become increasingly popular, since the mid 1980s [1.15].

The removal of damaged concrete and coatings prior to the refurbishing of structures has also grown into a significant business in the last ten years. It has now acquired its own term, that of **hydrodemolition**, and in some instances is the only method specified for concrete removal. The controlled removal of this material, and the uneven surface left after jet removal of a layer makes it easier to bond the repair layer to the original surface. As a result the finished surface has a greater integrity to subsequent wear than is the case where the damaged layer was removed by a mechanical grinding surface which left a smooth plane as the interface. In addition the highly localized cutting action reduces the vibrations on any internal steel reinforcing rods, which can be vibrated loose from the surrounding concrete where pneumatic chipping hammers are used.

At higher pressures, and lower flow rates, the waterjet cutting head will exert a small reaction force on the fixture holding it in position. This makes it an ideal tool to attach to the end of a robotic arm, for cutting applications. Such applications, originally confined to the cutting of relatively soft paper products, have gradually grown to include laminates, plastics, and, with a change in cutting fluid, food products.

In recent years the addition of abrasive to the waterjet stream has increased the range of products which can be cut to include glass and metals. The ability of waterjets to cut a metal surface without damage to the sidewalls of the cut have made it an ideal tool for cutting metals such as titanium and inconel. The flexibility of the cutting system have made it a very practical tool for cutting complex shapes, not only in metal but also in glass, and rock. Thus, the world map which comprises the centerpiece of the new Navy Memorial in Washington was carved from black and white granite slabs 5 cm thick using abrasive waterjets [1.16].

Waterjets continue to offer new, often unique opportunities to complete work otherwise almost impractical. This can range from the removal of rock to rescue a young child from a well in Texas [1.17] to the erosion of liver tissue in order to expose the blood vessels prior to their excision [1.18]. Many of these applications will be discussed in the following pages. As with any young technology, however, it will be exciting to see how the new applications which develop within the next few years. One recent example is the ability of waterjet systems to discriminately remove the layers of material overlying statues, exemplified by the cleaning, in 1993, of the Freedom statue from the top of the U.S. Capitol building in Washington, D.C. [1.19], others are, no doubt in the offing.

1.6 WHY ARE WATERJETS USED ?

When a plant manager looks at improving the operation of his plant by the introduction of new technologies and new methods, the advantages and disadvantages of any novel technology must be weighed in the balance. In early discussions with one mine operator his comment was that there had to be as yet unknown disadvantages to balance the obvious gains which might come about from the use of high pressure waterjets. After a number of years of industrial operation, some disadvantages of the tool are now apparent, but in many cases these have been significantly outweighed by the benefits derived from it.

In its earliest use as a mining tool, the safety of what is often a hazardous operation, was dramatically improved when hydraulic mining was introduced. In the mining of hydrocarbon ores and materials such as coal, fine dust particles easily become airborne and create both a respiratory and explosive risk. Waterjet mining occurs with the generation of almost no dust. Further, the fine mist created around the cutting zone makes it almost impossible to propagate a spark or gas ignition to the point that the operator is put at risk. The considerable range of the jet cutting equipment further has allowed the operator to stand back as far as 30 m from the working area. Thus, the miner and the equipment could be located in a strongly supported area and mine coal without the risk of immediate roof collapse. These benefits made the Sparwood Mine in British Columbia one of the safest mines in Canada in the 1970s.

The concentration of power within the waterjet stream and the ability to focus a large amount of energy over a small area significantly reduces the overall force both on the work piece and back on the nozzle holder. Further, the relatively simple directability of the jet stream over a wide range of angles with relatively little change in power, also differentiates it from mechanical cutting. These two features make it relatively easy to adapt high pressure waterjetting systems for use with industrial robots. The resulting jets, whether plain or laden with abrasive, can be used to cut both linear and intricate contours over a range of surfaces with little distortion of the final work surface. In addition, through the use of a second, counterbalancing jet at the back of a lance, the tool can be designed as a **zero-thrust cleaning** lance which has been found very effective by divers in underwater cleaning.

Where suitable precautions are taken, waterjet cutting equipment can be a quieter as well as a more productive tool. These benefits extend beyond the actual site and operation of the cutting process itself. For example, in process cutting operations, it is reported that the improved cleanliness of cut and reduction of waste generated, produces a lower interference with the transport rollers of the feed mechanisms. This improves process accuracy and reduces down time in the subsequent operational processes downstream of the jet cutting system.

In a number of the current applications of the technology, one of the benefits to the use of the waterjet system lies in the reduced wastage of material. The relatively thin slot cut through the target and the continued integrity of the cut surface material give two benefits. Firstly, it allows parts to be nested closer together in a cutting table layout, reducing the amount of material left as waste. Secondly, the edge quality is in many operations sufficiently high, that it can be left as a final surface. Parts are therefore cut to their final shape in the original cutting operation. This eliminates subsequent machining or grinding of the work piece to bring it to acceptable edge quality and tolerance.

Of course no tool will find universal application. At the present time there continue to be some disadvantages to waterjet use limiting its viability. Production rates are controlled by horsepower availability, practical cutting speeds, and the resistance of the target material. In a number of cases, these limitations provide a cost to the operation beyond that of other technologies. This is particularly true for a number of materials where traditional machining methods have evolved over a number of years. The relatively limited development of waterjet technology has not, as yet, evolved an answer to some of these problems.

High pressure waterjetting systems are still relatively expensive and therefore, for single small operations (as an extreme example, cutting a single slice of cake) they are not a practical tool. The use of high pressure water carries with it the risk of personal injury if the system is abused. Further, because of the higher precision and sensitivity of the equipment, operational and maintenance costs are a significant part of the process economics. Some of these underlying problems are currently being addressed as will be discussed in later chapters.

1.7 WHAT IS THIS BOOK PLANNING ON DOING?

As waterjetting has become more successful research groups are, increasingly, being asked to "just tell me what I need to cut my tomatoes (or whatever the product is)". This need should include a number of items, a pump pressure, a jet size, a cutting rate, and what, if anything should be done outside of conventional practice to improve the performance of the waterjets.

This text is in part an explanation as to why most of these are important to know. It is, at the same time, an explanation as to why, in many cases, it is better to find this out by actually cutting a number of tomatoes, than it would be sitting down for six months and running other tests on the tomatoes to find values to fit into an equation.

It sets out to show how waterjets can be used, and to discuss the likely range of pressures that might be tried first in different applications. A number of very talented individuals have worked, from time to time, in studies related to the jet cutting industry. Until very recently there was no central literature in which the reports of their work could be collected. Thus, much of the information gathered is not commonly known, and previous studies may be repeated because of this. This text sets out to be a vehicle for identifying some of the key studies that have been made in the past, identifying important considerations in the design and use of different systems, and hopefully, also providing a set of signposts to indicate where future work might profitably be directed.

It is important to make two points about units and their conversion at this point. Considerable discussion has taken place about the correct units to use in developing a new field of study. At the first International Symposium on Jet Cutting Technology, held in Coventry, UK in 1972, under the auspices of the then British Hydromechanics Research Association, it was decided to use metric notation with the use of

Megapascals (MPa) as the unit for reporting jet pressure. However, while conferees may choose the practical industry makes its own selection and moves on. Thus, in the United States, much use is still made of the old Imperial units (pounds per square inch or psi, for example). In an effort to compromise between the old and the new this text has used bar as the measure for jet pressure and liters/minute (lpm) as the units for flow volume.

Selecting these units has required that much of the work reported herein be converted so that the units conform to this standard. In making those conversions a certain approximation has been made. There are several reasons for this but three were considered paramount. The first is that many experiments were carried out at set increments (such as for example 10,000 psi) equivalent tests in metric units would be carried out at 70 MPa or 700 bar and thus where such selections appear to have been the choice close approximations of this type are made. The second reason for making this approximation is that pressures and flow rates fluctuate slightly during most equipment operation so that pressures were not accurately regulated. The third is that in many tests the coefficient of discharge and the amount of pressure loss in the line is not reported. Thus, values which are provided are often measured at the pump, where most pressure gages are located. The values given in the literature provide, therefore, only an approximate guide, and in some cases very approximate, to the actual pressures and flow rates of the jets issuing from the orifices of the equipment.

1.8 REFERENCES

In preparing the list of references to each chapter many of the citations are from the Biennial International Symposia on Jet Cutting Technology organized by British Hydromechanics Research Group of Cranfield, Bedford, UK. The other major source is the Biennial Water Jet Conferences organized by the Water Jet Technology Association headquartered in St. Louis, Mo. References to these volumes do not show this source fully, but the contribution of these organizations to the development of the technology has been a major one, and should be well and gratefully recognized by its proponents.

1.1 Wilkinson, A., The Ancient Egyptians, Publ. 1874, Vol. 2, p. 137.

1.2 Pliny (Caius Plinius Secundus) Natural History, Book 33.

1.3 Yufin, A.P., Hydromechanization, State Scientific Technical Press of Literature on Mining, Moscow, 1965.

1.4 Watson, W.B., Hydraulic Coal Mining with Particular Reference to New Zealand, M.Sc. Thesis, University of Birmingham, Birmingham, UK, 1958.

1.5 Horton, N., "No Blast, No Damnation," World Mining Equipment, Vol. 17, No. 2, February, 1993, pp. 22 - 26.

1.6 Longridge, C.C., Hydraulic Mining, Publ. Mining Journal, 1910.

1.7 Voitsekhovsky, B.V., "Jet Nozzle for Obtaining High Pulse Dynamic Pressure Heads," U.S. Patent No. 3,343,794, September 26, 1967.

1.8 Bryan ,E.L., "High Energy Jets as a new concept for wood machining," Forest Products Journal, Vol. 13, No. 8, August,1963, p. 305.

1.9 Walstad, O.M., and Noecker, P.W., "Development of high pressure pumps and associated equipment for fluid jet cutting," paper C3, 1st International Symposium on Jet Cutting Technology, Coventry, UK, April, 1972.

1.10 Field, J.E., "Stress Waves, Deformation, and Fracture Caused by Liquid Impact," Phil. Trans. Royal Society, London, Vol. 260A, July, 1966, pp. 86 - 93.

1.11 El-Saie, A.A., Investigation of Rock Slotting by High Pressure Water Jet for Use in Tunneling, Ph.D. Thesis, Mining Department, University of Missouri-Rolla, 1977.

1.12 Fairhurst, R.M., Heron, R.A., and Saunders, D.H., " ' Diajet' - a new abrasive water jet cutting technique," paper 40, 8th International Symposium on Jet Cutting Technology, Durham, UK, September, 1986, pp. 395 - 402.

1.13 Hood, M., A Study of Methods to Improve the Performance of Drag Bits used to cut Hard Rock, Chamber of Mines of South Africa Research Organization, Project GT2 NO2, Research Report No. 35/77, August, 1977, 135 pages.

1.14 Mazurkiewicz, M., Kubala, Z., and Chow, J., "Metal Machining with High Pressure Lubricoolant Jet - A new possibility, a new understanding," Journal of Engineering for Industry, February, 1989.

1.15 Anon, "Flowmole installs Gas Mains," Industrial Jetting Report, August, 1988, No. 62, p. 4.

1.16 Anon, "Waterjetting: The New Mapmaker's Tool ?" Industrial Jetting Report, February, 1988, p. 3.

1.17 Anon, "Baby in Well Miracle," Brisbane Sunday Telegraph, October 18, 1987, p. 1.

1.18 Uchino, J., Une, Y., Horie, T, Yokekawa, M., Kakita, A., and Sano, F., "Surgical Cutting of the Liver by Water Jet," Poster paper 1, 9th International Symposium on Jet Cutting Technology, Sendai, Japan, October, 1988, pp. 629 - 639.

1.19 Merk-Gould, L., Herskovitz, R., and Wilson, C., "Field Tests on Removing Corrosion from Outdoor Bronze Sculptures Using Medium Pressure Water," 1993, Fine Objects Conservation Inc., 3 Meeker Road, Westport, CT, 06880.

2 THE PARTS OF A HIGH PRESSURE SYSTEM

2.1 INTRODUCTION

When an operator chooses a high pressure system, the most important decision will normally be to choose the size and type of pump unit to be used. The second choice will be in terms of the delivery circuit through which the pump will send the water, under pressure, to the operating tool (Fig. 2.1). The tool will, in turn, direct this water to either one or several jet nozzles. At the nozzle the water flow is restricted to a small diameter and directed out, through an orifice at the work surface. The operator should decide what size of pump and parts to use, based upon the need for a given amount of delivered power at the working surface.

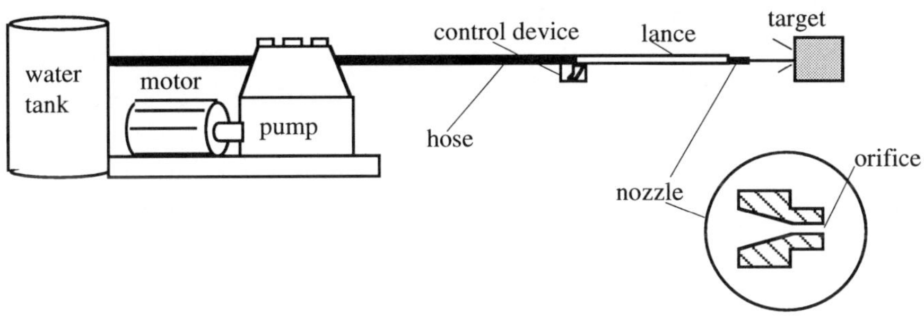

Figure 2.1 Basic components of a waterjet circuit.

The choice of the pump size is generally made in terms of the volume of water flowing through the nozzle and the pressure at which the pump delivers that water into the supply line. In many discussions with operators, it is clear that the pump operating pressure is considered to be the same value as the jet pressure when it reaches the target. At the same time the jet coming from the nozzle is considered to have the same power, regardless of nozzle shape, and to retain that power for a considerable distance from the tip of the nozzle. Both of these assumptions are generally wrong. To illustrate why this can be a problem, consider the stages through which the power, which is first supplied to the drive motor of the pump, is actually distributed.

To give a measure of this power change it is useful to have an agreed upon measure of the power of the jet. For practical reasons the power can be measured in terms of the effectiveness of the resulting jet which hits the work surface. The term used is called **specific energy** [2.1]. This is the amount of energy which is required to remove a unit volume of material from the work surface. The example offered is taken from an operation to cut coal from a coal seam, although it could equally well have been derived from the cleaning of ceramic waste from pipes, or cutting metal for automobile parts. To further illustrate the point, at each stage in the process the amount of energy available has been calculated in terms of a percentage of the power which was input to the motor.

2.2 DISTRIBUTION OF ENERGY

Consider the case where a 60 kW motor is used to power a high pressure pump, which will deliver a flow of 44 lpm at a jet pressure of 700 bar. The flow from the pump passes through 10 m of 4.76 mm internal diameter tubing to feed three nozzle orifices which are traversed over the working surface, which is at a distance of 30 cm from the nozzle. Each orifice is 1.4 mm in diameter. In order to provide an example of how energy is dissipated, consider the case when such a jet system is taking off 2.5 cm of material along a path which is 5 cm wide, and with the nozzle moving along that path at a speed of 60 m/min. Specific energy values can be calculated based upon the amount of energy available at each stage in the delivery line.

2.2.1 ENERGY INPUT TO THE MOTOR

The volume of material removed from the working surface each second is equal to the depth of cut multiplied by the width of cut and the distance which the nozzles have moved. In this text, unless otherwise stated these dimensions are given in mm. This can be calculated, for each second of operation as equal to:

Volume (v) = depth of cut (h) x width of cut (w) x distance moved (m)
= 25 x 50 x 1000 = 1,250,000 cubic mm = 1,250 cubic cm (cc).

Where: v is the volume of rock excavated in cubic mm
h is the depth of cut in mm
w is the width of the cut in mm
m is the distance cut/second in mm

The motor power of 60 kW is equivalent to 60,000 joules, so that the overall specific energy (the energy in joules required to remove 1 cc of material) can be calculated by simple division. The units most often reported are either Megajoules/cubic m or joules/cc. These are numerically equivalent. For this text joules/cc will be used, except where otherwise stated.

Specific Energy = power used/ volume removed
= 60,000 / 1250 = 48 joules/cc

This is the overall energy efficiency of the process in terms of the energy input. It is the most critical value since it is based on the quantity of energy which is entering the system, and for which the user must pay, and is related to the product achieved, for which the operator is paid. This is the initial energy value, which provides the 100% value against which subsequent levels of available energy will be compared.

2.2.2 INPUT POWER TO THE PUMP

If the electric motor is assumed to be 90% efficient, then the pump will only receive 54 kW of power from the motor. This is equivalent to 54,000 joules. However, with the same volume of material removed from the working surface, the specific energy value calculation gives:

Specific Energy = power available/ volume removed
= 54,000 / 1250 = 43.2 joules/cc

Available Energy = Energy Available/Energy Input
= 54,000/60,000 = 90%

2.2.3 POWER OUTPUT FROM THE PUMP TO THE WATER.

If the water delivered by the pump is flowing at a volume (Q) of 44 lpm and at a pressure (P) of 700 bar at the pump exit, then the contained power can be calculated (see section 2.7).

Water Power (K) = 1.666 x Flow Rate (Q) x Pressure (P) / 1,000
= 1.666 x 44 x 700 / 1,000 = 51.31 kW.

where: P is the Jet Pressure in bar
Q is the water volume flow rate in lpm
K is the water power in kW

Specific Energy = power available/volume removed
= 51,310 / 1250 = 41 joules/cc

Available Energy = Energy Available/Energy Input
= 51,310 / 60,000 = 85.51%

2.2.4 POWER OUTPUT TO THE JET AT THE NOZZLE.

The pressure loss in driving 44 lpm through 10 m of pipe with an internal diameter of 4.76 mm is roughly 25 bar for each m of tubing (see section 2.8.2). The pressure loss in turning the water to give the required divergent stream at the nozzle varies with angle and jet diameter. It is approximated, for this calculation to be 35 bar. The total pressure loss in the jet from leaving the pump to the point where it is discharged at the nozzle can thus be calculated.

Jet Pressure loss from the pump to the orifice = line loss + nozzle loss.
= 10 (m) x 25 (bar/m) + 35 = 285 bar.

Jet Pressure = Input Pressure - Pressure loss.
= 700 - 285 = 415 bar.

Water Power (K) = 16.66 x Flow Rate (lpm) x Pressure (bar) / 10,000
=16.66 x 44 x 415 / 10,000 = 30.4 kW.

Specific Energy = power available/ volume removed
= 30,420 / 1250 = 24.34 joules/cc

Available Energy = Energy Available / Energy Input
= 30,420 / 60,000 = 50.7%

Thus, in getting the water from the pump, through the line, and out of the nozzle, about half of the energy delivered to the pump has already been lost.

2.2.5 POWER DELIVERED AT THE WORKING SURFACE.

If the fluid flowing to the target is separated into three nozzles, then at 44 lpm, this will require that each nozzle diameter be 1.4 mm. This assumes that the coefficient of discharge is 1.0 and the jet pressure at the nozzle is 415 bar. (This point will be addressed in section 2.7.)

A stand-off distance of 30 cm is equivalent to 215 jet diameters. This is a little beyond the most effective cutting distance for a commercially available nozzle, since a normal jet will be at its most powerful only for a distance on the order of 150 to 200 nozzle diameters from the orifice (see section 2.7.3).

For a nozzle of this size and at this pressure, experiments have indicated that the pressure decay curve, in this distance range, can be approximated by the equation:

$$\text{impact pressure (bar)} = 389 \cdot e^{-0.0165 \cdot s}$$

where: s is the distance from the nozzle to the target surface in cm

Substituting values, this gives an impact pressure on the surface of 237 bar. This, however, is not a constant over the jet, but at this distance the pressure will be a bell shaped curve (see Fig. 2.25). By approximating the curve to a triangular function a simplified estimate of the amount of energy remaining in the jet can be made. This works out to be 8,100 joules. One can then recalculate the values for energy efficiency.

$$\text{Specific Energy} = \text{power available / volume removed}$$
$$= 8{,}100 / 1250 = 6.48 \text{ joules/cc}$$

$$\text{Available Energy} = \text{Energy Available / Energy Input}$$
$$= 8{,}100 / 60{,}000 = 13.5\%$$

The numbers which are given are for illustration purposes only (Fig. 2.2), and do not reflect the values which would be obtained with an optimized system design. They are however, somewhat typical of those which might be found in operational equipment which might be encountered in field use. The production values which have been used are taken from those found in actual experiments in the laboratory. The lesson to be learned from this preliminary calculation is that while it is only necessary to use some 6.5 joules of energy to remove one cc of material from the target surface, because of losses in the system, the power which is required to achieve this requires that 48 joules/cc be input to the pump motor. Thus, some 86.5% of the input energy is being dissipated before it reaches the target, and only 13.5% is available to do the work required.

Put another way eight times the amount of energy actually needed to remove the material must be used, in order to overcome system inefficiencies.

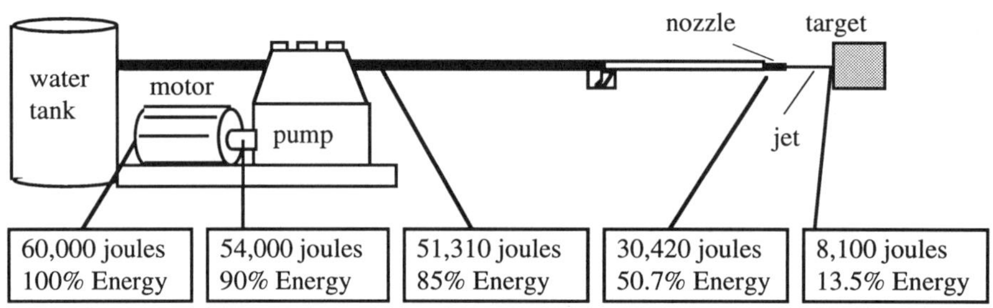

Figure 2.2 Relative levels of delivered energy, and input power percentages, at points along a high pressure system.

How then can the performance of the delivery system be improved, to retain more power in the jet, and thereby to improve its performance? To make that decision one must first examine each of the components which make up a high pressure delivery circuit, and see how each of these can be individually improved. One can then examine the best way of using the water coming out of the nozzle, and then use that information to design the best nozzle orifices and also the holder and drive system for those nozzles.

2.3 PUMP CHOICES

There are two types of pump which are commonly used in the waterjet industry, and a number of other, experimental units which have been tried. At lower pressures, pumps are normally piston driven, with a three piston or Triplex pump being the most common. Where the pressures are higher, and flow rates normally lower, then an intensifier system can be used. Each has its own peculiar advantages and disadvantages. The normal type of pump which one encounters at the lower pressure levels, i.e., below 1,400 bar, is one in which the water is delivered to the supply line by the action of oscillating pistons (Fig. 2.3). These pumps are called positive displacement, reciprocating pumps because the pistons normally move backwards and forwards at a constant speed. On each stroke a fixed volume of water is drawn into the cylinder above the piston, and then pumped out into the line. Thus, if the pump is operated at a constant speed, then each piston, and therefore the total pump, will provide a constant flow of water into the delivery line.

2.3.1 TRIPLEX AND POSITIVE DISPLACEMENT PUMPS

Figure 2.3 Sectioned view of the piston area of a triplex high pressure pump (courtesy of Paul Hammelmann Maschinenfabrik GmbH) [2.45].

There are several ways to control the volume of water which the pump will supply. It is possible, though not common, to vary the speed of the drive motor to the pump. This can change the oscillation speed of the pistons, and thus control the amount of water each pumps out every minute. An alternate method of achieving the same goal is to place a gearbox between the drive motor and the pump drive shaft. Thus, by changing gears, the speed of the pump drive shaft may be changed, and the volume coming from the pump altered accordingly.

This fixed volume of water flowing into the pump outlet manifold has both advantages and disadvantages to high pressure pump operations. The advantage lies in the steadiness of flow generated. However, there is a disadvantage in that if the nozzle should be blocked or partially blocked, then the water must still find a way out of the line. This will occur by either the water speeding up to get through the remaining area of the orifice, which will require a higher pressure exerted back on the pump pistons, or an alternate passage must be found. If such a passage in the form of a relief valve of some sort is not found then the water may make its own passage by bursting one of the hoses, or blowing apart one of the couplings in the line.

It is, therefore, important to ensure that proper safety valves be included in the line to protect the system and the operator against the consequences of a nozzle blockage. In their simplest form safety valves consist of a ball, held in place over an opening in the supply line, by a spring (Fig. 2.4). By adjusting the pressure in the spring, it is possible to set the valve so that it will open, when the line pressure exceeds a given level. When this pressure is reached, the force applied to the ball by the water exceeds that exerted by the spring, and the ball is pushed up out of its seat. This opens a gap between the ball and the seat, through which water can escape up through the valve. Safety valves can be located at several positions, and in a number of equipment layouts the practice has been to locate one unit at the pump, and a second relief valve near the operator position.

Because the flow coming from the pump is at a constant volume, there are several ways to control the pressure of the resulting water. The first is to provide a bypass circuit in the line, which diverts some of the water flow away from the nozzle. Generally flow through this line is controlled by an adjustable valve, and the water going down this line is returned to the pump reservoir for re-use (Fig. 2.5). The alternative method is to accept all the water into the delivery line to the nozzle and to size the nozzle according to the pressure required from the jet. The procedures to follow in this approach, and the potential pitfalls, are discussed in more detail later.

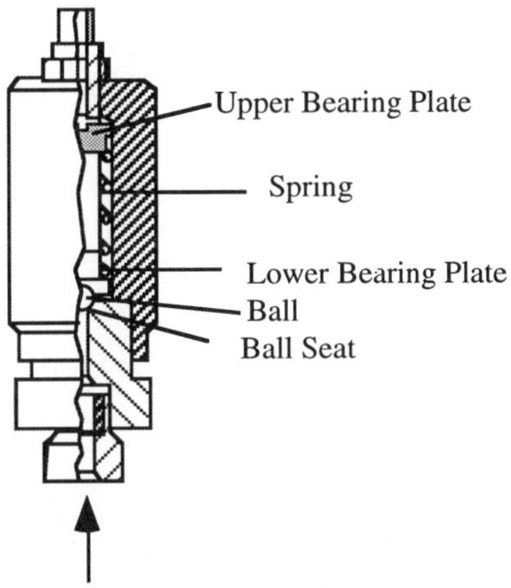

Figure 2.4 Schematic section through a safety valve, showing the main components of its construction.

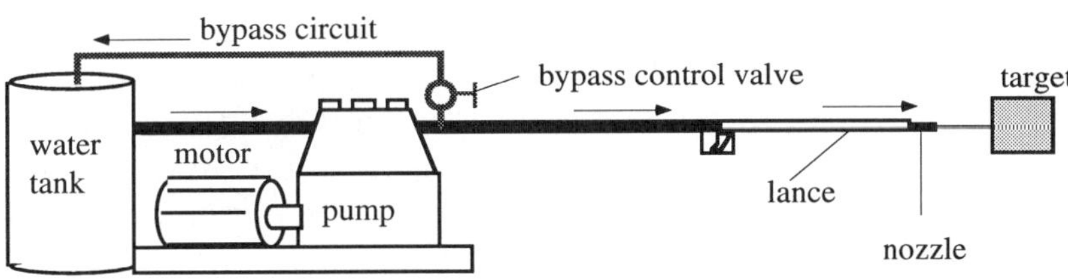

Figure 2.5 Typical arrangement of a bypass circuit.

Many of the drive motors connected to triplex pumps will run at a fixed speed. However, when the pump is used, different applications may require that the jet be operated at different pressures. In order to take full advantage of the power available from the main motor it is possible, with some models of pump, to change the diameter of the plungers and liner assemblies in the main pressure cylinders. This changes the volume of water pumped on each revolution of the pump, and a greater or lesser amount of water will be delivered to the nozzles. This will, in turn, change the pressure exerted on the plungers and, through the jet, on the target material for the same power delivered (Table 2.1).

Table 2.1 Variation in delivered flow rate and pressure for a given pump [2.2]

Plunger diameter (mm)	*Delivered volume* (liters/minute)*	*Allowed pressure** (bar)*
28.6	76	340
31.75	94	300
35	114	245
38.1	136	206
41.3	160	175
44.45	185	152

* at a Pump speed of 370 rpm, ** with a 45 kW integral electric motor.

Such changes will, for example, allow a 115 kW motor to drive a water pump to produce 175 lpm at 350 bar, 100 lpm at 700 bar, and 40 lpm at 1,750 bar. In fact these values will only be nominal since, at pressures above 700 bar a detectable compression of the water becomes apparent, which lowers the volume delivered from the pump to the nozzle. This will require an adjustment in the size of the nozzle used, but note that the water will expand once it reaches ambient pressure beyond the nozzle.

In order to allow an inexpensive method for coping with these changes some pumps come fitted with ceramic pistons and elastomeric seals between the piston and the cylinder walls to allow a less expensive, and quicker change in performance. These are generally easier to change than the other major design for the high pressure plungers where the plungers are honed to individually fit into each cylinder, giving a metal:metal seal on the piston wall. The latter systems perform well on initial installation, but, over the years, wear between the two surfaces will lead to some leakage and loss in overall pump volume to the nozzle assemblies. In this case the entire cylinder and piston assemblies have to be returned to the factory for refurbishing, a more timely and expensive process than replacing the worn seals of the alternate design.

Piston pump designs follow much the same basic designs, with only one or two exceptions. One such variation is to change the driving mechanism which moves the piston from a rotating crankshaft, as in the pumps described above, to a swinging swash plate. One example of this technique has recently been described by Japanese investigators seeking to develop a pump which could be used at depths of up to 6,500 m below the sea [2.3].

In order to overcome some of the complex problems associated with this location, while generating pressures of up to 700 bar above that of the surrounding fluid, some different approaches were required. For the design of the drive, the team chose to use an oscillating swash plate to alternatively oscillate the pistons (Fig. 2.6). The design was chosen to reduce inertial forces, and the torque required to start the unit. By using a symmetrical arrangement of five pistons for the unit, a compact pump could be designed, but required considerable care in selection of materials to hold up under the forces generated.

As materials and sealing technologies improve, so the capabilities of these systems have grown with them, to current working pressures which have reached 1,500 bar. This does not mean, however, that some basic common sense knowledge no longer need be applied to their use. De Santis [2.4] has addressed the method of operation of most triplex and quadruplex pumps stressing three aspects of pump operation. The most obvious, perhaps, is that the system must be designed for a given piston loading. This loading can be calculated by multiplying the design pressure of the unit by the area of the pistons. Obviously any time that this loading is exceeded, then the system will be over stressed, and failure can be anticipated.

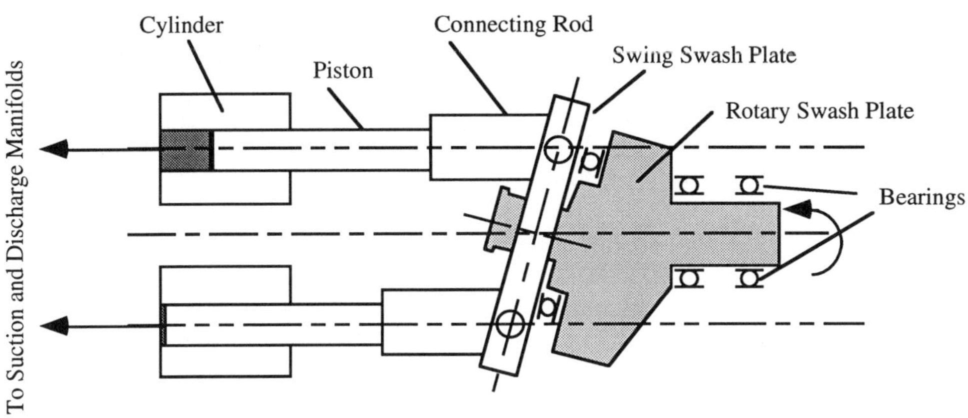

As the Rotary Swash Plate turns it also turns the Swinging Swash Plate, causing the connecting rods to oscillate back and forth, sequentially pulling in and discharging fluid through the pistons and cylinders.

Figure 2.6 Schematic of a pump using a swash plate drive [2.3].

The second point relates to the first but is not quite as obvious. In a typical triplex pump the pistons are equally arranged about a crankshaft at

120 degrees apart. This will not give an absolutely steady flow, since, as the pistons move backward and forward in the cylinders, water is drawn in, the inlet valve is closed, then the water is compressed by the piston motion, and it is only when it reaches a set pressure, that the delivery valve will open to push the water out to the manifold, and thence down the line to the nozzle (Fig. 2.7).

In the simple operation of the two valves which control the flow of water through this pressurizing cylinder, it is possible for several factors to complicate the operation. In normal operation the movement of each valve is controlled by the action of a spring, which will hold the valve closed, until the difference in fluid pressure across the valve causes it to open. Because the valve has a certain mass, and the springs have a limited strength, there will be a short break between the time that the flow stops into the cylinder, and the valve closes. Heron [2.5] has pointed out that this has two negative effects. Firstly the open valve will bypass some of the delivery water back into the supply line, reducing the total amount available, and secondly, when the valve does close, it will be with some pressure in the cylinder. The cutoff under pressure will induce a pressure spike into the system, which hastens fatigue and reduces the life of the component parts of the circuit. Similar effects can occur if the spring is too weak in the outlet line. Both problems can, to a large extent be resolved by increasing the pre-loads on the springs, but these values need to be properly calculated, as too large a load on the inlet valve may require additional supply pressure to open the valve on the inlet stroke.

De Santis (ibid.) plotted typical flow curves for triplex and quintuplex pumps which illustrate the flow capacity of the system at different stages in the pump operation (Figs. 2.8 and 2.9).

It can be seen that there is a considerable variation in the flow of water from the pump during both cycles, but that this is greater with the smaller number of plungers. On the other hand the smaller number of pistons lowers the pulsation frequency to the pump flow. At 400 rpm, the triplex would pulse the flow at 40 times/sec, while the quintuplex would give a lower pulsation, but at 66.6 times/sec. While this variation in flow, at pressure, must be anticipated in the delivery line, it also has a significant and often neglected effect on the suction side of the pump.

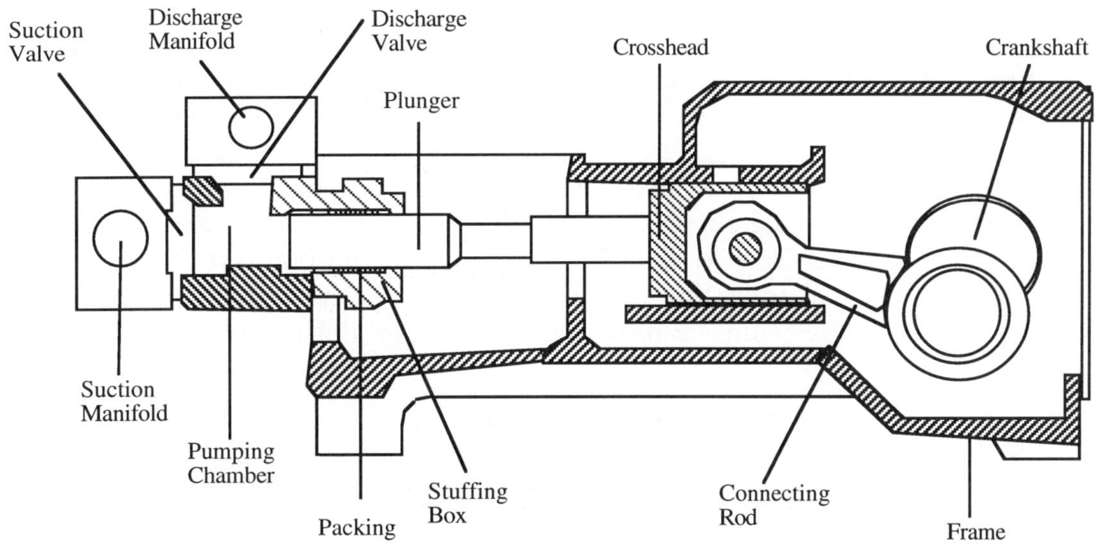

Figure 2.7 Sectioned view of a triplex pump, showing the valve locations [2.4].

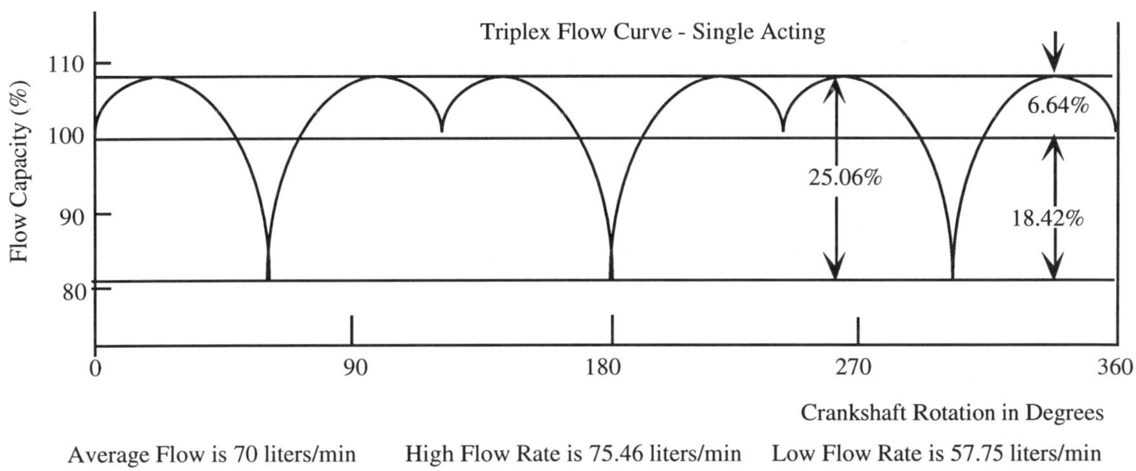

Figure 2.8 Flow variation during a single cycle of a triplex pump [2.4].

There are two numbers which are important in the delivery of water to the pump, the NPSHR and the NPSHA. These letters stand for the Net Positive Suction Head Required, and the Net Positive Suction Head Available. If the available head is less than the required head then the flow into the pump will cavitate. This has two serious effects. The first is that the flow to the nozzle will be reduced, but more critically the cavitation bubbles drawn into the pump will severely damage, and, frequently, in time will destroy the pump. The destructive ability of cavitating jets will be discussed in more detail in Chapter 10. As either the plunger size, or the pump rotational speed increase, so the volume of water drawn through the pump, and therefore also the required suction head will also increase. Without an adequate supply of fluid to meet this demand, this can lead to rapid onset of damage to the pump. Given that the pump will have a fixed diameter for the inlet line, the increase in demand may require that the water feed to the pump be provided at a certain, defined pressure.

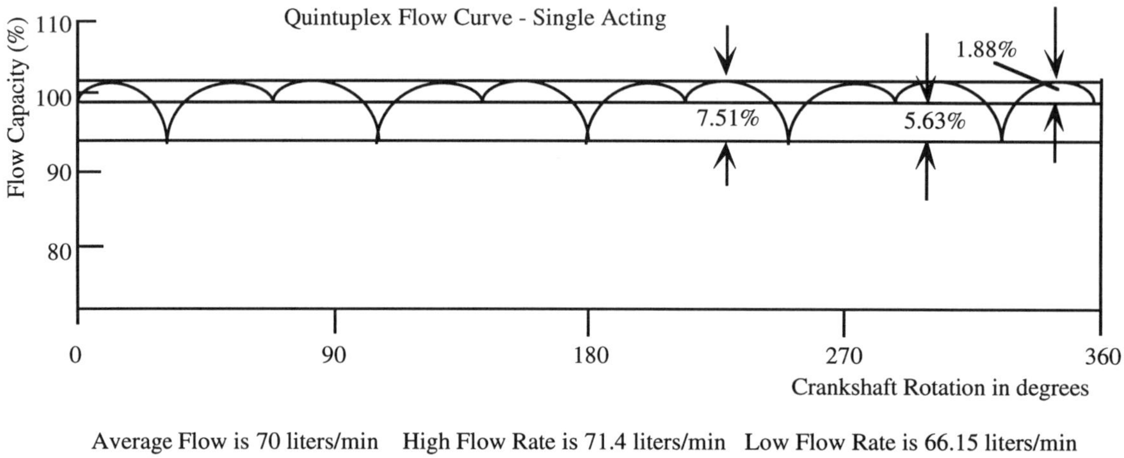

Figure 2.9 Flow variation during a single cycle of a quintuplex pump [2.4].

Operators may, on occasion, seek to obtain more output from a pump than its rated capacity. It is inadvisable to do this by running the pump above the recommended speed for several reasons. Initially one should recognize that while this increases the output of the pump it also increases the frequency of cycling the load and the power which is transmitted to the pistons through the pump parts. This overloading can exceed design specifications and induce fatigue failure in a relatively short time. Secondly the increased demand for flow has two contrasting effects on the supply line. It increases the demand for water, but at the same time it lowers the suction head available. While each of these in themselves can be a relatively small change, when they are combined they can be dangerous.

For example consider the change which occurs when a pump, normally rated at 400 rpm is driven at 500 rpm, for a 25% increase in output. At 400 rpm the NPSHR for a triplex pump supplied through a 3.175 cm diameter pipe from an open tank will be 0.55 bar. At 500 rpm, as the flow increases from 100 lpm to 125 lpm, the NPSHR rises to 0.62 bar, which is only a 12.9% change. However, under the same conditions the NPSHA, which begins at 0.79 bar with a 100 lpm demand, falls to 0.54 bar at 125 lpm. When the required suction head is thus contrasted with that available it can be seen that while, initially, there was a surplus of 45% this changes to a shortfall of 12% at the higher speed. The pump will cavitate, inadequate flow will reach the nozzle to provide full pump performance, and equipment lifetime will be markedly reduced.

An additional problem may arise, even if a pressurized tank or larger booster supply pump is used to supply water to the high pressure pump. The valve springs used to operate the inlet and outlet valves within the cylinders will have been manufactured to a required stiffness to respond to the slower cycle time of normal operation. Running the pump faster may cause the piston to cycle faster than the valve will close and this will thus induce shock loading of the system, as the valve will now close during the piston movement, and while the fluid is now under pressure.

Running the pump at a load above its rated pressure is likely to be a more common error. The problem may not appear as large to the operator as it actually is. Consider the case where an operator has inserted 3.175 cm diameter pistons into the pump in order to normally clean a surface with a jet pressure of 700 bar. The force on the piston will be:

$$\text{Force (kg)} = \text{piston area (sq cm)} \times \text{pressure (bar)}$$
$$= \pi \cdot r^2 \cdot P = 3.1412 \cdot (1.5875)^2 \cdot 700 = 5{,}540 \text{ kg}$$

where r is the piston radius in cm

During the operation of the pump, the operator notices that the jet is not cutting as fast as he would like, and thus he inserts a smaller nozzle orifice into the system. A smaller nozzle will increase the velocity of the water passing through it, and thus the required pressure to drive the water. If this change increases the pressure on the pistons to 1,000 bar, then the above calculation can be re-evaluated. At 700 bar, the pump loading through the piston, to the crankshaft, is 5,540 kg. At 1,000 bar, while the higher pressure may do the job faster, it will increase the force that the pump will see to 7,910 kg. This is a considerable design overload. It is one which may be the more common abuse of the pumping system, since to increase the nozzle pressure by this amount, providing the safety valves are also adjusted, may only require that the nozzle size be reduced from 1.00 mm to 0.914 mm. This small change may also occur when a nozzle is replaced, and when the replacement nozzle has a significantly lower coefficient of discharge than the nozzle being replaced. As a result the equivalent orifice size through which the water must pass will be reduced, and the system pressure will increase, without any conscious decision by the operator.

The increasing market for the application of high pressure equipment has seen considerable advances in the technology of the high pressure triplex systems. For example, in 1993 [2.42] Alkire described the development of a Direct-Drive system for an ultra-high pressure triplex pump capable of generating 26 lpm at a pressure of 2,800 bar. This development required the use of smaller high pressure components in the high pressure end of the pump, and, for greater flow, this required that the pump operate at a greater speed. It was by this means that the pump could be directly driven from a prime mover in the 1700 to 2100 rpm range. The pump which Alkire described also has an internal pressure-compensation device to ensure, within the pump capacity, that the pressure was maintained at the nozzle even where the nozzle size was below that which would use all the pump capacity. By internally and automatically diverting any excess flow within the pump, a greater control on system operation has been claimed.

2.3.2 INTENSIFIER DESIGNS

In a conventional high pressure pump the power is transmitted, through a crankshaft which forces the reciprocal motion of each piston, and in this way generates pressure in the fluid drawn into the space above the piston. This type of system works well at intermediate pressures, but as the pressure levels required mount, it becomes less reliable. An alternative approach has, therefore, been developed in which an intensifier design is used as a pumping unit.

In the basic design of such a unit (Fig. 2.10) oil is supplied at normal hydraulic operating pressures (in the range from 200 - 350 bar). The oil flows into a cylinder and applies pressure to the large surface area of the reciprocating piston within that cylinder. The pressure causes the piston to move, and in so doing, it drives water out of the volume ahead of its motion. This volume is however, held in a vessel of smaller diameter, so that with relative equal forces across the piston, much higher pressures can be generated in the delivery water.

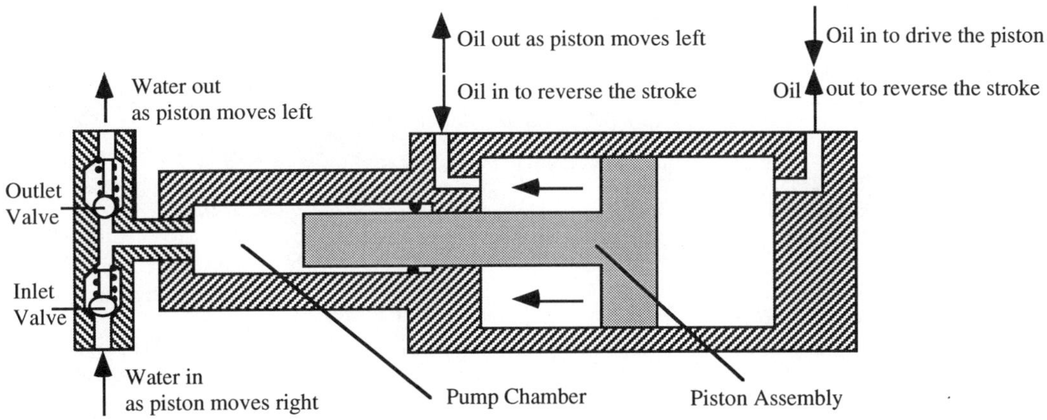

Figure 2.10 Schematic of the operation of an intensifier pump.

This can, perhaps, be most easily illustrated by example. If a piston is used where the oil pushes against a piston face with a 5 cm radius, then at a pressure of 350 bar, the oil will exert a force (= pressure x area) of (350 x π x 5^2) = 27,500 kg. If the piston size on the water side is only 1.58 cm in radius, then the pressure that will be exerted on the water is given by (force/ area = 27500/ (π x $1.58^{2)}$ = 3,500 bar. Thus by reducing the size of the piston from 10.0 cm in diameter on the low pressure side, to 3.16 cm diameter on the high pressure side, the pressure delivered can be increased by a factor of 10.

In order for the intensifier to give a quasi-continuous flow the units are generally made double acting (Fig. 2.11) so that, at the end of each stroke the cylinder is caused to change direction and then acts to pressurize the fluid in the opposing chamber, while allowing the inactive chamber to recharge. This alternate cycling can be achieved by alternately directing the flow of fluid to either side of the low pressure piston.

High pressure intensifiers have found a significant market, particularly in industrial cutting. They operate, however, most frequently at pressures at which metal behavior becomes of significant importance. Thus these pumps require that particular care be taken in the choice of material components, and in monitoring the quality of the water, or other fluid, which is used as the cutting fluid on the delivery size of the intensifier.

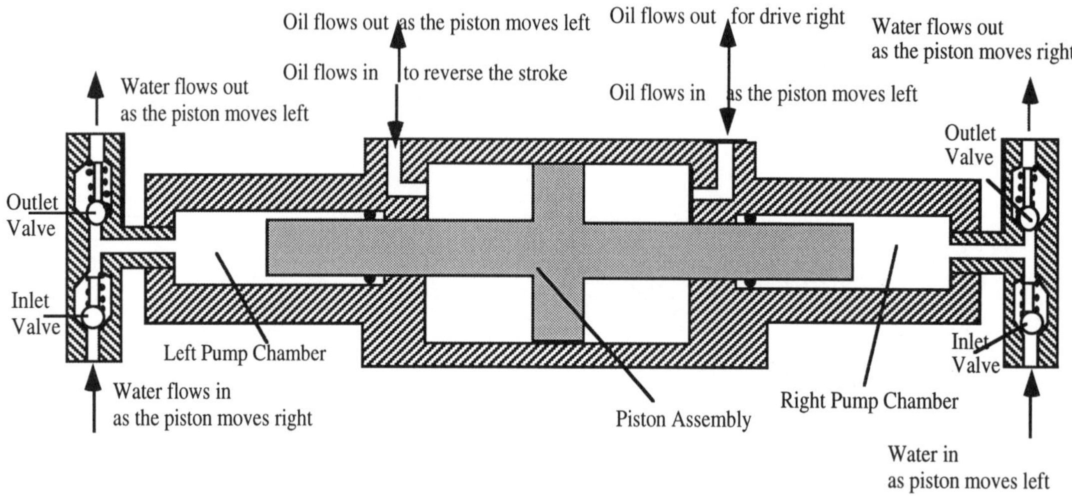

Figure 2.11 Schematic of the operational parts of a double acting intensifier.

An additional problem which arises with their use comes from the need to compress the fluid before pumping will begin on each stroke, given that the pump must also change direction before fluid delivery can start again. Singh and Benson [2.43] have noted that water is compressed by about 15% of its volume at 3,500 bar and thus, for intensifiers operating in that range, some 15% of each stroke must be completed before the pump begins to deliver fluid at pressure. There are two approaches which can be taken to solving this problem. The traditional one has been to include an accumulator or attenuator in the system (Fig. 2.12). Theoretical evaluations of accumulator design were discussed by Singh and Benson, while Chalmers [2.44] has made measurements of the actual performance of such systems.

Pressure and flow variations are highly undesirable in intensifier operations particularly in the high precision cutting applications for which they have been found very effective. Any variations in the jet stream can show up as a change in the contour of the cut and can lead to an unacceptable surface. In addition, as Chalmers points out, wide ranges of pressure fluctuation within the delivery system will lead to large cyclic loading, and a shortened fatigue life for the system. Sizing the attenuator to effectively reduce pressure fluctuations is critical to its effectiveness. Because the attenuator stores a volume of water at pressure, which it releases into the system as the pump pressure falls, it can even out some of the fluctuations (Fig. 2.13).

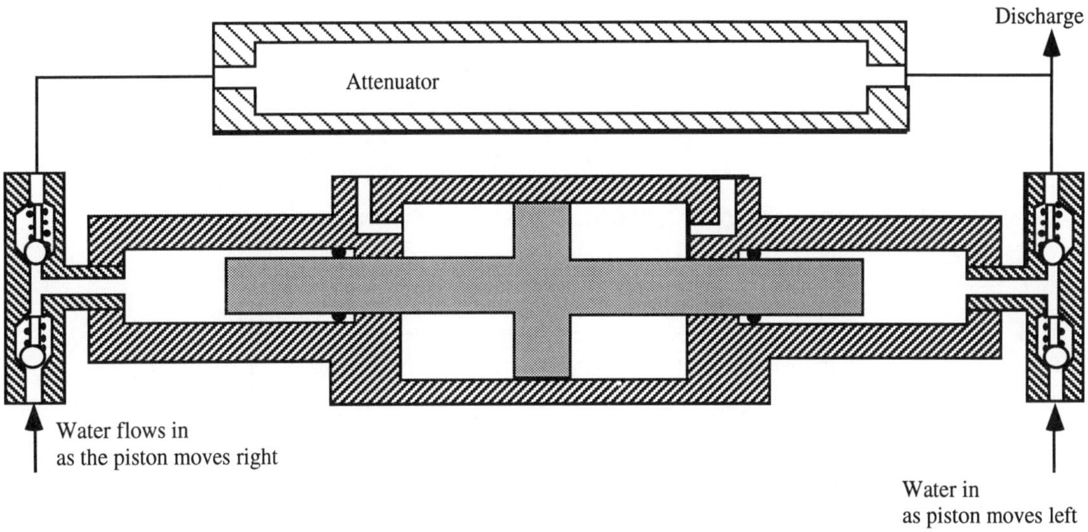

Figure 2.12 Hydraulic intensifier circuit fitted with an intensifier.

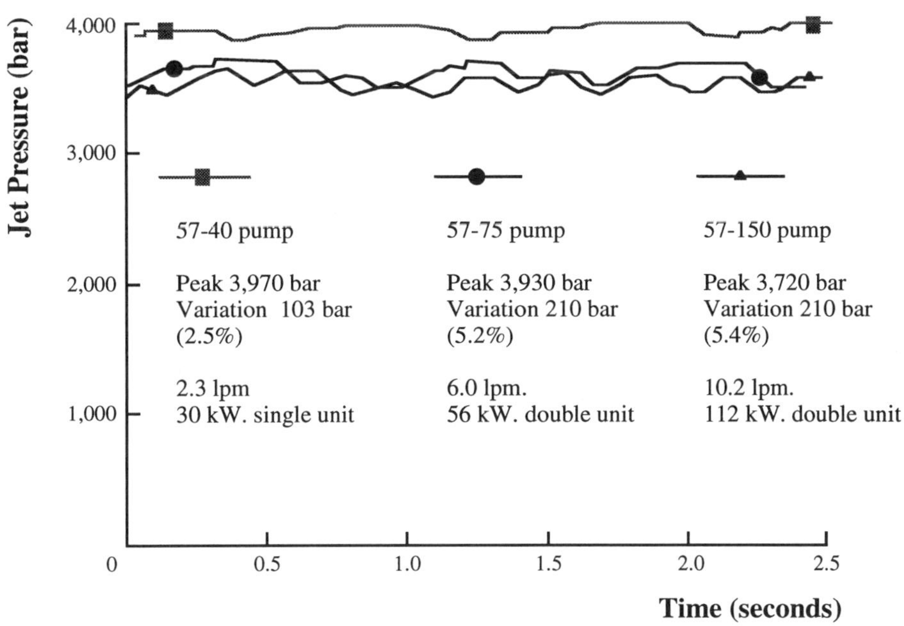

Figure 2.13 Variations in intensifier pressure output with an attenuator and for different pump sizes [2.44].

However the capacity of the attenuator must be sized to the delivery volume of the pump, and this can be seen from the data which Chalmers reported (Fig. 2.14). It is important to note that this system affects the overall efficiency of the operation and, in his paper, Chalmers found that the systems which he was studying had an operational efficiency of under 65% depending on the power supplied to the nozzle (Fig. 2.15). This value compares with one of 60 - 65% quoted by Alkire [2.42], and an efficiency of up to 95% for more conventional positive displacement pumps.

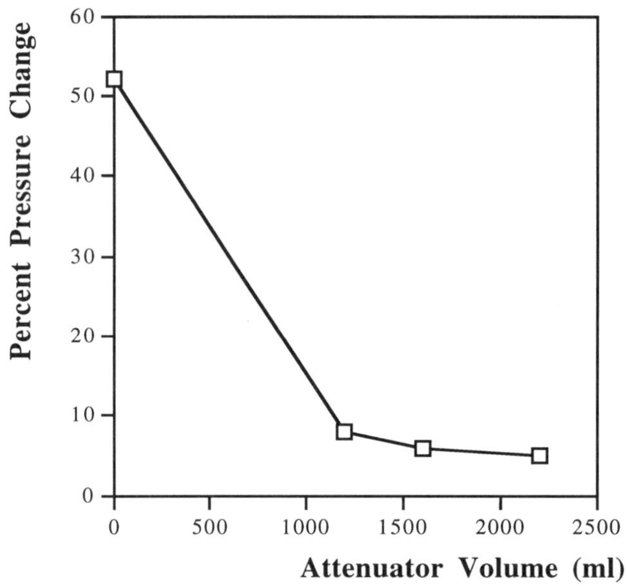

Figure 2.14 Effect of attenuator size on intensifier pressure fluctuations [2.44].

The alternative approach is to run two single intensifier pistons so that one has started its compression stroke, before the second has completed its delivery. Thus, as one piston stops, the other is already at pressure and continues the delivery (Fig. 2.16(a), 2.16(b), and 2.16(c)). Singh and Benson have described such a system [2.43] which does not require an attenuator. They further reported that the pressure fluctuations in the system were below that of a conventional dual acting intensifier (Fig. 2.17) and that in field operation the unit proved not only cheaper, but that the mean time to failure of the components tripled, reducing maintenance costs.

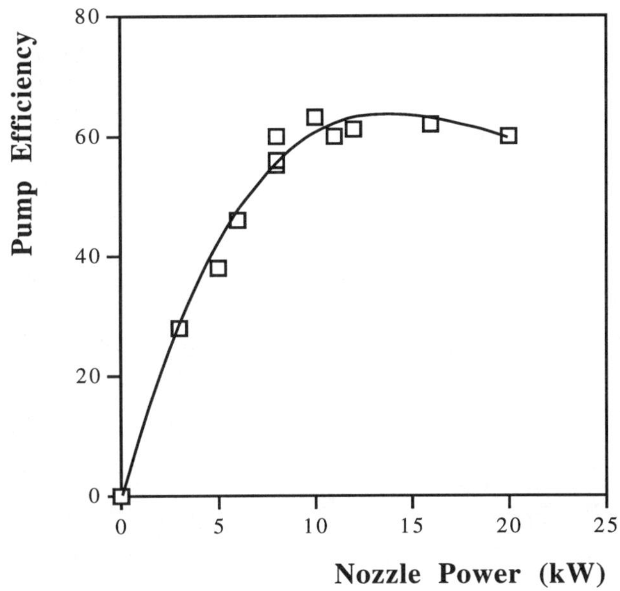

Figure 2.15 Intensifier efficiency as a function of nozzle power [2.44].

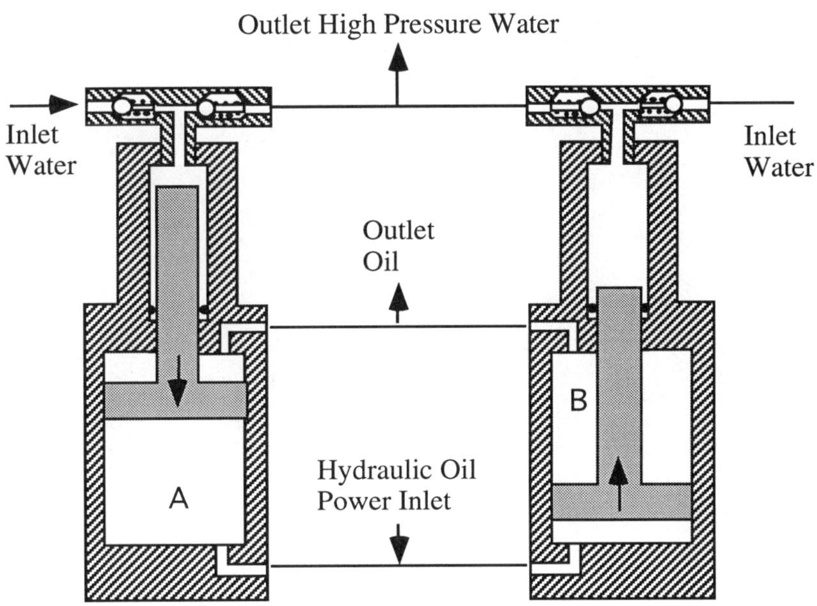

Figure 2.16(a) Sequential operation of a phased intensifier system [2.43]. Pistons A and B are independently controlled and at this stage A has started to retract, drawing water into the cylinder, while B is discharging high pressure water.

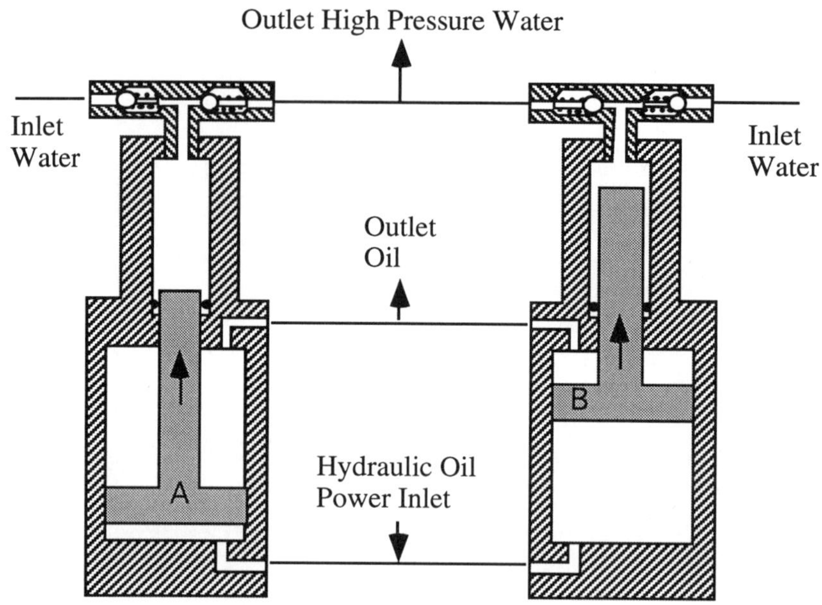

Figure 2.16(b) Operation of a phased intensifier system - piston A has started to move up, compressing water without discharge, B continues to discharge pressurized water.

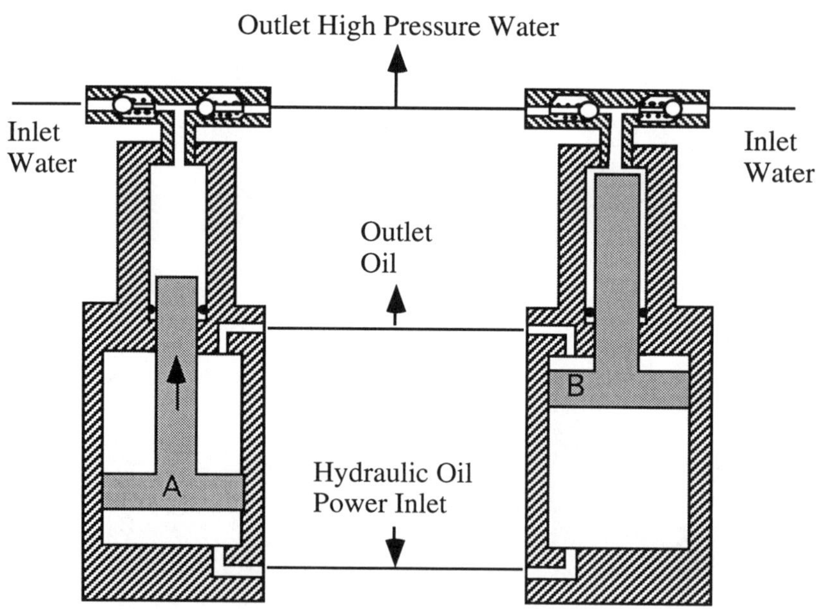

Figure 2.16(c) Sequential operation of a phased intensifier system [2.43]. Piston A has started to discharge, B has stopped and is about to retract.

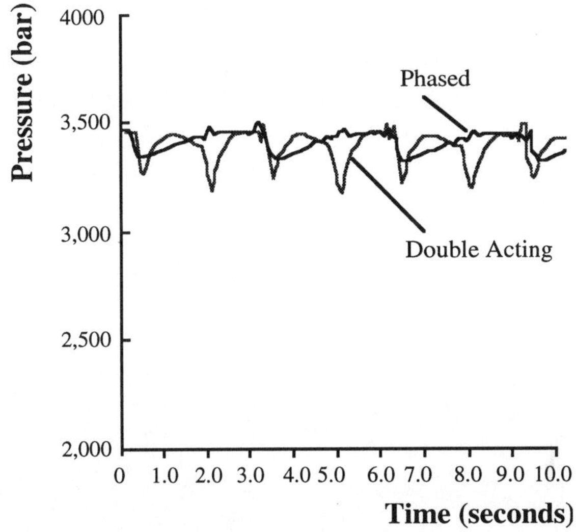

Figure 2.17 Reduction in pressure pulsation with a phased intensifier [2.43].

The increasing use of high pressure waterjets in industrial plants has led to a perception that intensifiers are large, heavy and expensive units. This need not be the case, and, for example, the phased systems discussed above are both smaller and lighter than their earlier counterparts. However, for very small and precise applications there is another alternative particularly when only relatively small water flows may be required (see, for example, section 4.9.1). For these low flow applications it is possible to obtain air driven intensifiers which use a compressed air feed to provide the power supply to the low pressure side of the intensifier. While such units may individually only generate flow rates on the order of 260 cc/min, they do this at pressure of 2,000 bar and above, and can be connected in parallel to provide greater flow volumes. With an initial cost of under $2,000 they provide a relatively inexpensive and small solution to some industrial application problems [2.6].

2.3.3 WATER CANNONS AND ULTRA-HIGH PRESSURE SYSTEMS

As pressure demands have increased beyond the normal operating limits of intensifiers, another method of generating pressure was required. Power requirements for continuous operation, particularly at larger flow rates become overwhelming at these pressures, and most equipment has, therefore, been designed to operate in a single shot, or interrupted mode.

The earliest equipment designed for this use, was built around the same idea as a military cannon. If a slug of water is placed in a barrel, and then a large volume of gas at high pressure is released behind it, then the water will issue from the "gun" at high speed (Fig. 2.18). In large measure this basic principle has continued for the generation of ultra-high pressures, although the method of generating the gas pressure has varied. Because the method of drive is based on military ideas, the devices have retained the sobriquet of "water cannons", and smaller versions have been locally referred to as "water derringers".

Figure 2.18 Modified 90 mm howitzer used as a water cannon at UMR.

The operation of the various designs can be exemplified by discussing the design and operation of such a "derringer" (Fig. 2.19). The device is built around a cylindrical steel body, which serves to contain the pressures generated, and to direct the gases and water produced. Within this cylinder a small chamber is located, which contains an aluminum piston. The piston is initially set at the bottom of the chamber, but located so that, as it moves upwards it will expose a vent hole drilled through the chamber and outer cylinder walls. A nozzle is attached to the center of the outer cylinder lid, and lowered until it engages a seal around the outer cylinder edge. A lower bottom plate, containing a small precision blasting cap and a calibrated amount of "Detasheet", an explosive, is attached so that the explosive chamber is located directly behind the aluminum piston. The assembly is held together with four retaining bolts which are tightened to a set pressure. The device is then located in the desired position, and a small quantity of water, on the order of 2 cc, is fed into the nozzle using a calibrated syringe. When the precision blasting cap is then electrically ignited, the resulting gases drive the piston forward, expelling the water through the nozzle. The driving gases vent through the side of the device, so that only the waterjet impacts on the target. Pressures of up to 70,000 bar can be generated in this manner [2.7].

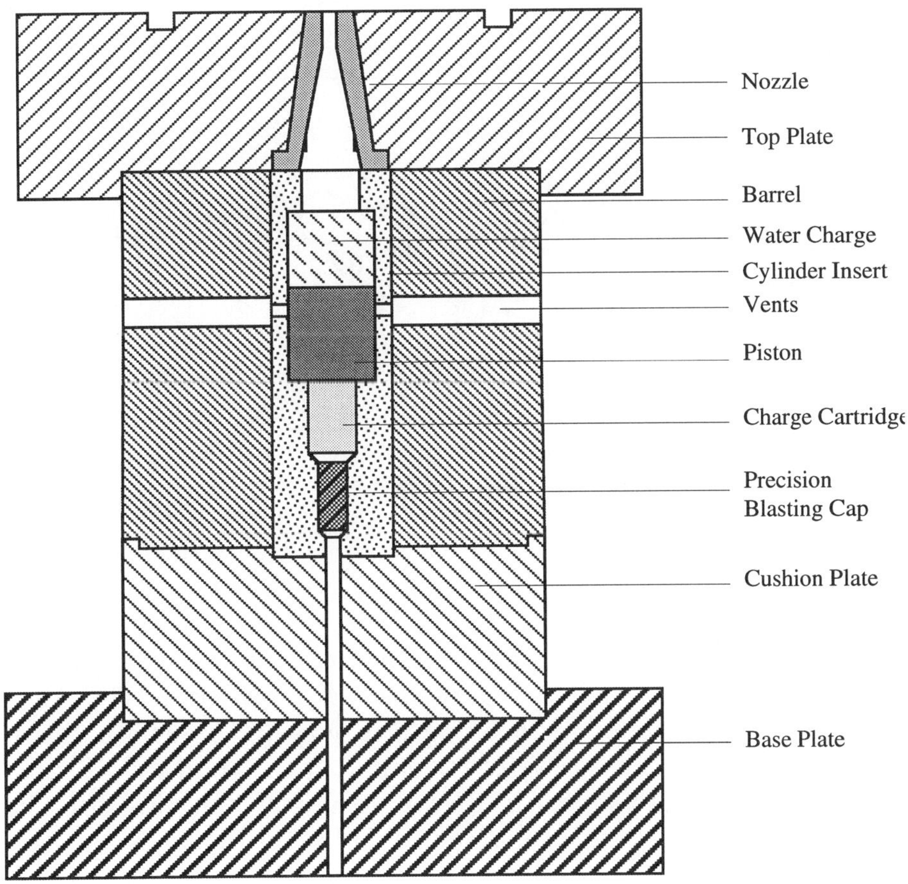

Figure 2.19 Section through a "water derringer" high pressure driver.

If the very highest pressures are required (on the order of 700,000 bar) then the gas driven devices become impractical. It has been found possible to generate "waterjets" to impact at such pressures using a lined explosive charge (Fig. 2.20). Instead of using a metal liner to the shaped charge, however, a thin film of water, held together with a small concentration of cerageenen, is laid over the explosive surface. When the charge is fired the water is driven into and along the axis of the charge, thereby accelerating and generating a very high velocity slug which can be aimed at a suitable target. Impact velocities for such a jet have been measured, at UMR, in excess of 10 km/sec [2.8]. Such systems have found, however, relatively little industrial application at this time. The mechanisms by which such devices work are discussed in more detail in Chapter 11. For those questioning the effect of the water under these conditions experiments have been carried out which show that the water does form a jet, which can punch a hole through a 25 mm thick steel plate at the pressures generated, within microseconds.

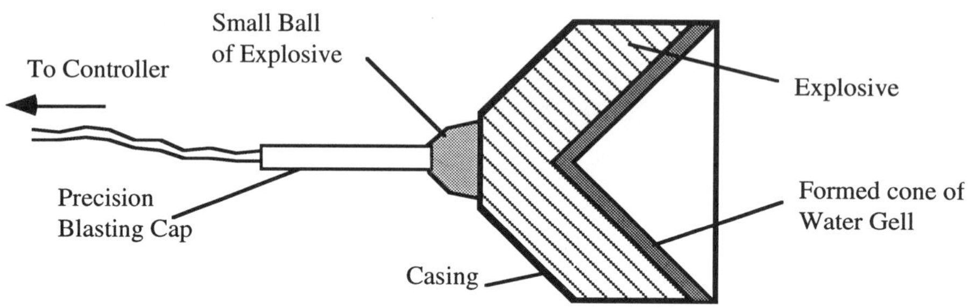

Figure 2.20 Components of a shaped charge waterjet driver.

2.4 POWER CALCULATIONS

The purchase of a high pressure waterjet system is normally decided based on the size of the pump needed to get the work required carried out in an acceptable amount of time. The pump size will be based on the power required to deliver the necessary flow volume at the required jet pressure. Hydraulic power is calculated for the pump alone and does not consider motor efficiency. At this stage it might also be pertinent to comment that when one purchases a pump, one will normally purchase it based on the reported, rather than the nominal volume of water delivered.

For example, a 120 kW pump can be anticipated, were it to use this full power, to deliver 102.9 lpm at 700 bar. However, when the pump is operated, and the flow of water out of the nozzle is measured, it is usually found that where the motor is running at 120 kW, that the delivered flow rate is somewhat less than 100 lpm. Typically the pump will operate to deliver somewhere between 80 and 90 lpm. The reason for this is likely to be two-fold. Firstly, there will be a number of small leaks, either deliberate or unintentional, in the system, (some pumps use this leakage as a lubricant) but more importantly there is an efficiency loss in transmitting power through the system. In the initial design and sizing of components, this system loss is often neglected.

Fluid power can be calculated based upon the following equation [2.9]:

$$\text{power (kW)} = \frac{1.666 \times \text{pressure (bar)} \times \text{flow rate (lpm)}}{1{,}000}$$

Note

In English units the equation is written as:

$$\text{Horsepower} = \frac{\text{Pressure (psi)} \times \text{Flow (gpm)}}{1{,}714}$$

2.5 PUMP SELECTION CRITERIA

Until 1976 the American Navy cleaned the water side of the boilers used to power its ships using a combination of chemical loosening (with a 10% solution of hydrochloric acid) and mechanical scrubbing. This effectively only removed about half of the fouling, and was not only time consuming, but also required frequent re-cleaning [2.10]. It might take, for example up to 150 hours to clean one boiler, with equipment costs of up to $2,000 each, and it would need to be done twice a year.

Following tests at the Naval Ship Engineering Center in Philadelphia, this procedure was changed to the use of high pressure waterjetting of the boilers. Not only did this reduce the time to clean an 80 bar boiler to 10 hours, it also improved cleaning efficiency. The cleaning, down to bare metal, was so effective that rusting occurred and it was necessary to add one kilogram of sodium nitrite to each 760 liters of the high pressure water in order to prevent this rusting. With this change, high pressure waterjetting was found to be a much more acceptable procedure than earlier methods for boiler cleaning. It reduced the set-up, operating, and take-down times; and it gave higher cleaning efficiency which reduced the required frequency of cleaning. As an example of cost, the 1972 cost for waterjet cleaning four destroyer boilers was given as $2,874, as opposed to

$8,000 mechanically. These results were sufficient to change the recommended procedures for cleaning boilers.

The U.S. Navy had then to decide which equipment to specify for purchase. Based upon preliminary testing the resulting procurement specified that a system be obtained that would provide a flow rate of 70 lpm at a jet pressure of 700 bar. A survey of the capabilities of the supply industry indicated that 23 companies could provide this equipment, and when they were contacted, six companies bid on the contract. Five companies carried out comparative testing, of nominally the same equipment, in cleaning sample tubes. The results indicate the reason for proper equipment sizing and selection.

Table 2.2 Relative cleaning effectiveness of nominally equivalent pumps [2.10]

Company	*Cleaning effectiveness*
A	92%
B	71%
C	70%
D	49%
E	36%

It is pertinent to comment that the equipment which the Navy purchased [2.10] performed at up to 98% efficiency, and when the equipment was retested in a later trial the four companies which supplied equipment were able to clean the tubes at between 82% and 98% efficiency.

During the course of the testing, the Naval investigators measured the pressure losses which occurred in the delivery line, while operating with a 700 bar pressure at the pump and delivering 43 lpm through the system. System losses were approximately 0.45 bar/m of 12.5 mm internal diameter hose, while, with a 7 m cleaning lance, 3.1 mm internal diameter, the pressure drop was 0.455 bar over the lance. This meant that the jet was issuing from the nozzle with a driving pressure of roughly 200 bar. Increasing the lance diameter to 4.0 mm increased the delivery jet pressure to 380 bar. In a subsequent test the lance internal diameter was increased to 6.25 mm, and this increased the jet pressure at the nozzle to 435 bar, at a flow rate of 70 lpm. The benefit of using larger feed lines to the nozzle assemblies was thus clearly demonstrated. The point will be discussed in section 2.8.2.

In section 2.7.4, reference will be made to work by Dr. Woodward [2.11], who has shown the considerable differences which can be found in nominally equivalent nozzles, all of which are sold to provide the same performance. It is much the same with the pump components of the system. The variation in performance extends, however, beyond just the actual performance of the pump but also affects other items which must be considered in estimating operating costs, such as the cost and lifetime of spare parts, maintenance and other running costs. For example, the pump which, at UMR, has consistently provided the greatest flow at its rated pressure and equivalent motor power, has also been the one which has had dramatically greater maintenance costs. Thus, in evaluating which components to install in an operation, it is important that the entire system performance be evaluated in terms of all the relevant standards of performance before deciding which system is most appropriate to obtain.

This was illustrated by Swan [2.12], who compared the operating costs of equipment from four suppliers, as compiled by his company over ten years. In addition to the purchase price of the equipment (generalized at roughly \$400 - \$625 per kW) costs for operational spares were also found to vary considerably. The following table is taken from that paper:

Table 2.3 Cost of operational spares per hour for pumps running at 420 bar [2.12]

Power	*Supplier or manufacturer*			
(kW)	A	B	C	D
30	\$2.00	5.00	6.00	4.50
50	\$2.50	6.00	6.50	5.00
110	\$3.00	9.00	11.00	6.00
165	\$5.00			

In selecting a pump, however, experience suggests that among the criteria used, the purchaser should determine the ease and cost of replacing the seals around the high pressure plungers. Pump designs exist which maintain the seal between the high pressure piston and the cylinder wall by lapping the two metal surfaces to a very close tolerance. This works well in the short term, but where the two interfaces are gradually worn, mainly by particles in the water, then it is necessary to return the complete piston and cylinder to the works for rebuilding. This may cost several thousand dollars and take several months. Recent units which use ceramic plungers with polycarbonate ring seals can be changed-out in a couple of hours. The operational maintenance costs of these pumps may, however, be higher.

Pump operations, and even the selection of which pump to purchase, may therefore also be mandated by the water quality which can be supplied to the unit. The costs of additional filters between the inlet supply and the pump may or may not be cost justified. Where the water is particularly dirty the rapid blockage of filters may suggest either that a settling system be installed, or that a pump be obtained which can cope with water of less purity. The increasing need to develop recycling systems for re-use of process water in order to comply with government regulations also will be a factor in assessing filtration system design.

The most vulnerable areas of the pump to abrasive wear are the valves and seats and the seals in the system. Wear may also be a significant item in the valves within the rest of the circuit, the bypass valve being particularly vulnerable if it is used to control flow, rather than just turn it on or off. Flitney [2.13] has traced the development of seals from the original soft packed glands to the use of elastomeric V ring packing. These have been changed to PTFE impregnated ramifiber packing, and reportedly, are successful in the range to 300 bar. Where more abrasion resistance is required, then the use of an aramid fiber based packing is potentially better. Clearance seals and diaphragm seals have also been used, particularly at higher pressures in order to resolve these problems.

Most pumps deliver water at a flow rate which is governed by the drive motor rpm. Where this is fixed, as with most electric motors, pressure control at the nozzle is adjusted by changing the flow volume feeding the nozzle. This is usually carried out by bypassing some of the water delivered by the pump, through a circuit which feeds back to the reservoir. This erodes the bypass valve and it may, therefore, be a wiser alternative choice, where possible, to control the volume of fluid flowing from the pump. If a diesel driven unit is used then, by placing a gear box between the pump and motor it is possible to change the pump rpm at constant motor speed, and this varies the flow rate to the nozzle without requiring a bypass line.

Choosing the right size of pump for an operation must include not only the output pressure and flow volume from the nozzle, but should add the pressure that will be required to overcome the line losses in delivering the water from the pump to the nozzle orifice. In addition, certain operational parameters must be considered in the system design. For example, in a cleaning operation, the operator of a hand-held lance can only withstand a certain amount of thrust (see below and Chapter 12). Thus, in designing a system the need for simultaneously using a number of guns must also be considered. If such a decision is made, then one may wish to include, within the circuit, flow control devices, so that the operation of one unit does not adversely affect the operators running other units fed with water from the same pump. This is recommended in some cases, because the change in volume fed to the different guns where one is shut off may otherwise cause a pressure surge in the line, which may cause an operator to lose balance under the unanticipated change in force from the gun.

The use of such devices will reduce the system operational efficiency. This is generally because the devices which control the pressure of the individual lances absorb a certain amount of energy in their operation. This arises because their general shape is a restricted orifice plate located within the feed line. The water passing through this orifice plate thus undergoes an acceleration as it passes through the orifice and a subsequent deceleration as it enters the larger area beyond it. With the additional induced turbulence that this creates, energy is lost to the system.

In deciding how many guns that a pump might supply one should consider both the pump delivered volume and pressure, and the reaction force which any one operator might experience in holding one of the cleaning guns. This reaction force can be computed from the equation:

$$\text{Backthrust (kg)} = 0.0227.\text{Q}.\sqrt{\text{P}}$$

Experiments have validated the accuracy of this equation, over the range of operating conditions likely to be found in most high pressure waterjetting applications [2.14]. At a flow of 20 lpm at 700 bar, the power consumed would be 23.3 kW. Using the above equation, as recommended by the Water Jet Technology Association [2.15] the reaction would be 12 kg. This is a relatively easily handleable force, although the Association has recommended that no one operator should be asked to handle more than 1/3 of their body weight in such a situation. Others have suggested that no operator should be asked to withstand a reaction force of greater than 23 kg [2.12].

This point is worth bearing in mind, since, as Swan has pointed out, operator practices and experience can make a considerable difference to the cost of an operation. For example while a new employee may at first only be able to clean a casting in 50 minutes, by the time he has cleaned some 60 castings he will have learned enough that the time to clean a casting may have dropped to 15 minutes. However the time which he will have taken at the beginning will also be extended because of greater fatigue due to the length of time the jet force has been resisted, and the inexperience of the operator's body in resisting this force.

Further, in comparing the costs of automated and manual cleaning, while it may have cost only $25 to clean a 1.5 m bundle of 100 tubes manually, as opposed to $78 automatically; the situation is reversed with larger jobs. Where the bundles are 8 m long, and contain 200 tubes, then manual cleaning will cost $2,240, while automated costs may be around $429. The cross-over in the example used appears to occur where the tube bundles are 3 m or more in length, or where there are more than 500 tubes in the bundle.

A typical cost estimate, calculated based upon the numbers Swan provided outlines the types of costs involved, although this experience is related to the European market, of the early 1980s.

Table 2.4 Cleaning costs for 100 castings using manual or automated equipment, 35 lpm at 700 bar, with a 60 kW pump [2.12]

Item		*Manual costs ($)*		*Automated costs ($)*
capital equipment		30,000		50,000
time taken (hours)		100		25
running costs		500		125
spares/ancillaries		25		62.50
labor	(2 workers)	1200	(1 worker)	150
amortization		300		125
total costs		2250		462.50
cost/part		22.50		4.62

Running a pump, even if properly maintained and operated does not come free of additional cost. Swan has provided some of these additional costs and other costs associated with jetting practice, as they related to operating equipment in the UK in the middle 1980's. Reference has been made earlier to the different costs which can be incurred for spares, as a function of supplier. Costs will also include fuel. Swan projected the following fuel usage, on the basis of field observation.

Table 2.5 Fuel usage as a function of power [2.12]

Power (kW)	*Fuel use in liters/hr*
30	7.1
55	11.83
110	23.66
165	47.32

It is not possible to give numbers which can be considered consistent for different pumps or conditions, since even the quality of the water used will have an effect on the lifetime of the parts of the system. Swan has, for example, graphed additional running costs incurred as a function of the pressure at which the system is operated, and the amount of fine material found in the water before it is pumped (Fig. 2.21).

In the latter case, then one should be able to establish, depending upon how dirty the water is, the economic benefit, of inserting cleaning filters between the water supply and the pump.

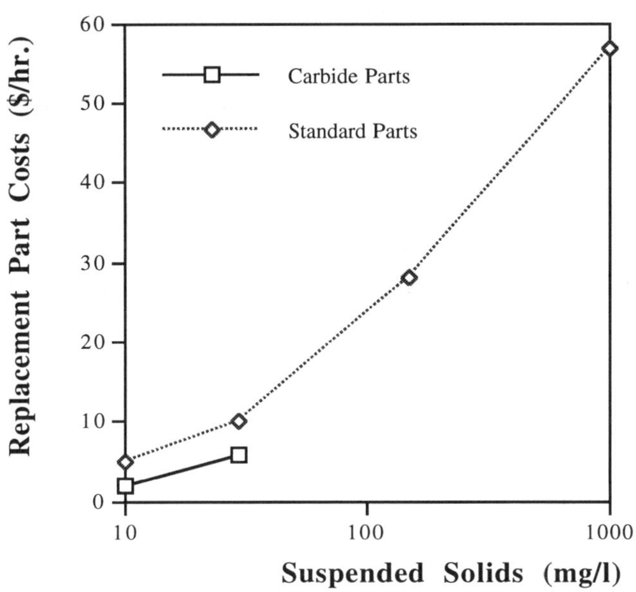

Figure 2.21 Effect of water cleanliness on maintenance costs [2.12].

2.6 TRANSMITTING THE ENERGY

The study by which the U.S. Navy established the best system for cleaning ships' boilers, pointed out that a large portion of the jet pressure could be lost just in delivering the water from the pump down the supply line to the nozzle. Specifically, with a flow of 48 lpm and a pump pressure of 700 bar, losses were approximately 0.45 bar/m of the 12.5 mm internal diameter hose, and, an additional 0.455 bar were lost in pumping the water down the 7 m cleaning lance which had an internal diameter of 3.1 mm This left roughly 200 bar pressure to drive the water out of the nozzle. This meant that significantly less than 1/3 of the input energy was being delivered to the target surface. Other data in the technical literature substantiate this type of loss.

In the case of the Navy study increasing the diameter of the lance to 4.5 mm reduced the losses to 320 bar, and with a lance diameter of 6.25 mm, the jet pressure was 435 bar, at a flow rate of 80 lpm. Even in this condition, which the Navy found satisfactory, over 1/3 of the input energy to the system was being lost to friction. The optimum sizing of components, and the choice of their shape, has a significant influence on system effectiveness. In some operations at UMR, for example, it has proved cost and system effective, to run two high pressure lines from the pump to the unit, in order to reduce such line losses. This choice was, in part, motivated by the additional cost and reduced mobility of using one single hose of large enough diameter to transmit the water with very low levels of pressure loss.

To establish the factors which influence the performance of the system, its component parts must be identified. Because they are the most common systems, the circuit which will be discussed is that of an operator who is using a high pressure waterjet system for a cleaning operation. Although the performance of the unit is ultimately controlled by the size of the pumping unit which is available, that choice is initially dictated by the pressure and volumes required at the working surface in order to achieve the desired work. Thus, an effective system design must begin at the nozzle.

2.7 NOZZLE CONSIDERATIONS

2.7.1 NOZZLE SIZE

Most of the pumps which are used in high pressure waterjetting applications operate on a positive displacement basis. This means that a fixed volume of water leaves the pump and enters the delivery line to the nozzle each minute. When all other passages are closed then this entire volume must pass through the orifice in the nozzle body. The jet acquires its final velocity as a result of having to pass through this very small hole in the nozzle. The pressure from the pump is that which is required to drive the total volume of water through this hole. Thus, with a pump which puts out a steady flow of water, the pressure of the jet which comes out of the end of the lance is controlled by the size of the hole (or orifice) in the end of the nozzle.

Where the flow is 20 lpm, and the required pressure is 350 bar, then we can calculate the size of the required orifice as follows:

One liter of water occupies 1,000 cc. Thus 20 liters will occupy 20,000 cc. If this must flow through the orifice in one minute, then one-sixtieth of this, or 333 cc, must pass each second. In an ideal world (reality comes later in the text) the velocity which a given pressure will generate is given by the relationship

$$V = 17.14 \bullet \sqrt{P}$$

Where V is the jet velocity in m/sec

Thus a jet pressure of 350 bar will give a jet with a velocity:

$$V = 17.14 \bullet \sqrt{35} = 320 \text{ m / sec}$$

The jet must thus move through the orifice we choose at 320 m/sec or 32000 cm/sec. Since we must move 333 cc each second this means that the area of the orifice must be 333/32000 = 0.01 sq cm.

Normally, however, we do not give the size of the nozzle orifices which we use in terms of their areas, but instead use the equivalent diameter. In this case that would be 1.12 mm. Thus in order to get a pressure of 350 bar from a pump which is putting out 20 lpm of water we would need to drive that water through a hole some 1.12 mm in diameter. In passing, and to illustrate the importance of carrying the calculation to the nearest 0.01 mm, consider a second calculation.

In rounding decimals we consider that from 1.05 to 1.15 mm diameter, we often round to 1.1 mm and may use this value to specify a diameter. At 1.05 mm diameter, the 20 lpm of water would need to move through the 0.866 square mm hole at a rate of 385 m/sec, while at 1.15 mm diameter, the same flow would move through a 1.03 square mm hole at a rate of 321 m/sec. When this is converted into a pressure equivalent the range is from 500 bar at the smaller size, to 350 bar at the larger. Thus very small changes in nozzle size can have a significant effect on the jet pressure, and, as we shall learn later, on jet performance. From this we learn, among other things, that making sure that the nozzles are in good condition is an important part of running a waterjetting operation.

The small size of the nozzles used, and the relatively great changes that flow velocity has on pressure can also be seen by recalculating the flow velocities as the volume is increased.

Table 2.6 Relative jet performance for different nozzle sizes

Nozzle dia. (mm)	*Nozzle area (sq cm)*	*Volume flow (lpm)*	*Fluid velocity (m/sec)*	*Jet pressure (bar)*
1	0.00785	20	424.47	612.7
1.05	0.00865	20	385.00	504.0
1.1	0.00950	20	350.8	418.5
1.15	0.01038	20	320.96	350.0
1	0.00785	21	445.69	675.5
1	0.00785	22	466.91	741.3
1	0.00785	23	488.14	810.3

It can be seen from the above table that increasing the flow from 20 to 23 lpm, or 15%, will increase the pump pressure required to deliver that volume through the same size orifice by approximately 30%.

As it happens the above calculation makes two assumptions which are not correct. The first is that the waterjet will issue from the nozzle at the same size as the orifice. In fact the water will converge to a smaller size than the orifice, either inside or just outside the nozzle body. This produces the smallest area, and thus the fastest velocity for the water. To calculate this change we need to multiply the area of the nozzle by the coefficient of discharge. Since this can vary from about 0.6 for a poor nozzle design to about 0.95 for a very good nozzle design this can have a significant effect on the results which are obtained with the nozzle.

Labus has documented, in a more comprehensive coverage of this topic [2.9], the calculations and equations which inter-relate the flow rate through a system and the pressure and pipe and jet diameters. Of particular interest are the examples of the coefficients of discharge which he provides for different generic nozzle shapes (Fig. 2.22). Values are given which correlate the volume which will pass as a portion of the flow where the jet would fill the entire orifice diameter (the Cd value), because the diameter is less the quantity which will flow will also be reduced.

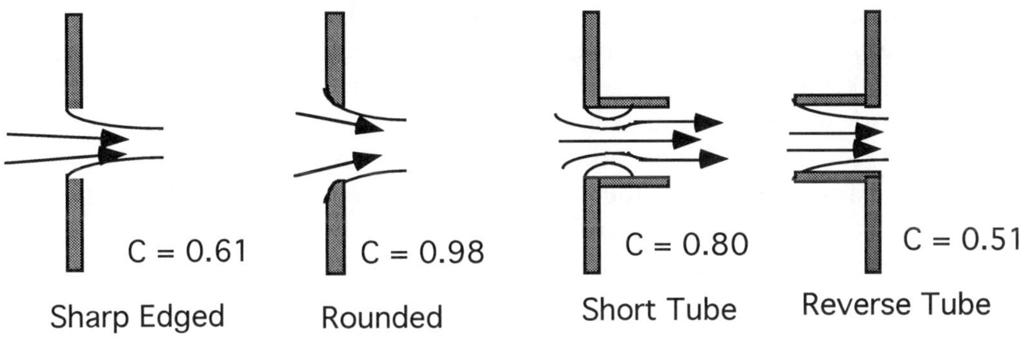

Figure 2.22 Typical orifice shapes and the nominal discharge coefficients [2.9].

In the earlier calculation, a 20 lpm flow through a 1.00 mm diameter nozzle required a jet pressure of 612.7 bar, with a coefficient of discharge of 1.0. If the nozzle was of good quality, then, with a coefficient of discharge of 0.95, the same flow would require a driving pressure of 680 bar. Alternatively the nozzle size would need to be increased to 1.026 mm in order to maintain the original design pressure. In contrast, however, with the very poor nozzle which has a coefficient of discharge of 0.6, the pressure required to drive this flow through the original 1.00 mm diameter nozzle would be 1,700 bar, or alternately the nozzle size would need to be increased to 1.291 mm to maintain the flow at the original pressure. The critical effect of the nozzle on system performance thus becomes evident, and it is pertinent to look into the changes which these parameters may bring to the nozzle performance in more detail.

2.7.2 NOZZLE PERFORMANCE

There are two parameters which are normally considered in the selection of a waterjetting system for a given operation., The first is the pressure which the system will produce, and the second is the volume flow rate which will be delivered. As discussed earlier these can then be combined together to predict the size of unit required by using the equation:

$$\text{power (kW)} = \frac{1.666 \times \text{pressure (P)} \times \text{flow rate (Q)}}{1{,}000}$$

Unfortunately life is not, however, that simple in reality and a number of other factors must also be included in the analysis. The word "unfortunately" is deliberately used since there are many occasions where the following considerations are not accounted for and the system may not appear to work at the level anticipated.

In order to most effectively size a system to get the performance desired, one should begin at the target surface, and initially determine the jet pressure required, **at that point**, to achieve the desired result. Once this minimum effective pressure has been established then, depending on circumstance, one can establish the flow rate needed. Two conflicting factors may arise at that point, the problems of using too much water, against the benefits which come, in terms of increased performance, at the higher flow rates.

It is interesting to note that, while the nozzle performance is the one aspect of the system which most frequently controls whether or not the waterjetting operation is a success, it is one which is quite commonly neglected. It should be borne in mind that the entire performance of a unit costing in the tens of thousands of dollars is entirely controlled by the $20 - $100 insert which accelerates the flow out of the end of the lance. And while detailed discussions may occur over which pump to buy, one often finds that nozzle selection is made strictly on the basis of which is the cheapest. Nozzle wear is also not given sufficient consideration and system performance can be considerably constrained where nozzles are not replaced regularly. This is not to say that the most expensive nozzles are necessarily the best, often manufacturing quality and materials, as well as design will vary between suppliers. The initial consideration, however, should be with the design of the nozzle.

2.7.3 NOZZLE DESIGN

The first major study of waterjet nozzle design, which may be considered important to this topic, was carried out by Rouse and his colleagues in 1951 [2.16]. This work, which was published in the ASCE journal of the time, looked at the best design of a nozzle to improve the performance of fire fighting monitors. This was followed, in Western Europe, by a study carried out in England by Leach and Walker in 1965 [2.17]. This latter study was carried out with the jets operated at higher pressures, particularly for use in cutting rock, and examined several nozzle shapes.

The most effective shape found in the latter study (Fig. 2.23) was that originally suggested by Nikonov and Shavlovskii [2.18], which was developed in Russia for use with hydraulic monitors. It is interesting that this study compared that design with the classical design which had been proposed by Rouse (Fig. 2.24).

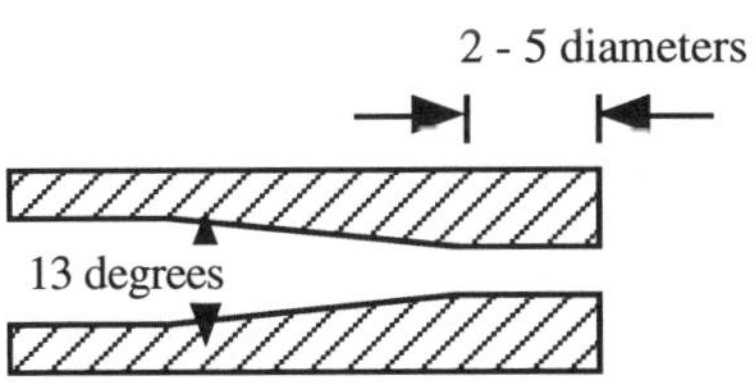

Figure 2.23 Nozzle design by Nikonov and Shavlovskii [2.18].

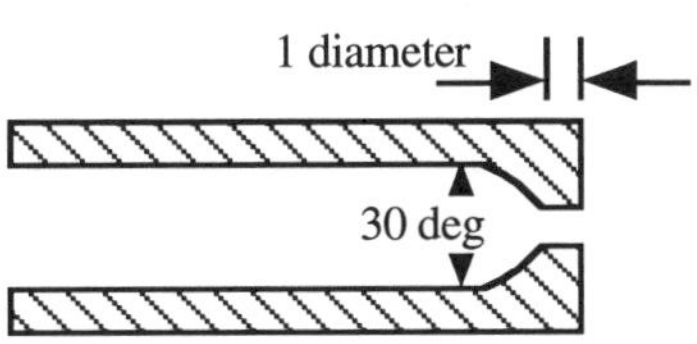

Figure 2.24 Nozzle design found to be best by Rouse [2.16].

The Rouse study had been considered the definitive one for nozzle design at lower pressures, but it empirically examined the waterjet nozzle shapes for fire fighting applications, and the design which it recommended has been found to be more sensitive than others to dirt in the water. The comparative evaluation which Leach and Walker carried out looked at the most effective pressure delivered by the jets as a function of the distance from the nozzle (Fig. 2.25).

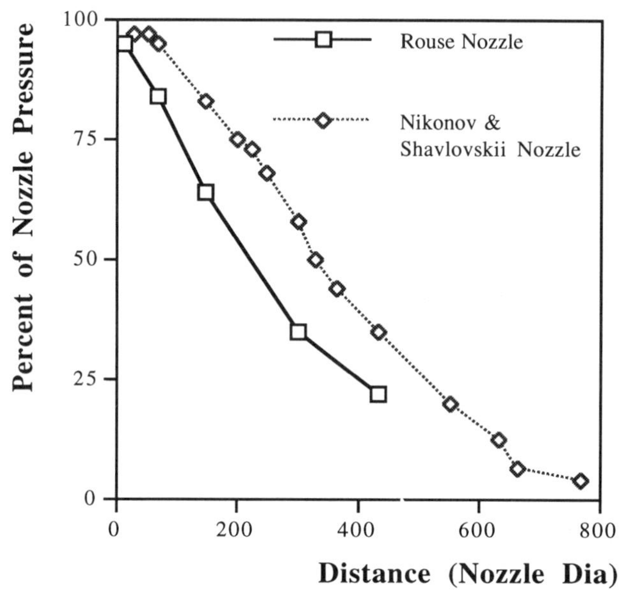

Figure 2.25 Comparison of the delivered pressure from two nozzle designs, as a function of stand-off distance from the nozzle [2.17].

The simplicity of the Nikonov design (Fig. 2.23), or as it has since been become known the Leach and Walker design (because of the study), lies in the two parts of the nozzle. The first of these is the fluid acceleration section with the channel linearly narrowed at a 13° included angle from the pipe diameter to the orifice size. The throat, or straight section, initially 2.5 times the nozzle diameter has recently been extended to 3.5 to 4 times diameter to allow for wear. Subsequent work in Russia indicated that the conic angle should be decreased as the jet velocity increases, and conversely can be increased to 18° to 20° at lower jet pressures. It is also found that a 2 to 4 diameter straight section can be used for optimum jet performance.

This initial work has been followed by other studies, by a number of investigators over the years. It is a common aspect of many new investigators in the field to begin by trying to "improve" the nozzle shape. Most particularly, investigators have chosen to look at various different geometric shapes for the interior of the nozzle. Thus, we have seen parabolic, exponential, and hyperbolic functions used to describe the internal structure of the jet. Many of these were developed for use at pressures up to 5,000 bar, but the majority of the nozzles which are in use are sold for conventional waterjet cutting and cleaning applications. The general conclusion, in terms of cost, ease of manufacture and performance is that the Leach and Walker recommended design is difficult to beat. At much higher pressures, it has been shown that a more complex nozzle shape can be more effective. However, such a design is much more expensive to manufacture. In 1980 one such nozzle design (Fig. 2.26) was manufactured for use at pressures of up to 7,000 bar, with equivalently high fluid flow rates. It cost $6,000 at that time.

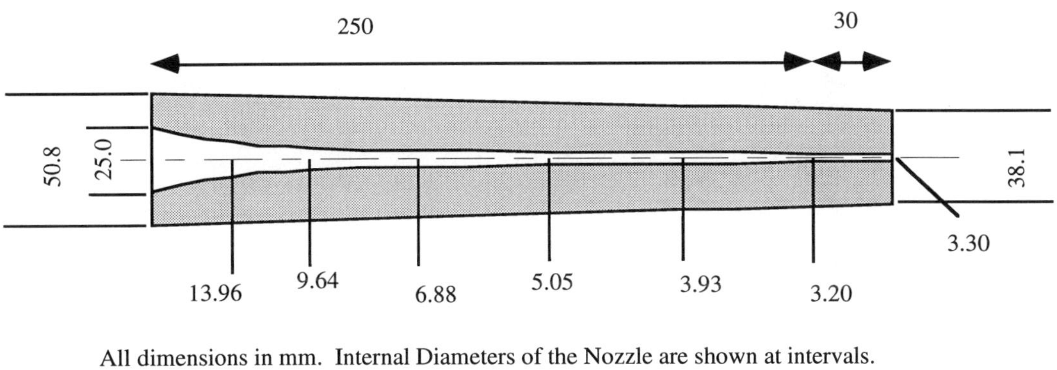

Figure 2.26 Nozzle design for use at very high pressure.

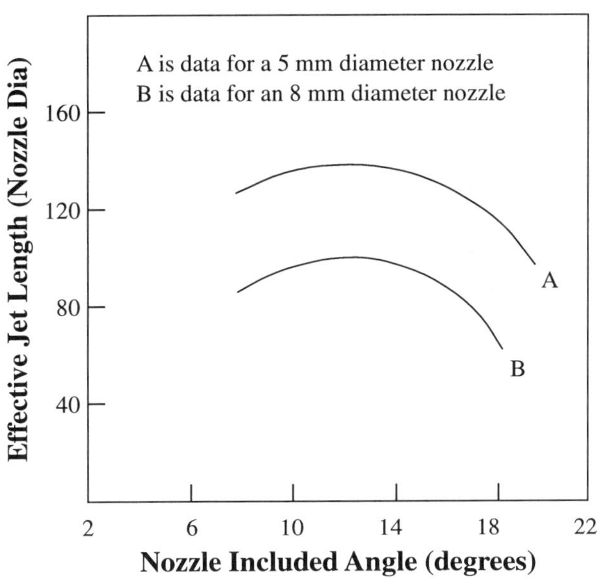

Figure 2.27 The effect of change in included angle on effective jet length [2.19].

There is often an assumption that all nozzles of the same shape and nominal outlet orifice size will produce jets of the same performance. The error in this assumption will be illustrated in the next section. A closer equivalence in performance can be achieved if certain precautions are taken in the manufacture and use of these nozzles. Early work on nozzle design was largely carried out in the then Soviet Union.

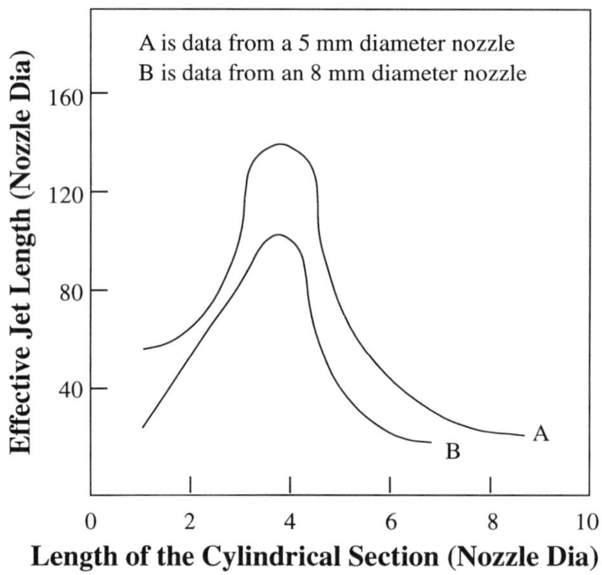

Figure 2.28 The effect of change in throat length on effective jet length [2.19].

The experiments which were reported [2.19] discuss the optimum geometry of the nozzle including both the angle of convergence of the cone (Fig. 2.27) and the length of the cylindrical throat section (Fig. 2.28). It is interesting, in light of other discussions throughout this work, that the maximum reported effective jet throw during those studies was less than 150 nozzle diameters.

One additional factor should be considered in the construction of the nozzles. This point is usually ignored in much of the work which has continued in nozzle design since the definitive Leach and Walker study. These investigators also examined the effect of shaping the corners of the nozzle as they were constructed. Although the change in performance is not great near the orifice, a significant effect can be seen (Fig. 2.29) at greater stand-off distances.

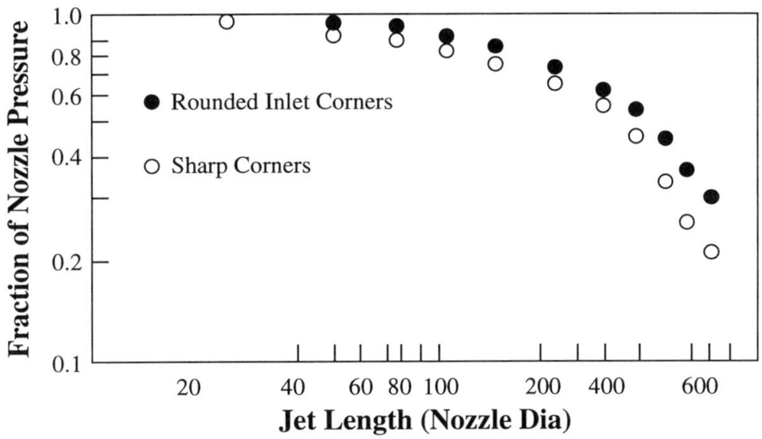

Figure 2.29 The effect of corner shape on effective jet length [2.19].

a)

b)

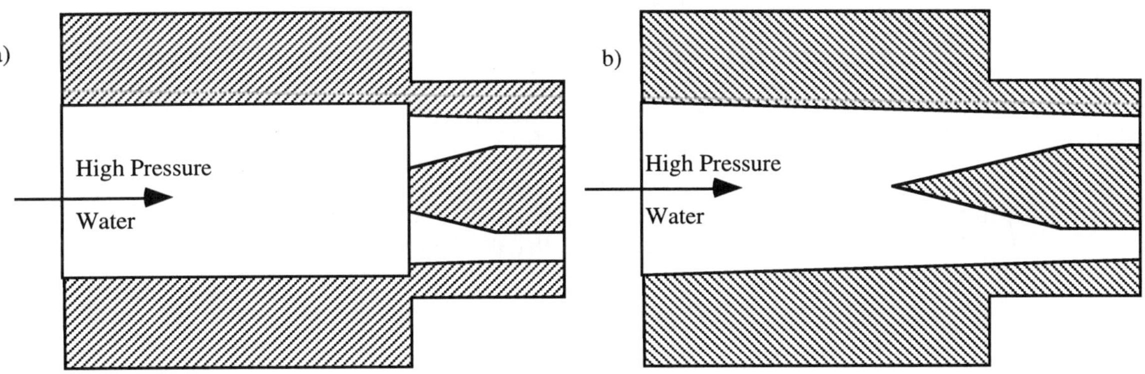

Figure 2.30 Comparison of a) "bad", and b) "good" nozzle designs for jet cutting with two jets [2.20].

The experience at UMR in this particular area is through the work of Barker and Selberg [2.20]. Their study looked at the improvements which could be achieved in jet performance, by simple fluid flow considerations. The two most important factors identified were the internal surface finish of the nozzle and the straightness of the flow into the orifice. This is particularly true where a multiple orifice jet is used and in these circumstances the authors showed that it is beneficial if the flow to the separate nozzles is divided before the fluid is accelerated (Fig. 2.30(a) and 2.30(b)). By such simple considerations, the effective throw of the jets can be considerably improved.

The majority of the work carried out at UMR, and elsewhere, indicates that the normal effective throw of a waterjet from a standard nozzle can be considered to be on the order of 200 jet diameters from the nozzle orifice, although for many nozzles considered "adequate" by their users, the effective cutting distance (at which pressure falls below half the nozzle pressure) is considerably less than this (Fig. 2.31). Where simple precautions are taken this distance is, however, attainable (Fig. 2.32).

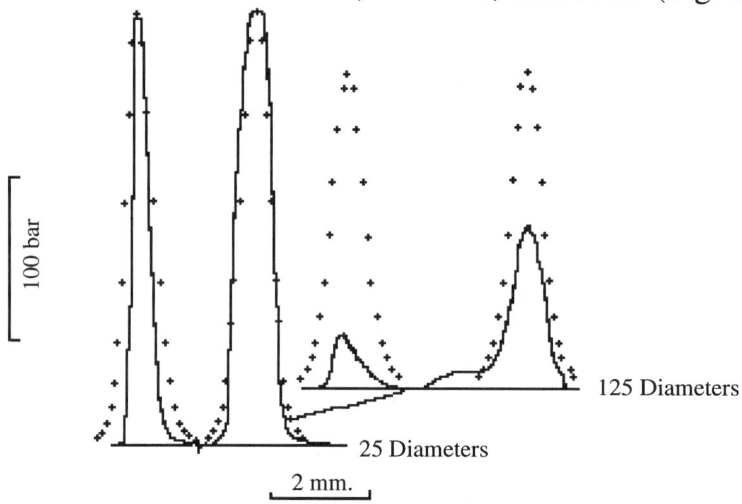

Figure 2.31 Pressure profiles with distance from a normal nozzle, jet diameters are 1.016 mm, jet pressure 280 bar, stand-off distance to the profiles is given in nozzle diameters [2.20].

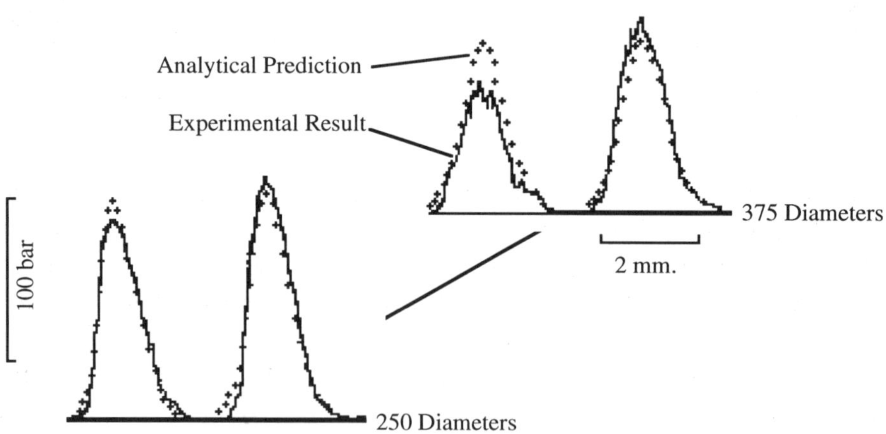

Figure 2.32 Pressure profiles with distance from a "good" nozzle, jet diameters are 1.016 mm, jet pressure 280 bar, stand-off distance to the profiles is given in nozzle diameters [2.20].

By removing any steps in the flow into the nozzle; by providing a straight section into the nozzle on the order of 100 pipe diameters or better; and by smoothing and aligning the passage to the nozzle, Barker and Selberg were able to extend the effective range of the jet out to about 2,000 diameters (Fig. 2.33).

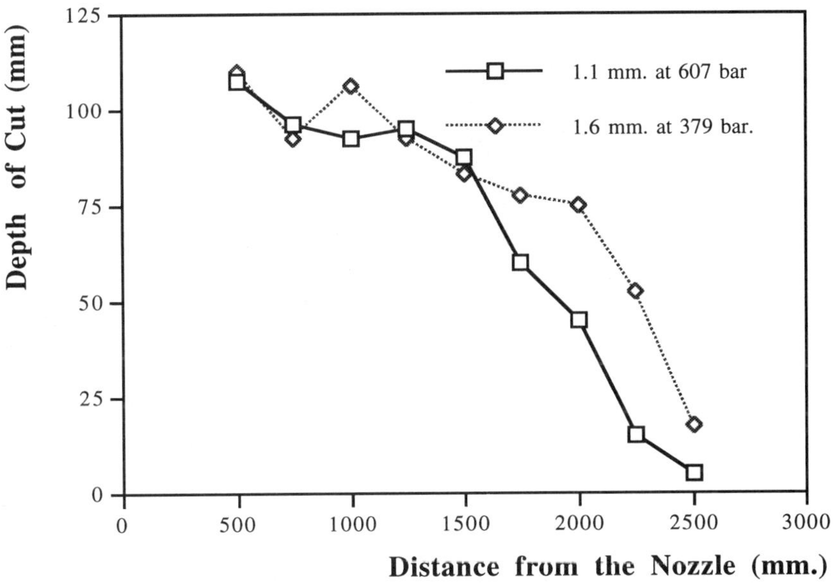

Figure 2.33 Volume of cut as a function of stand-off distance for a "very good" nozzle design [2.20].

In order to do this it should be mentioned that the cost of the nozzles was substantially increased. Prototype nozzles for this particular work cost on the order of $500 a piece in 1976 (bulk it is possible that their price would come down to somewhere around $200). The nozzle alignment was assured by inserting alignment pins between the nozzle and the supply line, ensuring that the flow line into the nozzle was not disturbed (Fig. 2.34).

In evaluating the methods used to manufacture improved nozzle bodies, a number of different possibilities have been examined. These have ranged from the use of ceramic nozzles through nickel plating and EDM (electron discharge machining) to electro-forming a nickel body around a removable mandrill. Of the different methods of manufacture this latter gives the best surface finish for the cost. The surface finish appeared to be considerably improved over that of the EDM process. Since Leach and Walker calculated that a 6 μinch internal surface finish was required for best performance, this was a deciding factor in the choice made at UMR.

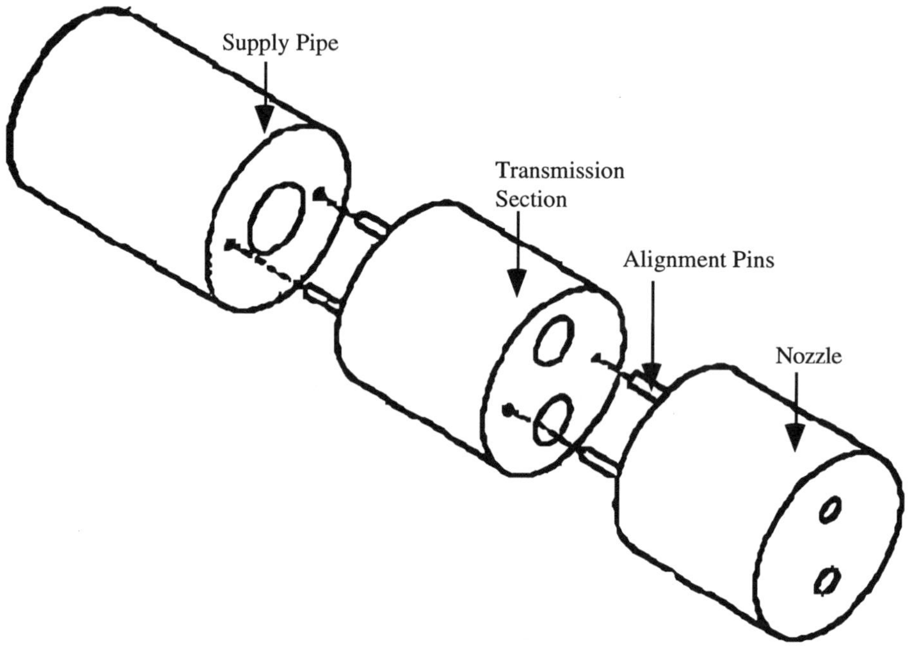

Figure 2.34 Method of ensuring nozzle alignment [2.20].

Ceramic nozzles appear to have a potential for future application and various carbide nozzles are now in use in the market. These nozzles are, however, more for use with abrasive laden streams where nozzle finish is not as critical, and operational wear is more important. The experience at UMR to date with ceramics, however, has been that these nozzles must be constrained in compression at all times due to their lower strength in tension. If this is not done then they fail relatively rapidly. Once this problem is addressed, these nozzles may hold some considerable promise.

The critical effect which nozzle conditions and size have on performance suggest that, with quite a varying quality control in the nozzles supplied, and with a possible rapid erosion in shape for nozzles that have been in operation for even a short time, that operators be aware of simple techniques which can be used to tell whether the nozzle is and continues to be effective.

2.7.4 NOZZLE EVALUATION TECHNIQUES

There are a number of different ways in which one can evaluate the best nozzle for an application, or by which one can compare the performance of nozzles. These range considerably in terms of cost, effectiveness, and the speed with which they can be carried out. While in many cases, the best evaluation is to put the different nozzles, in turn, on the end of a lance and try them, this is not always possible or accurate enough to give a true comparison.

There are three ways in which evaluations are normally reported. The more classical of the methods is the one most often used in scientific laboratories [2.17]. For this the nozzle is held in a clamp, with the jet directed normally horizontally. A small instrument which measures pressure is then slowly moved across the jet stream, usually at several stand-off distances (Fig. 2.35). The face of this transducer is generally protected with a cap of a harder metal which has a small hole drilled through to the transducer, so that only a small section of the jet stream is sampled at one time. From experience this hole should be on the order of 0.5 mm in diameter. With smaller holes it becomes difficult to make the cap, while larger holes do not give the accuracy required, given that the jet may be only 1 - 2 mm in diameter. Care is required to ensure that the transducer is traversed through the center of the jet at each location, at UMR this was usually checked by moving the instrument into the jet until a maximum reading was indicated, and then moving the instrument up and down vertically, to ensure that the profile obtained (Fig. 2.31) was through the jet axis. This was important since the readings were taken with the jet moving horizontally, and at greater distances the jet would drop a little.

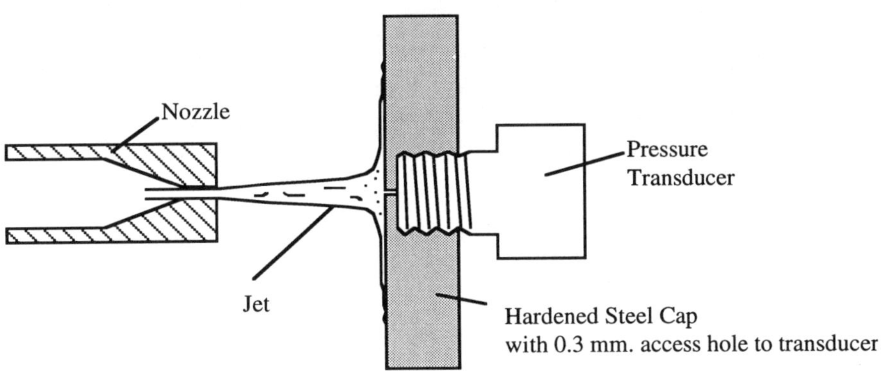

Figure 2.35 Profiling jet pressure across the jet stream [2.17].

This method is slow and requires both expensive equipment and considerable care, to be effective. It gives a very accurate understanding of the jet structure as the jet moves from the nozzle. A less qualitative method used, perhaps less frequently than it should, is to take a picture of the jet. Photographs must however also be taken carefully if they are to be effective. Normally a flash photograph is required, and depending on the speed of the jet, special short duration flash equipment may be needed. After all a jet moving at 200 m/sec will move 0.5 mm every millionth of a second. Thus the flash duration must be on the order of that time in order not to get a blurred picture. In taking the picture remember that the part of the jet of interest is the central core. This is often hidden from front view by the zone of small spray droplets (Fig. 2.36), so it is better to place the flash behind the jet (Fig. 2.37). To distribute the light along the stream, a sheet of ground glass works very well (Fig. 2.38).

For many cases, the simplest and easiest way of finding out if the jet works well is to use it to cut something. This method has the advantage that it can be set up quite easily and also used to decide when a nozzle is sufficiently worn that it needs to be replaced. In the evaluation of spray nozzle designs (which will be discussed later) the jet has relatively little power, and thus a soft target material should be used. In one such study the jet was directed to cut through Styrofoam blocks, placed at different distances from the nozzle. This allowed an estimate both of the jet force at that point, and the shape of the jet. A similar approach can be taken with other designs and pressures, tailoring the material being cut to the pressure and jet size, in order to still get meaningful results.

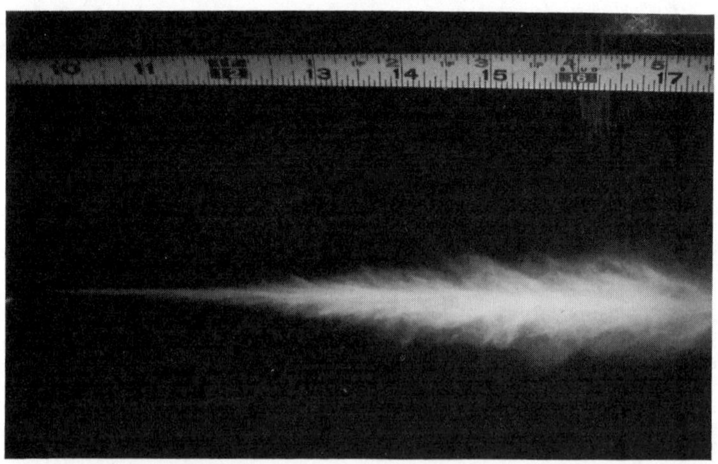

Figure 2.36 A waterjet at 2,100 bar issuing from a 0.25 mm diameter nozzle - the scale is in inches. The jet is lit from the front.

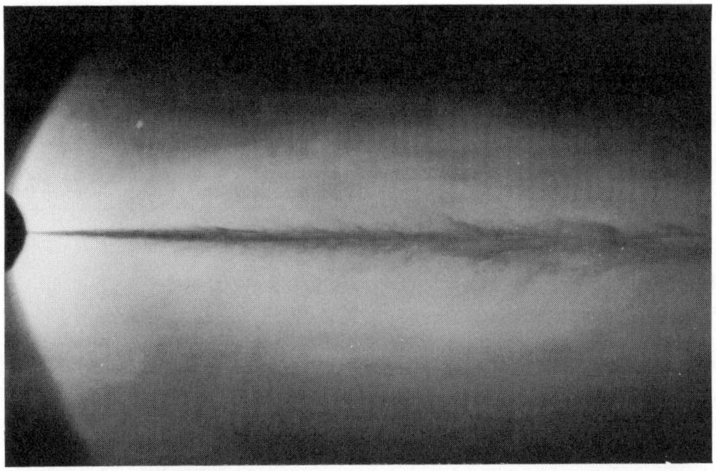

Figure 2.37 A waterjet at 2,100 bar issuing from a 0.25 mm diameter nozzle - the scale is the same as in Fig. 2.30. The jet is back lit.

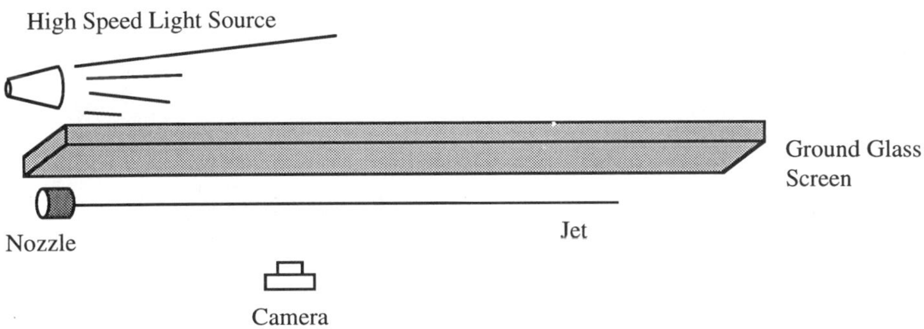

Figure 2.38 Plan view showing equipment layout for high speed jet photography.

In 1985 Woodward [2.11] published a review made by the Weatherford company of the different nozzle shapes then commercially available for cleaning applications. Some 50 different straight nozzles could be found, at that time, rated for operation at a nominal 38 lpm flow at 700 bar.

Eight of the nozzles were tested for cutting ability over a range of distance from the orifice. The study also included life expectancy (which ranged from 40 - 600 hours) actual flow rate achieved (which ranged from 34.87 - 44.66 lpm) at pressure, and included other significant factors which an operator should consider. These included cost, weight and nozzle size. The internal shapes of the nozzles (Fig. 2.39) and the variations in the cuts the different jets made (Fig. 2.40), can be compared with the data below.

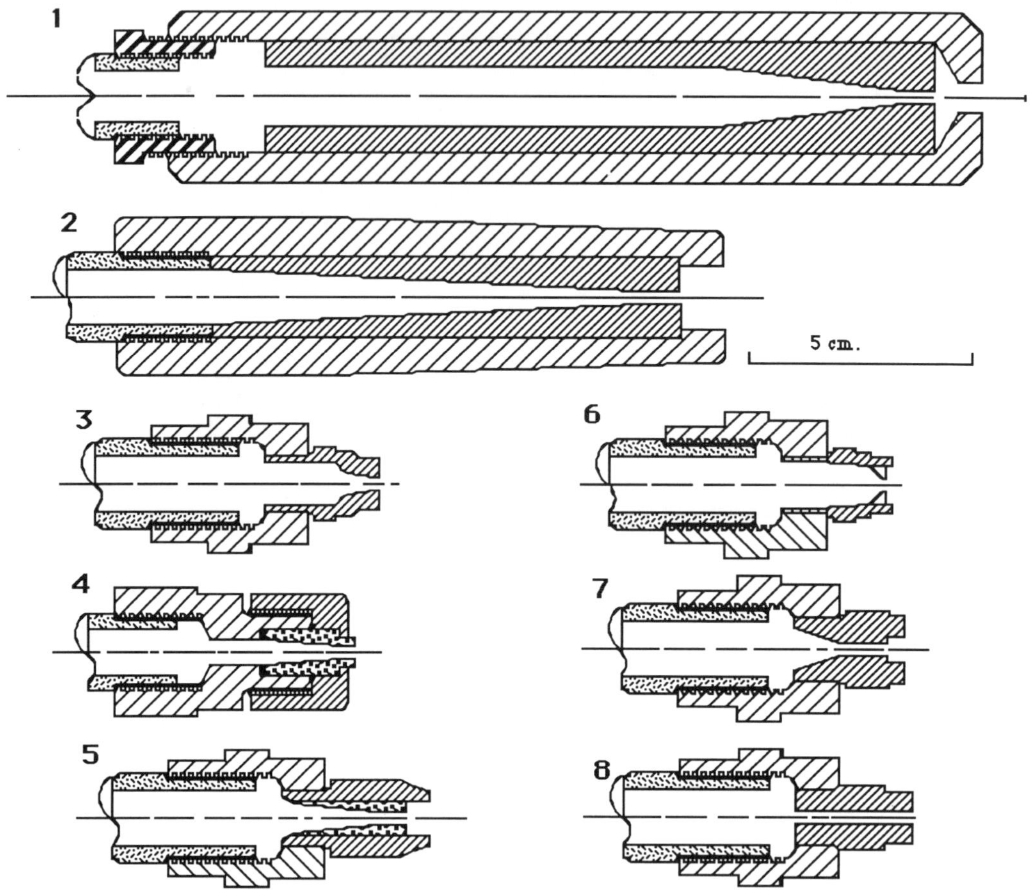

Figure 2.39 Sectioned view of the nozzle assemblies tested by Woodward [2.11].

To compare the performance of the jets coming from each of the nozzles, a series of target plates was set up, at 30 cm intervals away from each orifice. A jet at 700 bar was directed through the nozzle at each target for a period of 10 seconds. The size of the hole drilled in each plate was then measured using a steel rule, and this provided a measure of the path width and the effective jet distance, as shown (Fig. 2.40). The wide variations in result has led Woodward to recommend that the particular application be considered before deciding which of the jets would be most effective in a given situation. If the performance requirement was to clean, for example all the way through the shell side of a set of heat exchanger tubes, then nozzle #1 appears best, but for the general contractor, where the required jet throw is not as great, and the lance is hand-held, nozzle #5 would be a better choice.

Table 2.7 Comparative nozzle performance data [2.11]

Nozzle #	*Price (3/7/85) ($)*	*Weight (gm)*	*Life (hr.)*	*Power at 700 bar.(kW)*	*Material removed (cc/kW)*
1	250	1700	400	52	0.125
2	195	910	400	43	0.044
3	10	14.2	40	43	0.050
4	55	170	400	41	0.029
5	33	28.4	400	52	0.129
6	28	14.2	300	40	0.092
7	58	14.2	300	43	0.123
8	15	28.4	60	46	0.115

Two variations on this may be used by a general operator of the equipment to make sure that the nozzles are producing jets of the levels required. The first is to set up some standard measure of performance. For example, using a hand-held lance, the operator will often work with the lance held within 75 cm of the working surface. Thus, if a piece of plywood is set on a stand at a distance of 1 m from the nozzle, then the operator could check, before using the equipment, that the stream is at sufficient power at the target to be effective by seeing how long the stream would take to punch a hole through the wood.

Another approach is to set up a piece of wood, and plywood has some benefits for this test, so that it sits at a shallow angle to the jet axis (Fig. 2.41). The jet is powered up, and then moved laterally into and through the wood, but cutting along its plane. In this way, with the jet starting to cut at the far end of its range, a measure of the length over which the jet retains its power can be made. It has been interesting to find that, in a number of cases, this range is less than the 200 nozzle diameters usually used as the jet "effective range".

The choice of cutting conditions will often include both a selection of the optimum pressure for an operation as well as the size of the nozzle which will be most effective. The natural reaction is often to increase the jet pressure to the highest level which is can be obtained, in the hope that this will cut better. In fact this does not always occur. The reason is that when the jet pressure is increased, as discussed above, the jet velocity also goes up. In order to increase jet velocity, with the fixed volume of water delivered by the pump each minute, the area of the jet must be reduced. This has three potentially detrimental effects, which must be considered.

The first change that must be made is in the operation of the pump. If the pressure is increased for the same volume flow, then the horsepower demand on the pump will rise. If this exceeds the rated performance of the pump it will shorten its life and may create other problems as have been discussed in section 2.3.1. Alternately the volume through the pump may be reduced to maintain horsepower. If this occurs then an even greater reduction in the nozzle diameter is required, in order to maintain the pressure.

The smaller jet diameter at the higher pressure results in two disadvantages. The first is that the jet pressure falls away more rapidly as the jet moves away from the nozzle. This is because the jet is moving through virtually stationary air, and the shear at the interface tears off layers of water as fine droplets. The smaller the diameter of the jet, then the less material there is to wear away. At the same time the faster the jet is moving, relative to the air then the more rapidly the process occurs. The combination of these factors is such that, while a 1 mm diameter jet at a pressure of 700 bar may cut as far as 1 m from the nozzle, a jet at a pressure of 3,500 bar, flowing through a 0.125 mm orifice may be coherent for no more than 10 cm.

The second disadvantage relates to the way in which plain waterjets cut material. While this will be discussed in several of the later Chapters, findings at UMR have shown, over a range of materials, that for the same power increment once the critical pressure for cutting has been reached, cutting performance is improved more by increasing the jet diameter of the cutting jet than is achieved by a rise in pressure (see, for example, Figs. 6.22 and 6.23). In many industrial applications, however, higher volume flows may have significant disadvantages, and the very narrow cut of the higher pressures with low flow rates will, in those cases, argue for that combination, rather than the converse. It must always be borne in mind that the requirements of the specific application will always modify the general rule.

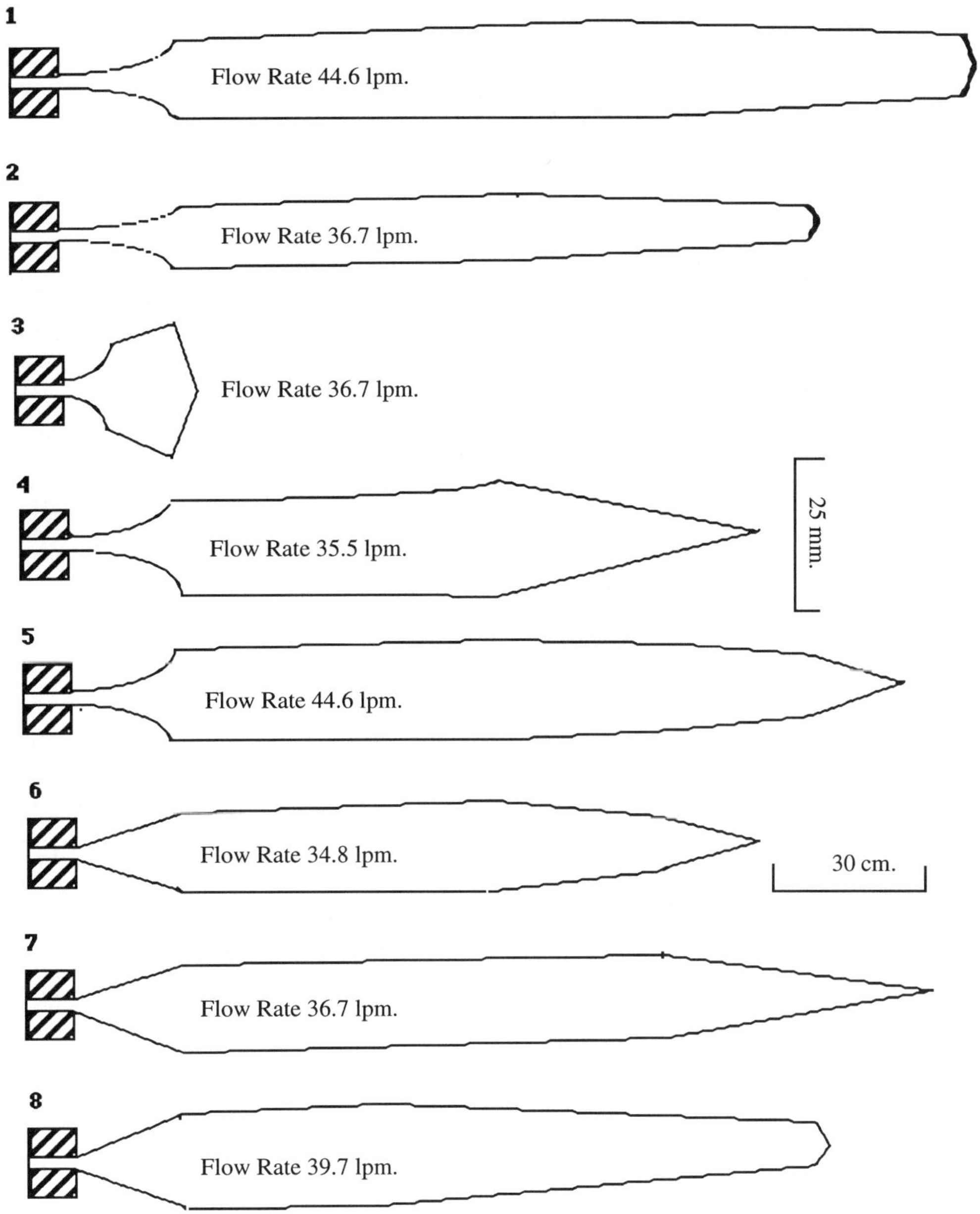

Figure 2.40 Cutting profiles achieved by the nozzles in Fig. 2.32, the measured flow rates are also given [2.11].

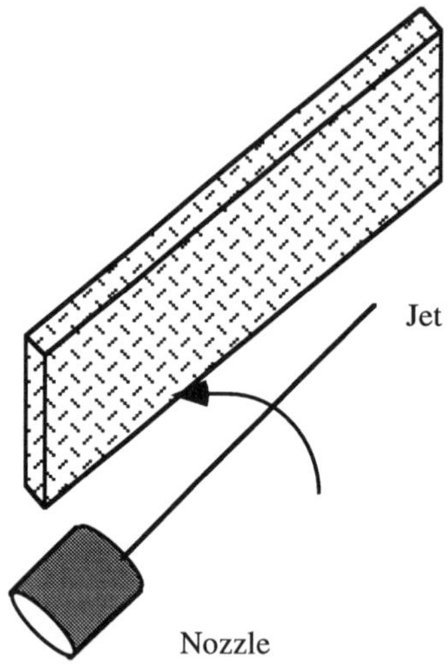

Figure 2.41 Schematic layout of equipment to find a jet effective range.

2.7.5 HIGH PRESSURE CUTTING NOZZLES

The majority of the research which has been undertaken in the design of nozzles has related to the lower pressure and higher volume flow rate applications. There is a significant difference between the data from these tests, and those where the nozzles have been designed for use at the higher pressures for cutting materials such as automobile carpets, plastic sheets, cardboard and paper (these applications and others are addressed in Chapter 4). In these applications the jet diameters must be smaller, to produce the low flow rates, at the higher driving pressures needed. These small sizes have historically made the normal methods of nozzle manufacture impractical, and alternate methods must be employed. Early in the development of the technology it was found that artificial sapphires, such as those used for watch mountings, were available in the right size and at a relatively low cost. While initial mounting techniques were quite crude, (Fig. 2.42), several studies have improved performance and seals [2.21].

More recently manufacturers have supplied the sapphire orifice already mounted a small fixture which is easier to mount in the fittings which come with the unit. This has become particularly important as the technology has moved from plain waterjet cutting to the inclusion of abrasive in the jet stream. Jet alignment in this latter case becomes critical to the performance of the combined stream and more elaborate methods have been developed for the combined nozzle assembly [2.22]. This topic will also be discussed in Chapter 11. It should be noted, however, that studies at the U.S. Bureau of Mines [2.22] have shown that the more abrupt transition in diameter used with the normal sapphire design produces a less effective cutting jet than does the normal tapered shape used for other applications (Fig. 2.43). The study indicated, however, that this difference disappeared at lower nozzle diameters (Fig. 2.44).

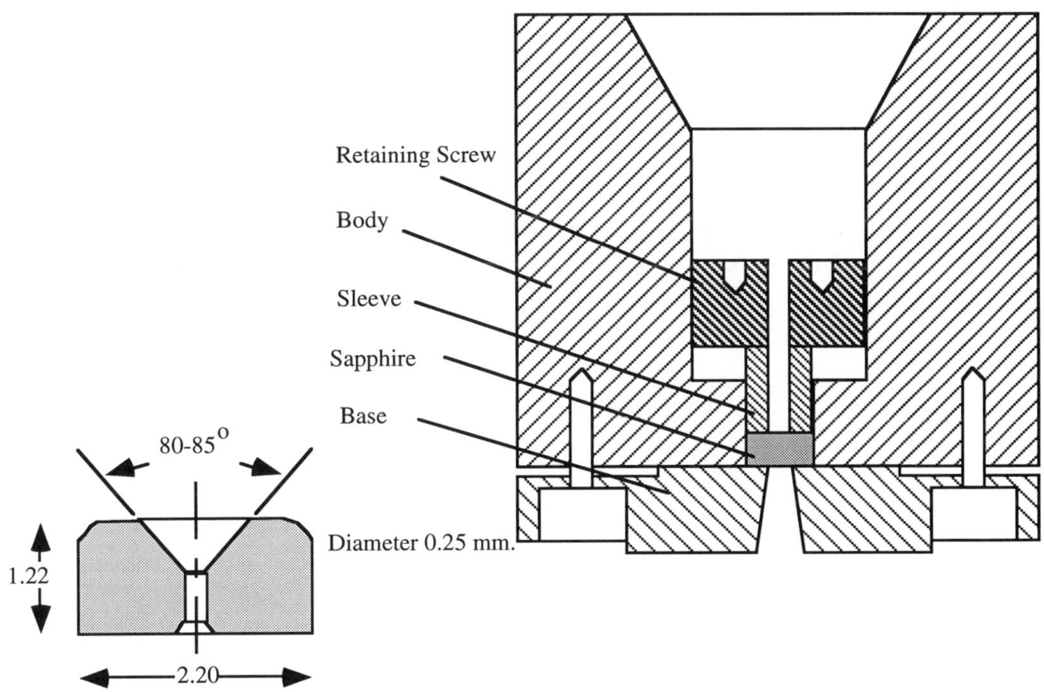

Figure 2.42 Profile of synthetic sapphire [2.21] and early method for attaching the nozzle to a waterjet pipe.

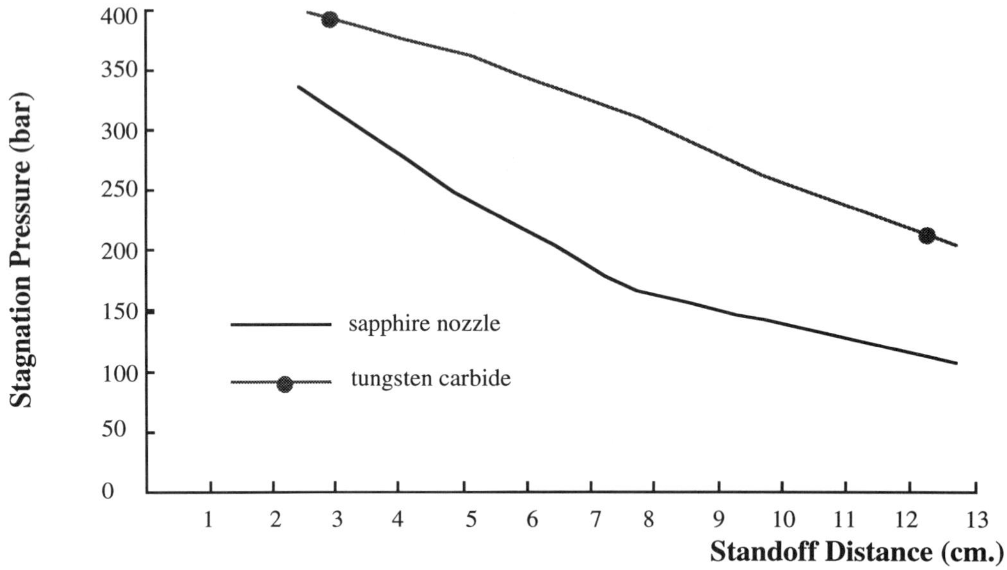

Figure 2.43 Measured jet pressure as a function of stand-off distance for a carbide and a sapphire nozzle, diameter 0.7 mm [2.22].

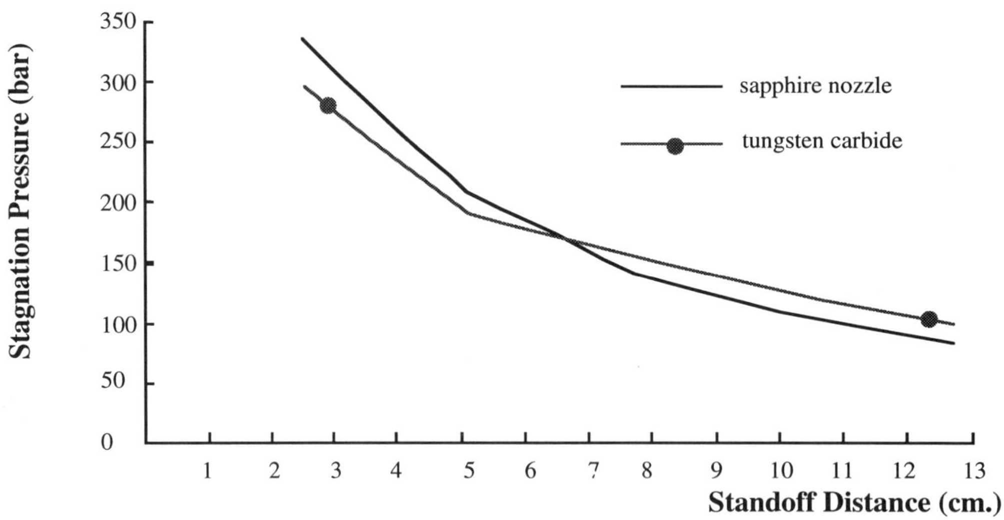

Figure 2.44 Measured jet pressure as a function of stand-off distance for a carbide and a sapphire nozzle, diameter 0.5 mm [2.22].

2.7.6 WATERJET CLEANING NOZZLES

Waterjet cleaning nozzles generally take a different shape from that found most useful for conventional cutting. Historically cutting jets have been round and issued from nozzles of the geometry described above. Cleaning jets, on the other hand, are normally designed to cover a wider area, with less potential for deep penetration, or damage to paint or other underlying surfaces. In such operations the use of a spray or fan jet nozzle is often preferred because these produce a jet which spreads to cover a strip of the working surface.

There are a number of different ways in which a spray nozzle may be made, but most are based on using a round orifice which has been bisected by a notch which runs across the face of the nozzle (Fig. 2.45). This method of manufacture has one important disadvantage which may not be normally recognized by the user. The edge of the orifice left by this method is very sharp, and quite thin. Under normal operating conditions this edge will wear away quite rapidly. This is particularly true where the nozzle is used with an untreated water supply which may contain a significant amount of abrasive material. The result is that the orifice wears into a less than optimum shape, and performance will rapidly degrade. Even when the nozzle is new, however, the shape can lead to unrecognized problems. The notched end of the nozzle will generate a waterjet which fans out into a sheet spreading at an angle, which depends on the nozzle shape. The shape also controls the thickness of the jet as it leaves the nozzle. As the jet moves out from the nozzle it starts to spread and will continue to spread with distance. Since a fixed volume of fluid exits from the nozzle at a given instant, then, as the sheet of water moves away from the nozzle, so it gets thinner.

For example, 20 liters of water, equivalent to 20,000 cc of water, issuing at 140 bar will be moving at approximately 168 m/sec. Thus, if the nozzle is initially 2 mm high and for simplicity assumed rectangular, then the initial width of the jet will be approximately 1 mm.

At a distance where the jet spray is 2.5 cm wide, the same volume of water will spread so that the thickness of the stream will reduce to 0.08 mm. At 10 cm from the nozzle, depending on the divergence angle, the stream will be further thinned to perhaps 0.02 mm. This will continue, with the jet getting wider but thinner, until the point where the sheet disrupts (Fig. 2.46). When this occurs, the fan sheet breaks up in two stages. At the point where initial penetration occurs, a large circular opening is created by the surface tension drawing the water backwards.

The creation of adjacent circles pulls the water into strings, which break into large droplets. These large droplets are virtually immediately broken aerodynamically into smaller droplets. It is, however, at this point that the waterjet fan may be at its most effective. Beyond the point where the water droplets become disrupted by aerodynamic forces into a very fine spray, the jet does very little but wet the surface. For a well designed fan spray the distance at which this jet break-up occurs can be as far as 10 cm from the nozzle.

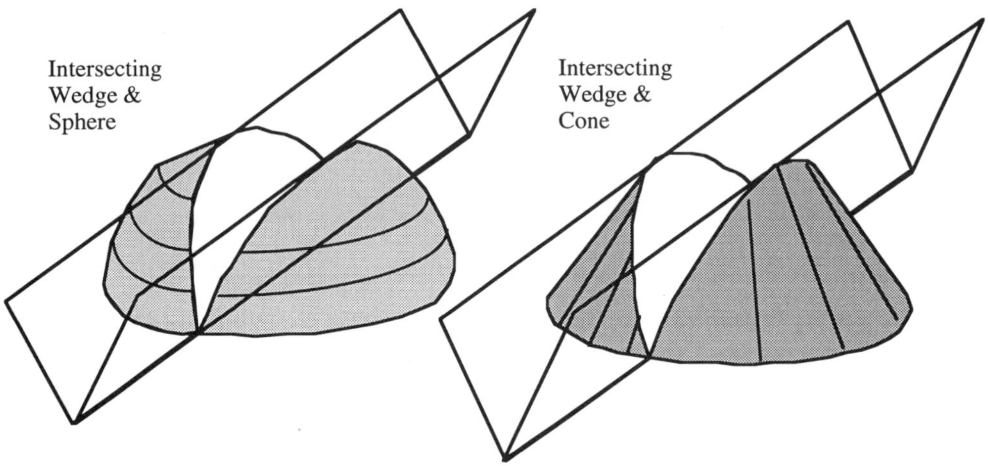

Figure 2.45 Methods of making a fan jet spray nozzle [2.23].

$$\text{Thickness} = \frac{20{,}000}{60 \bullet 16{,}800 \bullet 0.2} = 0.099 \text{ cm} = 0.99 \text{ mm}$$

Figure 2.46 Back-lit spray of water from a fan jet nozzle, showing the pattern of disruption.

2.7.7 DROPLET IMPACT

To explain why the droplets may be the most effective form of the jet, and at the same time to understand the limitations of this, it is necessary to refer to work which has been carried out at the University of Cambridge in England [2.24] and [2.25]. In this work, at the Cavendish Laboratory, the pressure was measured when a surface was impacted by a high velocity water droplet. It was found that, on the instant of impact, the surface of the impacting droplet is curved (Fig. 2.47).

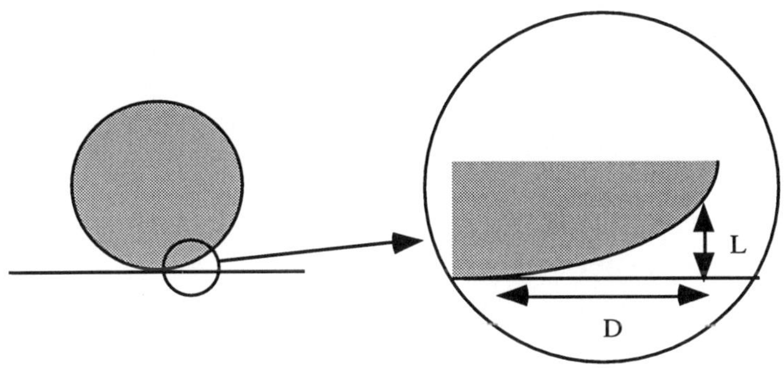

During the initial stages of impact L < D

Figure 2.47 Droplet impact upon a surface.

As this curved surface hits the target material the water that first makes contact tries to move out laterally to get away from the subsequently impacting water. Initially the radius of curvature of the impacting droplet is such that before it can escape a second ring of contact outside the initial contact zone collapses onto the surface. The water which has already impacted the surface therefore becomes trapped, and under increasing pressure from the arrival of the immediately overlying layer of water. This procedure continues outward with an ever increasing pressure build up at and under the surface, until the point is reached at which the water surface is no longer collapsing fast enough to contain the outflowing jet. At this point, a very high velocity lateral jet is created across the surface of a target, releasing the pressure buildup. Concurrently, the very high pressure peak which has been achieved on and within the surface of the material diminishes.

Normally when one considers water jet impact one talks of two different pressures. The first of these is the **jet stagnation pressure.**

One can establish this relationship relatively simply by an application of Bernoulli's Law which reduces the relationship simplistically to:

$$\text{impact pressure} = \frac{1}{2}\, r\, V^2$$

Where ρ is the density of the jet fluid in gm/cc.

However, one may on occasion get a much higher pressure especially on the instant of initial impact, such as in the case discussed above. This augmented pressure is usually referred to as the **water hammer pressure**.This increased pressure is created where a shock wave is induced into the system. The water hammer pressure is given by the equation:

$$P = r.\ C.V.$$

Where C is the sound speed in water.

Experimental measurements arising from out of the work at Cambridge has shown that pressures on impact can reach levels which are more than double the water hammer pressure [2.26].

The difference between these two pressures can be seen by the use of an example. A typical waterjet moving at 700 bar will have a velocity of approximately 400 m/sec. In contrast, the shock wave velocity (C) in water is 1470 m/sec. Thus, the pressure that will be generated on the surface from the stagnation pressure of the water impact will be 700 bar in contrast to a water hammer pressure which would develop an instantaneous pressure of approximately 5,600 bar with the same jet.

Work at Cambridge, however, has shown that on the perimeter of the water droplet, in the ring where the water is trapped initially by the impact, pressures of up to $2\frac{1}{2}$ times water hammer pressure may be achieved. Thus, with a 700 bar jet, a ring fracture may be induced by an instantaneous application of pressure which may reach, in this particular example, 14,000 bar. Thus one can find an augmentation in jet pressure on the order of 20 times stagnation pressure, at the instant of the droplet impact. The duration of this intensification is extremely short and the magnification does not occur where there is a soft layer on the surface since such a layer would not constrain the impacting water the way a dry target surface does.

This explanation describes the "knife" effect which can occur, where a fan is allowed to break up into droplets, just before it hits the surface. The stand-off distance range over which this droplet impact occurs, is however extremely narrow and, from observation, is likely to be less than 1 cm deep. It is a function of the jet geometry, the jet pressure, and the flow rate. For a well designed nozzle at a flow rate of 20 lpm and at a pressure of 150 bar it may be 10 to 11 cm from the nozzle orifice. It will not occur if the surface is wet or covered with soft dirt.

There is an additional problem which should be understood. Where a jet impacts on a surface, the very high pressure pulse may induce a very small fracture or flaw in that surface. This is very useful if one is trying to initiate penetration into materials such as glass or plastic, where there are no flaws to exploit initially. In the case of individual small droplets such as those created when a fan jet nozzle is used in cleaning, the depth of the flaws that may be created are extremely small because of the small droplet size. At the same time these small cracks cannot be effectively exploited by the pressure generated by the flow of fluid from the following droplet, because the droplet is so small. If one is traversing a cleaning water jet across the surface, then there is no subsequent water flow which will exploit these very fine fractures in the surface. So that it becomes extremely difficult to make much use of the water droplet impact phenomena just described. (This is a particular problem also with the water cannon approach to rock cutting, and will be further discussed in section 7.2.)

2.7.8 THE CHOICE BETWEEN ROUND AND FAN JET NOZZLES

The result of the rapid disruption of the stream from a fan jet nozzle is that a very inefficient jet cleaning system can be developed if sufficient care is not taken in the design and use of the nozzle assembly. And in this discussion it may be worth noting that most fan jet nozzles are often referred to as spraying nozzles. A **spray** is defined as a jet which is broken up [2.27]. The broken nature of the spray from a fan jet will rapidly reduce the efficiency of the jet generated. This can again be illustrated by way of example.

In evaluating a suite of nozzles to determine their relative cleaning potential, polystyrene foam blocks were rotated in front of the nozzle at different stand-off distances [2.28]. Very obviously if a waterjet cannot cut polystyrene foam, which can be cut at 0.7 bar, it will have some difficulty in cutting harder dirt and other encrusted material. The experiments, which used standard commercially available nozzles, showed that none of the fan jets were able to cut polystyrene adequately at a 22.5 cm stand-off distance. Most of the jets used were not able to cut the foam at a stand-off distance of more than 15 cm, and many no further away than 7.5 cm. This is because of the very rapid break up of the jet spray after it left the nozzle.

The reduction in cleaning effectiveness with distance can be overcome if the jet thickness is sustained with distance by using a round jet. In this case it is necessary to rotate the jet in order to cover the equivalent or greater width of the target, as would be covered with the fan jet. A 20 lpm flow flowing through a fan jet orifice, designed to spread a jet at 15° could, with a typical design, only cut polystyrene for a distance of 10 cm away from the nozzle. When an equivalent water flow, at the same pump pressure, was directed through a round orifice, however, and that nozzle rotated so that the resulting jet would cover the same 15 degree angle, the jet was able to cover a much greater area. The rotating jet would cover the same area as the fan orifice, at the same stand-off distance, since the jet was inclined out at the same angle. However, the greater cohesion of the round stream of water would retain its cutting energy so that the effective range was extended up to 90 cm away from the nozzle surface. If the relative areas swept by the jet are then computed it can be seen that the area covered in the second case is some 9 times greater than in the former (Fig. 2.48). By transposing this it can be seen that some 90% of the power and water required to clean a surface can be saved if spinning round jets are properly used to replace fan jet nozzles.

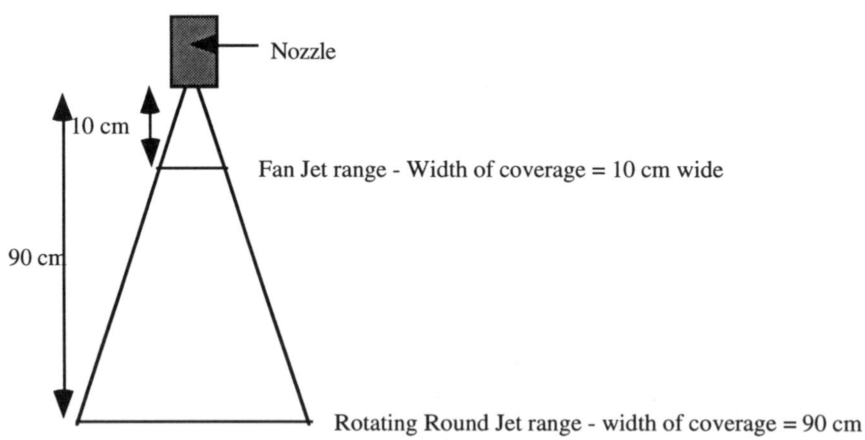

Figure 2.48 Gain in areal coverage when changing from a fan to a round jet.

2.8 SUPPLY LINE AND FITTINGS

Fluid from the pump will travel through a delivery line to the nozzle body. Typically there will be some form of mounting body into which the nozzle is attached, at the forward end, and into the back of which the supply line is connected. In manual cleaning operations this mounting body is known as a lance, and will comprise a number of important components. Usually in manual operations this will include a protective handle, a manually operated dump valve, which operates through a trigger, and a bracing mount. The dump valve passes the supply water through a valve into a large pipe which should be directed to carry the water away from the operator, but at low pressure. As the operator pulls the trigger, this closes a valve in that line and forces all the water down the lance to the nozzle. The small orifice in the nozzle then induces the pressure build-up, as the water must accelerate to get through it.

The dump circuit can be located to feed the water either into a return line, or even onto the ground, but should be directed to pose no hazard to the operator or other working personnel. It is important with such a system to ensure that, when two or more guns are operated from the same pump, simultaneously, that provision be made to isolate each lance from the active/dumped condition of the other. Safety is a major concern with water jet cleaning, and will be discussed in more detail in Chapter 12.

While the historic view of cleaning has developed around an operator manually sweeping a nozzle over a work surface, this provides a relatively slow, and often inefficient method for areal coverage, given the high speed at which a waterjet will move and remove material. One way of increasing the areal coverage is to increase the number of orifices within the nozzle body. While this may provide some advantage, particularly, for example when using a jet to clean the inner walls of a pipe, the poor quality of flow through a multiple orifice head can significantly reduce the effective pressure on the cutting surface. Recent improvements to the technology have therefore included small swivels which allow the jets to be rotated over the surface, giving a better areal coverage.

2.8.1 ROTATING NOZZLE ASSEMBLIES

Two different systems have developed to drive the nozzles in rotation. In the earlier development rotation was achieved by mounting a small motor on the cleaning lance and using this to rotate the nozzle.. This concept was also incorporated into a drive system for rotating a drilling head. There have been a number of different designs developed, once the original ability to create a swivel which would reliably operate at high pressure was developed. One of the advantages of using a drive motor is it controls the speed of the head and gives greater flexibility to the directions at which the jets can be pointed.

In the latest version of these swivels, however, a more innovative development has occurred [2.29]. If the operating jet nozzles are not only inclined outward from the axis of the lance, in order to give a better areal coverage, but are also directed tangent to the circle of their offset position, then as the jet issues from the nozzle, the reaction force it imposes on the nozzle body, will cause it to rotate (Fig. 2.49). Thus a much simpler and smaller fitting will be required to give a spinning jet system.

Gracey [2.30] has pointed out that the reaction force quoted earlier can also be used for this driving force, and that where the nozzles are offset from the axis of the lance the turning moments thus generated can be sufficient to turn the nozzle assembly (Fig. 2.50). With the proper design, the amount of power drawn from the jet can be very small, typically less than 75 Watts [2.29]. While this has considerable potential advantage, there are a number of cautions which should be borne in mind.

Figure 2.49 Self-rotating nozzle assembly using jet angle for drive.

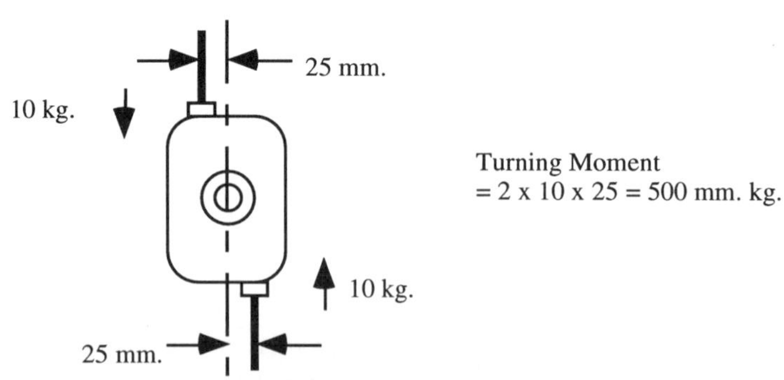

Figure 2.50 Self-rotating nozzle assembly using nozzle offset for drive [2.30].

Firstly one should remember that effective operation requires a certain jet residence time on the surface, so that too fast a rotation may be counter productive. This can be overcome by designing the swivel with some contained form of speed control. Hydraulic pumping, magnetic braking and viscous fluid systems have been marketed, and each has its advantage, depending on the speed range being used.

The second area of concern is with the flow path of the jet through the swivel and then out through the orifice. As with the flow into the nozzle body, this must be kept as smooth and gradual as possible. In one version of the self-rotating swivel, for example, the water is required to make two ninety degree turns immediately behind the nozzle orifice. The result of this will be to leave the jet with a degree of spin as it issues from the nozzle, which dramatically reduces the distance over which the jet will retain any effective cutting pressure.

Many swivels also are built with a relatively small internal bore. This diametrical reduction can be a source of considerable pressure loss, and where possible the system designer should size the system components, such as the swivel, to ensure that the passage does not narrow to the point of severely restricting the power of the jet after it leaves the nozzle. In some applications this is not a great problem since, for example where the jets are cleaning a pipe, the cutting range is very small. In other cases, however, this can be a problem and should be so recognized in choosing the most effective design of swivel for that use. An example would be in cleaning large tanks where the jets must cut at a distance from the nozzle. One advantage to this development is that the surfaces can be more reliably cleaned with a smaller number of nozzles. This in turn allows the orifices to be of a larger diameter, which gives a resulting better cutting and cleaning performance for the same power level.

Swivel reliability, particularly in the pressure range to 1,500 bar has improved considerably over the last ten years, and where, at one time only one or two of the available commercial units could be considered reliable over an extended operating shift, that is no longer the case. Swivel operational reliability has now been measured in the hundreds of hours. There is one caveat to that, however. One should, where possible, build the system so that no lateral loading is imposed upon the swivel. Generally this can be achieved by building bearing mountings just ahead of the swivel, so that any deviatory stresses are carried there, rather than in the swivel itself. Where a substantial lateral load is placed on the swivel, some models have been found to wear out in less than an hour of use. Nozzle designs now exist which can give not only two, but also three dimensional coverage of a volume, particularly useful for tank cleaning operations (see Chapter 3).

An alternative way of moving the nozzle to give the necessary areal coverage for deep slotting, yet avoiding the potential friction losses and complexities of a swivel drive is to use an orbiting design (Fig. 2.51). In this design the lance connection to the nozzle is flexible and, while non-rotating, it will sweep out a circular path. The drive mechanism is through a gear, eccentrically mounted on a rotating collar. As the collar is rotated, the nozzle travels a circular path and the jet sweeps a circular path on the target surface, cutting a path wide enough to give, for example, access to reinforcing steel in concrete so that it can be removed by other means.

Turbulence will be created at the entrance to the nozzle if the section of pipe leading into the nozzle is not straight. Historically it has been recommended that, where possible, the pipe immediately preceding the nozzle be straight, and aligned with the nozzle axis, for at least 100 diameters. Recent studies by the US Bureau of Mines [2.31] have shown that this distance can be reduced. Their work has shown (Fig. 2.52) that it is only necessary to use such a straight section over a distance of some 10 cm back from the nozzle for the jet quality to stabilize.

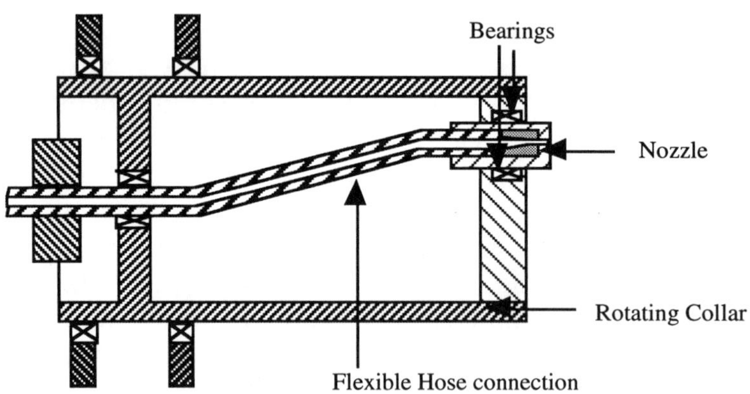

Figure 2.51 An orbiting nozzle schematic.

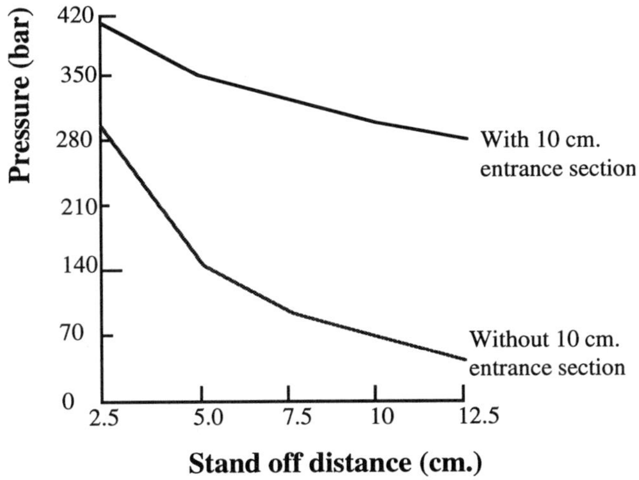

Figure 2.52 Entrance pipe straightness on resulting jet power [2.31].

2.8.2 FLOW STRAIGHTENERS

It is not always possible to achieve the long straight section of lead-in pipe which will produce the most effective jet. If the flow retains a high degree of turbulence, or spin, as it passes through the nozzle then the coherent jet length beyond the nozzle or effective length of throw, can be seriously reduced. Although, as referred to above, recent work has shown that the length required does not have to be as great as originally thought, yet it may still require more space than is available for many tools. Several investigators have looked at ways of shortening this length further. While applicable in a number of industries, the initial change came about for mining applications, where monitors were becoming too long to be effective. Investigators in Britain [2.32] were able to shorten the nozzle length by inserting flow straighteners behind the nozzle section (Fig. 2.53). Several designs were tested and the results indicated that a honeycomb structure gave the best performance. Subsequent studies with various nozzle designs have shown that while a straight length of pipe would give the best results, in terms of jet performance, where this option was not available, then flow straighteners would be an effective alternative.

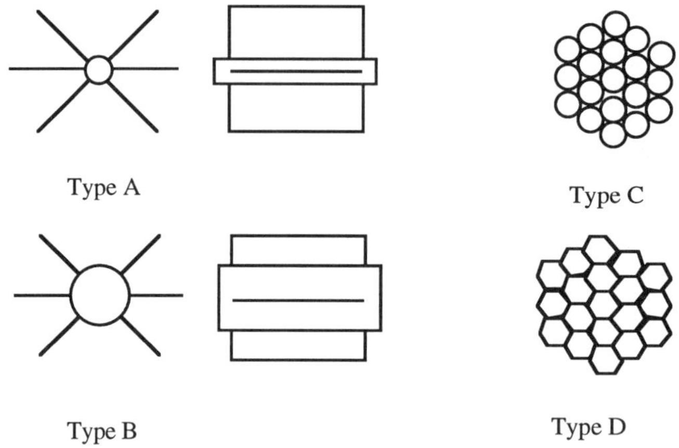

Figure 2.53 Flow straightener designs tried in the United Kingdom [2.32].

The use of flow straighteners is now widespread, particularly in those conditions where a lack of space does not allow the long entrance sections into the nozzle orifice, which produce the best jets. Various different geometries have been developed, for example, the nozzle designs used in China [2.33] differ in shape, as do the flow straightener shapes (Fig. 2.54). Flow straighteners of small size have also been found effective, in smaller sizes, in cleaning and other lower volume, higher pressure applications.

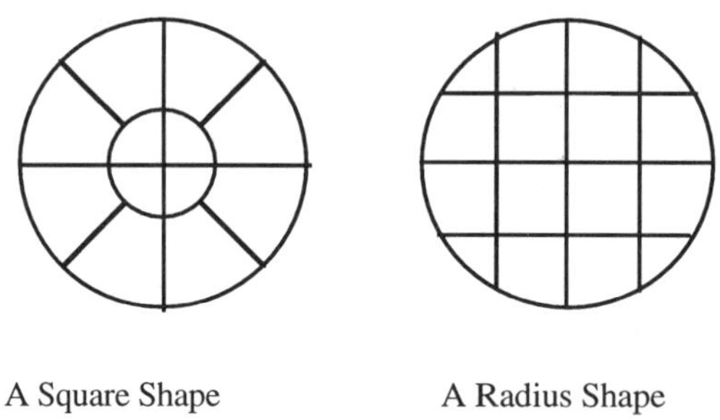

Figure 2.54 Flow straightener designs used in China [2.33].

An illustration of flow straightener effectiveness can be seen from experiments carried out by the U.S. Bureau of Mines in the use of a borehole miner for extracting small deposits of uranium. While the operation of the tool will be described in Chapter 6, the machine requires that the water flow down a vertical tube and then turn 90 degrees into a nozzle. From the nozzle the jet must be able to cut out the walls of a cavity for as far as 6 m from the nozzle. With the initial design, jets were not capable of reaching nearly this distance. Lohn at TRW, however [2.34], incorporated a flow curving and straightening section, in which the water was brought around the curve (Fig. 6.39). When this design was incorporated into the tool, it was able to achieve its designed cutting range.

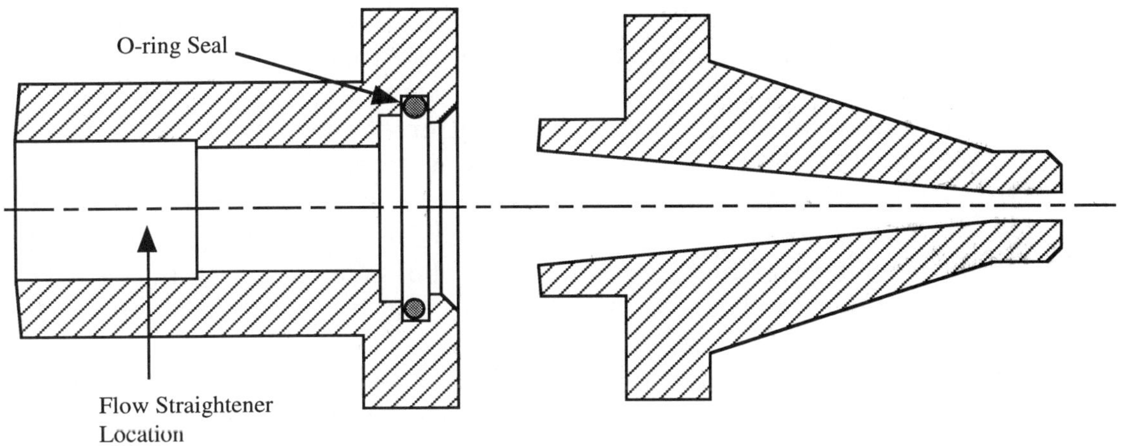

Figure 2.55 Position of the flow straightener behind a nozzle [2.19].

The exact position in which to locate the flow straightening device behind the nozzle, has not been studied by many sources. Shavlovskii [2.19] has suggested that where a length of 40 - 50 diameters of straight flow upstream of the nozzle cannot be achieved, that the use of the flow straightener can reduce the required length to only 10 - 15 diameters. The length of the linear plates of the stabilizer should be 2 - 4 times the diameter of the inner pipe. Shavlovskii further recommends that there should be a gap between the flow straightening device and the entrance to the nozzle of less than two pipe diameters (Fig. 2.55).

The inclusion of components in the system which allow movement of the cutting lance, or the provision of valves and fittings to improve safety and reliability all come at some cost. This does not refer just to the financial burden, but rather the additional friction losses which can be incurred in such a system. Frictional losses will also be generated in the line between the pump and the cleaning lance. To determine what these are, we must consider the various sources of loss.

2.8.3 COMPONENT PRESSURE LOSSES

In the discussion on system energy losses in Section 2.2 there were two major sources of loss in pressure. One was the loss in energy after the jet had left the nozzle (covered in Section 2.9 and Chapter 11) and the other was the loss in delivering the energy from the pump to the nozzle. Since almost every system will, over time, become uniquely configured for the job it is doing, it is important to recognize how much pressure may be lost if the components are not properly sized or assembled.

Energy may be lost through several different mechanisms. Where the flow is directed through sharp turns then some energy is required to change the direction of the water. Where the path undergoes a sharp contraction, or expansion, then additional energy may be lost in the turbulence patterns which are created. The greatest pressure loss will, however, often come because the inner pipe and hose are too small for the flow volume which they are transmitting.

Values for the pressure losses which are caused by the water flowing through valves and fittings should be provided by the manufacturer of the part, and should be correlated to the flow which is passing through the fitting. One way in which this information may be given is that the part will be given a resistance equivalent to a length of pipe of a given inner diameter. This makes it relatively easy to calculate the pressure loss for different volume flow through the part.

In calculating the pressure loss in driving the water through the line there are several methods which can be used. Labus [2.9] has suggested that the following equation can be used to determine pressure loss in the delivery line of most commercial operations where high pressure water is used:

$$\text{Pressure drop (bar/ m)} = \frac{0.597 \times Q^2}{100 \times D^5 \times R^{0.25}}$$

Where: D is the pipe diameter in cm
R is Reynolds number (given by 1116.5 (Q/D)

Consider a 1.25 cm internal diameter hose carrying a flow of 50 lpm. Reynolds Number for such a flow would be (1116.5 x (50/1.25)) or 44,660.

Pressure losses in the line would be 0.597 x 2500 / (100 x 3.052 x 14.53) or 0.33 bar/m. Running a 30 m hose from the pump to the cutting operation would thus generate a loss of approximately 10 bar. It is interesting to note that the measured pressure loss was slightly higher than this, 0.45 bar/m, in the Navy study cited earlier [2.10]. Such an additional loss may have been incurred in the fittings along the line or by a change in its internal condition.

Alternately the calculation may be replaced by reference to a nomogram. Versions of these have been developed by a number of companies and are relatively easy to use. However, given the availability of simple-to-use computers and analytical programs, these may be more accurate to use. It is important to evaluate such losses when carrying out design calculations before building a system. Consider, for example, water flowing at a rate of 80 lpm, through a 0.476 cm I.D. high pressure hose, the friction loss will be approximately 275 bar for every 3 m length of the pipe. A 15 m length of such pipe, would absorb 1,375 bar in passing the water from the pump to the nozzle. If the target material had a threshold pressure just below the pump pressure, then such a pressure loss may be sufficient to cause the waterjet to become no longer effective. (A calculation, using the formula suggested by Labus yields a pressure loss of 75 bar/m for a total loss of 1,125 bar over the same length.)

This illustration was used deliberately since an experiment which was carried out in a lead mine used just such a flow rate of 80 lpm. This experiment reached the stage where adding an additional 3 m of pipe to the system raised the line losses to the point that the nozzle pressure fell below that necessary to cut the rock. Another experiment, by another group (who will be allowed anonymity) was carried out where the supply line used was so small, for the flow, that over 75% of the energy developed at the pump was lost in delivering the water to the nozzle. This can be very important to an operation. In that particular case, for example, the jet pressure was on the order of 3,000 bar. Neither the experimental team, nor the client for the experiment, were aware that the pressure at the jet was only 875 bar. The resulting poor system performance led to a series of erroneous conclusions on the efficiency of jet cutting which have not yet been completely eradicated. They also reduced the performance of the system significantly below its operational capabilities.

The above nomograms were developed for low pressure flow and at higher velocity flows they may not be exactly correct. The data does, however, emphasize the need for trying to insure use of the maximum possible internal diameter of the hose or high pressure delivery tube, consonant with safe and effective operation. This is normally 1.25 cm internal diameter hose, although, depending upon the particular application, this may not always be the best option.

Where possible in designing fluid lines to a nozzle one should avoid step increases and decreases in flow diameter. This is not always possible, but sharp increases and decreases in the pipe flow diameter can lead to shock losses and create other energy losses which may reduce overall system performance.

2.8.4 CONSIDERATIONS OF SAFETY

Although this topic will be addressed in more detail in Chapter 12 at the end of this text, there are some factors which also relate to the design of the system. Three papers describe significant developments and considerations in this area.

One of the most prevalent methods of using high pressure waterjets is in cleaning surfaces. Most of the use comes with the jet delivered through a hand-held wand or lance. In 1991 Gracey [2.35] reviewed the development of these lances from their inception, which he dated back to the 1950s. From the initial flattened pipe used to hose down equipment at 140 - 200 bar, the lance sequentially acquired a nozzle, a handle, and, around 1963 Gracey places the introduction of the needle valve to control the pressure of the jet.

The design of the valves to control the flow to the gun has developed along three directions, the dump style, where the water reaches the gun, but is then bypassed to an open circuit when the trigger is released; the shut-off gun where the valve at the gun closes under pressure and retains pressure in the line - in which case an unloading or pressure regulating valve at the pump is required to redirect the flow from the gun; and the remote control gun, where the trigger opens and closes a valve further back in the circuit, so that water is only sent to the gun when the trigger is operated. In some applications these valves are mounted in a holder to allow the operator to control them with a foot, rather than using a hand-operated control.

Over the last ten years Gracey identifies three additional features which have been added to the guns. The first of these was a trigger guard, to prevent the accidental activation of the gun if it were dragged over the ground. The second feature was a safety catch, added to positively lock the trigger in a valve closed position, unless it is positively disengaged. The more recent addition has been the use of a second, larger sheathing hose, or safety shroud, fitted around the pressure hose feeding into the gun. With this design the operator is provided some protection should the supply hose rupture or leak.

The continued concern over user safety has also led some plants to require that the operator actively operate two trigger mechanisms to run the gun. This ensures that the operator has both hands holding the gun during its operation, and is considered to provide additional protection in case one of the valves malfunctioned.

At the same meeting Yie [2.36] discussed some of the problems associated with the use of conventional flow control valves, as the industry increasingly uses higher flow rates and pressures than the original designs were made to accommodate. One way of overcoming these problems, which Yie describes, is to use a pilot-operated design in which the pressure of the fluid is used to assist the operator in opening and closing the valve.

The discussion above on system losses has stressed the need to ensure that the inner diameter of the feed line to a nozzle be of sufficient size to allow the required water flow without significant pressure loss. This tube size cannot be taken in isolation, but also must be considered in terms of the thickness of the wall of the tube, to ensure that there is adequate strength in the material to hold the contained pressure. Fryer [2.37] has reviewed both common materials and the equations which have been used to analyze performance, both from a static and dynamic point of view. He points out the different approaches which may be taken to pipe wall design, and the concerns which must be considered in a situation where the pulsation in pressure within the line will lead to eventual fatigue of the equipment.

Until recently almost all fittings above 1,500 bar required that the water be delivered through rigid tubing, of the type Fryer described, however the advantages of using a flexible hose has led to their development. Several manufacturers now produce such hose, and the development of one such product has been described by Raghavan and Olsen [2.38]. In that paper the authors not only describe the basic hose construction, but also comment on the lifetimes which it can achieve. Where the hose is not subject to high stress it was found to last for more than 1,000 hours at operating pressures of 2,380 bar, while a lifetime on the order of 800 hours was reported for hoses which were required to flex in operation at pressures to 3,750 bar. It could be noted from the data presented that the pressure drop in the hose fittings was equivalent to roughly 3 m of additional hose length.

2.9 DELIVERING THE ENERGY

The majority of this chapter has been devoted to considerations in the layout of the equipment, and losses which can occur as the water passes though the system, from the pump to the nozzle. Such line losses can, as has been demonstrated, be very significant. It is thus of critical importance that care be taken in choosing the system components to minimize the energy losses before the water exits from the nozzle.

The work of the waterjet is not, however, completed when the jet leaves the nozzle. It is completed after the jet has impacted on the work surface, and has then left this working area. The difference lies in the distance that the jet has to travel to reach that surface, and the energy losses that will occur along the way.

Waterjets are rapidly eroded both in size and pressure, as they move away from the nozzle. This erosion of power is significantly increased where the jet is submerged, and becomes a major design consideration at depths below 150 m, or its pressure equivalent.

For this reason, it is important to size the cutting jet to operate at the lowest pressure, and the largest nozzle diameter consistent with the needs of the cutting operation. Studies of the decay in waterjet pressure away from the nozzle indicate that the pressure loss is exponential in nature.

A statistical regression on the pressure values measured, at fixed distances between 15 and 75 cm, and under pressures of 280 - 700 bar with three nozzle diameters (1.0, 1.25 and 1.5 mm diameter) indicated that the relationship was of the order:

$$\text{impact pressure} = \text{constant} \cdot \text{diameter}^{2.93} \cdot \text{pressure}^{0.438} \cdot \text{distance}^{-0.939}$$

In one example of this equation it was found that, at a distance of 50 cm from the nozzle, a 400 bar jet retained a maximum pressure of 140 bar.

The equation should not be considered as more than an indication of the form of the relationship, since the study showed that the values of the constants changed significantly as a function of all three variables. The reason that this is so is that the jet undergoes several different stages in its passage from the nozzle.

As the jet leaves the nozzle the pressure profile across the jet is almost a step function. There is a very rapid rise to full stagnation pressure, which is constant over the central core of the jet. Such a jet, incidentally, is not necessarily the most effective for cutting material.

As the jet moves away the pressure profile changes. The outer sheath of the water is slowed by the surrounding air, and this in turn slows the next inner layer. Within a short distance the pressure profile is more bell-shaped, and the jet often begins to cut much more effectively.

With increase in distance the jet begins to break up into segments, drops and fine mist. This break-up can occur in stages, the mist often obscures the internal structure of the jet. At the point where the jet first breaks into segments a number of investigators have shown that the jet performance is enhanced. This can be for several reasons. The one most often proposed is that the break up of the jet into large droplets is creating the enhanced water impact pressures discussed earlier. Studies have shown that this enhancement is reduced if the jet is hitting into a cavity filled with water. An alternate circumstance is the break-up of the jet into segments allows the water on the surface to escape before the next segment arrives. This removal of the cushion of water makes the arrival of the next segment more effective, since it does not suffer the counter flow of the immediately preceding water, which leaves the cavity almost at the same speed as the incoming jet. A number of investigators have sought to improve the jet structure on the way to the target. The most common method of doing this [2.39] has been to sheath the jet with a surrounding spray, either of slower moving water, or air (Fig. 2.56). The air sheath has been found quite effective in underwater applications, but has not found as many successful applications in an air environment. Because the need to improve jet performance has become a much greater issue in recent years, it will be discussed in more detail in Chapter 11.

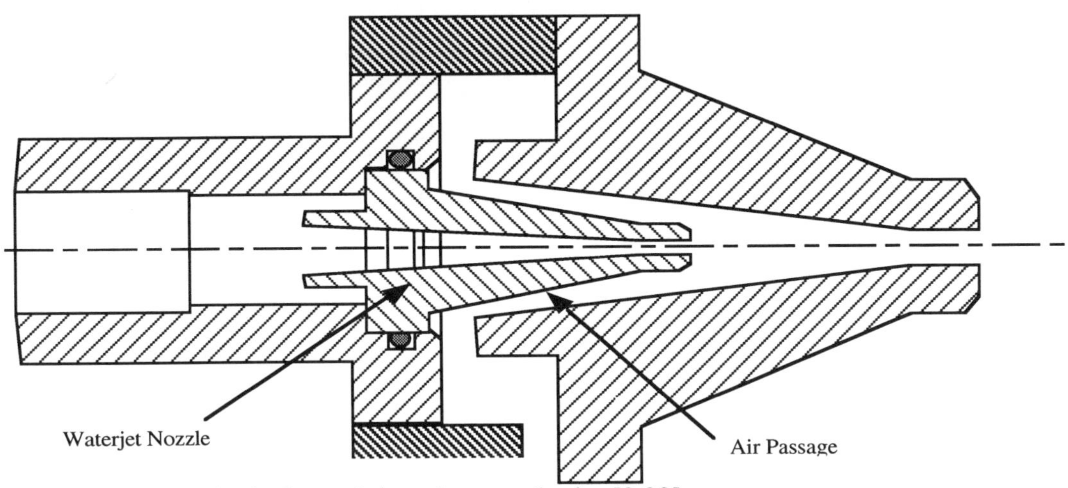

Figure 2.56 Method of entraining air around a jet [2.39].

2.10 USING THE ENERGY

Getting the energy to the working surface is the major objective of the system designer. But once there it is important to be sure that the waterjet is powerful enough to achieve its objective. Over the years a number of different pressures have been suggested as being necessary for cutting material. Unfortunately, in the tabulation of this data, it is rare to find an indication of several important data. For example, is the pressure measured at the pump, or the target? What is the flow volume and at what diameter is it hitting the target? How far is the target from the nozzle? Without the answers to these questions, the numbers which are proposed have only a limited value. Remember the study made by the Navy in which two thirds of the energy was dissipated before the water left the nozzle. Under those circumstances, while the pump was providing 700 bar, the cleaning operation only required 250 bar to be effective.

Waterjets work, in large measure, by exploiting the microfractures in the surface of the target. This was shown, in a classic demonstration by Field [2.40]. Griffith had shown [2.41], in the 1920s, that glass contains a number of small surface flaws, which strongly effect its strength. Field took such a piece of glass and etched half of it. He then fired a waterjet droplet onto the interface between the etched and the unetched surface. Where the glass contained no cracks because of the etching there were very few cracks or surface damage. Where the cracks had been left on the surface the glass was severely cracked under the water impact.

In designing a system for material removal, this knowledge must be included in the design consideration. Larger diameter jets encounter more surface flaws than small ones. A jet with a distinct pressure profile will remove more material than one of constant pressure, since the differential pressure will allow cracks to open and water to penetrate the surface and grow the cracks to failure. Surfaces with larger cracks are generally easier to penetrate than those with finer grains. Metals may not be penetrated until the jet pressure reaches levels of around 1,500 bar. It is toward the analysis and use of this information that the rest of this book is directed.

There are different considerations which hold depending on whether the surface is to be cleaned or cut. Waterjets at high pressure, low flow rate will cut through material in different ways than lower pressure, higher flow rate jets. Strategies to cut through materials such as concrete, where part of the material is too hard to cut, must be developed. These are described, but for simplicity the subjects are grouped according to their major function.

2.11 NOTES AND REFERENCES ON CHAPTER 2

Many of the references throughout this text will come from the Proceedings of the two major sets of Conferences which occur in the Waterjetting fields of application. These are the International Symposia on Jet Cutting Technology and the American Water Jet Conferences. The first of these has been organized, since 1972 by the British Hydromechanics Research Association, which is now BHR Group Ltd. Information on those Proceedings can be obtained from BHR Group, Cranfield, Bedford, UK. Unfortunately some of the earlier volumes are now out of print.

The American Conferences were initially held at host Universities, which also printed the Proceedings. The first such meeting was held at Colorado School of Mines; the second at the University of Missouri-Rolla; the third at the University of Pittsburgh; and the fourth at the University of California-Berkeley. The more recent Conferences have been run the Water Jet Technology Association, with its headquarters in St. Louis, MO. Information on the availability of the Proceedings may be obtained from that Association. Again, however, the earlier volumes of the series are now becoming out-of-print.

Other national meetings have now begun, based on the growth achieved with this new technology, and readers seeking more information are directed to the International Society for Water Jetting, who will be able to provide the addresses of local contacts for those meetings.

2.12 REFERENCES

2.1 Teale, R., "The Concept of Specific Energy in Rock Drilling," International Journal of Rock Mechanics and Mining Science, Vol. 2, No. 1, March, 1965, pp. 57 - 74.

2.2 Anon, Size 3 Triplex Instruction Manual, Kobe Inc., Huntington Park, CA, supplied to UMR, January, 1970.

2.3 Sugino, Y., Nagata, Y., Kanemitsu, M., Yoshimoto, S., Inokuma M., and Shimose, T., "Development of an Ultra-Pressure Marine Pump," Poster paper 5, 9th International Symposium on Jet Cutting Technology, Sendai, Japan, October, 1988, pp. 667 - 681.

2.4 De Santis, G.J., "Operational and Maintenance Misconceptions of High Pressure Power Pumps," 3rd American Water Jet Conference, Pittsburgh, PA, May, 1985, pp. 12 - 28.

2.5 Heron, R.A., "Interpretation of Dynamic Pressure Ripple Generated by Positive Displacement Pumps with Self-acting Valves", High Pressure Water Pumps, A Seminar, Inst. Mechanical Engineers, May 1, 1986.

2.6 Haskel Pump Catalog, 1990.

2.7 Summers, D.A., Tyler, L.J., Blaine, J.G., and Short, J.E. Jr., "The Growth of the WOMBAT," paper C4, 7th International Symposium on Jet Cutting Technology, Ottawa, Canada, June, 1984, pp. 153 - 162.

2.8 Summers, D.A., and Worsey, P.N., Development of PBX Washout Pilot Plant, UMR Final Report on Contract NOO164-84-C-0205, Naval Weapons Support Center, Crane, Indiana, January, 1987.

2.9 Labus, T.J., "Fluid Mechanics of Jets". Fluid Jet Technology, Fundamentals and Applications, A Short Course, Toronto, Canada, August, 1989.

2.10 Tursi, T.P. Jr., and Deleece, R.J. Jr, (!975) Development of Very High Pressure Waterjet for Cleaning Naval Boiler Tubes, Naval Ship Engineering Center, Philadelphia Division, Philadelphia, PA., 1975, p. 18.

2.11 Woodward, M.J., "An Experimental Comparison of Commercially Available, Steady, Straight Pattern, Water Jetting Nozzles", 3rd American Waterjet Conference, University of Pittsburgh, PA., 1985, pp. 29 - 43.

2.12 S.P.D. Swan, "Economic Considerations in Water Jet Cleaning," 2nd American Water Jet Conference, University of Missouri - Rolla, MO, May, 1983, pp. 433 - 439.

2.13 Flitney, R.K., "High Pressure Water Pump Seals," High Pressure Water Pumps, A Seminar, Inst. Mechanical Engineers, May 1, 1986.

2.14 Summers, D.A., Yao, J., and Wu, W-Z., "A Further Investigation of DIAjet Cutting," Chapter 11, 10th International Symposium on Jet Cutting Technology, Amsterdam, Netherlands, October, 1990, pp. 181 - 192.

2.15 Water Jet Technology Association, Recommended Practices for the Use of High Pressure, Manually Operated Water Jetting Equipment, 1987.

2.16 Rouse, H., Asce, M., Howe, J.H., and Metzler, D.E., Transactions of the American Society of Civil Engineers, Vol. 117, 1952, p. 1147.

2.17 Leach, S.J., and Walker, G.L., "Some Aspects of Rock Cutting by High Speed Water Jets," Phil. Trans. Royal Society, London, Vol. 260A, 1966, pp. 295 - 308.

2.18 Nikonov, G.P., and Shavlovskii, S., Gornyi Mashiny i Avtomatika, Nauchno- Tekh., Sb. Vol. 1, (18), p. 5.

2.19 Shavlovskii, S., "Hydrodynamics of High Pressure Fine Continuous Jets," paper A6, 1st International Symposium on Jet Cutting Technology, Coventry, UK, April, 1972, pp. A6-81 to A6-92.

2.20 Barker, C.R., and Selberg, B.P., "Water Jet Nozzle Performance Tests", paper A1, 4th International Symposium on Jet Cutting Technology, Canterbury, UK, April, 1978.

2.21 Saunders, D.H., "Some Factors Affecting Precision Jet Cutting," Paper F1, 3rd International Symposium on Jet Cutting Technology, Chicago, IL, May, 1976, pp. F1-1 to F1-14.

2.22 Singh, P.J. and Munoz ,J., "The Alignability of Jet Cutting Orifice and Nozzle Assemblies," 10th International Symposium on Jet Cutting Technology, Amsterdam, Netherlands, October, 1990, pp. 207 - 220.

2.23 Liljedahl, L.A., Effect of Fluid Properties and Nozzle Parameters on Drop Size Distributions from Fan Spray Nozzles, Ph.D. Thesis, Iowa State University, 1971, University Microfilms, Catalog No. 72-5224.

2.24 Field, J.E., Lesser, M.B., and Davies, P.N.H., "Theoretical and Experimental Studies of Two-Dimensional Liquid Impact," paper 2, 5th International Conference on Erosion by Liquid and Solid Impact, Cambridge, UK, September, 1979, pp. 2-1 to 2-8.

2.25. Lesser, M.B., and Field, J.E., "The Geometric Wave Theory of Liquid Impact," paper 17, 6th International Conference on Erosion by Liquid and Solid Impact, Cambridge, UK, September, 1983, pp. 17-1 to 17-9.

2.26 Rochester, M.C., and Brunton, J.H., "High Speed Impact of Liquid Jets on Solids," paper A1, 1st International Symposium on Jet Cutting Technology, Coventry, UK, April, 1972, pp. A1-1 to A1-24.

2.27 Anon, Websters Ninth New Collegiate Dictionary, Merriam-Webster Inc., Springfield, Mass, p.1141.

2.28 Summers, D.A., "The Effectiveness of Water Jet Sprays in Cleaning and the Mechanisms for Disintegration," paper C1, 6th International Symposium on Jet Cutting Technology, Guildford, UK, April, 1982, pp. 81 - 91.

2.29 Wolgamott, J.E., and Zink, G.P., "Self-Rotating Nozzle Heads," paper 47, 6th American Water Jet Conference, Houston, TX, August, 1991, pp. 603 - 612.

2.30 Gracey, M.T., "Industrial Applications for Rotating Nozzle Technology," paper 48, 5th American Water Jet Conference, Toronto, Canada, August, 1989, pp. 487 - 493.

2.31 Kovscek, P.D., Taylor, C.D., and Thimons, E.D., Techniques to Increase Water Pressure for Improved Water-Jet-Assisted Cutting, U.S. Bureau of Mines RI 9201, Report of Investigations, 1988, p 10.

2.32 Jenkins, R.W., "Hydraulic Mining" The National Coal Board Experimental Installation at Trelewis Drift Mine in the No 3 Area of the South Western Division, MSc Thesis, University of Wales, 1961.

2.33 Wang, F-D., "Status of Hydraulic Coal Mining in the People's Republic of China," Proceedings of the 2nd US Waterjet Conference, University of Missouri-Rolla, MO, pp. 263 - 268.

2.34 Lohn, P.D. and Brent, D.A., "Design and Test of an Inlet Nozzle Device," paper D1, 4th International Symposium on Jet Cutting Technology, Canterbury, UK, April, 1978.

2.35 Gracey, M.T., "Recent Developments in the High Pressure Waterblast Gun," paper 48, 6th American Water Jet Conference, Houston, TX, August, 1991, pp. 613 - 618.

2.36 Yie, G.G., "High Pressure Flow Control Valves," paper 43, 6th American Water Jet Conference, Houston, TX, August, 1991, pp. 575 - 588.

2.37 Fryer, D.M., "Evaluating Small Bore Tubing for Dynamic High Pressure Systems," paper 41, 5th American Water Jet Conference, Toronto, Canada, August, 1989, pp. 417 - 424.

2.38 Raghavan, C., and Olsen, J., "Development of a 7,000 bar Hose," paper 43, 5th American Water Jet Conference, Toronto, Canada, August, 1989, pp. 449 - 454.

2.39 Yahiro, T., and Yoshida, H., "On the Characteristics of High Speed Water in the Liquid and its Utilization of Induction Grouting Method," paper G4, 2nd International Symposium on Jet Cutting Technology, Cambridge, UK, April, 1974, pp. G4-41 - G4-63.

2.40 Field, J.E., "Stress Waves, Deformation, and Fracture Caused by Liquid Impact," Phil. Trans. Royal Society, London, Vol. 260A, July, 1966, pp. 86 - 93.

2.41 Griffith, A.A., "The phenomena of rupture and flow in solids," Phil. Trans. Royal Society, London, Vol. 221A, 1921, pp. 163 - 198.

2.42 Alkire, T.D., "Advances in Direct-Drive Pump Technology Brings the Competitive Edge Back to Ultra-High-Pressure Waterjets," paper 24, 7th American Water Jet Conference, Seattle, WA, August, 1993, pp. 351 - 362.

2.43 Singh, P.W., and Benson, D., "Development of Phased Intensifier for Waterjet Cutting," in Jet Cutting Technology, ed A. Lichtarowicz, Kluwer Academic Publishers, Proceedings of the 11th International Conference on Jet Cutting Technology, September, 1992, St. Andrews, Scotland, pp. 305 - 318.

2.44 Chalmers, E.J., "Pressure Fluctuation and Operating Efficiency of Intensifier Pumps," paper 22, 7th American Water Jet Conference, Seattle, WA, August, 1993, pp. 327 - 336.

2.45 Product Brochure, Hammelmann Corp., Dayton, Ohio, 1990.

3 THE USE OF WATERJETS IN CLEANING APPLICATIONS

3.1 INTRODUCTION

The use of water jets for cleaning can be considered as a separate process to its use in cutting since, in the first case, the object is to leave the main working surface intact, while in the latter case it is to cut through it. This difference in need gives a considerable advantage to the use of waterjetting over other techniques in a wide variety of cleaning cases. This is because the jet pressure and power may often be adjusted so that the water will have the power to remove the unwanted material covering the surface, but not enough power to damage the underlying surface or substrate.

Water has always been used as a cleaning medium but, until recently, the major reason for this has been because of its properties as a solvent and lubricant. Within the past 20 years the use of water at high pressure as a cleaning agent has become extensive throughout the world. While the range of applications has covered a wide variety of fields, one demonstration of the benefits from the use of high pressure waterjets for cleaning can be taken from the chemical industry.

Many large chemical plants use heat exchanger tubes to pass heat between different phases of the various processes. The heat exchanger itself consists of small diameter tubes, frequently as many as 700 - 900 in number, jacketed together in a container through which the second fluid will pass. As the exchanger works, the tubes, over a period of time, will experience a build up of precipitate from the material flowing through them. The tubes are initially of the order of 19 mm in diameter, and as the bore becomes smaller, so the effectiveness of the heat exchange process becomes reduced, requiring that the tubes be cleaned. Concurrently build-up of material on the outside of the tubes may also limit the effectiveness of the heat transfer. Access to remove this outer layer, known as that on the shell-side of the tube, is more difficult since it must be achieved through the "forest" of tubes which comprise the body of the heat exchanger.

Historically, several methods have been used to clean the tube, including driving rods down through the tubes; flame torches to burn out the deposits; a sand or grit blasting operation; and mechanical drilling. Of these, the first three can only be used in very particular types of materials, and drilling has been by far the most common method used. Drilling, however, requires that considerable care be taken by the operator to insure that the tube bundles themselves are not damaged, and it also consumes a large labor force and time. Ward [3.1] has given one example of a 990 tube bundle, some 3.6 m long, which took 450 worker-hours to clean (Fig. 3.1). At a waterjet pressure of 700 bar and a flow rate of 92 lpm, it required less than 24 man-hours to clean the same bundle using waterjets.

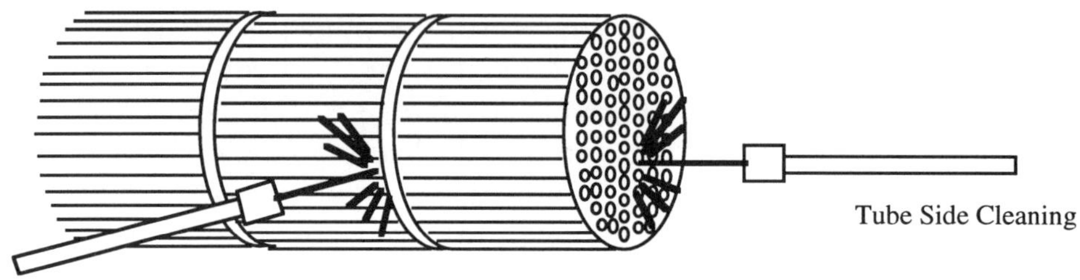

Figure 3.1 Cleaning directions for a heat exchanger bundle.

The surface prepared by high pressure jets is usually much cleaner than that found with the historical drilling or grit blasting techniques. This is because the water can penetrate into the small cracks and fissures within the surface and wash them free of contaminants without distorting the material. This is often not the case with the more conventional mechanical tools, where not only do the solid objects prove to be too large to get into the finer cracks, but they can also bend over some of the surface so that it traps corrosion pockets within the surface, lowering the time until the surface must be re-cleaned. The water further will flush into pockets outside the direct line of the jet, as it rebounds around internal part contours, and will carry the debris removed from the part.

As a result of this, the build-up of material, when the bundle is reintroduced into service, is much slower, increasing the service life of the heat exchanger before cleaning becomes necessary again by some 50% or more. This advantage is not restricted to heat exchanger tubes. For example, ships, hulls must be cleaned of animal and vegetable life, old paint and rust must be removed from most surfaces before repainting, and, in coke ovens the build-up of tar and other deposits from doors and vents, must be removed. Waterjet systems have been developed both on a hand operated basis and as semi-automatic systems for cleaning surfaces such as these.

Where clean water is used the surface is much less prone to further corrosion. It has been suggested that the reason for this is that the water is better able than mechanical tools, to remove the soluble sulfates and chlorides, the main causes of steel corrosion, at the same time that it is removing the rust and scale [3.2]. This is not a small advantage as shown by the size of the industry affected. By the early 1980s, some $600 million a year was being been spent by the construction industry in the United Kingdom alone on repainting and cleaning surfaces which have become corroded due to the failure of the initial protective coating.

In the historical cleaning of heat exchanger tube bundles the material which clogs the internal pipes of the exchanger is frequently very hard. To remove this material mechanically a striker bar could be driven down through the tube, fracturing and pulverizing this lining. However, should the bar be deflected, the force of the impacting tool could quite easily drive the bar through the relatively thin walls of the tube, damaging the vessel and requiring an expensive repair. It is also an inexact and often impossible tool to use in removing the fouling which may occur on the shell-side of the tubes.

In contrast, the waterjet pressure in a heat exchanger tube cleaning operation will frequently be on the order of 700 bar. This pressure, while high enough to erode and remove the mineral deposits, is considerably below that which would be needed to cut into the metal. Thus the cleaning can proceed with less worry about damaging the vessel. This ability of a waterjet system to be tailored to cut only the desired material, while leaving the remaining material undamaged is one of the major advantages of this technology, and it has a range of applications which go well beyond cleaning. In addition, the ability to use a flexible line to penetrate into the bundle, which can be achieved in cleaning some pressurized water reactors (PWRs) with the full jet pressure available at the end of the lance, makes the system more powerful for use in these installations than competing systems. The benefit gained from the cleaning is illustrated by one case where a plant lost 12 MWe of power, and an annual revenue of over $5 million, due to tube deposits [3.3].

A second illustration of the range of this practicality lies at the other extreme of waterjet use. The blockage of human arteries also occurs when an unwanted lining coats a tube which cannot be ruptured. Again by careful choice of jet pressure and size, this unwanted material can be broken out and aspirated away without damaging the surrounding arterial wall [3.4]. Other medical applications of this type will be discussed in Chapter 12.

Waterjet cleaning systems come in a variety of combinations, depending upon the particular need and the requirements of the application. At the lowest extreme the small pulsating water pick, used in dental hygiene and for debriding wounds in emergency rooms, allows a relatively gentle method of removing unwanted dirt without much pain or damage to the patient. At the other extreme the high cleaning rates, and low water usage required for a 3,500 bar system has led to the development of small cleaning guns which are used to remove paint and corrosion and to clean metal surfaces before repainting.

While many of the initial applications of waterjets in this field were somewhat mundane, the increasing complexity of some of the tasks has led to innovative solutions which can, with little effort, be transferred to other applications. For example, low pressure waterjets have been developed which are self-propelling and which move down sewer lines removing wall deposits. With time these have been modified so that smaller versions can be used in cleaning the pipes and tubes found in nuclear reactors. It is a relatively short step from this cleaning of existing pipes to the creation of new holes, i.e., the use of a modified system to drill through rock and soil. This Chapter will deal with cleaning applications, although some of the ways in which waterjets are used in other applications, described in other Chapters, may also be helpful for cleaning use.

3.2 EARLIER AND COMPETING METHODS OF CLEANING

Waterjet use is slowly replacing other methods for surface cleaning. This is particularly the case with the use of sandblasting, which has been, for many years, the most popular method of surface cleaning. In order to understand some of the benefits of waterjetting, and also to appreciate some of the ways in which the two systems might be combined, it is valuable to begin by reviewing the development of some of these alternative technologies.

There are two ways in which high velocity streams are used for cleaning which are not generally considered to be high pressure waterjetting practice. One is in the use of abrasive as the cleaning mechanism, with the particles being collected and driven down onto the surface with a high impact speed by a stream of pressurized air. The other is the use of steam as the cleaning agent. Since this latter leads into the use of water as the cleaning medium, the use of air-entrained abrasive will be briefly reviewed first.

3.2.1 AIRBORNE ABRASIVE CLEANING

The use of compressed air to accelerate a stream of abrasive particles down onto a surface, either for cleaning or cutting purposes, was apparently invented by B.C. Tilghman of Philadelphia in 1870. It is interesting to note that, although the original apparatus was designed for use with steam as the driving force, the patent which was awarded, number 2147 in the United Kingdom [3.5], described a system wherein the abrasive was carried and projected by means "of a jet of steam, air, water and other suitable gaseous or liquid medium." As Plaster points out [3.6] it also covered "any method or arrangement of jets or currents of steam, air, water etc." and provided for the situation whereby "the sand may be propelled by a current of air produced by suction, or a partial vacuum." The patent was thus a far reaching one, which foretold the development of many of the devices which have since become common to the cleaning industry. It is pertinent to note, given the history of such devices, that much of the initial market for abrasive jets was foreseen as being in the mining and cutting of rock. Thus the patent included the sentence "When a jet of water under heavy pressure is used, as in hydraulic mining, the addition of sand will cause it to cut away hard and close grained substances, upon which water alone would have little or no effect."

Until 1890 the main power to drive the jet of particles was provided by steam (Fig. 3.2), since this was the least expensive and most powerful system available. By the turn of the century this began to be replaced by compressed air, which has since been developed extensively, and is now the main carrier used in developing the abrasive stream.

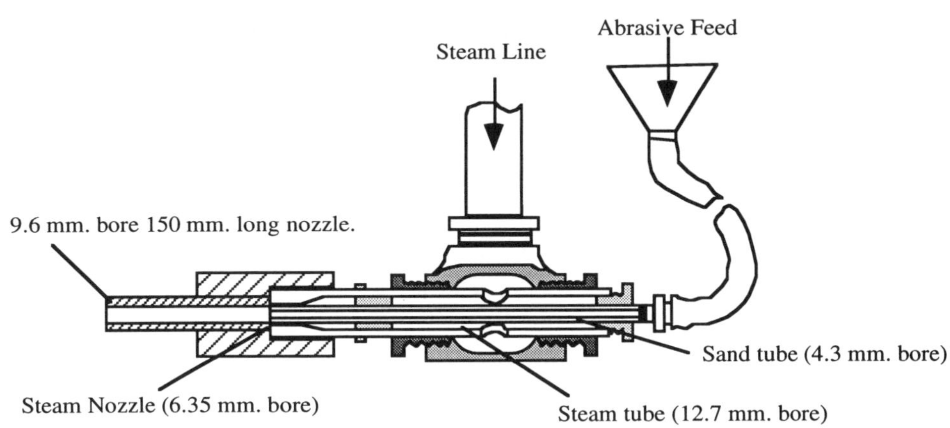

Figure 3.2 Original steam injected sand-blasting design [3.6].

The original machine was operated by steam at a pressure of up to 29 bar, since Tilghman had shown that system efficiency was better at higher pressures. Sand was fed, from the funnel, down through a pipe into a narrow (4.3 mm) tube centered in the steam tube, which was 12.7 mm in diameter. Steam entered this tube from ports, fed in turn from a supply line. Over the last 5.8 mm the bore of the outer tube was reduced to 6.35 mm leaving a narrow (0.37 mm) annulus about the steam pipe. The high velocity of the steam, as it then flowed out over the nozzle of the sand pipe was sufficient to draw sand into the stream. The resulting jet was collimated by a length of chilled iron pipe, some 150 mm long and 6.5 mm in diameter which acted as a collimating nozzle. It was found, experimentally, that putting flat plates out from the end of the nozzle, aligned with its edges, gave a better jet, with less lateral spreading when grooves or straight cuts were required. These were typically some 75 mm long.

Steam, however, tended to wet the sand, which would then adhere to the sides of the pipes, causing blockages, and also to the working surface Problems also arose with poor visibility, and unpleasantly hot and wet working conditions have also been cited [3.6]. Thus there was an incentive to change to air power, to replace the steam. By 1984 production rates for such systems of 0.36 sq m/minute could be achieved by one man working with a 9 kW compressor.

Abrasive cleaning has since reached sufficient market penetration that systems can be readily assembled for under $200 from components available at most local hardware stores. The low cost, and relative simplicity of the method have made it popular, and the use of abrasive has two significant advantages. Not only will the particles remove virtually any coating from a surface, they will also "tooth" that surface, by incising small cuts into it (Fig. 3.3). These rough patterns will provide a better anchorage to hold any subsequent paint or coating applied to the surface. The illustration shows the cuts made by individual particles in striking a steel surface. Under normal impact conditions, however, many particles will strike the surface so that it becomes indented by the particles and the individual impact cuts overlap. Under such conditions a much rougher surface is generated into which the coating can penetrate and anchor.

The smallest sand-blasting units operate on much the same principle as the largest units but frequently at lower air pressures, smaller particle sizes and with lower abrasive feed rates. Units can be purchased in which very fine abrasive, (typically a 27 to 50 micron sized range of aluminum oxide) is drawn from a small bottle held under the operating gun. Air at a

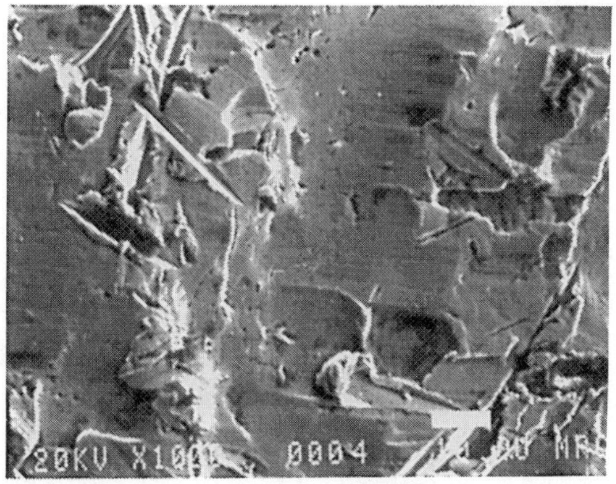

Figure 3.3 "Toothed" surface after abrasive particle impact.

pressure of 5 bar is fed to the gun from either a small reservoir or a small air compressor.

By controlling the type of abrasive, the quantity and the air pressure this very fine stream can be used either for cutting materials, such as, for example, silicon semi-conductors, or it can be used for cleaning surfaces. The abrasive gives sufficient power to the stream that it can be used to precisely deburr hypodermic needles, and it can be used to etch patterns into glass and other material for artistic purposes.

With an alternate abrasive, one of the softer minerals for example, it is possible to use this same system to delicately remove the smudges and dirt from ancient documents and to carefully clean statuary using fine crushed walnut shells (although this latter is now being replaced, on occasion, with waterjetting).

Nozzles for this type of small scale operation can be made from either sapphire or tungsten carbide. They can be either round or flat in shape and for the most delicate work, will typically be on the order of 0.13 - 0.8 mm in diameter. It is possible, using this system, to cut slots with a width of 0.013 mm in selected materials. Abrasive feed rates for such a system is on the order of 5 gm per minute. With an air pressure of 5 bar such a device will use approximately 0.008 cu m air/min to accelerate the particles to a speed of over 300 m/sec. Under such conditions the nozzle used will last on the order of 35 hours [3.7].

Where the air is supplied to the operating gun without abrasive (Fig. 3.4) the flow is controlled by an adjustable valve. Once past the valve, the air is accelerated through an initial nozzle into an enlarged chamber. Set in the side of the chamber is the feed line to the abrasive supply. The passage of the air through the mixing chamber generates a vacuum in the chamber, drawing abrasive up the feed line, and mixing it with the air. The combined flow is then collimated, and the air particles brought up to speed in the exit nozzle which directs the resulting stream down onto the target surface.

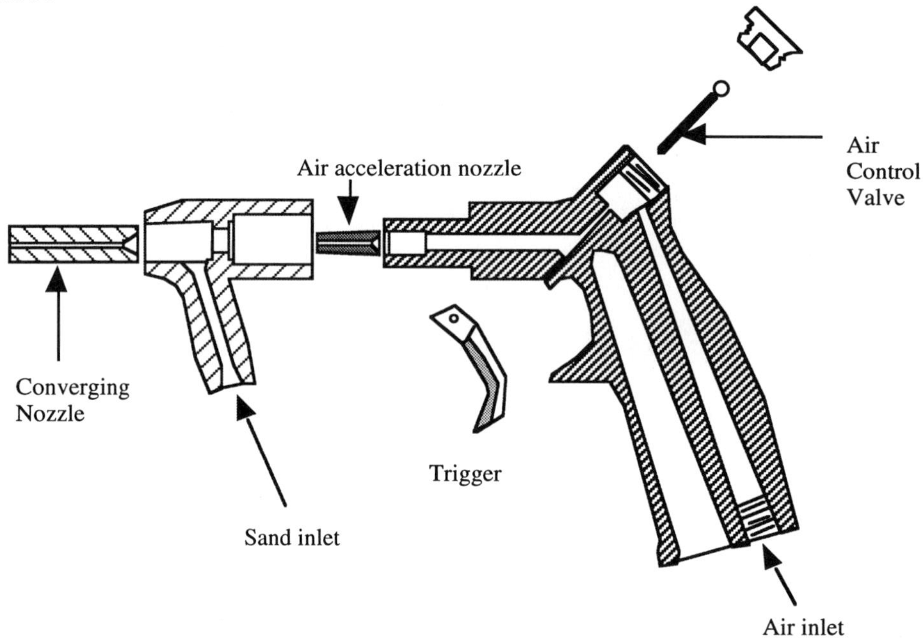

Figure 3.4 Schematic representation of an abrasive injection gun.

On a larger scale it is no longer as easy to retain the abrasive in a container attached to the cleaning gun. Under these circumstances it is possible to exchange the feed bottle for a hose which can be run to a nearby hopper from which it can draw abrasive into the feed line. Such a method of supply becomes cumbersome as the volume of material gets larger, and becomes less reliable as a means of providing a steady flow of abrasive to the operator. In addition this method of supply requires that two hoses, one for air and one for abrasive, must be attached to the cleaning gun. This increases the weight of the assembly the operator must support, and reduces mobility.

At the same time as this problem was being studied, concerns had arisen about the exchange of energy between the air and the abrasive. The mixing chamber is quite short, and although the collimating nozzle was lengthened to improve the energy transfer, there was still some acceleration of the particles occurring after they left the nozzle. This can be inferred from a study which showed that as the workpiece was moved away from the nozzle, for a short increment, the cleaning rate increased (Fig. 3.5).

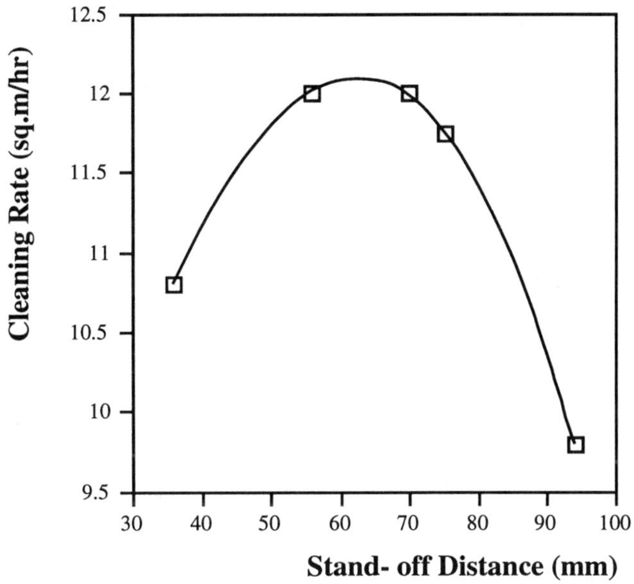

Figure 3.5 Effect of change in stand-off distance on cleaning rate [3.8].

There are two conventional ways in which this problem is overcome. Both feed the abrasive into the line from a supply hopper which can be pressurized. This has the potential benefit of increasing particle velocities by over 250% [3.6] and the development of such as device, known as a sand "pot" was foreseen in Tilghman's earliest patent. It should be noted, however, that where the sand is fed into the line at the nozzle, the systems are, in general, lighter and the process can be made more continuous, and will use less compressed air, than where a pressurized pot system is used.

The first idea is merely to move the mixing chamber to the bottom of the supply hopper. As the air passes the quantity of sand feeding into the line is controlled by a metering valve. However this process becomes much more accurate if the sand is pressure fed out of the hopper into the feed line (Fig. 3.6). Air pressure in this case acts both to push the sand out of the hopper and to aspirate it into the air stream. This provides a better mixing of the sand in the air and thus a better use of the resulting abrasive stream.

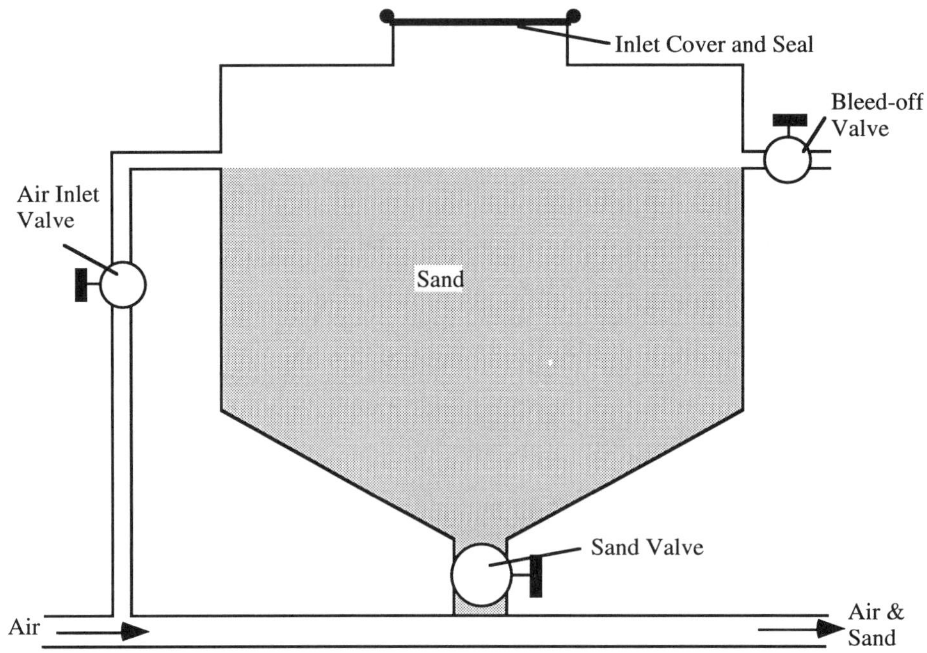

Figure 3.6 Early pressurized sand injection system [3.6].

One factor which must be borne in mind when using air:abrasive systems is the requirement that the feed of abrasive, particularly into the nozzle, be free of moisture, since any water present in the sand can lead to a build-up of material on the walls of the supply lines and the hopper so that flow is reduced and ultimately may be blocked. This is particularly the case with lower pressured systems which use vacuum to draw the sand into the mixing chamber, where the feed pressure is thus quite low. For these reasons it is important, even when pressurized systems are assembled, to make sure that only dry air is fed into the sand pot. Even moisture in the feed air will cause the sand to cake and block the feed out of the pot. Water should be removed by placing traps between the compressor and the unit (Fig. 3.7).

Large scale surface blasting has three separate purposes, for which it can normally be used. It can be used to remove loose material from a surface, a process which is normally known as brush-off blasting. It can be used for commercial cleaning of surfaces to remove contaminants from the surface, or it can be used more intensely to cut down through all the overlying layers, and to begin to penetrate the final surface to be left. It is this latter process which will leave the toothed surface mentioned above.

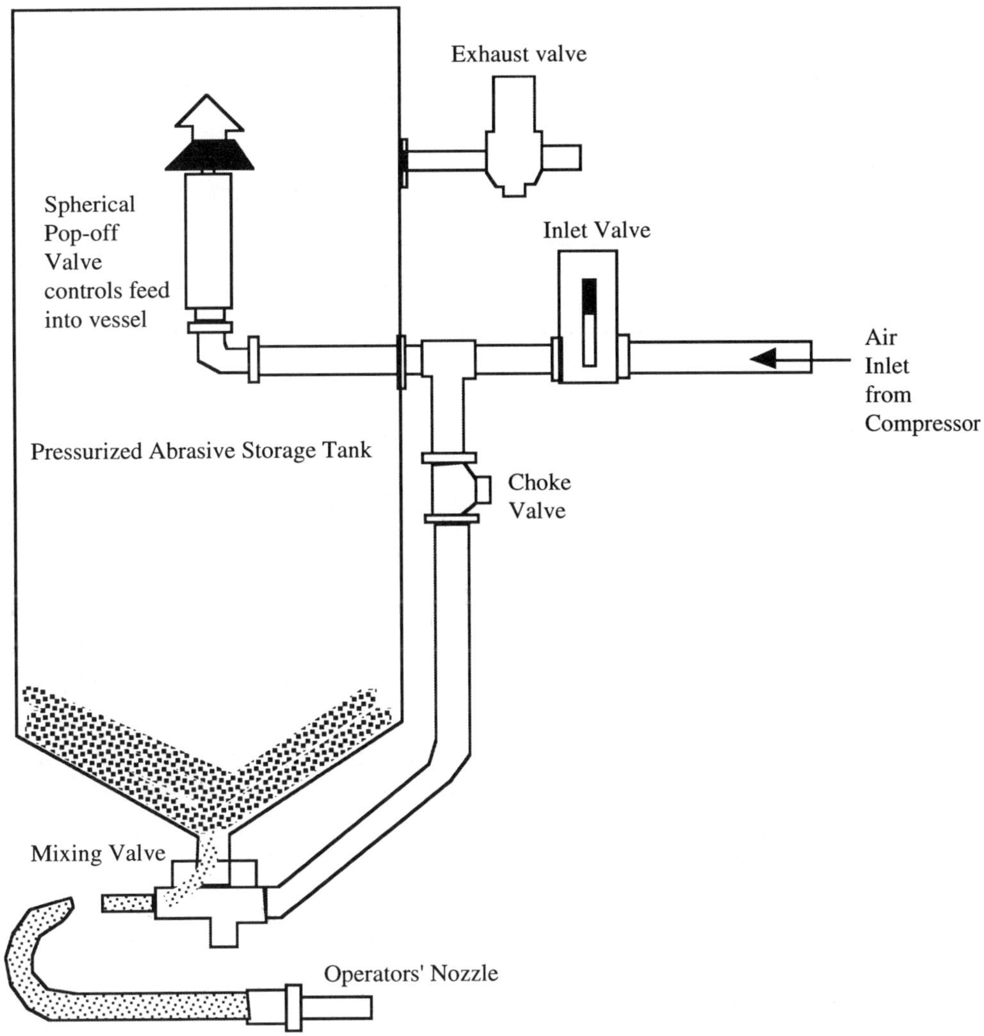

Figure 3.7 More modern pressurized sand mixing system [3.6].

In the normal use of abrasives for cleaning, two factors appear of most importance: the velocity and the concentration of the particles which impact on the surface. These parameters are, in turn, controlled by the pressure and volume of the air flow which is used, and the concentration of abrasive fed into the line. Normal commercial cleaning will use pressures of 5 to 8 bar depending upon the surface to be cleaned. Lower pressures will be used on surfaces such a brick or glass, while the higher pressure will be required to clean metallic surfaces. Relatively large volumes of air are required, typically 2 - 10 cu m/min, which must be supplied at a steady flow, both to ensure even mixing and application of the abrasive over the surface, but also to minimize holding problems for the operator of the equipment.

Table 3.1 Blasting pressures which may be used on various surfaces

Material	*Pressure (bar)*
sheet steel	2 - 4
glass, porcelain	2 - 4
wood carving	2 - 5.5
non-ferrous metals	2.7 - 3.4
marble carving	2.7 - 4
granite carving	4 - 6
heavy rolled steel	6 - 7
stucco buildings	6 - 7

The design of the cleaning system should also recognize that the use of an abrasive in the air line feed from the hopper to the nozzle will fill some of the volume in that line. Thus, in designing the size of line to feed from the hopper this additional volume must be included in the calculation. Where a 6 mm diameter nozzle is used at an air pressure of around 7.0 bar the volume of air required will be reduced by approximately 20% when a normal sand loading is mixed with the air [3.6].

3.2.2 AIR ABRASIVE NOZZLE DESIGN AND MANUFACTURE

There are four different parameters which are involved in the selection of a nozzle for abrasive cleaning. These are the nozzle length, the nozzle shape, the nozzle diameter and the nozzle material.

For the same volume of air flow, velocity increases as the cross-sectional area of the nozzle reduces. However, if sand is mixed in with that air, the driving force must be applied for a finite amount of time to properly accelerate the sand. The longer the nozzle is, within reason, the greater the energy transfer that will occur. Studies have shown [3.9] that where the nozzle length is doubled from 75 mm then the production from the nozzle will increase by 17%. Plaster [3.6] has reported on trials by Adlassing and John [3.10] which compared nozzle length, bore diameter and pressure on the removal of brass by quartz sand.

Table 3.2 Erosion rate as a function of nozzle length and pressure [3.10]

Air pressure (bar)	*Nozzle length (mm)*	*Brass removed (gm)*
2	50	6
2	150	11
4	50	20
4	150	33

When the effect of change in diameter was included, the relative levels of power could also be considered, based on the increased air volume required.

Table 3.3 Relative efficiency of brass removal [3.10]

Bore dia. (mm)	*Air pressure (bar)*	*Air volume (cu m/min)*	*Brass removed (gm)*	*Rel. efficiency*
6	2	1.13	6	26.54
13	2	4.58	16	17.46
6	4	1.89	13	17.19
13	4	7.78	46	14.78

It appears better to increase pressure than flow rate since, in doubling the pressure proportionately more effect is achieved than by increasing the bore. However, if the relative efficiency of the process is examined (by dividing the mass loss by the air pressure and the air volume) it can be seen that the smaller nozzle is clearly more efficient when operated at the lower pressure.

Short nozzles are, however, generally used where it is difficult to maneuver the longer nozzle, and where the operator stands closer to the surface. In the latter case the lower speed of the sand reduces the amount and velocity of the rebound which may hit that operator. Normal nozzle lengths lie in the range from 75 to 150 mm.

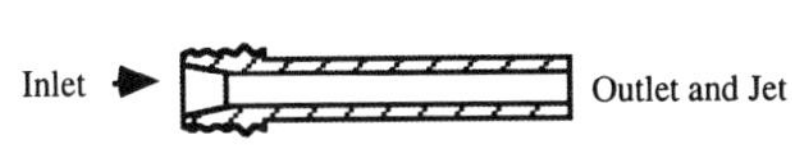

Figure 3.8 Early sand-blasting nozzle.

Until 1954, the nozzles used in sand-blasting operations consisted of a relatively short conic acceleration section followed by a long straight passage (Fig. 3.8). In that year, nozzles were developed in which a venturi section was inserted into the section (Fig. 3.9) and these have become known as laval nozzles. The addition creates a short throat for the nozzle followed by a diverging section of the nozzle.

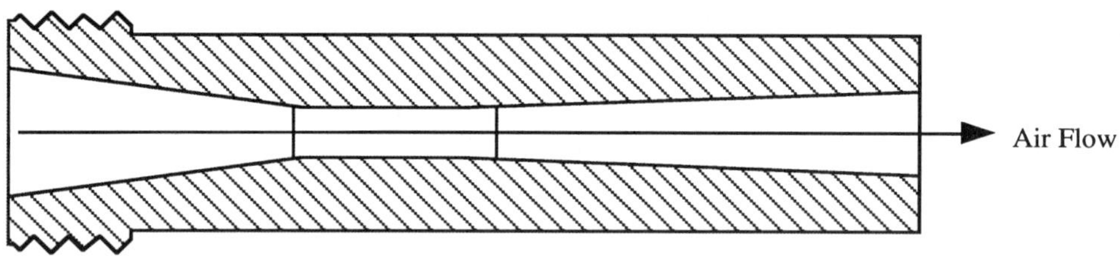

Figure 3.9 Venturi or laval design for an abrasive feed nozzle.

The advantage of this section is that, after the constant diameter, or throat section, the increasing diameter of the channel causes a drop in pressure in the nozzle. This, in turn, allows an acceleration of the air and abrasive and the velocity resulting is more than twice as high as it otherwise might reach. According to one report [3.11] velocity changes from 85 m/sec to 200 m/sec were measured, while another [3.9] reported an increase from 120 - 200 m/sec. In addition the area of target effectively cleaned has been reported as increasing by 30 - 40%, giving the range of values shown. Nozzle life is also reported to increase with this design change [3.6].

A specific patent on such a design (Fig. 3.10) was assigned to Albert in 1955 [3.12]. It should be noted that the transition from tapered to straight sections in the nozzle are carefully radiused. All dimensions are said to be critical. Air enters the nozzle at roughly 0.2 of the speed of sound, which it reaches in the throat. Pressure along the nozzle falls from 7 bar at the entry, to 3.5 bar in the throat, and to atmospheric (1 bar) at the nozzle exit.

It is reported that there is an average 30% increase in effectiveness in changing from a 150 mm long straight nozzle to a 200 mm venturi, with a resulting improvement ranging from 15 - 70% depending upon the target condition. With the aid of the venturi the pressure distribution across the nozzle is more even, such that the blast pattern becomes more effective. It is reported [3.13] that between 15 and 45% more work will be achieved with a venturi, with a reduction in abrasive of around 20%.

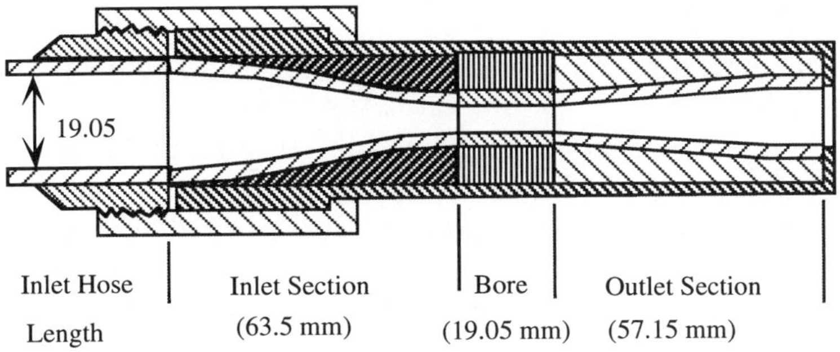

Inlet Diameter 19.05 mm, Bore Diameter 7.93 mm, Outlet Diameter 10.31 mm.

Figure 3.10 Albert's patented supersonic nozzle [3.6].

Table 3.4 Typical efficiency of some sand-blasting nozzles [3.13] (using as grit blasting sand)

Air pressure (bar)	*Area covered (sq cm/minute) with a nozzle diameter of:* 5 mm	6 mm	8 mm	10 mm
1.0	143	167	200	226
2.0	255	300	362	433
3.0	362	435	520	618
4.0	456	570	665	810
5.0	551	665	841	1000
6.0	643	800	1000	1140

In regard to the choice of nozzle to use, several factors are involved. A general recommendation appears to be that the venturi should be shaped with the exit diameter approximately twice that of the minimum bore. For a given volume of air flow, the smaller the nozzle diameter then the higher the air velocity must be. Higher pressure results in a higher particle velocity, which is generally found to be more efficient (Fig. 3.11).

Figures have been presented [3.9] which show that a reduction of 20% in pressure reduces effectiveness by 34%, while a pressure drop of 40% reduces effectiveness by 50%.

Nozzle size should be related to the grit and hose diameter, in order to ensure that there is no blockage of the line of the feed section into the nozzle. It is suggested that the optimum bore diameter be four times the largest grit size used, and that the nozzle be no smaller than a quarter of the inner hose diameter, for these reasons. Cleaning rate increases as approximately the square of the nozzle diameter. Thus doubling the diameter of the nozzle will quadruple the area of cleaning, provided that the same concentration of abrasive is kept, and the air velocity is maintained.

The life of the nozzle is a function of the abrasive which is used for cleaning. Under most conditions common sand is the cheapest and thus most effective abrasive. With sand, under normal conditions ceramic nozzles will last 1 - 2 hours, cast iron nozzles will last from 3 - 8 hours, tungsten carbide lasts approximately 300 hours, and norbide will last from 750 - 1000 hours, while boron carbide is claimed to be able to last up to 1500 hours. It is reported that the use of steel shot as the abrasive will increase most nozzle life by up to 2.5 times [3.9].

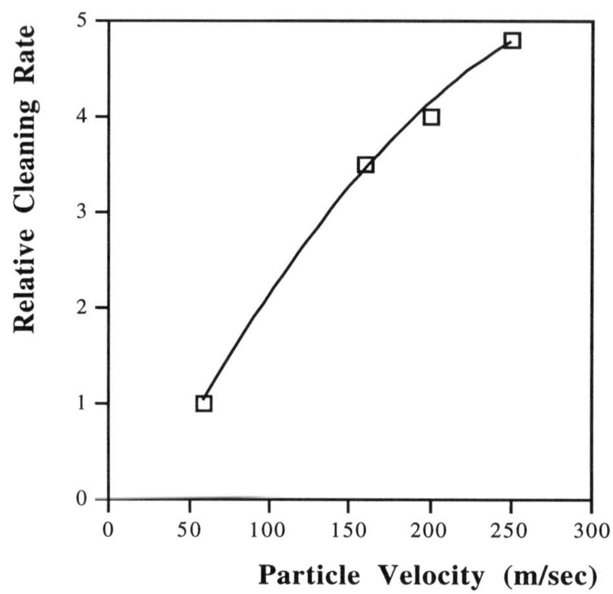

Figure 3.11 Improved cleaning rate with higher particle velocity [3.6].

Table 3.5 Sand consumption and production for a given nozzle type

Diameter (mm)	*Power (kW)*	*Air volume (cu m/hr)*	*Sand used (kg/hr)*	*Area clean (sq m/hr)*	*Relative efficiency*	
					m²/kW	*m²/kg*
					Power	*Abrasive*
2.38	1.5 - 2.25	10	45	0.8	0.4	0.018
3.18	2.25 - 3.75	25	67	1.6	0.53	0.023
4.0	3.75 - 7.5	40	90	3.2	0.64	0.035
4.76	7.5 - 10	65	135	4.8	0.55	0.035
6.35	15	130	225	6.4	0.43	0.028
7.93	30	200	360	9.6	0.32	0.027

3.2.3 CLEANING EFFECTIVENESS

It is important, in assessing the parameters to be used in cleaning a surface, to first establish "how clean is clean." For example, some foundry casts, such as manhole covers, are deliberately only cleaned of sand down to the black oxide layer [3.6]. Similarly, in cleaning some statues there is a benefit in not cleaning right down to bare metal. In contrast, in many surface cleaning cases, especially where the surface is being prepared for coating, the particles are expected to totally clean the surface down to fresh metal. If a single particle is not effective (and this may depend on the layer, the velocity, the angle of impact and the abrasive type) then the delivery system must be designed to ensure multiple impacts on the surface. Such impacts may, however, bend the surface to trap earlier particles, or residual surface contamination. Thus, for example, in preparing bath surfaces for enameling, it is reported [3.6] that changing abrasive type created a problem. Initially chilled iron grit was an effective abrasive, but this was changed to a mixture which included cut wire pellets, then unacceptable defects began to appear in the subsequent enamel coating. The wire appeared to give a "polished" appearance to the surface, and this was removed with a second application of iron grit, yet defects continued. It was only when the wire use was discontinued that a "zero" defect state was re-achieved. This suggests that it was a trapping of the occasional wire particle in the deformed surface which was creating the problem.

Such a residual contamination has alternatively been found to occur in cleaning stainless steel where the use of iron as an abrasive would leave an unacceptable corrodible film on the surface, not found when steel shot is used [3.6]. The role of individual particles in impact cleaning can be demonstrated by examining the thickness of scale left on a surface as a function of the amount of abrasive striking the surface (Fig. 3.12).

It is interesting to note that the scale removed by the end of the two second test shown was equivalent to that removed by a 45 second exposure of the surface to a 90° C sulfuric acid bath.

Resulting cleanliness is difficult to assess. One method has been to compare the relative pull strength of sprayed coatings, for example consider (Table 3.6) the results where new chilled iron grit was blasted across a surface before the sprayed on coating was applied.

These results are used purely for a specific illustration, not as a general rule, but they do suggest that if the cleaning is done at too high a pressure, that the impacts may "bend" the surface beyond the point where the most effective anchoring occurs.

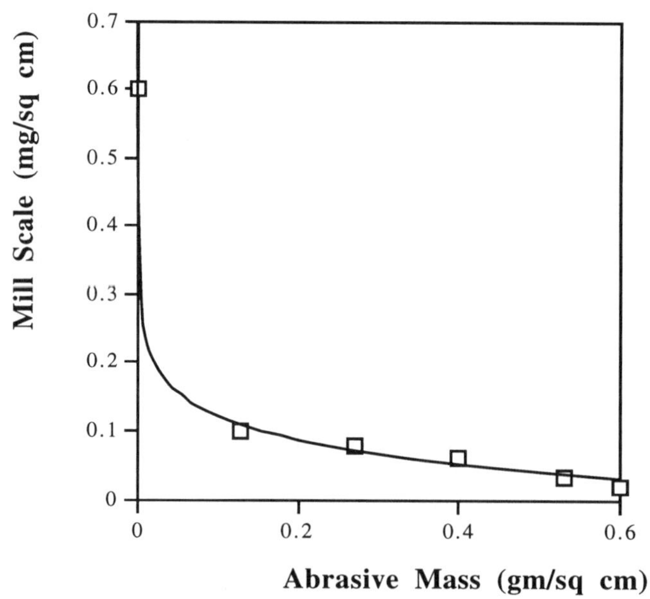

Figure 3.12 Residual scale as a function of abrasive impacting [3.14].

Table 3.6 Average pull strength of a coating as a function of abrasive cleaning pressure [3.15]

Air pressure (bar)	*Relative pull strength*
2.1	100
3.15	139
4.2	89

An alternative method of evaluating the cleaning effectiveness is to examine the final surface roughness. Again this is a function of the target material, abrasive type, and size and abrasive air jet pressure. Consider the effect of garnet particle size on surface finish, where a fine finish is required (Fig. 3.13).

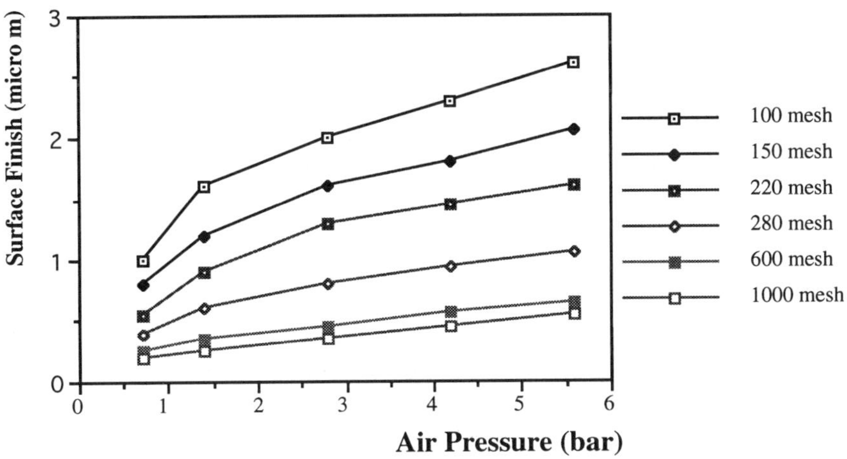

Figure 3.13 Effect of abrasive size and pressure on resulting surface finish [3.6].

3.2.4 ANGLE OF IMPACT

The optimum angle for cleaning effectiveness will vary with the surface, the coating to be removed and the type of abrasive being used. It is known, (and will be discussed in Chapter 11) that there is a significant effect on the optimum angle of impact, depending on whether the material under impact responds in a brittle or a ductile way. What may not be as widely recognized is an apparent equivalent effect due to abrasive type. Plaster [3.6] has quoted Neville [3.15] as finding that there is a significant variation in the optimum angle for material removal from steel, depending on the type of abrasive being used (Fig. 3.14).

In metal removal one can thus find that an impact angle of 45 degrees can remove as much as three times the amount of material removed with an impact angle of 90 degrees. However, it is reported that when removing a rust layer this changes, and as the layer of rust gets thicker, that the angle of impact should be brought closer to the perpendicular to the surface.

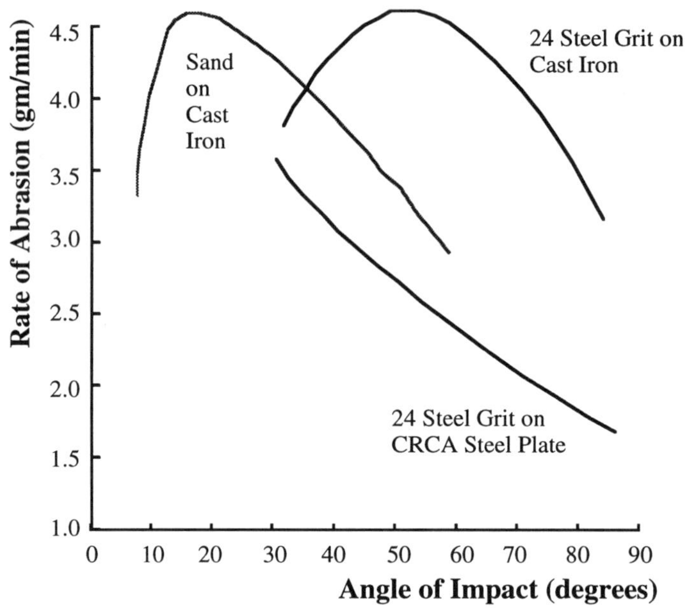

Figure 3.14 Optimum angle of cleaning varies with abrasive type [3.15].

3.2.5 AIRBORNE ABRASIVE MACHINING

While airborne abrasive have been used successfully for cleaning, their use in cutting applications is more limited. Kobayashi has recently studied this use for machining material [3.16]. The investigation studied the use of differing nozzle designs in improving the velocity of the exiting particles. Particle speed was determined by measuring the projected length of the particle image in a film exposed to the stream in a double exposure, with a 20 microsecond interval between the exposures.

Results of the experiments showed that with a plain air jet that there was an optimum performance of the nozzle which occurred when the exit orifice (Fig. 3.15) was greater, by 1.5 - 2.0 times the throat diameter (in this case 1 mm) (Fig. 3.16). Such an optimum value for the nozzle design was not confirmed when an airborne abrasive jet was used.

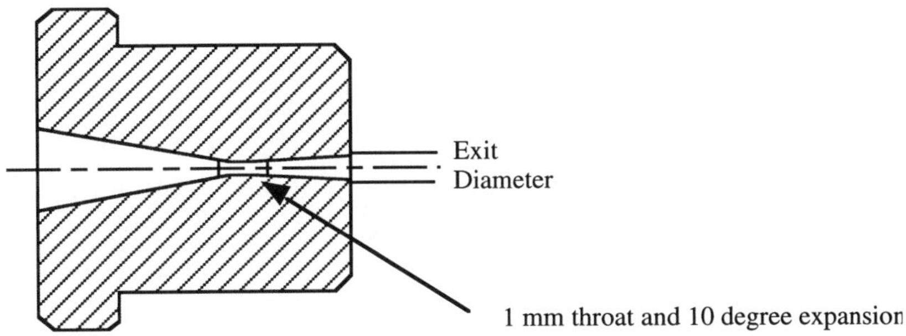

Figure 3.15 Design of air nozzle used by Kobayashi et al. [3.16].

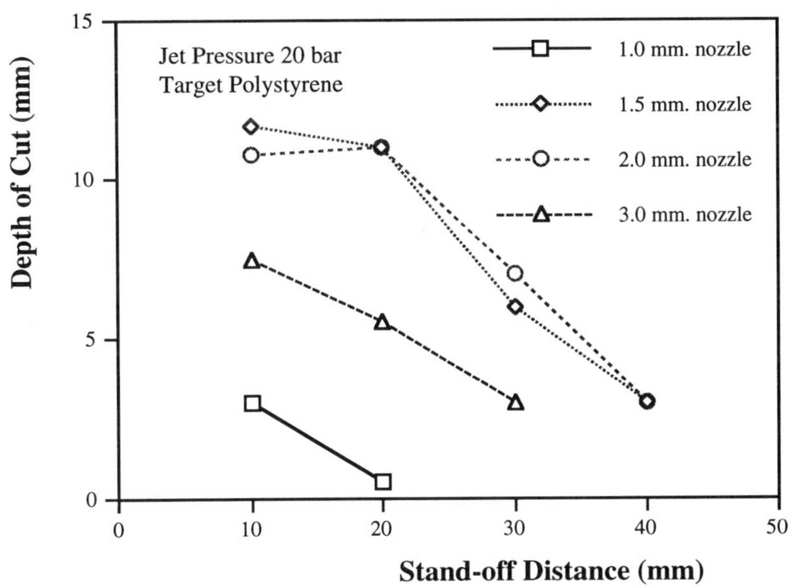

Figure 3.16 Depth of cut as a function of air abrasive nozzle diameter [3.16].

Tests were also carried out at higher pressures (up to 120 bar), and in contrast with results found at lower pressure with conventional abrasive jet cleaning, the results showed that the nozzles with the smaller diameters gave better results than those with a greater exit diameter (Fig. 3.17). The reason for this is postulated to be that the smaller nozzle diameters were able to accelerate the particles more efficiently. This was demonstrated by the velocity measurements which showed that the particles from the 2.0 mm diameter nozzle were moving at approximately twice the speed of the nozzles from the 4 mm diameter nozzle. It was important, however, to control the particle feed rate to achieve this optimization. While increased depth occurred as particle flow increased, beyond a flow of 200 gm/min at a pressure of 80 bar, through a 1 mm throat on the nozzle, (Fig. 3.18) it was found that the cutting depth was reduced due to inefficient particle acceleration under an overload condition. This indicated the significant effect of particle velocity on the erosion rate.

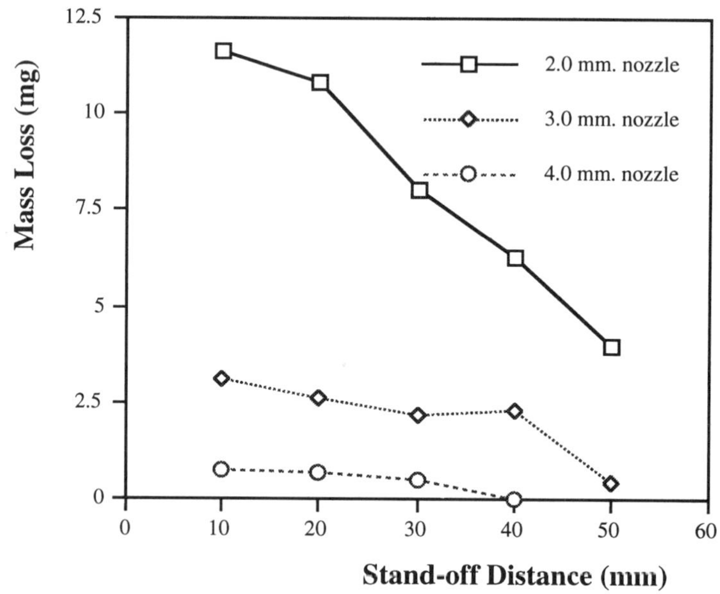

Figure 3.17 The effect of stream diameter and distance on the erosion of aluminum [3.16].

Studies on the effect of increasing jet pressure indicated that as pressure increased the optimum stand-off distance for material removal moved away from the nozzle. This suggests that the particles are still being accelerated after they leave the nozzle, but to an extent controlled by the pressure (Fig. 3.19).

This has considerable significance since the investigation showed that the mass loss due to particle impact on aluminum was related to the third power of the impact particle velocity. This confirms the need to adjust the particle feed and nozzle design conditions in order to optimize the resulting particle velocity before it reaches the target.

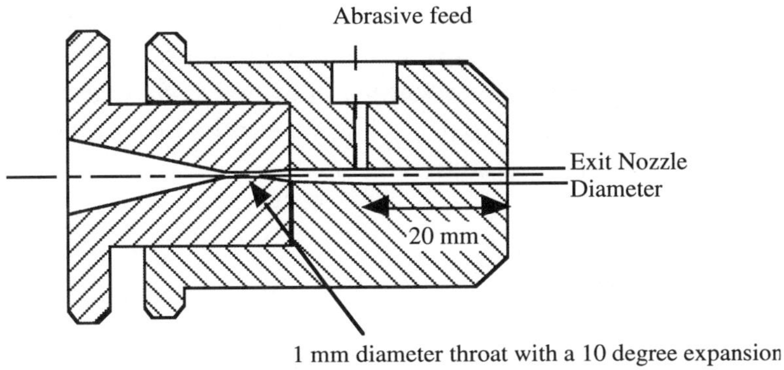

Figure 3.18 Kobayashi abrasive injection nozzle [3.16].

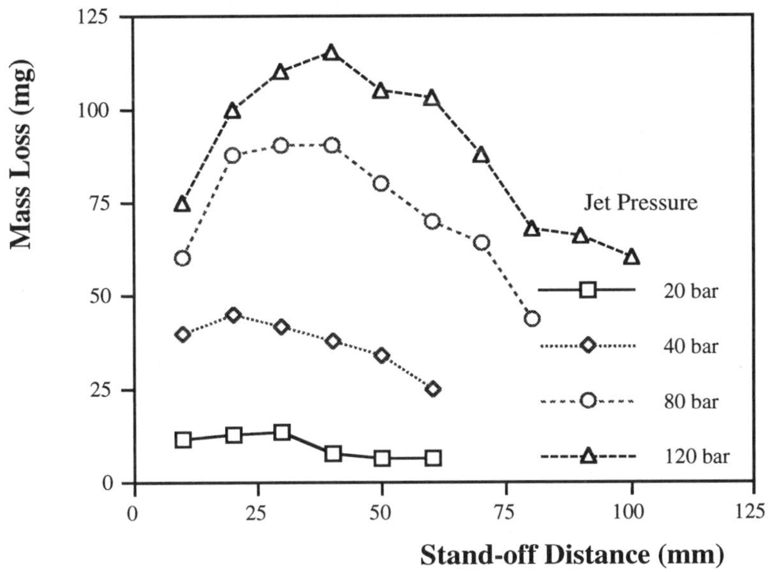

Figure 3.19 Change in mass loss with air jet pressure and distance [3.16].

3.2.6 ABRASIVE USE

Although Tilghman's generic patent was based on the use of sand, this was defined as "small grains or particles of any hard substance, of any degree of fineness, of which quartz sand is a type." In many applications chilled iron pellets are used extensively, cut wire pellets and steel grit and shot have also been developed. Plaster [3.6] reported that over a million tons of metallic abrasive was being used annually in cleaning by 1972. Glass beads have been used to clean tire molds, and ice particles have found applications in cleaning delicate parts. These applications, and others, will be discussed in subsequent Chapters.

3.2.7 STEAM POWERED SYSTEMS

While low pressure waterjets, fed largely by gravity, have been used in cleaning for centuries, one of the first cleaning applications under pressure occurred when steam was used, either alone, or with a water spray inclusion, at the beginning of this century. At its simplest level steam cleaning requires a boiler to heat the water, a delivery line, and a nozzle to accelerate the stream and to direct it onto the target surface.

Steam, and hot water have a considerable ability to enhance cleaning, particularly of surfaces which have a hydrocarbon contamination or where fats must be emulsified. Tests at UMR, for example, have shown that where water is heated to more than 85 °C, that a significantly improved cleaning rate can be achieved relative to that at lower temperatures. However, as with the delivery of most forms of power, transporting heat from the boiler to the target surface is not without the potential for considerable loss.

This loss, in a predominantly steam jet comes in two forms, firstly the radiant loss of heat through the delivery line itself, and secondly the velocity and heat loss as the resulting jet moves through the surrounding air. While some of the pipeline loss can be overcome by insulation, this is not as practical in the portable sections of the cleaning system. This loss must be considered and included in system design calculations in order to ensure that sufficient power reaches the surface.

Most steam cleaners have historically operated at a pressure of between 7 and 15 bar and use between 30 and 100 kg/hr of steam. At an inlet temperature of 10 °C, [3.17] a 50 kg system operating at 10 bar would require approximately 30 kW to raise the resulting steam to 100 °C. Within 30 cm of the nozzle the temperature of this steam will have dropped by 33% and by 1 m from the nozzle it will have fallen by 50% [3.18]. At this greater distance the steam velocity will have decreased to 16% of its original value and the jet will no longer retain any cleaning power. For this reason steam jets must be operated with the nozzle held relatively close to the surface to be cleaned. In addition an initial mechanical scrubbing of the surface is often required.

The mechanisms of jet disintegration have been discussed earlier, and result because the smaller and faster a droplet is moving through the air, then the faster it will be decelerated and broken up into smaller droplets [3.19]. This suggests that the power of a steam jet could be carriedfurther if the jet were to entrain larger water droplets as it left the nozzle.

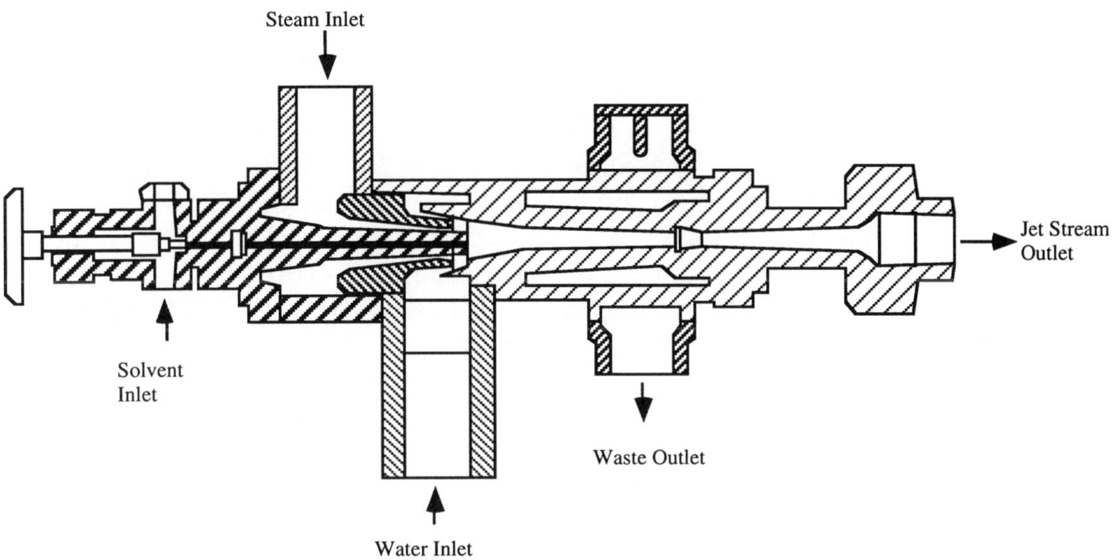

Figure 3.20 Schematic of a steam injection system.

This led initially to a concept for water injection through a venturi nozzle, marketed as a device called the hydraulic scrubber (Fig. 3.20). It is, perhaps appropriate to discuss the design of this nozzle in some detail, since the principle of its operation has been used in many devices since, to draw a variety of different materials into the fluid stream.

Steam is fed into the nozzle section through a large diameter pipe, and is then directed into a converging nozzle section, through which it accelerates. A central tube is fed through this nozzle with a tip located just beyond the exit section. At the exit the steam acceleration nozzle is slightly flared outward to permit the steam to expand, and thus to slightly reduce the pressure on the downstream side of the nozzle. This reduction in pressure is used to draw a detergent through the central tube, and mix it into the center of the cleaning spray. At the same time as this mixing is taking place the jet is passing though a converging chamber which transforms into a second acceleration nozzle. Around the original nozzle a feed pipe carries water which is also drawn into the converging chamber by the vacuum created in the chamber by the passage of the jet into the second nozzle. The water breaks into droplets as it is caught up in the jet stream and accelerated by the steam, through the converging nozzle. A final expansion section on the exit of the scrubber is used to give the jet its exit velocity and direct it down onto the target surface. This device, operated at 20 bar, has been found to clean surfaces some 30% faster than mechanical scrubbing and steam cleaning, and reduced the amount of detergent required to clean a meat processing plant by 20% [3.20].

One problem with the use of a heated fan jet, is that the heat will cause the water to partially "flash" as it leaves the nozzle. This has an immediate size reduction on the resulting spray. Short, for example, [3.21] has shown that while a normal atomizer will produce a spray with a Sauter mean droplet diameter of 250 microns, when heated to 100 ^oC, the average droplet diameter was 50 microns, in the size range from 20 - 120 microns. Studies by Liljedahl [3.19] have shown that, at 10 cm from the nozzle, droplets of 12 micron size had been decelerated to 3 m/sec while droplets which were 136 micron size were still traveling at 30 m/sec. Thus, the hotter jets will not reach as far from the nozzle, and will lose their power more rapidly.

These problems with the cutting range of the jet, and its rapid loss in power, ultimately led to a process where normal waterjet systems were redesigned to include a heating section. This can be as simple as passing the delivery line in a loop around the inside of a heating system on its way from the pump to the nozzle (Fig. 3.21). Commercial sales literature has suggested [3.22] that hot high pressure washers have three to four times the performance capacity of comparable steam cleaners.

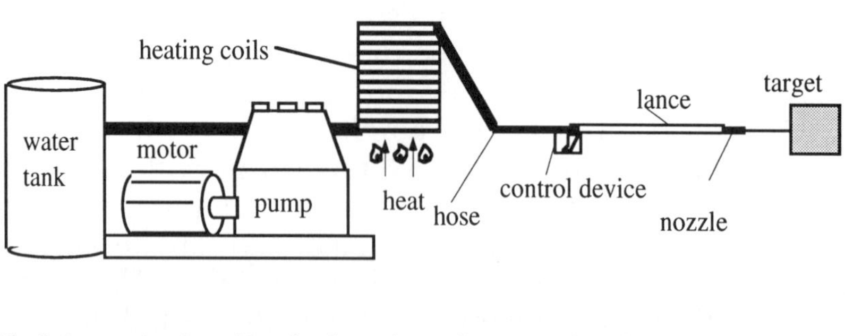

Figure 3.21 Schematic showing the location of a water heating coil.

Heated systems have continued to be used for the removal of grease and hydrocarbons. This is because of the ease with which these materials can be removed as their adhesion and viscosity are reduced with temperature rise. There are occasional problems where the units are used for sterilization purposes. If the temperature of the impacting jet falls too far below the boiling point then instead of removing the bacteria from the surface it is possible that they will be cultured and there are reports of increased contamination after a "hot" pressure wash, rather than a decrease as would be expected [3.23]. However, if the water temperature is increased above 85 °C then the surfaces are reported to be effectively cleaned [3.24]. These temperatures are important in cleaning areas where food is processed. Fish, for example, is sensitive to the use of high temperature which will reduce quality, and thus, while the clean-up after the fish is gone may use hot water, at 95 °C, one may find only cold water used in cleaning the fish tanks and food processing equipment. Other food processing plants such as dairies and beef processors have also found heat to be helpful [3.25] although temperatures used may be lower at 50 - 60 °C. Other precautions required in those applications may include the need for stainless steel fluid ends, water rather than oil cooling of the pistons, and the ability to pump drinking water grade fluid.

Significantly there are many operations, particularly at pressures below 350 bar, in which hot water has been found to give a significantly better result than cold. This is particularly true with oil and grease contaminated surfaces, cleaning car engines [3.26] where with a 23 lpm 100 bar unit will take half the time with 80 °C that it takes at ambient pressure. At the same time the operator reported that the use of chemicals, such as detergents in the water, was also cut by a third at the higher temperature. At higher pressures, 700 bar, a similar effect was reported in building cleaning [3.27] with the time taken to clean a petroleum station being cut from 4 - 6 hours down to 2 when the water was heated to 85 °C.

Many lower pressure cleaning systems use chemicals as part of the process. Particularly below 350 bar, heating the water has been found to be beneficial on a cost basis. Not only is the time required reduced, but the chemicals are also found to work more effectively so that the quantity of them used can be reduced. Sometimes the chemicals are applied first, and allowed to "work" on the surface for up to 20 minutes before the pressure washer is applied to rinse the surface clean [3.27] and other times the chemical is aspirated in with the pressurized water. Water at lower pressure can sometimes be used with the hotter system, and this can be critical, for example in the cleaning of historical buildings where the

100 - 130 bar conventional pressure might be too great and only 85 bar might be allowed to protect the fabric from erosion.

The economics of heating the water depend both on the material being removed, the pressure and flow of the delivery system, and the way in which the water is heated. Often the water is heated by running the pressurized line as a coil around a heat source. As a guide on the size of such [3.28] if that source were 100% efficient it might take 1.9 lphr of fuel oil to heat a flow of 13.25 lpm of water by 11 ^{o}C. Most boilers are however, less efficient, running from 65 - 95% depending on a variety of factors.

The requirement for more robust boilers and greater fuel to feed the larger flow rates common with higher pressure systems make their use less common. However, there can still be a considerable benefit to their use. For example, heating the water helps in the removal of plastic contaminants from gas collection tubes in chemical plants [3.29]. Bury, et al., have reported that the introduction of steam around a waterjet nozzle can improve the performance of a cleaning nozzle at a pressure of 350 bar by 12% in these circumstances. As well as providing an energy boost to the water, the role of the steam was believed to lower the adhesion between the plastic and the pipe wall. It is important, however, to ensure that the steam is supplied dry, requiring that the steam line to the lance be lagged, thereby making the process slightly more cumbersome.

Lufthansa has recently been reported to save over 10 million Dmark a year, by changing from the use of chemicals to jet cleaning at pressures of up to 500 bar. While a chemical pre-spray to swell the paint (using agents such as benzyl alcohol) is reported to be sometimes needed [3.30]. Working with WOMA a formula was developed to adjust the pressure and temperature of the wash water depending on the type of paint, its age and thickness. This has led to the construction of an automated facility in Hamburg where six rotating jets will strip the paint from an aircraft automatically and remotely, with the cost anticipated as being recovered in a year. Boeing tested the process to ensure that the treatment did not damage the skin of the aircraft and approved it for use in 1989. The process did not damage the skin, in contrast to the use of ice crystals and plastic granules, which have only been permitted to be used once on the skin of a plane, because of the deformation possible on the surface. Even at higher pressures (3,500 bar) water is reported [3.31] to cause less than a third of the distress to aircraft surfaces that is found with grit blasting.

It has also been reported that where the water is heated significantly above boiling temperatures, that an enhanced material removal was obtained. This was the subject of a discussion at a Cleaning Equipment Manufacturers' Association Meeting some years ago and a paper at the 9th International Symposium on Jet Cutting Technology [3.32] but has otherwise been little documented. It would appear that the jet performance is only improved over a relatively narrow temperature range, due to the more rapid disintegration of the jet due to flashing into steam at the nozzle. However, where the temperature provides a small amount of superheat, then the erosion rates by the jet can be doubled over that of cold water (Fig. 3.22).

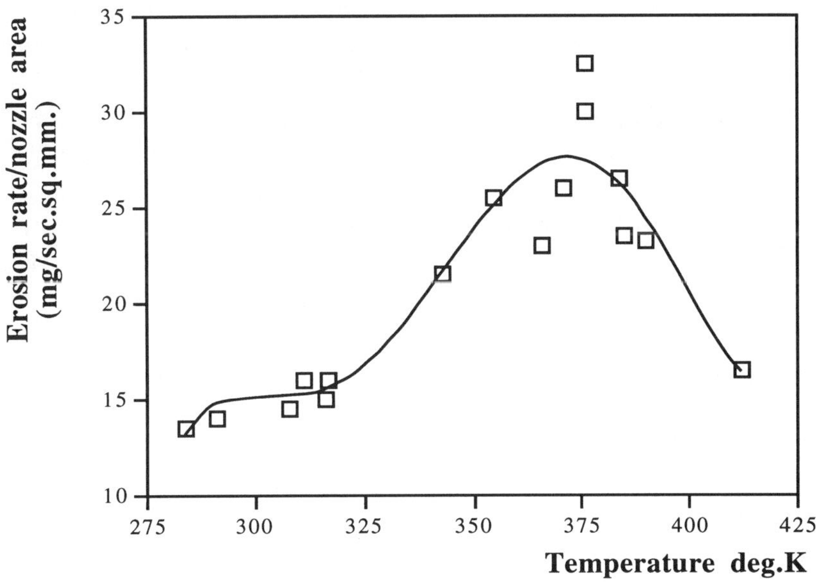

Figure 3.22 Improved cleaning effect with superheated water [3.32] - note that the interpolated curve has been changed from the original.

3.3 LOW PRESSURE WATERJET CLEANING

The use of waterjets for cleaning has largely developed over the 20th century, with most of the growth occurring after the Second World War. A major impetus to change has been as a replacement for chemical methods of cleaning [3.33] although, more recently, waterjets are also being seen as a means to replace sand and grit blasting and the use of needle guns and jackhammers [3.34].

There have been a number of different attempts to define the pressure ranges over which high pressure water is used. For this text, the upper limit on pressure for low pressure waterjet cleaning will be considered to be 350 bar. This is not a completely arbitrary number. It is chosen because, in cleaning applications, this is the pressure above which industrial manufacturers usually feel that too much energy is required to improve jet performance by heating the water. This is not always true, since there are specialized applications where heat is added to higher pressure systems, but stands as a general rule of thumb. Interestingly at these pressures it is more common to hear the hand-held lance called a wand, rather than the lance term which is the more popular term at higher pressures.

At 2 to 40 lpm, the major use of waterjets is largely in industrial and domestic cleaning. Systems are available from a number of sources, and products are increasingly becoming available for use by individual home owners. It has been suggested that there are as many as 40,000 contract cleaners in the United States [3.35] and while many of them do not yet use waterjetting systems the market for equipment alone is currently (1993) estimated as being on the order of $100 million a year with a potential for a 5 - 10% growth in the near term, both in market penetration and new applications [3.36]. As an example, a contractor with two assistants (one of whom cleans the gutters, while the other cleans the windows) currently can clean a home in about 90 minutes using a 40 lpm 200 bar unit for which he expects to get paid around $75 an hour [3.37].

At its most simple, a unit will consist of a prime mover, such as a motor, which drives a small pump either directly or through a fan belt. In the low end of this application many of these units are either driven by electric motors or small gasoline engines. Water from the pump is fed from the pump to a hose, which, in turn, carries it to the lance, on the end of which a nozzle is fixed (Fig. 3.23). A simple bypass circuit is usually included in the line to allow control of the volume of water going to the nozzle. A dump valve on the lance diverts the water through a large diameter pipe, thus keeping it at low pressure, until the trigger is activated. At that point a valve closes directing all the water coming into the lance through the nozzle. This causes it to accelerate and provides the pressure for operation.

It is not necessary to use either very high pressure, or great flow rates to find useful applications for such equipment. For example a 5 lpm unit working at 30 bar can effectively clean the coils of an air conditioning system. [3.38]. Higher pressures would be a disadvantage for this application since the fins on the cooling coils are quite fragile and could bend where any significant force is applied to them. Similarly higher pressures would not help in cleaning carpets, where additional force would drive the water too deeply into the fibers, that it could not easily be vacuumed out and could lead to mildew and early carpet failure. This type of unit can be powered by a 12 V DC pump.

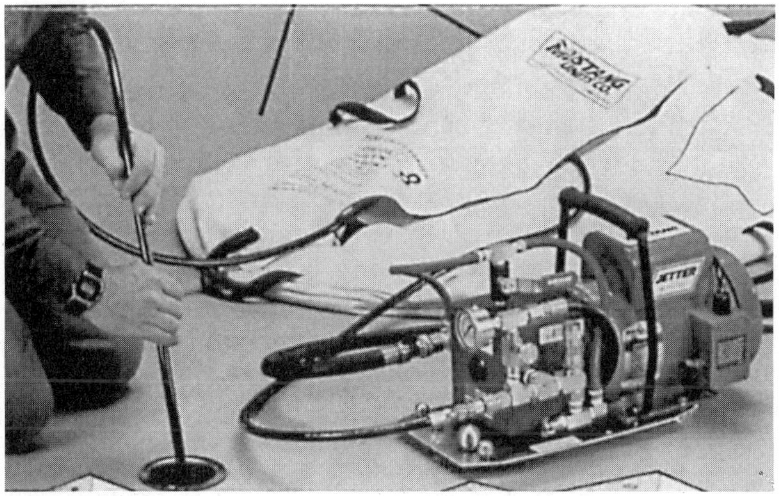

Figure 3.23 Simplified components of a jet cleaning system (courtesy of Mustang Units Company).

Building on this basic system, various solutions have been proposed to increase the performance of the equipment. Cleaning systems which use burners to heat the water have become popular, and, as discussed above, there is a significant advantage, particularly in the removal of hydrocarbon and similar dirt, if the water is heated to a high degree, although to be effective this would appear to require temperatures in excess of 50° C in some applications and 85 °C in those cases where there are grease and oil problems.

Alternatively chemical additives, usually detergents, are also a common practice. The most familiar is, perhaps, for use in washing cars. In the car wash industry, there are typically four different systems commonly available [3.39]. "Tunnel" car washes, where the car is moved through a frame tunnel and sequentially washed and cleaned by different sets of equipment, use 15 - 30 bar where the jets are used in conjunction with brushes and rotating cloth wipes. Where the tunnel does not use a contact brush then the rinsing jet pressures are usually increased to 70 bar. If the car stands in a bay and the equipment moves over it the jet pressure is higher yet, whether or not the system is "touchless." Self-service units are generally touchless and the customer uses a wand operated at 60 - 70 bar and chemicals to clean the car. As a side note soaps serve a dual function in the contact brush cleaning process. Not only do they help remove the dirt, they also serve to lubricate the movement of the brushes over the surface and to reduce the risk of the grit being retained in the brush and scratching the car.

Water recovery systems are becoming mandatory for use with car wash stations, as government regulators, such as the Environmental Protection Agency (EPA) in the United States become increasingly concerned about the run off of contaminated water. They also make economic sense since the cost of water, and the removed materials and sludge can prove to be a greater expense than the supply. In 1993 water reclamation systems were priced upwards from $10,000 depending on the volume flows. This has had a significant impact on the smallest of operations where it has been estimated that it could take up to 5 years to recover this cost. Sewer costs can be 300% of the basic water bill for such operations. For more sophisticated operations water collection, clean-up and recirculation systems have been priced at up to $50,000.

The addition of chemicals in the cleaning stream complicates clean-up, particularly where it is used as a direct feed into the waterjet supply to the nozzle. The chemicals are most commonly sold on the basis that they help in reducing the surface tension which holds the dirt onto the surface. It is interesting to carry out a simple calculation in this regard.

If the jet is traveling at some 150 m/sec and flows over the car surface for a distance of 5 cm, then the time the jet will be on the surface is on the order of 0.0003 seconds. This appears to be too short a time for the chemical to effectively achieve all that is required. Detergents and similar solvents should therefore be applied to the surface as a pre-soak before the jet cleaner is used at pressure, in order to give the solvent time to penetrate into the dirt layer and thus to achieve maximum effectiveness. This is the case with most tunnel and in-bay car wash systems where the detergent is applied first and the higher pressure water only used as a rinse later. (Similarly in cleaning buildings it is often the practice that a chemical "poultice" will be applied to the building first and then, after perhaps 20 minutes, this will be washed off by a high pressure water spray [3.40]).

There are, however, chemicals, most particularly of the long-chain polymer variety, which act instead by reducing the friction in the waterjet delivery line, and by seemingly "gluing" the water together allowing the jet to be cast a greater distance [3.41]. These can be quite beneficial in a number of applications, and, since they have such a range of uses, they will be discussed in depth in Chapter 11. Suffice it to say, at present, that they can considerably enhance cleaning performance, and may, on occasion, provide a viable alternative to heating the water.

As an alternate means of improving performance it is possible to aspirate abrasives into the jet stream. This is often carried out with the same type of aspiration system used in air blasting, and it is becoming more popular to use water-soluble abrasives for this as a means of simplifying clean-up. The topic is of sufficient complexity, however, that it will be discussed as a separate topic at the end of this chapter.

3.4 TUBES, PIPES, SEWERS, AND TANKS

Reference has been made in an earlier chapter to the benefits which the U.S. Navy found in changing to the use of high pressure waterjets for cleaning the tubes of naval ship boilers [3.42]. Such tubes are just a small part of the large numbers of pipes and similar fittings that must be cleaned on a regular basis. Some of these, such as boiler tubes, are quite small, others such as sewer lines, or road tunnels are rather larger. In either case the walls become coated with unwanted material which must be taken off.

3.4.1 CLEANING SMALL TUBES

It is relatively easy to take a section of high pressure tubing, seal off the end and drill small orifice holes through the wall of the tube to give a crude set of nozzles (Fig. 3.24). Such a system can quite easily be fed into a small tube, and the resulting jets, hitting the blocking material can erode and remove it. This basic system is simple, the jets do not have to cut very far from the orifice, so nozzle quality is less of a concern, and an array of small jets can be disposed to cover almost all the surface of the tube. It is, in fact, an advantage to have the orifices somewhat less than perfect in form, since the jets must cover the entire perimeter of the tube being penetrated, and with no rotation the jets otherwise will not directly cover this area.

Given the small size of some tubes such a tool may be all that is possible, but it does suffer some disadvantages. The small tube size can lead to very high pressure losses before the jets leave the tube, often more than 50% of the driving pressure [3.42]. Simple nozzles such as these have a poor quality control in many cases, and the jets can deteriorate very rapidly with orifice wear. It has been reported that such a lance may only survive one heat exchanger cleaning.

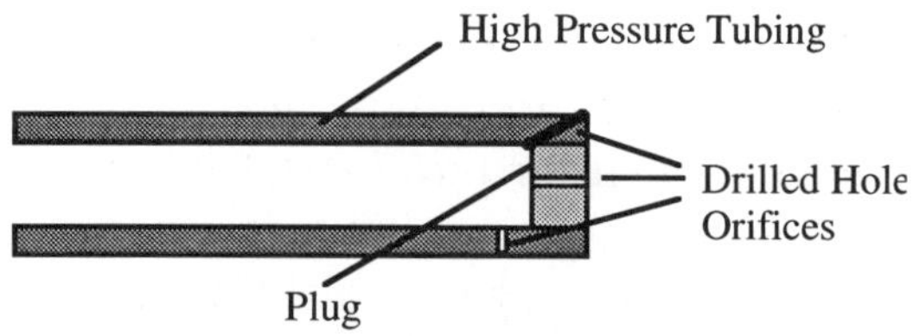

Figure 3.24 A simple nozzle for pipe cleaning.

From this simple beginning, there have been basically four types of pipe and tube cleaning nozzles developed [3.43].

There are four basic types of pipe cleaning nozzles (Fig. 3.24):

- forward cleaning, where the straight jets are oriented at up to 85 degrees from the lance axis but all point forward. These are usually used with a rigid lance.
- forward cleaning with front cutter, which have the same orientation as above but with one jet orifice pointed straight ahead.
- Retro-cleaning nozzles, in which the nozzles are pointed to the rear at up to 85 degrees from the centerline. These are usually used with flexible lances.
- Retro-cleaning with front cutter, again this is the same configuration as that immediately above but with a jet orifice pointed centrally forward.

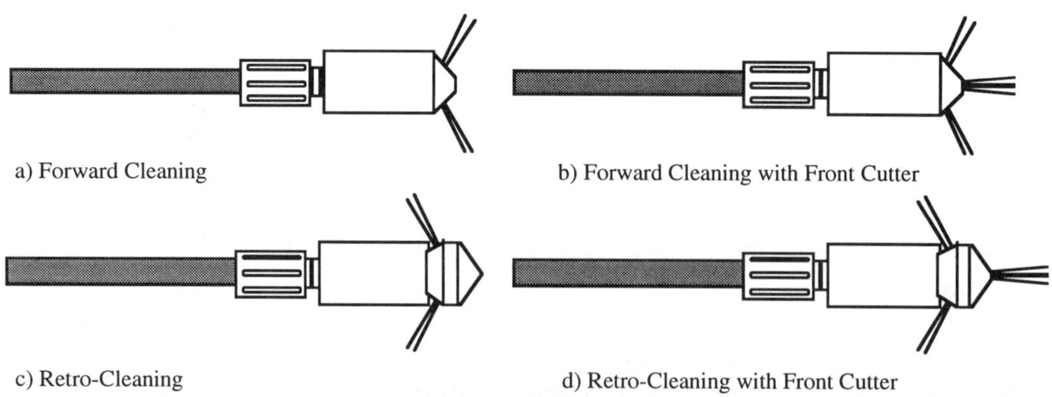

Figure 3.25 Basic types of tube cleaning nozzles [3.43].

From these simple divisions it is possible to come up with a wide variety of combinations to deal with different cleaning situations. Given the simplicity of the basic construction of the nozzle, it is readily possible to manufacture any desired combination of orifices around the central feed tube, providing that the number of nozzles, and their size is still within the capacity of the pump. As an illustration (Fig. 3.26) the patterns from the standard styles of a commercial manufacturer are illustrated.

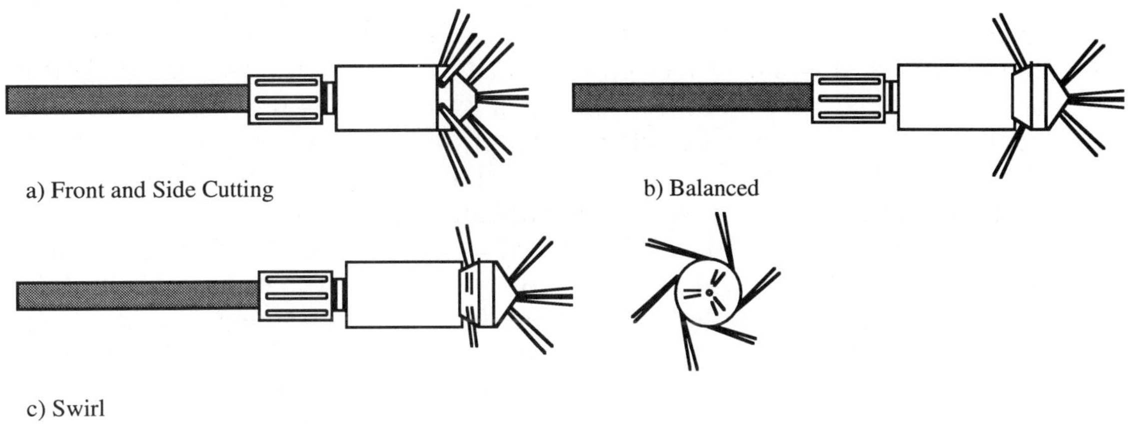

Figure 3.26 Combinations of orifices for tube cleaning [3.44].

The last of these has the advantage that it gives one method for trying to ensure that the jets cover the entire perimeter of the tube. This will happen since the jets will tend to follow along the wall of the tube after the initial impact, and will thus help peel off any material by penetrating along the tube:material interface. Manufacturers also try to cover the entire wall of the tube by adding up to 24 orifices to each nozzle. However, as well as requiring that the pump be capable of supplying these the operator must also be sure that with the large volume flow that they might require, that there is not too much friction loss in the line (see Chapter 2). If that were to happen then the nozzle pressure would be much less than the pump pressure and the efficiency of the cleaning would suffer. It is suggested [3.43] that wherever possible the least number of nozzles be used, but at the largest diameter to take advantage of the flow capacity of the pump. As an alternative approach some form of fan jet orifices can be used in the nozzle assembly.

Recent industrial development of small self rotating nozzles have overcome some of these disadvantages, although these devices can also absorb a considerable pressure loss over their length, further reducing the range of jet throw and effectiveness. The use of both rigid and flexible lances has given an increased freedom to maintenance personnel that had not existed with the old mechanical systems. Tubes that were not straight became cleanable if the rigid section of the lance is made short, and is attached to a flexible hose. The tube to be cleaned acts as a conduit to give rigidity to the hose, and guide it down the hole, while the hose flexibility allows it to negotiate bends.

It was still, however, difficult to push the hose around a complex pipe circuit, and to get enough thrust down a relatively soft hose, in order to keep the lance moving. Water power has the advantage, however, of being applicable in all directions. Thus, jets on the cleaning head could be directed, not only forward to clean out the obstacle, not only to the side to clean the walls, but also backwards along the tube (Fig. 3.25 and 3.26).

These **"retro-jets"** have two advantages, firstly they can be sized so that they provide a slightly higher reaction force to the head, than the jets pointing forward (this can be achieved by making the jets have a slightly larger surface area, without the need for more complex thrust value calculations). Secondly, they can assist in flushing the debris and spent water away from the cutting head. This is a fairly important consideration, particularly in small tubes where there is not much clearance between the nozzle body and the tube wall. This is why, incidentally, it is better not to have a perfectly circular nozzle assembly moving into a perfectly (at least when cleaned) circular tube. If there is not a passage for the debris to bypass the nozzle assembly then this may block ahead of the nozzle, or seal the passage way around it. This is particularly possible where the nozzle assembly is larger than the hose which is supporting it.

Several unfortunate results might occur as a result of this. The first may be that pressure builds up in the water ahead of the nozzle, reducing the effectiveness of the cutting jets to the point where they stop cutting. Alternately, if the blockage is a sudden one, the pressure ahead of the nozzle may get large enough to blow the assembly back down the tube. While this, in itself is dangerous to the operator feeding the nozzle into the tube, it is likely to be worse if the nozzle assembly is short, and the hose is immediately coupled, so that, in a larger tube, the nozzle can be bent back along its original passage way. Under this circumstance the nozzle is now impelled by its own reaction jets, as well as the pressure in the blocked tube, - as a result it may come back out of the tube, very fast and without warning, posing a very high risk to the operator. This is a situation which must be guarded against, and the WJTA and the British Association of High Pressure Waterjetting Contractors both recommend that a rigid lance section be placed immediately behind the nozzle assembly and that it is sufficiently long, relative to the tube bore being cleaned, (Fig. 3.27) that nozzle reversal is not possible [3.45] and [3.46].

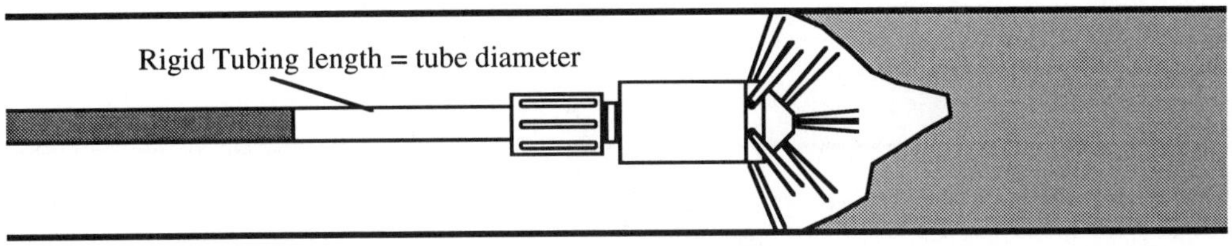

Figure 3.27 Rigid section of tubing in a line to prevent reversal.

It is especially important to ensure that the size of debris which is created by the jetting operation is able to escape past the nozzle assembly and out of the tube. Once the material has been broken from the wall of the pipe it becomes increasingly difficult to capture it within the jet stream long enough to break it further. For example, in the cleaning of scale from the pipes in a geothermal plant, Japanese investigators [3.47] first carried out a test at 392 bar to ensure that the cuttings could be broken to a small enough size, before they tried cleaning the calcium carbonate and silicate scale in place. As a result of their tests they found that the size of particles produced lay mainly in the range from 0.1 - 10 mm in size. As a result it was determined that the pipe could be cleaned by a 1.6 mm diameter nozzle. At a rotation speed of 50 rpm the jet cut through a 10 mm thick layer of calcium carbonate at a feed rate of 12 m/hr down the pipe. Thicker layers of silicate scale 20 - 30 mm thick were cleaned at 5 rpm and an advance rate of 18 m/hr except where the scale became harder, at which point the nozzle diameter was reduced to 1.2 mm, the pressure raised to 980 bar, and a cleaning speed of 16 m/hr was achieved.

The problem of flexibility within the tubes is an important issue as waterjet cleaning has become more popular for cleaning drains and narrow pipes which contain numerous bends and splits. There have been two significant developments which assist the lines in maneuvering through such passage ways [3.48]. At the same time there are other modifications which must also be made for the effective use of waterjetting equipment in cleaning drains and similar narrow lines. Back-flow prevention valves are also often required where such a unit connects to a fresh water supply.

Increasing the pulsation of the jet stream will help maneuver the head through the bends of the pipe, and adding a side jet to the assembly will also help in forcing the head around the corners within the pipe (Fig. 3.28).

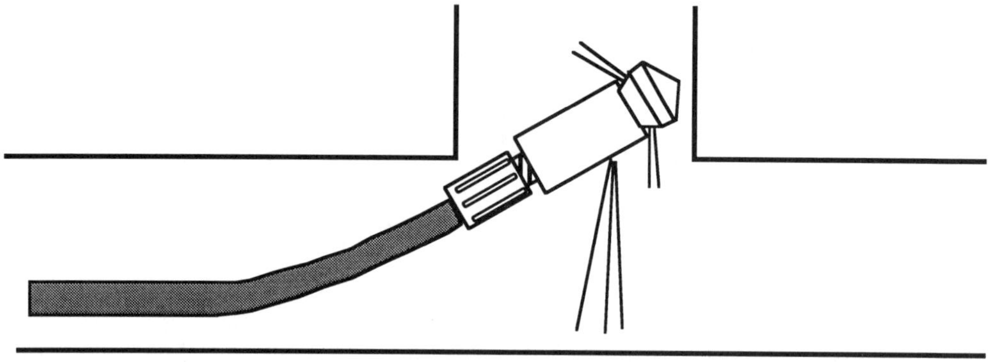

Figure 3.28 Cleaning a drain with a side jet to help make turns [3.48].

According to Henson [3.48] there are three reasons for pulsating the jet streams. The first is that this reduces the friction on the hose as it moves through the drain. The pulsation causes the hose to vibrate and this helps move it. This is particularly important in older sewer lines where the pressure is limited on the jets. The force to move the line in these cases comes from the reaction forces from the rearward pointing jets and if this is raised too high it might cause the jets to cut through the liner of the tube. Secondly pulsation helps the hose to maneuver around bends. The systems typically used for this application are around 12 kW in size, operating at 300 bar. At this pressure a fully pressurized hose is stiff, whereas under less pressure it becomes more flexible. Thus, pulsing the pressure will make the hose more limber to go around corners. The third benefit of pulsation is that when a blockage is reached the pulsation causes the head to beat on the blockage like a hammer, and will thus drive the head to beat its way through the blockage.

For drain applications the flexible requirement of the hose means that the rigid length of lance cannot be placed behind the nozzle assembly. The hose must be flexible, yet not damp out the pulsations, and it must also be light since it must almost float through any material in the line to reduce drag. One should be able to vary the pulsation rate and range in order to help drive the jetting tool, and some companies introduce equipment to enhance this pulsation and control it. These modifications, and the more immediate the need for the drain cleaning service mean that waterjet operators with this capability often charge more, perhaps as much as twice, as the rate for conventional waterjet cleaning at equivalent pressure.

3.4.2 MOLEING

A specialized technique has arisen for the use of tools of this type, known as **"moleing"**, because of the similarity to the animal. In this technique, as with drain cleaning, a flexible lance is inserted into a pipe, and uses reaction jets to pull itself along through the pipe, cleaning the surrounding walls as it progresses. The technique can be very effective, since it removes the operator fatigue that would be generated by physically holding, lifting and advancing a rigid lance. It also provides much of the same maneuverability as for the drain cleaner. However, because of the larger diameter tubes that are being cleaned it is important to introduce the rigid lance section behind the nozzle (Fig. 3.27) to prevent the flexible lance from reversing itself in the passage.

The procedure also carries with it another risk. In normal lance operations the operator is in control of the jet pressure through the trigger which controls the dump valve. Since he requires one hand to operate the trigger, and a second to steady the lance, at a point near the trigger, the operator's hands are fixed at a point behind the nozzle, and to that extent protected.

With a moleing tool, the operator needs both hands to feed the flexible hose into the tube behind the nozzle, and also to pull it out after the tube is clean. In order to control the jet pressure, therefore, the pressure controlling dump valve is frequently built into the circuit in such a way as to be foot controlled (Fig. 3.29). This itself poses no problem, except that, the operator may keep pressure on the jets as the lance is pulled out of the tube, in order to ensure better cleaning and that no residual debris is left in the tube. This creates the risk that, through a lack of warning or attention, the operator may also pull the nozzle assembly, with pressurized jets, through his hands, with a consequent severe injury. In order to minimize this risk, the hose should have both a highly visible, and physical structural marking sufficiently behind the nozzle assembly to warn the operator by sight and feel, before the nozzle exits the tube. The marker should be with some physical identifier, such as a metal ring or collar, since any color coding or other visible marker may become hidden beneath the coating that the hose may pick up as it moves through the line. These markers should also be maintained, since wear on them, as they move along the pipe, will also lead to them coming off the pipe or moving.

Because of the problems with the retro-jet system near the entrance to the tube, the operator should not use the moleing tool to start the cleaning of a tube. Rather a normal cleaning lance should clean the first 70 cm or so of the tube, so that the operator will not be exposed to the high speed fluid coming back along the pipe from the retro-jets on the nozzle.

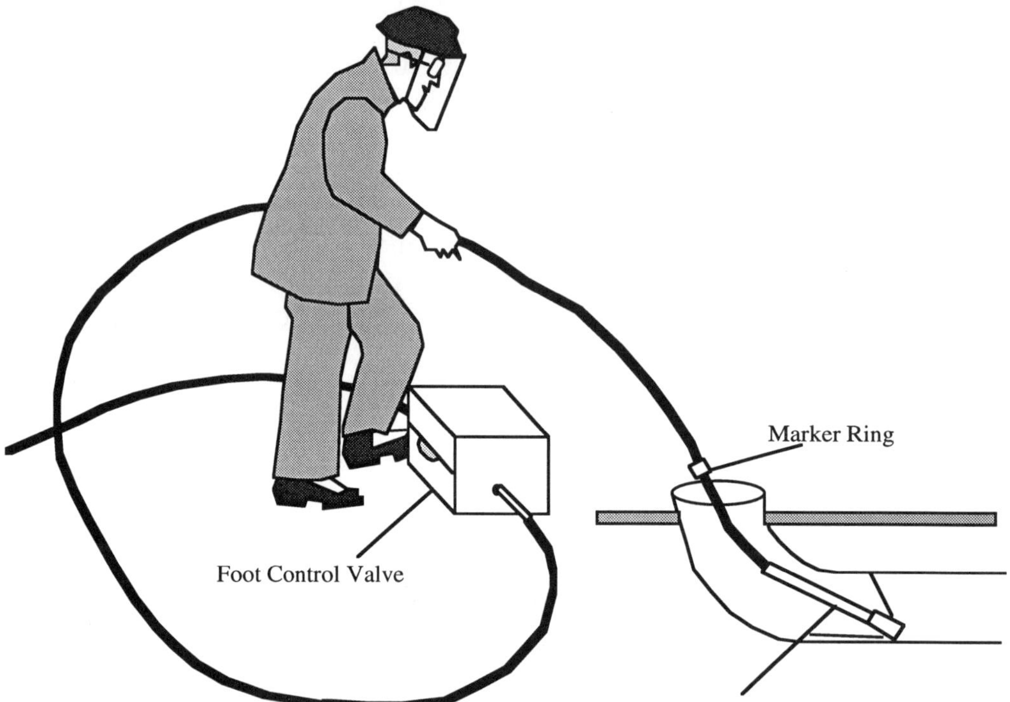

Figure 3.29 Moleing with a foot valve and hose rigid section and marker [3.49].

3.4.3 CLEANING LARGER DIAMETER TUBES

With time, the size of tubes which need to be cleaned by the jetting system increases. The simple nozzle array that would suffice to clean tube walls relatively close to the nozzle is much less effective as the target distance moves away. In order to hold the assembly equidistant from all the walls, and particularly from the floor where it may become submerged in the spent water and debris, a centering device may be used (Fig. 3.30). However, as the tube radius gets larger, having the jet nozzle in the center of the tube means that less and less power is reaching the walls. Under these circumstances, the nozzle orifices should be moved closer to the wall, by mounting them on the end of extension arms.

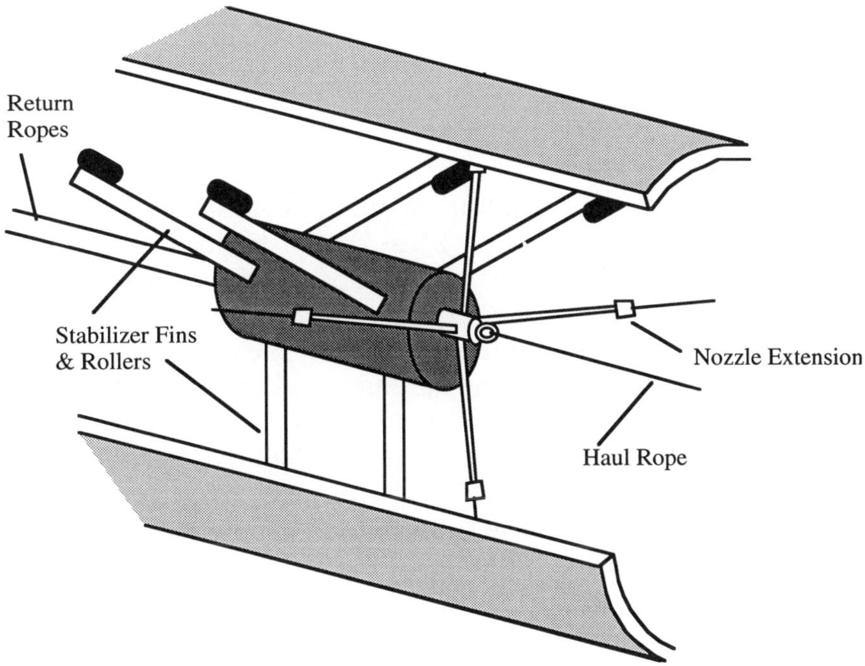

Figure 3.30 Centering device for large pipe cleaning [3.50].

An extension arm design should require some simple design considerations. As has been discussed above the water should undergo a process of increasing speed, as it moves through the line. Sudden decelerations as the water passes into larger cavities is not energy efficient. By the same token, running the water into very small diameter conduit is also extremely energy wasteful and should not be attempted unless no other alternative exists. Unfortunately the more efficient, larger diameter feed tubes are heavier and more easily unbalanced, leading to poor system performance. This makes it more critical, since there is less support from the tube itself, that the cutting nozzle assembly be designed so that it is balanced.

Balancing the jet forces does not necessarily mean that all the jets should be at the same size, and equi-angularly disposed around the distribution manifold. It does, however, mean that when the reaction forces from each of the jets is calculated, taking into account the different coefficients of discharge, and energy losses through change in jet direction, that there be no undesired out of balance forces on the nozzle. This is usually not as important along the axis of the nozzle, since reaction jets may be used to move the device forward. It is important for the forces perpendicular to the axis, since this can cause problems, should the device need to negotiate bends or obstacles.

3.4.4 SEWER CLEANING

Sewers are an interesting variation on tube cleaning, and the application brings with it different problems. Most of the build-up of sediment in a sewer will occur in the lower half of the tube [3.51]. Maximum efficiency is thus created where most of the jet energy is directed into this area. This argues against the use of simple rotating nozzles, since these will only spend a part of the time directed at the build-up and much of it directed at cleaning the overlying structure, which may be less important. Fixed nozzle arrays, and the use of fan jets may thus be advantageous. The array must, however, be designed to ensure full surface coverage, and to provide a motive thrust to the tool. These units, which can be more than 150 kW in size, operate at relatively low pressures - usually below 200 bar - and high flow rates (4,000 lpm) in order to remove the relatively thick but soft deposits normally encountered. Older sewer systems may have been penetrated by trees and pressure of up to 350 bar may be needed in those cases to cut the wood, although where the sewer line is old, precautions should be taken to ensure that the pressure is not so high as to dislodge the brick lining to the pipe. The recent advent of low pressure direct injection of abrasive systems also has a potential application in solving some of these problems.

The cleaning of sewers underlines one recognizable problem in waterjet cleaning. The process of cleaning the surface will result in the creation of debris and spent water. Where tubes are small the exit water may carry enough energy to carry the debris away with it, particularly over small lengths of sewer. In larger tubes, however, where the water does not fill the volume, the speed of the water will fall, and the debris will rapidly fall out of suspension. The debris disposal must therefore be controlled so that it does not pile up sufficiently to block the free movement of the tool in and out of the pipe, nor must it block the pipe after the job is finished. The normal way for ensuring this is to provide extraction points at suitable manholes, where the material can be removed, and this does not then become an after-treatment problem. It is easier, in moving material, if the jet can break it into small particles since these will be carried by the water over a greater distance, and they pose less risk of jamming between the tool and the wall and causing a blockage. Vacuum systems are often built into sewer cleaning equipment so that the solids can be removed more easily from the sewers and pipes [3.52].

3.5 PARAMETER CONTROL FOR EFFECTIVE CLEANING

The development of reliable, high pressure swivels, allowed the use of rotating nozzles, on a broad commercial scale. This has required a change in the design of traversing cleaning systems to ensure that the jets cover all the surface to be cleaned, preferably with a significant overlap to allow for individual nozzle inefficiencies. Zublin has addressed this point in a paper given in Rolla in 1983 [3.53].

The situation he described considers the use of a waterjet system in cleaning the deep end of an oil well. At depth the waterjet must cut through a layer of water and oil to reach the wall of the liner to be cleaned. This fluid is at a back pressure which normally restricts the jet cutting range to about ten nozzle diameters. Because of turbulence and other effects around the nozzle, jet cleaning in this condition will clean a path which is roughly eight times the width of the jet at the nozzle. The traverse speed of the jet nozzle over the surface is given by the equation:

$$V_t = ((R \cdot D_h/19.09)^2 + (V_v/60)^2)^{0.5}$$

where:

V_t is the traverse speed in cm/sec

R is the rotational speed in rpm

D_h is the inner diameter of the pipe to be cleaned (cm)

V_v is the vertical feed of the cleaning nozzle down the tube (cm/min)

This gives an effective speed that the jet is moving over the surface. However, this is the inverse of the important value. What is required is the effective amount of time that the jet should impact on the surface to ensure that the surface has been cleaned. In a continuous feed operation multiple passes are rarely an easy option, and thus the system should be designed to clean the surface with a single pass. The path can be adjusted, however, so that there is an overlap between adjacent tracks over the surface. Zublin defines this as an **overlap factor**, which counts the number of times a point on the surface is impacted by a jet. An alternative value is the **%age overlap**, which counts how much of the previous track is covered on the following one. Thus a value of 2 for the overlap factor would be equivalent to a 50 %age overlap.

There are three factors which need to be considered at this point. These are the amount of power reaching the surface, the power required to remove the material, and the overlap factor. Within the relatively short stand-off distances which are effective underwater, the power remaining in the jet can be predicted, as a function of distance from the orifice, from the equation:

$$\% \text{ power } (E_f) = 213/ (X/d)^3$$

where plain water is used, and

$$\% \text{ power } (E_f) = 2130/ (X/d)^3$$

where a long chain polymer, such as Superwater, is added to the cutting fluid.

These equations can then all be combined to give a recommended linear traverse speed of the jet over the surface in the form:

$$V_t = K \cdot D^2 \cdot p^{1.5}\, N \cdot E_f / \rho^{1.5} \cdot CE$$

where:

K is a constant
D is nozzle diameter (mm)
P is jet pressure (bar)
N is the number of passes required/cm
ρ is fluid density (which may vary in the bottom of an oil well)
CE is the relative energy required to remove different materials

Zublin's original equation was in the pound-foot units, to simplify calculations the unit conversions have been incorporated into the relative energy values, which are tabulated below. This has been standardized to a constant value of 1.0.

The calculation can be made in the following stages. Given the nozzle diameter of 0.838 mm and a jet pressure of 517 bar, pumping water, to clean a pipe of 10 cm diameter, where the nozzle is held 6.35 mm from the wall of the tube. The overlap factor is presumed to be 2 for the calculation. At this stand-off distance we presume that the path width is equal to six nozzle diameters or 0.5 cm, which gives a value for N of 4, if we are overlapping twice. If the material being removed is calcium carbonate, then this gives a traverse velocity of 7.9 cm/sec. This, in turn, translates into a cleaning rate of 119 sq cm/min or 0.71 sq m/hr. This latter value is derived by dividing the traverse rate by the number of passes per cm, and converting the value into an areal rate per minute. It should be noted that this does not include the use of polymers. Given the significant enhancement which Zublin reported from their use (which will be discussed in a later chapter) it is potentially possible to increase the areal cleaning rate by a factor of 10 when Superwater is used in this type of an underwater environment.

Table 3.7 Relative energy requirements for different tube contaminants [3.53]

Material	*Relative energy required*
barium sulfate	2.598
silicates	2.226
calcium carbonate	2.041
calcium sulfate	1.670
carbonate-sulfate-silica complexes	1.410
water scales and hydrocarbon complexes	1.187
coal tar	1.113
coke	0.928
waxes	0.742
paraffins	0.445
sludges	0.371
thixotropic materials such as mud	0.297
non-thixotropic materials	0.186

Finally one can take the areal cleaning rate given above and translate this into a feed speed for the system. With a pipe having a 10 cm diameter, if the jet is traversing over the surface at 7.9 cm/sec, then the head will be rotating at a speed of approximately 4 sec/rev or 15 rpm (this is derived by dividing the perimeter of the pipe = $\pi \cdot D$ = 31.42 cm by the traverse speed).

If the head is making four passes per cm., then the lance will be fed into the liner at a speed of roughly 3.75 cm/min. This rate may appear relatively slow to operators working on the surface, where cleaning rates in the orders of m/min are achieved. They should bear in mind, however, that this operation is being carried out at a considerable depth below the surface, where jet effectiveness is markedly reduced by the back pressure of fluid in the well.

The above calculation also only gives the predicted performance for a single nozzle, however, as in most cleaning applications, it is likely that a multiple orifice nozzle will be used. The calculation can be modified to adjust for this change. In underwater applications, however, even more so than on the surface, considerable benefit may arise from using flow through a smaller number of larger diameter nozzles than the reverse. This is because of the rapid decrease in jet effectiveness with stand-off distance which occurs in an underwater environment, where jet throws of much greater than ten nozzle diameters may not be effective. Thus smaller diameter jets lose power much more rapidly underwater than when projected through air at the surface.

3.6 CLEANING TANKS AND LARGE SURFACES

3.6.1 REMOTE OPERATION

Most pipes and sewers are readily accessible from their ends, and tools can easily be made to fit into the space available and to move so that the jets cover all the surface to be cleaned. There are many installations, however, where this relatively easy shape may not exist. Typical examples of this are railroad cars, and chemical treatment tanks. Access to these containers is generally through relatively small manholes, and the device which is used to clean must cover a large area once in place. Special rotary cleaning units have been developed for this application, exemplified by a device described by Bardrick [3.54].

A frequent problem in chemical plants is that the walls of the reaction vessels in which the chemicals form useful product become coated with that product. This has several effects. First it reduces the heat transfer into the process, secondly it reduces the volume of material which can be treated in a given batch, and thirdly, where the same vessel is used for different processes, it can lead to cross contamination between batches. As the build-up of product grows, it becomes uneconomic to continue using the vessel until it has been removed. Historically this has been achieved either by using a chemical treatment or by sending in laborers to manually remove the material. Where the processes use heat, it is extremely expensive to shut the vessel down, allow it to cool, clean it, reheat it and then start to re-use it. However, there are practical limits to manual labor use in hot tanks, so that large periods of downtime may be required for a cleaning. This has made management schedule such operations only at infrequent intervals, where it had become a necessity.

Bardick points out that it is more effective to insert a device between successive batches of chemical and to routinely clean the walls of the vessel at frequent intervals. The device used, however, must be easy to insert, non-contaminating of the process, and directed so that it will cover and clean all the walls of the tank, and around any tubes and fixtures within the vessel.

To achieve this level of cleaning, the device must be controlled from outside the vessel, and, since there are areas where the operator cannot make a visual inspection, it must be designed so that it will automatically follow a path to clean all the surfaces. Cleaning should remove all the material down to the skin of the tank, since it is more difficult for chemical to build up on a clean skin than it is to grow onto material already deposited on the wall.

Bardick illustrates the relative costs of different methods of cleaning reactors, from which the following figures can be derived:

Table 3.8 Annual comparative costs for cleaning chemical reactors [3.54]

chemical cleaning of two reactors	\$26,000
manual cleaning of reactor	\$16,000
waterjet cleaning of reactor	\$ 4,000

The costs given are detailed in the paper and were given in English pounds from 1978, but are illustrative of some of the gains that can arise from the use of waterjetting. These costs do not include the enhanced output from a quicker cleaning operation, improved batch production, and more efficient reaction chemistry because of the cleaner vessel. At the time Bardick estimated that these hidden gains might reach some \$80,000 a year. However this evaluation of benefit did not cover the additional costs for getting rid of the chemicals used in the chemical cleaning process. When that also was included an additional \$20,000 savings was estimated. Given the more stringent controls which are now being imposed for chemical disposal in both the United States and Europe, such a savings may now be significantly greater.

Cleaning the vessel between successive batches has an additional advantage. When the coating is first deposited it will likely be softer and quite thin, and thus more easy to remove. Where it builds up into a thicker layer, the stronger underlying material will take more energy to dislodge and it will require more time for the jets to penetrate through the material to reach the wall of the tank. Thus, overall process costs will be reduced with more frequent cleaning. The shorter duration cleaning cycle will also reduce the temperature losses within the tank between process cycles, giving an additional savings to the operation.

In order to clean the vessel on a regular basis a special tool can be designed to traverse through the vessel automatically cleaning it between batches. The design of such a device is almost always a custom operation, since the shape of the vessel and the internal fixtures will change with the application. The path of the tool will be limited, and thus the range of the jets, in turn controlled by pressure diameter and nozzle entrant flow becomes critical to its performance. Rotary units usually work best, with the nozzle assemblies moved to travel as close to the wall as can be arranged. Storage of the unit between runs also requires some design effort, since the tool must be kept out of the process vessel, and the storage unit should not provide a path for any noxious or otherwise hazardous vapors to escape the vessel and enter the workplace of the operator.

Note that in the design shown (Fig. 3.31) only one nozzle assembly is shown. In many operations a multiple array of nozzles may be used and the assemblies may be driven by some mechanism. This is in contrast to the system shown where the reaction forces from the jets will cause the head to spin around the high pressure swivel.

Where there is a lot of hardware within a tank then it is no longer possible to freely move a cleaning head around the tank to clean all the surfaces. An extreme example of this occurs in the cleaning of sludge from within nuclear power station steam generators [3.3]. In these installations sludge deposits can lead to corrosion damage and to a loss in heat transfer across the tube walls. The deposits can grow to a thickness of more than 0.25 mm. Access to the tube walls is, however, limited since the tubes are stacked close together. A main "blowdown" lane is provided through the stack, some 90 mm wide and this can be used to insert tools which must then negotiate through the less than 10 mm gaps between the tubes to reach the deposits.

Foster Miller have developed a tool known as CECIL (Consolidated Edison Combined Inspection and Lancing) which can be used to send a nozzle array down the narrow gap out to a distance of up to 6 m cleaning the tubes. Depending on the nature of the deposits to be removed, the lance can be fitted either with medium pressure nozzles to remove loose sludge, with fan jet nozzles to remove scale, or with round jet nozzles to remove hard deposits. Because of the narrow gap, the supply line diameters are small for the device, limiting flow volumes, and thus the nozzles must be brought relatively close to the surfaces of the harder deposits to be effective. The process is expected to save between $5 and $7 million over the chemical cleaning alternative. That process is so expensive because of the need to store, treat and dispose of the chemicals involved, whereas the water can be simply filtered and recycled at much lower cost, reducing also the volume of material to be disposed of.

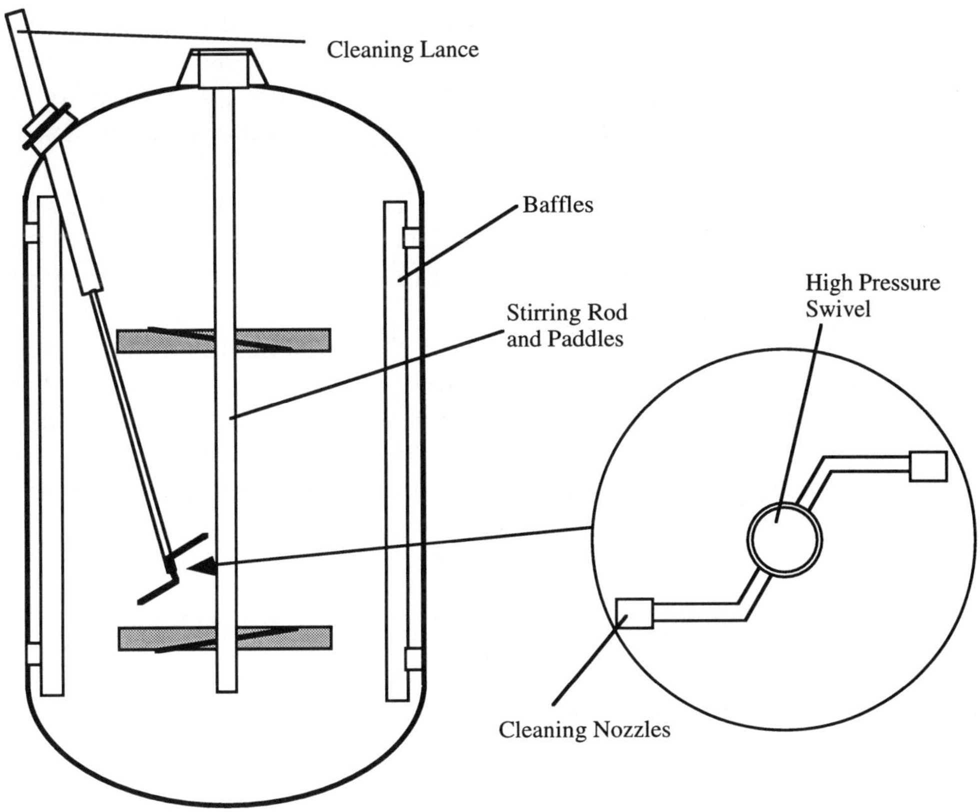

Figure 3.31 Simplified configuration for an automatic tank cleaner [3.54].

The problems of radioactive material removal and disposal have become more visible with the public awareness of the condition of the storage tanks used for the historic containment of this material. Historically much of the radioactive waste generated has been stored underground in large, single shell, storage tanks. This name was given from the construction of the tanks which used a single layer of steel to make the wall, although the gap between the tank and the surrounding ground was also filled with concrete. With the concerns for tank leakage and for more secure containment in stronger tanks, as well as to provide for treatment and volume reduction, this waste must now be removed from its original storage.

Depending on the material in the tank this may, or may not, require the use of very large pressures. For example, the removal of radioactive sludge from storage tanks at Oak Ridge National Laboratory [3.55] required the use of a relatively low pressure, bentonite-carrying jet. Approximately 1.5 million liters of radioactive sludge had been collected in 6 storage tanks during operations at Oak Ridge and this had to be removed for disposal. Bentonite was added to the low pressure jets to ensure that once the sludge was in suspension, that it would remain there until it reached its final destination. The jet had to penetrate the sludge to a depth of between 30 cm and 3 m. Most of the material removed was less than 10 μm in size. Larger particles could be crushed to below 20 mm and these would then remain in suspension at concentrations of up to 25% by weight in a 2.5% by weight bentonite suspension.

In order to achieve effective sluicing of the material over the range of stand-offs and with the amount of consolidation which had occurred, it was found advantageous to raise the jet pressure by reducing the nozzle diameter from 1.6 - 1.3 mm. The resulting stream provided a more effective pulverizing tool. Although the process was relatively slow, (it took several months of operating time) and some problems were encountered, the process (Fig. 3.32) was successfully completed using remotely operated equipment.

While this method is effective where the sludge is relatively soft and easily disintegrated, it becomes less effective where the sludge is harder. As a result two different approaches have been developed, particularly for use at the Hanford site in Richland, Washington [3.56] and [3.57].

At Hanford radioactive waste is stored in a total of 149 underground tanks which have an individual volume that can range up to 3,785 cu m. The waste is found in a variety of forms, ranging from a liquid, through sludges of varying consistency to a hard "saltcake" [3.56]. Because there is some risk that the integrity of the outer containment of the tanks has failed, it was considered important that no additional fluid be introduced into the tanks during waste removal. Further, given the radioactive nature of the waste, the material would be extracted using remotely operated equipment to dislodge and convey the material out of the tank. Target production rates of approximately 0.1 cu m/min were set for material removal over and above that required for cutting the waste.

An evaluation of a variety of different techniques suggested that the low reaction forces available with high pressure jetting made this the most likely candidate for the dislodging tool, with either hydraulic or pneumatic transport being used to remove the material from the tank. The constraint on liquid volumes to be used led to the evaluation of both medium pressure (circa 700 bar) mining systems and high pressure (circa 3,500 bar) cutting systems for the dislodging of the material.

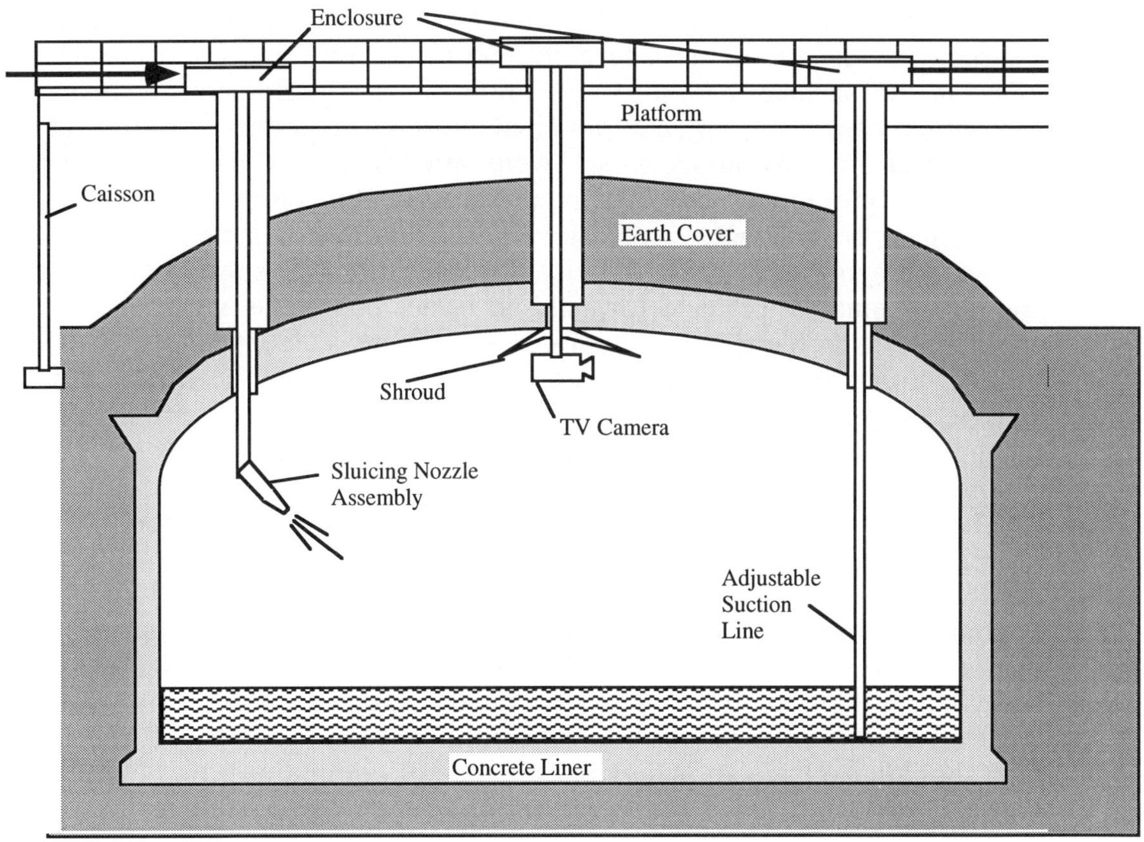

Figure 3.32 Method for removing radioactive waste from underground storage tanks at Oak Ridge [3.55].

In either case the cutting head would be moved over the surface at a rate which would remove between 2 and 5 cm of material at a pass, over a designated width, and at a rate required to give the production rate needed.

In order to contain all the cutting fluid the cutting head would operate from within a shroud [3.57] which would act to contain all the rebounding water and debris from the cut, so that, while still mobilized by the jetting action, it could be fed into the exhausting piping which would carry it out of the tank. At the beginning of 1994 both approaches were being evaluated to determine their relative applicability to deal with the varying conditions likely to be found in the different tanks on site.

3.6.2 HAND-HELD EQUIPMENT

With further increase in size of a facility it becomes practical, and often desirable, to have a manual operator enter the facility and use a hand lance to clean the surface. Manual operation of the equipment, if properly performed, can give a much better final surface, since the operator can see and correct for any areas which were not effectively cleaned. Certain ancillary requirements also accompany an operator into a tank, however. There must be adequate ventilation, good lighting, a sound footing and adequate bracing to support the cleaning operation. Disposal of the water and debris must be planned for, and no fumes should be generated which could pose a problem. The operator should be adequately protected by way of proper clothing and protective gear, and also by adequate training and understanding of the job before it is undertaken. It should be recognized that spraying in a confined chamber may create breathing conditions which may not be healthy.

The operator should understand the operating parameters of the equipment which is being operated. For example, if a fan nozzle is being used with an effective range of 10 cm then the operator should know that holding the nozzle 15 cm from the surface will not remove material. The operator should understand the behavior of the material being removed, and what particular problems it may raise. Since waterjetting is largely a team effort this suite of knowledge should be understood by all the team, who should also have an understood method of communication between the various members, since the noise inside a tank being cleaned may rise above 100 dB.

There is a significant change in operator conditions where one goes from the old method of mechanical drilling and the use of jackhammers to using waterjets to remove hard materials. This is the concept of backthrust, or the force required to hold a water jet nozzle in position.

The back thrust from a linearly directed jet can be calculated from the equation:

$$\text{Backthrust (kg)} = 0.258\ Q\ (P)^{0.5}$$

where :

Q is the flow rate in lpm
P is the jet pressure measured in bar

This reactive force value must be considered in the design of any holding fixture which will retain the piece being washed, or which an operator is required to manipulate. The Water Jet Technology Association recommends that no person be asked to withstand a reaction force greater than 1/3 of their body weight [3.45]. This force is, however, considerably less than that which would be required with a mechanical method of cleaning, particularly those which use compressed air as the method of providing power to the tool.

When the waterjet is directed through a hand-held lance, this limits the maximum pressures and nozzle diameters which can be used, since the average operator is only capable of holding about one third of their body weight. As Torpey has pointed out [3.58], this means that the maximum horsepower which the operator can handle runs in the region of 15 - 20 kW, with the maximum level probably being at the order of 30 kW. Such a system, however, provided that the driving pressure of the jet is between 500 and 700 bar, is sufficient to clean most industrial surfaces. Where this is not in itself the case, there are certain enhancement techniques which can be applied to the jet (Chapter 10) such that the system does cut the material.

In choosing the system of jetting to apply the operator should give considerable thought to the response of the target surface [3.59]. For example if the material is brittle, so that it shatters upon jet impact, then a round jet which carries high levels of energy to the surface, and concentrates it in a small area, may be most effective. Pardey has noted that calcium carbonate can be removed at rates in excess of 2 m/min from tubes, provided that a jet pressure over 500 bar is used in the jet. In contrast where the material is more resilient, and the round jet will cut a thin slot through it, then it may be more effective to use a fan jet. For example, in removing a layer of vinyl chloride from the walls of a reactor vessel one might use a 400 bar fan jet to cut into the interface between the layer and the wall, and then to peel back the coating. Changing the method of removal will usually also require that the operator change the angle at which the jet attacks the surface. While aiming perpendicular to the surface will usually be most effective in a shattering application, where the operator is trying to peel away the material then an angle of attack between 30 and 60 degrees might be more effective.

In 1978 David Odds [3.60] described one of the more graphic illustrations of the benefits of jet cleaning which arise from suitable design of waterjet systems. By balancing the reactive force from the cleaning jet with a second jet of larger diameter, and lower pressure, a tool was built for use by divers in cleaning the legs of the oil rigs in the North Sea. The paper cited the case of the diver who first used a jet cleaning unit underwater on a rig. A few minutes after the trial began, the diver halted the test, came to the surface and without a word, walked over to the jackhammer unit previously used and with an insulting farewell, threw it over the side.

In cleaning ship structures, waterjets are usually fanned rather than circular to cover a larger surface. Such fan jets, operating at pressures of approximately 200 - 350 bar, have also been used for cleaning ships' holds, particularly of the crude oil residue between the different loads which oil tankers must carry. It must be pointed out, however, that the use of such a technique is one which requires considerable care. Explosions have occurred where petroleum vapor build-up has been allowed inside the tank before cleaning began [3.61]. In such circumstances, the mist around the waterjet can acquire an electrical potential sufficient that sparks are generated, igniting the surrounding petroleum gas and blowing the ship apart.

More recently, as the need for an efficient, large scale system for cleaning structures such as the hulls of ships, has been defined, the use of cavitation bubbles [3.62] and, on occasion, the induction of abrasive particles into the water [3.63], has been found to help in increasing the areal coverage of the cleaning unit, without increase in fluid horsepower. These two subsystems can be used to clean materials which might otherwise require higher jet pressures and thus more powerful units. They can also find application in other ranges of the cutting and cleaning industry, and will thus be discussed in much more detail later (Chapter 10). There is, of course, a price to pay for this benefit, and it is most easily seen in the increased complexity which can develop in the equipment required.

To give an example of the benefits which can accrue with these additives comparative cleaning rates are available for cleaning surface areas found in such applications as removing the barnacles from ships' hulls.

Many of the "improved" versions of jets which have been suggested for use in the lower pressure ranges, will also have application at higher pressures, though not always with the same level of success as at lower pressures. For example, the power requirements to heat flow rates above 40 lpm become too expensive for the gain that can be achieved. While as one cavitates jet flows above 1,000 bar then the risk of destroying the nozzle becomes more prominent.

Table 3.9 Effective cleaning rates for ship hulls

Material removed	*Cleaning rate in sq ft/hr*			
	Dry sandblasting	Wet sandblasting	High pressure sand/water jets	Cavitating jets
light scale	160	192	450-725	900
heavy scale	135-140	150	300-550	
heavy encrusted	-	-	200-400	

One way in which more power can be transmitted to the face is to redesign the nozzle system to include retro-jets. In this manner an operator can handle an input power of up to 30 kW, increasing productivity. Torpey points out, a 6 m tube bundle containing some two hundred tubes can and has been cleaned in under 18 minutes. In another example the Partek Corporation found that a tube bundle which took 300 manhours to clean using hand tools and chemicals, could be completely cleaned by two men within an hour [3.64].

Other approaches are, on occasion, required to overcome more intractable problems. For example, as described earlier, conventional waterjets, even at pressures of 700 bar, were unable to adequately clean the gas collection lines of ICI polymer plants [3.29]. The lines, which included stems and valves ranging in diameter from 50 - 250 mm, were cheaper to replace rather than try to clean them effectively. Since, however, the material blocking the pipes was heat softenable, the plant modified the waterjet assembly to include a venturi section which drew steam along the surface of the waterjet. It then proved possible for two men to clean pipe sections in a period of 45 minutes which would have taken at least 1 man-day to clean previously. It also proved possible to clean the bends and valve assemblies of the pipes without any great difficulty.

Many applications of waterjets for cleaning in industrial structures can be cited. Waterjets are used to clean structural surfaces of deposits in chemical plants, refineries, and petrochemical processes and, in the building and construction industry, to remove concrete, and clean off aggregate and other material prior to reworking the surface. The flexibility of the waterjet system and its ability to clean materials from areas difficult of access for a number of reasons have meant that the waterjet cleaning industry has grown considerably in the thirty years or more since high pressure waterjet cleaning was begun.

Waterjets have applications in areas normally inaccessible not only by reason of geometry but by reason of environment. Thus, for example, in the steel industry steel billets initially start out with a thick deposit of scale around them. While much of this is removed by movement of the billet itself, some will adhere to the steel, raising the potential for damage in equipment downstream. In order to remove this material, waterjets at pressures of 100 - 200 bar are directed onto the billet, removing the scale by a combination of thermal and mechanical actions.

In a similar operation, the National Research Council of Canada developed a system for cleaning out the gas collector pipe which leads from large coke ovens [3.65]. This pipe rapidly becomes clogged with materials from the coking process and must frequently be cleaned. Normally this is done by a hydraulically driven mechanical device, but this does not completely clean the surface, and build-up of material occurs relatively quickly once the process is restarted. Working at a pressure of 700 bar and at a flow rate of 81.4 lpm, two nozzles of diameter 1.6 mm were rotated at a speed of 300 rpm and completely cleaned the pipe in a period of some 30 seconds at a feed rate of 8.5 mm/rev. As a subsidiary advantage to the system, Brierley pointed out that not only was the service period of the coking oven extended, but the cleaner pipe surface reduced the operating pressures within the coke oven.

3.7 ADVENT OF HIGHER PRESSURE

One of the major advantages of the high pressure jet cleaning systems has been the relatively low reaction forces which are generated from a cleaning lance. However, when the deposits to be cleaned are harder, then the operator must increasingly use higher pressures. Under normal conditions the higher pressures continued to be used with significant flow rates, and the thrust levels experienced by the operators became significant, and the maneuverability of the systems became less.

The advent of very high pressure systems for industrial cutting has provided an alternate approach. By significantly increasing the jet pressure it is possible to reduce the volume of water used to around 3 lpm. At a pressure of 2,500 bar, this gives a reaction force of 3.89 kg, easily handleable by most operators. Concurrently the very high pressure of the jet allows it to be used for a wide variety of otherwise difficult cleaning tasks. Most of the guns which have been developed for this use are relatively small and contain some mechanism for rotating the nozzle over a circular path on the surface.

While some devices use a swivel for this purpose another method is to use an orbiting path for the nozzle, as described by Barker [3.66] and subsequently used by Steele [3.67]. In this system the high pressure lance is gimbal mounted to a support at the hose end. The free end is then eccentrically fitted in a bearing mount on a circular plate, which can rotate about the tool axis (Fig. 3.33). The plate is rotated by a small motor and as it turns so it moves the nozzle in a circular path. The nozzle is allowed to rotate in the plate mount by the bearing so that, while the jet sweeps out a circular path, the lance itself does not rotate about the bearing. This design removes the problem of swivel reliability and pressure losses which were a major concern early in the decade, but which have since been ameliorated by equipment improvements.

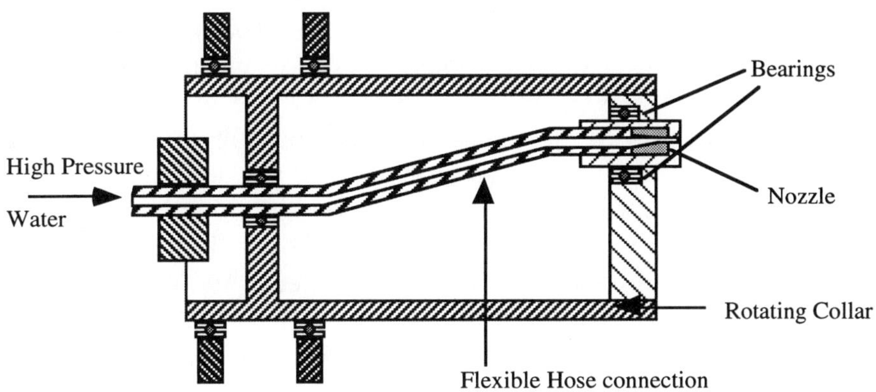

Figure 3.33 Orbital drive for a rotating waterjet device [3.66].

3.7.1 ROUND AND FAN NOZZLES REVISITED

In the discussion of nozzle selection in Chapter 2, a strong recommendation arose for the use of rotating round jet systems as a viable alternative to the use of a fan jet system. It is important to understand, however, that this is a generalized recommendation, and that there are circumstances where it is not appropriate, i.e., in narrow pipe cleaning the use of a rotary device may not be practical. In other circumstances the properly engineered use of a fan nozzle system may provide a better end result. This is particularly true, as has been discussed above, where the target material responds more to peeling action rather than to shattering by the jet.

One illustration of this use was given by Saunders and Barton [3.68]. They compared the use of a rotating nozzle system, operated at 700 bar using two 1.5 mm diameter nozzles with that achieved using an array of fan jet nozzles. The target material was a special four-coat epoxy paint used in the shipping of radioactive materials, a material more responsive to peeling. Threshold pressure testing of this material indicated that while it required 275 bar to initiate damage to the paint with the round jets, only 155 bar was required when the fan jets were used. When the rotary round jets were applied to the surface they only marked the surface of the paint, without significant material removal.

In contrast when the fan jets were used at a pressure of 620 bar, and at a stand-off distance of 90 mm, between 50% and 100% of the paint could be removed over a track width of up to 44 mm. The study investigated, in some detail, the performance of the fan jet nozzles. Individual nozzle performances were studied and, as a result, the design of manifolds for multiple nozzles could be evaluated. These were then compared experimentally.

The cleaning action of the jets was such that it took three passes at 52.8 cm/min to remove paint over the central 50% of the jet impact path width and leave a bare metal surface. Below this speed a wider swath of material was removed and at higher speeds paint was left over the full section of the path. Path width could be increased by a change in the nozzle design, although their finding reflected the results of a UMR study that the angle of effective jetting is approximately 7.5° greater than that of the nominal description of the nozzle.

As a result of the tests carried out, the authors developed a prediction equation for the maximum effective traverse speed required to remove material from the surface. This they expressed as:

$$T = 1.6 \times 10^{-6} \times P^{2.7} \times S^{-3.8} \times Q^{2.4} \times (\text{Tan } A/2)^{-4.7}$$

where

T is the traverse velocity in mm/sec
P is the jet pressure in bar

S is the stand-off distance in mm
Q is the water flow rate in lpm
A is the actual included angle of divergence of the fan jet

It is interesting to note that their study of manifold design provided a corroboration of the results reported earlier from the Bureau of Mines on the need for a straight section into the nozzle body [3.69]. The BHRA study showed that 15 and 30 mm straight sections ahead of the nozzle gave roughly 50% of the performance of a single lance lead in to a nozzle, whereas with a straight section 75 mm long the performance of the jets matched that from the single jet with a long lead in. The investigation also showed that it is important to stagger the nozzles, or otherwise align them so that the jets from one nozzle do not physically interfere with the jet from another. Where such interference does occur it significantly reduces the effectiveness of the jets. The overlap of nozzle paths is required in a manifold design to accommodate variations in stand-off distance, and to allow for a change of up to 15° in the actual angle of the jet issuing from nominally similar nozzles upon replacement.

The control of the impacting pressure also plays an important part in cleaning where the underlying surface is not strong, or where the cleaning action must remove the contamination from the surface without damaging the original surface condition. One example of this is in the cleaning of historic bronze statues. In recent years the manual removal of surface debris from these monuments has been replaced by other means, in an attempt to speed-up the process, while minimizing surface damage. One technique examined has been that using airborne abrasive cleaning with crushed walnut shells, because of the reduced damage to the underlying surface [3.70]. Similar materials have been examined by NASA for use in refurbishing Solid Rocket Booster (SRB) components [3.71]. The results from the testing at NASA illustrate the reasons for the popularity of this abrasive (Table 3.10). As two additional benefits to its use, NASA reported that walnut hulls were the cheapest of the abrasives to buy, and, since they were biodegradable, the least expensive to dispose of.

Walnut hulls, however, tend to crush the surface of the target, and may drive the corrosion into any pits created on the surface. Thus although the surface may appear cleaner it may not be perfectly refurbished. From that point of view the use of high pressure waterjets may be more beneficial, given that they share some of the cost benefits of the abrasive. However, experiments at UMR have shown that antique bronze is not greatly resistant to waterjet action and can be damaged if incorrect pressures are used

[3.72]. The pressures must be controlled when cleaning valuable monuments to within a relatively narrow pressure range [3.73]. It is a wise precaution to set the jet pressure below that at which surface damage can be created, and to use a fan jet nozzle for such applications. This reduces the risk of damaging the underlying coating and, by adjusting the stand-off distance between the nozzle and the workpiece, allows the corrosion coating to be removed down to the layer of the surface which is desired (Fig. 3.34).

Table 3.10 Stress and warpage of surfaces impacted by abrasive [3.71]

Abrasive	*Surface stress (bar)*	*Surface warpage (0.01 m)*	*Roughness (μm)*
garnet 25	-	1.1	5.05
silicon carbide 30/60	1,605	1.18	4.19
aluminum oxide 36	1,755	0.94	4.17
silica sand 40/70	2,075	1.3	3.55
garnet 80	1,254	0.81	3.11
aluminum oxide 80	2,302	0.62	2.06
silica sand 80/90	2,250	0.80	2.03
walnut hulls 12/20	no stress detected	0.049	1.06

Note. The abrasive was directed at the surface using a standard air injection system, using air pressure of 2.75 - 5.5 bar. The surface will warp in a concave manner under the peening action of the abrasive, while this is a function of the amount and energy of abrasive impact, the table indicates the amount of curvature generated.

Where fan jets are used they will also provide a more rapid cleaning of the surface, while making it possible for the operator to get into the various crevices created by the sculptor when creating the work of art.

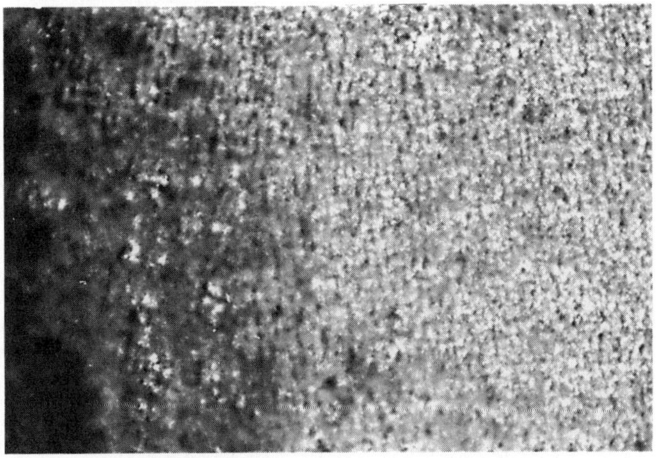

Figure 3.34 Removal of corrosion layer from bronze while leaving the surface coating intact. (The left side of the material is in its original form, layers are removed to the right with the final surface being at the point where metal is being removed unacceptably.)

Figure 3.35 Linda Merk-Gould, Fine Objects Conservation, Inc., cleaning the Freedom Statue from the top of the U.S. Capitol Building.

3.7.2 CLEANING RADIO-ACTIVE PIPES

One problem which arises in cleaning surfaces comes from the question "How clean is clean?" As was just discussed in regard to cleaning statues, it is not always required that all the coatings be removed from a surface for the cleaning to be effective. On the other hand were one to be concerned with cleaning toxic chemicals from a surface, or removing radioactive material, then the standard of cleanliness required would be much more stringent. Assessing how clean a surface has become is also not always easy. When a car is cleaned with water it can appear "clean and shiny" when wet, but will still dry to show a layer of road film, if the jetting job was not carried out properly. Examination of walnut blasted surfaces may, to a superficial examination appear to show that the crushed walnut hulls cleaned a surface better than did a waterjet. In both cases a close examination with a microscope would show the true result, but it is a rare individual who can bring a powerful microscope to a cleaning operation and be able to use it.

Measuring the amount of cleaning of a surface is more difficult in some cases, than measuring the effect of a jet in cutting. This is because the relative measures of success can be less clearly defined. However, it is a necessary part of planning to decide how to measure the success of the operation. This means that one must properly consider the results needed when one sizes and designs a waterjet cleaning system. Sanders and Bond illustrate an approach to the problem [3.74]. The particular problem that they addressed related to the removal of radioactive deposits from the piping in a nuclear reactor. This process, while it has inherent problems beyond that of the normal cleaning operation, also has one advantage. That is that, by measuring the dosage, it is possible to obtain a reasonable measure of the amount of material which has been removed from the surface.

Because the cleaning is carried out from within the filled pipes the initial tests of different cleaning configurations also took place submerged. Preliminary tests with round jet nozzles indicated that they were not very effective, and as a result two different 15° nozzle bodies were compared in the testing. In addition a comparison was made between sample surfaces which had been left untreated and surfaces which had been pre-treated with a reagent. In contrast with many cleaning operations the amount of precipitate to be removed in the program is relatively thin, approximately 5 - 10 μm for the treated samples and 25 - 35 μm for the untreated surfaces. The results of the test program showed that cleaning was most effectively achieved when the operator made use of the cavitation cloud formed as the cleaning jet exited into the surrounding fluid.

Because of this discovery, the nozzle designs used were specifically chosen to enhance turbulence in the nozzle region, and used a 180° contraction angle and a very short nozzle straight section. The optimum stand-off was found to be on the order of 14 nozzle diameters (25.0 mm). This was related to the distance over which the cavitation bubbles survived, and gave a path width of 12 - 14 mm. An equivalent round jet would give a path width of only 2 - 3 mm. It was found that, because cavitation was being used as a material removal mechanism, that underlying metal could be removed from the surface at a pressure of 560 bar, although this undesirable removal could be reduced by increasing the traverse speed of the lance through the pipe.

This example is a useful one to study, not only because of the methodical way the engineers determined the best individual nozzle that could be used in the process, but because they carried the process one step further. The pipe being cleaned was 55 mm in diameter. This provided a circumference of (π x diameter) 172.9 mm. If each nozzle cleaned a path 13 mm wide then this would require a total of 13 nozzles (172.9/13) even if one discounted the need for some overlapping percentage. At a pressure of 207 bar, the flow required for each nozzle would be 30 lpm. Thus, the total flow to the cleaning head would be 13 x 30 = 390 lpm.

Allowing the supply line and lance to have an internal diameter of 12.7 mm and using a 90 m required length creates a significant problem with the amount of friction created in delivering such a flow. If one refers to the equation which Labus proposed for flow through a pipe:

$$\text{Pressure drop (bar/m)} = \frac{0.597 \times Q^2}{100 \times D^5 \times R^{0.25}}$$

where:

Q is the flow in lpm
D is the pipe diameter in cm
R is Reynolds number (given by 1116.5 (Q/D)

Reynolds Number for such a flow would be (1,116.5 x (390/1.27)) or 343,000. The pressure drop would be 11.35 bar/m or 1,020 bar over the length of line anticipated. If the delivery pressure required was 200 bar, this would require a pump producing over 1,250 bar, obviously excessive.

The solution proposed was to lower the flow to 100 lpm which would give a pressure drop on the line of 140 bar. This flow would however only supply 3 nozzles of the required size, and thus a rotating head, or multiple pass system was then required to give the required complete coverage of the pipe area.

3.7.3 RUNWAY CLEANING

One of the fascinating stories of the waterjet cleaning industry is that of Bob White, and his wife Donna. In 1972 Mr. White unsuccessfully tried to clean the rubber deposits left from the wheels of landing aircraft from the runway at McClellan Air Force Base. Although able to remove material at only 5.5 sq m/hr, he was convinced, particularly after seeing that the state of the art system was a chemical treatment, that waterjetting was the answer. Through a combination of loans from a variety of sources to the tune of $180,000 he built a four-pump, 24 nozzle spray bar system, initially operating at 700 bar and went into business. By driving himself through the night and thus being able to underbid the competition he took the first 25 of 28 jobs which were bid, and by the end of that year he had paid off his loans. By 1977 he had 5 rigs on the road around the country and was anticipating his first million dollar year.

After the first year on the road it was found, for most applications, that the jets removed the rubber better at 350 - 400 bar than at 700 bar and there was the added risk that at the higher pressure the water would either damage or polish the underlying concrete, depending on its quality and that of the aggregate contained. By supplying the flow from each pump to a six nozzle section of the spray bar it was possible to isolate a section should it have a problem, while still operating the rest of the units. The rig used spray bars fitted with 36 nozzles of hardened steel, 1.58 mm orifice diameters with a 30 degree spread, required to give 25 mm of coverage at the 19 mm stand-off used. Other available equipment used 2,000 kW pump units to supply 1.98 mm diameter jets held 25 mm apart, and with up to 96 nozzles on the spray bar. Such a unit could clean paint build-up over 4.75 mm thick at a rate of 1,200 sq m/hr, at a cost of $0.64/sq m. Lowering the nozzle diameter to 1.0 mm resulted in a bar 2.4 m long which when held 37.5 mm above the surface, at 500 bar would allow cleaning at the rate of 18 m/min, for a combined rate of 3,700 - 4,600 sq m/hr. The cost was estimated at $0.376 to $0.50 per sq m.

In 1974 White was winning contracts at $0.45 per sq m, and going below $0.32 per sq m to get others. He had at that time 34 competitors, and yet grossed $280,000 that year. In 1975 he was bidding runway cleaning at less than $0.22 per sq m for larger areas. Nozzle life was on the order of 50 hours for the stainless steel tips, and 250 hours for the steel holders. In January of 1978 Bob White discovered he had cancer and although successfully treated by 1980 he had sold off his rigs to highway painting contractors while he himself had turned to publishing [3.75].

3.8 ABRASIVE JET CLEANING SYSTEMS

The use of very high pressures can improve the ability of waterjets to clean material, but they can also be more expensive and demanding in their use. An alternative way of removing stubborn material and giving a white metal surface has been available for many years. This has been the use of an abrasive laden stream to act as a cleaning tool.

Air-laden abrasive streams have been used by industry for many years. Air-laden abrasive cleaning, or sand-blasting, is a widely available system, and at relatively low air pressures metal can be cleaned down to a bare surface and "toothed". This toothed surface caused by abrasion of the particle on impact can be controlled by the design of the system, and is used to provide a better bonding surface when a paint or other coating is to be applied to the surface.

Even at low air pressures abrasive streams can cut through most materials. This, however, puts them at somewhat of a disadvantage since prolonged exposure of any surface to the abrasive jet will cause penetration. These tools require somewhat greater care in their application since they can, therefore, damage the substrate below the layer of material which must be cleaned off.

Air driven streams can be made sufficiently gentle that they are used to clean dirt from ancient documents, for example. The controls on abrasive blast cleaning through pressure and air volume controls have meant that it could be used as an effective tool in a number of applications. Yet a stream containing 5 g/min, mixed with 0.16 liters/sec of air at 5 bar, will move at 335 m/sec. When this stream strikes the target some of the abrasive rebounds into the surrounding air, either as entire particles or broken into a smaller and thus more dangerous particle size range. However, in America the Clean Air Act of 1970 created a National Primary Ambient Air Quality Standard which defined an acceptable limit of 75 microgram/cu m annual geometric mean, and a 360 microgram/cu m daily maximum particulate count for respirable dust. These particulate counts are far surpassed in the air surrounding conventional sand-blasting, which thus is increasingly being found unacceptable. In contrast wet sand-blasting in a high pressure waterjet stream can provide airborne dust levels which meet the standard [3.76].

Dry sand cleaning will give cleaning rates of up to 1,140 sq cm/min at a pressure of 6 bar when directed through a 10 mm nozzle diameter. Nozzle geometries control the performance to an extent, as does the type of abrasive which is used (see the discussion at the beginning of this chapter). However, the controls on the performance of the abrasive, and for the waterjetting stream change when the two systems are combined. (The subject is reviewed in more detail in Chapter 10).

There are currently four ways of adding abrasive to waterjetting systems for cleaning purposes. The first of these is the conventional method used by most of the air-abrasive companies. A ring main fitting (Fig. 3.36) is attached to the end of the conventional dry blast nozzle and will inject a water spray into the air-sand stream to reduce rebound and dust generation. An alternative approach to this is to feed a water:sand slurry to the nozzle, where it is accelerated by injection into a moving air stream. In this second alternative one would normally use a finer mesh of sand than that used with the ring main system.

In contrast the water can be directed from a straight feed into a nozzle, and thus into a mixing chamber where sand is aspirated into the jet from the side (Fig. 3.38). This technique, which is also used at much higher pressures, has several advantages, but requires some care in the design of the mixing chamber and the water and sand feed lines.

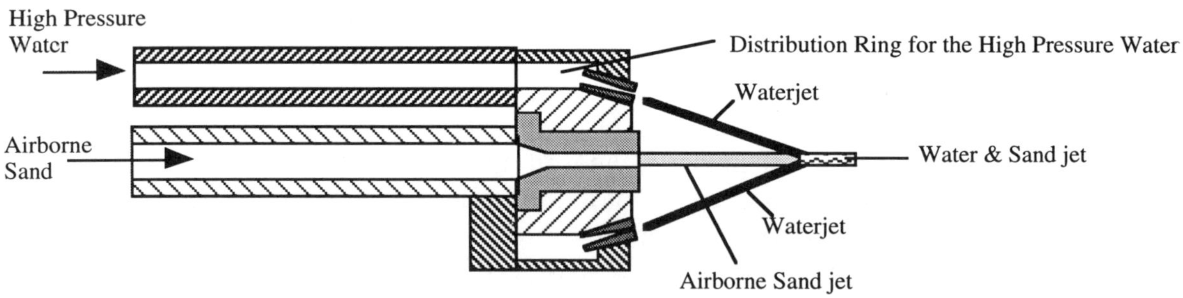

Figure 3.36 A ring main application of waterjets around an abrasive-laden air jet.

In contrast, most pressure washer cleaners have developed modified assemblies which can be attached to conventional water cleaning lances but which retain the higher jet pressures. Two ways of achieving this can be illustrated, one of which feeds water around the abrasive injection line (Fig. 3.37). This is relatively inefficient because of poor water flow into the mixing chamber, but it ensures the sand is in the middle of the stream.

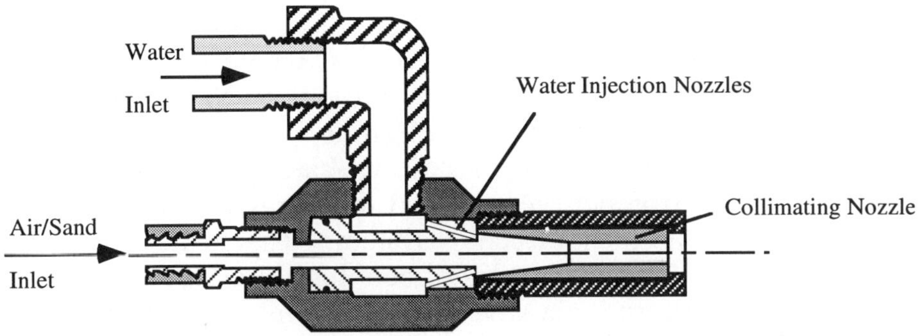

Figure 3.37 Wet abrasive blasting nozzle with water added to an airborne sand jet [3.77].

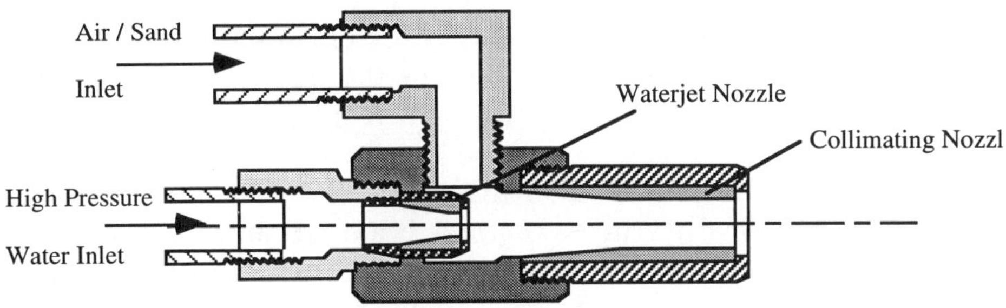

Figure 3.38 Wet abrasive blasting nozzle with sand added to the waterjet [3.77].

Comparative cleaning rates of the several systems were recently investigated by Appleman and Bruno [3.77]. Their investigation was carried out in the field and evaluated ten different wet blast units in their effectiveness on cleaning rusted, pitted, mill-scaled and painted steel surfaces. The comparisons were made, where possible between an air-abrasive cleaning unit (the dry blast system), an air/water/sand unit (the wet blast system) and the use of high pressure waterjet units with abrasive injection (high pressure units). Tests with a high pressure unit operating at 1,400 bar and a flow rate of 32 lpm were very fatiguing to the operator and were reportedly unable to remove tight mill scale. A high pressure waterjet system alone was therefore not judged practical as a result of those trials.

The results of the tests showed that where a ring main was used to direct waterjets into an accelerated air abrasive stream that cleaning rates of 90 - 200% of that of an equivalent air blast system could be achieved. Where the water was mixed into the air:abrasive stream as it reached the final acceleration nozzle (Fig. 3.36) a cleaning rate of 70 - 140% of the air blast system could be achieved. Slurry blast systems, where the air was used to accelerate a sand:water mixture only provided 20 - 80% of the performance of the equivalent dry blast cleaning.

These figures, and the conclusions drawn are, of interest, not only because of the comparison between dry and wet abrasive, but in light of the design of apparently the most effective system. This would appear to be the system in which a ring of nozzles directs high pressure water jets into the otherwise dry airborne abrasive stream from a conventional sand-blasting gun. This result should be considered in light of results reported by Woodward on waterjet combination with abrasive cleaning [3.63] in which a comparative evaluation was made of the performance of a nozzle which combined a fan jet and a dry abrasive stream into a large surface area cleaner (Fig. 3.39). This novel design of cleaning nozzle was found, when properly applied to significantly lower the cost of cleaning surfaces over that of dry sandblast, but also to significantly increase the cleaning rate.

The test program, however, also illustrated the dangers of making an absolute comparison of differing systems without running a proper series of comparative tests. In his study Dr. Woodward compared changes in flow volume, flow pressure, abrasive feed rate, and nozzle stand-off distance in terms of their effect on the 1987 cost for cleaning steel plates.

His results indicated that while there were relatively small changes in cost with abrasive feed rate, and stand-off distance, there were significant changes when the pressure and flow rate were varied. For example in changing the pressure from 70 bar to 280 bar the cleaning cost dropped from $6.32/sq m (area cleaned 11.8 sq m/hr) to $1.56/ sq m (area cleaned 47 sq m/hr). There was a much less significant drop in cost benefit by increasing the pressure beyond that level. By the same measure increasing the flow volume from 8 lpm to 30 lpm at a pressure of 350 bar lowered the cleaning cost from $2.92/sq m (cleaning 26 sq m/hr) to $1.21/ sq m (cleaning 65 sq m/hr). Again there was little significant gain found in increasing the jet volume above this value. When this is compared to a cleaning cost of $1.92 for dry abrasive blasting, which was achieving a cleaning rate of 19.5 sq m/hr, it can be seen that, with proper tuning, a waterjet abrasive cleaning nozzle can be the more effective tool.

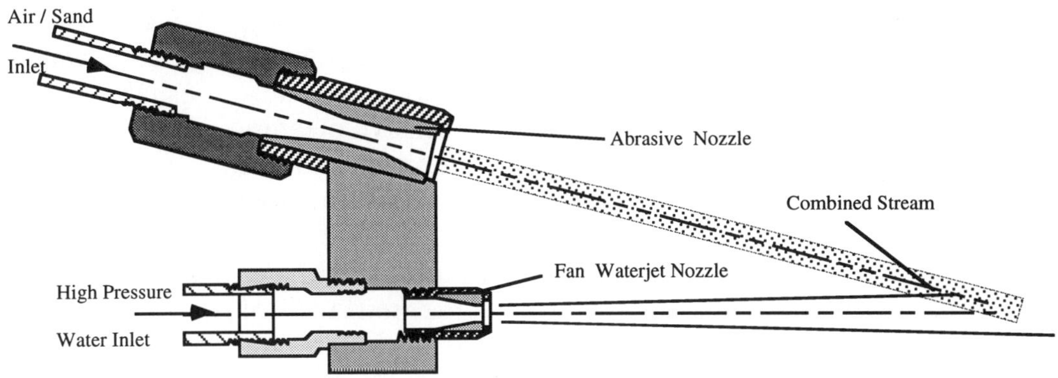

Figure 3.39 Schematic of an improved wet abrasive nozzle design by Woodward [3.63].

A related cost which must increasingly be considered is that of clean-up. Dry abrasive blasting required about 32.6 kg/sq m of sand, while the wet process reduced this amount of disposable material to 20.9 kg/sq m. Significantly higher cleaning rates were also achieved when the nozzle was properly tuned, and combined with others in a cleaning boom.

The results from this paper stress the importance of ensuring that a system is operating under the correct conditions for the task to be carried out. It illustrates the problems which may arise in field comparisons of equipment, where the unit is not properly adjusted for the conditions, and where, by example, costs might be as high as 5.25 times the minimum where the unit is improperly sized or operated.

3.8.1 SOLUBLE ABRASIVES

The increased costs for disposal of the debris from cleaning operations has led to a search for different approaches to the reduction of this volume. Where water alone is used as the removal agent this can be separated out by filtration and re-used, so that only the coating is disposed of. However, as discussed above, the use of abrasives can be very effective in cleaning metal surfaces. The normal separation of coatings from the abrasive can be expensive and time consuming, yet without it the volume of material to be disposed of becomes high.

One solution to the problem is to change the abrasive used to one which can dissolve in water. Such abrasives are now becoming more common, since, after application they dissolve in the wash water, which can again be easily separated from the removed coating. The chemicals chosen for the abrasive are such that they can be disposed of in the normal water flow, rather than needing to be collected, drummed and disposed of to a waste storage site. These biodegradable abrasives are also, usually, less aggressive than those used in conventional abrasive cleaning. Thus they are more likely to be used for removing stains, light build-ups of corrosion and for the removal of graffiti where normal methods would be too aggressive [3.78].

By 1993 there were at least seven biodegradable abrasives on the market with capabilities allowing, for example, the removal of 0.1 mm thick urethane coatings at an areal rate of 1,350 - 2,250 sq cm per hour, using only 0.7 - 7 bar pressure in a flow of 19 lpm of water and carrying 25 - 110 kg of abrasive per hour. At a higher pressure it was reported possible, using a modified baking soda abrasive, to strip the paint from classic cars without needing to mask the glass or chrome. The resulting surface was cited as being equivalent to a 400-grit sanded finish.

Questions have arisen as to the relative effectiveness of these abrasives, and the condition for their use, given that they tend to be more expensive than conventional abrasive. (This is in part since they cannot be re-used.) Woodward [3.79] has evaluated the conditions for using this abrasive in terms of its ability to remove paint from aluminum siding.

He carried out a series of tests to evaluate changes in jet pressure (from 41 - 895 bar), water flow rate (from 4.9 - 22.3 lpm) and abrasive feed rate (between 0.45 and 4.5 kg/min) on the productivity of the stream. The results indicated that cleaning rate is linearly related to the jet pressure and varies with the square root of water flow. As a result Woodward recommends that cleaning be carried out at higher jet pressure with lower water and abrasive flow rates. This recommendation is based, in part, on the likelihood that the cleaning will be carried out using a manually operated lance and this combination, as well as cleaning well will be less tiring to the operator. Production rates of up to 60 sq m/hr were reported from the tests. It should be noted that, for most effective use of the abrasive it should not be introduced into the jet stream until just before impact.

3.8.2 SLURRY ABRASIVE CLEANING

An alternate way of using abrasive is to introduce it into the feed line between the pump and the nozzle. While this system is discussed in more detail in later sections, the tool has been found quite effective in cleaning.

For example in China it has been reported that cleaning rate with conventional mixing is able to double performance over sand-blasting, reaching 13.2 sq m/hr, and in contrast with just plain high pressure water which can only achieve 5.4 sq m/hr [3.80]. Sand was fed into the stream at the rate of 2.5 kg/min and was in the 0.3 - 0.5 mm size.

In contrast, using a direct method of abrasive injection, it was found possible to lower the operating pressure of the system to 100 bar and still be able to achieve cleaning rates of up to 16.8 sq m/hr [3.81]. The abrasive size range was again below 0.5 mm, but the concentration was increased to 30% by weight of the water flow (i.e., 10 kg/min with a flow of 30 lpm of water). It was reported that this gave optimal results in terms of abrasive concentration in the water.

3.9 CLEANING ADDITIVES

Although the use of chemicals in the water has been discussed earlier, there is one additional point which must be considered. This is particularly the case in cleaning metal surfaces. Water has a small ability itself as a chemical agent in cleaning. Its major role, however is to mechanically remove unwanted material from a surface. When the jet is particularly active it can remove not only the desired waste, but also any protective film of oxide which has built up on the surface. Under such conditions it is possible for the surface to almost immediately rust over. This rusting may be sufficiently active and intense as to provide an unacceptable level of corrosion to the surface. In order to counter this the U.S. Navy used to mix 0.45 kg of sodium nitride with every 350 liters of water used in the cleaning process, to protect the metal in the ships' boilers from further corrosion [3.42].

On occasion, where surfaces are being cleaned which are of a critical nature, the waterjet may be replaced by an oil jet which does not pose the problem. At lower jet pressures a detergent has frequently been added to the waterjet in order to enhance the cleaning of parts between manufacturing processes. However, at higher jet pressures this chemical became both more expensive and less effective as the jet force was sufficient to remove the material.

Chemical additives are thus used for several purposes. Those dealing with the enhancement of jet performance has been partially discussed already, and will be referred to again in Chapter 10. The use of chemicals to inhibit corrosion is, however, something which is related to the cleaning operation.

The subject was discussed at the BHRA meeting in Durham in 1986, by Hall [3.82]. Conventional corrosion protection is through application of a thin film of oil, but this is removed by any waterjet cleaning of the surface. There are two conditions which must, therefore, be addressed. When hot surfaces are cleaned the water may evaporate from the surface relatively rapidly after the wash. The parts are then dry and may remain relatively rust free for several days. On the other hand if the parts remain wet, then rusting can be a problem within a few minutes.

Additives must be placed in the water which will prevent both situations from giving a rusty product. Four additional considerations must also be borne in mind. The additive must not cause the fluid to foam, since this would have several bad effects; it must be of such a nature that bacteria, fungi and other health threatening organisms do not easily proliferate; it must not provide any hazard to the operator; and it must be easily disposed of when too contaminated for additional use.

Protection cannot come from a thick protective layer since only the evaporated concentration from the thin film of water on the surface will remain to provide protection. These safety considerations argue against the continued use of nitrites (ibid.).

It is interesting to note that both Hall and Swan [3.83] have reported that the problem of bacteria is resolvable within the operation of the pump. The shearing and other forces generated in the pump and nozzle have a strongly destructive effect on the bacteria count. Thus a regular trip through a pump will provide some significant control on bacterial growth, which can further be inhibited by chemical additives.

Hall points out that the chemistry of an additive system will vary with the equipment being used and the purpose. Aluminum parts may be subject to unacceptable corrosion and pitting when an electrolyte is used to inhibit target corrosion. This was, in one case, caused by an adjacent stainless steel part in the pump and could only partially be compensated for.

As a general rule inhibitors which can be used include the inorganic electrolytes which are said to aid cleaning and provide a relatively stable biological environment. This stability can be improved with boron salts. If a polyelectrolyte is used in a mildly acidic form it will increase the dispersion of the dirt in the wash water, thus acting to reduce redeposition on the surface. It will also react with any fatty soils to create a soap. The polyphosphate is particularly useful since it also has the advantage of generating relative little foaming.

3.10 UNDERWATER APPLICATIONS

From the first symposium which BHRA hosted on waterjet technology, the advantages which waterjet cleaning brings to operations in a marine, or underwater environment have been recognized. It was not until later symposia, however, that the combination of forward-pointing, high pressure jets, and balancing larger diameter, lower pressure retro-jets for divers' use became more clearly advantageous. The marine environment is a particularly corrosive one, and also includes the potential for relatively large animal and plant growth on surfaces, all of which must be removed both to maintain equipment performance, and also to allow proper inspection of the surfaces.

Many of the problems which waterjets have tackled are equivalent to those encountered in above water situations. The marine environment does pose particular problems, however. The cost of cleaning ships is also not cheap. As Pardey pointed out in 1976 [3.84], drydocking and cleaning a very large crude oil carrier can cost more than fifty thousand pounds. For the largest carriers this is not a realistic option, given the lack of facilities and the risk of potential damage to the ship structure. For this reason, underwater cleaning became important. Underwater operations often make it difficult to observe the actual cleaning process, since removed solids can hang in the water, which may not have been clear originally. Thus effective use of waterjetting systems require that the changing conditions which occur in waterjet usage, particularly with increasing depth, be understood.

Pardey (ibid.) illustrated some of these factors by carrying out comparative tests of equivalent nozzles in air, and with the jet submerged. Cutting both polystyrene and concrete he found that whereas a jet would cut more than 250 mm in air, with reduced influence of stand-off distance and there was almost a linear decrease in effectiveness with distance. Increasing the jet horsepower by raising the pressure was not found to be as effective as increasing the jet flow. This effect has been illustrated also by work at Rolla [3.85]. In order to examine the effect of water conditions on jets issuing from various nozzles tests were carried out in which a jet was traversed over a sloping surface. Thus, over the course of the test, effective cutting distance of the jet could be observed, as a function of pressure in the water. It rapidly became clear, that very small jets can become ineffective over relatively short distances from the nozzle (Table 3.11).

Table 3.11 Effective cutting distance of a waterjet in pressurized fluid [3.85]

Nozzle dia (mm)	*0*	*35*	*Back pressure (bar)* *70*	*105*	*140*	*17.5*
0.75	28.7	16.0	16.0	7.85	-	-
1.00	47.5	22.35	22.35	11.18	-	-
1.63	69.85		31.75	23.88	20.81	17.52

Cutting depths are given in mm. The operating jet pressure was 700 bar, and the tests were carried out on berea sandstone.

One observation by a number of observers has been that the increase in turbulence around an underwater jet can significantly improve the performance of the jetting unit. The major factor in this enhancement is the generation of cavitation bubbles, and their use. While the mechanisms will be dealt with more in Chapter 10, there are two salient points to bear in mind. Firstly the pressure of the surrounding fluid has a significant effect on the effectiveness of the cavitation stream. Small changes in the pressure of the water, as a diver goes deeper, for example, will significantly lower the cutting range of the cavitation, and may thus significantly adjust the effectiveness of the process. Concurrently the noise generated by a cavitating flow is quite high and intense. Given that the operating diver has his head in the water near the noise source, it is important that he wear sufficient protection to guard against damage from this hazard.

As the areas which can be cleaned underwater have become more established, and larger areas have become the norm, the difficulty in operating with a diving crew who are limited in the amount of time that they can spend on the job, and who may have difficulty in observing how well the jets are cutting in certain areas has led to a change. As with a number of other industries, the advent of automated or semi-automated equipment has also, increasingly made itself evident in the submarine environment. Much of the equipment must, however, be specially designed for the contours of the individual surfaces to be cleaned. It must take into consideration the depth at which the cleaning will take place, the thickness of the deposit to be removed, since this will control the ease with which the equipment moves over the surface. It will also control the stand-off distance of the jet from the final surface, a much more critical factor than in an air environment. The use of fan jets becomes more of a question than in surface areal cleaning, given the reduction in power with stand-off, but it can be an advantage in shallow operations where the increase in turbulence can enhance the supplementary cleaning which comes with cavitation.

3.11 LESSONS FOR USE

Cleaning is currently the largest application of waterjets in industry. The wide variety of items that must be cleaned of a varied suite of materials make it impossible to provide one common answer to every problem. Some lessons are, however, universal. The logical approach to a problem wherein the first tests should determine the optimum nozzle shape, flow rate and **operating pressure at the nozzle**, should be relatively obvious. Configuring a number of orifices to give optimum coverage of a surface is less easy. Coating thickness may require either multiple passes of the cleaning head, or that adjacent nozzles overlap. The effect of stand-off distance in removing thick coatings, and the ability of the jet to effectively reach the underlying surface, particularly underwater, must be established.

Once the flow and pressure have been established, it is important to ensure that there is not too great a line loss between the pump and the cleaning head, so as to bring the operating jets below the optimum values required, at least without conscious decision. Once a job has started, proper maintenance and continued effective performance requires that the nozzle condition be monitored at regular intervals. The mere observation of a jet, and the conclusion "it looks good" does not meet this requirement.

Because waterjetting appears to be a simple operation, pumping water under pressure from a unit, through a nozzle onto a surface, does not mean that it cannot be a very inefficient one. As the industry becomes more competitive, the winners and survivors will finally be those who best engineer their systems to minimize system losses and maximize the delivery of energy from the jet to the target material. Sometimes this will also require different strategies to optimize material removal. It is sometimes easier and much more effective to get under a layer of contamination and peel it off, than it is to try and slice the material off in a 90° attack. The business, in reality, is a lot more complex, than it may appear.

3.12 REFERENCES

3.1 Ward, G.M., "Safety Considerations arising from operational experience with high pressure jet cleaning," 1st International Symposium on Jet Cutting Technology, Coventry, UK, paper F1, April, 1972, pp. F1-1 - F1-24.

3.2 Brierley, W.H., Workshop on the Application of High Pressure Waterjet Cutting Technology, Rolla, MO., November, 1975.

3.3 Ashton, A.T., Gay, J., "SID seeks sludge and foreign bodies in steam generator upper heads while . . . ", Nuclear Engineering International, March, 1993, pp. 20 - 23.

3.4 Siegel, R.J., Fishbein, M.C., Forrester, J., Moore, K., DeCastro, E., Daykhovsky, L., and DonMichael, T.A., "Ultrasonic Plaque Ablation - A New Method for Recanalization of Partially or Totally Occluded Arteries," Circulation, Vol. 78, No. 6, December, 1988, pp. 1443 -1448.

3.5 Tilghman, B.C., Patent 2147, United Kingdom, August, 1870.

3.6 Plaster, H.J., Blast Cleaning and Allied Processes, Industrial Newspapers Ltd., 17/19 John Adam House, Adelphi, London UK, 1972, 2 vols.

3.7 Anon, S.S. White Industrial Abrasive Unit, S.S. White Division of Pennwalt Corporation, Piscataway, NJ, Bulletin 7706A.

3.8 Remmelts, J., Optimum Conditions for Blast Cleaning Steel Pipe, Report 94c, Netherlands Ship Research Center, 1968.

3.9 Anon, Clemco Presents Blast Off, Publication 114-0476, Stock Number 09294, Clemco Industries, 1657 Rollins Road, Burlingame, CA 94010.

3.10 Adlassing, K., and John, W., "The Properties Of Nozzles And Blasting Materials In Compressed Air Blasting Units," Gieberei, 1960.

3.11 Anon, Tetrabore Sandblasting Nozzles, ESK catalog supplied by Abrading Machinery and Supply Co., Chicago, IL, 1981.

3.12 Albert, G.D., Patent 722464, 1955.

3.13 Anon, Sand-All Sandblaster, Catalog 777, from Ace Enterprises Inc., Miami, FL, 1981.

3.14 Zeiler, W., and Schmithals, P.U., Properties and Applications of Cut Wire, quoted in Ref. 3.6.

3.15 Neville, F.W., quoted in Reference 3.6.

3.16 Kobayashi, R., Fukunishi, Y., and Ishikawa, T., "Machining of Solid Materials by High Speed Air Jet," paper K1, 10th International Symposium on Jet Cutting Technology, Amsterdam, Holland, October, 1990, pp. 281 - 291.

3.17 Anon, Electric Steam Boilers, Chromalox Industrial Heating Products, Pittsburgh, PA, Bulletin FX920M.

3.18 Axelson, W., "Why Clean Equipment?" Heavy Duty Equipment Maintenance, July, 1976, An Irving Cloud Publication.

3.19 Liljedahl, L.A., Effect of Fluid Properties and Nozzle Parameters on Drop Size Distributions from Fan Spray Nozzles, Ph.D. thesis, Iowa State University, 1971, University Microfilms, Catalog Number 72 - 52224.

3.20 Trauberman, L., "Saves with Total Cleaning Cost," Food Engineering, June, 1970.

3.21 Short, W.L., Some Properties of Sprays Formed by the Disintegration of a Superheated Liquid Jet, Ph.D. thesis, University of Michigan, 1963, University Microfilms, Catalog Number 64 - 8208.

3.22 Anon, Hot, High Pressure Washer, Siebring Manufacturing Company, George, IA, Catalog TCS860 10M.

3.23 McIntyre, J., "Fish Processing: Pressure Washing Just for the Halibut," Cleaner Times, Advantage Publishing, 17319 Crystal Valley Road, Little Rock, Arkansas, Vol. 5, No. 2, February, 1993, pp. 6 - 10.

3.24 Maasburg, W., Cleaning in the 80's, Water Jet Contractors Meeting, Sheffield, UK, 1980.

3.25 Mora, J., "In Plant Cleaning Systems - The Six Principles for Successful in-Plant System Design," Cleaner Times, Advantage Publishing, 17319 Crystal Valley Road, Little Rock, Arkansas, Vol. 5, No. 8, August, 1993, pp. 7 - 10.

3.26 Tilton, L., "Accident Leads to New Business for North Carolina Pressure Cleaner," Pressure Concepts, Advantage Publishing, 17319 Crystal Valley Road, Little Rock, Arkansas, Vol. 1, No. 3, May, 1993, pp. 32 - 33.

3.27 Anon, "Cleaning House: Pressure Washing and Residential Exteriors," Cleaner Times, Vol. 5, No. 3, March, 1993, pp. 6 - 11.

3.28 Henderson, J.C., "Understanding and Measuring Boiler Efficiency," Cleaner Times, Vol. 5, No. 3, March, 1993, pp. 20 - 24.

3.29 Bury, L., Housden, R.K., Stephensen, R., and Thompson, C.R., "Steam-Boosted Waterjet Cleaning at 300 bar," paper F1, 2nd International Symposium on Jet Cutting Technology, Cambridge, UK, April, 1974.

3.30 Scharwat, F., "Alternative Methods of Reducing the Environmental and Human Health Risks in Paint Stripping," Cleaner Times, Vol. 4, No. 12, December, 1992, pp. 16 - 17, 22.

3.31 Thorpe, M.L., "Ultra High Pressure Water jet Cleaning and Stripping," Cleaner Times, Vol. 4, No. 4, April, 1992, pp. 24 - 25.

3.32 Uranishi, K., "Effect of Superheat on High Speed Water Jet," paper J4, 9th International Symposium on Jet Cutting Technology, Sendai, Japan, October, 1988, pp. 479 - 494.

3.33 Bloch, C., Knox, J., Nesbitt, D., and Heist, C.H., "Historical Perspective of Industrial Cleaning Business," Industrial Cleaning Contractor, Vol. 1 No. 5, July/August, 1993, pp. 23 - 29.

3.34 Johnson, S., "Using High Pressure Water to Grow," Industrial Cleaning Contractor, Vol. 1, No. 2, April, 1993, pp. 19 - 21.

3.35 Hinderliter, R.W., in Industrial Cleaning Contractor, Vol. 1, No. 2, April, 1993.

3.36 Anon, "High Pressure Water Jetting Systems - Part 2," Industrial Cleaning Contractor, Vol. 1, No. 4, June, 1993, pp. 20 - 23.

3.37 Tilton, L., "Yes He Does Windows," Pressure Concepts, Advantage Publishing, 17319 Crystal Valley Road, Little Rock, Arkansas, Vol. 1, No. 6, August, 1993, pp. 24 - 26.

3.38 Cook, J., "Mega Market Potential for Mini Power Cleaning Equipment," Cleaner Times, Vol. 4, No. 9, September, 1992, pp. 32 - 34.

3.39 Osburn, D., "Car Wash: Surviving Despite Tough Times," Cleaner Times, Vol. 4, No. 9, September, 1992, pp. 6 - 21.

3.40 Tilton, L., "Cleaning the Mormon Temple," Pressure Concepts, Advantage Publishing, 17319 Crystal Valley Road, Little Rock, Arkansas, Vol. 1, No. 3, May, 1993, pp. 6 - 9.

3.41 Howells, W.G., "Polymer Blasting - A Chemists Point of View," 2nd U.S. Water Jet Conference, Rolla, Mo., May, 1983, pp. 442 - 447.

3.42 Tursi, T.P., Jr., and DeLeece, R.J., Jr, Development of Very High Pressure Waterjet for Cleaning Naval Boiler Tubes, Naval Ship Engineering Center, Philadelphia Division, Philadelphia, PA., 1975, p. 18.

3.43 Leach, M., "Industrial Applications for High Pressure Water," Cleaner Times, Vol. 4., No. 8, August, 1992, pp. 18 - 21.

3.44 Anon, Jet and Line Moles, Arthur Products Co., 620 E Smith Rd., Medina, OH, 44256, Catalog 84, 1984, 16 pp.

3.45 Summers, D.A., et al., Recommended Practices for the Use of Manually Operated High Pressure Water Jetting Equipment, Water Jet Technology Association, St. Louis, Mo, 1994.

3.46 Anon, Code of Practice for the Use of High Pressure Waterjetting Equipment, Association of High Pressure Water Jetting Contractors, London, UK, December, 1982, 25 pages.

3.47 Matsuki, K., Okamura, K., and Sugimoto, F., "Removal of Geothermal Scales with High Pressure Waterjets," Industrial Jetting Report, May, 1988.

3.48 Anon, "Increasing Profit Margins with Sewer and Drain Waterjetters," Cleaner Times, Vol. 5, No. 10, October, 1993, pp. 8 - 13.

3.49 Anon, High-Pressure Water Blasting, Construction Safety Association of Ontario, Toronto, Canada, April, 1980, 12 pages.

3.50 Anon, Industrial Jetting Report, December, 1987.

3.51 Ewald, M.H., "The 'Skipjack' Sewer Cleaning Nozzle," 2nd U.S. Water Jet Conference, University of Missouri-Rolla, Rolla, MO, May, 1983, pp. 123 - 126.

3.52 Lancaster, J., "Combination Cleaners: Cleaning Up Additional Profits," Pressure Concepts, Advantage Publishing, 17319 Crystal Valley Road, Little Rock, Arkansas, Vol. 1, No. 5, July, 1993, pp. 4 - 8.

3.53 Zublin, C.W., "Water Jet Cleaning Speeds - Theoretical Determinations," 2nd U.S. Water Jet Conference, University of Missouri-Rolla, Rolla, MO, May, 1983, pp. 159 - 166.

3.54 Bardick, I.D., "An approach to remote controlled jet cleaning of chemical reaction vessels to gain the maximum benefit from the technique," paper H7, 4th International Symposium on Jet Cutting Technology, Canterbury, UK, April, 1978, pp. H7-87 - H7-94.

3.55 Weeren, H.O., Lasher, L.C., and McDaniel, E.W., "Cleanout of Waste Storage Tanks at Oak Ridge National Laboratory," Annual Meeting, Materials Research Society, 1984, Boston, Mass.

3.56 Bamberger, A.A., and Steele, D.E., "Developing a Scarifier to Retrieve Radioactive Waste from Hanford Single-Shell Tanks," paper 55, 7th American Water Jet Conference, Seattle, Washington, August, 1993, pp. 737 - 746.

3.57 Summers, D.A., Mann, M., and Galecki, G., A Mining Strategy to Remove Radioactive Waste from Underground Storage Tanks, Report to Battelle Pacific Northwest Laboratories, December 1993, University of Missouri-Rolla.

3.58 Torpey, P., "Some Experiences in the Manufacture and Application of High Pressure Water Cleaning Equipment," paper D1, 1st International Symposium on Jet Cutting Technology, Coventry, UK, April, 1972, pp. D1-1 - D1-11.

3.59 Anon, Guide to Water Jetting Techniques, F.A. Hughes Company, Ltd., August, 1977.

3.60 Odds, D., "Water Jetting under the North Sea," paper H3, 4th International Symposium on Jet Cutting Technology, Canterbury, UK, April, 1978, pp. H3-39 - H3-46.

3.61 Reif, R.B., and Hawk, S.A., Review and Evaluation of the Literature on Electrostatic Generation in Tank Cleaning, Final Report on Work Order 733842.1, to the U.S. Coast Guard, Department of Transportation, from Battelle Columbus Laboratories, October, 1973, 41 pages.

3.62 Conn, A.F., Rudy, S.L., and Mehta, G.D., "Development of a Cavijet System for Removing Marine Fouling and Rust," paper G4, 3rd International Symposium on Jet Cutting Technology, Chicago, IL, May, 1976, pp. G4-31 - G4-44.

3.63 Woodward, M.J., and Judson, R.S., "The Development of a High Production Abrasive Water Jet Nozzle System," 4th US Water Jet Conference, Berkeley, CA, August, 1987, pp. 137 - 146.

3.64 Wolgamott, J., "Plant/Construction Applications," Section 7, Fluid Jet Technology - Fundamentals and Applications, Water Jet Technology Association, 1991.

3.65 Brierley, W.H., and Vijay, M.M.,"Rotating High Pressure Water Jet Gooseneck Cleaner," paper G3, 3rd International Symposium on Jet Cutting Technology, Chicago, IL, May, 1976, pp. G3-23 - G3-29.

3.66 Barker, C.R., High Pressure Fluid Jet Cutting and Drilling Apparatus, U.S. Patent #4,369,850, January, 1983.

3.67 Steele, C., Newsletter, U.S. Water Jet Technology Association, Vol. 2, 1986, p. 1.

3.68 Saunders, D.H., and Barton, R.E.P., "The Use Of Fan-Shaped Water Jets In Preference To Straight Jets To Remove A Paint Coating," paper 37, 8th International Symposium on Jet Cutting Technology, Durham, UK, September, 1986, pp. 353 - 362.

3.69 Kovscek, P.D., Taylor, C.D., and Thiomons, E.D., Techniques to Increase Water Pressure for Improved Water-Jet-Assisted Cutting, U.S. Bureau of Mines Report of Investigations, RI 9201, 1988, 10 pages.

3.70 Merk-Gould, L., Herskovitz, R., and Wilson, C., "Field Tests on Removing Corrosion from Outdoor Bronze Sculptures Using Medium Pressure Water," 1993, Fine Objects Conservation Inc., 3 Meeker Road, Westport, CT, 06880.

3.71 Colberg, W.R., Gordon, G.H., and Jackson, C.H., Refurbishment of SRB Aluminum Components by Walnut Hull Blast Removal of Protective Coatings, NASA Technical Memorandum TM - 82499, July, 1982, 13 pages.

3.72 Summers, D.A., et al., The Removal of Corrosion Products by Pressure Washing, Final Report to Fine Arts Conservation, Inc., by The University of Missouri-Rolla, May, 1993.

3.73 Anon, "Freedom Statue to get Face Lift," St. Louis Post Dispatch, May 6, 1993, page C1.

3.74 Sanders, M.J., and Bond, R.D., "The Use Of High Pressure Water Jetting To Remove The Corrosion Deposits From Samples Of The WSGHWR Primary Circuit Pipework," paper C1, 7th International Symposium on Jet Cutting Technology, Ottawa, Canada , June, 1984, pp. 99 - 118.

3.75 White, R., The "Duck" books and other correspondence, 1979, Robert White Inc., P.O. Box 1928, Cocoa, FL 32922.

3.76 Taylor, S.A., and Judson, R.S., Use of Very High Pressure Water Jet Cleaning in Marine Maintenance, Weatherford/AAI, 10950 Old Katy Road, P.O. Box 55343, Houston, TX, 77055, 16 January 1976.

3.77 Appleman, B. R., and Bruno, J.A., Jr., "Evaluation of Wet Blast Cleaning Units," Journal of Protective Coatings and Linings, August, 1985, pp. 34 - 42.

3.78 Gracey, M.T., "Another Way to Use Biodegradable Media," Cleaner Times, Vol. 5, No. 8, pp. 30 - 32.

3.79 Woodward, M.J., "Water Soluble Abrasives," paper 28, 7th American Water Jet Conference, Seattle, Washington, August, 1993, pp. 397 - 403.

3.80 Xue, S., Huang, W., Zhao, S., and Shi, D-J, "Equipment and Test Research of High Pressure Water Jet for Rust Removal," paper 48, 7th American Water Jet Conference, Seattle, Washington, August, 1993, pp. 653 - 662.

3.81 Liu, B-l., Beihua, J., and Zhang, D., "The Premajet Derusting and the Abrasive Recovery System," paper 46, 7th American Water Jet Conference, Seattle, Washington, August, 1993, pp. 629 - 641.

3.82 Hall, P.G., "Additives for High Pressure Cleaning," paper 34, 8th International Symposium on Jet Cutting Technology, Durham, UK, September, 1986, pp. 331 - 334.

3.83 Swan, S.P.D., conversation at the 5th American Water Jet Conference, Toronto, Canada, August, 1989.

3.84 Pardey, P.H., "Underwater application of continuous water jets", paper G1, 3rd International Symposium on Jet Cutting Technology, Chicago, IL, May, 1976, pp. G1-1 - G1-12.

3.85 Summers, D.A., and Sebastian, Z., The Reaming out of Geothermal Excavations, Final Report to Sandia Laboratories on Contract 13-3246, April, 1980, 155 pages.

4 HIGH PRESSURE INDUSTRIAL WATERJET CUTTING

4.1 INTRODUCTION

The development of waterjets, as a tool for the cutting of geotechnical applications, had been progressing at a readily slow rate, when an entirely different venue was opened up for the commercial use of high pressure jet technology. This application, for industrial cutting, has since grown to dominate the very high pressure use of waterjets. With the subsequent addition of abrasive particles, it has become a method of choice for the cutting of expensive metals and glass. And because of the low reaction forcc which it imparts to the holding fixture, it has become a method easily adapted for use with robotic equipment.

4.2 BACKGROUND

Credit for the development of the industrial cutting application rightly belongs to Dr. Norman Franz. Based upon his observation that steams leaks were found by noting where they cut through a straw detector, he conceived the idea of using high pressure waterjet systems for cutting through wood products [4.1]. After guiding a Ph.D. student through a study of the possibility [4.2] Dr. Franz worked with the McCartney Manufacturing Company of Baxter Springs, KS (now a part of Ingersoll-Rand) and the first commercial unit was installed at Alton Box Board of Alton, IL on November 15, 1971 [4.3]. The initial results obtained with this equipment were significantly better than had been anticipated, additional units were installed, and a new industry was in the making.

The initial equipment consisted of two, 150 mm stroke, intensifiers, a polymer mixer, an accumulator and a panel control board, located some 18 m from the work station. Water was supplied to the nozzle at a pressure of 2,800 bar, at a flow rate of 350 l/hr. The cutting nozzle was mounted on a vertical stand, so that it could be moved up and down the target surface. The target itself, a 1.25 cm thick cardboard tube, was rotated around a vertical axis in front of the jet so that, in combination, a profile cut could be achieved. In its original application the unit cut contoured forms for use in the furniture industry but the basic approach taken has since been proven in a wide range of other applications.

It is important to note that,even the in original installation, an automatic polymer mixing system was included in the package. While the subject of jet enhancement by polymers will be addressed in more detail in Chapter 10, it is important to note the two advantages sought for the use of the polymer, enhanced cutting capability, and the reduction in the wetting of the target material have otherwise remained as problems for the industry until the present time. A study by Franz about the time of the first installation [4.4] showed that a concentration of 0.4% by weight of the long chain polymer, Polyox WSR-301 increased cutting depth, at a target distance of 3.2 mm from the nozzle, by a factor of 3. The variation in the water retention on the sides of the cut was a function of jet traverse speed and jet pressure. It dropped from a high of 0.45 g/cm for a jet pressure of 1,380 bar, and a traverse speed of 91.5 m/sec to a low of almost no detectable water at a pressure of 2,750 bar and a feed rate of 213.4 m/sec when the polymer was added [4.5].

The first products to be commercially cut were a thick paper tube, and the advantage of the waterjet tool in this use, brought its application to a number of industries. The most immediately obvious advantage is illustrated by the cutting of cardboard. With a conventional system which uses a mechanical knife for the cut, the speed of cut is limited. In addition, it is not possible, because of the force on the workpiece, to cut the material without compressing the flute (Fig. 4.1). While this has some disadvantage when cutting across the flute, it is also a serious problem when cutting along the flute of the structure. With the waterjet cutting, the very high speed of the water as it traverses through the piece, virtually eliminates sidewall compression, and the part is left without significant water penetration. It is this high quality edge cut, achievable with this tool, that has brought it further penetration into the market place.

The move toward broader industry application was hastened by the entry of a second company, Flow Industries, into the market. And although a number of other investigators and companies have worked in the business since their involvement, it is still to this company, and its sister organizations in Kent, WA., that the major credit must go for moving the application of waterjets into so many industrial applications within the last fifteen years.

Figure 4.1 Comparative cutting of cardboard by a waterjet (bottom) and mechanical tool.

4.3 EQUIPMENT FOR INDUSTRIAL APPLICATIONS

4.3.1 ACTIVATION VALVE AND CATCHER

In order to integrate waterjets into industry two additional pieces of equipment were required. These were an automatic on-off valve at the nozzle, and a catcher. The valve is required to ensure that when the jet is not actively required for cutting, that it is not in action. This allows the nozzle to move over the part without damage to unwanted areas, and it minimizes undesirable part wetting. It is important that this valve function is as nearly instantaneous as possible since this will control the accuracy of the edge cut and the precision in cutting the internal contours of pieces. The valve is placed immediately upstream of the nozzle so that there is no drainage from the nozzle after it activates.

Many waterjet cutting installations also now include a catcher on the other side of the part from the jet nozzle [4.6] (Fig 4.2). The catcher has several roles. Firstly it captures the spent water and cut material. Waterjet cutting is thus one of the best methods for reducing dust and atmospheric pollution from the cutting process. Secondly by retaining the jet near the workpiece it significantly reduces the noise that would be generated if the jet were to be allowed to expend itself into the air beyond the piece. Thirdly, by retaining the water the possibility of damage caused by the jet to adjacent surfaces, and to possible individuals in the vicinity is minimized, provided the catcher is kept in good order.

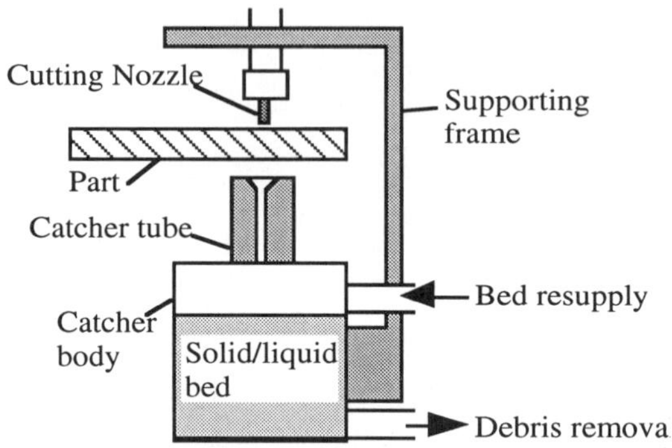

Figure 4.2 Catcher arrangement under a trimming waterjet nozzle.

One disadvantage which this useful tool incurs is that of weight, particularly when abrasive waterjets are used. While the normal weight of the nozzle and support structure is on the order of 3.6 kg, if an abrasive system is used, this can increase to up to 20 kg. This brings the total weight at the end of its arm close to the operational limit for some robots, and this may therefore have a negative effect on the speed and accuracy of the device at its upper levels [4.7].

4.3.2 INLET WATER TREATMENT

When high pressure waterjet cutting systems were first installed in factories it was anticipated that the equipment would be able to operate for a considerable period of time without needing a great deal of maintenance and upkeep. However, as equipment began to be installed in factories around the United States, a particular and unexpected problem arose from the water. The problem is that most of the water in the United States comes from individual wells, and it is not all of the same quality. The standards which the water must meet, may make the water acceptable for drinking and normal use, but can sometimes give problems in equipment where the water is subject to pressures above 3,500 bar and where it is moving in excess of 700 m/sec. For these reasons the original 5 micron fiber cartridge filter fitted to the first machine has not always been adequate to protect the pump and jet delivery system, even though, in that installation, no erosion of the seals and shafts had occurred after the first six months of operation.

In these circumstances, slight changes in water chemistry can significantly affect the erosion resistance of some of the components within the pump mechanism. By the same token, the presence of small amounts of solid material in the jet fluid can cause an accelerated wear of either the nozzle body or of other points within the circuit where jet velocities are high. Alternately, as the jet leaves the nozzle some of the dissolved solids may precipitate out of the fluid onto the nozzle bore. This will lead to a gradual blockage of the nozzle and a dramatic reduction in its effective performance. The problem of water quality has been recognized and addressed by the vendors of the equipment (for example [4.8]). As a result it has been found necessary to set standards in order to operate high pressure waterjet cutting systems effectively.

Table 4.1 Water quality required for an Ingersoll-Rand Waterjet Cutting System [4.9]

total dissolved solids (TDS)	<500 mg/liter
total hardness	<25 mg/liter
fluoride	<250 mg/liter
iron	<.2 mg/liter
manganese	<.1 mg/liter
turbidity	<5 NTU
free chlorine	<1 mg/liter
water pH	6.5 to 8.5

Unfortunately, in a number of localities this water quality will not initially be met. While at the most common problem areas it is where the water contains too much calcium carbonate. This will frequently precipitate out during the cutting process and may, therefore, cause a blockage of the nozzle. For example, in one independent set of tests which they had performed, Flow Systems had shown that with tap water it is possible to have the nozzle totally blocked within 34 hours [4.8].

There are several different ways of treating the water. This ranges from a softening process, familiar to most home owners, to the complexity of reverse osmosis or a separate circuit for deionizing the water. Softening the water involves passing the water over a resin bed, which replaces the calcium with sodium. The resin bed is then regenerated in a separate cycle, and, with the replenishment of salt to provide the sodium, the process can continue. Reverse osmosis involves passing the water through a very fine membrane, which effectively acts as a super filter operating at the molecular level. In this way water can pass through the membrane, while the undesired materials are left behind. In a de-ionizing circuit the process is somewhat similar to softening, except that the process is more comprehensive and occurs with an ion exchange, such that the negative ions such as the chlorides and sulfates are replaced with hydroxyl ions, while the positive ions such as sodium and calcium are replaced with hydrogen. The system is regenerated with sulfuric acid and caustic soda. Flow compared the results from these treatments on water samples [4.8].

Table 4.2 Effect of water treatment on water content [4.8]

	Original tap water	*Softened water*	*RO water*	*De-ionized*
alkalinity	324 mg/liter	364	10	<1
calcium	23.7 mg/liter	<0.01	0.14	<0.01
chloride	38 mg/liter	56	<1	<1
hardness	140 mg/liter	<1	<1	<1
magnesium	15.4 mg/liter	<1	<0.01	<0.01
pH	7.12	7.83	6.87	6.96
sodium	212 mg/liter	241	4.02	<0.01
sulfate	158 mg/liter	163	<1	<1
Total dissolved solids	637 mg/liter	652	14	<1

It can be seen from the above table that the use of a deionizing circuit is the most effective of the different methods of treatment for water. In the study in which Flow funded, it was found that using this technique orifice life could be achieved which was consistently in excess of 200 hours. With reversed osmosis the average lifetime of the orifice was approximately 190 hours, while conventional softened water gave a nozzle lifetime average of only 78 hours, in contrast to the use of the tap water where the average was 34 hours.

Before the water undergoes treatment the particulate matter in the water must be removed to insure it does not either affect the water treatment circuit or cause more rapid erosion of the nozzle. For this reason the water is pre-filtered by being passed through successively finer screens, typical sizes being 10 micron, 1 micron, and 0.5 micron filter series. The use of a deionizing unit is not always recommended since the resulting water can increase corrosion of the components of the pump and delivery system. Reverse osmosis with specific filters for use where the hardness, iron or magnesium content of the feed water poses a problem, provides an alternate treatment. When the water coming into a waterjet system is particularly turbid a depth filter may be used, otherwise the filter sizes should be adjusted for the conditions encountered. The treatment of water is not necessarily a simple inexpensive process. For example, Cutshall [4.10] priced a recommended deionizing water treatment plant for a pump operating 8 hours a day at $15,000 in 1986.

4.4 NOZZLE AND PART POSITIONING-ROBOTIC USE

The first high pressure waterjet cutting installation contained many of the features which have since become common in industrial applications [4.3]. The controlled jet position, achieved by a combination of nozzle and part positioning, in order to achieve a complex contour, has become the feature of many installed systems. Simplification of the relative movement by separating the motion in the vertical "X" direction which was provided by the movement of the nozzle, from the horizontal "Y" motion which was provided by the movement of the table and/or part. This made it easier to cut the complex shape required in the original plant and this has been translated into similar approaches to solving similar cutting path problems in current manufacturing. The separation makes it simpler, and thus potentially less expensive, to install the connections for the high pressure cutting system and to give the system the flexibility which it needs. The original unit operated at a cutting speed of approximately 500 cm/min at which the jet nozzle was moved around the cardboard tubes.

Moving the nozzle of the jet system, when the internal piping leading to it may contain water at a pressure of up to 4,000 bar requires some considerable engineering consideration. At the time that the original installations were made, high pressure swivels capable of allowing free joint rotation in the supply line to the nozzle were not available in reliable form. In order to allow free movement of the nozzle across the workpiece, therefore, a flexible feed was required. Fortunately the waterjet volume required in high pressure industrial use is quite small. Normal operating conditions require approximately 4 lpm of water, and this can be supplied through high pressure tubing which is 14.28 mm in outer diameter, 2.38 mm in internal diameter. This tubing is quite flexible, and can thus be engineered to provide the flexibility required for the task. This is normally carried out in one of two ways. If the space available is sufficient, then the tube leading to the nozzle mount can be made long enough and of sufficient curvature to allow the relative nozzle movement over the limits of its traverse path, without applying any additional stress to the connectors at each end of the supply line. This was the method used for the first installation.

Where space is at more of a premium, then the motion of the nozzle may be allowed by turning the tubing into a series of coils (Fig. 4.3). These can be wrapped around the mounting frame of the nozzle support bracket, or left free-standing above the nozzle assembly. The purpose of the coil is to provide relative motion of the nozzle block in two directions. In the coil shape there is the equivalent of a spring, which will allow vertical motion of the underlying nozzle by expanding or compressing the coils. At the same time there is a limited ability to rotate the nozzle perpendicular to the axis of the coil, with a lateral motion of approximately 9 degrees on either side of the at rest position for each circle of the coil (Fig. 4.4). Obviously this flexibility varies with the pipe diameter and the tightness of the coil generated, the value given is for a 6.35 mm diameter tube wrapped in a 20 cm diameter circle [4.12].

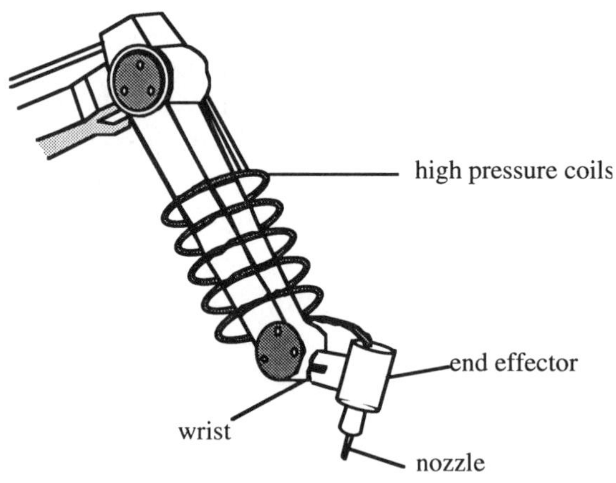

Figure 4.3 Coil spring assembly to allow nozzle movement [after 4.11].

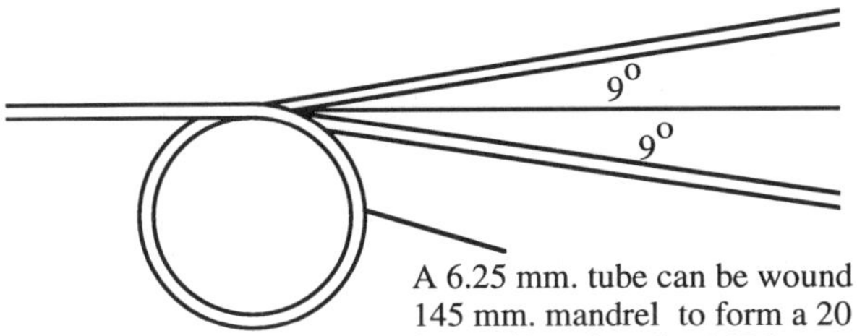

Figure 4.4 Deflection allowed by a single coil [after 4.12].

In the initial installation the light weight, approximately 0.4 kg, of the nozzle assembly made it relatively easy to move the nozzle using initially a manually driven rock and pinion drive, while the table rotation was achieved using a peripheral carriage drive. As confidence with the system grew the manual operation was changed to an automated drive. Servo-controlled motors were installed for both part rotation and nozzle manipulation. Motion of the two could then be coordinated from an optical tracing system, working from a part blueprint. Although the machine could operate at cutting speeds of up to 600 cm/min two factors made this an impractical speed. The first was the edge quality that was achieved. At the highest cutting speeds a "ripple" appears along the edge of the cut over the width being sliced. This is unacceptable for outer edge fabrication, and thus requires a slower speed. The problem will be addressed further in the discussion on abrasive jet cutting. The other consideration lies in the problem of turns. Even though the head was very light, it, and its supporting structure, still had considerable momentum, and thus at the higher speeds it becomes more difficult to make tight turns at high rates of cutting. These factors became more important as the industry developed.

In the first layout the points where acceleration or deceleration was required could be given by placing "Match marks" on the drawing which the tracer was following [4.13]. More recently this can be accomplished by identification of the points from within the computer programming which drives the unit.

The use of high pressure waterjets has grown considerably from that first installation. As it has developed, however, two different approaches can be differentiated in the way in which the waterjets have become useful to the cutting industry. One is the cutting of parts from sheets of material fed into the machine, and the other is the trimming of parts already made by other processes. The first requirement will often require a simpler drive unit, which can be built around the X-Y table concept described above, while the second frequently requires the more complex movements in three dimensions which call for positioning of the cutting nozzle under the control of a multi-axis robot.

4.4.1 ROBOTS AND THEIR CONTROL

In order to understand the differences in approach it is useful to have some basic idea as to some of the operating parameters of robotic systems. An industrial robot has been defined as "a reprogrammable, multi-functional manipulator designed to move materials, parts, tools or specialized devices through variable programmed motions for the performance of a variety of tasks" [4.14]. Often the tool which is used to carry out the purpose of the robot is called an **end effector**, since it is attached, through an **interface plate**, or **wrist**, to the end of the positioning limb or **arm** of the robotic device (Fig. 4.3).

A robot can be mounted so that the main manipulator arm mount can be moved in either or both the X and Y directions (Fig. 4.5), or it can be mounted on a fixed platform or pedestal, with a reach which extends in a circular, or partially circular path around the station (Fig. 4.6). Although there are several variations on this basic distinction, for simplicity they will not be discussed, but rather the robots will be grouped into the first class, **gantry robots,** and the second, **pedestal robots**.

Note that the control cables for the end effector and cross-feed on the gantry are carried in cable guides which bend over each other to allow flexibility while minimizing space needed. The ability of the robot arm to move the cutting lance is normally described by its **degrees of freedom**. This is the number of different axes around or along which the end effector can be moved. For example, a robot arm which moved the nozzle in only one linear direction (say the X axis of the gantry) would have one degree of freedom. This movement would occur if the nozzle was moved across the table using the cross-feed mechanism on the gantry robot. If the nozzle had to be moved in two perpendicular directions (the X and Y axes) then both the cross-feed and the longitudinal feeds would be used on the table. These two motions could be used to cut shapes (see Fig. 4.9), for example out of paper or leather, and the nozzle is said to have two degrees of freedom. If the nozzle is held in a fixture which can also raise the assembly in the direction perpendicular to the first two (the Z direction) giving a height control, then there are three degrees of freedom.

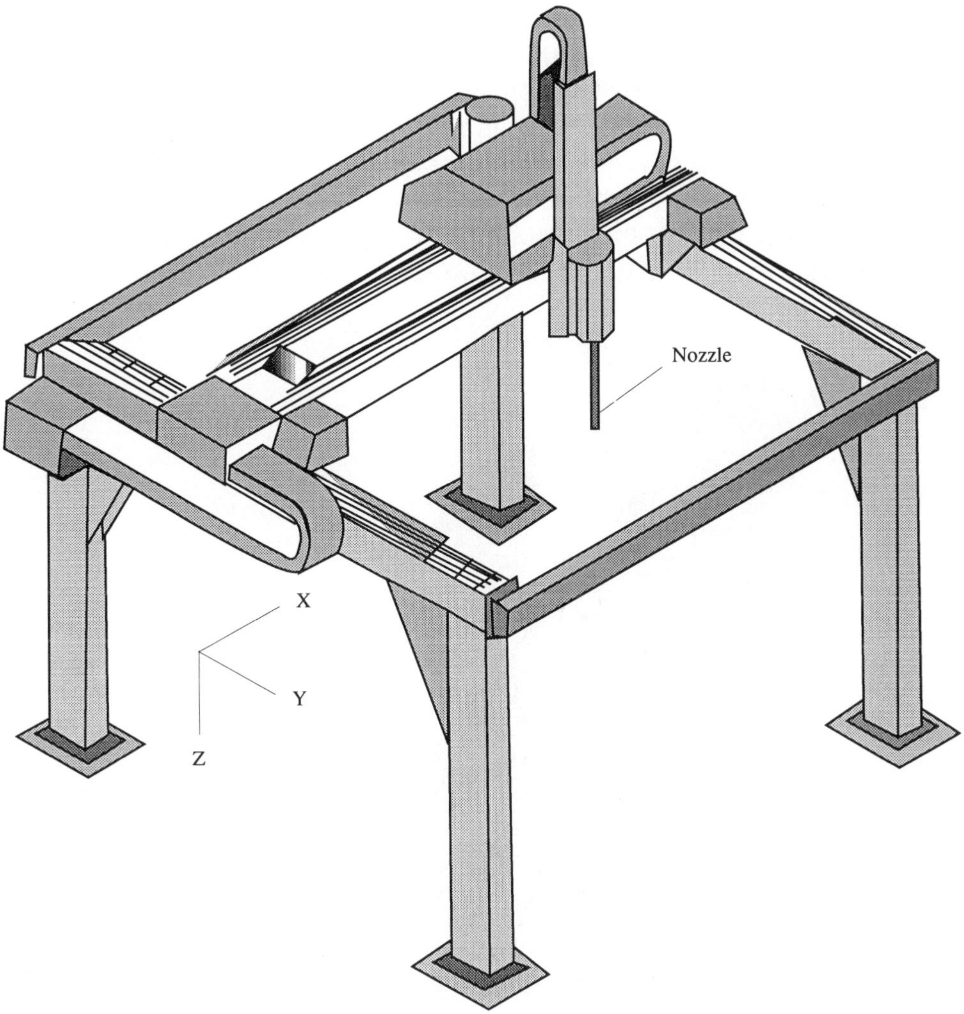

Figure 4.5 Schematic representation of a gantry mounted robot.

The above feeds allow the nozzle the ability to move through three-dimensional space, but in order to be effective the jet must be pointed to cut perpendicularly through material beneath it. This requires that the nozzle be mounted on a wrist which can rotate. Rotations can, however, come from several sources to move in different planes. It is perhaps easier to make this distinction using a pedestal robot as an example. Turning around the supporting arm is called the **roll** motion, and gives a fourth degree of freedom to the nozzle. A fifth degree of freedom is found in the **yaw**, or rotation in the horizontal plane, while the sixth degree of freedom comes from rotation in the vertical plane, or the **pitch** (Fig. 4.7). The addition of each degree of freedom adds more expense to the robot, makes the writing of the control program more complex, but allows a considerable increase in the performance capabilities. Three of the degrees of freedom can be included in the motions of the wrist, but when all six are required for the pedestal robot, this requires additional motion from rotation at the elbow of the arm, and rotation and extension from the shoulder as well as the turning or **sweep** of the arm itself around a central mount within the pedestal.

Figure 4.6 Artists rendition of a pedestal mounted robot.

In designing the path which a waterjet nozzle must follow as the end effector on the arm, it must be remembered that the cutting point is the jet itself, which may be cutting several centimeters beyond the nozzle orifice. This means that pivoting of the jet, or turning it in three dimensions around a part corner, must relate to a center of rotation which lies below the arm, rather than being at the arm tip. This requirement makes programming the motion a little more complex. This is particularly true when setting up an arm to move along a linear cut, when one is using a pedestal robot. One must also consider the placement of the catcher under the part, since to be effective this has to move with the same accuracy and positioning as the nozzle.

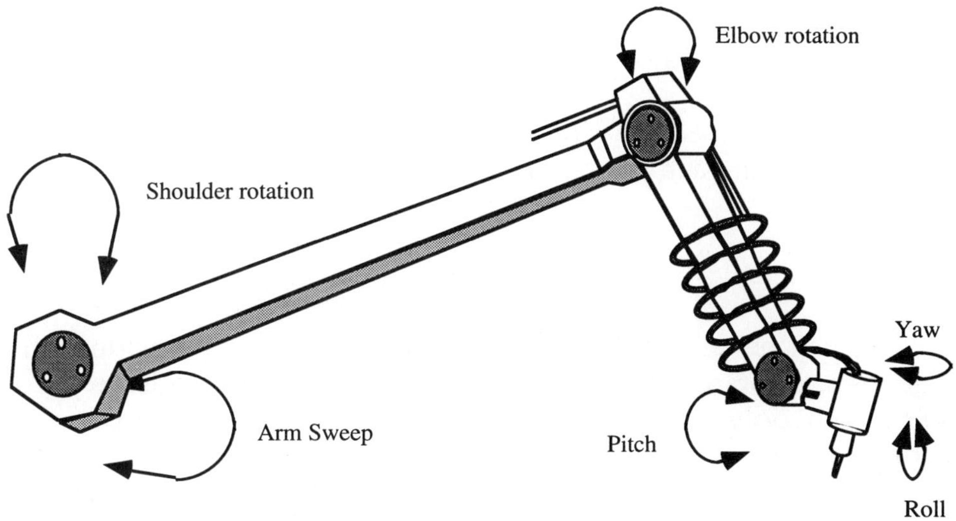

Figure 4.7 Schematic showing 6 degrees of freedom for a pedestal-mounted robot arm.

4.4.2 ACCURACY AND PRECISION

In the early days of the marriage between robots and waterjets, the applications and abilities were sufficiently novel that relatively wide tolerances could be accepted in the cutting abilities of the waterjet. The advantages from their use, in many installations allowed some tolerance variation with room temperature, season of the year, and rigidity of equipment mounting. With their continued use, and spread into the aerospace and automobile industries, tolerance requirements have increased significantly [4.15]. Current requirements are on the order of ± 0.5 mm although they are moving toward ± 0.1 mm and better. This poses particular problems when abrasive waterjets are used, as will be discussed later. It also poses problems with part accuracy in robotic control. For example, even with that tolerance, if the accuracy of three axes is controlled to ± 0.5 mm, the combined errors in accuracy will be cumulative. Errors arise both from an inaccuracy in the exact initial starting position (**repeatability**) and in the accuracy with which the manipulator is moved along the line (**Run out** or **accuracy**).

Schubert [4.16] for example has shown that the errors can be combined through the relationship:

Total error =

$$((X_{pos} + Y_{x\ run} + Z_{x\ run})^2 + (Y_{pos} + X_{y\ run} + Z_{y\ run})^2 + (Z_{pos} + X_{z\ run} + Y_{z\ run})^2))^{0.5}$$

and if this is illustrated by representative values, in 0.1 mm, where the error is 0.05 mm in each direction:

Total error =

$$((1.5 + 1.5 + 1.5)^2 + (1.5 + 1.0 + 1.5)^2 + (1.0 + 1.0 + 1.0)^2)^{0.5} = 6.73$$

Thus, it can be seen that, without proper monitoring instrumentation and attention, errors can grow into the cutting operation which exceed the tolerance limitations anticipated. For this reason, as accuracy has become more of an issue so more accurate positioning of the parts is required. Where the parts are mounted on the frame manually, high positioning accuracy is more expensive to achieve, and in these circumstances it is often cheaper if the computer identifies the exact part location, through target marks on the pieces, before cutting. Once the position is identified, then accuracy in cutting requires that the position of the nozzle be continuously monitored, either by external instrumentation, such as a laser unit, or by internal controls, such as those available with some stepping motor assemblies.

The problems of changes in the cutting pattern, and thus the width and angle of the cut, will be addressed later in the chapter, since they relate more to abrasive jet cutting than to plain jet cutting. Current accuracies which are reportedly available (ibid.) are on the order of ± 0.25 mm, - 0.16 mm in drilling 190 holes in a part each 2.49 mm in diameter. This drive for increased accuracy requires that the initial location and mounting of the robotic system be carried out with the care normally required when installing a machine tool facility, demanding that greater attention be paid to the foundation of the structure. The positioning accuracy also depends on the size of part being worked on, so that it is possible [4.15] to find accuracies in cutting of 0.04 mm in parts which fit within an X-Y-Z coordinate space some 0.9 by 1.5 by 3.0 m in size.

One way of overcoming this problem can be used if the part being treated only requires an external trim. Under these circumstances the part can be mounted on a fixture, which will not only hold it in place, but also provides an underlying or overlying template against which the nozzle attachment can be held as it traces the edge. This can give a simpler and accurate contour cut. This aspect of waterjet application is also one that can still be performed manually, since in its original form, before waterjets, the trimming was performed with a routing tool. Adams [4.17] has described how this can be replaced with a hand-operated waterjet device (Fig. 4.8) which the operator can move around the part edge.

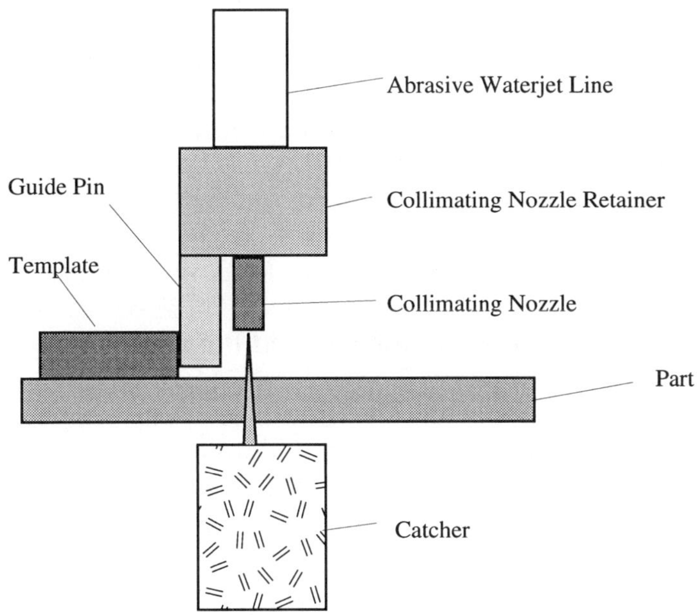

Figure 4.8 Hand-held waterjet reamer schematic.

4.4.3 CASE STUDIES OF X-Y CUTTING

Several different installations have been described in the literature which exemplify the types of robotic control and how they can be used. Printed circuit boards, now commonly called printed wiring boards, require a precise cut around the assembled piece, and sometimes internal cutting as well. These circuits are often mounted on boards made up of several layers of material. If the cutting is not carried out carefully in these materials, the jet can penetrate the layers and cause delamination of the board, making it unusable. It has been found that backing the jet up to some 6 - 25 mm above the board surface can overcome this problem

[4.17]. The reason for this improvement relates to a change in the internal structure of the jet which will be discussed in a later chapter.

The waterjet cutting of these boards has the advantage of significantly reducing or eliminating tool wear delamination, operator fatigue, and the shock, bending, dust and other problems of manual treatment. In one installation [4.18] panels 2.38 mm thick were cut at a speed of up to 3 m/min using an X-Y table and Allen Bradley Model 8400 machine control. The jet cut the parts at a pressure of 2,900 bar and with a diameter of 0.23 mm. Nozzle life, using a reverse osmosis water treatment was 330 hours for the sapphire on average, and the system paid for itself in five months due to a reduced number of reject boards and reduced routing tool replacement costs. It concurrently doubled production. By 1988 the size of this market alone was estimated [4.19] to be on the order of $3 million, with custom equipment being priced in the $120,000 - $350,000 range depending upon requirements.

To put that figure in perspective, the total market size for industrial waterjet cutting was estimated at $200 million in 1990, of which some $70 million was for the waterjet components and the remainder for the robotics and installation equipment [4.19].

The cutting of circuit boards illustrates one application of the technology, in which the jets cut around defined profiles at the edge of parts. In other applications shapes are cut from plain sheets of material. Use of a waterjet system carries a number of advantages in this case. The width of the cut is very small, and the low force on the piece means that parts can be cut very close to one another. For most efficient operation this requires that an intermediate step is required in the process. This is referred to as **nesting**, and is the process of fitting the parts together into the workpiece, in such a way as to maximize the number of parts that can be cut from it, reducing wastage in the process.

Three examples can be given of the use of this procedure. The first is in the cutting of gasket material [4.21]. Trane Company manufactures refrigeration and air conditioning units for which it uses closed cell neoprene as an insulating material. This is supplied in sheets measuring 91.5 by 122 cm in thicknesses of 9.6 and 19.05 mm. The company used a Calma CAD/CAM system to "nest" the parts which must be cut out, on the available sheet. The process is carried out in three stages, laying out the parts, assigning the motion commands to the cutting manipulator, and the illustration of the "nest" and how the nozzle will follow the cutting path. Part descriptions retained in a file are used to optimize the layout and reduce wastage. Given the strength of the material being cut it is possible to use a common cut between two surfaces to liberate both, and this way to eliminate any wasted material between the two.

During the second stage of the program, the cutting pattern is transferred to an NC tape for machine use. The technician running the program selects the cutting pattern to minimize the travel time between cuts. The program is designed to select the most effective cutting speed for the material and thickness being cut. The program also recognizes and adjusts for the length of line or turning radius being cut. This is important to ensure that the jet cuts straight through the material and does not cut on a slope which would give an unacceptable edge. Even though the arm can move the nozzle at speeds of up to 25 m/min the short lengths of cut mean that the head cannot accelerate/decelerate for very long and make a proper cut. For this reason Trane recommended the following cutting speeds:

Table 4.3 Speeds for cutting short lengths:radii in neoprene [4.21]

Length of line or radius (mm)	*Cutting speed (cm/min)*
0.00 - 12.5	190
12.6 - 100	380
>100 mm	2540

Once the program has been run and checked, it is transferred to the tape and sent to the machine for operation. A similar process is used at the plant for cutting fiberglass parts and gaskets, except that a commercial nesting program was purchased by the company from Optimization, Inc. for that purpose.

It is reported that the initial part programming took 1700 worker-hours, in the first year, with updating estimated at 850 man-hours a year.

Once the unit had been installed other advantages and uses became apparent. The method of manufacture made it possible to develop a better "Just-in-time" manufacturing plan, with consequent reduction in overhead. Further the equipment could be used to make panels from Styrofoam which previously took 4.16 minutes each to make by hand, and which the unit could cut out at the rate of 14 every 4.6 minutes. The internal rate of return to the company was reported at 51%.

The initial savings on the first year of operation are listed as follows:

Table 4.4 Labor benefits and material savings in waterjet cutting [4.21]

Material type	*Manual cutting cost*	*Waterjet cutting cost*	*Material savings*	*Total savings*
neoprene	$181,053	$73,159	$16,372	$124,266
fiberglass	$10,487	$5,185	negligible	$5,302
gaskets	$10,644	$2,129	negligible	$8,515
Total savings				$138,083

A similar process has been developed for the cutting of shoe components from feed stock. Cutting of shoe parts was an early objective of the industrial user. Initial programs to develop the technology were carried out in both America [4.22] and Europe [4.23]. One of the problems arising in the cutting of the parts was the need to eliminate splash back from the underside of the cut. The material being cut must be supported under the jet, and yet there must be a path for the water to escape. An initial idea was to support the sheets of material on thin piano wire at discrete intervals. However, while partially effective, there was still some splash back into the parts when the jet hit the wire. It must also be borne in mind that the sheets can be quite heavy. Thus the support system under the sheet might have to support perhaps 36 kg over a 3 square meter area, with a fine enough mesh that parts cannot fall through. Tests by SATRA [4.23] showed that while 0.25 mm wire was relatively quiet when the jet hit it, it had a lifetime of only a few hundred cycles, at 0.5 mm the noise was greater, but the lifetime was in the low thousands of cycles, while at 0.75 mm the noise level was very high, and although there was some increase in life there was a noticeable splash back on the underside of the cut. This wetting of the underside was not evident in the smaller wire sizes. More recently this problem has been solved by using a "wrinkle" [4.24].

The wrinkle (Fig. 4.9) is a means of providing an X-Y motion, and support for the material being cut, without having a surface under the part which will reflect the water back up. Simply put the sheet to be cut lies on top of two conveyor belts. These each fold over a roller, and return, one on each side of the cutting nozzle. This leaves a gap which is approximately 5 cm wide, under which the catcher can travel to pick up and remove the spent water and debris from the cut. The nozzle thus continues to give the X and Y motions, but the support is provided by the underlying rollers which move along the fabric belt with the nozzle, riding on the fabric belt, while the catcher can move across the sample in the space left between the rollers.

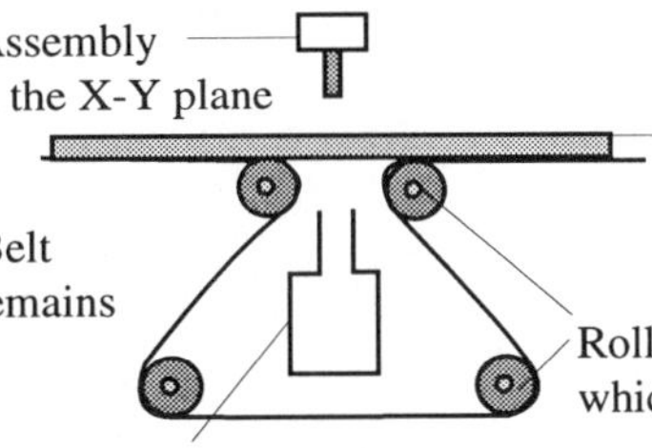

Figure 4.9 Method of using a wrinkle to support the part and move a catcher under a jet.

The third example comes from metal cutting [4.25], and compares the performance of an abrasive jet cutter with that of die cutting. Conventional die cutting of metal parts which are 20 to 25 mm thick would require that the parts be spaced approximately 12.5 cm apart so that only six parts could be cut from a bar which is some 75 cm long and 15.8 cm wide. In contrast (Fig. 4.10) the low reaction force and narrow cut width allow the same parts to be more closely nested on the bar. It is possible to now cut 13 parts per bar, which not only reduces changeout time, it also reduces the amount of material committed to each part from 3.6 to 1.6 kg.

Some manufacturers have found that it is better, in terms of the long term operation of the machines, to reduce operating pressure to around 3,000 bar, rather than running at maximum pump performance [4.24] and [4.26]. Although this slightly reduces output, it pays for itself in terms of operational and maintenance cost reductions, which may be as great as 40%. Part of the reason for this comes from the fatigue induced with the rapid on-and-off cycling of the system. As Pilkington has pointed out [4.23] the machine switches on at the beginning of a part cut, it traverses around the part (say 82 cm in perimeter) with an acceleration of 0.5 g and a maximum speed of 40 cm/sec and then stops at its starting point. The jet switches off, after an operating time of perhaps 3 seconds, and the head indexes to make the next cut. If a period of 0.2 seconds is allowed, before the jet starts to move for pressure build-up, and 0.3 seconds at the end of the cut for pressure decay, before moving, allowing 1.5 seconds for movement to the next piece, gives a 5 second cycle time. This translates into an average of 400 to 600 parts/machine each hour or a total for a factory with several machines of perhaps 20,000 parts in an 8 hour shift. This cyclic loading of the system can induce fatigue in the components, and will be addressed later, in Chapter 12.

Waterjet Cut - 1.6 kg/part 13 parts per bar

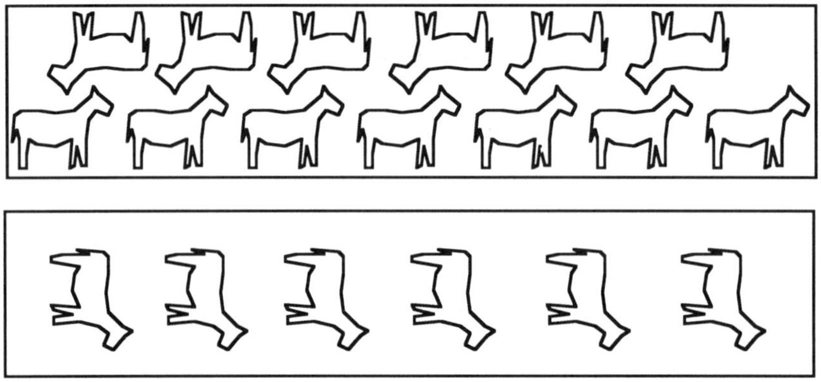

Die Cut - 3.6 kg/part 6 parts per bar

Figure 4.10 Layout of metal parts on a bar, nested for waterjet cutting, and spaced for die cutting (after [4.25]).

It is important to ensure, when cutting softer fabrics and material, that the piece being cut is flat, without ripples, and some manufacturers use a "foot" which fits around the nozzle assembly and holds the material flat, as the jet passes over it. Production from such cutting systems, for preparing shoe parts for example, can be controlled by three people operating two machines, one to load and operate the 2 machines, the other two to unload one machine each. This latter job is more time consuming since the individual parts must be picked out of the sheet [4.24]. Waterjet cutting and effective nesting can save as much as 18% of the material in shoe cutting since the parts do not require as great a spacing for die cutting in that material as they do in metal. Given that the 1976 cost of material for making ladies shoes approximated $860 an hour [4.23], this gives an annual saving of perhaps $270,000, with an 8 hour day and 220 working days a year.

The production advantages gained from waterjets can be increased by cutting more than one layer at once, however, it is a function of the part thickness as to how many plies may be cut at one time. For example a 4.75 mm rubber is better cut in 2-ply since this will produce a cleaner and faster cut that a 4-ply cut. In contrast, insole material which is 2.375 mm thick can be cut in 6 or 8 plies, and leather is generally cut only one layer at a time. Part of the reason for the latter is that leather has unpredictable blemishes and shapes and thus no two overlapping pieces are likely to have the same cutting pattern.

It is interesting to note that perhaps the most widely used application of high pressure waterjets at the present time is in the cutting of diapers. It was reported by the mid 1980s [4.26] that over 400 high pressure waterjet cutting installations had been made in 43 countries in which high pressure waterjets were used to cut the contours for such products. One such installation is at Braco Inc. in NJ. [4.27]. In 1989 this plant used 5 waterjet nozzles to cut the materials for the contoured diapers, producing up to 240,000 garments in a two-shift 16 hour day. In order to ensure that the plant had no problem with nozzle blockage or nozzle wear, the water conductivity is monitored continuously. Although an acceptable standard is 20,000 ohms, still well above the recommended limit. In this way nozzle life could be kept at around 180 - 200 hours of operation, an economic advantage given the $225.00 cost of the five-nozzle set.

4.5 MULTI-DIMENSIONAL CUTTING

The definition of a robot given earlier implied that these machines have abilities which go beyond simple manipulation of a waterjet nozzle. One example of the usefulness of this new manufacturing tool is at the Chevrolet Motor Division, Adrian Manufacturing Plant, where 10 robots work on two lines to trim and stack the plastic fuel tank shields for Chevrolet trucks [4.28]. The system is sufficiently flexible that it can cut the shields so that they can be mounted on either the right or the left hand side of the vehicle. The 4.95 mm thick polyethylene shields are first molded to shape and then automatically transferred to the anchor hole drilling and trimming station. Historically, the part had passed through a line of machine tools which were used to drill the holes and to trim the edges of the piece as well as dividing it into two separate pieces. Such a system required consistent and regular maintenance of the cutting dies and frequent die changes to replace worn parts. It also required a significant amount of floor space. With the change to a waterjet cutting system a significant reduction in the amount of space required was achieved.

The part is allowed to cool after manufacture and then attached to a traveling fixture which carries it through the waterjet station (Fig. 4.11). At the first location, two robots (A & B), equipped with waterjet nozzles, cut the internal holes and slots in the shield body. The piece then moves to the second station where the edges of the shield are trimmed and it is divided into two identical parts by a second pair of robots (C & D) which each cut a path approximately 3.5 m long around the perimeter of the shield. The nozzles move around the part at a stand-off distance of approximately 1.25 cm, at a maximum speed of 25 cm/sec. Finally, the fifth robot (E) at the end of the station picks up the completed shield and moves it to a shipping basket mounted beside the production line. Scraps from the cutting process fall onto the underlying conveyor and are carried to a scrap bin for disposal.

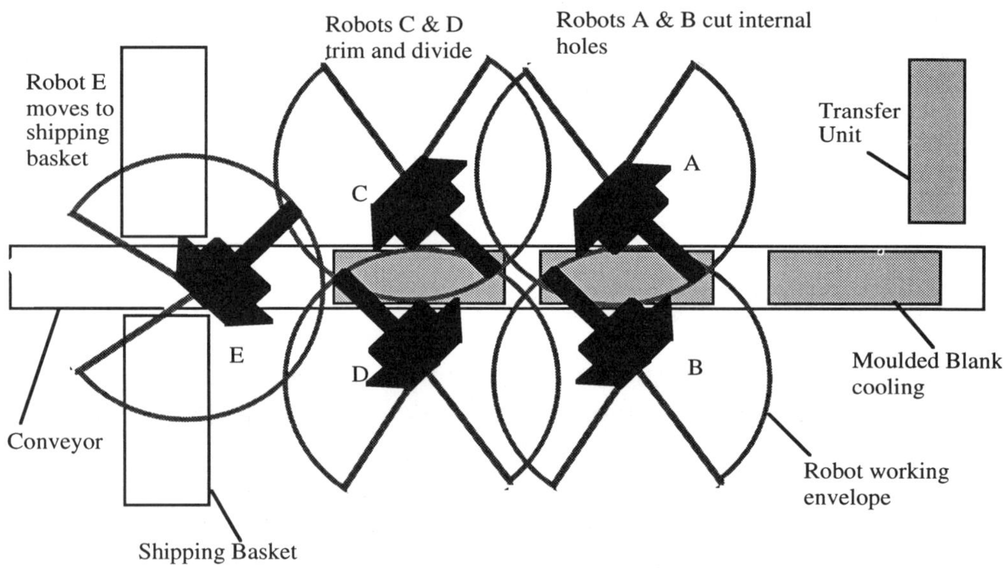

Figure 4.11 Layout of a workstation for cutting gas tank shields [4.29].

The total production time for a pair of shields is 64 seconds which matches the thermal formal time required to make the part. Repeatability in manufacturing for this process is better than ± 0.5 mm and this installation was the first application of the six-axis GMF robots to waterjet cutting. Because five robots work together to carry out the different parts of the cutting process, it was necessary to stack them close enough that their operational envelopes overlapped. For that reason, it was important to write into the software programming a number of safety interlocks to ensure that the robots would not come into contact with one another.

The waterjets for this operation worked at a pressure of 3,850 bar and issued through 0.2 mm diameter nozzles. The waterjets nozzles were also fitted with a catcher, located under the part, which captured the water and the cut material after passing through the polyethylene. It is interesting to note that the single pumping system which supplied water to the nozzles used less than 60 l. of water each hour. Approximately 70% of this water evaporated during the cutting process, leaving only residual water to be disposed of to waste.

4.5.1 CONSIDERATIONS IN THE USE OF INDUSTRIAL WATERJET CUTTING

The waterjet cutting process, under remote or robotic control, can bring a significant number of cost savings including the following:

- initial cost reduction - the robots cost less than the stamping dies and transfer devices which would otherwise be used in a manual process.

- the quality of cut achieved with the waterjets was superior to a manual cutting and trimming of the parts.

- only one operator was required to oversee the complete line.

- the waterjet cutting system uses replenishable water and therefore the down time while tools were replaced and sharpened was virtually eliminated.

- because the robot programming could be achieved with relatively simple software changes, the time required to change cutting parts is limited only to that required to change the mounting fixtures.

- less floor space is required.

Among the disadvantages of the system initially identified but ameliorated was water quality, since initially the minerals in the supply water had a negative effect on nozzle life. The problems caused by overlapping between the robot work envelopes had also to be solved, and a significant amount of time was required to program and interface all the equipment in the line.

One unanticipated problem was that though the waterjet cutting is relatively dust free, since the cuttings are drawn into the jet, and down through the cut, these small particles from the cutting collect and float upon the waste water. If this is not removed, they can be swept back into the air by ventilation currents.

It is interesting to note that there is some heat created, since it is often referred to as a cold cutting process. For example Imanaka reported [4.30] at the first International Conference that a waterjet striking copper would generate a temperature increase related to the jet pressure (Fig. 4.12). The pressures used in that study went as high as 10,000 bar, to obtain a 200° C temperature rise. There was more than a 20° C drop in temperature when the jet was traversing, rather than attempting to pierce the target.

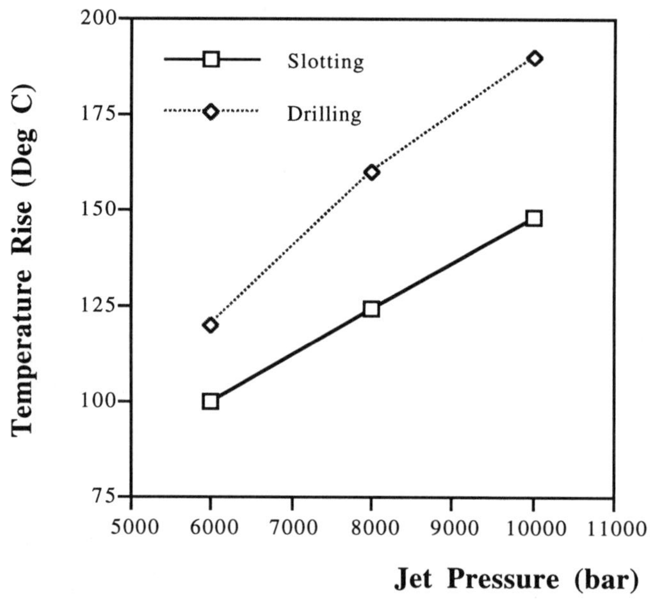

Figure 4.12 Temperature rise in cutting and drilling copper [4.30].

A related study was carried out at the University of Wisconsin-Madison [4.31] in which a jet at a pressure of 7,000 bar and a diameter of 0.25 mm was directed at targets of PVC, polyethylene and steel. The jet used was a di-2-ethyl hexyl sebacate (Plexol) in order to overcome boiling problems with the water. An earlier study on a hardened steel was reported to generate a temperature rise of 260° C, which was considered to lie close to the stagnation temperature of the jet. In the study cited, temperature rises of up to 175° C were reported, with a rise time of 0.25 seconds, as the jet cut within 0.5 mm of a thermocouple in PVC (Fig. 4.13). Temperatures in the polyethylene and the steel were some 80° C lower. There was a rapid decay in the measured temperature rise as the thermocouples were placed further from the jet path.

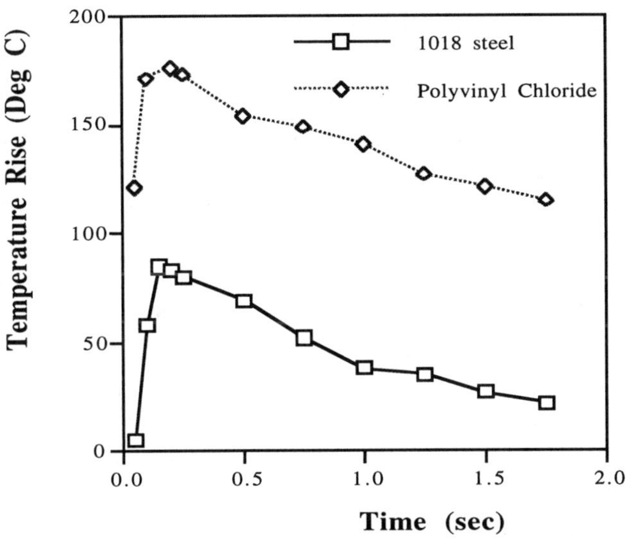

Figure 4.13 Temperature rise in drilling PVC and steel [4.31].

A more recent study by Worsey [4.32] determined the relatively localized nature of the heating event, and its very transient nature, due to both the rapid removal of material from the work site, and the rapidity with which the cutting process itself occurs.

4.5.2 EXPLOSIVE MATERIAL PROCESSING

Although some temperature rise has been detected when a waterjet impacts upon the surface of the target material, at normal operating pressures this is relatively insignificant and makes the process relatively safe for use in hazardous environments. One example of this is in the removal of propellant and explosives (often collectively known as **energetic**

material) from military munitions and rocket motors. Historically, these materials were bonded together by waxed based materials which could be liquefied by steam, making it straightforward, therefore, to melt the explosive and remove it from the casing. With the advent of plastic bonded explosives where the bonding agent may be a lexan or similar plastic, this option is no longer available. Since there continues to be a need to remove explosives and propellants from these casings, a new method had to be established for this process.

One program to carry this out has been the subject of an investigation at the University of Missouri-Rolla, funded by the Naval Weapons Support Center in Crane, Indiana [4.33]. Although a number of programs had examined the use of waterjets for removing propellant from missile casings, a process which has become a standard procedure for certain propellant ranges, this was one of the first studies of a similar means for removing the explosive. This program began by establishing the relatively safety of the operation. This was carried out by subjecting samples of a suite of explosive materials to waterjet impact at increasingly high velocities. Initially, the waterjets were generated using a high pressure pump. As higher pressures were required, these were first fully achieved using a small derringer type similar to that described in Chapter 2, and finally a shaped waterjet charge was manufactured and used to produce impact velocities in excess of 10 km/sec. As a result of these experiments it was shown that the use of high pressure waterjets under normal operating pressures was a relatively safe means of removing the potentially explosive material from the casing. This was demonstrated since the pressure required to induce explosive reaction was found to be significantly higher than the pressure required to remove each of the explosives from the required casing.

A remotely-controlled, computer operated device was built and established in an underground facility for such washouts. The controlling computer was programmed to move a high pressure lance within the internal geometry of the explosive container in such a way as to totally remove the explosive contained therein. In order to make the waterjets more efficient, the nozzle was rotated during its movement. It was decided to move the nozzle rather than rotating the casing, a practice which has been established in other installations, [4.34] because in this way it made it possible to more easily cope with internal tubes and pipes which may be found in some types of munition. This robotic device, given the acronym WOMBAT, (Fig. 4.14) was controlled from outside the facility and had 4 degrees of freedom, forward and backward, left and right, up and down, and nozzle rotation (the roll motion).

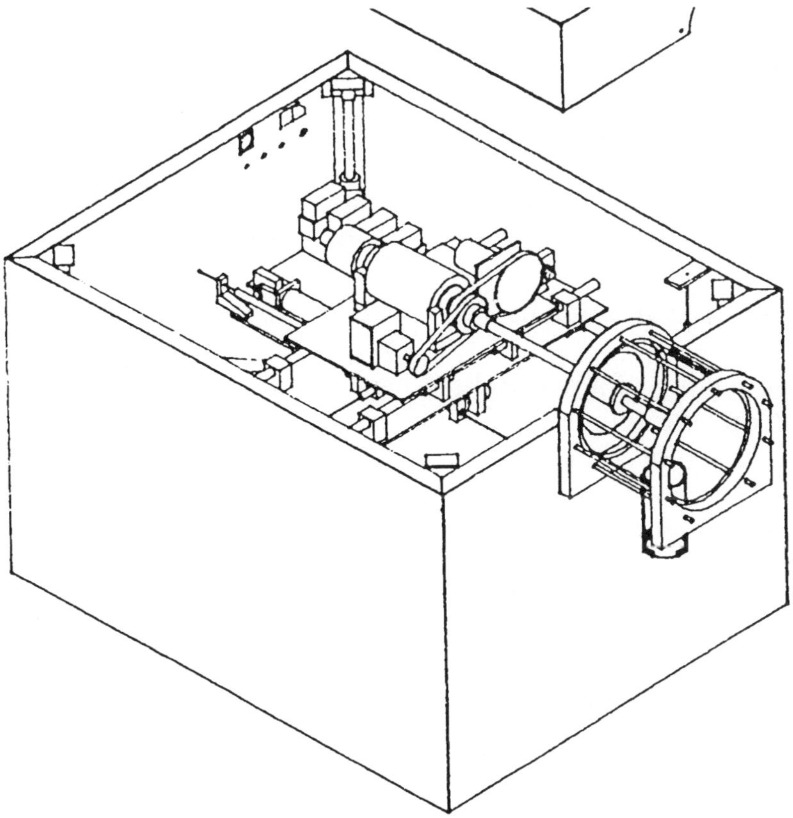

Figure 4.14 Schematic of the assembly of a WOMBAT (by E. Hammerand).

It is particularly important with explosive material to insure that the nozzle does not contact the explosive (since in a subsequent event, a violent impact of this type did induce an explosive reaction), but also for the integrity of the casing to insure that the cutting arm did not impact on any part of the casing itself. Resolving this particular problem requires that safeguards be built into the controlling software to control nozzle position at all times. This must include the possibility of power loss, and the resulting movement of the nozzle. The programming must recognize that the movement of the nozzle to an initial start position or on return to a home position, could, if not guarded against, move the nozzle assembly through the position of an internal component of the casing.

Experiments with the device to establish its safety show that under normal operating conditions, this system would perform with no discernible risk of explosive reaction. However, it was observed that when some of the explosive was subjected to cavitation attack, even under relatively low ambient fluid pressures that a reaction could be detected from the surface of the explosive [4.35]. For this reason, it is recommended that cavitation not be used as a method of explosive or propellant removal and that the cutting and removal of such reactive materials be undertaken in air rather than under water so as to obviate the risk of cavitation induction and the potential for an explosive reaction.

4.5.3 CUT EDGE QUALITY

The edge quality of a waterjet cut has been studied by several investigators, almost from the time that waterjet cutting began as an industrial tool. At the University of Hanover, for example, a study compared the edge cut quality achieved in cutting thermoplastics [4.36]. Cuts were made using diamond disks and circular saws as well as the waterjets. As with the abrasive jet cutting the cut quality created by the waterjet can be distinguished by a relatively smooth cut in the upper zone, and a rougher striated lower section (Fig. 4.15). In cutting thermoplastics an unacceptable "fringe" of melted material can also be created on the lower edge of the cut. The investigators found that increasing pressure and nozzle diameter improved quality, while increasing traverse speed and sample thickness reduced it.

Figure 4.15 Typical cut made by a high pressure waterjet showing the smooth and "rippled" section.

A similar problem has been addressed in the cutting of pack foam [4.37]. Conventional technologies for the assembly of missile packing foam has required that the 56 cm thickness be cut in 5 cm layers, which are then assembled. By using a 2,800 bar waterjet at a diameter of 0.25 mm it was possible to drive a single cut all the way through the 56 cm foam at a traverse speed of 2 mm/sec. The depth of cut over which the required edge quality was maintained varied between 25 and 37 cm, in part as a function of the addition of a long chain polymer to the water. The use of the additive Superwater gave a considerable improvement in the quality control which could be achieved in making the cut.

Similarly Howells has shown [4.38] that it is possible to lower the pressure required to cut shoe material from 3,750 to 3,400 bar and concurrently achieve a better quality of edge to the cut, where the long chain polymer is added to the cutting water (Fig. 4.16). It is also reported that the nozzle life using a diamond orifice was extended from 720 to over 2,000 hours and that the plant was able to sustain a 15% increase in production rate.

Figure 4.16(a) Shoe component edge cuts made with plain water. Jet pressure 3,750 bar, 0.2 mm diameter nozzle. (Courtesy of Berkeley Chemical Research, Inc. Photographs: Photo Design, David Hayashida, Oakland, CA.)

Figure 4.16(b) Shoe component edge cuts made with 0.1% Superwater [4.38]. Jet pressure 3,400 bar, 0.2 mm diameter nozzle. (Courtesy of Berkeley Chemical Research, Inc. Photographs: Photo Design, David Hayashida, Oakland, CA.)

High pressure waterjet systems have thus exceeded the potential benefits for their use envisioned in the first development papers [4.39] In that paper some advantage was foreseen in the ability of waterjets to trim the edges of parts. In Sweden, for example, insulation materials are trimmed in fiberglass. As with many other applications this reduces dust and develops only a small kerf in the material. Savings are quoted at 1200% [4.40].

4.6 TECHNIQUES FOR CUTTING HARDER MATERIALS

The market for waterjet cutting was growing at a relatively slow rate because of the limited number of materials which a plain waterjet could cut. However, in 1984 the first commercial marketing of the Paser™ System, by Flow Industries and the Hydroabrasive™ nozzle by Ingersoll-Rand marked the advent of abrasive waterjet cutting into the industry and increased the range of application of this tool significantly. By 1986, the U.S. market for high technology cutting machines was on the order of $164 million [4.41].

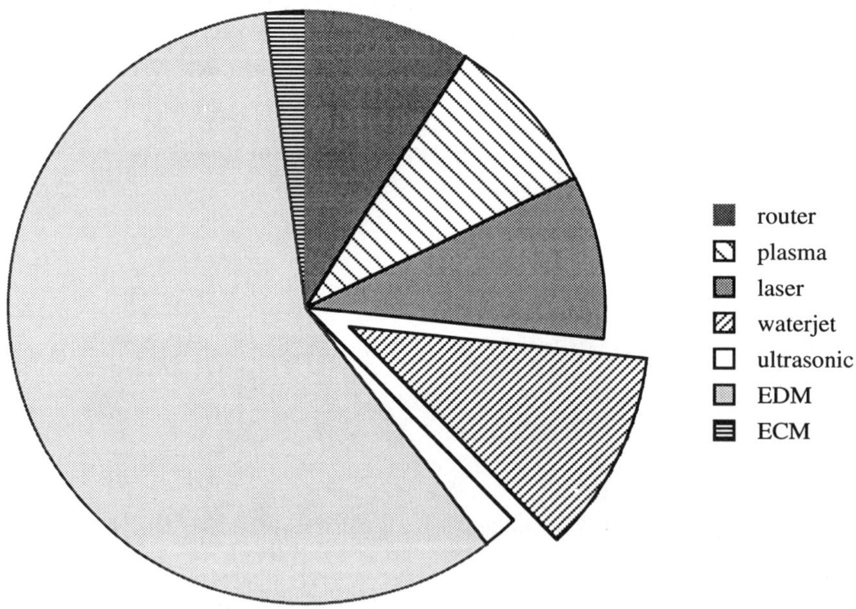

Figure 4.17 1986 market for non-traditional machining technologies (after [4.41]).

Of this market, waterjets had already captured 11% (Fig. 4.17) in 1986 and by 1990 sales of $70 million a year were being quoted [4.42]. Abrasive waterjet cutting was able to build upon the technology already established for robotically controlled plain waterjet cutting, but with the advantage that the tool was now able to cut through materials of virtually any strength, most particularly those materials such as inconel, titanium, glass, and ceramics which are difficult to cut to exact contour by conventional machining technology. The market for the industry therefore grew significantly.

The injection of abrasive into the waterjet system was originally developed along the same lines as that which had been used in the cleaning industry to create a water abrasive jet for cleaning purposes (compare Fig. 4.18 with Fig. 3.38).

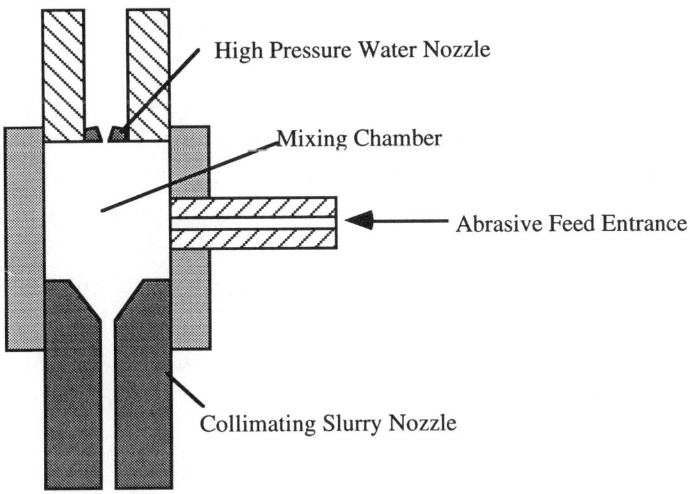

Figure 4.18 Sketch of conventional abrasive injection into a high pressure stream.

However the systems that are conventionally used for abrasive jet cutting are operated at much higher jet pressures and with significantly less water than are used for the cleaning applications. Thus, in most installations, the basic components of the system are very similar to that of the plain, very high pressure waterjet cutting systems which have been described above. However, with the addition of the aspirating nozzle assembly to the front of the high pressure line, and the feeding of an abrasive particulate flow into that chamber, the two combine to produce an abrasive laden waterjet flow through the second collimating nozzle, with much stronger and more powerful jet results.

4.6.1 NON-TRADITIONAL MACHINING

To understand some of the advantages of this new tool it is pertinent to examine some of the alternate methods which are now used for cutting complex shapes into materials which are often difficult to machine. The requirements for many operations requires that high surface tolerances be achieved with high repeatability and accuracy. In the areas of surface cutting of materials competing technologies are often thermal, starting with oxy-acetylene cutting this has progressed to the use of high temperature flames, the plasma torch, and the use of lasers.

4.6.2 LASER MACHINING

Laser machining and the use of electron beams has become popular in the same time frame as the use of high pressure waterjets. Typically, cutting lasers employ either Neodymium doped Yttrium-Aluminum-Garnet or Nd:yag lasers, or CO_2 lasers for many of their cutting processes. Typically an Nd:yag laser will generate a power on the order of 6.5×10^6 watts per square cm. Because these lasers operate in a shorter wave length, they can achieve a better coupling, and more effective heat transfer to the workpiece for the same input power than can the CO_2. The Nd:yag laser will typically produce a cut with less microcracking and a smaller heat affected zone around the cutting area. On the other hand the CO_2 laser can put out a much higher level of power. Typically, while a typical Nd:yag laser will operate at 400 watts, the CO_2 laser may be obtained at power levels in the range from 500 to 5,000 watts [4.43], which allows for cutting thicker pieces of material and faster cutting rates (Figs. 4.19 and 4.20).

Nd:Yag lasers can drill holes up to 0.5 mm in diameter while the CO_2 laser can drill holes to about 2.0 mm in diameter, depending on the material. Typically, a laser drilled hole will taper approximately 1% over the drilled depth. This taper can, however, be controlled to a degree by machine performance. Edge quality and laser cutting is typically on the order of 3 mm. However, a relatively simple basic laser system can cost approximately $250,000. The minimum kerf which can normally be cut with a laser beam is on the order of 0.4 mm. An additional problem with the use of the laser is that it will damage optical surfaces and it does require some gas to assist in the cutting process [4.43]. Engel and Labus [4.42] compared the operational cost of a CO_2 laser with that of an abrasive waterjetting system and estimated that, in 1990 dollars, that a laser system would cost \$72.72/hour to operate, while an abrasive waterjet system would cost \$111.30/hour.

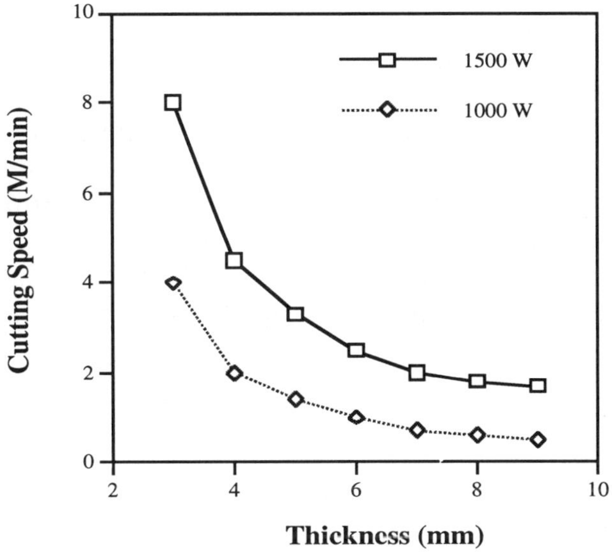

Figure 4.19 Typical cutting rates for a CO_2 laser, through carbon steel [4.43].

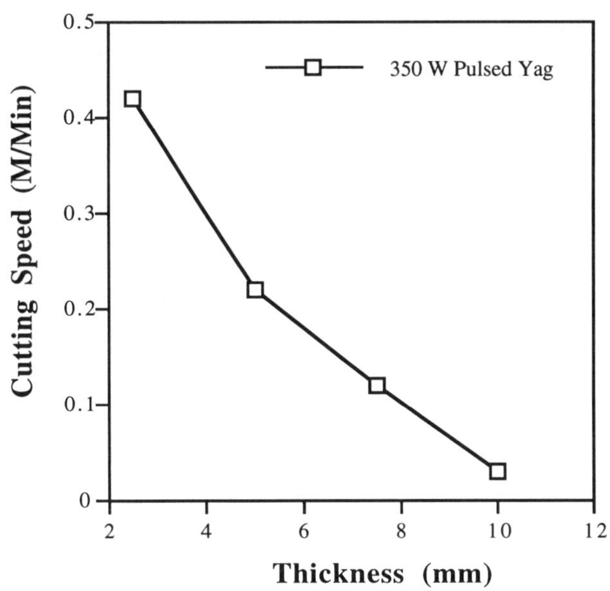

Figure 4.20 Typical cutting performance of a pulsed Nd:yag laser in steel [4.43].

Laser removal rates on the order of 5 x 10^{-3} cc/sec have been measured for a 560 watt carbon dioxide laser generating an average surface roughness of 3 - 5 mm. However, in machining ceramics it is reported that the strength of the part may be reduced by up to 30% when laser machining is used because of the microcracking and thermal weakening which can be introduced into ceramic parts [4.44]. This disadvantage does not occur with abrasive waterjet cutting. Wightman [4.45] gives an interesting illustration of this. The company LAI had been trying to provide dental implants made of titanium for the Department of Defense. Using lasers to cut the material, however, left microcracks in the metal, which provided nucleation sites for colonies of bacteria in the mouth. Changing to abrasive jet cutting eliminated the problem and allowed LAI to regain the contract.

4.6.3 PLASMA TORCH

Normally a plasma torch operating at 220 V and drawing between 20 and 50 amps and using compressed air as the gas, at 5 bar pressure, can cut mild steel 1 cm thick at a rate of 65 cm/min [4.46].

Plasma cutting of 3 cm thick Inconel 718 can be achieved at rates of up to 20 times faster than with an abrasive waterjet, particularly if the abrasive jet speed is reduced to maintain a reduced taper on the cut, and relatively little roughness on the lower edge. However, the waterjet system does not leave a heat affected zone which must otherwise be milled off. Plasma has a cut width of 0.5 - 1.75 cm in comparison with the .75 - 1.5 mm wide slot with the abrasive waterjet. The combined material savings can thus be as high as 20%. Operating costs for a abrasive waterjet cutting job shop in Los Angeles were around $250/hour in 1990. The development of more resistant nozzle materials has allowed a 28 m long cut through 21 mm thick steel to be made in 15 hours without changing the nozzle. (It took 20 hours to write the controlling computer program.) The kerf width was 1.5 mm with a surface finish of 1.5 mm [4.47].

4.6.4 ELECTRIC DISCHARGE MACHINING

In cutting shapes in metals electric discharge machining (EDM) has a significant presence. This technology cuts and drills metals by sparks generated between an electrode, and the target material in a bath filled with a dielectric fluid. The sparks erode a cavity in the surface, and the shaped electrode is fed into this gap, machining in this manner a mirror image of the electrode. The technology is now changing so that a simpler electrode can be used to cut the shapes required, with the electrode moving under computer control to establish the required cavity. Spark settings are adjustable to control the surface finish required. The market for this tool is currently estimated to be growing by about 18% a year, with one firm alone selling $20 million of equipment in the U.S. in 1988 [4.48]. Metal removal rates vary with the quality of surface required. For example to achieve a 4 mm surface finish might require a 25 amp circuit which would erode material at a rate of 2.6 to 6.5 cc/hr with between 9 and 19% electrode wear. With a faster roughing cut, at 8 mm finish, rates of 8.8 to 14.7 cc/hr can be achieved [4.49].

4.6.5 ELECTROCHEMICAL MACHINING

A modification to the idea of EDM is Electrochemical Machining (ECM) where the tool acts as the cathode and the workpiece as the anode in an aqueous salt electrolyte bath. A typical power supply might require between 200 and 40,000 amps at between 7 and 25 volts DC [4.50]. Penetration rates of up to 50 cm/min are quoted in broaching the interior of gun barrels. Transmission die cutting was achieved to an accuracy of 1 mm in 10 hours which was 1/8 the time taken by EDM. A modified version of this process in which the tool is rotated has been given the name of Rotolytic finishing. This can deburr an aluminum part for a turbocharger in 30 seconds [4.51]. The process required an 8 volt DV, 250 amp, power supply and worked in a 15% sodium nitrate bath removing burrs up to 0.5 mm high with the tool spinning at 1600 rpm.

4.6.6 ULTRASONIC MACHINING

Ultrasonic Machining (USM) is a technique where a slurry nozzle feeds grit between a shaped tool and the target material (Fig. 4.21). The shaping tool is mounted on the end of a sonic transducer which vibrates at approximately 20,000 cycles/sec over a 0.025 to 0.050 mm range.

Ultrasonic Machining is limited usually to working with ceramic pieces which are less than 100 mm in diameter. Typically either silicon or boron carbide particles in the 200 to 80 mesh size (0.066 - 0.011 mm diameter) are used. Cavities can be cut to depths of 65 mm with an accuracy of $\pm$ 0.01 mm or better. Volume removal rates are on the order of 0.4 - 1.3 cc/min. When this tool is used for drilling ceramics penetration rates from

0.25 - 1.5 mm/min can be achieved with a surface finish of down to 0.4 mm [4.49].

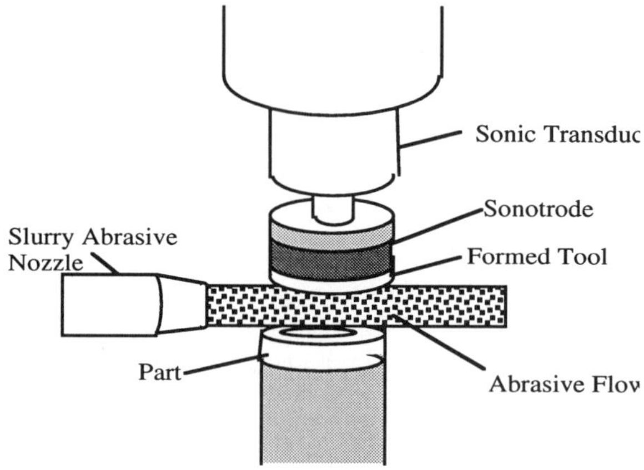

Figure 4.21 Schematic of an ultrasonic machining process (after [4.49]).

4.6.7 ABRASIVE FLOW MACHINING

When very small holes need to be deburred or drilled, abrasive flow machining (AFM) may be used. In this, a fluid of relatively high viscosity carries small abrasive particles, in the size range from .005 mm - 1.5 mm. The particles can be of aluminum oxide, silicone or boron carbide or diamond dust, depending on the target material. The fluid is forced through the holes that need to be finished. This procedure on which the method is based is that as the fluid is forced through the passage, it becomes more viscous and the abrasive particles thereby are spread uniformly around the passage walls. Then, as the fluid migrates upwards, the particles evenly polish the passage surface. This procedure can produce surface finishes of around 0.05 mm and can debur holes as small as .02 cm to tolerances ± 50 mm. The procedure works at pressures of between 35 and 100 bar with an abrasive concentration of about 66% by weight in the viscous fluid. Flow rates are on the order of 1 liter per stroke, which can take between 1 and 3 minutes to occur. The lower the viscosity of the fluid the higher the surface roughness. Machines are available to operate at pressures between 7 and 220 bar with flow rates exceeding 308 lpm. Tooling is used to control the direction of the abrasive flow and to accelerate it in the areas where abrasion is required. Larger abrasives cut faster but the smaller sizes produce higher surface finishes in smaller holes.

4.6.8 COMPARISONS OF PERFORMANCE

Carlson and Huntley [4.43] give examples comparing the performance of the different cutting techniques. For example, in preparing a stator assembly from a casting made of inconel alloy it was necessary to machine a slot 0.2 mm to 0.3 mm through an I-Beam section varying in thickness from 1.14 mm to 2.41 mm. It was also required that the slot be cut at an angle of 5° to the vertical axis. Using a traveling wire EDM and a 0.2 mm electrode, the slot cut was made well within the permitted tolerances. There was a recast layer approximately 0.02 mm thick with no microcracking. However, it took 5.75 hours to do each part.

Using an Nd:yag laser, was not initially possible in this application since the standard kerf cut by of the machine was 0.3 - 0.4 mm wide. By adjusting machine parameters it was possible to reduce the slot width to the required size, but at a much reduced operating speed. Under those conditions it was possible to make a part, but it took 3 hours per unit. Waterjet cutting of this particular part was not possible, since the minimum slot width achieved was 0.71 mm. Because EDM has a lower cost rate per hours, it was competitive with laser cutting for that particular operation.

In the second example, an exhaust augmentor flap was required to be trimmed after braising from Waspalloy. The critical requirements was to trim the boundary within ± 750 mm of the blueprint nominal dimensions. The recast layer could not exceed 500 mm, microcracking had to be held within 37.5 mm and a surface finish of 6 mm was required.

Because of the long length around the perimeter, EDM would not work in this particular process, therefore the choice was made between laser processing and waterjet cutting. In the initial evaluation, a 600 watt CO_2 laser was tried but failed due to excessive recast and microcracking of the surface.

By reducing the feed rate to 12.5 cm per minute an Nd:yag laser was able to meet the metallurgical requirements but the cutting time was somewhat longer. Using an abrasive waterjet system, the abrasive water was able to trim the part, at a speed of 3.75 m/min with acceptable edge quality. Because there was no thermal deformation and the edge finish was better than 3 mm, this made the process very attractive. However, using a traverse speed of 3.75 m/min made it difficult to hold tolerance around contours and the speed was therefore reduced to 1.25 m/min. It was found during this cutting trial that the slurry flow rate had to be monitored very closely in order to make sure that the kerf remained at approximately constant width.

Abrasive waterjet systems are slower than laser cutting for metal thicknesses up to 6 mm but come to have an advantage in being able to cut through thicker metals with the same basic set-up. Part of the reason for this is that as the parts get thicker the Heat Affected Zone (HAZ) increases in size to the point where it leaves an unacceptable edge to the product. Thus Engel and Labus suggest that in thicknesses above 6 mm that abrasive waterjets can be a more competitive alternative to lasers [4.42].

4.7 ABRASIVE WATERJET CUTTING

4.7.1 NOZZLE DESIGN

There have been a number of different designs proposed to improve the basic shape and concept developed for the abrasive injection. One such [4.52] brought the abrasive flow into the center of the jet by focusing converging smaller jet streams within the venturi chamber and drawing the abrasive material through the central common focus (Fig. 4.22).

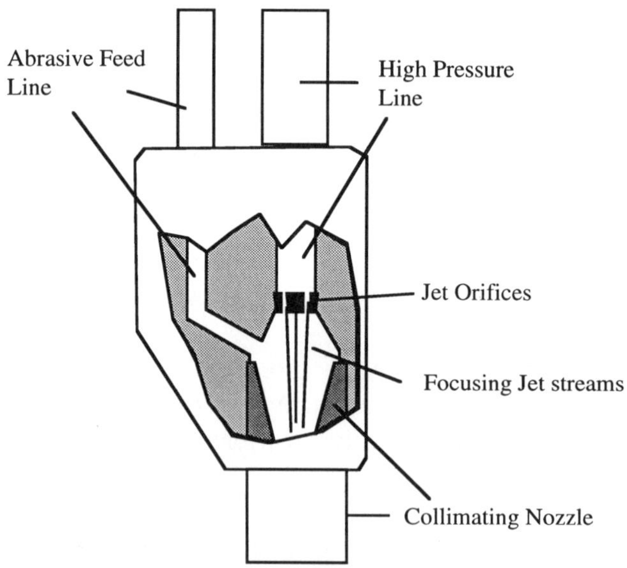

Figure 4.22 Nozzle design for abrasive jet cutting (after [4.52]).

Other designs have also been suggested from time to time in order to improve the performance of the system. However, regardless of the particular way in which the abrasive and the waterjet mix (a topic discussed in more detail later in this chapter and in Chapter 10), nevertheless the new tool was very readily able to cut through virtually all materials in its path. In its earliest conception, the abrasive jet rapidly eroded the nozzle used to collimate the flow after the abrasive mixing. Yet, even with the relatively short life of this nozzle, the economics of the cutting operation and the ability to cut target materials which otherwise could not be shaped with sufficient accuracy and care, made the tool very rapidly a market player in the high technology arena.

4.7.2 ABRASIVE JET CASE STUDIES

The magnitude of its accomplishment is perhaps illustrated in part by the ability of an abrasive waterjet to cut through a 254 mm thick titanium block. This cut which was made by the Cudini Corporation of Romulus, Michigan [4.53]. The titanium was required for a part used in the aerospace industry. Normal cutting of the titanium using traditional methods would include pickling to remove the oxides, heat scales, and other damage generated by a thermal cut. However, since the abrasive waterjet would in effect be a "cold cut", this pickling would not be necessary. The cut was made at a feed rate of 2 mm/min and used a flow of 81 kg of garnet/hour of cutting. Although this test did successfully show that the titanium could be cut, it would take as long as 75 hours to complete one part for the contractor and thus, this had not become a major market opportunity at this time.

Less dramatically, but economically more significant, Rockwell International installed an abrasive jet system to cut parts for the B-1B bomber. Using both a pin-router type of installation and an X-Y table for cutting parts from sheet stock, cost savings of up to 50% were estimated. These arose from the elimination of the inking, and scribing parts of the manufacturing process, the reduction in filing, after cutting, and the replacement of the hand-sawing process. The company cut 1.6 mm thick titanium at a speed of 30 cm/min, using approximately 0.68 kg of garnet a minute [4.54]. The system is integrated with a Thermwood CNC controlled gantry robot.

The use of the abrasive in the waterjet flow does introduce one significant variable which must be considered in the construction of the robots and support tables, relative to those used for plain waterjet cutting.

Because of the rebound of some abrasive and the ability of crushed particles of abrasive to become airborne, the moving parts, joints, and sliding paths for the robotic system must be contained and protected from the deposition of the abrasive dust. Similarly parts underneath those being cut must be protected from the impact of the abrasive waterjet flow. Because the energy is contained in the solid particles, rather than the malleable water, the energy is retained over longer distances.

This, in turn, can make the waterjet dangerous over a much greater distance than that of the conventional plain waterjet system. For this reason, the ways are often covered in corrugated sleeves. In contrast to the relatively open work cells of the plain waterjet system, abrasive waterjet cutting will often take place in cutting rooms which are barred to human access during the cutting operation. Where X-Y tables are used for cutting parts the operation may be submerged in a water bath, since in this way it is possible to capture the cutting jet and reduce its energy without the noise and inconvenience of the more conventional catcher. The bath is usually filled with up to 15 cm of gravel to protect the metal on the bottom of the tank.

4.7.3 OPERATIONAL CONSIDERATIONS

In a typical abrasive waterjet cutting operation the abrasive is likely to be a fine grade of garnet (80 - 100 mesh is reported as the most common [4.55]) and the jet cuts a slot typically 0.75 to 2.5 mm in width with a surface roughness of 2 - 6 mm. Low priced systems can be obtained for as little as $55,000 for a 20 hp unit with a simple X-Y table controller, while more complex systems can cost over $1 million. In October of 1989 ASI Robotics claimed to dominate the 5 axis gantry waterjet robotic market with 95% of the instrument panel (40 units) and 85% of the automotive waterjet installations (including 40 units cutting car and truck carpets). They had installed 179 systems of which only 9 were abrasive waterjet units.

Abrasive waterjets can cut through Kevlar, at speeds of up to 250 cm/min for a saving of 40% in labor and 20% material over manual methods, as reported for its use in Boeing Military Aircraft Co. in Wichita KS. Using a highly precise alignment Jet Edge have developed a machining facility for turbine blades which uses abrasive waterjet to cut to a tolerance of + 7.5 mm a tolerance which requires laser gaging [4.56].

Edge quality is an increasingly important attribute in cutting precision parts. Hashish has reported [4.57] on tests carried out on 1.29 mm thick metal sheets to determine the effect of different jet cutting conditions on waviness, burrs and kerf width. Using 60 mesh garnet at a pressure of 2.400 bar five abrasive flow rates were tested, over a range of traverse speeds (Fig. 4.23). Hashish has divided the cutting zone into two parts, the cutting wear zone and the deformation wear zone (see Fig. 4.15). Where cutting is restricted to the former (which will be discussed in Chapter 10) then there are few and small burrs, when the speeds are greater and the second regime is discerned then burrs are more prominent.

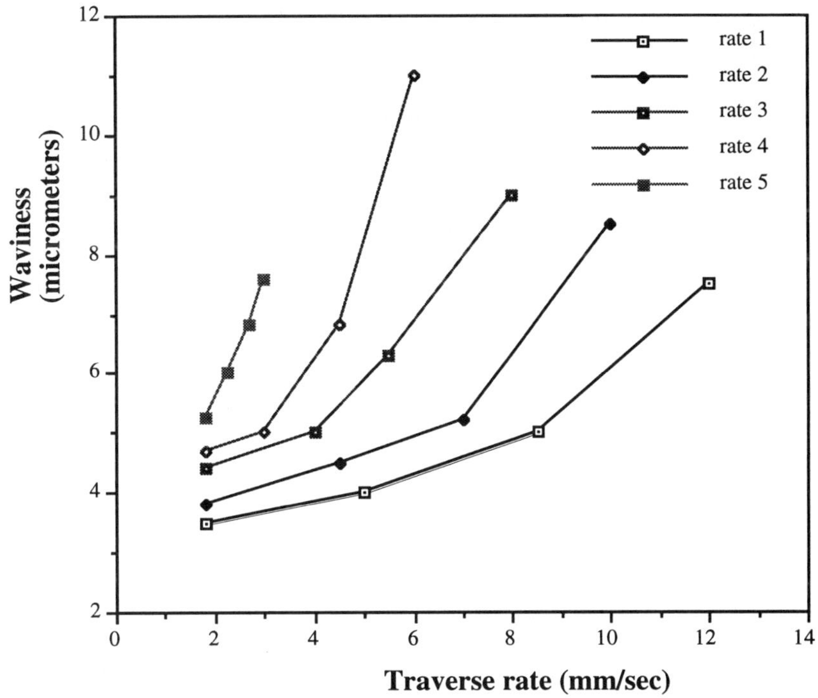

Figure 4.23 Effect of speed and abrasive flow rate on surface waviness [4.57].

Cutting material with abrasive waterjets is a somewhat more complex operation than just using plain water. The reason for this is that, in thicker materials, the cutting edge is not usually straight down under the impact point but curves back along the cut. This makes it somewhat more difficult to program a cutting arm to follow complex contours. The speed of traverse must be more precisely controlled to ensure that, as the jet goes into, and around a corner, that this cutting curve has caught up with the top edge of the cut over the full thickness being cut. This ensures that the correct radius is cut without overcutting to the outside of the turn on the underside of the part. Two additional points must be considered in programming a cut.

When the abrasive waterjet is programmed to pierce into the material and start a cut from a blind hole, the impact point will be surrounded by an eroded zone and the hole diameter will commonly be larger than the slot width cut on traverse. To remove these blemishes from the final cut edge - which may be at net final shape - the piercing can be started inside the material to be removed, and then brought around a curve onto the line of the required cut (Fig. 4.24). In this way, as the jet cuts back to this point, there is no change in edge quality. One may also find this cutting into the path used when a nozzle is changed on a high precision part.

Figure 4.24 Path used to cut an internal contour without edge effects.

The second problem relates to edge quality. Where the jet is traversed slowly and the target is relatively narrow, a smooth cut across the target surface can be obtained. As the jet traverse speed increases, or the target becomes thicker, this causes a change in even-ness of the cutting process. As a result waves begin to appear in the cut, which increase in magnitude with depth (Fig. 4.15). While these are acceptable in some processes, where the surface will be subsequently treated, in many operations these are unacceptable and so, the cut speed is limited to reduce their presence. This may also introduce an unanticipated problem [4.58]. When a robot arm must move at very low speeds, below around 1 m/min, the motor drive circuits are no longer always functioning at maximum efficiency, and thus may "hunt" around the location point, which can cause an oscillation in the motion, and generate surface roughness. Totally stopping the jet will also cause an increase in the kerf angle, and create erosion around the point, and thus should be minimized.

Program update speeds should also be recognized in the setting of the path, and the speed of cut. Cutshall (ibid.) points out that if the cutting head is moving at 60 cm/min and has an update rate of 50 ms, then the head will move 0.5 mm between updates. But if the location points in the program are 0.25 mm apart, for example, then the robot might pause, or even back up, after each step since it has overshot the mark. This would result in a poor and inefficient cutting. The point spacing should thus be chosen to resolve this, or a faster update speed must be built into the programmer. This may require a greater memory capacity, and when purchasing a unit, it is important to determine, ahead of time, that sufficient memory for path storage, will exist. As a warning, let it be noted that in one operation a robot was purchased with more than 1 Mb of memory. It was not, however, until late in the programming that it was discovered that the robot needed about 960 kb of this for its own internal operations, leaving only 40 kb for path storage, an inadequate volume.

The abrasive waterjet allows the same types of cutting as with the plain waterjet, except that metal and ceramic pieces may also be cut. Small and large parts may be nested for cutting under X-Y table conditions, although in this case, where the sheet is large, the cutting may take place with the workpiece under, or on the surface of a water catcher tank. The problem of part support is not normally as great with these materials, since they have inherent strength, however, when cutting small parts a method must be found to stop the pieces from falling down into the catcher tank. This can be accomplished, where possible with the part shape and need, by leaving small detents or tabs around the piece. Thus, the cut is not completely made by the jet, but a small bridge of material is left to hold the part in place (Fig. 4.25). When the sheet is removed at the end of the cutting program, the individual parts can then be broken off, by snapping these dents usually by hand.

An illustration of the steps required in cutting parts is illustrated by the Lockheed procedure used to manufacture an **intercoastal** [4.59]. Intercoastals are flanged structural ribs contoured to fit between the inside surface of an aircraft skin and the underlying support beam. An aircraft tail structure will need fourteen of these, each uniquely shaped, and to provide spares for four units, a total of seventy pieces were to be made. The parts were to be cut from 1.8 mm thick sheets of 15% by volume silicon-carbide whisker in 2124 aluminum. The manufacturing process was carried out in two parts. Firstly two tooling holes were cut in the sheet, and then each profile was cut. The tooling holes were used to locate the part in the final assembly, and thus had to be cut to 6.4 mm diameter with a tolerance of ± 51 mm.

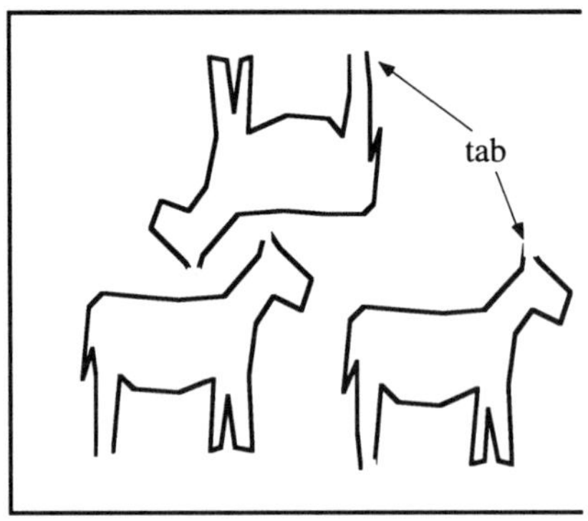

Figure 4.25 The leaving of remnant tabs or detents to hold parts on a sheet. Tabs are exaggerated and shown at the ear and rear foot of the cut out.

Once the nesting pattern had been established using a CADCAM program, this was transferred to a CNC tape which controlled the cutting robot. The sheet was held down within the frame of a 4-axis robot by placing lead weights on it. The cutting nozzle was fixed over the starting point, which Ginburg calls the **touch-off point**. After initiating the program the abrasive nozzle is first moved to the center of the first hole at no flow, the jet is then activated and pierces through the material. The nozzle is then moved in a spiral cut until it reaches the required hole perimeter (see Fig. 4.24) at which point it moves along the path which cuts the hole perimeter at 1.3 mm/sec. At the end of the cut the jet is shut down and the nozzle is moved to the next hole where it repeats the procedure. The perimeter of the entire rib was then cut at a speed of 6.4 mm/sec. It took 40 minutes to cut the entire set of ribs for one assembly and, after using the abrasive waterjet cutter, the parts required neither trimming nor deburring.

One additional aspect of this program is worth noting. The parts were cut to a high tolerance. At different cutting speeds the jet will cut a different kerf width, and thus the path of the nozzle must be adjusted to ensure that the edge left is at the right tolerance. This can only realistically be determined by experiment with the nozzle and parts being cut. Ginburg gives the following example:

Table 4.5 Kerf width in a metal-matrix composite [4.59]

(Jet pressure 3,150 bar; grit 80 mesh at .22 kg/min; stand-off 19 mm; nozzle 0.74 mm; orifice 0.178 mm; target 2.4 mm thick)

Jet speed (mm/s)	*Kerf width (mm)*
12.7	0.61
11.4	0.635
10.2	0.686
8.89	0.711
7.62	0.737
6.35	0.787
5.08	0.838
3.81	0.889
2.54	0.991
1.27	1.067

In a related paper Faticanti [4.60] has reported than an 8 ply, 2 mm thick boron reinforced 1100 aluminum can be cut at a speed of 21.2 mm/sec with a surface roughness of about 3 mm, although it requires a secondary deburring stage. Silicon carbide:aluminum can be cut in the 2 to 10 mm/sec range but the amount of burring is controlled by the process speed. Since the paper was written Faticanti has reportedly been able to improve the cutting ability to 32 plies without burring or the need for secondary finishing.

Titanium composites (which costs about $4,500/kg) have been cut without needing secondary finishing, with the speed controlled by the number of plies and thickness.

A diamond wheel will cut an 8 ply composite at a rate of 0.338 mm/sec, while wire EDM averages 0.0035 mm/sec. On the other hand set-up time for the diamond wheel is much shorter than that for the waterjet system on a single operation, which made it important in this instance to ensure that the entire cutting operation could be achieved with a single set-up, where possible. The diamond wheel cost $95 and lasted over a cutting distance of

30 m while the abrasive cost was on the order of $13.20 for a 30 m long cut.

Table 4.6 Cutting speed for titanium aluminide composites [4.60]

Composite # of plies	*Thickness (mm)*	*Cutting rate (mm/s)*
4	0.81	10.6
8	1.62	6.35
12	2.43	4.23
24	4.98	2.12
32	6.50	0.847

As another example of the differences in cost between traditional machining and abrasive waterjet cutting, Zaring has costed the cutting of a turbine wheel from titanium [4.61]:

Table 4.7 Relative time for cutting turbine wheel [4.61]

	Conventional method	*Abrasive waterjet*
rough cut hours	1400	48
cost/hr	$30.00	$30.00
total	$42,000	$1,440
cost savings		$40,560
productivity increase		29 times

Hashish has reported that there is also a slight hardness change where waterjets cut metal [4.62].

Table 4.8 Surface hardness of different metals after cutting [4.62]

Metal	*Base hardness*	*Hardness of cutting*
titanium	R_C 34.0	R_C 34.3
6061-T6 aluminum	R_B 54.8	R_B 58.7
magnesium	R_B 54.1	R_B 58.4
A-572 carbon steel	R_B 82.8	R_B 84.5
tool steel	R_B 91.5	R_B 93.3

The problems of cutting to tolerance, which are critical to the gain in performance which can come from initially cutting to net final shape, are compounded by the fact that the operating nozzle which collimates the abrasive jet will undergo wear. Early data suggested [4.63] that the nozzle diameter would grow on the order of 0.1 mm each hour, under a flow of 0.22 kg of 60 mesh garnet at a jet pressure of 2,200 bar. Since that time there has been a change in the materials from which the collimating nozzle is made, and it has been reported that nozzle lifetimes have been extended from a base of 1 - 3 hours to 40 - 60 hours for intensive use and up to 75 hours for lighter work in thinner materials [4.64]. The relevance of this must be recognized in the growing move toward the cutting of larger target shapes, from harder materials. In these applications, the change from a worn to a new nozzle may bring with it a change in the width of the kerf which, if not compensated for, will change the surface finish. Thus, longer life for the nozzle can minimize this required change and justify the investment in the more expensive product.

As an example of this with the longer lasting nozzles it was possible [4.47] to make a spiral cut in a 22.2 mm thick steel plate which had an effective length of 28 m with the entire cut made in 15 hr with a single nozzle.

Cutting costs for the use of an abrasive, or a plain waterjet system, do not, however, end with the cost of the abrasive or the nozzle. To give some perspective to the types of costs which are involved Labus [4.41] tabulated typical costs at the end of the 1980s.

Table 4.9 Operating costs for a 30 kW intensifier [4.41].

waterjet nozzle diameter (mm)	.178	.356
operating pressure (bar)	3,150	3,150
power costs ($/KW-hr)	0.10	0.10
number of nozzles	1	1
flow/nozzle (liters/hr)	46.1	184.6
abrasive nozzle diameter (mm)	.76	1.60
abrasive flow (kg/min	.23	0.91
abrasive cost ($/kg)	0.33	0.33
carbide nozzle life-hrs	6	3
carbide nozzle cost ($)	27.50	27.50
power required/hr (KW)	6.18	24.72
power cost ($/hr)	0.62	2.47
hourly cost of replacement parts		
check valve (1500 hrs life at $40.00)	0.03	0.03
sapphire orifice (150 hrs at $15)	0.10	0.10
high pressure seal (1000 hrs at $20)	0.02	0.02
pump filters (500 hrs at $21 for 3)	0.04	0.04
operating/maintenance costs		
parts & power	$0.81	$2.66
abrasive	$4.50	$18.00
carbides	$4.58	$9.17
total hourly cost	$9.89	$29.83

Tubing costs are $167/m and a nozzle assembly costs $2,000.
A pump costs $65,000 and an abrasive injection system $75,000 installed [4.65].

4.7.4 GLASS CUTTING

The cutting of glass has provided a number of new applications for waterjet systems, and allowed the creation of sculptures and artwork not previously achievable [4.66]. The use of abrasives themselves is not new in this application. Air entrained abrasives have been used for fine detail etching and cutting for a number of years [4.67]. Typical jets, in those installations, were on the order of 0.125 mm to 2 mm in size and were used for fine cutting or frosting areas of glass. A gas pressure of 5 bar and a volume flow rate of 35 liters/hour have been used for the cutting which gives surface finishes in the 0.15 - 1.5 mm range. Material removal rates are on the order of 16 cu mm/min, the equivalent of a slot some 0.5 mm wide, and 0.25 mm deep being cut to a length of 128 mm in a minute. While this gives some idea of the precision which can be achieved, it is too slow an operational rate for many concerns, and the additional power available with a high pressure waterjet has thus found a use.

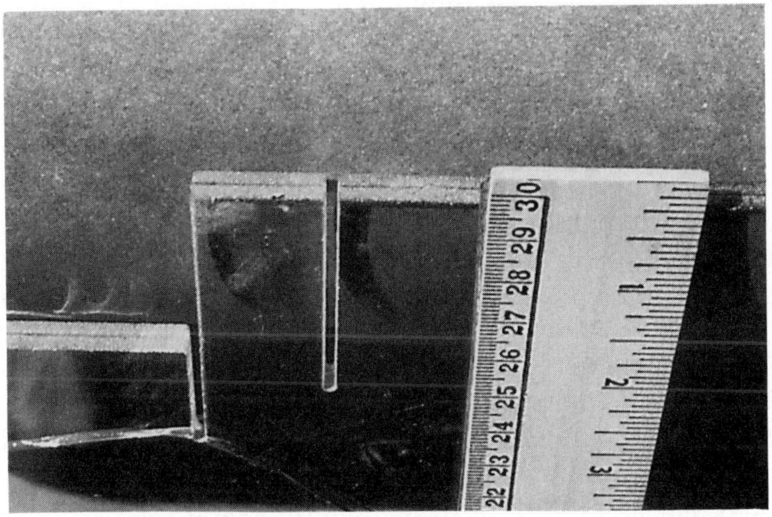

Figure 4.26 Abrasive jet cut through two laminated layers of glass.

The use of an abrasive laden waterjet has an additional advantage in glass cutting. The ability of the waterjet to cut through all the materials it encounters, within reason, means that it can be used to cut through composite stacks of material, one of which is laminated glass. A particular problem in the conventional cutting of safety glass arises out of its construction as two thin panes of glass held together by an interlayer of plastic. In conventional cutting, where the cutting plane is first scribed on the surface with a diamond wheel, and then impacted to create the edge, this interlayer can create up to 30% wastage. With an abrasive jet which cuts through all three layers with no deviation (Fig. 4.26) this problem is immediately overcome. Such laminated plate glass can be cut either with internal or external contours, with the same performance requirements, particularly in thicker materials that have been addressed above.

Glass does however require that additional care be taken in handling the workpiece [4.68]. If the jet is at all diffuse it can create a frosting along the edge of the cut, which is usually unacceptable. This does have a slight advantage in that the slight edging on the cut does reduce the potential for injury to personnel handling the piece. The presence around the workpiece of stray particles of abrasive can lead to scratching of the piece, either before or after cutting, just through the handling process. Because of its brittle nature, the piece also must be carefully supported, since any bending may induce stresses, around the cutting point, which will cause it to act as a stress concentrator, and lead to catastrophic crack growth. It is this potential for exacerbating the individual cracks created during the initial "stationary jet" part of the process which has led, in part, to the recommendation that where possible in piercing glass, that the jet pressure used be on the order of 700 to 800 bar and that the jet not be raised to actual cutting pressure until the hole has penetrated. The result of crack pressurization may not, in this case be total plate failure, but localized chipping of the surface may occur [4.63].

This consideration of the stresses generated in the part is particularly important when heat treated and tempered glasses are considered as candidate materials. Under normal conditions these materials are designed so that they will shatter upon cracking. This is achieved by building internal stresses, perhaps on the order of 500 bar, into the glass. Thus, the onset of any crack can cause sufficient concentration of stress at the tip that total failure of the sample will occur. For these reasons it is not currently recommended that jet cutting of these materials be tried.

4.7.5 ABRASIVE WATERJET MILLING

The precision with which a waterjet can now be controlled to remove material can now, also, be used for internal milling within parts. A considerable effort was carried out by Hashish in determining the basic parameters required for such an operation [4.69].

Milling requires that the cutting tool hold a very accurate control over the depth of cut achieved. This is not immediately easy to achieve since the abrasive waterjet system will tend, under normal conditions, to cut a slot to a varying depth. Hashish has found that control of jet angle and traverse speed can control the irregularity of the surface, as does the distance from the surface to the target. In all cases there are optimum conditions for the material being cut. In order to reduce the variation of depth which occurs with change in head direction, and the resulting change in the time that the jet is playing on the surface, it is possible to have the jet run outside the target area, onto a covering, resistive plate. This will allow the speed and position changes to occur "off specimen" and thus give an even feed, and cut, where it is needed.

In contrast with mechanical material removal processes it is not helpful to use a coarse particle to do the rough cutting, and then to finish the surface with a finer grit. Since the material removal process is different, the result, with an abrasive jet, is that the finer particle finish will, if anything, exacerbate the roughness created in the surface from the original coarse abrasive feed. It is better to accept the slower material removal rate and to use fine abrasive from the beginning since this will produce the better result [4.69]. In glass material removal rates on the order of 16 cc/min can be obtained [4.70].

In assessing the parameters which control milling investigators at Hanover [4.71] suggest that the work be carried out with the highest jet pressure and the largest jet nozzle, and with an abrasive concentration of 35% of the water mass flow rate. Based on optimizing water and energy use, they suggest that a nozzle diameter of between 0.2 and 0.25 mm be used.

4.8 INTERMEDIATE PRESSURE ABRASIVE JET CUTTING

The ability of abrasive laden waterjets to cut through materials is not restricted to waterjets at high pressure. Systems are available in both the United States and the United Kingdom, in which waterjets at lower pressures have been used for abrasive jet cutting. Jordan [4.72] reported that carbon steel plate, 2.5 cm thick, can be cut at a speed of 5 cm/min using approximately 0.9 kg of garnet abrasive at 1,400 bar, while 1.25 cm thick plate can be cut at 15 cm/min using 1.14 kg of abrasive/min. The cost of the system is given as between $400 and $650 per kW for the pump, and some $7,500 for the abrasive injection system.

At the Welding Institute in England, a unit is operated at 1,000 bar at a flow rate of 25 lpm, in which a flint abrasive is added at up to 3 kg/min [4.73]. When cutting softer materials the flint, which costs about $190 a tonne, is sized at 0.1 to 0.4 mm, while a size range of 0.4 to 1.4 mm is used when cutting tin, harder materials, and metals. The Institute first looked at feeding the abrasive in a slurry form, but the mixing was found to be less efficient, while the carrier water had to be accelerated, with a loss in delivery efficiency. Dry feed of the abrasive, in a similar form to that of the higher pressure systems was therefore adopted. The cost of the system was reported as being comparable with that of the BHRA DIAjet system reported below. Performance of the system has been reported for several materials as in the attached table:

Table 4.10 Abrasive waterjet cutting below 1,000 bar with conventional injection [4.73]

Material	*Thickness (mm)*	*Cutting speed (mm/min)*	*Water pressure (bar)*
mild steel	3	210	750
steel armor	8	40	750
mild steel	10	32	690
C-Mn steel	12	50	750
C-Mn steel	25	25	750
C-Mn steel	30	20	750
C-Mn steel	50	15	750
stainless steel	3	200	750
stainless steel	8	60	700
stainless steel	10	35	690
stainless steel	30	15	700
aluminum	3	500	900
aluminum	3	350	690
aluminum alloy	6	250	690
aluminum alloy	10	125	690
aluminum alloy	12	130	740
aluminum alloy	25	70	900
aluminum armor	25	70	700
aluminum armor	50	50	700
copper	3	150	750
titanium	12	36	690
titanium	25	25	700
titanium composite	9	20	700
reinforced concrete	70	20	690
reinforced concrete	120	25	720
ceramic	4	200	700
ceramic tile	8	300	700
abrasive disc	3	660	700
marble	25	50	700
kevlar	4	130	720
kevlar + 5 mm ceramic	35	125	700
carbon fiber composite	2	1000	690
carbon fiber composite	8	350	690

The system used 2 kg/min of #5 Flintag (0.4 - 1.4 mm diameter), flowing through a 3 mm diameter nozzle at 2 mm stand-off distance.

The study has shown that the abrasive waterjet was particularly effective in cutting metals which had been surface hardened, and those which are adversely effected by heating.

4.8.1 ABRASIVE SUSPENSION JETS

Recent work at Southwest Research Institute has returned more to the concept of the earlier abrasive jet cutting systems [4.74], in which a polymer-enhanced water is used to suspend abrasive particles which are then fed through the nozzle onto the target. The jet system has been referred to as a suspension jet and was initially used to cut 6.35 mm samples to demonstrate its capability. At a pressure of 510 bar, and at a traverse speed of 10.2 cm/min the jet cut through glass, with a kerf width of 0.79 mm. When the pressure was increased to 1,000 bar, the kerf width increased to 3.2 mm. A comparison of the potentials of the system can be seen from its comparison with conventional abrasive jet cutting (ibid.).

Table 4.11 Conventional and suspended abrasive jet cutting of 6.35 mm steel [4.74]

	Conventional jet	*Suspended jet*
pump pressure (bar)	2,070	520
traverse speed (mm/min)	10.2	5.1
power (kcal/min)	65.7	9.4
power/mm - (kcal/mm/min)	0.64	0.18
abrasive used (gm/min)	217	81.6
abrasive used/mm - gm/mm	2.6	1.6
abrasive size	60-80 mesh	75-106 mm
water use (liters per mm)	0.013	0.015
kerf width (mm)	1.6	<0.79

Hashish has reported [4.75] that, on an equivalent pressure basis, the abrasive suspension jets can cut to more than double the range of the conventional abrasive cutting systems (Fig. 4.27). In order to create these jets a suspension of abrasive in polymer thickened water is often first prepared, and then placed on the downstream side of a plunger in a separating cylinder which isolates the cutting fluid from the pressurizing fluid. The conventional high or ultra-high pressure pump can then be used to pressurize the fluid, without the risks to the pump associated with driving the fluid through its cylinders.

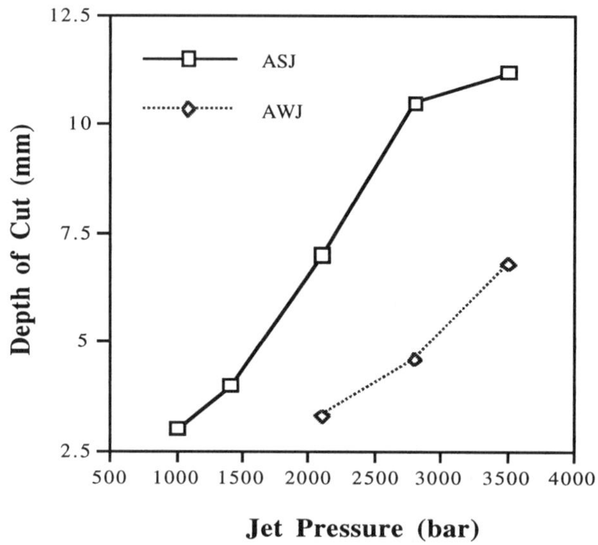

Figure 4.27 Comparison of conventional (AWJ) and suspended abrasive (ASJ) waterjet cutting using 80 mesh garnet, 0.229 mm nozzle diameter, a stand-off distance of 1.5 mm, and a traverse speed of 17 mm/s [4.75].

4.9 DIRECT INJECTION OF ABRASIVE

This idea of adding the abrasive to the waterjet before it exits the primary nozzle has also been investigated elsewhere. Beginning with a Masters thesis at Cranfield [4.76] investigators at BHRA developed a system which has since become known as the DIAjet system [4.77]. This design overcomes some of the disadvantages of the conventional method of entraining abrasive into a waterjet stream. Instead of mixing the abrasive in a chamber, following the acceleration of the water, the abrasive is metered into the flow of water from the pressurizing pump to the nozzle. This is achieved (Fig. 4.28) by first adding the abrasive to a tank which can be pressurized to the delivery pressure, and then connecting this tank into a loop attached to the delivery line. The connections are made through valves, and flow through the loop is controlled by a metering valve. In this way the concentration of abrasive feeding into the line can be controlled.

Initial experiments were reported in 1986 which showed that the system was capable of cutting through a 13 mm thick mild steel plate at a rate of 51 mm/min when copper slag was used as the abrasive. Subsequently this product has been marketed, with the primary product a unit which operates at 350 bar, with a flow rate of some 70 lpm. This unit has been developed for commercial industrial cutting, and has been used for the cutting of titanium and armor plate. It is able to achieve the same level of performance as conventional abrasive jet systems, but at significantly less pressure. The relevance of this can be seen since it is possible to purchase pumps to deliver this volume and flow for less than $10,000. This, in itself is a savings of $55,000 over the cost of the pump cited above, but additional savings come from the less expensive piping and fittings which are also required for the system to operate. Potential disadvantages to the system lie in the increased volume of water which the system uses, and the concomitant increase in abrasive quantity since the two systems use roughly equivalent concentrations of abrasive in the water.

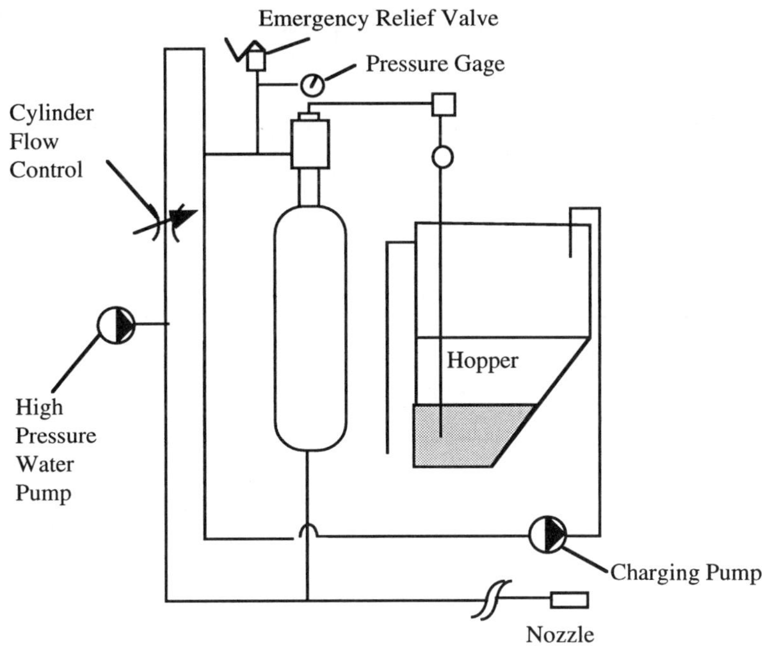

Figure 4.28 Schematic representation of the DIAjet flow circuit [4.78].

4.9.1 DIAJET CASE STUDY

The commercial application of this system was first described in 1988 [4.79] although actually carried out in 1986. Work at the Philips-Imperial Petroleum Oil refinery required that the internal metal lining of a distillation column be removed. Given that there was a risk of gas presence, the method for removing the 2 mm liner plates and the supporting structure from the 5.64 and 3.35 m diameter sections of the column had to be capable of operating in an explosive environment. At a pressure of 275 bar the jet could cut through the 20 mm steel support brackets at a speed of 150 mm/min using copper slag as the abrasive. The cutting was varied out at 325 bar at which pressure the nozzles lasted between seven and twenty-five hours.

The work was completed within the time schedule anticipated. Subsequent to this development the inventors have gone on to develop a higher pressure version of the system, with an improved cutting capability. This system was launched commercially in the summer of 1990 [4.80]. Representative performance characteristics for the old and new systems are given below. The characteristics of the system, in so far as they have been investigated, will be discussed in Chapter 7.

Table 4.12 Comparative performance of 350 and 700 bar DIAjet systems

Performance given is for the cutting of 13 mm thick steel

	345 bar	*690 bar*
speed of cut (mm/min)	300	150
kerf width (mm)	2.8	0.75
water flow (lpm)	85	9
abrasive feed (kg/min)	10	0.75
hydrokinetic power (kW)	50	10
surface generated (sq mm/kg)	390	2600
surface generated (sq mm/kW)	78	195
surface generated (sq mm/liter of water)	46	216

4.10 FUTURE TECHNOLOGY

It is interesting to note that the potential for this new technology has not gone unnoticed. In 1989 the American National Center for Manufacturing Science sought to determine the needs for industry in the area of waterjets for the next five years. Based upon their survey they sought two major waterjet developments, as well as improved support robotics. One of these was a higher pressure system, capable of cutting through materials faster and with a cleaner cut, aiming the operating pressure of the system toward the 7,000 bar range. At the same time the potential benefits from abrasive injection before the nozzle are sufficient that the second major effort required was to develop a method for applying a direct injection method of abrasive addition in order to provide a more precise method of cutting material. The Center made the award for this work to Quest, the original Flow Research, in collaboration with a number of universities in early 1990 [4.71].

4.10.1 WATERJET MACHINING

As with other cutting tools it is possible to apply waterjets, and particularly abrasive waterjets to the three dimensional cutting of surfaces, and particularly to the internal milling and machine turning processes. The most promising avenue of application for this technology is in the machining of ceramic materials which are difficult and expensive to shape by other means. By locating an abrasive waterjet in the position of a conventional machine cutting tool it has been found possible to use the jet as a cutting tool in carving required shapes. Although not widely commercialized yet, this technique has been able to machine internal cavities and bring parts close to near-net shape condition [4.82]. Rods of aluminum and magnesium boron carbide were turned from an initial 25.4 mm to 6 mm diameter. Although the rate for the latter was 38% slower than for the aluminum conventional machining would have required some 5 - 10 times as long. The primary measures of performance were the ability of the jet to reduce the part to the required size, and to achieve a satisfactory surface finish. It was reported that the surface waviness could be repeated within 60 - 90 mm given the variations in jet pressure, feed rate, and abrasive feed rate which were occurring during the cutting process. Additional passes were found capable of reducing the surface waviness from 122 - 64 mm.

It was reported that by changing particle size, abrasive type, and the number of passes the surface roughness could be reduced, however, the waviness of the surface was less easily controlled and required a better control on some of the process parameters.

4.10.2 WATERJET ASSISTED MACHINE TOOL CUTTING

Similarly a new application of waterjets to conventional material removal is only just developing from initial laboratory testing [4.73]. It has been shown that the addition of a high pressure waterjet to the leading edge of a machine tool not only improves the surface quality of the material being cut, it also reduces the force on the tool, increases tool life, and instantaneous output and produces small, cool chips of metal which are easier to handle (Figs. 4.29 and 4.30).

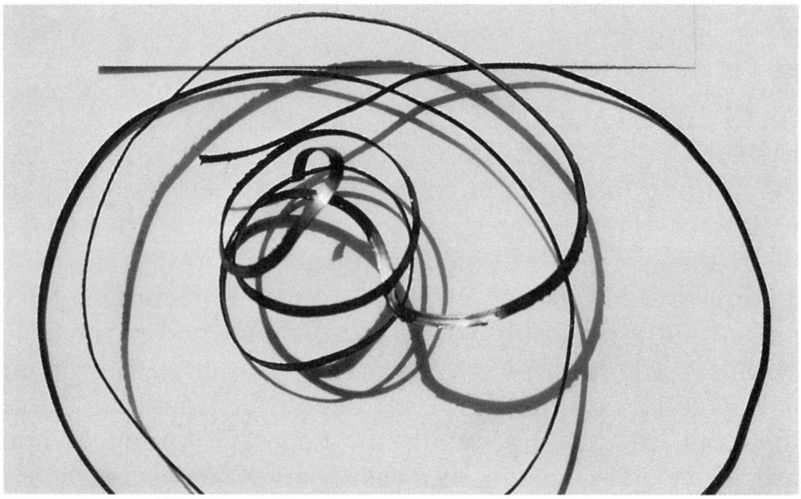

Figure 4.29 Strips of metal removed by machining without waterjet cooling.

Figure 4.30 Small shiny chips removed with waterjet assisted cooling in the machine tool metal interface.

The process has recently [4.84] been extended for use in metal grinding. In this instance the jet action is somewhat different, since it serves to continuously clean the stone being used. reducing the blinding of the stone as it becomes impregnated with chips of metal. Preliminary experiments have shown that the process reduces the need to redress the wheel, and could thus reduce grinding costs by somewhere in the region of 12%. This indicates yet another potential avenue for waterjet applications into areas as yet only barely delineated. This technology has recently also been investigated at another University [4.85], with very promising results. Schoenig and his colleagues report that the addition of 3,500 bar waterjets to the cutting interface increased tool length 1000%; increased cutting speed from 1.5 to 2.5 times; and fractured the chips (as shown above).

With this type of innovation and some of those others described above still being developed, the applications of waterjets in industry therefore, continue to look very promising.

4.11 NOTES - CHAPTER FOUR

Many of the references in this chapter came from a series of Clinics held by the Society of Manufacturing Engineers on the topic of High Pressure Waterjet Cutting. These are referred to for simplicity by the initial SME and the title, location and date of the meeting. Further information can be obtained from SME.

4.12 REFERENCES

4.1 Miller, R.K., Waterjet Cutting, Technology and Industrial Applications, SEAI Technical Publications, P.O. Box 590, Madison, GA 30650, 1985.

4.2 Bryan, E.L., (1964), Ph.D. dissertation, Wood Technology Department, University of Michigan, Ann Arbor.

4.3 Walstad, O.M., Noecker, P.W., "Development of high pressure pumps and associated equipment for fluid jet cutting," paper C3, 1st International Sympo- sium on Jet Cutting Technology, Coventry, UK, April, 1972, pp. C321 - C328.

4.4 Franz, N.C., "Fluid additives for improving high velocity jet cutting," paper A7, 1st International Symposium on Jet Cutting Technology, Coventry, UK, April, 1972, pp. A7 93 - A7-104.

4.5 Szymani, R., A study of corrugated board cutting by high velocity liquid jet, MS thesis, Univ. of British Columbia, September, 1970.

4.6 Davis, D.C., "Robotic Abrasive Water Jet Cutting of Aerostructure Components," Automated Waterjet Cutting Processes, SME, Detroit Michigan, May, 1988.

4.7 Ambrosis, J., and Jones, E.P., "Integraticn of Industrial Robots with Flow Systems, Inc. Waterjets," Flow Engineering Technical Report TR-1020 at the Robotic Waterjet Cutting Seminar, SME, Napierville, IL, June, 1986.

4.8 Johnston, C.E., "Effect of Water Quality on the Waternife™," Engineering Technical Report TR-1031 by Flow Systems, at the Robotic Waterjet Cutting Seminar, SME, Napierville, IL, June, 1986.

4.9 Wightman, D.F., "Waterjets on the cutting edge of machining," Conf. on Flexible Manufacturing Systems, paper MS86-171, Society of Manufacturing Engineers, March 3-6, 1986, Chicago, IL.

4.10 Cutshall, David W., "Waterjet Cutting in a Robotic Environment," Automated Waterjet Cutting Processes, SME, May 10, 1988, Fairlane, Michigan.

4.11 Cover Plate, Robotic Waterjet Cutting Seminar, SME, Napierville, IL, June, 1986.

4.12 Rogers, C., "Flow Systems, Supplemental information," Robotic Waterjet Cutting Seminar, SME, Napierville, IL, July, 1987.

4.13 Walstad, O.M., "Commercial Utilization of the McCartney Fluid jet cutting concept," paper E2, 2nd International Symposium on Jet Cutting Technology, Cambridge, UK, April, 1974, pp. E2 11 - E2 18.

4.14 Lane, J.D., "Robot Basics," Robotic Waterjet Cutting Seminar, SME, Napierville, IL, June, 1986.

4.15 Woolman R., "Precision Multi-axis Waterjet Machining of Aerospace Materials," Effective Waterjet Machining, SME, Los Angeles, CA, September, 1991.

4.16 Schubert, G.C., "High Tolerence Abrasivejet Machining, Waterjet Cutting West, SME, Los Angeles, CA, November, 1989.

4.17 Adams, R.B., "Waterjet Machining of Composites," Proceedings of Composites in Manufacturing - 5, Los Angeles, January, 1986 paper EM86-113, SME.

4.18 Snider, D.E., "Waterjet Cutting in a Production Environment," 5th American Waterjet Conference, Aug. '89, Toronto Canada, pp. 231 - 236.

4.19 Anon, "Waterjets Cut Circuit Boards," High Technology Business, Vol. 8, No. 4, April, 1988, p. 9.

4.20 Savanick, G., "From the Presidents Desk," Jet News, December, 1990, Water Jet Technology Association, St. Louis, MO.

4.21 Ouellette, D.E., "Ingersoll-Rand Waterjet Cutting using CAD/CAM Programming," Robotic Waterjet Cutting, June, 1987, Napierville, IL.

4.22 Leslie, E.N., "Application of the Water Jet to Automated Cutting in the Shoe Industry," paper F3, 3rd International Symposium on Jet Cutting Technology, Chicago, IL, May, 1976, pp. F3 23 - F3 36.

4.23 Pilkington, D.J., Fletcher, J.V., "Some aspects of applying jet cutting to the production of footwear components," paper F4, 3rd International Symposium on Jet Cutting Technology, Chicago, IL, May, 1976, pp. F4 37 - F4 46.

4.24 Gossage, A. (Lodge & Shipley, 3055 Colerain Avenue, Cincinnati, Ohio) personal communication 14 September 1990.

4.25 Kiehl, R.E., "Waterjets-Their Place in Metals Processing," Successful Applications of Waterjet Cutting Systems, SME, March, '91, Cleveland, OH.

4.26 Flow Systems Inc. Corporate Report 1986.

4.27 Allen, L.G., "Clean water keeps water knife cutting cleanly," Automation, September, 1989, Penton Publishing, Cleveland Ohio 44114.

4.28 Ingersoll, F.V., "Waterjet Cutting of Thermo Formed Plastic Parts," Automated Waterjet Cutting Processes, SME, Southfield, MI, May, 1990.

4.29 Ostby, K., "Robot Waterjet Cutting," paper MS84-359 Robots Eight Conference, SME, June 4-7, 1984, Detroit, Michigan.

4.30 Imanaka, O., Fujino, S., Shinohara, K., and Kawate, Y., "Experimental study of machining characteristics by liquid jets of high power density up to 10^8 Wcm^{-2}," paper G3, 1st International Symposium on Jet Cutting Technology, Coventry, UK, April, 1972, pp. G3 25 - G3 35.

4.31 Neusen, K.F., and Schramm, S.W., "Jet Induced Target Material Temperature Increases during Jet Cutting," paper E4, 4th International Symposium on Jet Cutting Technology, Canterbury, UK, pp. E4 45 - E4 52.

4.32 Worsey P.N. Derringer Manual, UMR June 1993.

4.33 Summers, D.A., "Considerations in the design of a waterjet device for reclamation of missile casing," International Waterjet Symposium, Beijing China, September, 1987, paper 6-7. Sponsored by the Water Jet Technology Association in collaboration with the Ministry of Coal Industry and the China Science and Technology Exchange Center.

4.34 Information to David Summers on Halliburton and Aerojet rocket motor cleanout facilities.

4.35 Summers, D.A., Tyler, L.J., Blaine, J.G., Fossey, R.D., Short, J., and Craig, L., "Considerations in the design of a waterjet device for reclamation of missile casings," 4th American Water Jet Conference, Berkeley, CA, August, 1987, pp. 51 - 57.

4.36 Haferkamp, H., Louis, H., and Schikorr, W., "Precise Cutting of High Performance Thermoplastics," 7th International Symposium on Jet Cutting Technology, Ottawa ,Canada, June, 1984, pp. 353 - 368.

4.37 Niehaus, F.A., Rock, A.O., Yazici, S., and Summers, D.A., The use of High Pressure Waterjets as a Means of Cutting Foam, Report NWSC/CR/RDTN-245, Ordnance Development Department, Naval Weapons Support Center, Crane, IN, September, 1987.

4.38 Howells, G., Information provided to David Summers.

4.39 Kurko, M.C., and Chadwick, R.F., "High Pressure Jet Machining," National Conference on Fluid Power, Chicago, IL, October, 1971.

4.40 Haylock, R., "Waterjet Cutting," Automech Australia '85, SME MM85-651, Melbourne, Australia, July, 1985.

4.41 Labus, T.J., "Section 8", Fluid Jet Technology Fundamentals and Applications - A Short Course, 5th American Waterjet Conference, Toronto, Canada, August 28, 1989, pp. 145 - 168.

4.42 Engel, S.L., and Labus, T.J., "Industrial Applications ", Section 8, Fluid Jet Technology Fundamentals and Applications - A Short Course, 6th American Waterjet Conference, Houston, TX, August, 1991.

4.43 Carlson, L.D., and Huntley, D.T., "The Advantages of High Energy Beam Processing Over Conventional Methods," paper MS89-810, Non-Traditional Machining Conference, October, 1989, Orlando, FL.

4.44 Kennedy, W.J., Skaar, Eric C., "Improving the Machining of Ceramics," paper MS89-813, Non-Traditional Machining Conference, October, 1989, Orlando, FL.

4.45 Wightman, D., "Hydroabrasive Cutting," Automated Waterjet Cutting Processes, May, 1990, Southfield, Michigan.

4.46 Anon, Sales literature on the System 100, Harris Calorific Division, Emerson Electric Co., St. Louis, MO.

4.47 Anon, "Abrasive-waterjet jobshop thrives," American Machinist, October, 1989, pp. 80 - 81.

4.48 Rajurkar, K.P., "Adaptive control systems for EDM," SME paper EE89-801, Non-Traditional Machining Conference, October, 1989, Orlando, FL.

4.49 Clouser, H.A., "High Speed EDM Electrode Forming," SME paper EE89-802, Non-Traditional Machining Conference, October, 1989, Orlando, FL.

4.50 Risko, D.G., "Electrochemical Machining," SME paper EE89-820, Non-Traditional Machining Conference, October, 1989, Orlando, FL.

4.51 Vishnitsky, V., "Rotolytic Finishing," paper EE89-804, Non-Traditional Machining Conference, October, 1989, Orlando, FL.

4.52 Yie, G.G., "Cutting Hard Rock With Abrasive-Entrained Waterjet at Moderate Pressure," 2nd U.S. Water Jet Conference, Rolla, MO, May, 1983, pp. 407 - 422.

4.53 Behringer-Ploskonkaca, C.A., "Waterjet Cutting - A Technology Afloat on a Sea of Potential," Manufacturing Engineering, November, 1987, Vol. 99, No. 5, pp. 37 - 41.

4.54 Anon, "Giving Water the power of a laser beam," Compressed Air Magazine, 1985.

4.55 Mason, F., "Water and sand cut it," American Machinist, October, 1989, pp. 84 - 95.

4.56 Corcoran, M.J., "Automated Production to very tight tolerances using an abrasive-jet - two new approaches," SME Automated Waterjet Cutting Processes, May, 1989, Southfield, MI.

4.57 Hashish, M., "Characteristics of surfaces machined with abrasive waterjets," ASME Winter Annual Meeting, MD, Vol. 16, ASME.

4.58 Cutshall, D.W., "Waterjet Cutting in a Robotic Environment," SME Automated Waterjet Cutting Processes, Detroit, MI, May, 1988.

4.59 Ginburg, D.M., "Abrasive Waterjet Cutting of Aerospace Materials," SME Waterjet Cutting West, November, 1989, Los Angeles, CA.

4.60 Faticanti, R., "Waterjet Cutting Of Continuous Fiber Reinforced Metal Matrix Composites," SME Waterjet Cutting West, November, 1989, Los Angeles, CA.

4.61 Zaring, K., "Abrasive Waterjet Cutting Hardware and Applications," SME Automated Waterjet Cutting Processes, Southfield, MI, May, 1990.

4.62 Hashish, M., "Data Trends in Abrasive Waterjet Machining," SME Automated Waterjet Cutting Processes, Southfield, MI, May, 1989.

4.63 Anon, "Design criteria for PASER abrasive jet cutting and piercing," SME Robotic Waterjet Cutting, Detroit, MI, 1987.

4.64 Kiehl, R., "System Integration," SME Automated Waterjet Cutting Processes, Southfield, MI, May, 1990.

4.65 Wightman, D.F., "Waterjets on the Cutting Edge of Machining," SME MS86-171, Flexible Manufacturing Systems, Chicago, IL.

4.66 Henry, S.J., "Glass Cutting Today With Abrasive Waterjet Systems - From Job Shop To Computer Integrated Systems," SME Waterjet Cutting West, Los Angeles, CA.

4.67 Lavoie, F.J., "Abrasive Jet Machining," Machine Design, September 6, 1973, pp. 135 - 139.

4.68 Patell, F.A., Scott, D.A., and Saari, K., "Job Shopping with Waterjet/Abrasive Waterjet," paper 23, 5th American Waterjet Conference, Toronto, Canada, pp. 237 - 244.

4.69 Hashish, M., "Abrasive Jets", Section 4, Fluid Jet Technology Fundamentals and Applications - A Short Course, 5th American Waterjet Conference, Toronto, Canada, August, 1989, pp. 49 - 100.

4.70 Anon, Glass Cutting Solutions, Flow Systems, Brochure 10/87 FS-040.

4.71 Laurinat, A., Louis, H., and Meier-Wiechert, G., "A Model for Milling with Abrasive Water Jets," 7th American Water Jet Conference, August, 1993, Seattle, WA.

4.72 Jordan, R., "Water Jet Cutting at 20,000 psi," SME Automated Waterjet Cutting Processes, May, 1988, Detroit, MI.

4.73 Harris, I.D., "Abrasive Water Jet Cutting And Its Applications At The Welding Institute," Welding Institute Research Bulletin, Vol. 29, February, 1988, pp. 42 - 49.

4.74 Hollinger, R.H., Perry, W.D., and Swanson, R.K., "Precision Cutting With A Low Pressure Coherent Abrasive Suspension Jet," paper 24, 5th American Water Jet Conference, Toronto, Canada, August, 1989.

4.75 Hashish, M., "Cutting with High-Pressure Abrasive Suspension Jets," paper 33, 6th American Water Jet Conference, Houston, TX, August, 1991, pp. 439 - 455.

4.76 Fairhurst, R.M., Abrasive Water Jet Cutting, MSc thesis, Cranfield Institute of Technology, January, 1982.

4.77 Fairhurst, R.M., Heron, R.A., and Saunders, D.H., "Diajet" - A New Abrasive Waterjet Cutting Technique," 8th International Symposiumon Jet Cutting Technology, Durham, UK, September, 1986, pp. 395 - 402.

4.78 Yazici, S., Abrasive Jet Cutting and Drilling of Rock, Ph.D. Dissertation, University of Missouri-Rolla, 1989.

4.79 Fairhurst, R.M., and Roff, M.F., "A field application of the DIAjet Abrasive Water Jet Cutting Technique," paper H2, 9th International Symposium on Jet Cutting Technology, Sendai, Japan, pp. 399 - 409.

4.80 Miller, D.S., and Bloomfield, E.J., "The Future for Abrasive Jet Cutting," paper 42, 6th American Water Jet Conference, August, 1991, Houston, TX, pp. 561 - 574.

4.81 Anon, "Major R & D Program on Manufacturing Technology," WJTA Jet News, August, 1990, p. 3.

4.82 Hashish, M., "Turning, Milling and Drilling with Abrasive Waterjets," Paper C2, 9th International Symposium on Jet Cutting Technology, Sendai, Japan, 1988.

4.83 Mazurkiewicz, M., Kabala, Z., and Chow, J., "Metal Machining With High Pressure Water Cooling Assistance - A New Possibility," ASME Journal of Engineering for Industry, Vol. 111, February, 1989.

4.84 Borkowski, J., and Mazurkiewicz, M., "Aluminum Grinding With High Pressure Water Jet Assistance," paper 25, 5th American Water Jet Conference, Toronto, Canada, August, 1991, pp. 253 - 261.

4.85 Schoenig, F.C., Khan, A.K., Atherton, A., and Lindeke, R.R., "Machining of Titanium with Water Jet Assistance through the Insert," 7th American Water Jet Conference, Seattle, WA, August, 1993, pp. 801 - 812.

5 WATERJETS IN CIVIL ENGINEERING APPLICATIONS

5.1 INTRODUCTION

It is difficult to divide the use of waterjets in mining applications from those applications where the technique has, instead had a civil engineering purpose. This is because the two disciplines have many features in common, particularly in the excavation and removal of geotechnical material. Because of this dilemma it was decided to group the cutting of soil and concrete, and the procedures more commonly considered to be of a civil engineering use into one chapter, and those more clearly related to mining into a second. To make the mining distinction, the materials to be mined are considered to be rock, with the weakest category as a rule being considered to be coal.

Making this distinction has raised one unanticipated problem, since the earliest use of **hydraulic mining,** as the practice of using waterjets in excavation in known, was to remove soil. For this reason, and since the results correlate better with the studies of soil and weak rock removal, of more practical use now to civil engineers, this somewhat historical section has been included in the civil engineering, rather than the mining engineering section.

5.2 EARTH MOVING APPLICATIONS

The destruction of earth following any heavy rain, or during any severe storm at high tide continues to demonstrate the great power water has in moving earth. It is a power often applied with water under little pressure, yet results have a tremendous effect. This capability is not always recognized in modern applications, where use of higher pressures is argued for, without recognition of the capabilities waterjets have demonstrated at lower pressures in earlier and even current times.

The power that water has to break and remove material has been applied for many centuries; the Ancient Egyptians were familiar with the practice that is now known as "booming" [5.1]. In this technology water is first impounded in a form of reservoir and then, when a sufficient head has developed, a trap would be opened directing the water down a prepared channel to the valuable deposit. This form of ground sluicing was given the name of **booming,** because of the boom when the water first hit the sluice [5.2]. The technique is referred to by Pliny the Elder in his book on

"Natural Science" written in A.D. 42. The book describes the mining of gold deposits in the mountains around Leon in what is now northern Spain. The gold ore was weakened by fires before being hit with the water [5.3]. The water not only broke out the valuable mineral, but also, because of the way the water eroded the ore from the solid, separated it into individual grains. These grains were then carried down through a prepared channel, to the point that the valuable material could be separated out and collected.

It may be possible that a variation of this technique was known in earlier, Greek times. The reason for this comes from the way in which the gold was separated out and collected after the water has hit the deposit. In practice, the stream of water flowing down through the flume was directed along a series of trenches carved across the ore body, and it was during this that the water picked up the ore particles and carried them away from the deposit and into the trough. As the water entered the trough, the original turbulence and speed would be lost so that the heavier gold particles would settle to the bottom of the flow, and any roughness on the bottom of the channel would capture the particles. The Romans placed fibrous plants along the bottom of the flume to trap the heavy gold mineral [5.3] In earlier times the fibrous silex material was replaced by sheepskin and other hides. Strabos, writing in 7 B.C. has been reported [5.4] as describing this practice as being used as early as 4,000 B.C. in the country of Saones. It is easy, therefore, to deduce that it was the story of these sheepskins, laden with gold particles, which brought Jason and his Argonauts to Euxine, in around 1,200 B.C. in search of the Golden Fleece. The practice has been found to work so well that it has remained in use into the early twentieth century, when Spanish miners were still running the slurry form hydraulic mining over ox hides in order to collect the gold particles [5.5].

5.2.1 SOIL REMOVAL IN CALIFORNIA DURING THE GOLD RUSH

The Great California Gold Rush began when gold was found some 70 miles northeast of Sacramento in 1849. So rapid was the exploration and exploitation of the gold-bearing streams that the majority of the "easy" gold had been mined by 1851. The ore that was left lay, often in a sandy material, so that the practice of booming was initially resurrected. By 1850 it was costing $1.00 per miner's inch/day and this would cost, to run a single sluice, approximately $8.00/day [5.2]. At that time water was simply channeled to flow over the deposit and required that a man move the sluice physically up to the ore. In 1852, Edward D. Matthison (1823 -

1903) was almost buried, when the bank he was sluicing fell in. He apparently then sought a technique that would allow him to stand back from the working area. He joined with a hydraulic engineer named Chabot and changed the system so that water was first collected in a nail keg at the top of the 9 m high bank of weak sandy rock. Water was then fed through a 12 m length of 100 mm diameter, rawhide hose which Chabot built. To the end of this he fitted a trumpet shaped sheet brass nozzle, approximately 90 cm long and 4 cm in diameter, fitted inside in a wooden jacket. This initial test took place in February 1852 at the American Hill Mine near Nevada City [5.6].

By the time of the public demonstration of the technology in June 1853, written up in the Alta Californian paper [5.7], they had switched to a canvas hose and a solid metal nozzle, made by a blacksmith named Miller. Some indication of the productivity of the system can be indicated by the fact that they were, at 75 cents/miner's inch of water, paying a water bill of $153.00 a week. The operation nevertheless yielded the four partners of the program $50.00 a day in profit. It is perhaps pertinent to mention that a miner's inch is the volume of water flowing through a vertical 2.5 cm square hole under a defined head, typically on the order of 15 cm. To convert it to more modern units. 40 miner's inches are approximately equivalent to 1 cubic foot/second or 1.62 cu m/min, depending on your miner. (In terms of the above calculation, therefore a bill of $153 translates into a volume of 204 miners inches, or 8.25 cu m/min. Not a small operation).

The success of the monitors was such that over the period from 1871 to 1880, $121 million worth of gold was produced from the area around the Malakoff, and there were 425 individual companies at work. A typical monitor would produce 3-1/2 cu m per 24 hour inch day although it might at times get up to 10 cu m. The comparative economics of different methods of mining, at that time can be estimated from the following figures. A company which would pay a miner $4.00/day, gave the cost of cleaning a cu m of earth at $8.00 using panning, $4.00 using a rocker, and $1.00 using a long tom, while a sluice box cost $0.50. A miner would manually clean roughly 0.5 cu m/day using the conventional pan [5.8].

In 1855 the goose neck swivel was developed [5.10]. This was a flexible iron joint of two iron elbows which would allow horizontal movement as well as vertical movement of the monitor. The initial development was just for the horizontal movement, vertical movement coming shortly thereafter, and by 1870 a Giant, comparable with the modem machine came into use (Fig. 5.1). It is noted that the idea of putting a deflection plate in the water to swivel the monitor was credited to Dave Stokes a pit foreman in the Malakoff, who cleaned his shovel by placing his blade against the side of a jet of water, and who noticed that this caused the monitor to turn on the goose neck [5.11]. This in time led to the patent that was awarded to Superintendent Henry Perkins for this device. It was, also, the forerunner of the current device used in watering lawns.

Figure 5.1 Early hydraulic monitor.

Unfortunately, the success of the hydraulic mining operations in eastern California led to considerable controversy. The disaggregation of the rock by the waterjet action separated the individual grains of material and made them easier for the water to move. The heavier gold particles rapidly settled out, as was required, but this still left the finer particles of lighter clay and sand, suspended in the water. This fine silt was carried in suspension by the relatively rapidly moving streams of water, until it reached the navigable streams of the state. Here it would settle out. In the Yuba, Beam and American rivers it was estimated that, by 1890, some 155 million cubic m of debris had been deposited. This raised the beds of the rivers, relative to the banks, so that the channel of the river became smaller. Thus, when there was a heavy rain, the river would no longer fit in the channel and would overflow its banks, inundating the surrounding farmland and causing serious flooding. The flood of 1880, for example, flooded 43,000 acres at a cost of $2.5 million. This led to legislature and court action, the most famous of which was tried before Judge Sawyer in California. In 1884 he set a ruling which prohibited the dumping of debris into the navigable streams. At the time, it was said that at one blow the decision reduced the visible assets of the state of California by $100 million [5.11]. In fact if one compares the productivity over the period, one notes that the value of the gold produced from the diggings went from $121 million in the years between 1871 and 1880, to $38 million in the years 1881 to 1890.

5.2.2 THE EXTRACTION OF CLAY AND MONITOR DESIGN

Hydraulic mining practice rapidly spread from California, by 1862 it was in use in Idaho, (see Fig. 1.2) and subsequently it has been used in Alaska, Russia, China, Europe and the Yukon, where it is still practiced. The use of low pressure water to disaggregate and move loosely bonded material such as soil, China clay, or weak deposits of sand, has been developed around the world since that time. One example of this can be found in the English China Clay operations in Southern England, and has been described by Davies and Jackson [5.12].

In order to improve operational economics, Davies examined this process from two points of view. The over-riding parameter of performance was to minimize the amount of energy required to mine a unit volume of the clay. This in turn was controlled by the optimization of the nozzle design to improve the efficiency with which the water was delivered to the clay surface, and then to optimize the use of that water in creating a "**wash density**". This value is the amount of suspended material carried in the water as it flows away from the point where the water has eroded the bank. Once removed this slurry must be separated, chemically treated and dewatered to recover the valuable clay minerals, thus the higher the initial slurry density the lower these processing costs will be. Typical slurry densities vary from 1,005 kg m^{-3}, for hard deposits, to 1,200 kg m^{-3} for material which has been ripped and therefore weakened by a bulldozer.

Davies work showed that nozzle performance was optimized when the nozzle was kept within 100 diameters of the surface. The nature of the mining operation, where a soil/rock wall is eroded, from the bottom, mandated, from a safety standpoint, that such an operation be carried out remotely. This makes production a little more difficult to optimize, given that an operator at the machine might find it easier to steer the monitor to take advantage of cracks and fissures in the rock surface. By driving the water into these the rock can be brought down in larger sections and slurried nearer the nozzle where the energy is significantly higher.

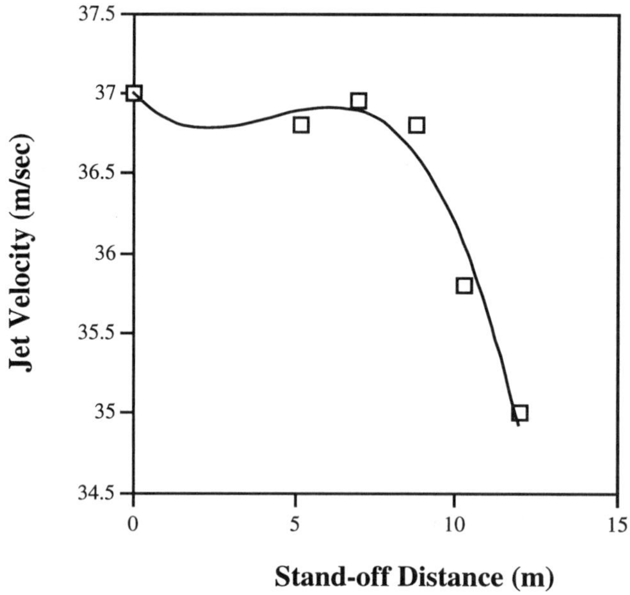

Figure 5.2 Break-up length of a waterjet from a monitor, shown by velocity change (after [5.13]).

A laboratory study indicated that of three initial nozzle designs tested (a taper, a design based on the Rouse recommendation and a nozzle design with a long throat) a tapered nozzle gave the most coherent jets, even at diameters of 2.5 to 3.18 cm [5.13]. This validates the conclusions from the Leach and Walker study [5.14] on nozzle design but at significantly larger diameters. The study did show, however, that the jet would retain over 90% of its initial pressure up to the point that it began to break into droplets, at which time the water very rapidly decelerated (Fig. 5.2).

This break-up in the jet structure had a marked effect on the amount of material which the monitor mined, as reflected in the density of the water flowing away from the operation (Fig. 5.3).

For tapered nozzles an equation was suggested for predicting jet pressure, as a function of distance. A least squares fit provided that the jet pressure was given by:

$$\text{Jet pressure} = \text{Initial pressure} \;.\; 240.\; (x/d_o)^{-1.19}$$

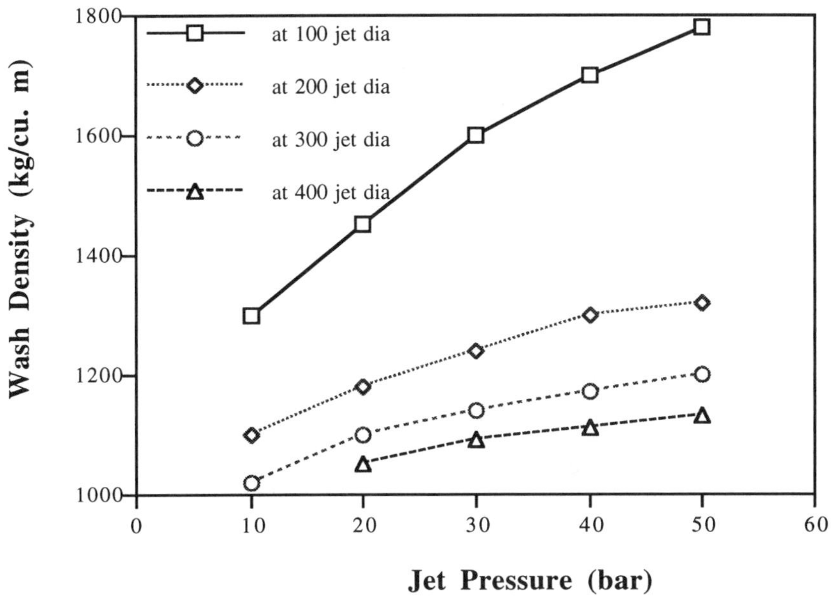

Figure 5.3 Production of material as a function of pressure and stand-off distance (after [5.12]).

where the target is more than 100 diameters from the nozzle. At this distance the mean jet pressure was found to be 25.7% of the maximum. The effective diameter of the jet increased with stand-off distance, following an approximate form given by:

$$\text{jet diameter} = (1 + 0.0086(x/d_o))$$

where the stand-off distance was less than 125 nozzle diameters, at greater distances the relationship was derived:

$$\text{jet diameter} = (-0.27 + 0.021(x/d_o))$$

Material removal rates in the original material were linearly tied to impact pressure, once a given threshold impact pressure (0.356 ± 0.09 bar) was achieved (Fig. 5.4). The value of the threshold pressure varied with the nature of the rock, depending upon whether it was in its original state, whether it had been blasted or whether it has been ripped and stacked by a bulldozer. In the latter two cases production rates were not tied to impact pressure.

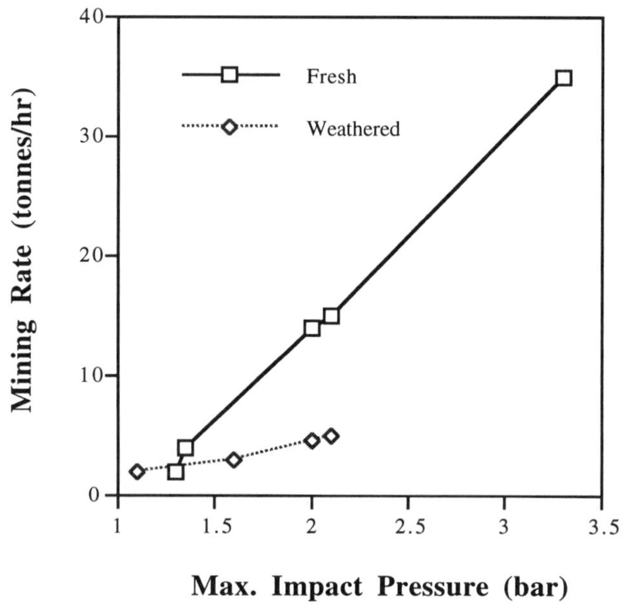

Figure 5.4 Material removed as a function of jet pressure (after [5.15]).

The above experimental data suggested that performance could be significantly improved where the stand-off distance was reduced. When this was tested in the quarry, the lowering of the distance from the nozzle to the clay bank from the conventional stand-off distance to one within one hundred nozzle diameters lowered specific energy by five to tenfold [5.12] to a value ranging between 1 MJ tonne^{-1} and 4 MJ tonne^{-1}. Such a conclusion required, when mining in a virgin area that nozzle diameters be increased from 25 mm, to 60 mm. It was suggested that a more intermittent operation be undertaken rather than a continuous jetting. Production was also controlled by the ability to remove blocks of material too hard and insufficiently changed into clay to be useful.

5.2.3 SOVIET EXPERIENCE IN SOIL REMOVAL

Similar conclusions have been reported where larger diameter jets were used for mining soil-like materials hydraulically in the Soviet Union [5.16]. The criteria for the definition of **effective jet length** was considered to be the distance over which the jet pressure was more than 50% that at the nozzle.

Table 5.1 Effective jet length (m) as a function of diameter and pressure (after [5.16])

Nozzle diameter (mm)	*Jet pressure (m of water column)*			
	75	*90*	*100*	*110*
60	1.5	1.52	1.54	1.65
80	2.75	3.20	3.50	3.80
90	3.50	4.00	4.50	4.90
100	4.50	5.20	5.50	6.00

A study of the pressure required to excavate different soils from surface quarries indicated that the following impact pressures would be required:

Table 5.2 Effective pressures for mining various soil types [5.16]

light soil and sandy loam	0.63 bar
medium soil and medium loam	2.25 bar
heavy soil and loam	3.90 bar
very heavy soil or sandy clay	5.80 bar
firm clay	8.00 bar
very firm material such as argillaceous shale	10.00 bar

As with the study in the clay, output is considerably affected by the diameter of the nozzle, and statistics on this performance were given as follows:

Table 5.3 Specific flow (Q)* and pressures (in meters of head) required for soil removal (after [5.16])

Soil type	*Height of bench (m)*					
	3-5		*5-15*		*< 15 m*	
	Q	*P*	*Q*	*P*	*Q*	*P*
preloosened, not bedded	5	30	4.5	40	3.5	50
fine-grained sand	5	30	3.5	40	3.5	50
powdered sand	5	30	3.5	40	3.5	50
light sandy loam	6	30	5.4	40	4	50
friable loess	6	40	5.4	50	4	60
peat	6	40	5.4	50	4	60
medium grained sand	6	30	5.4	40	4	50
variously grained sand	6	30	5.4	40	4	50
medium sandy loam	7	40	6.3	50	5	60
light loam	7	50	6.3	60	5	70
compact loess	7	60	6.3	70	5	80
coarse grained sand	7	30	6.3	40	5	50
heavy sandy loess	7	50	6.3	60	5	70
medium and heavy loams	9	70	8.1	80	7	90
lean flowing clay	9	70	8.1	80	7	90
sandy gravel	12	40	10.8	50	9	60
semi-fat clays	12	80	10.8	100	9	120

*Specific flow is the volume of water (cu m) required to remove 1 cu m of soil.

As an example of the typical output from such a monitor, working in a bench less than 5 m high, the monitor might use 100 liters/sec to mine 336 cu m a shift in light sandy soil, but would only produce 96 cu m/shift in sandy soil where there is more than 40% gravel in the soil.

Overall production from a monitor is controlled by the distance the monitor must move between operating points. In tighter ground this may be as close as at 5 m intervals, while in softer ground the move may be as great as 15 m. It is recommended that the slurry gathering sluice be moved with the monitor and the operating motion of the monitor be such that it works from the section closest to the sluice outwards. It is important to work with the cut bank at as high a slope as possible to keep the mined material within the maximum operating potential of the jet, and to allow a more efficient use of gravity in helping to bring down and break the material.

Hydraulic mining started in Russia well before the October revolution [5.17]. In 1890 it was used in the placer gold deposits at the Lena mine near Lake Baykal, where it was reported to lower production costs by a factor of 2.5. In 1914 waterjets were used by Klasson for the processing of peat for the Electroperedacha Electric Power Station (now the R.E. Klasson Hydroelectric Power Station). After the first World War it is reported (ibid.) that one third of all industrial peat in the Soviet Union was extracted by hydraulic mining. By 1931 the use of high pressure waterjets in construction was first used as a means of excavating for the Dneprostry Dam and then for the excavation of the Moscow Canal beginning in 1934, although it was only used to excavate about 5% of the total volume.

Following the Second World War the magnitude of the hydraulic operations increased, with over 60% of the excavation for the hydraulic projects on the Don, Volga and Dneiper rivers being through use of waterjet removal. Over 100 million cu m of mainly sandy soil was mined, moved and selectively redeposited in this work. It was found easier to work with the sandy type of soil since it has little cohesion and can thus be easily broken into particles, which can be transported by the water. Cohesive clay is more difficult to mine and move, since the clay does not disintegrate as easily, but must be liquefied into a fine slurry by closely adjacent passes, since large lumps of clay cannot be easily picked up and transported. Clay also has damaging effects on the cutters and mud pumps of the mining and transport system.

Contrasting conditions must be evaluated in the use of the monitors, since safety constraints suggest that the monitor should be no closer to the face, than the height of the cliff. However, since the mining of benches less than 3 m tall is not economic and the jet energy dissipates after traveling 3 - 4 m, through the air, this argues against the need to get the monitor as close to the face as possible in order to get most effective mining. The introduction of remotely operated monitors close to the face allowed a reduction of pressure by a factor of two in mining cohesive soils, reduced water consumption by 55 - 80% and power required by 65 - 78%.

As with the experience in the English China clay operation, removal of mined material can slow the work. The Soviet solution was to add a second monitor with a larger, but less powerful jet, which sluiced away the material. Presoaking the face also improved productivity. Where faces are presaturated the height of ground which can be worked can be increased. Since the face is mined by first undercutting the bank with a horizontal slot at the base of the slope, and this will use up to 50% of all the energy needed to mine the bank, this can significantly improve productivity.

5.2.4 TUNNELING IN SOFT MATERIAL

The process of hydraulic erosion of soft rock is not confined to surface extraction of material. The sandstone rock underlying Minneapolis, MN is a soft material which can be mined by a pressure of approximately 17 to 30 bar, through a nozzle measuring between 0.9 and 1.8 cm in diameter [5.18]. This capability is used in driving the tunnels under the city. Special trains are assembled which carry the equipment to pressurize the water and feed it through nozzles to the cutting face of the tunnel. The streams of water are first used to cut the rock out into blocks which are dropped onto the floor of the tunnel. Once the blocks are on the floor they are then broken into particles under the waterjet impact, and then the water and fine material are collected and pumping away. Where the rock is slightly harder it may be initially blasted from the solid, and then broken into a slurry for pumping out of the tunnel. In normal operations however the jets can cut into the sandstone, eroding it away in small enough pieces to be immediately slurried. Advance rates of up to three 1.5 m rounds can be achieved in an eight-hour shift.

5.2.5 REDUCED FLOW RATE STUDIES

In the United States, recent studies on the removal of soil-like materials have largely been carried out working at higher pressures and significantly smaller jet diameters than those referred to above [5.19]. A study by the US Army Corps of Engineers examined the ability of waterjets to cut soil at pressures of up to 550 bar, through a 1.0 mm diameter nozzle. Stand-off distance was held constant at 150 nozzle diameters, which, as the discussion above indicates, may have been beyond the critical distance for that nozzle. The study concentrated on determining the effects of soil properties on the jet penetrability and cutting action.

An initial experiment (Fig. 5.5) at 70 bar showed that with no jet movement it was possible to penetrate between 15 and 62.5 cm into various soils within a 20 second exposure. The amount of penetration was found to vary with the degree of cohesion of the soil (the greater depth being achieved in soil that had none). As traverse speed increased the depth of cut declined at varying rates (Fig. 5.6). It was observed, again as a function of soil type, that the zone of jet influence around the traverse line had a width ranging from 30 - 65 mm.

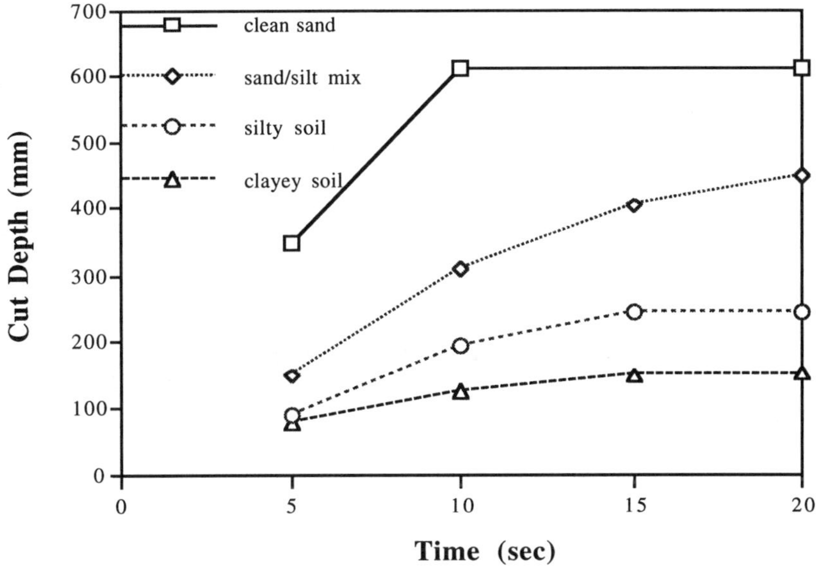

Figure 5.5 Depth of jet penetration into soil, as a function of time (after [5.19]).

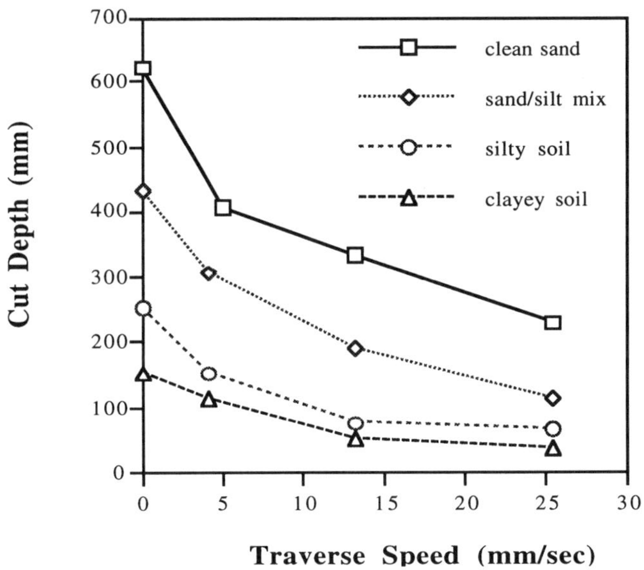

Figure 5.6 Cutting depth in soil as a function of traverse velocity (after [5.19]).

This penetration was considered to increase with an increase in the amount of water that was already in the soil. Thus the more saturated the soil was the easier it was for erosion to occur. An experiment to quantify this indicated that an increase of 50% in soil saturation increased the depth of penetration by up to 150% (from 25 to 62.5 cm). However, this, again, was found to be a function of soil type (Fig. 5.7) with finer sands and clays being more sensitive to the effects of density and permeability than the medium to coarser sand samples which were tested.

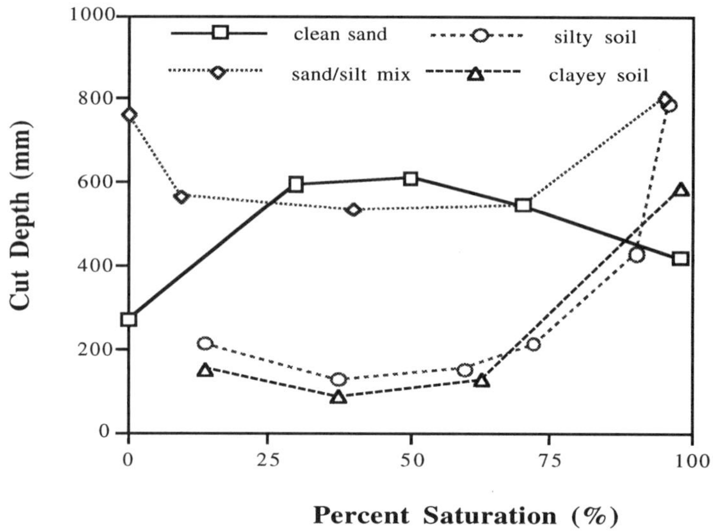

Figure 5.7 Effect of soil saturation on cutting depth (after [5.19]).

5.3 THE MECHANISM OF SOIL REMOVAL

The lessons that can be learned from the above case histories and experiments is not always clear to those who are studying the subject today. Where the jet encounters large grained, highly porous material, the jet can penetrate through the cracks and gaps in the material into a significant volume surrounding the impact point. This penetration of the water will either increase the pressure and complete the saturation of any water that is already present, as in saturated soils, or it will flow to saturate the soil. In so doing it surrounds individual grains. Once the soil is saturated, then the continued impact of water on the surface will pressurize that already in the soil. This will cause it to try to expand and to move. In doing this it causes "lift-off" of individual grains, and once separated can be carried away by the water flow. This is **erosion.** The water can also concentrate stresses around the hole, causing the growth of large cracks in the material.

A simple experiment can be used to illustrate this principle. If an air rifle is loaded with a small pellet, and the gun is fired at a block of Plexiglas, no damage is visible to the naked eye. If, however, a small hole is drilled in the block first (Fig. 5.8), and this is filled with water, then the impact of the pellet at the top of the column will cause cracking in the block, radiating from the hole (Fig. 5.9). It is interesting to observe that very small notches on the side of the hole can be used to direct the growth of cracks away from the original hole (see the left of the two drilled holes in Fig. 5.9).

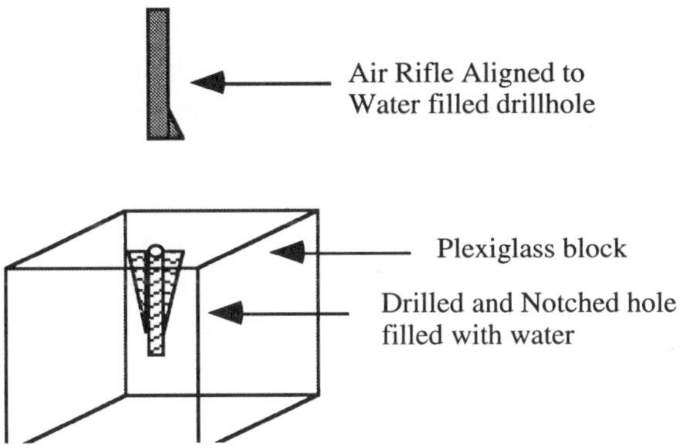

Figure 5.8 Schematic showing experimental layout for Plexiglas impact.

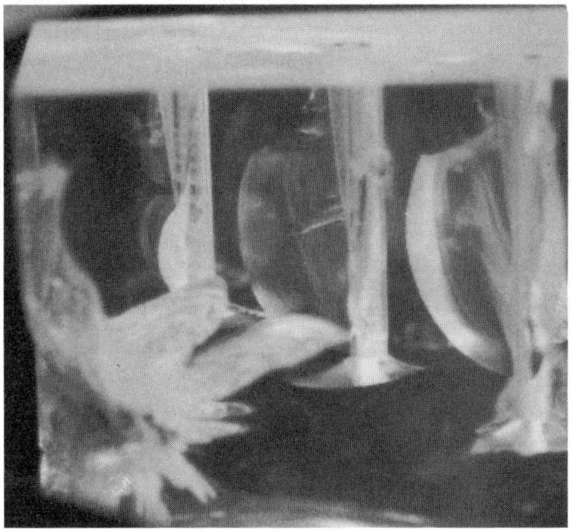

Figure 5.9 Result of pellet impact, showing crack growth from holes.

This method of material removal works well when the grains of the rock are large and the water can easily penetrate into the material. Such a process becomes less effective in tighter soils, where there is some cohesion to resist the uplifting effect of the water. Further, Griffith [5.20] has shown that cracks grow more easily from longer original flaws than they do from short ones. Brace, in turn, has shown that the dominant crack size in rocks (which are a form of cohesive soil) is related to the grain size [5.21]. Thus it is easier to grow cracks in large grained material, than it is in small grained material, because of the greater length of cracks in the former. It is these structural features of the material being cut that effect the ability of the jet to penetrate and thus to remove material.

This can be illustrated by considering what happens when a soil is cut by a waterjet, as the viscosity of the water is changed. As the concentration of a long chain polymeric additive such as Polyethylene Oxide, for example, is increased in a waterjet, the cohesion of the water increases. (To the point that water can be induced to run uphill if it contains sufficient additive.) The cohesion of the jet will initially improve the energy transfer between the nozzle and the soil surface and reduce the friction losses in the delivery system (a point further addressed in Chapter 10). However, if the concentration is further increased a point is reached when the cutting ability of the water starts to reduce (Table 5.4). This occurs when the water becomes sufficiently viscous that it finds it more difficult to penetrate into the gaps between the grains of the sand.

The effectiveness of the jet and the polymer can be changed by the degree of compaction of the soil. For example consider a jet at 2,100 bar, diameter 0.3 mm cutting into clay at a 25 cm stand-off distance.

Even though the pressure is significantly higher in this latter case, and the nozzle is held much closer to the surface, the depth of cut is much less. The reason for this apparent change is that the narrow, polymerized jet cuts a very narrow slot which collapses in on itself, reducing the apparent depth of cut. One must determine the object of the exercise therefore in designing the most effective jet for cutting different materials. In this particular instance the actual initial depth of cut, before the slot collapsed in the loose clay with the polymer was some 50 mm deeper than that given in the table above. This point will be resurrected at later stages in this work.

Table 5.4 Slot dimensions for traverse tests on soils with varying concentration of polymeric additive (after [5.22])

Type of soil	*Loam*		*Sand*		*Clay*	
traverse speed (cm/s)	15	90	15	90	15	90
Water						
depth of cut (mm)	58.4	53.3	38.1	45.7	66.0	40.6
width of cut (mm)	104	96.5	102	99.1	99.1	107
500 ppm Nalco BX 254						
depth of cut (mm)	40.6	30.5	41.9	44.5	39.4	47.0
width of cut (mm)	112	102	107	127	117	81.3
1000 ppm Nalco BX 254						
depth of cut (cm)	27.9	22.9	40.6	39.4	31.8	19.0
width of cut (cm)	96.5	66.0	137	94.0	107	96.5

Moisture content: clay 17.1%; Loam 13.1%; Sand 3.2%
Jet pressure 175 bar, Nozzle Diameter 1.5 mm; Stand-off distance 1.67 m

Table 5.5 Depth of cut into clay and the effect of compaction on jet performance (after [5.22])

	Compact clay	*Loose clay*
water	12.5 mm deep	50 mm deep
500 ppm Nalco 625	25.4 mm deep	18.7 mm deep

5.3.1 FROZEN SOIL AND THE CHANGE IN CUTTING MECHANISM

The basic structure of the soil may apparently stay the same, but the cutting characteristics of the material change significantly if the soil is now frozen [5.23]. Working with a waterjet on the ice can result in several modes of failure depending on the pressure. The presence of the ice in the soil changes the material erosion characteristic from an easy movement of grains, to a required chipping and fracturing action to break the ice. The ice will chip under jet attack, and, with sufficient pressure and flow rate, the water can penetrate into weakness planes and break out relatively large pieces. At higher pressures the jet will melt the ice and create a relatively narrow slot through the material. This latter case is the most difficult to deal with since the narrowness of the cut can cause almost immediate refreezing. This would argue that intermediate pressures and flow rates are more effective than higher pressures in cutting through ice, since the wider slots do not refreeze as rapidly.

Experiments in cutting frozen soil require, however, some additional design changes if the waterjet cutting is going to be effective. The presence of larger pebbles in the matrix, which can easily be moved when the soil is relatively loose, become more of a problem in frozen soil where the matrix is more coherent. The material becomes more like a concrete and the jet path must follow a pattern which cuts at least twice the maximum particle width, if the slot being cut is advanced down into the material. Such a width will remove particles ahead of the nozzle body so that it can advance.

The Bureau of Mines has examined the use of higher flow rates, up to 151 lpm, at a pressure of 310 bar in order to examine the potential for driving tunnels in the gravel and for mining purposes [5.24]. Interestingly the least efficient cutting, at approximately 240 joules/cc occurred with sample blocks in a laboratory setting. When tests were carried out at a lower pressure but in the actual tunnel then the specific cutting energy fell to 157 joules/cc at pressures below 290 bar. In an open pit experiment the waterjets were able to mine material with energy levels as low as 14.4 joules/cc. It is interesting to comment that a man with a pick can, with knowledge, mine relatively soft material with a specific energy consumption of 4 joules/cc.

These figures are of interest because of the considerable amount of excavation that must take place in Arctic regions. The major supporter of work in this area was, for a period, the US Army Cold Regions Research Engineering Laboratory (USACRREL) [5.25]. Their study began [5.26] with a contract through Foster-Miller to Exotech, a company who used an ultra-high pressure water cannon as the excavating tool. The decision to use this device was based on the Foster-Miller conclusion that continuous jets provided low energy effectiveness, high power demands and large water requirements. Water cannons on the other hand generated a short, high intensity pulse of water, of limited volume but high power density. (At the time that work was being carried out it was concurrently being concluded [5.27] that in the cutting of limestone it was much less efficient to cut with a water cannon than it was to cut the rock with a continuous jet - the reason for this will be discussed in Chapter 7).

Using the Exotech machine water was fired through a 5.59 mm diameter nozzle at a jet velocity of 610 m/sec yielding an excavation specific energy on the order of 7.6 to 43.4 MJ/cu m (this unit is equivalent to the same number of joules/cc). Smaller diameter continuous jets cut the frozen soil at significantly higher values, even though the material was quite fine grained.

Subsequent work confirmed that continuous jets were likely to require more energy for material removal than did the pulsed jet. The work at that time did not, however, examine the use of adjacent slots for kerf removal (see Chapter 9) or the change in attack mechanism to erode the softer matrix from around the hard pebbles, which will be discussed later in this chapter when talking about concrete removal. Thus the conclusions which were drawn must be taken within the limited context of the comparisons which were made. However, despite the higher horsepower, Mellor concluded that continuous jetting units would ultimately be cheaper to obtain. From the data obtained he then developed a predictive equation for the depth of cut to be obtained as:

$$D = D_o [1 - \exp(-K_2 \cdot P_o^2 . d/v)]$$

where:

- D is the depth of penetration
- D_o is the penetration depth at zero traverse speed
- K_2 is a material constant
- P_o is the jet pressure
- d is the nozzle diameter
- v is the traverse velocity

From this Mellor derived an optimum cutting condition for which

$$D = 0.44(1000d - s)$$

predicated on the assumption that the maximum depth of cut which a nozzle will achieve is 1,000 times its diameter, based on jet throw, and this is then reduced by the stand-off distance(s) of the nozzle from the surface.

It is worthy of comment that in 1972 Mellor predicted the use of waterjet assisted mechanical excavators, although these were not tried.

5.3.2 CUTTING AND REMOVAL OF ICE

A continuing interest has developed over the years in using high pressure waterjets to remove ice, or to cut through it. The interest has arisen from a number of different perspectives, aimed at resolving sometimes quite disparate problems. Regrettably as yet there has been no consistent application of the technology. The developments undertaken do illustrate some of the potentials for the tool, if they can be improved.

A perhaps unique problem which ice introduces into waterjet cutting is more pertinent to the use of smaller diameter, higher pressure jets, rather than to larger diameter, lower pressure systems. At high pressure the jets will cut, or may melt, a slot in the ice which is sufficiently narrow that, within a relatively short period of time, the slot may refreeze and close. While there is an application, described below, where this is helpful, in the more general case this is one disadvantage of the technique.

As part of the study of frozen gravel referred to above [5.25] USACRREL funded a small study of ice cutting at UMR [5.28]. The tests, carried out at relatively low pressures indicated a steady increase in penetration with pressure (Fig. 5.10).

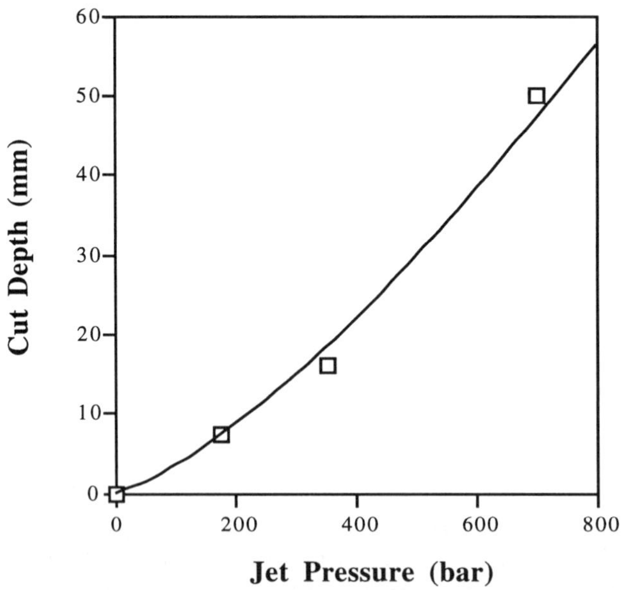

Figure 5.10 Low jet pressure on depth of cut in ice [5.28].

In part as a result of this work a study was carried out on floating ice at higher jet pressures. Because the floating ice failed, dumping the equipment in the water, the second half of the test took place using blocks of ice drawn under the nozzle. Results (Fig. 5.11) indicated that the jets were able to cut to a significant depth, with such cutting being better accomplished with a single pass rather than multiple jet passes, which at higher pass numbers gave only half the depth for equivalent residence time.

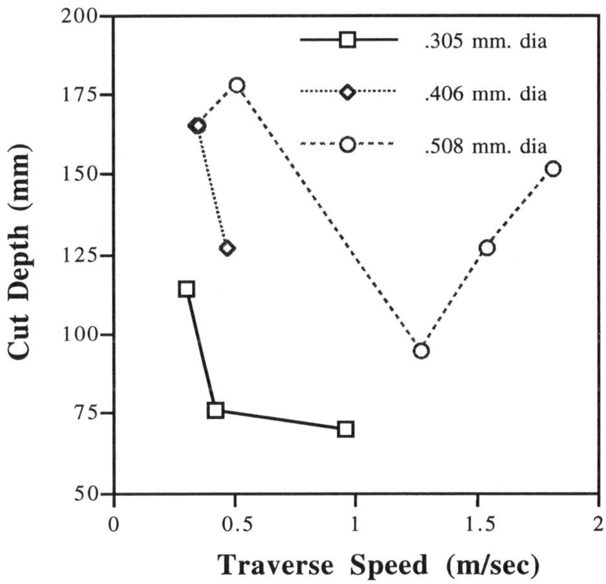

Figure 5.11 High jet pressure effect on depths of cut (after [5.29]) jet at 0.508 mm diameter was operated at 6,900 bar, the others at 4,140 bar.

Cutting depths of up to 2,000 diameters, significantly better than the 1,000 diameters predicted in the earlier work, were achieved with the potential for cutting to depths of 3,000 diameters being established. It was clear, particularly with the larger diameter data, however, that there was considerably more scatter and variation in the data than is found in cutting other material.

The data was analyzed to determine if waterjets could assist river icebreakers which need to cut ice up to 0.6 m thick at a traveling speed of 2.57 m/sec, although a cut of 0.46 m at a speed of 1.54 m/sec would be possibly acceptable. Using the equations derived, and based upon an extrapolation of those results it was determined that a 1,400 bar jet issuing from three 5 mm diameter nozzles would be required to meet the required slots ahead of the breaker. This would require some 4,500 kW. By comparison a towboat on the river uses about 2,250 kW, and 3 mechanical saws to give the same cutting would require 1,500 kW while lasers might need 5,700 - 11,000 kW. These power levels were considered excessive and the idea was discontinued. Waterjets were, however, considered to be of advantage in assisting in emplacing pipelines in the Arctic. In light of this set of results it is interesting to note earlier papers from the Soviet Union [5.30, 5.31 and 5.32].

Although the data obtained was less broadly based than that of the American study, the investigators found that waterjets at a pressure of 300 to 400 bar could satisfactorily cut the ice [3.30] and that, as a result it was possible to design an ice-breaking ship capable of dealing with ice floes up to 0.5 m thick [3.31]. The design used the high pressure waterjets to cut the ice ahead of the ship into strips, which were then pushed under the ice pack, leaving a free channel for the ship. Interestingly the study suggested that the jet effective length should be considered to be only 400 to 500 jet diameters, rather than the longer cutting depths suggested from the American study.

Tests of a thermal lance, similar to those found in granite quarrying [3.32] found that while this also could be used to cut through the ice, the penetration rate was only 10 to 12 cm of penetration in 1.5 minutes. This is significantly slower than that reported above for waterjet cutting. This is even though the flame temperature reached 2,000 ^{o}C in the lance, and the gas velocity was in the 1,200 to 1,500 m/sec range. The maximum rate of material removal was 69 cc/sec but the large width of the cutting nozzle made a much wider cut than that required with the waterjet system.

The work carried out in America [3.29] had also examined the potential of high pressure jets for removing the ice from pavement and concluded that the power and energy requirements as well as the water quantities involved would be also be too high for this to be practical (Fig. 5.12).

Investigators at the NRC, initially funded by USACRREL, began their own study and modified the equation which Mellor had proposed to become [5.33]:

$$D = [-50 + 15.58(\frac{H_p}{\sqrt{v}})]$$

where H_p is the power contained in the jet in kW.

D is the depth in mm

v is the traverse velocity in m/sec

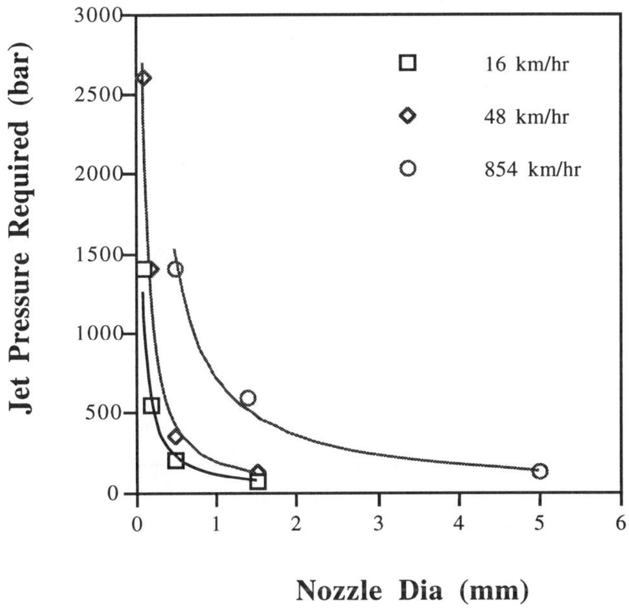

Figure 5.12 Predicted performance in the removal of ice from pavements [5.29].

Their interest was continued toward the de-icing of lock walls in winter, an offshoot perhaps of pavement cleaning. The experiments again took place as a cooperative investigation with USACRREL [5.34]. Commercial high pressure waterjet cleaning units were used in these trials which established that while waterjets could, in fact, clean the ice from the lock walls they were more expensive and slower than mechanical saws and chemical means which were more traditionally used. Depths of cut achieved in the tests (Fig. 5.13) were however quite significant for the conditions of test. Nozzle diameters of around 2.2 mm at a jet pressure between 600 and 700 bar were used in the tests.

The consequence of this relatively negative set of results has been the virtual stopping of the study of waterjet use in cutting ice. Vijay [5.35] subsequently examined the use of high pressure waterjets as a means of drilling ice and frozen ground. He found that the tendency of the ice to spall under impact inhibited drilling and there was considerable scatter in the data. It was concluded that a rotational speed of 400 rpm was most effective, using jet pressures of 400 bar and a dual orifice nozzle to drill the holes.

The increased size of the data base did, however, indicate that a better prediction for the penetration could be obtained if the equation was further modified to:

$$d = 1.3\left[\left(\frac{H_p}{\sqrt{v}}\right)^{0.55}\right]$$

where d is in cm,
H_p is in kW,
and v is in km/hr

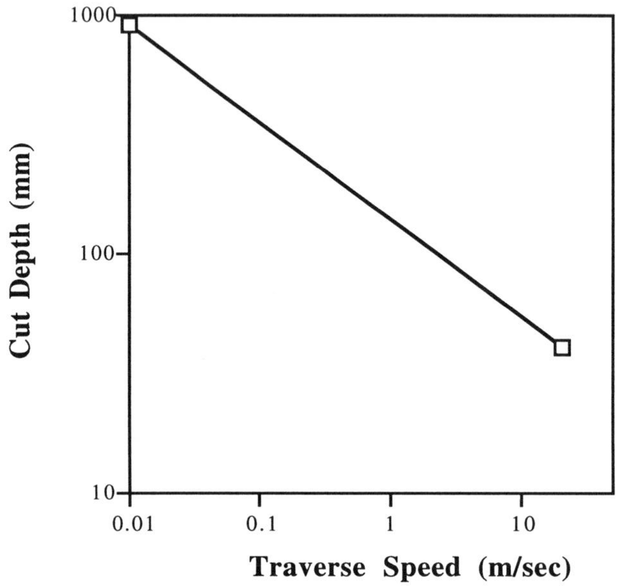

Figure 5.13 Summarized trend for ice cutting from the walls of river locks (after [5.29]).

However, despite these results, and recent word of further investigations in the Soviet Union, there appears to be little continuing work in using waterjets in this area. The Strategic Highways Research Program in the United States has, however, become concerned at the 10 million tons of salt used on the highways of the United States for de-icing, and initiated a program to find better ways of removing ice from roads.

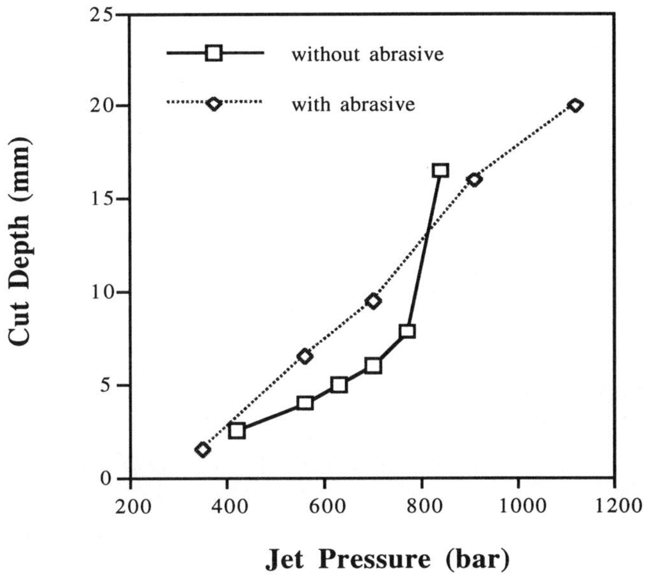

Figure 5.14 Effect of abrasive injection on jet cutting of ice (after [5.36]).

It has therefore initiated a program at Michigan Technical University to examine the use of abrasive laden waterjets as a means of improved removal of ice. The program was in its initial stages in 1991, but the results to date appeared to have some practical interest [5.36]. The addition of the abrasive increases the depth of cut (Fig. 5.14), but only at lower pressures. The work is, as yet, only in its preliminary stages, so that a full explanation for the reasons for the curve have yet to be given. The investigation has, however, revealed that the addition of 20% ethylene glycol inhibits the slot from refreezing closed.

Ice is a difficult material to quantify, since its characteristics change significantly with the way it forms on a surface and how it is treated after that. A simple experiment was carried out to determine the validity of some of the conclusions cited above.

A self-rotating waterjet fitted with two nozzles, 0.925 mm in diameter, was used to remove a sheet of ice some 50 mm thick at the UMR High Pressure Waterjet Laboratory (Fig. 5.15). It was found relatively easy to remove the surface ice, down to a layer which varied between 10 and 12 mm thick. This lower layer of ice was the first to be laid down during the storm, and was firmly bonded to the underlying concrete. Thus while the overlying material could be removed at pressures of around 200 bar, the jet pressure had to be increased to approximately double that value to cut down to and clear the underlying concrete. This latter value appears to agree both with Soviet and Canadian results reported above.

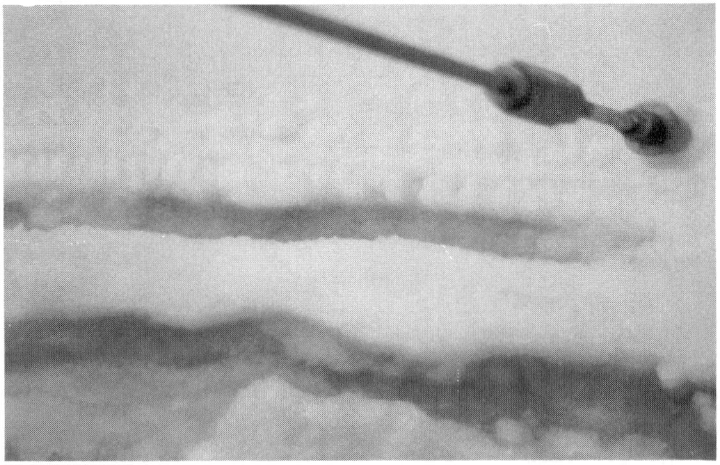

Figure 5.15 Cutting ice with a self-rotating waterjet assembly.

An attack perpendicular to the surface did not remove as much material as when the lance was held at an incline to the surface. This was, perhaps, because the method of most effective removal had two parts. The first was to delineate free edges around a block of ice adhering to the concrete. Once this free surface had been established then the jet was directed in under the sheet along the concrete interface. In this way blocks of ice could be freed from the surface and removed more effectively than by individually eroding all the particles of ice which had been deposited. There was a significant improvement in cutting performance (observed only qualifiedly) at jet pressures above 450 bar, similar in scope to that reported by the Michigan investigators. Conceivably this difference may be related to the warmer temperature (only slightly below freezing) at which the UMR experiment was carried out.

One interesting example of waterjetting where the refreezing of the hole works advantageously is illustrated by some experiments carried out by researchers from Flow Research. Considerable interest has developed in providing fresh water to Arabia by dragging large icebergs to the coast, and there allowing them to melt into reservoirs. In order to tow the bergs, however, some means of attachment is required. At the same time icebergs are a considerable threat to shipping off the east coast of Canada. These icebergs can weigh up to 20 million tonnes, and can roll so that attaching a cable to allow them to be towed is a dangerous task, best carried out by a remotely operated vehicle (ROV).

Studies had suggested that the average iceberg in the Labrador Sea has an average weight of 900,000 tonnes, a drag coefficient of 0.55 and a keel area of 36,000 sq m. Thus a tugboat capable of a 50-tonne bollard pull would be able to move this berg at an average speed of 0.14 knots. Pulling such a mountain would require a substantial anchor, and it was this need which led to the use of the high pressure waterjet lance. An anchoring tool was designed some 10 cm in diameter and 4.5 m long, which would drill a hole into the ice, and then freeze the anchor in place using liquid carbon dioxide. Lab tests indicated that a four nozzle array would be able to drill the ice, at a pressure of 1,400 bar and at a speed of 0.5 m/min. The low thrust levels required for the drill meant that it could be manipulated by an ROV, and a field experiment was accordingly undertaken.

Experiments were carried out, on icebergs off the coast of Newfoundland using a slightly smaller tool for the initial tests. Because of the difficulty in holding the ROV against the ice wall drilling rates of only 10 - 15 cm/min were achieved on the first trial and the tool, which desirably should be 4.5 m into the ice, could not totally penetrate the berg. A first test, with a hole some 1.8 m deep had an anchor length of 1.2 m and was pulled at 30 tonnes for 75 minutes before the anchorage failed. The second was 2.7 m deep with a 2.1 m anchor length. Failure was at 60 tonnes after just under 3 hours of loading at increasing pull. The tests indicated the practicality of using waterjets to drill anchorage into icebergs for the installation of haulage cables, which could then be used to move these ice mountains [5.37].

5.4 APPLICATIONS OF WATERJETS IN SOIL EXCAVATION

5.4.1 EXCAVATION OF THE BAR-LEV LINE

An interesting example of the speed and effectiveness with which waterjets can remove relatively soft soil formations arose during the 1973 war between Egypt and Israel. The Israeli defensive positions at the Suez Canal were mounted behind an earthen and sand barrier known as the Bar-Lev line. Egyptian intelligence had determined [5.38] that the Israeli Army had assumed it would take 24 hours for this barrier to be breached, and a total of 48 hours for the Egyptian tank forces to successfully penetrate the line. The response time of the military units was planned accordingly.

This time estimate was based upon the time assumed to be required to make a breach 6.8 m wide through the barrier, an excavation which must move 55 cu m of material. A total of 60 such holes would be required for effective penetration. Using either bulldozers or explosives it would have taken 10-12 hours to create a hole. The Egyptian Army, using an idea proposed by a junior officer, used high pressure water monitors and made the holes in 3 to 5 hours, gaining a considerable initial military advantage by this means.

5.4.2 SOIL EXCAVATION AND BARRIER WALLS

The United States is one of many countries faced with problems of containing potentially leaky contaminated waste dumps. An innovative solution to sealing such sites was discussed at the Waterjet Conference in Rolla, in 1983 [5.39]. Once a site has been filled with contaminated material, the risk is always that the containment will fail, and that spilled material will migrate into the surrounding water, from which it can pass into the drinking supply.

The idea behind the method is to break the waste site free of the surrounding solid, and then to float this "block" upwards in a bath of injected slurry material, such as soil bentonite and water. The slurry material would then form an impermeable barrier around this site. It must be confessed that this idea was greeted with some skepticism by the current author when it was initially proposed. It was therefore fitting that the results of the successful practical demonstration of the project were given at a Conference for which he acted as host and editor.

The role of the waterjet, or slurry jet in this instance, was to create the initial notched holes under the site which could then be enlarged to create an underlying free surface beneath the waste dump (Fig. 5.16). A cost comparison of conventional slurry walls (discussed in 5.4.3) and this technique indicated a range of conditions where the cost of cutting down to an impermeable rock with the grout walls would be too expensive (Fig. 5.17).

The initial intent, as shown above, was to create a perimeter separation using a jetting technique, and then to pump grout into the injection wells and thus create a horizontal fracture, which would isolate the site. However, in the first field test it was decided that the injection of the grout would be sufficient to create the fracture and to lift the ground without sidewall cutting. This required a careful preparation of the underlying fracture, and this was created by pressurizing notches cut into the ground from the bottom of seven cased wells drilled down to the required level. To make the notches some 6.9 m below the surface a jet containing a 2% bentonite was used to cut slots some 0.6 m diameter into the surrounding ground. The jet issued through a 12.7 mm diameter nozzle, and, by keeping the hole filled with compressed air during the cutting, notch depths could be increased to 1.2 m and then gradually out to a distance of 3.9 m radially from the well, the planes intersecting to cover the floor of the projected block.

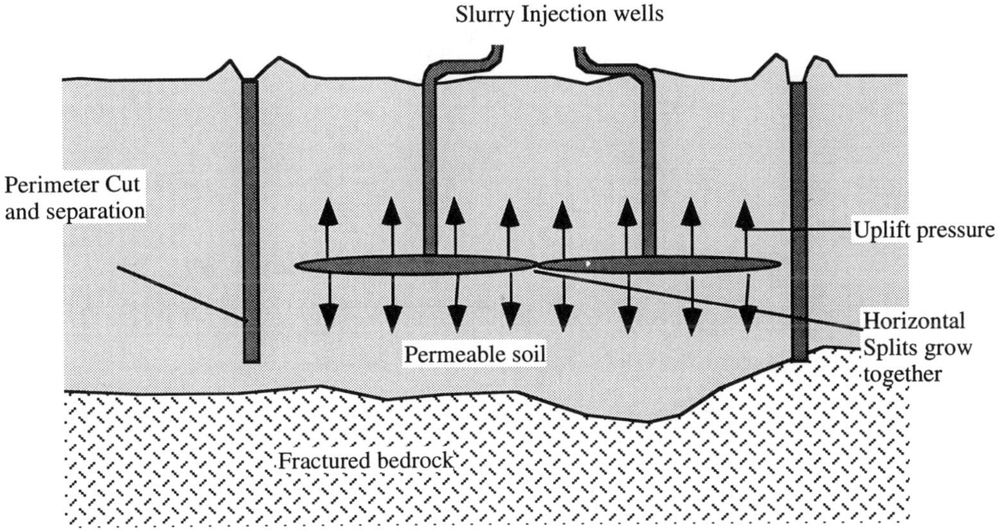

Figure 5.16 Method for block movement over waterjetted slots [5.39].

Jet pressures were quite small for this operation, the highest pressure of 10 bar, was required to cut an initial 1.8 m diameter notch in the central injection well. The notch was then propagated by pumping the slurry into the hole at a pressure of 1.7 bar to grow the notch until it intersected the adjacent notches. Once the underlying plane had been established some 54 cu m of gout was pumped under the site, raising the block on the "sea of slurry" by a distance of 280 mm. The jet notching, crack growth, and block displacement had been successfully demonstrated in a relatively weak material.

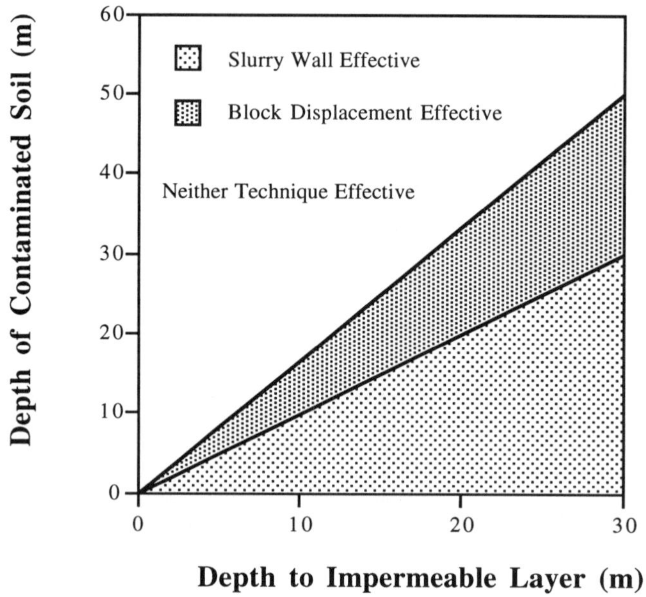

Figure 5.17 Cost effectiveness of block displacement versus a slurry wall [5.39].

5.4.3 GROUT WALLS AND ARTIFICIAL PILE CONSTRUCTION

Creating artificial walls by lifting a block of material over a fill of slurry is one way of establishing barriers. A more simple way is to merely install a grout wall or curtain around or under the required location. Such a wall can form an effective barrier, particularly when it is needed to stop water percolating through the ground. Depending on whether the wall must be rigid or flexible, a variety of different mixes can be used to fill the slot, ranging stiff cement, to a very flexible, but weak bentonite mixture.

The first moves to create such a barrier, using jetting, were reported at the second international waterjet meeting in Cambridge [5.40], although the initial trial of the method was in 1962, and the first Japanese patent was obtained in 1970 [5.41]. In conventional grouting around an excavation site holes are drilled into the soil. Thin cement grout is then pumped into the ground, under the hope that this will permeate the ground evenly around each borehole and create a set of adjacent cylinders of solidified material which will join together to create a barrier to the penetration of water or other fluids. While this can be a very effective method of grouting [5.42] it can also be ineffective should the grout find a weak channel along which to flow. In this latter case cylinders may not form, or may be misshaped so that they create an incomplete barrier around the injection points.

To overcome this problem Japanese investigators first tried cutting a channel through the ground with a high pressure waterjet and filling this with thin cement. This did not prove completely effective, in its original form since the waterjet is required to cut inside a hole which has been filled with a bentonite slurry in order to keep it stable during the grouting process. The thicker bentonite fluid reduced the range of operation of the jet system (Fig. 5.18). To overcome this problem the waterjet was sheathed with a surrounding column of air (Fig. 5.19). The waterjet nozzle had a diameter of 2 mm, and this was surrounded by an air jet issuing from an annular slit 1 mm wide. The jet was operated at a pressure of up to 700 bar, at a maximum flow of 56 lpm. Air pressure was 7 bar. The effect of increasing air volume can be seen on the length over which the jet retains 50% of its pressure (Fig. 5.18).

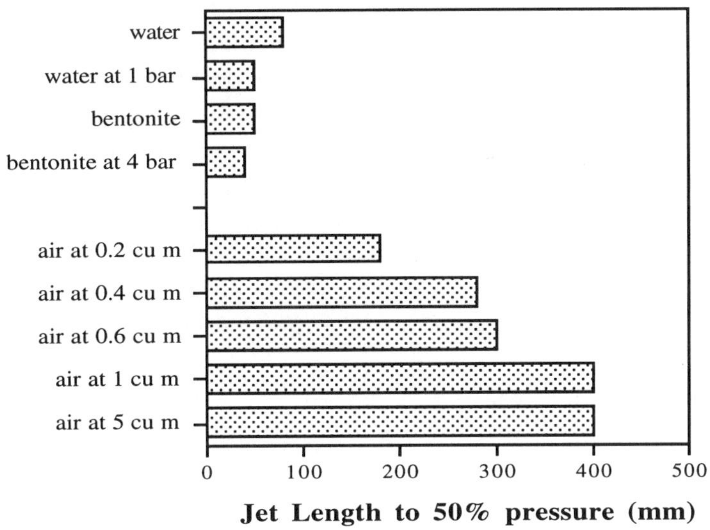

Figure 5.18 Effect of ambient fluid condition on a 250 bar, 2 mm jet with varying volumes of air sheath and discharging into different fluids at possible hydrostatic pressure [5.40].

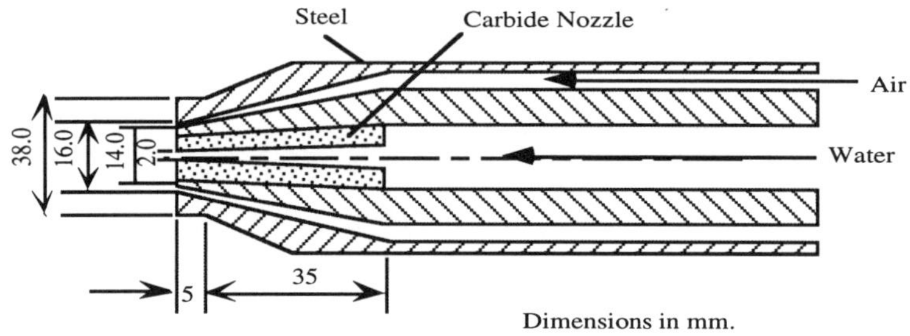

Figure 5.19 Nozzle used with an air sheath for induction grouting [5.43].

This procedure was initially given the name of **induction grouting.** In the original method for installation of such a wall, holes some 10 cm in diameter were first drilled along the projected line of the grout curtain and filled with a bentonite slurry (Fig. 5.20). A waterjet monitor with the nozzle directed perpendicular to the axis of the hole and towards the adjacent hole, was then lowered to the bottom of a hole and raised to pressure. The resulting jet would cut through the soil and make the connection to the next adjacent hole. The penetration through the soil could be detected, by the appearance of air bubbles coming up through the bentonite in the receiving hole. Once the connection had been made, the monitor was slowly raised, cutting a slot through the intervening material. Either a cement or a bentonite grout was immediately injected below the cutting nozzle, filling the slot cut and creating the required impermeable barrier. The jet system was then moved over to the next hole and the procedure repeated. On average an area of 11.3 sq m/hr could be achieved with the equipment.

In more recent times, there have been other approaches to the solving this problem, and one such has recently been developed by Brown and Root, with a more positive method for developing the wall as a continuous feature along the deserted path (Fig. 5.21).

The procedure was found to work well in grouting of sand, silt and very weak rock, to a depth of 45 m. Studies of water permeability through the ground indicated that the grout curtain had reduced permeability by to 1/10,000 of its original value.

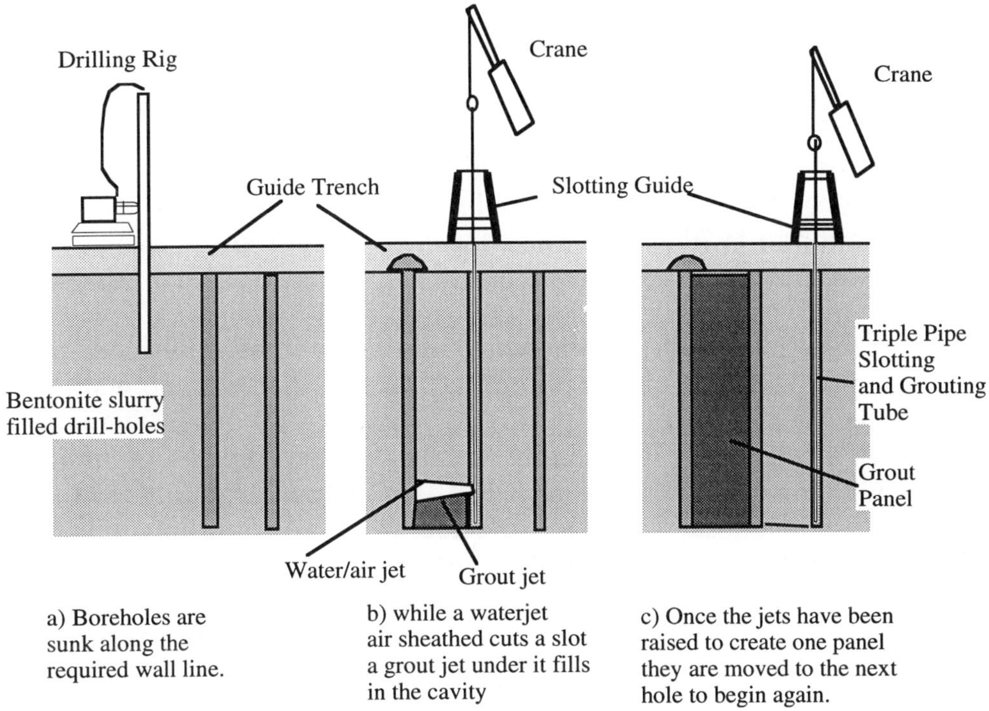

Figure 5.20 Schematic of the method of induction grouting [5.43].

Figure 5.21 Successful grout wall, 30 cm thick, installed and then excavated to show its effect [5.44].

The excavation of soil is also frequently required when it is necessary to dig out space in which to build the foundations under future buildings. In many urban environments, however, the space surrounding a building site is already heavily built-up. This limits the amount of earth that can be removed on site before the adjacent structures are threatened. In order to overcome this problem Japanese investigators developed a new method for pile construction in 1970 [5.45]. The Japanese market alone was estimated at over $25 million with an equivalent market being identified outside that country.

The process involved is somewhat similar to that described above, except that a cylindrical column is generated rather than a plane wall panel. In this procedure a hole is initially drilled, some 8 - 20 cm in diameter, in order to provide access for the jetting nozzle and feed pipe. Depending on the material being cut either a waterjet or light cement grout is then used as a reaming tool (Fig. 5.22). It is projected from a nozzle lowered to the bottom of the hole, and rotated at a speed of 5 - 10 rpm. A typical jet, some 2 mm in diameter, will cut out soil to a diameter of 0.8 to 2.5 m at a pressure of 400 bar. A jet pump can be used to extract this material as it flows to the center of the hole, leaving a void into which a cement grout is pumped. Alternately the following cement jet can mix with the broken up soil to create a cement column which has the required strength characteristics. This grout column forms a pile which can then be used to act as a foundation for the building. Where larger columns are required, the jet throw can, as illustrated above (Fig. 5.18), be enhanced by surrounding the jet with an air collar. This procedure is, however, more expensive and complex, and should be economically justified before use. The procedure has been used down to 50 m in Japan and to 27 m in Europe where it was introduced in 1979.

It is an indication of the ready acceptability of the technique that it was used in over 98 projects in Japan alone between 1972 and 1974 [5.43]. In some instances the initial hole is drilled by a combination of waterjets, pointing both down and radially outward from the pipe. In soil a pressure of 20 bar can be successfully used for this part of the process. Reaming can be carried out at various pressures depending upon the soil type and whether or not the waterjet is supported by a surrounding air sheath. Where it is not, a pressure of 200 - 250 bar has been suggested [5.43]. By using a concrete milk to form the jet, the resulting agitation of the soil by and in the jet stream creates a low strength concrete, or soilcrete, with sufficient strength to act as a load bearing member. Where no air sheath is used, however, the size of the cavity is restricted, and columns must be placed with in 40 - 50 cm in order to ensure that an impermeable barrier is created.

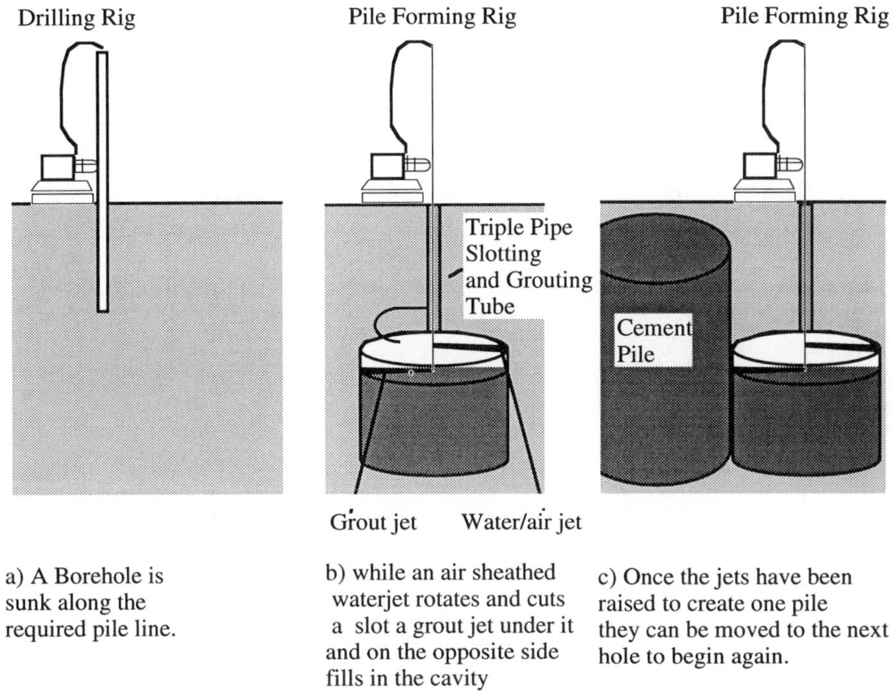

Figure 5.22 Schematic representation of excavating and grouting a pile (after [5.45]).

The air sheath around the jet increases the range of the jet considerably. For example, in 1982 Yahiro reported [5.45] on the further development of the method, now referred to as the **Column Jet Pile Method.** Operating at a pressure of 300 - 500 bar and a flow of between 50 and 70 lpm a single 1.6 mm diameter nozzle, assisted by an annular air nozzle 1 mm wide, could create a pile up to 3.5 m in diameter in sandy soil, and 2.5 m in diameter in clayey soil. An air flow of 4 cu m/min at 7 bar was required to support the waterjet. Once the original 15 cm diameter hole had been driven and the reaming nozzle assembly introduced into the hole it was rotated at 5 - 10 rpm and could be raised at a speed of 5 - 10 cm/min. Obviously the faster the nozzle was raised, the smaller the diameter of the resulting column. For example by slowing the rotation to 5 rpm, and reducing the rate of elevation to 5 cm/min it was possible to ream a cavity, and thus install, a 3.2 m diameter pile in clayey soil.

At one of the early trials of the method with a jet pressure of 400 bar an average of 1.5 piled holes could be drilled to a depth of 28 m a day, while it took 6 hours to complete the drilling and grouting of each hole. Some of the performance parameters for the project were was follows:

Table 5.6 Performance parameters for column jet grouting [5.45]

number of piles installed	286
total length of piling	5,859 m
quantity of cement in each hole	1.94 t/m
quantity of bentonite used to stabilize hole	0.005 t/m
volume of soil removed	3.08 cu m/m
labor	
-drilling	0.34 persons/m
-jetting and grouting	0.22 persons/m
-mixing and supply	0.18 persons/m
-general labor	0.58 persons/m
-electrical work	0.03 persons/m
-machine operator	0.16 persons/m
electric power	42 kW hr/m
water	3.9 cu m/m
light oil	4.9 lpm

The procedure can also be used to renovate the foundations of existing old buildings. For example, Baumann reported [5.46] that the 600 year old church at Hofgeismar in Germany was in danger of collapse since its underlying wood foundation had rotted. A carefully planned series of concrete columns were injected under the walls and support columns during 1983 and 1984, and were able to restabilize the church. The speed of the reaming and grouting process are best achieved if the grouting is carried out at the same time as the jet cutting. To make this possible a triple tube assembly (Fig. 5.22) has been used to feed high pressure water to the cutting nozzle, air to the surrounding sheath for the waterjet, and cement to the opposing jet which fills the void.

The cement jet has a diameter of 10 - 12 mm and is set below the level of the waterjet cutting nozzle. Grout which has been premixed is injected through this line to fill the cavity left by the jet cutting.

It is possible to use the technique in other ways to create barrier walls. A primary injection cut using a cement grout, for example, and operating under the same conditions but without nozzle rotation, can inject a barrier wall into soil, some 35 cm thick. Such a barrier will give a relatively high compressive strength of 120 - 180 bar in sandy soil, with a permeability of 10 - 9 to 10 - 11 m/sec. Such a process has acquired the name **panel jetting.**

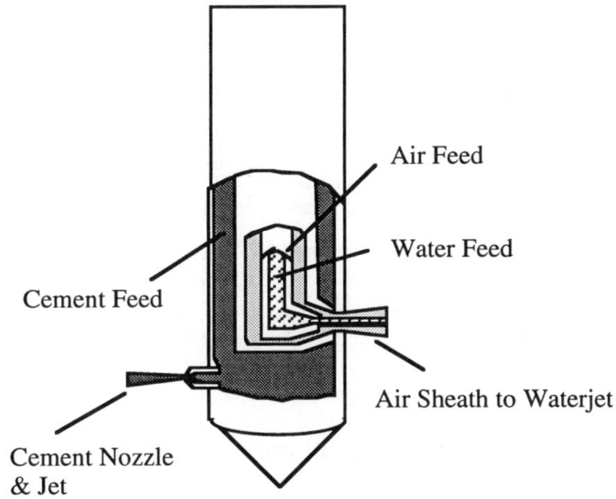

Figure 5.23 Combined jet reaming and grouting nozzle configuration [5.46].

Experiments in the optimum location of the three nozzles, the waterjet, the air jet and the grouting nozzle have indicated [5.41] that the optimum gap between the air sheath and the waterjet should be between 1 and 1.5 mm, while the grouting nozzle should be pointed slightly downwards and at a greater distance below the waterjet nozzle than the combined thickness of the disturbed soil layer from the waterjet and the thickness of the grout stream. The experiments were required in China before the technique could be used to insert grout columns into the failing area of a tailings dam. A total of 160 piles were installed, 1.6 m apart, and to a maximum depth of 19 m. The project took 6 months and used an average of 0.71 t/m of concrete at a total cost of 193 yuan/m to successfully stabilize the area and stop water leakage.

The flexibility which this method allows for reinforcing failing structures is illustrated by a case study from the United Kingdom [5.47]. A tunnel in the suburban Glasgow British Rail system had been driven over a 25 m wide glacial hollow in the rock, at a distance of 100 m inside the entrance, or portal of the tunnel. This section of the tunnel was therefore lying on a layer of saturated silt and boulder clay some 4.5 m thick into which it was slowly subsiding and this was causing the track to sink. Over a period of eight 29-hour weekend periods a series of two rows of 2 m diameter columns were drilled and jet grouted into place, at 2 m intervals along the affected length of track (Fig. 5.24).

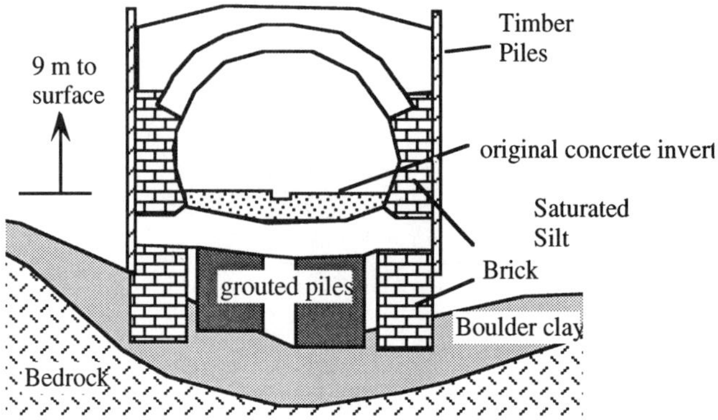

Figure 5.24 Layout of grouted concrete columns under a Glasgow rail tunnel [5.47].

The concrete injected fully stabilized the floor, or **invert,** of the tunnel but it was found that the work had to be interrupted on occasion to allow **low pressure conventional** grouting of some gaps which had developed in the original wooden mattress on which the first concrete invert had been laid, some 100 years before. British Rail estimated that the cost of the operation was some 20% of the only alternative, which would have been to close the tunnel and completely rebuild the tunnel section.

A second case history from Gateshead in the UK [5.47], where jet grouting was used to consolidate the soil ahead of a tunneling machine, illustrates one area of difficulty with the technology. The excavation combined jet grouting with use of a bentonite shield tunneling machine, a combination successfully used over 400 times by the Japanese. However, it was found that the jet grouting did not give complete ground consolidation in all areas ahead of the machine. Investigation revealed that the problem arises in soil containing relatively large boulders of rock. These have the strength to deflect the cutting and grouting jets, and thus of "protecting" the area behind them from treatment. This is normally a problem when the particles are significantly larger than the jet, and are found close to the nozzle. As a result the areas not covered by the jet are not consolidated, and may collapse into the tunnel when it is driven past them. Nevertheless, despite the additional costs required for the additional remedial work to overcome the problems thus encountered, the work was completed within the $170/ cu m tender price.

The application of this technology is not restricted to vertical installation. Using the same technology, but under the trade name Rodinjet, grout columns have been installed in horizontal and inclined patterns for both ground stabilization and as a water barrier [5.48]. Generally the same nozzle configurations can be used, with either a grout jet being used to drill and ream the hole through the same column or a triple nozzle assembly with an air sheathed waterjet used for creating the void, and a third nozzle then injecting the grout (Fig. 5.25). The ball check valve in the single nozzle arrangement is used to change the direction of jet flow from vertical to horizontal.

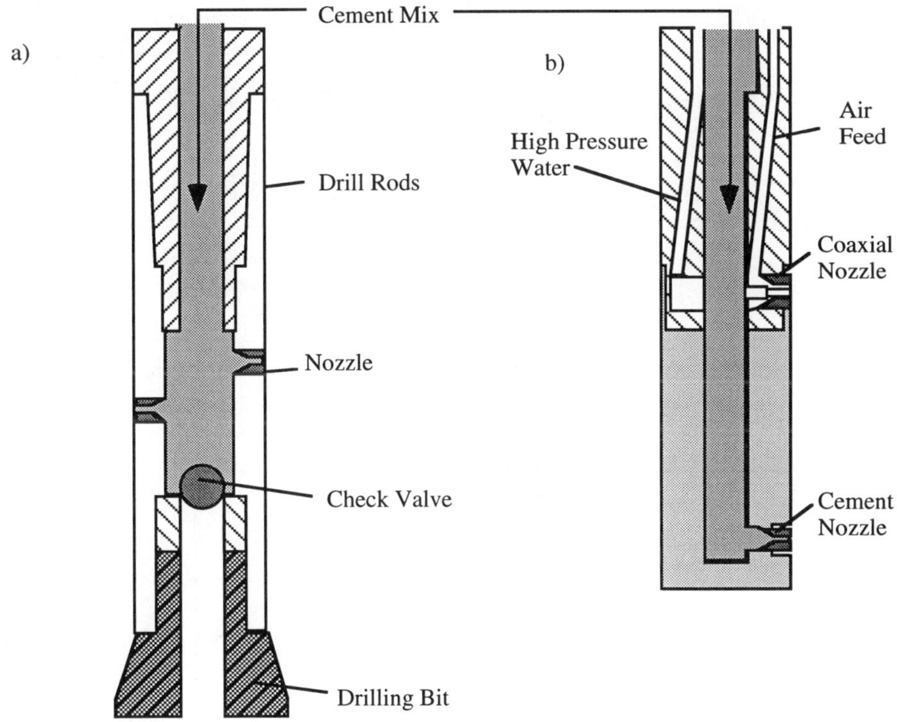

Figure 5.25 Jetting nozzle designs for horizontal installations a) Rodinjet-1 b) Rodinjet-3 [5.48].

In the initial installations for which this technique was used in Italy, the grouting was used to stabilize the ground above an excavation for an underground railway station in Milan. Holes were drilled using a conventional rotary percussive drill along the line of proposed excavation. Nine holes were drilled at each location, each some 9 m long and 0.4 m apart, at an upward angle of 13° (Fig. 5.26). The holes were then jet grouted and once set, the concrete columns created permitted safe excavation under the thin layers of fine sand and silt found in the area.

The technology has now become widely adopted and recent experiments [5.49] have sought to increase the jetting diameter and lower the cost.

Results of a factorial experiment varying jet pressure and flow rate indicated that this can be better achieved by increasing flow rate than from by raising jet pressure (Fig. 5.27).

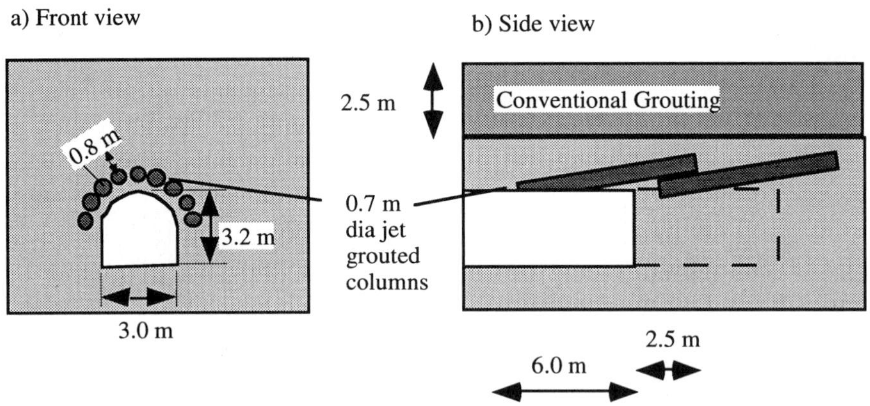

Figure 5.26 Jet grout stabilization pattern over the Milan underground [5.48].

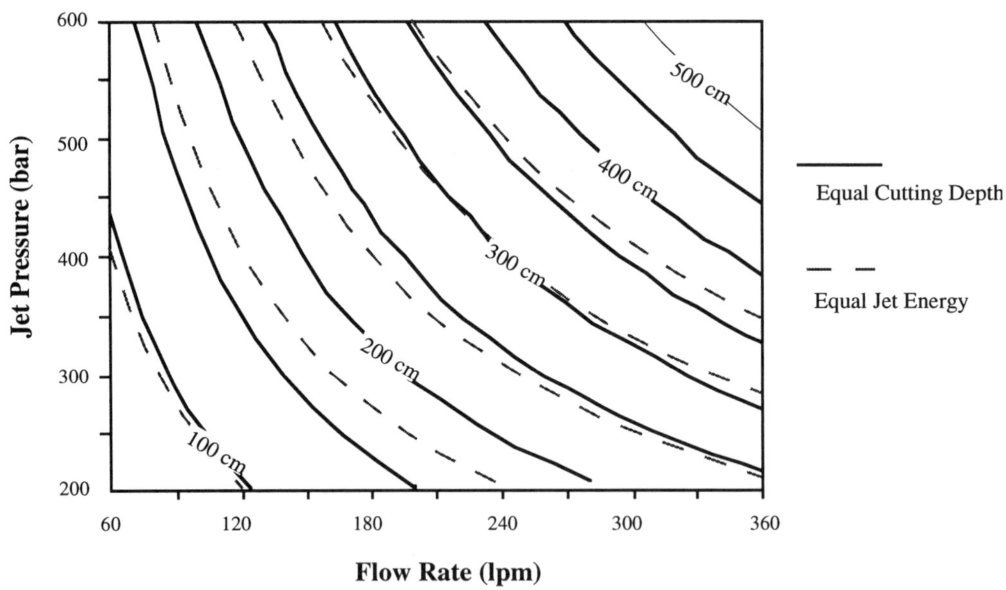

Figure 5.27 Effect of jet pressure and flow rate on depth of cut in soil (after [5.49]).

The procedure itself has also been modified [5.50] to improve the strength of the installed pile (Fig. 5.28). Simplistically the procedure proposed is to use a rotating waterjet lance to drill a hole down to the required depth. The jet pressure is then raised and the assembly rotated at a slower speed as it is slowly retracted over the lower section of the hole. This reams the bottom of the hole over a defined interval. The nozzle assembly is now removed and the hole filled with a cement milk. A special metal pile is then driven down into the excavation. On the bottom of the pile is a special fiber shell which can then be inflated by concrete injected through an internal pipe within the pile. The spreading of this fibrous shell into the bulb of concrete creates a much stronger structure at the bottom of the pile.

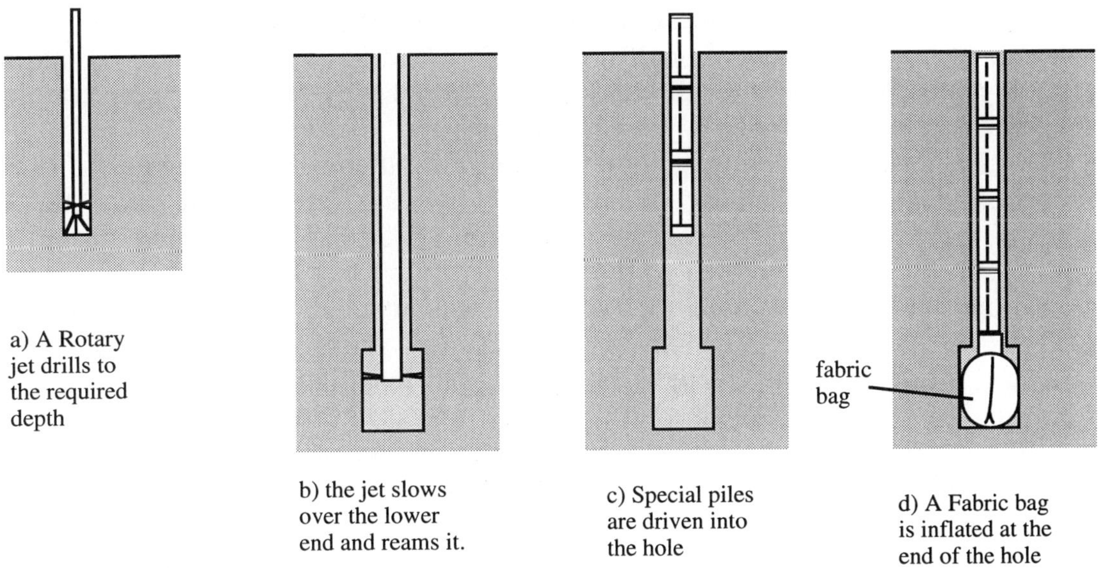

Figure 5.28 Improvement to piling using an enlarged anchor [5.50].

Interestingly the procedure for drilling and reaming the hole has now been computer controlled so that rotation speed and advance rate are adjusted to take into consideration the type of ground through which the drill is advancing. Based upon this feed rate the amount of concrete being injected into the cavity is also controlled. Through these precautions it is reported possible to install the piles within 25 cm of the boundary of the site.

The initial hole is drilled by a mechanical tool and reamed by a waterjet of either 3.2 or 4.0 mm diameter which rotates at a 200 mm diameter cutting the path required for the future injection of the pile. Some

1,600,000 m of piling had been installed by this method by 1988. The ground is normally cut out to a diameter of 40 cm, using a jet pressure of 200 bar. The drill is advanced into the ground at an average speed of about 1 m/min.

Improved methods of conventional pile injection have normally just added waterjet nozzles to the leading edge of the pipe or metal sheet being driven into the ground [5.51]. A jet pressure of 300 - 500 bar at a flow rate of up to 120 lpm is used in this application.. The water has two effects, it removes material ahead of the pile, allowing it to slide into the ground at a faster rate, and the outward flow of the spent water along the edge of the pile reduces the friction on the pile:soil contact, again lowering the installation force required. The method reduces pile driving time to a quarter of that conventionally required, while lowering ground vibrations by about 50%. As a result costs can be reduced by 50% over conventional pile driving, and by 15% over boring assisted pile driving.

Because pile driving becomes more difficult as the soil changes to rock, the advantage of using the waterjet to ease penetration into harder structures becomes of even greater benefit. Studies of the best design of nozzle to use have indicated that a fan jet might be of greater benefit than a round jet [5.52]. Where such a nozzle was attached to a pile and operated at 600 bar, (Fig. 5.29) that up to 50 sq m of piling could be installed in a day, on average, with roughly a doubling of installation performance. The rate however increased from 0.95 times as fast in a sand which is normally easy to penetrate, to a level of 5.0 times in mud stone which is normally very difficult. The fan jet was a 30° fan, operating at a flow of 0.05 cu m/min.

It has been suggested that pulsating such a waterjet would reduce the amount of energy required by two-thirds and the procedure would become noiseless. A system to develop such a method was discussed in 1978, [5.53] and a machine was to be built in Japan to prove this improvement to the technology.

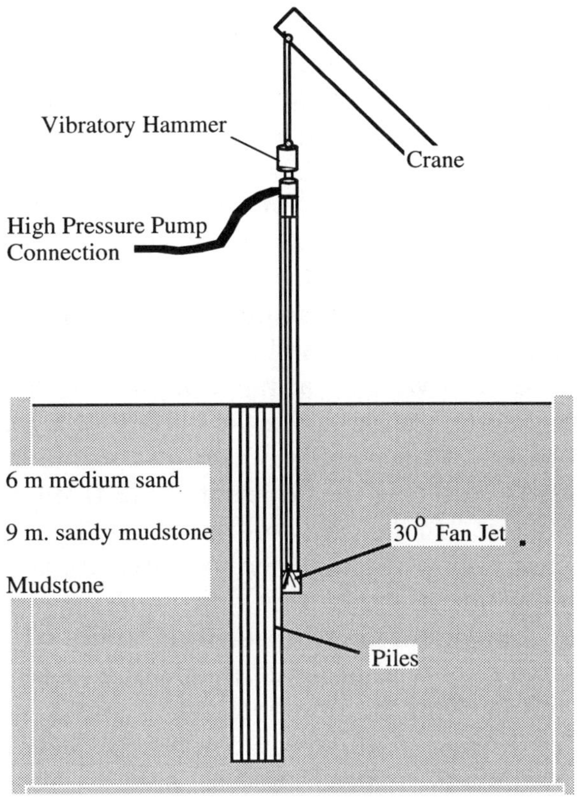

Figure 5.29 Attachment of a waterjet to a pile installation [5.52].

5.4.4 HORIZONTAL HOLE DRILLING

Experiments in the cutting of soil have followed a different path in more recent years. As the urban environment spreads, with changing demands for power and utilities, and these need repair, it has become increasingly expensive and disruptive to cut through the concrete and asphalt paving to access power and service lines located under roads. Similarly it becomes disruptive to trench through established yards and gardens to install new utility lines. It is a major, but hidden problem, to repair the underground infrastructure much of which is now quite old. In the United States there are over 4.5 million km of underground pipes, and a similar length in Western Europe [5.54], with a world-wide market in 1993 of $5.8 billion for excavation.

One of the first approaches to using waterjets in this field arose, initially under funding from EPRI, when investigators at Flow Industries sought to design a way in which waterjets could be used to cut along existing cables in order to free them from the surrounding soil, so that they could be withdrawn and replaced [5.55]. Under normal circumstances the cost for cable replacement could exceed $300/m, in 1982 (ibid.). By adjusting the jet operating pressure to less than 350 bar, waterjets could be tailored to cut through the soil ahead of the tool, without damaging the adjacent cable and utility lines. A flow rate of approximately 15 lpm was used and initial tests were carried out in both loose sand and consolidated clay. The 15 cm long tool was capable of advancing at rates of over 1 m/min, and had the ability to either advance or retreat down the cable. In a loamy soil the tool could advance at a speed of up to 60 cm/min in a stable hole. As a result, the jets from the head could free the adjacent cable from the soil, so that it could be easily extracted. In sandier soil, however, hole stability could be a problem, in one case collapsing onto the tool after an advance of 20 m.

To overcome this problem the cutting fluid in the jets was changed from water to bentonite, a standard oil well drilling mud which stabilizes the hole by building a thin clay wall around the opening as the water permeates through into the sand beyond the borehole wall. Field tests of this system in extracting cable showed that advance rates of between 60 and 120 cm/min could be achieved. The tool was able to work effectively in loosening cables over a distance of 90 m, and that the cables could then be extracted with a force of less than 100 kg. In cutting the unconsolidated material it was not necessary to rotate the head, but when the tool was used in a clay ground, rotation was required in order to cut a clear path for the head through the ground. This is because the relatively narrow slot which the small jets cut into the clay would leave ridges of material protruding from the face between the jet cuts. These ribs would block the advance of the tool, and required that the tool be rotated so that the jets would cover the entire face of the hole, and remove all the material ahead of the device.

Reichman found that, because of the small nozzle size, and the presence of mud in the hole, a jet could only cut about 2.5 cm ahead of the nozzle. By using a suitable arrangement of nozzles, however, it was possible to still achieve a hole diameter of 12.5 cm with the 15 lpm flow rate. The jets had, however, to be designed not to concentrate on the hole perimeter since, in softer soils, this would enlarge the hole too much. The paper analyzed operating costs and estimated that the use of the tool would cost approximately $30/m, in contrast with a minimum cost of $75 for conventional trenching and replacement.

That particular tool did not become popular, since among other reasons it was found that when the cable had been originally installed, excess length was often just dropped into the trench and left. Thus the tool could not easily follow and free a cable which wound around or overlapped itself. Instead, a free driving hole drilling tool has become more popular, and several companies now market jet assisted cutting tools for horizontal, and quasi-horizontal drilling for trenchless utility repair and installation. This overall subject has become the object of special conferences and, in 1992 a new magazine was published dealing with this topic [5.56].

Although the market is still dominated by mechanical excavation devices, there has been a significant development of waterjet assisted tools which are beginning to successfully penetrate the industry. In large measure these new tools combine waterjet cutting with mechanical means to excavate and remove the material, and also to steer the tool as it penetrates the ground. A procedure has evolved in which a pilot hole is first drilled through the soil, using 350 bar jets issuing from a series of small orifices disposed around the cutting head. In several of these cutting heads the jets act with the mechanical action of the wedge shape on the head to remove and carry back the soil from the face [5.57]. Once the pilot hole has been drilled then the cable is attached to the back of a reaming tool. As the cutting system is retracted through the hole, enlarging it as it goes, the cable is pulled into the hole behind it, thus completing the location of the cable without disturbing the ground surface between the small pit dug for the insertion point of the tool, or the second small hole at the far end of the operation, where the tool head is changed and the cable is attached.

The practicality of this new tool can be illustrated by example. In 1988 the British Gas Board wished to lay a new main pipe in Milton Keynes. The pipe had to be installed over a 7 km length and the line had to be installed with five crossings of a busy main arterial road. The Flowmole Guidedril was used to install the polyethylene pipe in a period of 10 weeks, half the time anticipated for the project, achieving more than 75 m of installed pipe in a day. The drill was able to install the pipe without surface disruption, and avoiding cables already installed in and around the proposed route [5.58]. Flowmole is now known as UTILX.

One advantage of waterjets which as previously been discussed is the low reaction force which they can exert on an operator, while applying high pressure levels to the cutting surface. This is a particular advantage when holes must be drilled in relatively inaccessible locations, and where the use of conventional equipment would be extremely expensive due to the problems of access. For example, it is necessary to install tieback anchors and drainage holes into soil banks which show the risk of failure. Such soil slopes can often be high and difficult to get too. The relatively small, hand-held waterjet drills which have now been developed [5.59] can be carried to the site by a man working in mountain climbing gear. However the original 14 kg manual drill could only drill to depths of between 1.5 and 6 m and was limited in the strength of material it could penetrate.

A second, larger unit was therefore fabricated which was track mounted, and this could be used to drill not only deeper holes in harder material, but could also drill diameters up to 10 cm in diameter, drilled to depths of 18 m. A small air motor was required to drive the rotation, and feed was by use of a hand crank. Subsequent versions of the drill have been automated, and can drill on a rack 7.5 m long to install holes up to 15 cm in diameter. The unit normally operates using a 700 bar pressure. The largest drill incorporates three nozzles and uses carbide blades to ream the hole to the required size. Depending on the diameter of the nozzles the flow of up to 70 lpm is sufficient to clear the hole of debris. Before the hole is used, and the design was developed for earth anchors for civil engineering structures, the hole must be blown clear of dirt and water using a compressed air blast.

5.4.5 JOINT AND GAP CLEANING

The drilling of long holes through relatively weak soil, and the ability of the jet head to turn within these confines, points out some of the advantages of this light and to an extent delicate tool. Another illustration of the unique advantage which waterjets bring to construction practices is in the cleaning out of clay and other material from weakness planes in rock construction walls. In large scale construction it is common to find that the rock at the excavation site contains joints and that these can be filled with clay or other weak material. These can be a threat to the stability of the overlying structure, or can be weakness planes in the bed of a dam site, which can, if not treated cause massive failures, such as at the Vaiont Dam in the Alps. It is difficult to remove this soft material from the depth of the joints conventionally and to sufficiently clean their surfaces that a grout injection can strengthen the structure, and seal off any potential water path. High pressure waterjets (Fig. 5.30) can, however, remove this clay at relatively low pressure, and can reach back, beyond the range of normal tools to remove the clay and to clean the joint surfaces, even where the gap is quite thin. It is suggested, however, that the faces of the cut be rinsed with low pressure water after the main jet action (Fig. 5.31). This is because the high pressure jet will continue to bring material back across the face from the back of the cleaned area where it is used.

Figure 5.30 Waterjet removal of clay seam at the St. Louis theater excavation.

Figure 5.31 Cleaned joint, after low pressure wash down.

For example in the construction of the Feitsui Dam in Taiwan [5.60] waterjets were used to clean out joints and drill holes were located to ensure that adequate coverage of the seams removed all the potentially dangerous material. Once the clay seams had been removed from joints in the surrounding stronger sandstone and siltstone, the seams were injected with an non-shrink mortar grout. The project was of considerable importance given that the dam, which will have a crest height of 510 m holds a reservoir of water of 6 million cu m. In order to locate and clean the seams, access tunnels were driven into the sides of the valley across which the dam would stretch. The tunnels were driven 8 m apart up both banks. Because of the narrow gage of the seams, and their undulation within the rock waterjets were found to be the only method capable of cleaning out the material [5.61]. Where the seams were more than 1 cm thick a 200 bar jet flowing 50 - 100 lpm was found to be most effective, but for the narrower seams a 2,400 bar jet at 10 - 15 lpm was able to clean the material out more effectively.

Because of the risk of rock collapse only portions of the seam could be removed at one time. Thus segments 3 - 5 m in length were cleaned out, a similar length left to act as support and a third length cleaned. Once these intermittent lengths had been cleaned and grouted, the grout was allowed to reach a specified strength, and then the intervening lengths of the seam were cleaned out and grouted. In order to assist in cleaning the thicker seams a compressed air line was attached to the high pressure hose. This left only the coarsest particles in the seam, and these would mix with the cement then injected and assist in providing a proper seal.

Where the high pressure jets were used in the thinnest crevices, cleaning holes had to be drilled only 20 cm apart due to the small range of the jet, and the reduced carrying power of the jet due to the lower volumes of water which it used. It was possible using these methods to clean the seam at an average rate of between 0.322 and 0.46 sq m/hr.

A similar problem was encountered during the construction of the second underground movie theater in the Jefferson National Expansion Memorial under the Gateway Arch in St. Louis [5.62]. As the underground room was being excavated layers of clay were found which posed a considerable threat to the stability of the west wall of the theater. The clay was removed to the depth of 1 m using a waterjet at a pressure of 100 - 200 bar, with a hand-held lance, again removing segments of only 1.5 m in length at a time. These were backfilled with concrete, which was allowed to set and acquire strength before the next segment was removed.

In this way the entire section which could potentially cause problems was treated.

Similar sized gaps are found in concrete roads and bridges and are inserted into the roadbeds to allow the road to expand and contract due to changes in the temperature. In 1974 the Texas Highway Department funded a study at the Texas Transportation Institute to find ways of removing the debris and material which is used to fill these joints, so that they could be refurbished [5.63]. A high pressure waterjet lance was found to be "far superior to any other that they had tried." It was usually not necessary to feed the lance or nozzle into the slot, but all the debris could be removed with the nozzle held at the surface of the bridge. A single round jet was found most effective, at a pressure of 700 bar, with the nozzle mounted on a small cart. The cart removed the strain from the operator who maneuvers the device along the joint. An attempt to use the jets to widen the joints into the concrete on either side was less successful. The technique was further examined by the Corps of Engineers [5.64] for cutting the joints in fresh concrete. The technique was found to be as effective, where 700 bar waterjets were used, as wet sand-blasting and it required less clean-up. The application of waterjets in cutting concrete will, however, be addressed more fully later in this chapter.

5.4.6 SOIL AND TURF PENETRATION

Perhaps one of the more interesting of the recent developments in waterjet applications to soil is the Water Jet Turf Aerator recently marketed by the Toro company [5.65]. The device which is self-powered, can be steered over the greens of golf courses. As it traverses the course it shoots small jets of water, at approximately 200 bar pressure, into the ground. These jets of short duration, penetrate to depths of 10 to 20 cm and allow not only aeration of the course, but also the potential for injection of fertilizer or herbicides. Because of the small holes drilled by the jets, the course is left apparently undamaged and thus, in contrast to more conventional aeration, the procedure can take place between rounds of play.

This ability of the waterjet to easily penetrate soil has been put to a different use by investigators at Southwest Research Institute. Tasked with finding a way to locate underground pipes for the gas industry a small low volume high pressure waterjet was fired into the ground at intervals. When it hit a buried obstacle it created a noise. By calibrating an acoustic monitor it is possible to characterize the signal measured and identify when the jet is hitting a pipe (Fig. 5.32). The tool has been able to detect both plastic and steel pipe and normally penetrates to the possible pipe depth in less than a minute. It has proven to be 88 percent reliable [5.66] and provides a depth as well as material characterization. For the device to work a 700 bar jet is attached to a probing lance. A small jet on the end of the lance uses a flow of 0.3 lpm and the tool is pushed into the ground, at a rate of 1.2 m/min by the operator until it penetrates to the depth of the pipe (about a meter). The noise of impact which the jet makes on the pipe is transmitted up the body of the probe and read by a vibration sensor mounted on the handle.

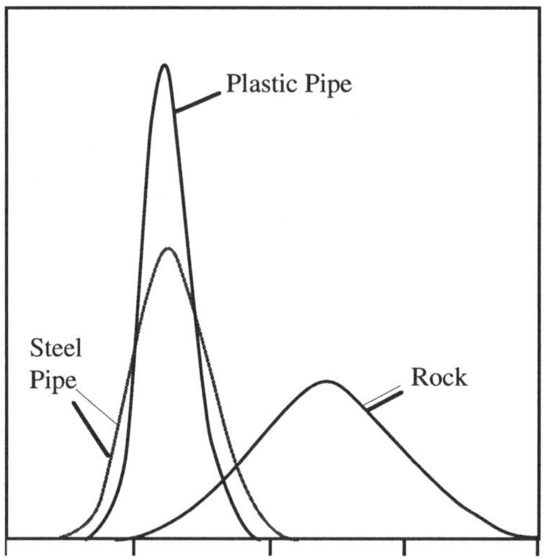

Figure 5.32 Representative spectra of acoustic signals from jet impact on pipes and rock [5.66].

One additional potential use for waterjets has recently been suggested, following the recent conflicts in Kuwait, the Falkland Islands and elsewhere [5.67]. It has been suggested that jets at a pressure of 700 bar would be able to expose the land mines laid in those conflicts, and that, once exposed, the mine could be cut apart using higher pressure (1,750 to 4,200 bar) jets. Projected performance for such a system would be that it could clear a 10 cm deep swath some 2.4 m wide at an advance rate of 0.6 m/min. The power required would use some 90 kW. While such a scheme might work in clearing fields after a conflict, during such an event the amount of time available for mine clearance is significantly less, and the depth to clear the mines from the ground is often greater so that additional power and alternative approaches will often be necessary. The use of waterjets is not, however, precluded, and in other segments of this work, reference is made to the use of high pressure waterjets in both cutting through munitions and in removing the explosive contained within them.

5.4.7 WATERJET ASSISTED EARTH PLOW

As an alternate means for installing power cables in soft ground, where again, the need is for a deeper cut, the Electric Power Research Institute (EPRI) worked with Flow Industries to assess the use of waterjet assisted jets to a conventional cable plow, used for cable and utility installation. Experiments were carried out with a variety of combinations of waterjet nozzles attached to a plow blade, set to operate at 0.75 m into the ground. The jets were inclined to cut ahead of the plow blade (Fig. 5.33) and with jets directed either forward at the sides or in combination on the plow blade. The results were compared with those obtained where the blade is vibrated as it is pulled through the ground [5.68]. It was found that significant reductions in force could be achieved when waterjets were added to the plow, with the combined side and forward jets working better than forward jets alone (Table 5.7). Jet pressure was 840 bar, and a flow of 15 lpm. The jet nozzles were located on the lower half of the blade only, to reduce water requirements [5.69].

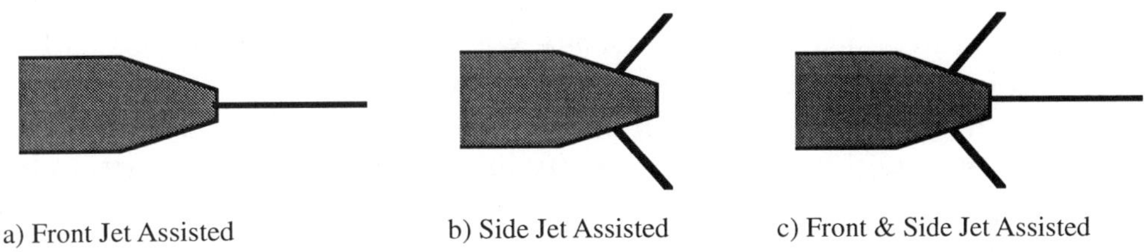

Figure 5.33 Configuration of waterjets to assist an earth cable plow [5.68].

Table 5.7 Results of improvements to a drag plow (after [5.68])

Method	*Percentage improvement*
waterjets to the front	40
waterjets to the side	33
waterjets to front and side	44
vibrating plow	38
waterjet assisted vibrating plow	117 in damp soil 48 in hard soil

5.5 CONCRETE CUTTING AND HYDRODEMOLITION

It is an interesting comment on the dangers of prediction to review, with over 20 years of hind-sight a paper presented on the use of waterjets for concrete removal at the first BHRA symposium in Coventry [5.70]. At that time the authors, who worked for a major British construction company, had evaluated the likely performance of waterjets as a potential new tool for the removal of concrete. Relatively poor results were obtained in terms of the energy required to meet a proposed performance criterion. In evaluating the likely performance which a machine would have to meet to be useful, it was concluded that the machine would need to operate at a pressure of at least 3,800 bar, and have a pump power of 370 kW. Since such a unit would cost, in 1972, around $250,000 - a cost too high for most construction companies - the technology was therefore considered to be an impractical one for use in the construction industry.

In May of 1987 an article in the magazine Civil Engineering [5.71] commented that, within two years of its introduction, the State of Indiana Highway Department had included waterjetting in its standard specifications as a means of concrete repair. Such a turn around surely deserves comment. Not only perhaps specifically for this application, but because it illustrates just one of a number of situations where waterjet use was initially written off, on the basis of experimental results, but where it has since found a practical use. (Incidentally tunneling is another one such - but one where the applications are only now becoming apparent). It is appropriate, before discussing the topic in more detail, that the original prediction on cost was closer than some of the other estimates.

5.5.1 THE INITIAL STUDY

Concrete is a material of widely varying properties and composition. Simplistically it is obtained by mixing quantities of a cement, a sand and small rock pieces known as aggregate, with a specified amount of water. The mixture is poured into place and the water reacts with the cement to transform it into a solid material. This solid acts as a matrix to hold the sand and aggregate together in the solid form which is referred to as concrete. The mixture can be made up of varying quantities of each of the ingredients. The ingredients themselves will change from location to location, and this includes the chemistry of the water used in the process. It is especially important to note that the size and rock type which makes up the aggregate will change. This last change is, for many tools, the most critical one.

Consider how a diamond wheel cuts a slot through a concrete slab. The blade must cut through all the material in front of the blade, so that it can move forward. The advance is thus controlled by the ability of the blade to cut, to the required depth, through the hardest aggregate which is present, along the line of passage of the blade. Given that there are places where quartzite is the aggregate and that this material can have a strength of up to 4,200 bar, and the size of the problem becomes apparent.

Initially McCurrich and Browne [5.70] tested waterjet cutting at a pressure of 700 bar using a variety of nozzles to find if this tool could be used for slotting or drilling concrete. They also compared specific energies of cutting for the waterjet and comparative existing techniques (Table 5.8).

Table 5.8 Early comparative specific energies for removing concrete [5.70]

Technology	*Specific energy (GJ/cu m)*
pneumatic hammer	0.26 - 1.6
rotary percussive drill	0.21 - 0.84
drop hammer	0.001 - 0.003
thermic lance	33 - 134
plasma arc	0.87 - 4.3
diamond saw	1.10 - 4.50
waterjets	180 - 4,000

In evaluating the use of high pressure waterjets it was concluded that a minimum pressure of 2,200 bar would be required to cut the aggregate, which at that time was considered necessary in order to create a practical tool. The report also concluded that a jet pressure of 3,800 bar would be likely to be more effective. By the same token it was considered that a minimum nozzle diameter should be on the order of 1 mm. Traverse rates should be on the order of 1 m/sec which was considered optimum, with depths of penetration designed to be 0.3 mm or more.

5.5.2 ESTABLISHING THE PARAMETERS

This discouraging report slowed, but did not stop the development of the technology. By the second international conference, just two years later, there were three papers specifically directed at concrete cutting. Olsen [5.72] was concerned with the cutting of thin slots in concrete, to replace the diamond blade discussed above. In many applications where these thin cuts are required, such as for skid reduction surfaces and expansion joints, slot depths of 25 - 75 mm are considered adequate. Making use of a Flow Industries intensifier capable of generating 4,130 bar, studies were made of how changes in the jet pressure, flow rate and traverse speed affected the cutting ability of the waterjet stream. At lower pressures the earlier conclusions, that waterjets could effectively remove the sand and cement, but could not cut the aggregate, were substantiated.

Even at the full jet pressure it was, however, not possible to guarantee that the aggregate would all be cut. This was particularly true at higher traverse speeds (2.5 m/sec) where no aggregate was cut. Olsen therefore recommended that the optimum traverse speed be reduced to 250 mm/sec, since at this speed almost all the aggregate was cut, and the edge quality was acceptable. However, in the requirement for a greater cutting depth it was considered important, by both investigators, that the jet advance into the slot being cut. This requires a slot wider than the nozzle or holder. Since Olsen's nozzle was 0.305 mm in width, and the holder required a path 50 mm wide, the jet must traverse an oscillating path to cut wide enough clearance. This, in turn, reduced the forward advance speed, to around 30 mm/sec. Rather than use the high power suggested in the earlier paper, however, Olsen suggested mounting a small intensifier operating on a 20:1 ratio on a back hoe, which has a normal 18 kW hydraulic system which will produce 130 bar. This lower pressure would feed the low pressure side of the intensifier, which would thus be able to generate a jet pressure of 2,600 bar, albeit with only 1/20th the flow rate provided from the system hydraulics. This size of unit would be competitive with diamond saw applications. For larger field applications a 600 kW unit was alternately proposed, using a multiple array of 0.3 mm nozzles.

A similar series of experiments was carried out at this time in Canada [5.73]. Pressures were increased to 4,830 bar for these tests, using a 0.178 mm diameter nozzle. The study confirmed that a high feed rate was necessary to obtain better cutting efficiency, with multiple fast passes appearing more effective than one single slower pass. At this pressure and a feed rate of 5.1 cm/sec it was possible to cut approximately 5 mm deep into concrete. In order to obtain a wider cut it was possible to remove the intervening rib to a distance of some 7 nozzle diameters. Because this still produced some spalling of the edge of the cut, an undesirable feature where the jets are used to cut anti-skid grooves, a procedure to eliminate this was developed. This, in essence comprised a method of compressing the material along the edge of the cut (Fig. 5.34). This proved effective in leaving a clean edge to the jet cut.

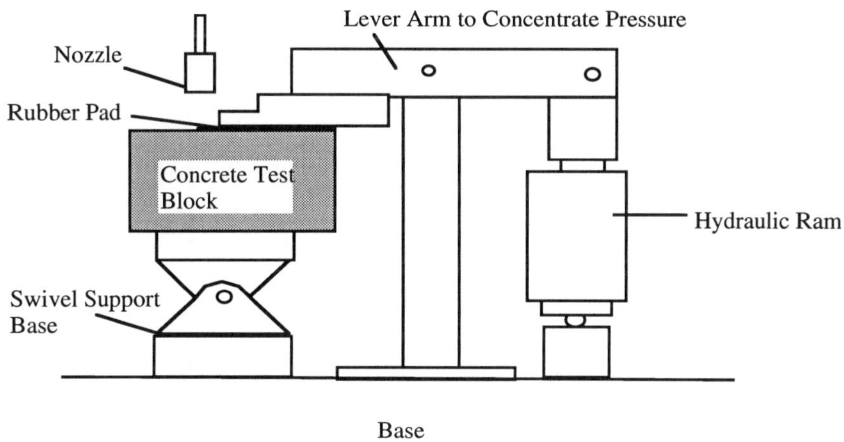

Figure 5.34 Method of compressing the cut surface to eliminate cut edge spalling [5.73].

It is worthy of comment that none of the above studies had commented upon work carried out by the Bureau of Mines on the same topic. Most likely this was because the work was only briefly referred to at the first conference [5.74] and because it dealt with large volume removal, rather than precision cutting. The study, described in more detail in a report [5.75] used the lower pressure, higher volume monitors which the Bureau had been using for conventional hydraulic mining trials, in order to cut concrete blocks. Jet pressures ranged from 234 - 345 bar and significant slots were cut into the blocks. While the average specific energy of removal was 1,630 joules/cc (1.63 GJ/cu m), values measured lay between 0.33 and 405 GJ/cu m. More significantly, even at these lower pressures the slots were open, with both aggregate and cement removed.

This point was noted by Japanese investigators [5.76] who noted that the Bureau study had been with jets of larger diameter. Thus in evaluating lower jet pressures, up to 500 bar, the effects of nozzle diameter and spacing between adjacent cuts were pursued in more detail. Although the specific energies of cutting still remained quite high in the results of this Japanese work, they did find (Fig. 5.35) that specific energy of cutting decreased with increase in nozzle diameter, and also with an increase in spacing between adjacent cuts, as long as the intervening rib of material could be removed. The most efficient removal of this rib of material appeared to occur when the two adjacent jets were spaced at a distance of between 9 and 18 nozzle diameters apart (Fig. 5.36).

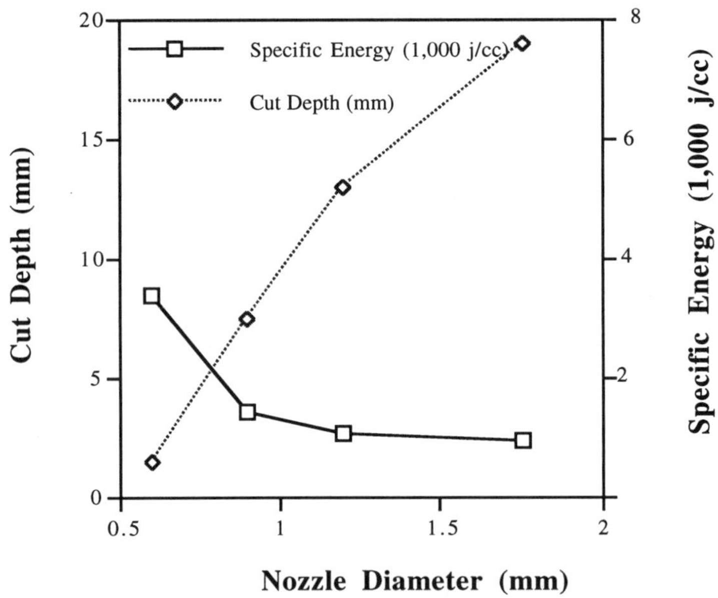

Figure 5.35 Effect of nozzle diameter on the depth of cut and specific energy of concrete removal (after [5.76]).

The machine that this investigation suggests is most practical is thus a 140 kW machine, operating at a pressure of 1,000 bar with a 2 mm nozzle diameter.

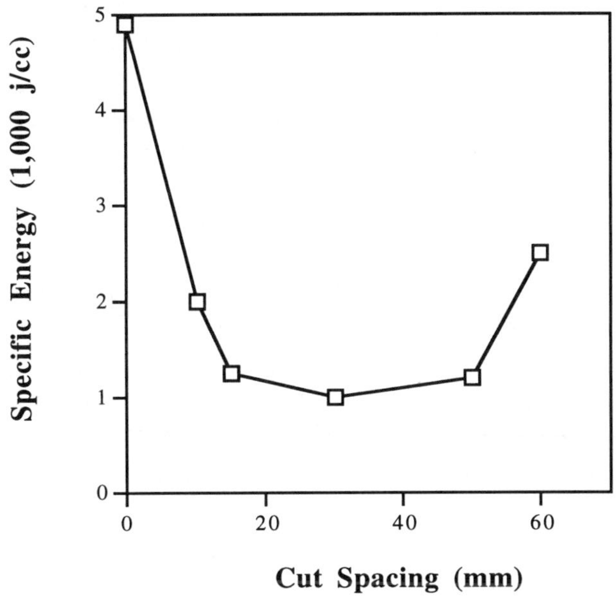

Figure 5.36 Effect of slot spacing on specific energy of concrete removal [5.76].

A clear difference in the approaches to concrete removal was becoming apparent in much the same manner as was occurring in the approach to cutting of rock, as will be discussed in Chapter 7. There was the attempt to use increasingly large pressure impulse jets for massive concrete fracture, then there was the use of very high pressure, but small volume continuous jets for cutting thin slots in concrete, and finally there was the lower pressure higher-volume continuous jet approach directed more toward slot cutting and the volume removal of material. Ultimately a fourth option, that of adding abrasive to the water became available and that too has found a place in the technology. Because of the differences in approach and the resulting applications each of these options will now be discussed separately, even though for most of the decade of the 1980s all four options were in development and use concurrently.

5.5.3 HIGH PRESSURE SINGLE PULSE UNITS

Development of high pressure single shot or high pressure pulsing units was aimed at replacing the drop hammer and pneumatic pavement breaker. To be effective such a unit must not only be able to break concrete effectively on a single pulse, giving instantaneous productivity, but should also be able to recycle quite quickly and maintain consistent operation during a day, giving high levels of overall production.

Early experiments with high pressure water cannon types of equipment had been able to generate extremely high impulse pressures (see Chapter 2). They were, however, restricted to a lengthy recharge operation, so that quasi-continuous operation became more difficult. Two developments began to change this operation. Exotech, in Maryland developed a pulsating unit which could be mounted on a backhoe (and which was used in early ice cutting tests q.v.) [5.77]. Concurrently Burns at the University of Waterloo began work with the Gas Research Institute to find a way of developing a pulsed jet machine. The first experiments concerned the development of a unit which would use the power from a gasoline-air combustion to drive the piston of the intensifier [5.78]. Because of the need to retain a high jet pressure throughout the stroke of the piston, in order for the jet to be most effective, this meant that the gas in the chamber must still be at high temperature and pressure at the end of the stroke, when it would be vented to atmosphere, ready for the next cycle. As a result a considerable percentage of the energy created by the combustion was lost after the cycle, making this first design quite inefficient.

A second machine was therefore constructed, using the sudden release of compressed nitrogen, as the driving power to the piston (Fig. 5.37), [5.79]. Two units were constructed, capable of generating jet pulses of 328 cc to a maximum pressure of 6,900 bar. This jet could be discharged through a nozzle with a diameter of between 1.0 and 2.5 mm. By using gas as the driving mechanism the unit could be recycled (Fig. 5.38) at a rate of 8 shots a minute, with a maximum frequency, for the early design, of 10 shots a minute. Historically it has been this problem, that of designing a unit to fire at a relatively rapid rate, with the resulting shock loading that this imposes on the supply pipelines, which has limited the commercial development of this product for the market.

A field unit was constructed to fit on a trailer and field experiments carried out. At a peak pressure of 4,140 bar the pulsed jet consistently fractured 20 by 20 by 12.5 cm blocks of concrete from a single central impact, using a 2 mm diameter nozzle [5.80]. This was not possible with a 1 mm diameter jet. When larger blocks (40 by 40 by 15 cm) were struck these could be fractured by pulses from a 2 mm nozzle at 3,800 bar, either by individual blows breaking to the sides of the block, or when secondary impacts broke to the initial crater and to fractures developed from the initial pulse. At 4,200 bar jet pressure the blocks were consistently fractured.

When tests were carried out on large slabs but near an edge it was confirmed that this increased the amount of material that could be removed, and a single shot to the center of an area delineated by saw cuts 15 cm apart would remove the central volume of concrete to the depths of the saw cut. This could not be consistently achieved at a greater cut spacing. The effectiveness of the concrete removal did not appear influenced by the presence of rebar in the concrete. Jet action did penetrate along the interface between the concrete and the steel, breaking the concrete from the reinforcing.

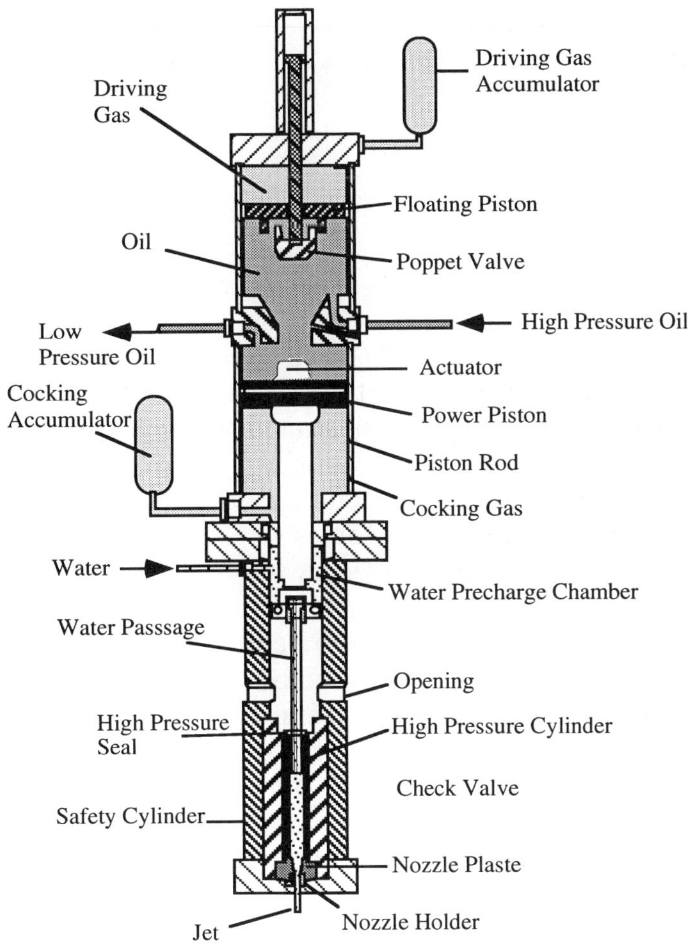

Figure 5.37 Design of a gas driven pulsation unit for concrete breaking [5.79].

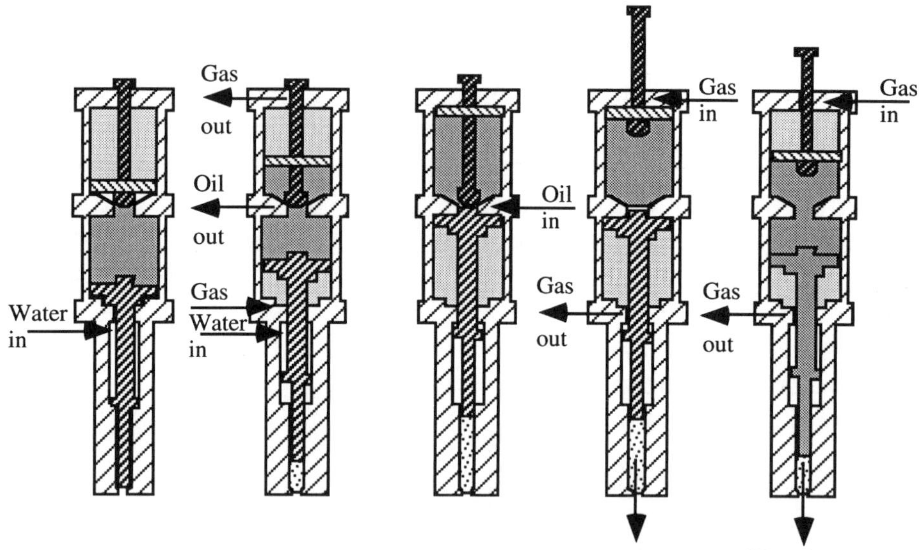

Figure 5.38 Schematic of operating cycle of the pulsation unit [5.79].

This ability of the jet to penetrate along interfaces and delaminate materials was also successfully demonstrated where an asphalt overlay had been applied to the concrete. For example 7.5 cm thick asphalt overlays could be removed from concrete by cutting two saw cuts 25 - 30 cm apart and firing the pulsed jet along the center line between them. The asphalt layer could then be lifted and removed by hand, even though it showed few fractures (ibid.).

The studies showed that for the jet to be effective in breaking concrete it had to penetrate over 50% of the concrete thickness. This required a relatively large diameter jet with a sustained pressure over the duration of the pulse. The system worked best when working within 8 cm of existing edges, but could be designed to work to earlier cavities created by the equipment. Subsequent experimentation confirmed these conclusions [5.81] and led to the recommendation for a change in the design of the equipment.

The concept of a pulsed jet device was further examined by Pater [5.82]. A device was manufactured and attached to an impact breaker framework. Experiments showed that exponential nozzles had a much longer operational life, on the order of 85,000 cycles, in contrast to conventional conical nozzles. The noise generated by the device was a concern. Levels had earlier been measured at around 115 db in the working area of the Canadian device [5.80]. The technology from Dravo was transferred to Briggs Technology, and the driver for the water pressure changed to a Blow Down design (Fig. 5.39) which could achieve several shots a second ([5.83]. Seal problems became evident in the operation of the equipment, however, and the technology is not currently being aggressively pursued.

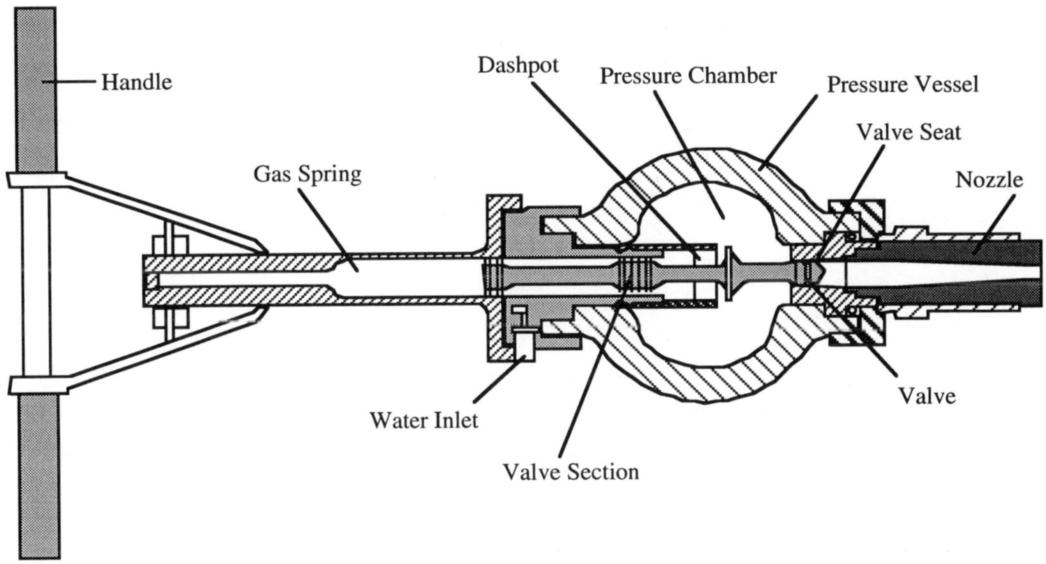

Figure 5.39 Blow down water cannon [5.83].

5.5.4 CONTINUOUS HIGH PRESSURE LOW FLOW SYSTEMS

The work which Olsen had begun at Flow Research had an interesting consequence. The islands around Seattle are connected by bridges with concrete decks. These must be occasionally repaired, a procedure which, historically is carried out by first using jackhammers to chip out the damaged concrete. This is manually intensive and can be quite slow. On one such bridge the contractor was falling significantly behind deadline using this method and approached Flow to determine if they could help. By using an intensifier system it proved possible to use the high pressure waterjets to remove the damaged concrete, and allow the concrete removal and repair to be completed on time. This early demonstration of the technology was probably the first such application for waterjets. It did not create an instant business because the cost benefit to waterjet use, while advantageous in rapidly clearing a bridge in an intensively traveled area, did not easily translate, in those days, into a normally competitive system.

Flow Research were not the only early investigators of this approach. Under funding from the National Science Foundation [5.84] tests were carried out by IIT Research Institute in Chicago in both the laboratory and field examining the use of jets operating in the pressure range between 4,000 and 7,000 bar. Because of the high pressures used, the flow volumes available only allowed the use of nozzles ranging between 0.4 and 0.6 mm in size. The study yielded a predictive equation for the depth of cut which might be achieved, in terms of nozzle diameter:

$$\frac{h}{d} = K_1 \left[\frac{d}{s} \cdot \frac{P}{\sigma_c} \sqrt{\frac{V_j}{V_t}} \right]^{\alpha}$$

The paper is also of interest, however, in that it compared both the costs of cutting concrete, and the rates of excavation achievable by waterjets and conventional methods (Fig. 5.40). Given that the data given was for the lowest waterjet pressure and smallest diameter tested, the authors suggest that increasing the jet operating parameters would have made the figure comparison even more impressive. A cost comparison was made of means for cutting an expansion joint in concrete. A wheel cutter would require 7.5 minutes to cut a slot to a depth of 22.8 cm, over a length of 3.66 m. Using a 4,140 bar jet, through a 0.4 mm orifice, it was projected that a waterjet system would take 4.64 minutes. The teeth on the wheel will normally cut between 2 and 16 joints, for an operational life of perhaps 2 hours. There are 108 teeth on the wheel, and in 1978 they cost $1.33 each. In contrast to the $23,000 capital cost for the cutter, a waterjet system was estimated to cost $66,000. However the jet nozzles, the main wear part, would last up to 800 hours. The authors suggest, based on these figures, that the benefit of using the waterjets would come after 299 hours of operation.

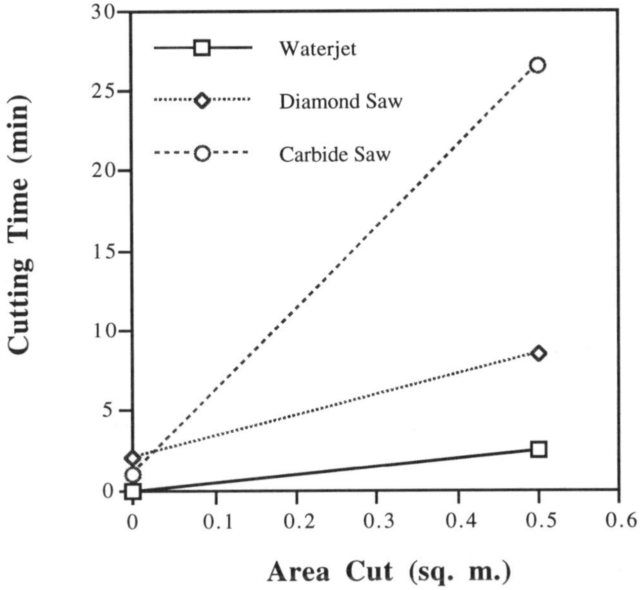

Figure 5.40 Comparative concrete cutting times for a 4,140 bar waterjet, a diamond saw and a carbide saw [5.84].

Although the work at IIT Research Institute continued with a field study after the initial laboratory data was obtained, that work was directed more at removing damaged concrete. It is thus, more relevant to first review the work continuing at Flow Research [5.85]. Working with the Electric Power Research Institute (EPRI) Flow had developed a machine to cut trenches in concrete. The system was designed based upon tests in cutting a concrete made up with a granite aggregate. The choice of the pressure was determined based on determination of **the threshold pressure** for this aggregate material. This is defined as the minimum pressure required to penetrate the material, and this was established as 1,000 bar. Because of some conflicting data which will be discussed later in the text, it should be pointed out that this result was obtained by cutting the concrete at different pressures (Fig. 5.41). A quite small nozzle was used for these tests, as well as a high traverse speed of 50 mm/ sec. In addition it should be noted that the data shown was the depth achieved after a total of ten passes. The data was extrapolated back to a zero depth of penetration, and it was reported that no penetration of the aggregate occurred at that pressure. Based upon a statement "From an energy stand-point, three times the threshold pressure is (the) most efficient pressure" it was then concluded that the machine should operate at a pressure of 3,000 bar.

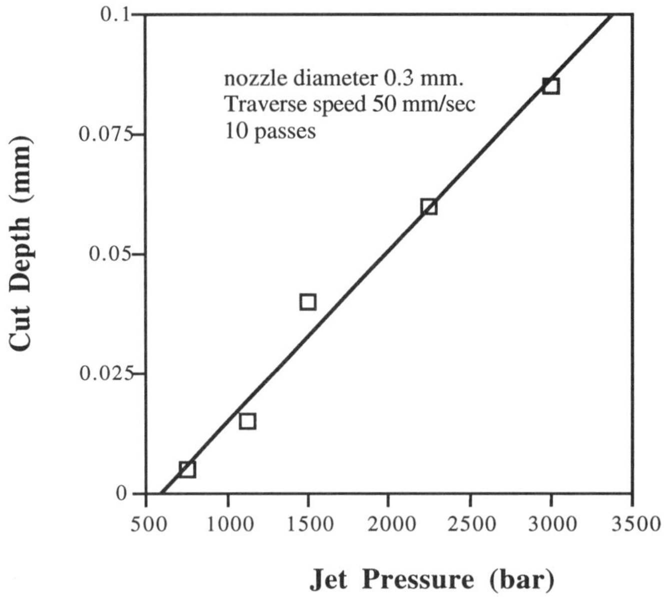

Figure 5.41 Depth of cut in concrete as a function of jet pressure [5.85].

Jet performance is considerably effected by the traverse speed of the nozzle over the surface. Under most test conditions the jet is not traversed fast enough to find an optimum value for the cutting rate in terms of the area of cut created in a second. For example, at IIT [5.84] the study had only carried out tests at speeds of up to 400 mm/sec at which point the curves of area created as a function of traverse rate were still increasing. The Flow study suggested that, for a 0.3 mm diameter jet this increase peaked at a traverse rate of 760 mm/sec (Fig. 5.42). The area of cut created is determined by multiplying the depth of cut by the traverse speed.

Although the jet is cutting efficiently at this speed the depth of cut achieved is quite small and thus, to cut through a 220 mm thick concrete slab some cutting strategy is required. The Flow approach was to cut through the concrete by using three nozzles to make a series of successive passes over the concrete surface, and then to insert a smaller nozzle into the trench thus created and cut through the remainder of the concrete (Fig. 5.43). The three jet assembly had the two outer nozzles inclined out by one degree, in order to cut a slot of constant width. The jets were 12.5 mm apart, and the material between the slots was thus thin enough to break off, and be removed by the jet action. The resulting slot was then wide enough to allow a single nozzle, and holder, to penetrate to the bottom of the cut for the final slotting.

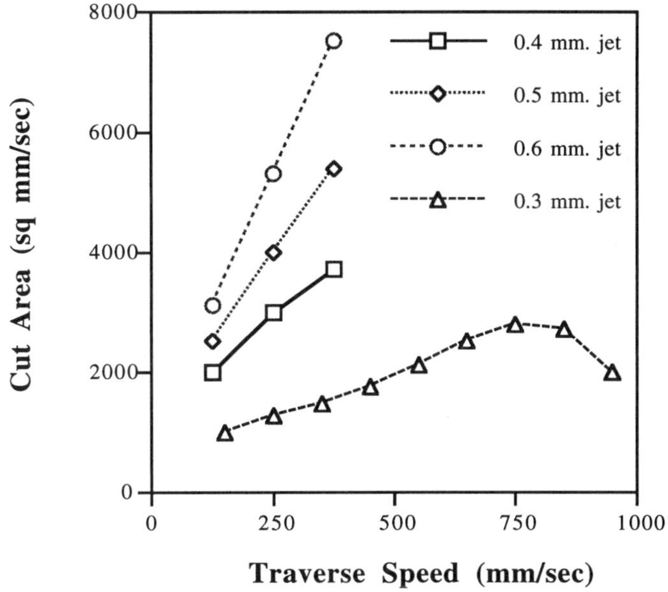

Figure 5.42 Area of cut created as a function of traverse speed [5.84 & 5.85]. The 0.3 mm data was obtained at 3,000 bar, the remainder at 5,510 bar. Stand-off distance was 12.7 and 12.5 mm respectively.

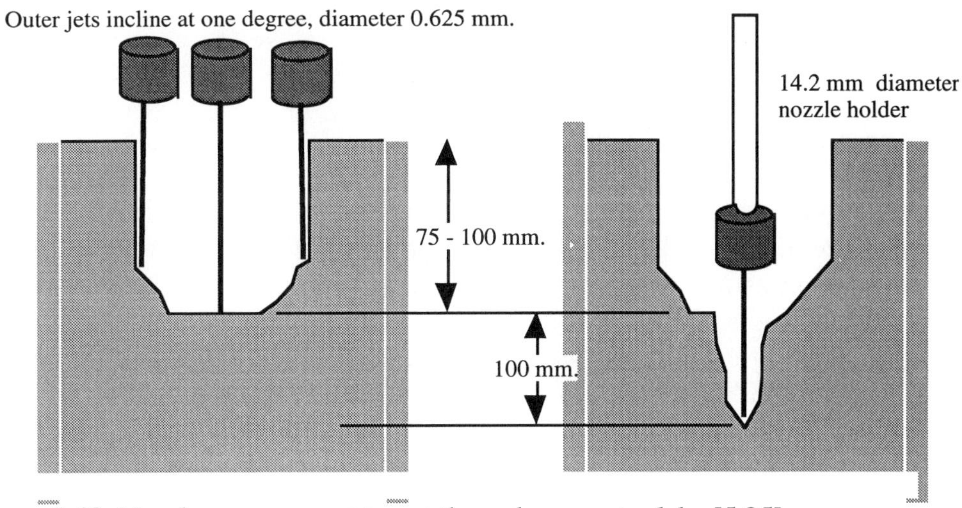

Figure 5.43 Nozzle arrangement to cut through concrete slabs [5.85].

This moving the nozzle into the slot was necessary since the efficiency of cutting dropped rapidly with stand-off distance (Fig. 5.44), and thus as the number of passes is increased, and the distance to the cut surface increased, the relative gain which can be obtained from each cut became less.

Part of the reason for this is that the jet has only a given finite length. This effective length is reduced with higher jet pressures and smaller jet diameters and can also be diminished by entrant flow conditions into the nozzle, and very high lateral nozzle speeds.

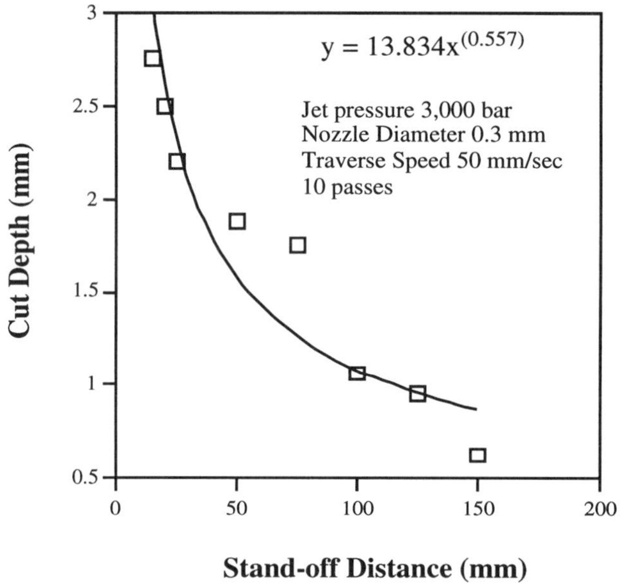

Figure 5.44 Effect of stand-off distance on depth of cut [5.68].

Based upon this concept for concrete slotting a mobile equipment trailer was built containing a pump capable of producing a jet at a flow rate of 30.3 lpm at a pressure of 3,660 bar, with a power of 187.5 kW. In 1980, and based on the sale of 10 units a year, such a system was priced at $215,000. A single operator was priced at $20 an hour, and maintenance was estimated at $20 an hour. The price for cutting concrete then becomes a function of how much can be cut each year, since this must absorb the depreciation of the machine. At 1,000 hours of use a year, and assuming that the machine can cut at 9 m/hr, a price of $10/m is reached, half that of a conventional trenching tool. Such an overall average rate, however, assumed that the highway consisted of both asphalt and concrete, and the jets could cut asphalt much more rapidly than they could cut concrete. The equipment was trailer mounted and tested in field operation by a number of utilities.

Advantages anticipated for the unit included:

- lower, less offensive noise levels from both the equipment and the cutting.
- physically easier to operate.
- creates very little dust.
- does not stress the concrete around the excavation.
- one operator is required who can run the tool from a weatherproof cab.
- does not vibrate, and thus separate, the reinforcing rod from the concrete.

Concurrently work with the IIT Research Institute had switched to a slightly different application. While the utilities are interested in cutting through concrete for access, most highway construction involves the removal of damaged concrete and its replacement. The equipment which had been developed for cutting trenches in concrete was therefore tried in this application on a bridge deck in Cicero, Illinois [5.86]. The effectiveness of different operating pressures was evaluated by tests at four pressure levels from 700 to 2,800 bar, with the highest pressure being found most effective. A different approach was taken in order to achieve a significant penetration depth. Rather than traversing an array of fixed nozzles over the surface, the equipment used a dual orifice nozzle which was rotated as it traveled. This removed all the material in a trench ahead of the nozzle. The traverse speed was standardized at 50 mm/sec, using two nozzles each 0.5 mm in diameter, and rotating the nozzles at 600 rpm. The combined rotation and traversing of the nozzle assembly thus swept out a path over the area to be removed. It proved possible to remove a layer of concrete 63.5 mm deep over an area of 1.13 sq m in an hour using the unit. In comparison Klarcrete removes 0.57 sq m/hr, and a manually operated jack hammer 0.46 sq m/hr. The waterjet requires an 82 kW system, the Klarcrete a 193 kW system, and the jackhammer a 74.5 kW system.

It is already interesting to comment that the cost figures which McCurrich and Browne [5.70] had estimated were being reached, but that the development, being funded by the combined power utilities, and by the Federal Government was not precluded because of this. In fact such price levels have been sustained (but now in 1990s dollars rather than those of 1972) and equipment is quite widely being used.

The apparent success of waterjets as a method of removing concrete, and particularly damaged concrete led to an investigation of the technique by the US Army Corps of Engineers [5.87]. The study was particularly directed at the rapid repair of bomb damaged concrete for runways. This required that an acceptable device be capable of cutting through a slab of concrete, 30 cm thick, at a rate of at least 120 m/hr. Thus the areal cutting rate required should be at least 36 sq m/hr.

Because this is a repair need, the quality of the edge cut was important. Because of these requirements an evaluation of competitive methods in 1983 indicated that the diamond saw remained the most effective technology (Table 5.8), although further work on plain and abrasive laden waterjets was recommended.

Table 5.9 Comparative performance of equipment for trenching concrete in 1983 [5.87]

Technology	*Type of action*	*Performance (sq m/hr)*	*Status in 1983*
carbide saw	cut	5 to 27	commercially available
diamond saw	cut	2 to 32	commercially available
waterjet cutting	cut	0.35 to 22	R&D/commercially available
waterjet and pick	cut	up to 18	R & D
abrasive waterjet	cut	34	R & D
concrete saw/impact	cut/break	4.75 to 30	R & D
powder torch	melt/cut	0.75	R & D
thermal lance	melt/cut	0.75	R & D
powder lance	melt/cut	0.75	R & D
fuel oil/comp air	melt/cut	10	R & D
laser	cut	0.1 to 35	mainly theoretical

The study also had an experimental component, and in July of 1981 a demonstration was held in Vicksburg [5.88]. At this time a different viewpoint was introduced into the discussion. The approach which had been the basis for equipment design until this time had been that the jets must cut through all the components of the concrete, including both aggregate and cement paste. However the cement paste is significantly weaker than the aggregate components, and can thus be much more easily removed. A design philosophy was therefore suggested in which the waterjets be used to erode the cement and not directly attack the aggregate [5.89]. While this would leave some aggregate in the cut supported by cement outside the cut, the removal of support for the rest of the aggregate would let it fall out, and be washed away by the water.

Such an approach has a more effective use in clearing larger areas of concrete, and is particularly suited for use in removing damaged concrete. Because of the technical requirements, it required greater fluid flows that the systems proposed to date, but did not require as great a pressure. In the demonstration it proved possible to cut a slot in blast fractured, but otherwise unweakened concrete at a jet pressure of 800 bar with two jets angled at 22.5 degrees from the vertical. The jets cut a slot some 5 cm wide at an advance rate of 0.7 m/min to a depth of over 5 cm. Progress was slowed by the very large size of aggregate in the concrete at the demonstration site.

This approach was also successfully used in cutting through several reinforced concrete walls at the University of Missouri - Rolla using a jet pressure of 800 bar, and a flow rate of 80 lpm. The flow was directed through a single nozzle which was moved in an orbital path along the outlines of the vertical trench to be cut in the walls (Fig. 5.45). A slot some 5 cm wide was cut through the walls, to a depth of up to 0.6 m as the cuts were completed around the outlines of doors required in building refurbishment on the campus.

Figure 5.45 Slot cut to make a door at the UMR Rock Mechanics Facility.

5.5.5 CONTINUOUS, MEDIUM PRESSURE, MEDIUM FLOW SYSTEMS

The use of higher flow rates, and lower pressures to remove the cement paste, from around the large aggregate pebbles in the concrete, requires consideration of additional parameters in the design of equipment. Given the relative weakness of the cement paste, the pressures need can be much lower, but the power and range required to cut under and around the aggregate requires that the jet cut further, and that it be angled to range under the aggregate (Fig. 5.46). Thus the angle at which the jet is directed to the vertical becomes important in the system design (Fig. 5.47).

a)

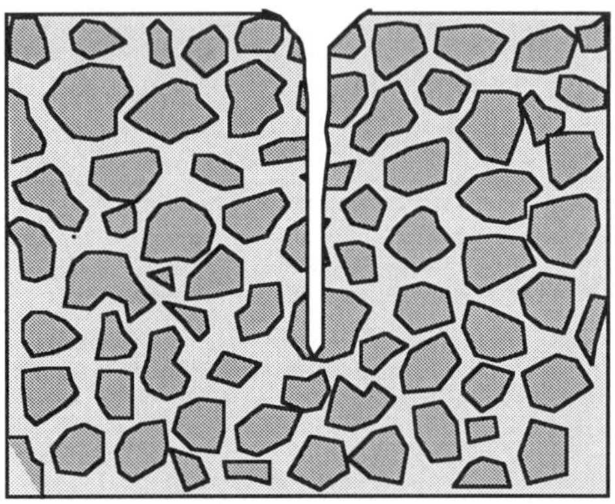

b)

c)

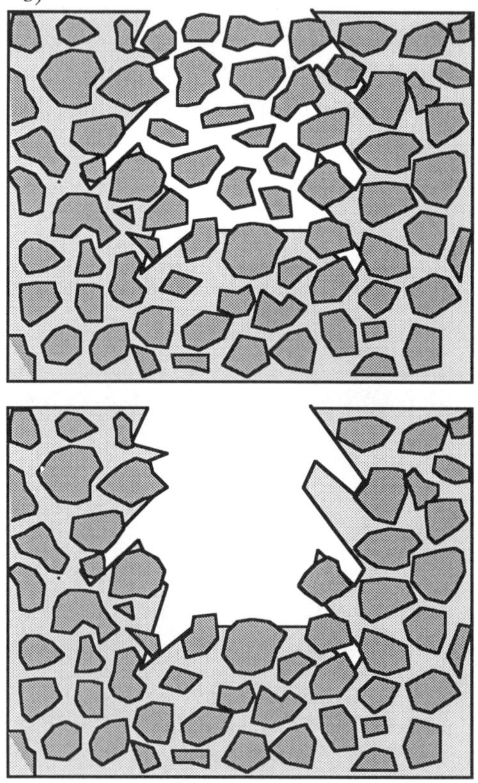

The ultra high pressure jet cut (a) cuts through all the aggregate and cement and leaves a narrow cut, it is constrained by its ability to cut the aggregate.

The medium pressure, higher flow rate jet (b) cuts only the cement paste but washes it out from around all the aggregate. This aggregate falls out (c) leaving a much wider cut. The cutting ability is constrained by the ability to access and remove the much lower strength cement paste.

Figure 5.46 Comparison of high pressure and moderate pressure cutting philosophies.

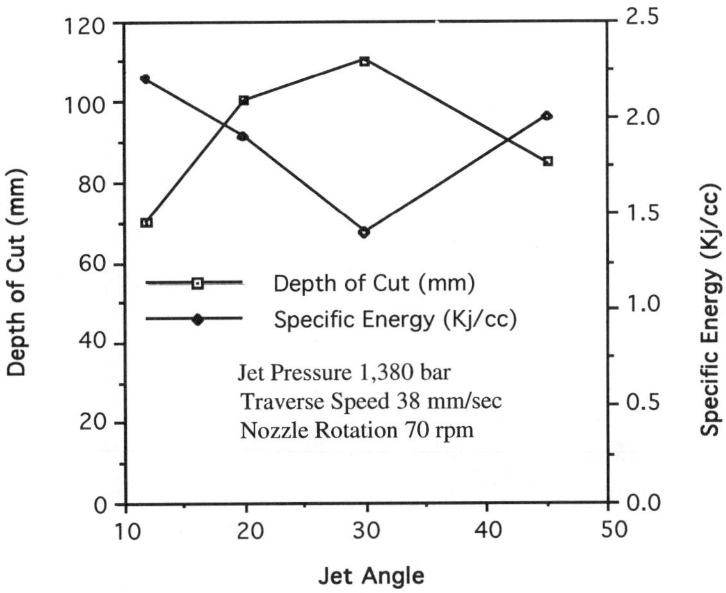

Figure 5.47 Effect of jet angle on concrete removal efficiency [5.90].

While this technique was being developed in America, a similar approach had become commercially successful in Europe. Developed with the technical name of **Hydrodemolition,** a research effort was initiated by the Italian company FIP in 1979 [5.91]. A prototype of the equipment was successfully demonstrated on the Viadotto del Lago in November of that year, and the first commercial equipment was available in 1980. The developed commercial version of the equipment was first tried in Sweden in 1984, and demonstrated in a Toronto parking garage in 1985. The equipment was given the name "The Shrimp". It operates at a pressure of between 850 and 1,000 bar. Concurrently with the later stages of the development of this equipment, competing firms were also introducing similar equipment.

This approach has proved to be very effective for removing deteriorated concrete. The reason for this, as will be discussed in Chapter 11, is that waterjets work by penetrating and extending cracks in the material. Deteriorated concrete is material with a greater number of cracks, and cracks which have grown in length. By using the power of the waterjet to fill and pressurize these cracks it is possible to get them to grow and spall off the damaged material. At the same time the pressure in the jet is set below that required for crack growth in the healthier underlying concrete. In this way the tool has some "inherent intelligence" in that, by the nature of the way it works, it can selectively remove damaged concrete while leaving the stronger healthy concrete in place.

Figure 5.48 Equipment for removing concrete from surfaces.

A number of competing devices are now available, and illustrative examples of their performance, may be the best way of explaining how they work. In most cases the basic form of the equipment is the same. A trailer houses the high pressure cutting equipment drive and pumps, and this feeds water to a boom on which the cutting equipment is placed (Fig. 5.48). The high pressure nozzle moves along a guide rail within a shrouded channel, on the end of the boom. The shroud acts to contain the resulting spray and debris, which can be collected after the excavation has been completed. The jets will effectively remove concrete from above and below reinforcing rods and leaves the aggregate protruding from the concrete on the surface. This makes it easier to bond the patching concrete to the surface, and does not leave a thin flat plane as a joint, which might fail prematurely (Fig. 5.49)

Figure 5.49 Concrete surface after waterjet removal of the top layer.

5.5.6 CASE HISTORIES

The technology is not restricted to removing concrete from the upper surfaces of flat roadbed and bridge decks. In 1984, for example, the road tunnel under the river Tyne in Northern England needed some 200 sq m of damaged concrete to be removed to a depth of 20 mm behind the reinforcing rebar [5.92]. The small cutting head required of a waterjet system, and its easy maneuverability led to the choice of this technique for removing the concrete. However problems encountered included poor visibility and the limited access space available, so that the total excavation period which SDL Pressure Jetting Ltd required for the project was some 7 weeks, including the time to twice dismantle and reassemble the special equipment built for the work.

In April 1985 [5.93] one of the interstate highway, I-190 bridges over the Niagara River near Grand Island, NY was closed for repair. Some 4,500 sq m of the bridge needed partial removal and replacement of the concrete, and 630 sq m needed total concrete replacement. The FIP equipment was operated at a pressure of 800 to 1,000 bar, at a flow rate of 240 lpm. The cutting nozzle swept out a path 1.9 m long with the nozzle oscillating to cut a channel from 2.5 - 7.5 cm wide, spaced some 10 - 15 cm apart due to the forward motion of the machine. The jet action thus removed the intervening material, and fully exposed and cleaned the upper course of the reinforcing steel. Each highway lane required two passes by the unit to completely cover the width. While normal perform-ance of the unit is around 13.5 sq m/hr, the harder concrete at the site reduced the rate to 4 sq m/hr when cutting to an average depth of 7.5 cm. The jet removed over 90% of the material, but some concrete remained to be removed by chipping hammers. This was largely concrete in the zone within 2.5 cm under the reinforcing, which the contract mandated be removed, even if undamaged and well bonded to the rebar. The unit could not also reach and remove the curbs and adjacent concrete. In 1985 this equipment also worked on the repair of the Lincoln Memorial Bridge in Washington, D.C.

Atlas Copco have studied the potential use of waterjets in construction and mining for a number of years, and have developed a "Con-Jet" machine for the removal of deteriorated concrete. This unit is built in three parts, a power unit, a remote control unit, and the boom mounted cutting arm, which contains the waterjet nozzle. Interestingly this unit was also tried on the repair of a Grand Island Bridge of I-190 in New York in 1985 [5.94]. The unit was operated at 1,200 bar and used 96 lpm of water. The unit took about 20 minutes to set up and the remote operation allowed the operator to stand up to 78 m from the action. Some 7 m from the unit the noise level was reported as only being 75 decibels. In order to speed operation on the bridge the contractor used two power units to supply a second nozzle on the cutting arm which was then above to remove between 90 and 180 sq m/day of concrete working 24 hours, and cutting to a depth of 7.5 to 15 cm. The concrete debris was then picked up by a vacuum truck. Using this process the bridge was repaired in some four months. This is the more remarkable since the contract was initially bid based on an estimated damaged area of 1260 sq m, but when examined it was found that some 6,750 sq m had to be treated.

The unit was reportedly able to scarify concrete at a rate of 8 sq m/hr in removing the top 6 mm of concrete, and at a speed of 7.2 sq m/hr when removing the top 12 mm of concrete. A 1985 price of $500,000 was given for the unit and support equipment, and operating costs were adjudged the same as for conventional methods, some $90 to $110 per sq m.

A comparison was made by the Indiana Department of Transportation (DOT) between this technique and conventional jackhammer operations in repairing two bridges in that state [5.95]. The jackhammer operation required some 7 persons 12 days to remove 317.2 sq m while the hydro-demolition unit took 6 days to remove 467 sq m. After two years of operation it was possible (ibid.) to obtain operating costs for the Atlas Copco system (Table 5.10) in 1987 dollars.

Table 5.10 Comparison of jackhammer and hydrodemolition [5.95]

Item	lane 1	lane 2	lane 3	lane 4	lane 5	lane 6
Method	hammer	waterjet	hammer	waterjet	hammer	waterjet
Days	12	6	17	12	20	3
Person/machine hours	756	39.5	1071	128	728	18
Repair size (sq m)	317.2	467.4	365	799	139	105.8
Av depth (mm)	88.4	52.3	94.5	61.2	95.5	153.4
Repair rate	0.42	11.8	0.34	6.24	0.19	5.87

Table 5.11 Operating costs for a hydrodemolition unit for a 10 hour day [5.95]

Item	*Cost ($)*
fuel - 0.013 l/sec	96.00
grease and lube	2.00
oil	0.50
filters	5.00
traction system (robot)	4.00
pump maintenance (8 year)	55.00
engine and radiator	8.00
pistons	20.00
cylinders	30.00
5 meter hoses	15
nozzles	50.00
miscellaneous	2.00
Total maintenance	**287.50**
machine operator	262.50
maintenance person	30.00
Total labor cost	**292.50**
Total operating cost	580.00
equipment cost amortization	945.32
average production	72 sq m/day
Cost/sq m	**$21.18**

The Equipment cost was derived based upon a delivered purchase price of $436,300 and a use of the equipment for 90 days a year.

This can be compared with the use of the Flow equipment which was demonstrated on the Dearborn Street interchange bridge in Seattle [5.96]. This equipment operated at 1,700 bar with a flow rate of between 57 and 167 lpm. Improved performance was achieved both by increasing the flow rate to increase jet power, and by keeping the horsepower constant and increasing the jet pressure (Fig. 5.50).

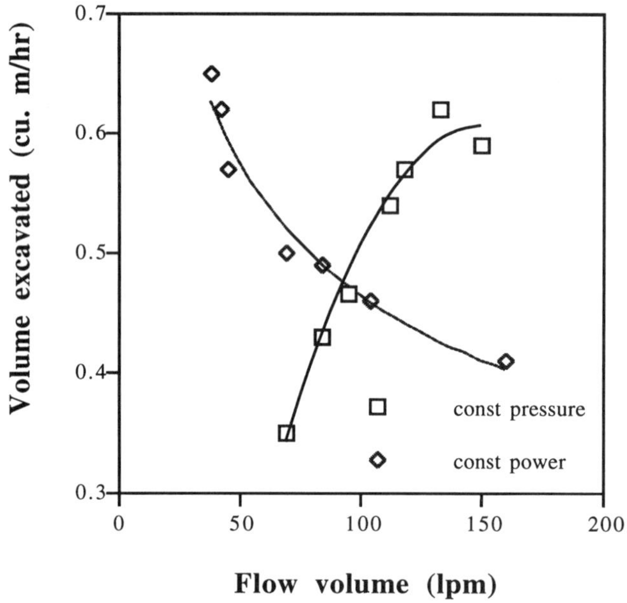

Figure 5.50 Change in volume removed with flow rate (after [5.78].

Schmid estimated that equipment maintenance and fuel would cost $5.00 an hour for a jackhammer, and that where labor costs were less than $12.00 an hour that the use of the waterjets would be economic using the Flow system. The paper also reported that the shear strength of a waterjet cut and repaired surface was some 2.3 times higher than a jackhammer repaired surface, and the pull-off strength was some 3.1 times higher.

The English problem is a little different to that in the United States since the major damage to bridges often occurs underneath the bridge, rather than on the more easily accessible deck. By changing to a 2,000 bar system, at a flow of 12 lpm it proved possible to use hand-held equipment to access the work area, and to clean the damaged concrete from the rebar at a rate three to four times faster than that achievable at 1,000 bar [5.97]. In relation to the original cost estimate for equipment, a 1990 price ranged from \$40,000 to \$300,000 and this brought the equipment into the range affordable by a number of small companies.

5.5.7 INJECTION OF ABRASIVE FOR CONCRETE CUTTING

The continued need for a faster method of cutting slots in concrete led both EPRI and GRI to fund a study at Flow to examine the potential for abrasive injection into the waterjet stream as a more effective method of cutting concrete. This method had the advantage of also being able to cut through any rebar present in the concrete, and of achieving total penetration through the aggregate and cement to the bottom of the slab on a single pass, without the need to feed the nozzle into the slot.

Initial results [5.97] suggested that the use of garnet was more effective than either silica sand or silicon carbide. The optimum feed rate was at an abrasive flow of 68 gm/sec although a 38 gm/sec feed was only 11% less effective. The first trials showed that it was possible to completely cut a 380 mm thick slab of concrete at an advance rate of 0.4 mm/sec using a jet at a pressure of 2,415 bar, a waterjet diameter of 0.635 mm, and an abrasive feed rate of 72 gm/sec. In these experiments it was possible to simultaneously cut a reinforcing bar 10 mm in diameter, located 75 mm below the top surface of the concrete, but a second rod, some 127 mm lower was only partially cut at a feed rate of 1.3 mm/sec. The nozzle design used in the initial study was refined [5.98] and it then proved possible to cut through a 250 mm thick concrete slab and two 18 mm diameter courses of reinforcing (at 75 mm and 175 mm into the slab) at a feed rate of just over 25 mm/min using approximately half the amount of abrasive initially required.

It was then estimated (1983) that hourly costs for cutting such a slab would be approximately \$12.00/hr for equipment, \$9.00/hr for abrasive, and \$4.00/hr for other expendables.

In order to improve on the initial performance data presented, Japanese investigators have also studied the problem of abrasive use in cutting through concrete [5.99]. Tests were carried out at flow rates of up to 15 lpm at a maximum of 2,940 bar. An initial comparison of the relative effectiveness of different abrasive indicated (Fig. 5.51) that iron grit had considerable potential as an alternate abrasive.

Iron grit was found more effective than aluminum oxide from two points of view. Firstly as the depth of cut increases the greater density particle carries a greater proportion of its energy to the bottom of the cut, and secondly the harder and more brittle aluminum oxide tends to shatter on impact, while the iron retains its structure. This allows the recycling of the particles with subsequent savings in material costs. An optimum feed rate of 25 kg/ min for a jet at 1,960 bar was found optimum for the iron grit, under which condition it could cut through a slab 100 mm thick, at a traverse speed of 200 mm/min. In comparing the effective life of ceramic and tungsten carbide mixing nozzles, it was found that the wear rate of the tungsten carbide nozzle was significantly greater when iron grit was used rather than aluminum oxide, but that this relationship reversed when the nozzle was made of a ceramic material (Fig. 5.52).

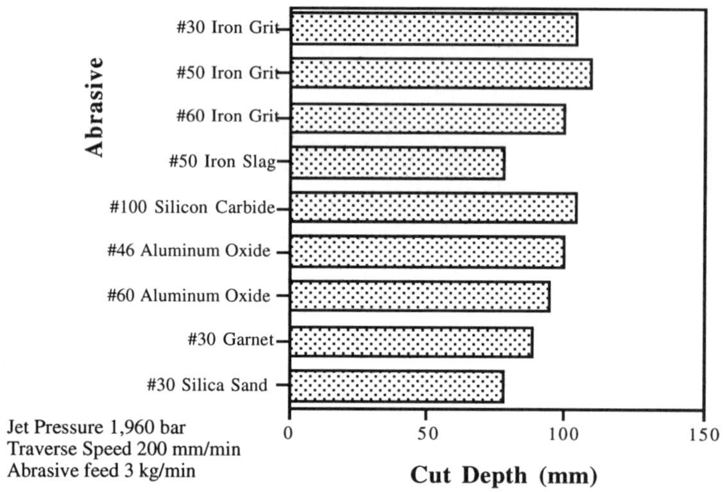

Figure 5.51 Effect of different abrasive materials on cutting concrete [5.99].

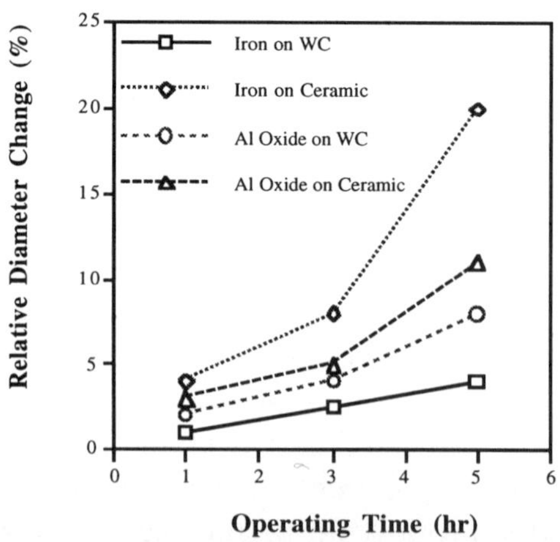

Figure 5.52 Effect of abrasive and collimating nozzle material on nozzle life [5.99].

Because iron grit is of greater density than garnet, the commonly used abrasive for jet cutting, a smaller number of particles are involved in the cutting process for equivalent mass flow rates. However Yie, using a nozzle design with a multiple jet flow into the mixing chamber, (which will be discussed more in a subsequent chapter) found that equivalent depths of cut could be achieved for the same mass flow rates [5.100]. With that nozzle design optimum abrasive feed rates were on the order of 1.0 kg/min. For example, operating at a jet pressure of 1,000 bar, it was possible to cut through approximately 180 mm of concrete at a feed rate of 100 mm/min. If Yie's results are compared with those of the Japanese [5.99] at a traverse speed of 200 mm/min he reported being able to cut through 100 mm of a concrete made up with a granite aggregate. This is with a jet of approximately half the pressure and containing half the abrasive used for the Japanese result. It was also possible to cut to depths of more than 500 mm with this system.

The relatively small and easy maneuverability of the waterjet head has been advantageously used before, and these particular advantages still hold true for abrasive waterjet cutting. The increased cutting performance of the jets, and the ability to cut through both concrete and metal has found several applications of particular interest.

In Japan there was a need to cut through the concrete walls between adjacent apartments while remodeling buildings [5.101]. A comparison with impact breakers, hydraulic breakers, and diamond saws indicated that abrasive waterjets had a greater potential for effective cutting. A special cutting rig was therefore constructed (Fig. 5.53).

Because of the cutting through the wall a special protective unit was required on the far side of the wall and this was achieved with a 45 degree inclined deflection plate made of silicon nitride (Fig. 5.54). The jet cut through the wall at a pressure of 1,960 bar and a flow rate of 13 lpm 45 kg/min of garnet was mixed with the waterjet and cut the 180 mm thick reinforced concrete wall at a rate of 1.3 mm/sec. It was particularly important in this application to reduce the operating noise level, and a rubber buffer was held in place around the cutting equipment by applying a small vacuum to the enclosure. In this way the noise was reduced to 77 db in the operating area and 58 db in the surrounding rooms (Fig. 5.55). Operating costs for the system was estimated to be 32,900 yen/m.

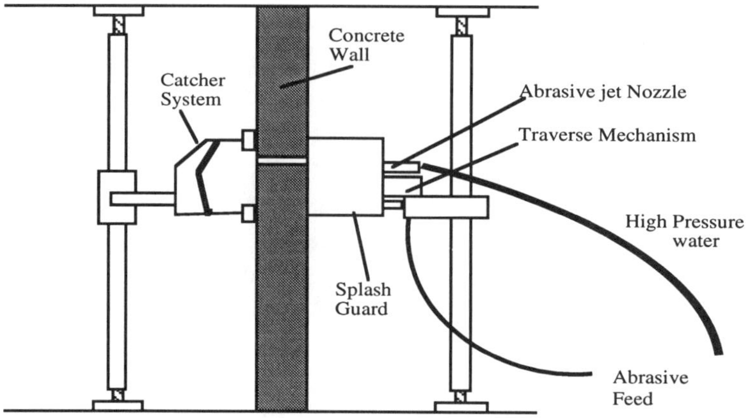

Figure 5.53 Equipment layout for cutting apartment walls [5.101].

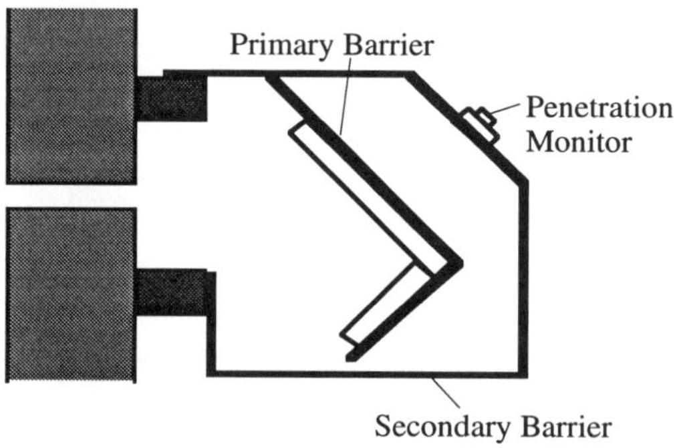

Figure 5.54 Detail of catcher used to prevent damage behind the wall [5.101].

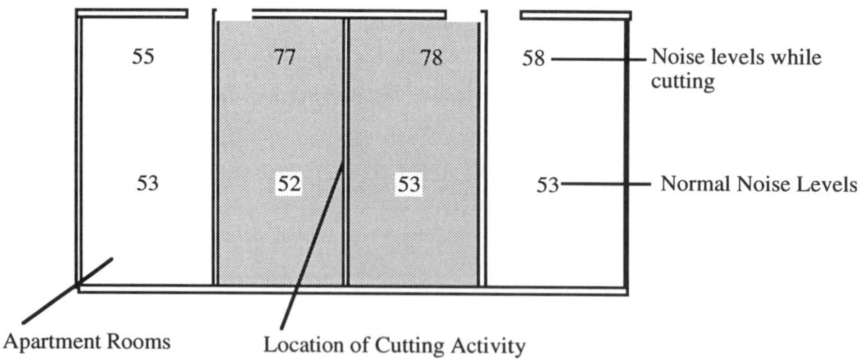

Figure 5.55 Noise levels around an abrasive jet cutting apartment walls [5.101].

A similar system was used to renovate a hospital in a suburb of Tokyo [5.102]. In this operation a jet pressure of 2,450 bar was used with a flow rate of 11.4 lpm and a feed of 2.2 kg/min of garnet to the cutting nozzle. The wall measured between 160 and 200 mm thick, and could be cut at a rate of 50 mm/min. Noise levels near the machine were 86 db, and 51 db in the nearby hallway. This compared with a noise level of 110 db which the cutter created, without the use of the noise suppression equipment.

A second application of abrasive cutting systems is developing in their use for dismantling nuclear reactors. A preliminary study of this use has been reported [5.103] in which an abrasive waterjet lance was rotated and cut into a concrete block in a deep trenching experiment. In this application, where the collection and disposal of contaminated materials has a significant role in process economics, it was concluded that at a disposal cost of $600/cu m, recycling steel shot as the cutting abrasive became an economic advantage at a recovery rate of 80% of the material for reuse. A field demonstration of the equipment was carried out at the West Valley Nuclear facility in 1987 [5.104]. A kerf 1.2 m deep could be cut into the concrete at a rate of 0.46 sq m/hr.

The problem of contaminated spoil disposal increases where abrasive is used with the waterjets and provides an opportunity where cavitating waterjets may have an advantage.

The results of the development of the abrasive jet cutting system as a means of cutting access holes for use by the power utility companies was reported in 1985. A 2,400 bar system was marketed [5.105] which was reported to be capable of cutting through a 150 mm thick concrete slab at a feed rate of between 6 and 1,000 mm/min. It would take the system 4 minutes to cut the perimeter of a 250 mm diameter circle in 150 mm thick concrete.

Although systems have been developed, and commercially used which can cut through reinforced concrete at a single pass, the amount of effort to cut through the material becomes constrained by the relatively slow feed rate required to ensure cutting of the steel. For this rate it has been recommended [5.106] that, where possible the concrete be cut first, and then a return cut made to remove the rebar. In this way the greater energy demand for the cutting of the rebar, up to twenty-eight times that needed to cut the concrete, will not be wasted when the rebar is not present.

More recently the advent of the DIAjet system has allowed concrete cutting systems to be developed which operate at much lower pressures, typically on the order of 350 bar [5.107]. One problem with these systems in some of the applications just described is that they generally use more water and abrasive, which becomes considerably more difficult to manage within dwellings and must accordingly be planned for, particularly where possible to ensure recycling of both water and abrasive. Nevertheless such systems can operate with very light and simple equipment and can cut through not only the walls of the opening, but also adjacent layers of reinforcing steel (Fig. 5.56). The improved cutting rates and the reductions in the requirements for water and abrasive suggest that higher pressure systems are likely to be more beneficial, regardless of method of abrasive injection, in such building modifications.

Figure 5.56 Wall section cut with a traversing direct abrasive injection system.

5.5.8 CUTTING CONCRETE WITH CAVITATION

The demand for an economic means of cutting access holes for utility companies did not stop with the development of the abrasive jet cutting system. The relatively high cost of the abrasive suggested that an alternate system, with cavitation as the jet enhancing mechanism, might have a potential use. Reported costs for abrasive jet cutting were $35.00 to cut a perimeter length of 0.6 to 1.2 m around a hole 200 mm thick using a jet at a pressure of 2,000 bar. Conn [5.108] reported that a similar slot could be cut at a cost, in 1986 dollars, of $18.70 using a center body cavitating waterjet at a pressure of only 700 bar. This is in comparison with a cost of $42.47 for conventional cutting of the same hole.

A simple demonstration of this technique was carried out in Baltimore to repair a leaking gas main. The trial was delayed by problems of gaining water (the adjacent two hydrants were broken) and the water supplied was very muddy. The cavitating nozzle was repeatedly passed over the slot to be cut, and required a total of 60 passes, at a traverse speed of 229 mm/sec to cut through the 89 mm thick asphalt, including a 25 mm topcoat, and underlying 150 to 250 mm thick concrete. It took a total of 13 minutes including several equipment moves, to cut around the 2.74 m long perimeter of the concrete block. An additional ten minutes was required to break out and remove the central concrete. This is in contrast to a total of 45 minutes which would have been required for a jackhammer operation. A further discussion of this and other uses of cavitating jet systems will be given in Chapter 10.

5.6 NOTES AND REFERENCES TO CHAPTER 5

The development of waterjets in civil engineering applications relates very closely to their development for mining applications. When this chapter was first begun, the two applications were combined and jointly described. Unfortunately this proved to be an unwieldy way of writing, and a partially artificial distinction has thus had to be made. Since many of the tools can be jointly used, it is recommended that all the chapters which deal with the cutting and cleaning of geotechnical materials be reviewed, since much of the information will transfer from one area to another.

5.7 REFERENCES

5.1 Wilkinson, The Ancient Egyptians, 1874, Vol. 2, p. 137.

5.2 Kelley, R.L., "Forgotten Giant - The Hydraulic Gold Mining Industry of California," Pacific Historical Review, Vol. 23, November, 1954, pp. 343 - 356.

5.3 Pliny (Caius Plinius Secundus), Natural History, A.D. 42, Book 33,

5.4 MacDonald, E.H., Alluvial Mining, Chapman & Hall, 1983, 508 pp.

5.5 Longridge, C.C., Hydraulic Mining, publ. Mining Journal, 1910.

5.6 Anon, Hutching's California Magazine, July, 1857.

5.7 Anon, Daily Alta Californian, June, 1853.

5.8 May, P.R., Origins of Hydraulic Mining in California, Holmes Book Co., Oakland, CA, 1970.

5.9 Kelley, R.L., Gold vs. Grain, Glendale, Arthur H. Clark, Co., 1959.

5.10 Egenhoff, R.L., The Hydraulic Mining Situation in California, 1931.

5.11 Kallenberger, Memories of a Gold Digger, Garden Grove, CA., 1970.

5.12 Davies, T.W., and Jackson, M.K., "Optimization of Nozzle Flow/head Requirements for China Clay Mining," paper F1, 7th International Symposium on Jet Cutting Technology, Ottawa, Canada, June, 1984, pp. 293 - 314.

5.13 Davies, T.W., Metcalfe, R.A., and Jackson, M.K., "The Anatomy and Impact Characteristics of Large Scale Waterjets," paper A2, 5th International Symposium on Jet Cutting Technology, Hanover, Germany, June, 1980, pp. 15 - 32.

5.14 Leach, S.J., and Walker, G.L., "Some Aspects of Rock Cutting by High Speed Water Jets," Phil. Trans. Royal Society, London, Vol. 260A, 1966, pp. 295 - 308.

5.15 Davies, T.W., and Jackson, M.K., "Optimum conditions for the hydraulic mining of china clay," paper 3.4, 1st U.S. Water Jet Symposium, April, 1981, Golden, CO.

5.16 Okrimenko, V.A., Hydro-Monitor Operator in Coal Mines and Pits. State Scientific Technical Press of Literature on Mining, Moscow, 1962, 264 pp. (Translation U.S. Army Foreign Science and Technology Center, Document AD 820634, 1967).

5.17 Yufin, A.P., Hydromechanization, State Scientific Technical Press of Literature on Mining, Moscow, 1965 (Translation U.S. Army Foreign Science and Technology Center Translation J-1061, 1967).

5.18 Nelson, C., "Present Tunneling Techniques - St Anthony Falls Storm Sewer Project," Workshop on the Application of High Pressure Water Jet Cutting Technology, Rolla, MO, November, 1975, pp. 359 - 373.

5.19 Atmatzidis, D.K., and Ferrin, F.R., "Laboratory Investigation of Soil Cutting by a Water Jet," 2nd U.S. Water Jet Conference, Rolla, MO, May, 1983, pp. 101 - 110.

5.20 Griffith, A.A., "The Theory of Rupture," 1st International Congress on Applied Mathematics, Delft, Holland, 1924, pp. 55 - 63.

5.21 Brace, W.F., "Dependence of Fracture Strength of Rocks on Grain Size," 4th Symposium on Rock Mechanics, Univ. of Pennsylvania, 1961, pp. 99 - 103.

5.22 Summers, D.A., and Zakin, J.L., The Structure of High Speed Fluid Jets and Their Use in Cutting Various Soil and Material Types, Final Report on Contract DAAK002-74-C-0006, U.S. Army Mobility Command, R & D Procurement Office, Fort Belvoir, VA., 1975.

5.23 Summers, D.A., The Water Jet Cutting of Frozen Gravel, Final Report on DACA 89-71-1452, U.S. Army Cold Regions R & E Laboratory, Hanover, NH, September, 1971.

5.24 Frank, J.N., Fogelson, D.E., and Chester, J.W., "Hydraulic Mining in the USA," paper E4, 1st International Symposium on Jet Cutting Technology, Coventry, UK., April, 1972, pp. E4 45 - E4 60.

5.25 Mellor, M., "Jet Cutting in Frozen Ground," paper G2, 1st International Symposium on Jet Cutting Technology, Coventry, UK., April, 1972, pp. G2 13 - G2 24.

5.26 Foster-Miller Associates, Inc., Fundamental Concepts for the Rapid Disengagement of Frozen Soil, USACRREL Technical Reports, 233 and 234, 1971.

5.27 Summers, D.A., and Henry, R.L., "The effect of change in energy and momentum levels on the rock removal rate in Indiana Limestone," paper B5, 1st International Symposium on Jet Cutting Technology, Coventry, UK., April, 1972, pp. B5 77 - B5 88.

5.28 Summers, D.A., The Water Jet Cutting Of Frozen Gravel, report for USACRREL Hanover, NH., 1971, unpubl.

5.29 Mellor, M., "Cutting Ice with Continuous jets," paper G5, 2nd International Symposium on Jet Cutting Technology, Cambridge, England, April, 1974, pp. G5 65 - G5 76.

5.30 Shvaishtein, Z.I., "The Use of Steady High-Pressure Jets to Cut Ice," in Studies in Ice Physics and Ice Engineering, Ed. G.N. Yakovlev, Israel Program for Scientific Translation, Jerusalem, 1973, pp. 148 - 155.

5.31 Peschanskii, I.S., and Shvaistein, Z.I., Description and Operating Principle of an Ice-Cutting Vessel, Leningrad, Izdatel'stvo Arkt. i Antarkt. Instituta, 1966.

5.32 Yakovlev, G.N., "The Use of a High Speed Gas Jet to Break Ice," in Studies in Ice Physics and Ice Engineering, Ed. G.N. Yakovlev, Israel Program for Scientific Translation, Jerusalem, 1973, pp. 135 - 147.

5.33 Brierley, W.H., Workshop on the Application of High Pressure Water Jet Cutting Technology, UMR, 1975, pp. 171 - 188.

5.34 Calkins, D., and Mellor, M., "Investigation of Water Jets for Lock Wall Deicing," paper G2, 3rd International Symposium on Jet Cutting Technology, Chicago, USA, May, 1976, pp. G2 13 - G2 22.

5.35 Vijay, M.M., Grattan-Bellew, P.E., and Sinha, N.K., "Drilling and Slotting of Ice and Permafrost with Rotating High Pressure Waterjets," paper 17, 8th International Symposium on Jet Cutting Technology, Durham, UK., September, 1986, pp. 177 - 188.

5.36 Ohadi, M.M., Haase, R.A., and Whipple, R.L., "A basic study of ice detachment from road surface via high pressure abrasive liquid jets," paper 23, 10th International Symposium on Jet Cutting Technology, Amsterdam, NL., October, 1990, pp. 335 - 348.

5.37 Kolle, J.J., "Moving Ice Mountain," Mechanical Engineering, February, 1990, pp. 48 - 53.

5.38 London Sunday Times, December 16, 1973, p. 33.

5.39 Brunsing, T., "Jet Notching used in the Construction of a Chemical Waste Isolation Barrier," 2nd U.S. Waterjet Conference, Rolla, May, 1983, pp. 299 - 305.

5.40 Yahiro, T., Yoshida, H., "On the Characteristics of High Speed Water Jets and Their Utilization in the Induction Grouting Method," paper G4, 2nd International Symposium on Jet Cutting Technology, Cambridge, UK., April, 1974, pp. G4-41 - G4-63.

5.41 Du, JiaHong, and Shi, HaiYu, "Research on how to apply the High Pressure Water Efflux Technology to Leak Prevention or Water Cut-off Project of Tailing Dam," International Water Jet Symposium, Beijing, China, September, 1987, p. 5-57 - 5-66.

5.42 Bowen, R., Grouting in Engineering Practice, Applied Science Publishers, Ltd., 1975, 187 pp.

5.43 Nissan Freeze Co., Ltd., CCP Jet Grouting Method for Consolidating and Stabilizing Soft Ground, 1974, N.I.T. Co., Ltd., 5-21-18, Higashigotanda, Shinagawa-Ku, Tokyo, Japan.

5.44 Isaac, R., Communication with David Summers, May 1994.

5.45 Yahiro, T., Yoshida, H., and Nishi, K., "Soil Improvement Utilizing a High Speed Waterjet and Air Jet," paper J2, 6th International Symposium on Water Jet Technology, Surrey, UK., April, 1982, pp. 397 - 428.

5.46 Baumann, V., "On the Execution of Soil Improvement Method Utilizing Water Jet," paper III-3, International Symposium on Water Jet Technology, Tokyo, Japan, December, 1984.

5.47 Coomber, D.B., "Tunneling and Soil Stabilization by Jet Grouting," Tunneling '85, Proceedings of the 4th International Symposium, Institution of Mining and Metallurgy, March, 1985, pp. 277 - 283.

5.48 Tornaghi, R., Perelli, and Cippo A., "Soil Improvement by Jet Grouting for the Solution of Tunneling Problems," Tunneling '85, Proceedings of the 4th International Symposium, Institution of Mining and Metallurgy, March, 1985, pp. 265 - 276.

5.49 Yoshida, H., Shibazaki, M., Kubo, H., Jimbo, S., and Sakakibara, Mr., "The Effect of Pressure and Flow Rate on Cutting Soil Utilizing Waterjet for Wider Application," 5th American Water Jet Conference, Toronto, Canada, August, 1989, pp. 297 - 305.

5.50 Horiguchi, T., and Kajihara, K., "Development of a Piling System of Precast Piles by Applying Rotary Jet," paper L4, 9th International Symposium on Water Jet Technology, Sendai, Japan, October, 1988, pp. 591 - 610.

5.51 Anon, "Sheet Pile installation with waterjet," Industrial Jetting Report, May, 1987, No. 47, p. 3.

5.52 Yahiro, T., Yoshida, H., and Nishi, K., "Sheet Piles Driving and H steel Pulling by High Speed Waterjets," paper D4, 5th International Symposium on Water Jet Technology, Hanover, Germany, June, 1980, pp. 237 - 250.

5.53 Hoshino, K., Hagihara, M., Shikata, S., and Yamatani, Y., "750 kW Water Jet Pump for Pile Driving," paper H5, 4th International Symposium on Water Jet Technology, Canterbury, UK., April, 1978, pp. H5 63 - H5 66.

5.54 Kramer, S.R., "The Global Business of Trenchless Technology," Trenchless Technology, Vol. 2, No. 3, May/June, 1993, pp. 46 - 47, 57.

5.55 Reichman, J.M., Kelley, D.P., and Marvin, M., "The Further Development of an Underground Cable Following Tool," 2nd US Water Jet Conference, Rolla, MO., May, 1983, pp. 307 - 314.

5.56 Trenchless Technology, Trenchless Technology, Inc., P.O. Box 190, Peninsula, OH., 44264.

5.57 Woodward, M.J., "Construction Applications," in Fluid Jet Technology - Fundamentals and Applications, Water Jet Technology Association, 1993.

5.58 Anon, "Flowmole installs Gas Mains," Industrial Jetting Report, August, 1988, No. 62, p. 4.

5.59 Mahoney, R.D., and Carville, C.A., "Recent Developments and applications using waterjetting for tieback anchors and hillside dewatering," paper 11, Proceedings 5th American Water Jet Conference, Toronto, Canada, 1989, pp. 111 - 119.

5.60 Hann, P., and Reinhardt, W.G., "Thin arch gets a new foundation," Engineering News Record, February 20, 1986, pp. 26 - 27.

5.61 Yung, C., "Seam Treatment in the Abutments of Fei-Tsui Arch Dam," paper 26, 8th International Symposium on Water Jet Technology, Durham, UK., September, 1986, pp. 259 - 263.

5.62 Summers, D.A., Yao, J., Blaine, J.G., Fossey, R.D., and Tyler, L.J., "Low Pressure Abrasive Waterjet Use for Precision Drilling and Cutting of Rock," 11th International Symposium on Jet Cutting Technology, St. Andrews, Scotland, September, 1992, pp. 233 - 251.

5.63 Moore, W.M., Debris Removal from Concrete Bridge Deck Joints, Texas Transportation Institute, for Federal Highway Administration, September, 1974, PB 240626.

5.64 Tynes, W.O., and McCleese, W.F., Investigation of Methods of Preparing Horizontal Construction Joints in Concrete Report 4 - Evaluation of High-Pressure Water Jet and Joint Preparation Procedures, Technical Report 6-518, U.S. Army Corps of Engineers, Vicksburg, MS., August, 1973, AD 766694.

5.65 Anon, "Water Jet Turf Aerator," WJTA Jet News, August, 1990, p. 1, Water Jet Technology Association, 818 Olive Street, St. Louis, MO., 63101.

5.66 Anon, "Waterjet Probe Identifies Buried Pipe More Accurately," Gas Research Digest, Fall, 1988, pp. 25 - 26.

5.67 Ashley, S., "Flushing Out Land Mines," Mechanical Engineering, Vol. 114, No. 3, March, 1992, p. 122.

5.68 Hashish, M., Reichman, J., Cheung, J., and Nelson, T., "Development of a Waterjet-Assisted Cable Plow," 1st U.S. Water Jet Symposium, Golden, CO., April, 1981, pp. IV-1.1 - IV-1.15.

5.69 Reichman, J., Yie, G., and Hashish, M., "The Application of Waterjet Cutting to Underground Utilities Installation," 1st U.S. Water Jet Symposium, Golden, CO., April, 1981, pp. IV-3.1 - IV-3.11.

5.70 McCurrich, L.H., and Browne, R.D., "Application of Water Jet Cutting Technology to Cement Grouts and Concrete," paper G7, 1st International Symposium on Jet Cutting Technology, Coventry, UK., April, 1972, pp. G7-69 - G7-91.

5.71 Godfrey, K.A., Jr., "Water Jets: Concrete Yes, Tunneling Maybe," Civil Engineering, May, 1987, pp. 78 - 81.

5.72 Olsen, J.H., "Jet Slotting of Concrete," paper G1, 2nd International Symposium on Jet Cutting Technology, Cambridge, UK., April, 1974, pp. G1-1 - G1-10.

5.73 Norsworthy, A.G., Mohaupt, U.H., and Burns, D.J., "Concrete Slotting with Continuous Water Jets at Pressures up to 483 MPa," paper G3, 2nd International Symposium on Jet Cutting Technology, Cambridge, UK., pp. G3-31 - G3-39.

5.74 Frank, J.N., Fogelson, D.E., and Chester, J.W., "Hydraulic Mining in the U.S.A.," paper E4, 1st International Symposium on Jet Cutting Technology, Coventry, UK., April, 1972, pp. E4-45 - E4-60.

5.75 Frank, J.N., and Chester, J.W., Fragmentation of Concrete with Hydraulic Jets, BuMines Report of Investigation 7572, 1971, 32 pp.

5.76 Hamada, H., Fukuda, T., and Sijoh, A., "Basic Study of Concrete Cutting by High Pressure Continuous Water Jets," paper G2, 2nd International Symposium on Jet Cutting Technology, Cambridge, UK., April, 1974, pp. G2-11 - G2-30.

5.77 Clipp, L.L., and Cooley, W.C., Development, Test and Evaluation of an Advanced Design Experimental Pneumatic Powered Water Cannon, Exotech Inc., Report TR-RD-040 to the U.S. Department of Transportation, March, 1969, 62 pp.

5.78 Mellors, W., Mohaupt, U.H., and Burns, D.J., "Dynamic Response and Optimization of a Pulsed Water Jet Machine of the Pressure Extrusion Type," paper B4, 3rd International Symposium on Jet Cutting Technology, Chicago, IL., May, 1976, pp. B4-47 - B4-12.

5.79 Mohaupt, U.H., Burns, D.J., Yie, G.G., and Mellors, W., "Design and Dynamic Response of a Pulse-Jet Pavement Breaker," paper D2, 4th International Symposium on Jet Cutting Technology, Canterbury, UK., April, 1978, pp. D2-17 - D2-28.

5.80 Yie, G.G., Burns, D.J., and Mohaupt, U.H., "Performance of a High Pressure Pulsed Water Jet Device for Fracturing Concrete Pavement," 4th International Symposium on Jet Cutting Technology, Canterbury, UK., April, 1978, pp. H6-67 - H6-86.

5.81 Vallve, F.X., Mohaupt, U.H., Kalbfleisch, J.G., and Burns, D.J., "Relationship between Jet Penetration in Concrete and Design Parameters of a Pulsed Water Jet Machine," paper D2, 5th International Symposium on Jet Cutting Technology, Hanover, Germany, June, 1980, pp. 215 - 228.

5.82 Pater, L.L., "Experiments with a Cumulation Pulsed Jet Device," paper B3, 7th International Symposium on Jet Cutting Technology, Ottawa, Canada, June, 1984, pp. 83 - 90.

5.83 Pater, L.L., "The Blow Down Water Cannon: A Novel Method for Powering the Cumulation Nozzle," paper 20, 8th International Symposium on Jet Cutting Technology, Durham, UK., September, 1986, pp. 203 - 210.

5.84 Hilaria, J.A., Labus, T.J., "Highway Maintenance Application of Jet Cutting Technology," paper G1, 4th International Symposium on Jet Cutting Technology, Canterbury, UK., April, 1978, pp. G1-1 - G1-8.

5.85 Reichman, J.M., Kirby, M.J., and Rodenbaugh, T.J., "The Development of a Water Jet Cutting System for Trenching in Concrete," paper D1, 5th International Symposium on Jet Cutting Technology, Hanover FDG, June, 1980, pp. 201 - 214.

5.86 Hilaris, J.A., and Bortz, S.A., "Field Study for Highway Maintenance Application of Jet Cutting Technology," paper D3, 5th International Symposium on Jet Cutting Technology, Hanover, FDG, June, 1980, pp. 97 - 104.

5.87 Styron, C.R. III, "A review of methods of cutting concrete," 2nd U.S. Waterjet Conference, Rolla, MO., May, 1983, pp. 281 - 286.

5.88 Anon, "Water Jet Cutting," The Military Engineer, September-October 1982, Vol. 74, No. 482, pp. 382 - 383.

5.89 Summers, D.A., and Raether, R.J., "Comparative Use of Intermediate Pressure Water Jets for Slotting and Removing Concrete," paper J1, 6th International Symposium on Jet Cutting Technology, Guildford, UK., June, 1980, pp. 37 - 395.

5.90 Puchala, R.J., Lechem, A.S., and Hawrylewicz, B.M., "Mass Concrete Removal by High Pressure Waterjet," paper 22, 8th International Symposium on Jet Cutting Technology, Durham, UK., September, 1986, pp. 219 - 229.

5.91 Medeot, R., History, Theory and Practice of Hydrodemolition, FIP Industirale S.p.a., Report NT 722/86, December, 1986.

5.92 Anon, "Tyne Tunnel Repairs," Industrial Jetting Report, February, 1985, pp. 2 - 3.

5.93 Anon, "Hydrodemolition speeds bridge deck reconstruction," Roads and Bridges Magazine, March, 1986.

5.94 Anon, "Waterblasting robot helps complete bridge repair 14 months early," Concrete Construction, Vol. 30, No. 9, pp. 785 - 786.

5.95 Nittinger, R.J., "Hydro Demolition - Technology for Productivity and Profits for America," Proceedings 4th American Waterjet Conference, Berkeley, CA., August, 1987, pp. 65 - 71.

5.96 Schmid, R.F., "High Pressure Hydromilling of Concrete Surfaces," paper 15, 5th American Waterjet Conference, August, 1989, Toronto, Canada, pp. 157 - 163.

5.97 Hashish, M., "The Application of Abrasive Jets to Concrete Cutting," paper K2, 6th International Symposium on Jet Cutting Technology, Guildford, UK., June, 1980, pp. 447 - 464.

5.98 Hashish, M., "Cutting with Abrasive Waterjets," 2nd U.S. Waterjet Conference, Rolla, MO., May, 1983, pp. 391 - 403.

5.99 Nakaya, M., Kitagawa, T., and Satake, S., "Concrete Cutting with Abrasive Waterjets," 7th International Symposium on Jet Cutting Technology, Ottawa, Canada, June, 1984, pp. 281 - 292.

5.100 Yie, G.G., "Cutting Hard Materials with Abrasive Entrained Waterjet - A Progress Report," paper P1, 7th International Symposium on Jet Cutting Techology, Ottawa, Canada, June, 1984, pp. 481 - 492.

5.101 Yamada, B., Yahiro, T., and Ishibashi, J., "On the Development and Application of a Method of Remodeling Utilizing an Abrasive Jet System," paper E1, 9th International Symposium on Jet Cutting Technology, Sendai, Japan, October, 1988, pp. 203 - 216.

5.102 Sashida, K., Sasaki, K., Kamoshida, B., "Waterjets; Construction Applications," paper L1, 9th International Symposium on Jet Cutting Technology, Sendai, Japan, October, 1988, pp. 561 - 570.

5.103 Hashish, M., Echert, D., and Marvin, M., "Development of Abrasive Waterjet Concrete Deep Kerf Tool for Nuclear Facility Decommissioning," International Water Jet Symposium, Beijing, China, September, 1987, pp. 4.11 - 4.33.

5.104 Echert, D.C., Hashish, M., and Marvin, M., "Abrasive Waterjet and Waterjet Techniques for Decontaminating and Decommissioning Nuclear Facilities," 4th American Waterjet Conference, Berkeley, CA., August, 1987, pp. 73 - 81.

5.105 Anon, "Circle Cutter for Concrete and Steel," Industrial Jetting Report, No. 25, June, 1985, pp. 6 - 7.

5.106 Arasawa, H., Matsumoto, K., Yamaguchi, S., and Sumita, K., "Controlled Cutting of Concrete Structure with Abrasive Waterjet," 8th International Symposium on Jet Cutting Technology, Durham, UK., September, 1986, pp. 211 - 218.

5.107 Fossey, R.D., Blaine, J.G., and Summers, D.A., "The Feasibility of Commercial DIAjet Use," 11th International Symposium on Jet Cutting Technology, St. Andrews, Scotland, September, 1992, pp. 255 - 265.

5.108 Conn, A.F., "Rapid Cutting of Pavement with Cavitating Waterjets," 8th International Symposium on Jet Cutting Technology, Durham, UK, September, 1986, pp. 231 - 240.

6 THE USE OF WATERJETS IN MINING

6.1 INTRODUCTION

Chapter 5 described the development of waterjets from their initial application in removing the soil cover over valuable mineral deposits, through their use in excavating veins of softer valuable mineral, to mining such materials as china clay and gold bearing sands. Some lessons learned in that development will have equal validity in mining softer rocks discussed in this chapter. They will apply where large volume low pressure flows, such as those commonly used in mining coal, will be discussed.

The evolution from the erosion and washing away of softer material to the cutting and excavation of harder rock has no clear boundary or time frame. Rather as different applications have been seen, technology has evolved. In order to simplify the discussion of this evolution, yet at the same time to highlight meaningful lessons and technical developments this chapter will examine the trend based initially upon the large studies carried out into coal mining, and then, from these branch out into the developments in harder rock excavation. The concept of waterjet assisted cutting evolved from harder rock application, and is only now becoming applied in coal mining machines. This subject will be discussed in Chapter 9 in a single section, rather than as two separate parts. A separate chapter will be devoted to coal and rock drilling since that technology has also evolved along a common development.

6.2 COAL MINING DEVELOPMENT

6.2.1 SOVIET DEVELOPMENTS

Initial development of hydraulicking was very successful as a method of mining soft gold bearing ores and other valuable deposits and rapidly spread around the world from the U.S. It was reported, under the colloquial name of "Hushing" in the North of England [6.1]. It is in its development for use in coal mining that the technology evolved in Russia. The importation of hydraulic monitors to Russia in 1867 for mining gold deposits around Lake Baykal had, by 1890, reduced operating costs by 60%. The technology was given some publicity, and attempts were made to make it of practical use in extracting other materials. In 1914 the technique was tried for using peat, and between the wars had reached a level where 30% of all Soviet peat was mined hydraulically. The first hydraulic mine was established to mine gold near Lake Baykal in 1939.

The first underground trials of a waterjet system took place in a coal mine shaft in the Donets Basin (Donbass) in 1915. V.S. Muchnik wrote his thesis on water jet cutting of coal in 1935 [6.2], but during this time the main use of the waterjets was in the flushing away of broken coal which had previously been broken from the solid using explosives.

Commercial mining of coal hydraulically began in 1952 at the Tyrganskie-Uklony Mine in the Kuznetsk Basin (Kuzbass) where production was planned at 500 tonnes/shift, using water pressures of up to 42 bar to wash away coal which had been pre-blasted. By the end of the year output/shift had reached 600 tonnes, more than twice conventional levels in the area. It has since been found that jet pressures need to be no more than 20 bar for this operation, with water flow rates of 3 - 6 cu m/tonne, and a power consumption varying from 1.24 kWhr/tonne for a 32 mm nozzle, to 0.74 kWhr/tonne for the 50 mm nozzle. The water is generally supplied through nozzle diameters of up to 50 mm, since, not only is it more efficient, but the jet can be kept effective for distances of over 20 m [6.3].

A second mine was converted in 1953, again initially using water jets to wash away pre-blasted coal. In 1957 the pressure of the water was raised sufficiently that pre-blasting was no longer required and the practice has largely been discontinued for coal mining. (As an aside the practice is still widespread for extracting gold from the stopes of the mines in South Africa [6.4]. Ultimately nine major hydraulic mines were developed in the USSR, of which the smallest (No. 4 Ordzhonikidzeugol) produced 245,000 tonnes in 1976, while the largest (Yubileynaya) mined 3.5 million tonnes (only some of which was hydraulically obtained). In addition there were five hydraulic sections in mines which otherwise were operated conventionally [6.3, 6.5]. The total hydraulic production had risen from 15,000 tonnes in 1952 to 8.9 million tonnes by 1979.

The early hydraulic monitors, such as the RGM-1m, were hand operated (Fig. 6.1), with large wheels and deflectors being employed to help the operator move the monitor, but these machines were replaced by remotely operated, hydraulic ram steered, hydraulic mining units [6.6]. This change provided a large degree of extra safety for the operator, who could thus control the machine, with hydraulic rams from a distance of up to 10 m.

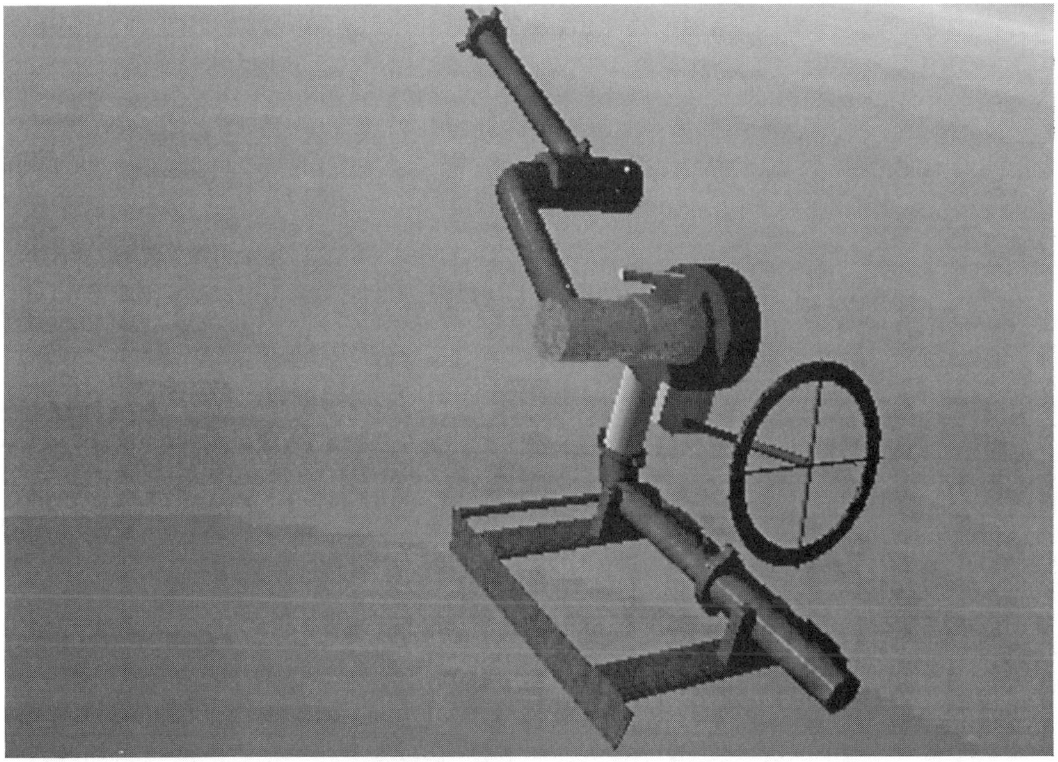

Figure 6.1 RGM-1m hydromonitor.

The most common unit became the GMDTs - 3M monitors which could operate in coal seams which had a height greater than 75 cm, and which dipped at more that 6 degrees (Fig. 6.2). These monitors were able to mine coal at more than 50 tonnes/hour, and operated with a jet pressure of approximately 100 bar, using water at rates of 150 cu m/hour, to mine coal with a hardness of approximately .82 on the Protody'akonov scale. They weighed 170 kg, without the oil reservoir.

The other highly popular monitor in the Soviet Union has been the 12GD-2 monitor which was developed for use in thicker coal seams, in which, mining under the same operating conditions, it could extract up to 70 tonnes/hr although requiring some 400 cu m of water to do so. This design resembled in many ways that of the GMDTs - 3M. However, these units were somewhat larger than the earlier machines used in the preliminary experiments and the pressure was raised from the original 70 bar and 115 cu m/hr. The use of hydraulic controls allows the operator to be sited up to 50 m from the monitor [6.7]. The major feature of this design has been the improved design of the joints which allow monitor motion in the vertical and horizontal planes. Comparing the illustrations (Figs. 6.1 and 6.2) one can see the number of turns which the water had to go around in order to reach the nozzle with the earlier design. By using a straight through flow design, the hydraulic resistance built up by the passage of fluid through the monitor is reduced, and this increases jet performance since it allows more of the original driving pressure to be retained by the fluid [6.8]. The monitor weights 300 kg, and, with a jet pressure of 120 bar, and a flow of 400 cu m/hr it can mine up to 100 tonnes/hr of coal which has a Protodyakonov hardness of 0.8 - 1.0. Face productivity with this machine has reached 115 tonnes/person shift.

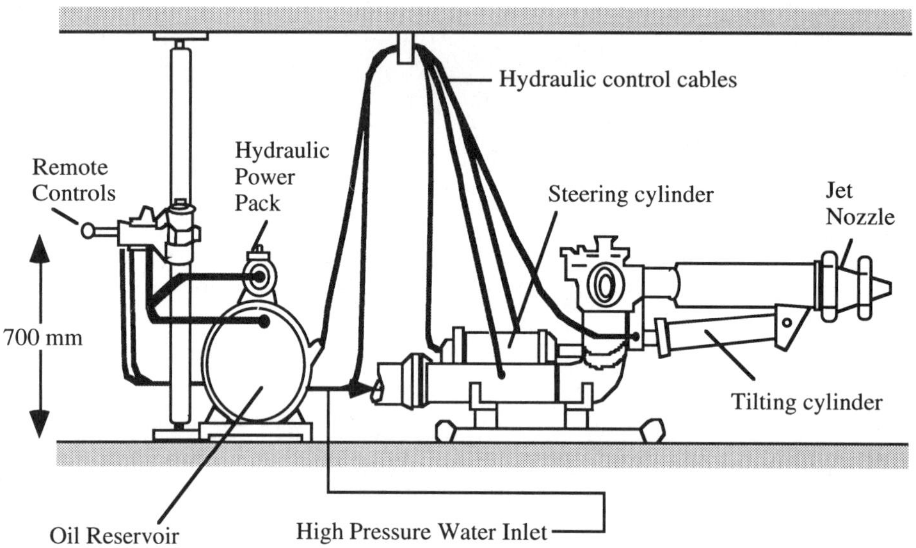

Figure 6.2 GMDTs - 3M hydromonitor.

An estimate of the productivity of hydromonitors has been proposed by N.F. Tsyapko, who suggested the relationship:

$$productivity = \frac{1.802 \times d^2 \times P^{3/2}}{10^{9/2} \times E_t}$$

where d is the diameter of the nozzle in mm;

P is the pressure of the water nozzle in bar;

E_t is the total power consumed in the hydraulic extraction. Where E_t is given by the expression:

$$E_t = 0.2\,E_1 + 0.8\,E_3 \text{ kw hpt}$$

$$E_1 = \frac{4.3 \times 10^3 F_t^{5/2}}{P}$$

$$E_3 = \frac{24F_t^2}{P^2}$$

where F_t is the coal hardness, equal to the product of the Protodyakonov number (F) and the jointing factor of the coal, which in the Kuznetsk Basin is on the order of 17. The Protodyakonov number is a method which measures coal friability by determining the amount of comminution of a standard sample under drop weight impact. It has been evaluated for use as a method of correlating waterjet resistance in other materials [6.9] with limited success. The role of rock properties in the evaluation of jet performance is further addressed in Chapter 11.

It should be pointed out that these extraction rates are for full area extraction rather than being achieved in the driving of headings. Within the confined space of an entry, the productivity achieved by the monitor will only be some 60 - 70% of that achieved in full extraction. Recent trends in the Soviet Union have been toward reducing the nozzle diameters to between 6 and 9 mm while increasing the jet pressure to between 280 and 500 bar.

The methods of mining used in extracting these seams varied as a function of the thickness of the coal, the stability of the overlying rock and the angle at which the seam dipped. Okrimenko [6.10] has divided the methods of mining into three sections:

- Where the seam is of average depth, and dips at a gentle angle long pillars are extracted along the strike of the seam, with the development headings being driven up along the dip, and subsequent developments being in strips along the dip (Fig. 6.3). The development of the section begins with an outline of a wing of the panel from the main access drives. This is typically 600 - 1,000 m long and up to 200 m deep. The coal is extracted by running drifts up the slope, some 14 - 16 m apart, and then removing the coal in alternate wings on either side of the drift as the drift is retreated. The extraction drift is 5 - 6 m wide and will be driven half way to the next drift. If the coal is friable it is possible to take advantage of this and extend the size of the individual area extracted in a lift. A **lift** is the term sometimes used to denote the amount of coal mined during one cycle of the mining process, before the equipment is moved or another step is taken.

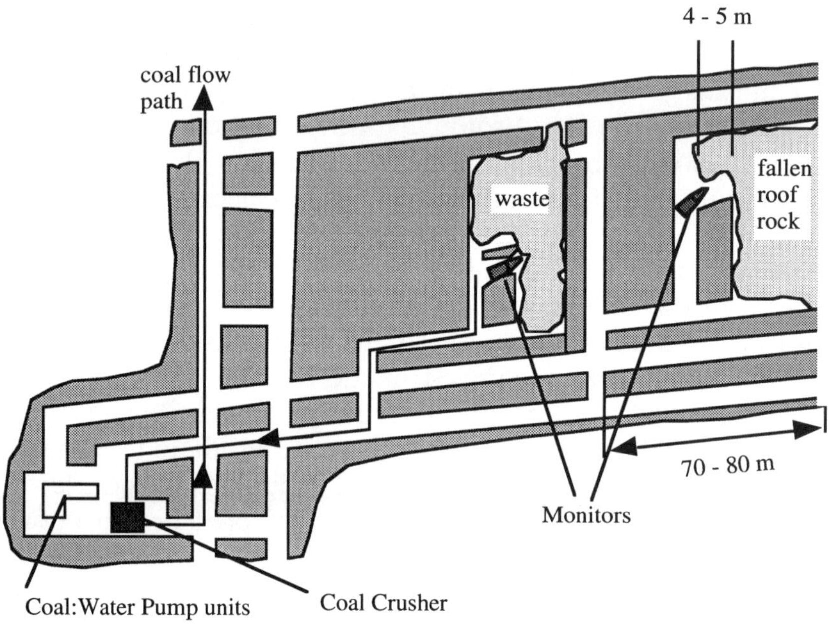

Figure 6.3 Method of mining gently dipping seams of average thickness (after [6.10]).

• Where the seam was deeper, yet still gently dipping, such as in those seams found in the Tomusinakaya-1/2 shaft in the Kuznetsk Coal Basin (the Kuzbas), a room and pillar method of excavation, without the need for support of the excavated area, has been developed. (Fig. 6.4).

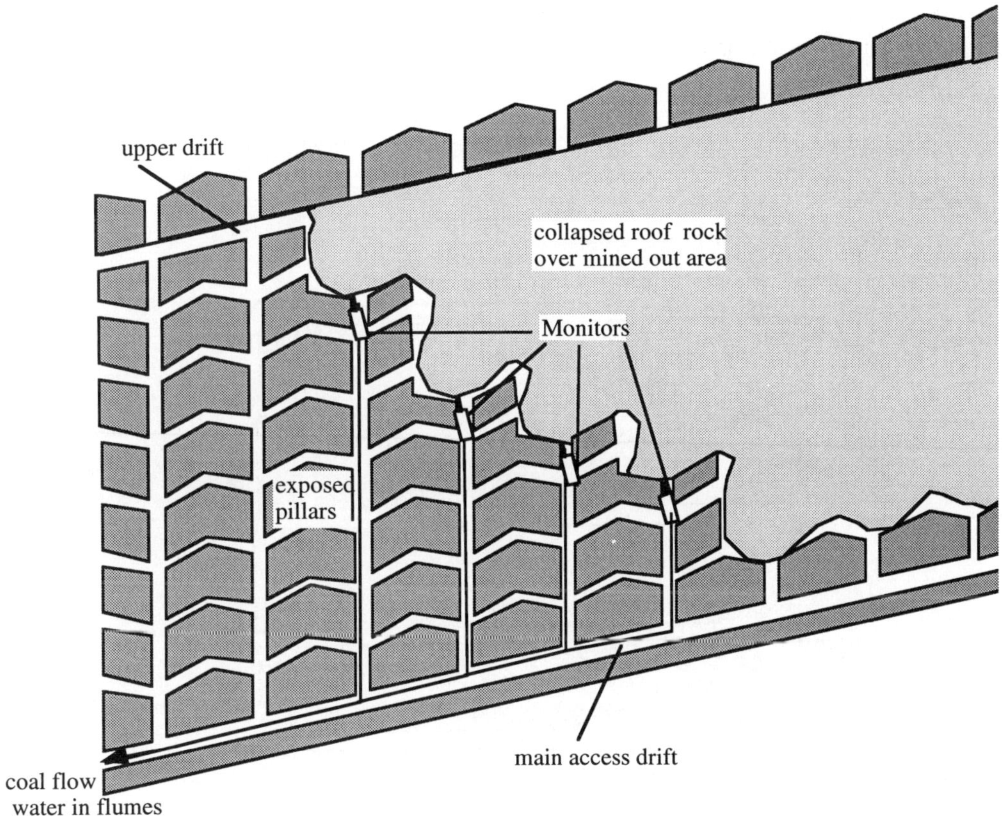

Figure 6.4 Method of mining gently dipping, thick seams (after [6.10]).

• Where the seams were steeply inclined then a substage excavation system with hydraulic fragmentation has been adopted, again without the need to support the excavated area (Fig. 6.5). The seam is extracted from a series of sublevels driven 2.5 m wide and from 6 - 15 m apart. These sublevels or access entries are normally driven at a sufficient angle to the horizontal (known as the "**strike**" in mining terms) that the coal can be collected, with the spent water in a flume, and then carried together, under gravity, back to the main access tunnel. This tunnel, or drift, has been driven either at, or close to, the direction at which the coal slopes at its maximum angle (known as the "**dip**"). Thus gravity will, again, work to carry the coal, in the spent water, down the flume to the main tunnels driven to the entry shaft. By maintaining this tunnel at a gradient, similar to that of the initial extraction tunnel, it may then be possible to continue the gravity feed of the material all the way to the shaft. At that point the coal can be fed into a pumping station, which will send it through a pipeline up to the surface.

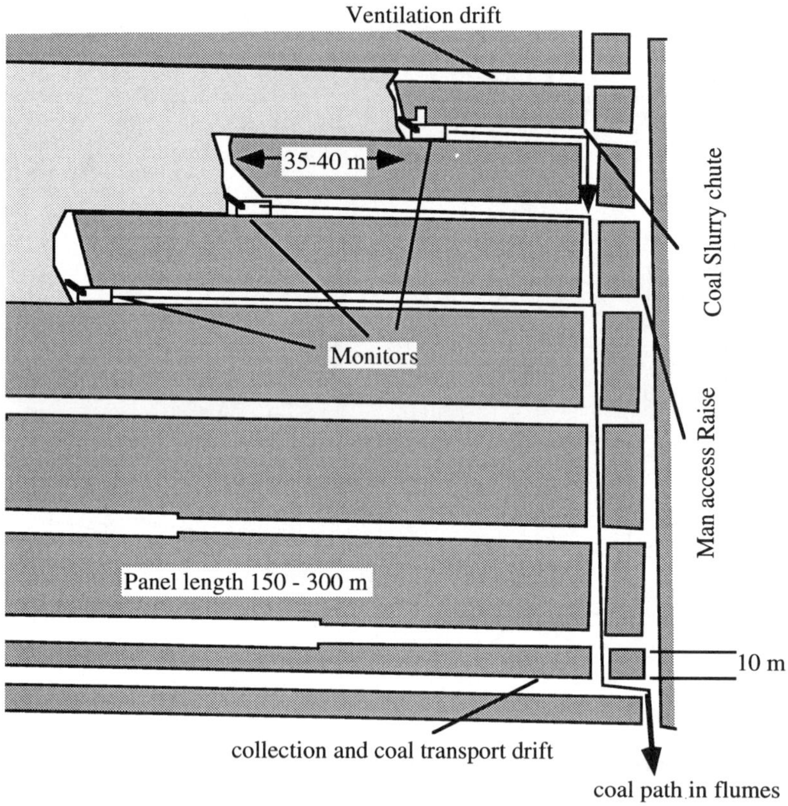

Figure 6.5 Method of mining steeply dipping seams (after [6.10]).

Extraction occurs in descending order along the sublevels, with the lower level extraction following at least 15 m behind the upper. In mining each section of the panel, the coal is removed from the protection of the unextracted tunnel section, with the coal being mined out from the bottom of the seam, and moving the extraction upwards and towards the monitor by directing the jet in an oscillating series of passes.

Cooley, in his survey of mining methods, found that a fourth method of mining, that of pillar extraction either with open or closed "lifts" could be worked (Fig. 6.6).

The choice as to which method to use would depend on the relative strength of the roof overlying the coal, as well as on the condition of the coal itself, the underlying floor, and the degree at which the seam was dipping. The depth of the seam will influence the pressure on the pillar and thus will also control the way in which it breaks as the pillar size reduces during extraction.

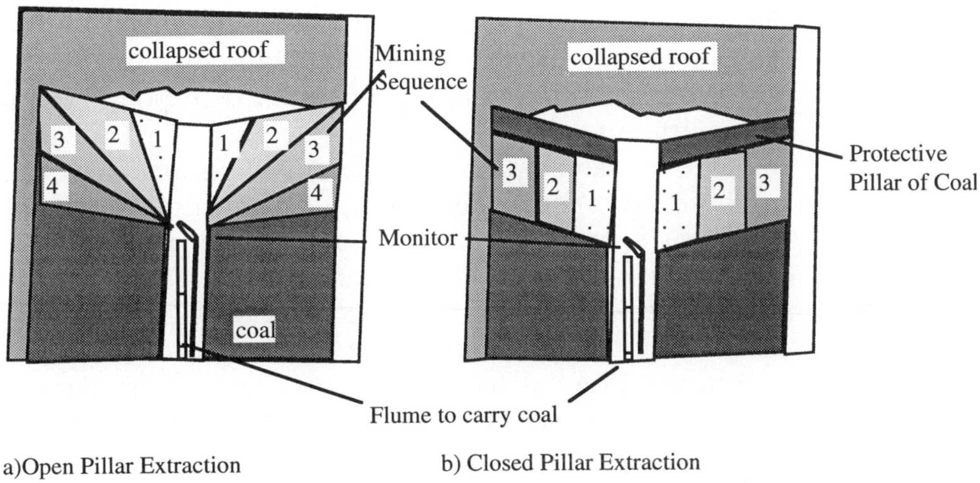

Figure 6.6 Pillar extraction by open and closed lifts [6.5].

The procedure to extract the coal from a heading was [6.8] to first cut a deep horizontal cut at the bottom of the seam, and then to raise the angle of the monitor by 5 - 10 degrees and then to make a second pass. This pass would not only cut a second slot, but it would also collapse the intervening rib of coal, providing it was not too thick. The process was then repeated moving up the face until the entire section was extracted. Because the face would not break evenly, and greater stand-off distance reduced the performance of the jets, self propelled monitors were developed such as the 12GP-2. These would operate at 120 bar pressure, 400 cu m/hr flow and could mine 120 tonnes/hr. This would give a face output of 139 tonnes/person shift.

The production of coal from hydraulically mined sections was considerably greater than that which could be achieved by conventional means, and the mining costs in 1957 were almost half those of conventional mining [6.11].

In a subsequent visit in 1978 Mills reported [6.7] on discussions in which the intent was expressed of increasing hydraulic mining production from the Kuzbass to 25 million tons a year by 1993. He reported that monitor pressures of 50 to 150 bar could mine to a depth of up to 20 m in plowable coal, although this would be halved in cases where the cleat was on bord (this is a mining term and relates to the direction of the vertical weakness planes or "**cleat**" relative to the direction of the tunnel). At one mine a team of 60 men produced an average of 1,200 tons of coal a day for over five years, from a 1.1 m thick seam at a gradient of 10 degrees with one heading machine and one hydromonitor. Costs were cited as ranging from 5.1 - 11.6 rubles/ton, with the comment that the large amount of fines in some of the coal reaching the preparation plant was responsible for a significant part of the difference in cost.

As the technology has developed so different monitors have been developed [6.3]. These are increasingly being designed for use under remotely control. The jet pressure required to mine the coal will vary from seam to seam, but is generally in the range from 100 - 160 bar, although there is a trend to increase pressure to levels of 500 bar. This latter change requires that the monitor nozzle sizes be reduced to between 6 and 9 mm. The additional pressure is obtained by using piston pumps to act as pressure boosters. Similarly the flow volume through the monitors is being increased by increasing the size of the flow channels.

The GMDTs - 3M machine, for example, will typically mine 50 tons/hr at a jet pressure of 120 bar, and a flow rate of 150 cu m/hr. In contrast the 12GD-2 operating at the same pressure, but a flow rate of up to 400 cu m/hr may mine 70 tons/hr from somewhat thicker seams. Mills, for example, reported that one machine mined 3,400 tons of coal in 48 hours, including 16 hours required for moving and reconnecting the monitor [6.7]. A further significant increase can be achieved where the pressure at the monitor is increased, to as high as 160 bar using conventional pumps. Once broken from the solid the coal is then washed into flumes and carried, under gravity feed, to the shaft where it can be pumped to the surface. Some crushing of the coal is required for transportation, so that while over 65% of the coal may be larger than 13 mm in size when it is mined, by the time that it reaches the preparation plant only 20% of the coal is still this large. At the same time the size fraction below 1.25 mm increases from 11.5% at the face to almost 30% at the plant.

The water is ultimately recirculated back to the monitor. During the coal mining and transportation losses may be as high as 20 - 30%. In earlier systems the operating hydraulics for these units has been based on the use of the pressurized process water. This has not proved reliable and more recently there has been a change to the use of oil-based hydraulic controls. Because of the fine grains of coal and rock which are not extracted before the water is recirculated, frequent maintenance is required, with a nozzle life of 300 - 500 hours and a monitor life of two to four years.

The water required for flume transport is less, by an order of magnitude, than the 70 to 250 cu m/hr of water required to carry coal along the floor of the seam to a collection point. Since the excess water increases the problems of coal recovery and water treatment, this method of washing the coal to the collection point has become less common, and the water and coal are entrained in flumes to concentrate the power of the water.

The GVD series of hydromonitors, (Fig. 6.7) were developed for remote operation in thin seams, and were designed to be able to mine a strip of coal measuring 1 m wide, 10 - 12 m long and over the entire seam section in a period of 10 - 12 minutes, at a jet pressure of 100 bar, and with a nozzle diameter of 19 - 24 mm. In this way the machine could advance up into the seam, away from the entry tunnel, and mine coal at rates of up to 76 tons/hr from seams 0.6 to 1.5 m thick, more than doubling the productivity of the labor force (Fig. 6.8).

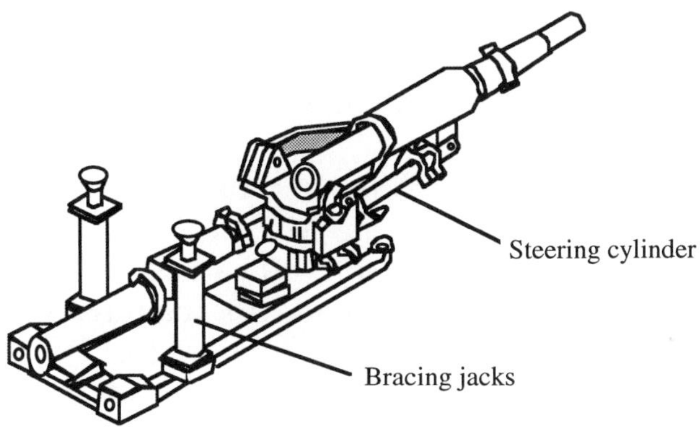

Figure 6.7 GVD-3 hydromonitor.

The monitor is launched from the lower entry in a seam, and mines up dip with the coal being washed down the mined out passage by the flow of the spent water from the monitor. It is captured in the main access tunnel by a flume, not shown in the sketch, and then carried in a series of flumes out of the mine, in much the same way as for the other methods of hydraulic mining.

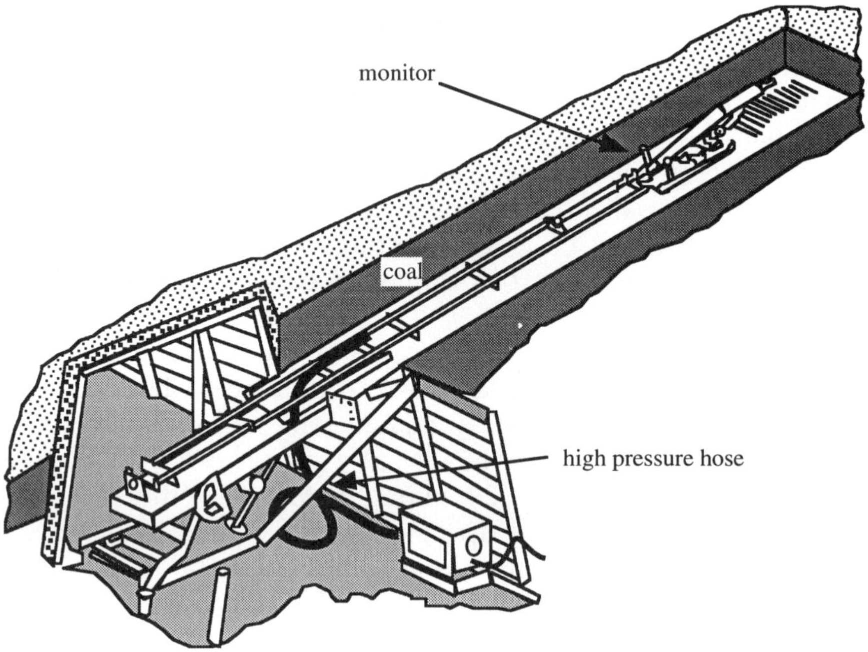

Figure 6.8 Method of mining with the GVD series hydromonitors.

6.2.2 TRIALS IN THE UNITED KINGDOM

A search for new technology following the Second World War, led investigators from the United Kingdom to visit Russia in May 1956 [6.11]. As a result of their trip an experiment was carried out, at Trelewis Drift Mine in Wales, to find ways of improving the cutting ability of waterjets and to see if it would be practical in British mines [6.12]. The experiments were directed both at cutting the coal from the solid, and providing a method of moving the broken coal in flumes away from the face. Two methods of mining were investigated, both driving headings and the extraction of pillars on either side of the heading. The tests were carried out in the Brithdir seam, approximately 175 cm thick, at a gradient of 1 in 14.7. Both roof and floor were normally a hard shale. High pressure water was carried some 900 m from the surface to the test area and supplied to the monitor at a pressure of between 42 and 80 bar. Early experiments showed that the operator had to be kept back from the monitor in order to protect him from splash back which could throw coal particles up to 12 m from the face. For this reason a remote monitor was constructed (Fig. 6.9). It was found advantageous to use a honeycomb flow straightener within the monitor to improve jet throw (Fig. 6.10). Among the nozzle designs tested that which closely approximated the design recommended by Leach and Walker (Fig. 6.11) was found to be the most effective.

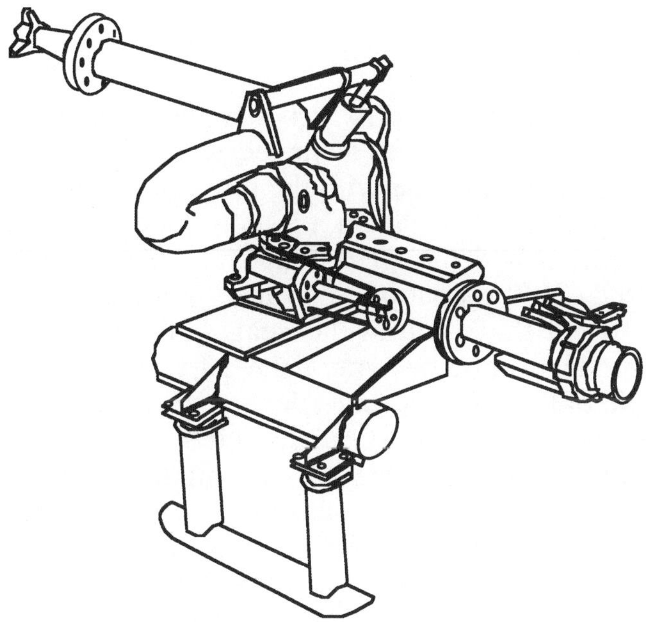

Figure 6.9 Hydromonitor used in the Trelewis Drift experiments [6.12].

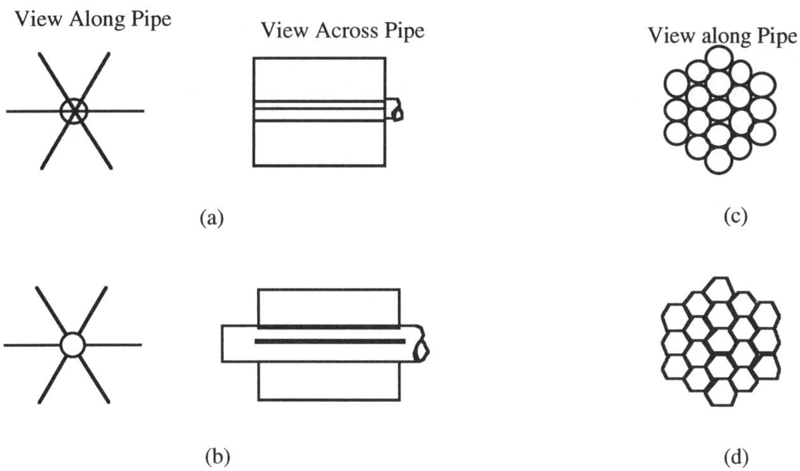

Figure 6.10 Flow straightener designs used in Wales [6.9].

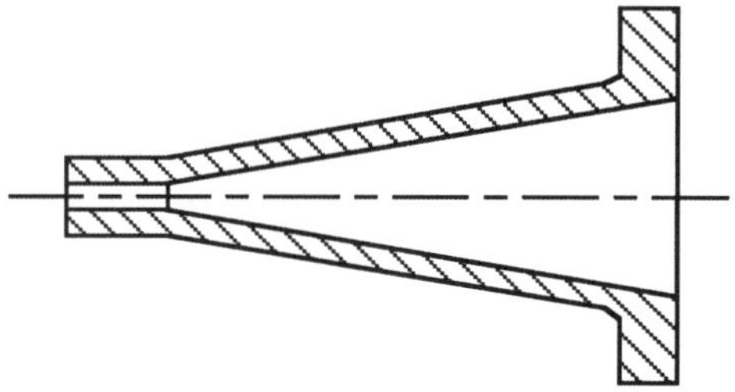

Figure 6.11 Nozzle design used at Trelewis Drift [6.9].

Various experiments with entry drivage indicated that the maximum production, some 25 tonnes/hr was achieved when the monitor was used to drive a heading some 1.9 m wide with the monitor set up so that it was first started with the jet some 2 m from the coal face. Over a period of 40 minutes the jet cut to a depth of 3.6 m over the seam height, but was somewhat limited in production by the presence of a stone band approximately 10 cm thick within the seam. It was confirmed that the optimum coal production arose where the jet is used to impact the coal between the angle of the face and butt cleats. Widening the entry after the initial pass produced coal of up to 60 cm cubes, too difficult to load from the floor. For most of these operations the higher pressure waterjet was used to move the coal, but it was found necessary to supplement that flow with a higher volume low pressure feed, to achieve effective coal transportation [6.13]. This may have been because the coal and water were not immediately collected and concentrated within a flume or channel cut in the floor.

Side pillar stripping (Fig. 6.6) was tested in the retreat of these headings, using a variety of jet traverse patterns but testing was restricted by frequent falls of roof rock, in one case burying the monitor. A maximum production rate achieved was on the order of 40 tonnes/hr. Some 13 men were employed in these tests, including both surface and underground employees. With the coal conditions at the mine a production rate of 45 tonnes/7.2 hour shift was estimated for a monitor during drivage, and up to 137.5 tonnes/shift in production. Capital costs at the time were some 66,833 pounds. Power requirements were 10.55 kWhr/tonne for extraction and 16.1 kWhr/tonne for the overall system. The overall cost was estimated to give a minimum potential cost of approximately fifty pence/tonne in 1963.

Despite these figures, based upon the limited areas of favorable geology and the higher productivity achieved using other machines, the ultimate conclusion of that study was that in the established mine workings that occur in most of Britain, where the seams were relatively flat, that hydraulic mining would not be an answer.

Subsequent to their experiments with the use of lower pressure waterjets experiments were carried out by the Safety in Mines Research Establishment (SMRE). These led to the development of a mining plow with a series of waterjets embedded in the body (Fig. 6.12). The concept on which the design was based was that the jets would be injected into the coal ahead of the machine at pressures of up to 5,000 bar. The jets would weaken and fracture the coal ahead of the wedge, and thus make it possible for the plow to take a deeper bite [6.13].

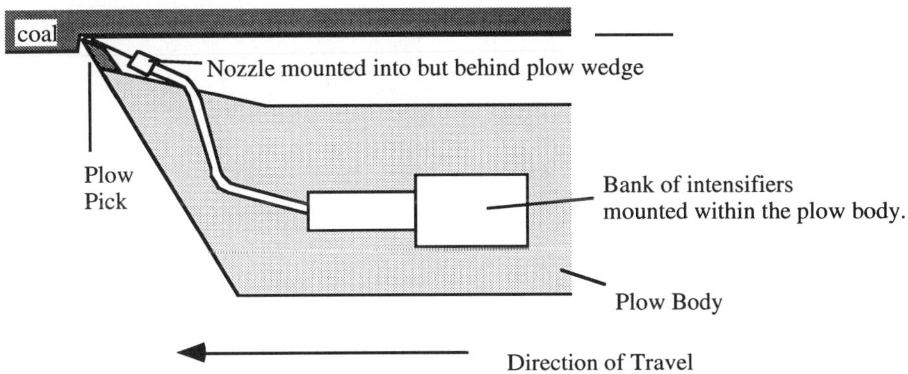

Figure 6.12 Plan view of British waterjet assisted coal plow (after [6.13]).

A machine was built to determine the practicality of the design and tested in an open pit in a coal seam approximately 1.15 m thick. The plow body used for the tests had a 30 degree face angle to the line of advance. For the test series only one nozzle was used, 1 mm in diameter, and a single pulse of 120 ml was fired at one time. Some 73 shots were fired in the test series and were a mixed success. One of the problems encountered was that if the coal fractured along the line of the jet injection then the water pressure was relieved before the pressure could fracture and weaken the entire coal body ahead of the wedge. Pressures of up to 5,000 bar were used in the test series and higher pressures gave better results in general.

Because of the limited nature of the trials it is not possible to completely extrapolate the potential for this design from the data obtained. It is interesting to note, however, given other comments on the relative inefficiency of jet cutting that excavation efficiencies of 4 joules/cc were averaged, compatible with that of conventional coal cutters. Cutting forces were reduced in 30% of the trials, and a minimum cutting energy of 1 joule/cc could be achieved. A typical thrust curve showed the effect of the jet impact on the power required to drive the plow shows an instantaneous drop in the required load of over 60% (Fig. 6.13).

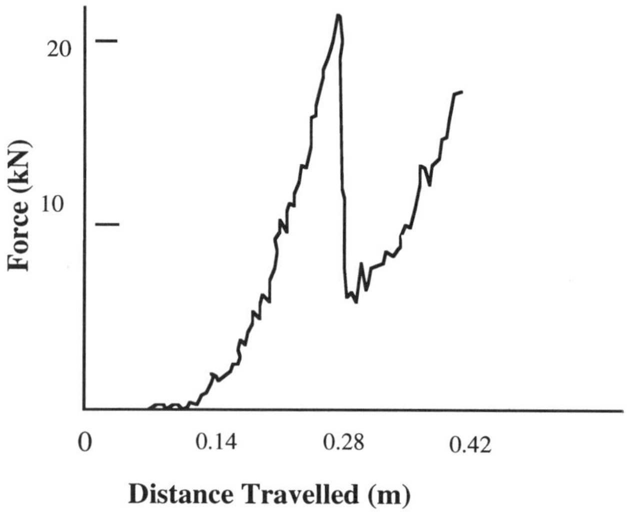

Figure 6.13 Typical load displacement curve for the SMRE waterjet plow with jet firing (after [6.13]).

This was instantaneously effective and did give some relief to the plow forces after the unit had fired, but it still provided a high cyclic load to the plow body and did not give the reduction in forces, or allow the greater depth of cut to be taken by the plow teeth, in contrast to the potential of competing systems developed in both the United States and Europe. For this, among other reasons, the program has since been discontinued, in this form, although further work has been carried out with waterjet assisted mining machines in the UK (Chapter 9).

6.2.3 DEVELOPMENTS IN THE UNITED STATES

The mining of coal in the United States was preceded by its development as a means for mining gilsonite. Gilsonite is a friable hydrocarbon, which is extremely explosive when it is in a dust cloud. Thus conventional mechanical mining has not been a safe method of extraction. Beginning in 1949 the nearly vertical veins were mined by taking horizontal cuts along the deposit. Jets at a pressure of 140 bar were used, at a flow rate of 310 lpm to mine out benches of material. The cut gilsonite was washed into flumes and gravity used to carry it to a pumping station [6.14]. A less expensive method, using a vertical drill and ream concept, was adopted in 1961 [6.15].

In this second method. a series of access holes, 15 - 20 cm in diameter was first drilled at 6 m intervals along the vein, from the surface and leading down to a tunnel located at the bottom of the vein. A pair of 9.5 mm diameter nozzles were then fitted to the end of a special pipe, loaded in the holes, and water supplied to the nozzles at a pressure of 150 bar (Fig. 6.14). By rotating the assembly at 1 rpm, and raising it at a speed of 1.2 - 3.7 m/hr the hydrocarbon could be remotely mined with a greater degree of safety and lower cost.

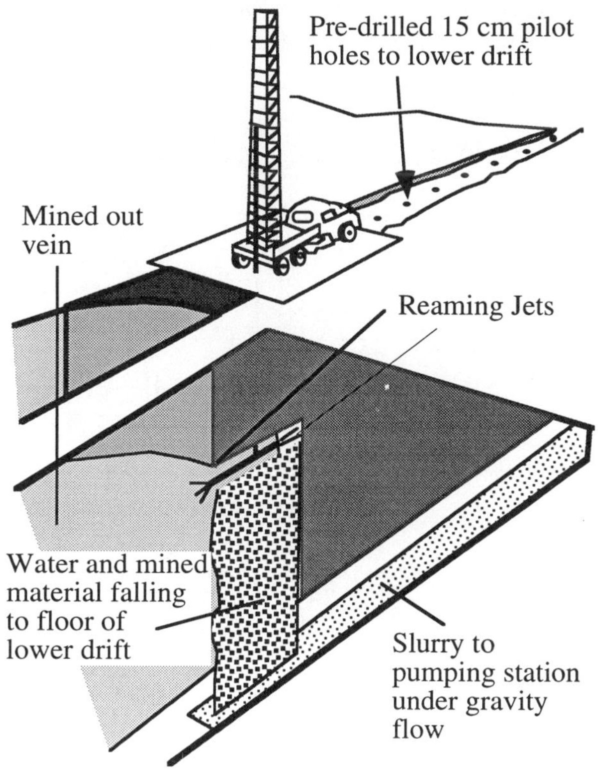

Figure 6.14 Method for mining gilsonite.

In the same time frame as trials began in England, the U.S. Bureau of Mines also was also seeking new ways of improving coal mining production.

Again, as a result of reports emanating from the Soviet Union, the potential for the use of hydraulic mining was investigated. Trials were initially carried out using hand-held monitors to cut slots in a dipping bituminous coal seam [6.16], and led up to a final test series, in anthracite, with a remotely controlled monitor directed from a cab, in which the monitor head was able to move both laterally and vertically across the coal face [6.17].

In the early trials of this equipment, tests were carried out in relatively steeply dipping seams. The Russian practice has been to fix the monitor and use the jet to reach out and mine large sections of coal which could be as far as up to 12 m away from the monitor. Enough water was used, however, to flume the coal away from the working face, and then through flumes, through tunnels which ran down the dip, and thence away from the working area. In the Bureau tests, in contrast, the monitor was mounted on a horizontal bar, relatively close to the coal face (Fig. 6.15).

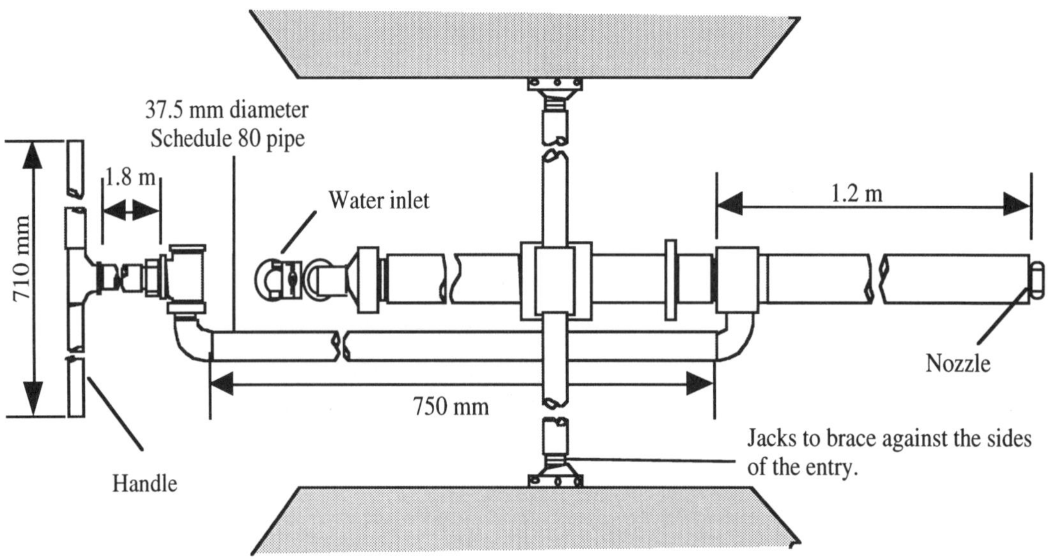

Figure 6.15 Early Bureau of Mines experimental arrangement (after [6.16]).

This produced mining rates of up to 1.3 tonnes/min over a 3.4 m cutting width. The monitor was then attached to a loading machine. At a pressure of 280 bar, and a flow rate of approximately 800 lpm the 9.5 mm diameter jet could mine up to 1.8 tonnes/min. This rate declined rapidly as the nozzle moved away from the face. Production was maximized at a traverse speed of 60 cm/sec at which speed the jet was cutting a slot up to 30 cm deep into the face [6.18]. This production rate, significantly higher than that from the UK, was sufficient to encourage further studies.

In an experiment at the Thompson Creek mine in Colorado a cutting lance was attached to a rigid prop and operated at 130 bar and a flow rate of 151 lpm. Over a period of 61 production shifts, the three-man face crew mined 2,567 tonnes and advanced the face 107 m. Productivity was 35% more than for conventional mining, and the elimination of explosive reduced supply costs by 74% [6.16].

A new approach was then tried using what was called the Water Jet Miner waterjet lances mounted on a self-advancing chock system so that it could advance along and up the face (Fig. 6.16). While it must now remain conjecture, since the research group is dispersed, it has been reported that many of the problems which slowed production may have arisen from the difficulty of manipulating this equipment. Certainly during mining, production rates of over 1.2 tonnes/min were achieved, but the unit only operated for 12% of the available face time and productivity levels of 4.1 tonnes a person shift were below the conventional level of 6.8 tonnes.

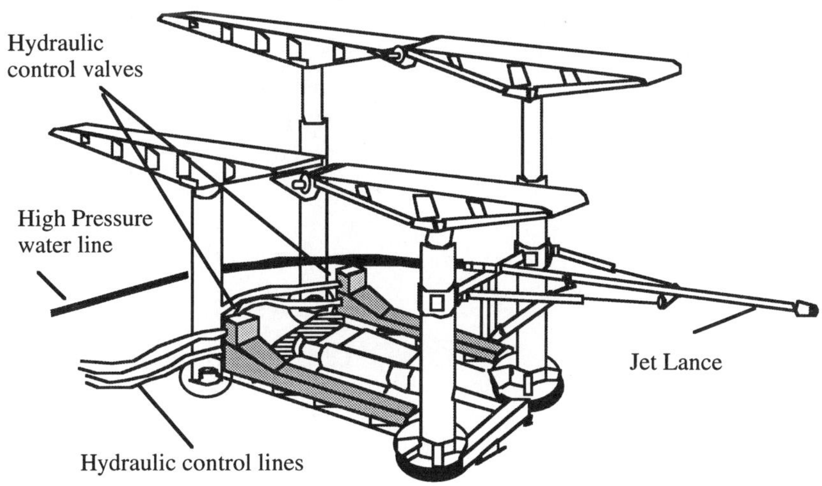

Figure 6.16 Schematic of the U.S. Bureau of Mines waterjet miner [6.16].

A final experiment was carried out in the anthracite mines near Sugar Notch, PA. A special mining machine was constructed which operated at 345 bar, with a flow rate of 1,135 lpm through nozzles ranging from 10 to 12.7 mm in diameter. With approximately 25% operating time, the machine mined at an average of 0.79 tonnes/min with an average power requirement of 15.6 kWhr/tonne. The collapse of the anthracite market closed this experiment and the government study of conventional low pressure hydraulic mining of coal fell into abeyance in the United States.

However, in part as a result of the success of the work which was, at the time continuing in Canada, a study was begun on the use of hydraulic mining near Newcastle in Colorado [6.20]. A limited number of experiments were carried out in the mine development. Three different mechanisms were identified for the action of the jets on the coal. In the first the jets drill into the coal with relatively low efficiencies of material removal (Fig. 6.17). Once penetration is achieved then the coal is pressurized and this will induce growth of existing fractures in the coal around the impact point, at somewhat higher efficiency, while the final mode in which large slabs of coal can be removed requires even less energy although needing more face development before it can be achieved, since a significant free surface must exist for the coal to slab into.

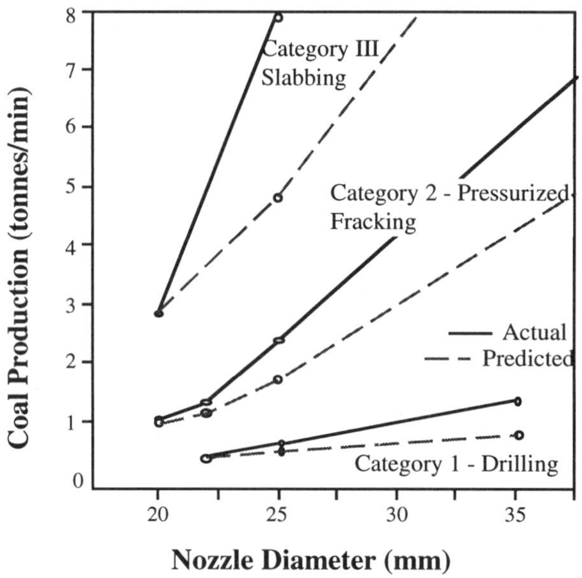

Figure 6.17 Coal removed by different mechanisms of failure [6.20].

The results of the testing indicated that production should be around 23 tonnes/person-shift, and that lower initial costs for mining and transportation equipment would be required. It was suggested that the technology would require approximately half the initial investment for equipment and mine development, relative to that needed to develop a mine for conventional mining. These figures can be compared with equivalent values reported by the Russians for the cutting and extraction of coal using similar units [6.3]. Unfortunately the reduced size of the international coal market has limited the development of this property.

6.2.3.1 HIGHER PRESSURE COAL MINING RESEARCH

In 1973 the U.S. Bureau of Mines began a search for improved methods of mining, following two serious mine accidents, each with multiple fatalities. At that time hydraulic mining had been found to have a number of drawbacks, as it related to most American coal deposits. In contrast to the Soviet Union, most American coal lies in relatively flat deposits. Thus, not only can water not be used to help gravity transport, but if not removed, it will collect on the mine floor, which is often a clay-like material, weakening it and causing a number of problems to continued mining operations. Thus, while the idea may have had merit, the existing, low pressure, high volume systems had several disadvantages.

Private companies in the United States had been carrying out some of their own research in coal cutting with waterjets, and thus at the time the Bureau of Mines expressed their interest in the work, different approaches were suggested. As a result, three initial contracts to examine these different approaches to the use of higher pressures for jet mining of coal were funded.

6.2.3.2 4,500 BAR SYSTEM FOR ROOM AND PILLAR COAL MINING

The highest pressure to be suggested for the mining system was that proposed by a consortium under the major direction of Bendix Research Laboratories [6.21]. The consortium included Joy Manufacturing Company, Continental Oil Company and G.L. Judy Associates. The consortium worked with data previously acquired by Bendix and Continental Oil Corporation to (in the words of the contract):

> "Develop a design for a practical, hydraulic coal mining machine to produce coal at 6 tons per minute in seams 5 - 8 ft high, using water at cutting pressures up to 65,000 psi. Operators safety, visibility exposure to machine noise and equipment permissibility are important design parameters."

(Which translates into the need to use waterjets at a pressure of approximately 4,500 bar).

In light of subsequent developments it is interesting to note some of the findings from the data analysis. A statistical analysis of the data from the tests carried out by the two companies indicated that the depth of cut for a waterjet into coal could be predicted from the equation:

$$\text{Depth of Cut} = \text{constant x (Jet Pressure)/(traverse speed)}^{0.37}$$

This correlates well with the subsequent equation which has been empirically developed at UMR for general analysis of jet cutting data of the form

$$\text{Depth of Cut} = \text{constant x (Jet Pressure)x (Nozzle Diameter)}^{1.5}\text{/(traverse speed)}^{0.33}$$

This relationship will be discussed further in Chapter 11. It is interesting to draw attention to this however, not only for the coincidence of the structure, but also because the Bendix equation did not include nozzle diameter in the relationship. As will be discussed shortly, this was a critical absence.

In addition to deriving the parametric relationship, the study showed that it was possible to easily remove the remaining rib of coal between two successive passes of the jet, if this rib were less than two-thirds the depth of the cut.

The design which the group sought to develop was based upon the modification of a Joy 12CM continuous coal mining machine. In order for the machine to be most effective it was decided that eight individual intensifiers (made by Bendix) would be mounted on a cutting arm and used to cut vertical slots in the face as the arm was raised or lowered. The intervening ribs of coal could, simultaneously be broken out by having a set of wheels or cutters follow along the jet path, wedging the ribs of coal down to the floor. Here the coal fragments would be picked up by the gathering arms of the underframe of the machine as it was advanced. A series of primary and secondary nozzles was included in the design (for a total of 23 nozzles) only one set of which operated at a time (primary on the downward stroke, secondary on the upward). Seven of the nozzles were left as an auxiliary array and were not normally to be used in operation and were to be capped off. The nozzles were arrayed to leave a pitch or rib between adjacent cuts of between 18 and 22.5 cm.

Because the addition of the intensifiers to the boom would give it a considerable weight, and since, for optimal efficiency, the boom should move at a design speed of up to 0.4 m/sec, additional power was required to move the traversing boom. Limitations on the overall power available to the machine became a constraint in the design, and a 900 kW limit was placed on the power which could be used. This was then separated into the following categories of use:

Table 6.1 Power distribution for the Bendix hydraulic mining unit [6.21]

hydraulic power to intensifiers (2 @ 260 kW pumps and motors)	520 kW
low pressure hydraulic supply	130 kW
water precharge pump	15 kW
mechanical cutter drive	105 kW
gathering head motor	45 kW
additional traversing pump motor	37.5 kW
traction motors	52.5 kW
total power required	905 kW

Note that the hydraulic power was used to power 8 intensifiers at 40 kW each.

The intensifiers, it should be noted, were rated at only 48% efficiency when the relative efficiencies of the input power systems were correlated with the size and pressure of the jets produced. The water flow to the intensifiers was on the order of 100 lpm. In order to maintain the power levels at the overall limit which had been established, some compromises had to be made, one of which was to lower the system pressure from the original concept in which the waterjets operated at 4,500 bar pressure to one in which the waterjets operated at around 2,400 bar.

Even with these conditions it was not anticipated that the machine could mine more than 2 tonnes/min or only 20 - 25% of the production of an existing JCM. At the same time the power which the jets would require was over twice that used in a conventional machine. Given the constraints of time in terms of developing a different and more radical approach, the conclusion from that study was that waterjet cutting of coal using a very high pressure unit did not justify additional development, and the program concluded.

6.2.3.3 3,300 BAR SYSTEM FOR ROOM AND PILLAR COAL MINING

The second program which the Bureau of Mines funded was to a group which included IIT Research Institute (IITRI) and the Goodman Equipment Corporation [6.22] and [6.23]. The basic conditions for the contract were the same as those identified for the Bendix operation, although starting with the lower pressure. (Interestingly the starting pressure for this contract was the same as the pressure at which the Bendix group ended up).

The machine was designed around the use of a Goodman 968 loader and used four intensifiers rather than the eight of the Bendix design. Each intensifier, designed by IITRI, was rated to produce 27.3 lpm of water at 3,330 bar, with the flow supplying two 0.6 mm diameter nozzles. As with the Bendix design these were mounted on a boom, however, in this case the boom was designed so that the jets would sweep the face in a series of horizontal passes, as the boom moved up the face. In this way the ribs of coal left on the face would be removed by the adjacent jet action, and no mechanical cutting would be required. The IITRI design broke the power requirements into:

Table 6.2 Power distribution for the IITRI design [6.22]

Intensifier Operations	746 kW
Loader functions	134 kW
Total Power Requirement	880 kW

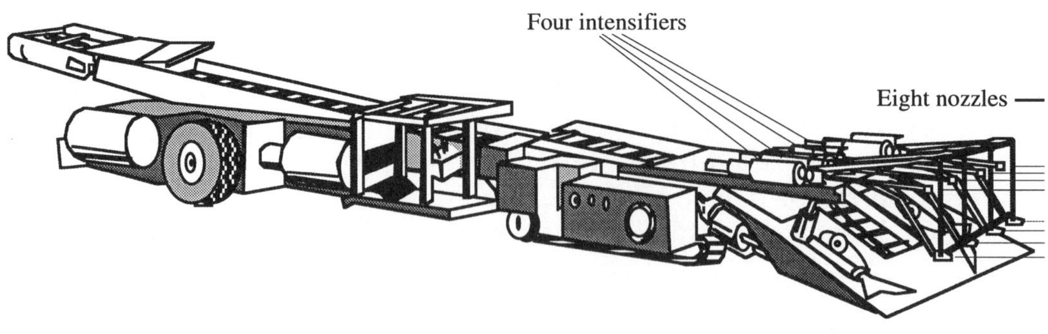

Figure 6.18 Schematic of the IIT Research Institute waterjet mining machine [6.22].

The design was based upon field tests carried out in a surface mine by IITRI, during which dust levels generated by the machine were measured to be on the order of 813 mg/tonne, with a specific energy of cutting of 38.7 joules/cc, considerably higher than that reported for conventional mechanical machines.

Although the performance predicted for this machine was on the order of 6 tonnes/min, within the range achieved by the equivalent mechanical machine this design still required more than double the power of existing machines, without any great benefit appearing to accrue from its use. Thus this program also was discontinued.

It is interesting to note, however, that by changing the cutting pattern the designers were able, for an equivalent sized machine, to predict the ability to mine three times as much coal as that predicted from the Bendix design.

6.2.3.4 700 BAR SYSTEM FOR LONGWALL COAL MINING

The third program which the Bureau of Mines funded took a different approach to the problem. The most productive mining system in general use is longwall mining. In this system of mining the machine is set up within a long tunnel and mines the coal by progressively peeling slices of coal from the side wall of the tunnel, as it moves along it. The roof over the machine is held up by hydraulic **chocks** or **shields**. These are adjacent cylinders, which press steel plates between the roof and the floor, and which then move forward into the space left by the machine, after its passage, allowing the roof behind them to collapse (Fig. 6.19). Typically two different machines are used to mine coal in this system. One of these is called a **shearer**, and uses one or two large drums, fitted with picks, which rotate against the coal grinding the material from the solid. This machine will usually move at speeds of up to 6 m/min taking a cut width of around 67 cm. The second machine consists of a series of larger teeth, held in a wedge shaped frame, and pressed hard against the coal seam by a series of jacks, acting against the roof-supporting cylinders. These machines, or **plows**, wedge the coal from the solid. They typically move at speeds of up to 100 m/min, but take a slice of coal which may range from 2.5 - 10 cm thick.

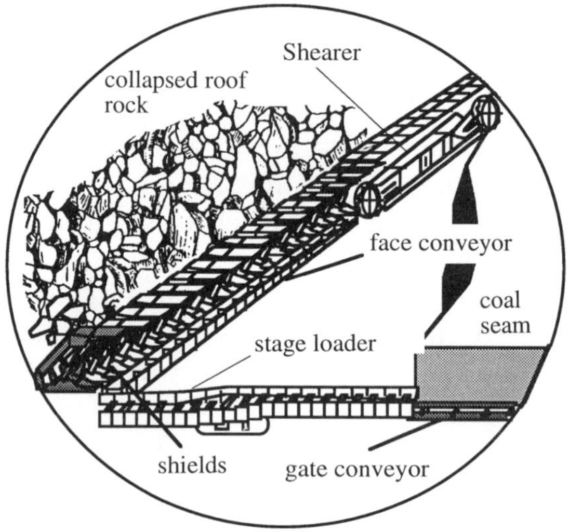

Figure 6.19 Section through a longwall face showing equipment used.

Historically a third machine, known as the Meco-Moore Cutter Loader had been used in British coal mines, and extracted coal by making two horizontal cuts, one at the bottom of the seam, and one in the center, while a third cut was simultaneously made at the back of the slice being taken. These cuts were made with a mechanical sawing system, and together extracted a slice of coal at one time which could be up to 2.1 m thick. The third contract from the Bureau of Mines was to the University of Missouri-Rolla to adapt this concept, and that of a plow into a machine, now known as the Hydrominer, which would replace the mechanical cutters of the Meco-Moore concept with waterjets, and use the shape of a plow to load the coal from the coal face, onto the adjacent conveyor [6.24]. In essence the machine consists of a series of oscillating waterjets which cut at the back, top and bottom of a wedge shaped plow which, when dragged along the coal, peels off a layer up to 1 m thick, and to the height of the coal seam (Figs. 6.20, 6.21).

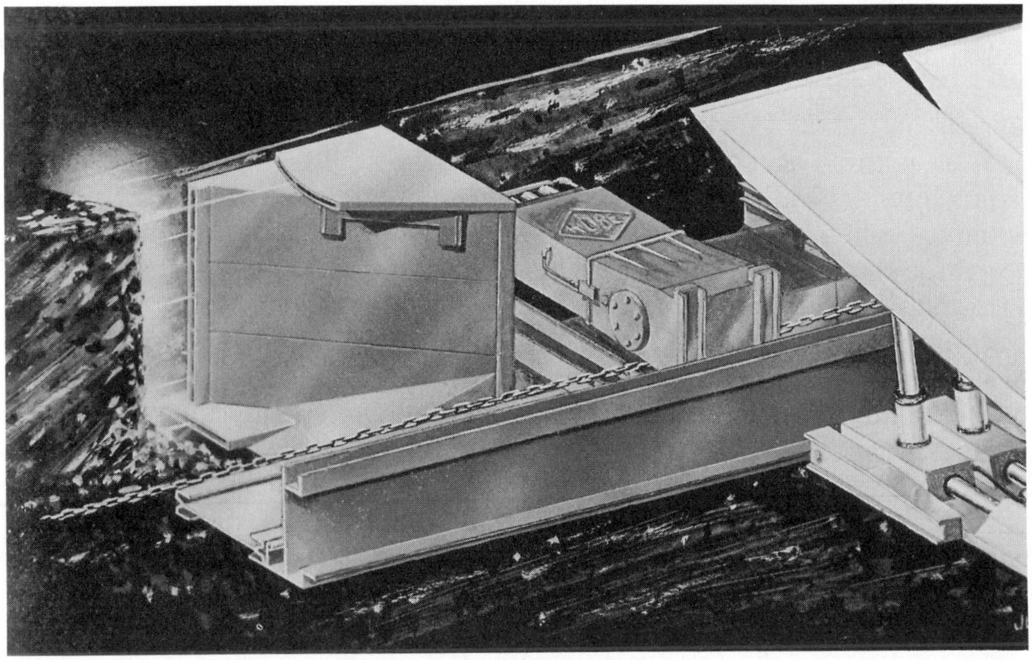

Figure 6.20 Artist's drawing of the first hydrominer.

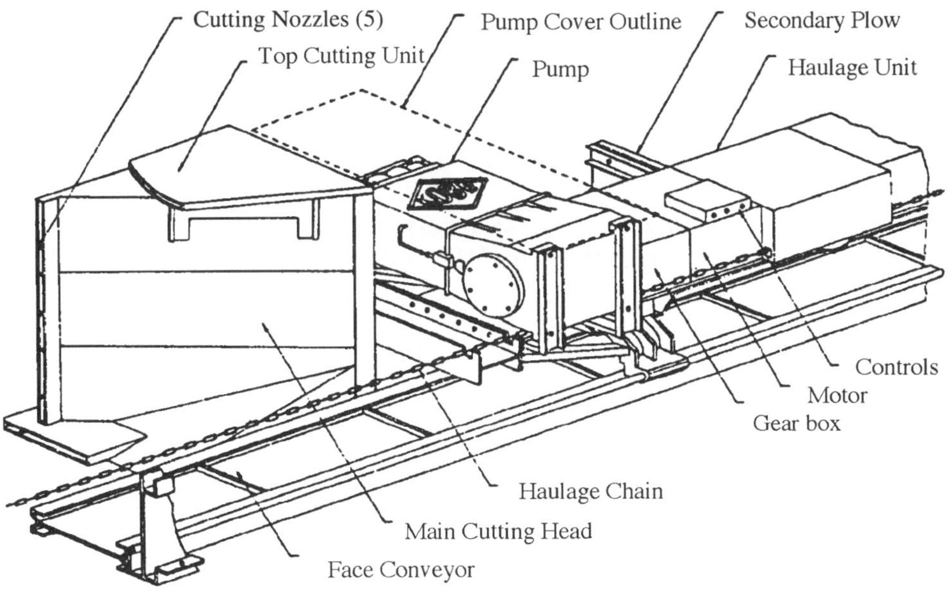

Figure 6.21 Sketch showing the component parts of the UMR hydrominer.

In contrast with the earlier designs, this concept which partly evolved during the contract, incorporated a different approach to the use of waterjets on the coal face. Previously the intent had been to use waterjets alone, or with adjacent cutters, to break all the coal from the solid. With this machine the waterjets were to be used to create deep slots which could then be exploited by mechanical action to ease the mining of the coal.

Several changes to this approach were made as the machine was built and tested. Firstly, the work of Read et al. [6.21] had shown that depth of penetration increased with increased jet pressure, and with increased jet power. However, in that analysis they had included jet flow rate in the jet power calculation without recognition of the greater importance that an increase in the flow, rather than an increase in pressure, would make to the efficiency of the mining machine.

As discussed earlier horizontal mining practice can only sustain the presence of limited quantities of water. However the evidence from hydraulic mining is that waterjets could mine coal at pressures of 70 to 100 bar. To thus use pressures of 3,500 bar seemed a little excessive as a way to reduce the amount of water used. Conventionally up to 200 lpm of water are used on existing mining equipment to suppress the dust generated by the cutting picks. This water can normally be readily coped with in the mine. This level thus provided a standard against which to design a machine.

A series of tests were carried out examining the relative roles of jet pressure and flow rate, as expressed by nozzle diameter. The results, (Figs. 6.22 and 6.23) were normalized to determine the relative effect of increasing either parameter. As the criteria of performance the energy in the jet, the volume of material removed and the specific energy of cutting were calculated, and normalized. The curves which resulted showed that it was much more effective to increase the flow rate of the jets than it was to increase jet pressure, once the threshold pressure of the material had been exceeded.

By reducing the jet pressure to 700 bar, the power required for the waterjets dropped to under 225 kW. This brought the machine within the power spectrum of existing equipment, and thus made it a more competitive system.

In order to cut the slots ahead of the unit wide enough so that the nozzles and leading edge of the Hydrominer could feed into them, a second change in the design of the nozzles was made. Single jet cuts have the disadvantage that, when they are used to make multiple cuts over the same line, it is difficult to align the jet with the previous cut, and thus energy is lost in recutting the slot, and in the pressure loss to the edge of the cut. This is particularly true where the nozzle is being moved into the slot at the same time. By changing to a dual orifice design, (two jets are inclined outward from a common axis), this problem can be overcome.

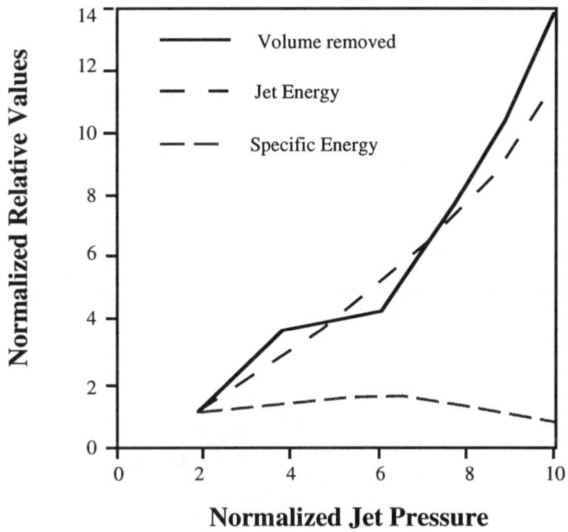

Figure 6.22 Effect of increase in jet pressure on the jet energy, volume removed and specific energy of cutting.

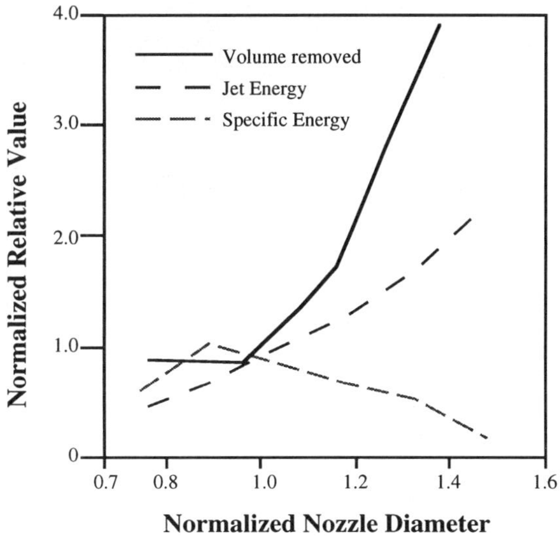

Figure 6.23 Effect of increase in nozzle diameter on the jet energy, volume removed and specific energy of cutting.

Experiments showed that where this dual orifice design, (Fig. 6.24) was used to cut coal, then the two jets would also remove the intervening material to a width of over 5 cm, without the need for any additional cutting equipment to move the coal. The resulting slot (Fig. 6.25) is therefore wide enough that it will allow the jets to travel through air on the next pass of the two jets until they reach the back of the existing cut. This significantly increases the penetration of the rock and reduces the amount of energy which is otherwise lost in multiple pass cutting of a surface.

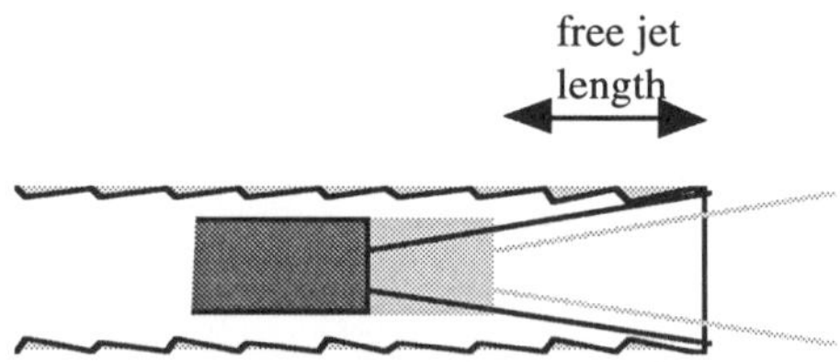

Figure 6.24 Dual orifice nozzle cutting schematic, showing the gain on the second pass.

Figure 6.25 Slot cut by a dual jet nozzle in coal, showing pyrite within the cut.

For this design to be effective, however, the jet must move forward into the cut between passes. Where this occurs, which happens when the machine is moving down the coal face, it proved possible, in the field, to cut into the coal to depths of over 50 cm ahead of the Hydrominer blade (Fig. 6.26). This gave a considerable leverage to the wedging action of the plow, significantly reducing the forces necessary to break the coal from the solid.

Figure 6.26 Cut made ahead of the blade by the first generation hydrominer.

A third modification to the system was also required. Nozzle arrays in the alternate approaches had been tied to the booms of the cutting heads. This reduced the speed at which the jets could traverse across the surface, and lowered their efficiency. By using rotary unions and individual lances, to which the nozzles were mounted, it was possible to separate the motion of the jets from the motion of the head. Thus only the very light nozzle assemblies had to be oscillated. This reduced the power required for the motion to about a kilowatt and allowed cutting traverse speeds of over 1 m/sec, at which speeds the jets were much more efficient. Calculated extraction efficiencies for the resulting machine were predicted to be on the order of 4 joules/cc within the range of existing equipment and, partially as a result, a second phase to this contract was undertaken [6.25].

A prototype of the cutting head was constructed and tested in a surface mine during the summer of 1976. These trials validated the basic concept, indicating also that because of the relatively deep cut made by the jets ahead of the plow, relatively little force was required to break the coal from the solid. This finding held true even when the thickness of coal peeled from the face exceeded one meter. The machine was able to achieve advance rates of over 3 m/min. A second cutting head was constructed, following the results of the first trial (Fig. 6.24), but was never successfully accepted.

Figure 6.27 Second generation Hydrominer on test.

The following advantages were found from the first set of trials:

- The Hydrominer mines coal with effectively no dust.
- The size of product is on average larger than that from a shearer with substantially less fine coal produced. By choice of jet angle in the nozzles this size could be controlled to minimize large, i.e., plus 15 cm in coal.
- The haulage forces required to move the unit are lower than those required to pull a shearer down the face under equivalent conditions. Concurrently, there is much less vibration of the cutting unit as it advances since the actual coal cutting occurs ahead of the Hydrominer.
- The Hydrominer is able to cut webs of increasing depth, in the tests from 0.5 - 1.0 m, with little increase in unit horsepower. The head,

although designed for this operation to mine to a height of only 75 cm of coal, was able to successfully cut coal to the maximum seam height of 1.3 m.

- The cutting head was able to penetrate, without loss in speed and no wear on the machine, pyrite layers 5 cm thick.
- Coal coming from the Hydrominer was infused with water so that little dust will be generated by later degradation of the coal as it is transported out of the mine.

Following these tests a German equipment manufacturer built a similar design of equipment, although it did not include the dual orifice cutting jets or the undercutting arm. Both these features are important, since the first allows the greater cutting depth while the second breaks the solid beam of coal ahead of the plow, relieving it of stress, and making it considerably easier to mine [6.26]. It is this feature, for example, which makes the machine more productive than the Polish concept which was developed along the same lines [6.27]. The Polish machine was able to advance at a speed of up to 5 m/min with a pressure of 300 bar, and a flow rate of 360 lpm. Jet diameters were, however larger, at 4 - 5 mm and the machine could mine only at rates to 250 tonne/hr. The German machine, in contrast, was able to mine large coal at advance rates of up to 24 m/min while taking an average 0.4 m bite of coal from a seam of coal 1.6 to 1.7 m thick [6.28]. The machine produced large coal, with 21% less fines over that of a shearer. Flow rates were up to 226 lpm, and the coal had a moisture content of 4.6% in contrast with the 8% from a comparable shearer face.

Regrettably, although this design thus proved successful a market survey of the industry suggested that only ten machines could be sold each year, and the program was canceled.

6.2.3.5 700 BAR SYSTEM FOR ROOM & PILLAR COAL MINING

In a continuing effort to find improved methods for mining, the Bureau of Mines has funded at least two additional approaches to the application of waterjets to room and pillar mining. In 1976 Ingersoll-Rand Research, Inc. was funded to evaluate different methods of using a low pressure, high volume machine for use in conventional mines. The resulting report [6.29] examined four concepts for mounting a monitor on a variety of different carriages and suggested that the technique could be economically advantageous. The program did not, however, continue.

A second program was funded through an inter-agency agreement with the National Aeronautics and Space Administration, and was initially conducted at the Jet Propulsion Laboratory [6.30]. This program followed a series of developments at that laboratory which had sought innovative methods for extracting coal from the underground. One such method was the use of a 30 - 40% mixture of water and carbon dioxide which would dissolve some of the constituents of the coal, and weaken the rest. After proving this concept in the laboratory a method for using this in a remote method of mining known as the "Coal Worm" was proposed [6.31]. The method was conceived either to use the mixed fluid, or as a more conventional means, using waterjets alone, to mine the coal.

This concept, which was ongoing at the same time that vertical borehole mining (Section 6.3) was being developed, approached the problem of remote mining in a slightly different manner. Following the vertical drilling of an access hole, long horizontal holes are drilled up down the dip of the seam. A sump and slurry pump are then located at the end of the hole, through a second hole drilled from the surface. The horizontal drilling head changes to a second jetting mode in which it is reaming coal over a square section around the borehole, as it now retreats up the seam and back to the original borehole (Fig. 6.28). The resulting slurry flows down the gallery which the worm passage has created, to the sump pump, from where it is pumped to the surface. By then drilling adjacent cross cuts perpendicular to the initial drive, a panel of coal can then be similarly extracted without the need for underground access of personnel (Fig. 6.29). This concept requires the ability to turn sharp corners with a waterjet drill - a concept proven at UMR later in the decade [6.32].

That program did not continue, but it led into another, based upon a modification to existing room and pillar mining practice. A system of mining was developed which included six mining units, given the acronym SAMS (Semi-Autonomous Mining System) each capable of mining over 1.5 tons/minute was conceived [6.30]. The mining machine upon which the concept was built was designed in collaboration with faculty from the University of Missouri-Rolla. Following completion of the original concept a proposal to the Department of Energy, Energy Related Invention program was approved, and the University proceeded to build a prototype machine [6.33].

The basic concept relied upon the same idea as the Hydrominer, namely that a waterjet system would first cut relieving slots into the coal face ahead of the machine, and then that wedges would break the outlined blocks of coal into the free space (Fig. 6.30). The UMR machine, given the acronym

RAPIERS, was built to take advantage of existing used mining machine components. Thus the main frame of the unit was the loading apron from a continuous miner (in the same way as the frame for the Bendix and IIT machines had been).

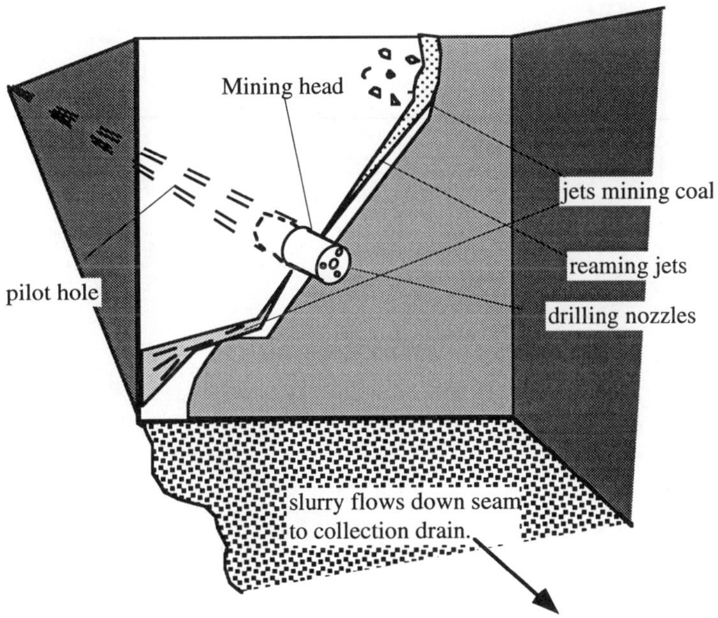

Figure 6.28 The coal worm concept [6.31].

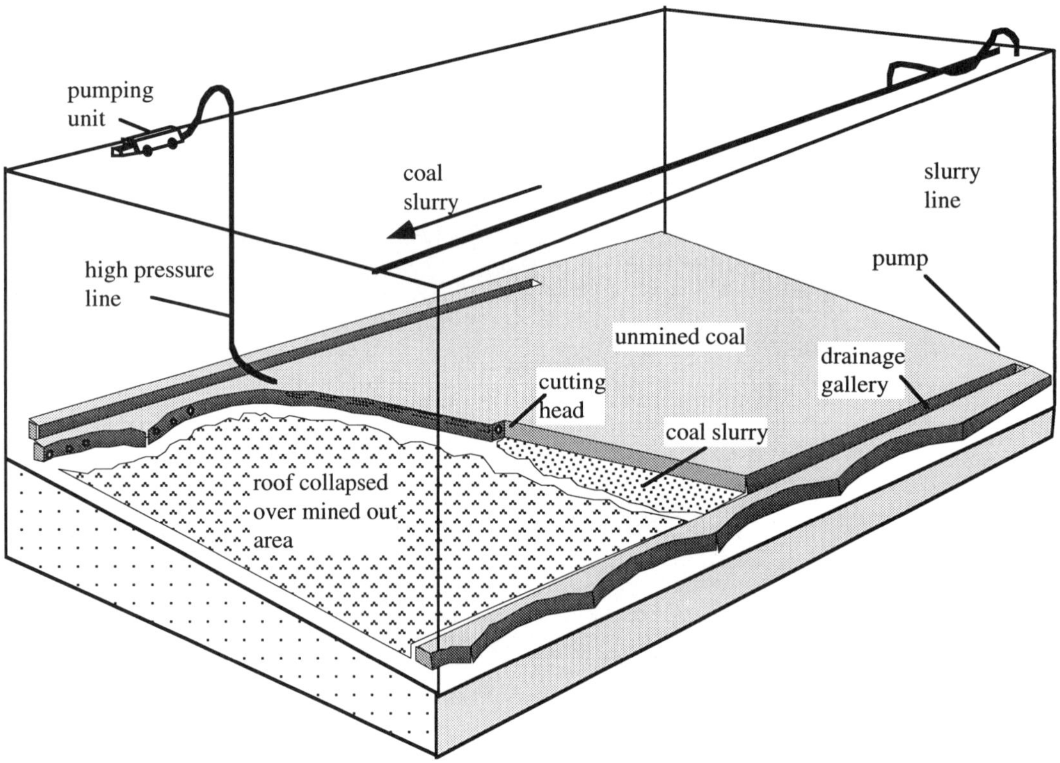

Figure 6.29 The coal worm mining plan [6.31].

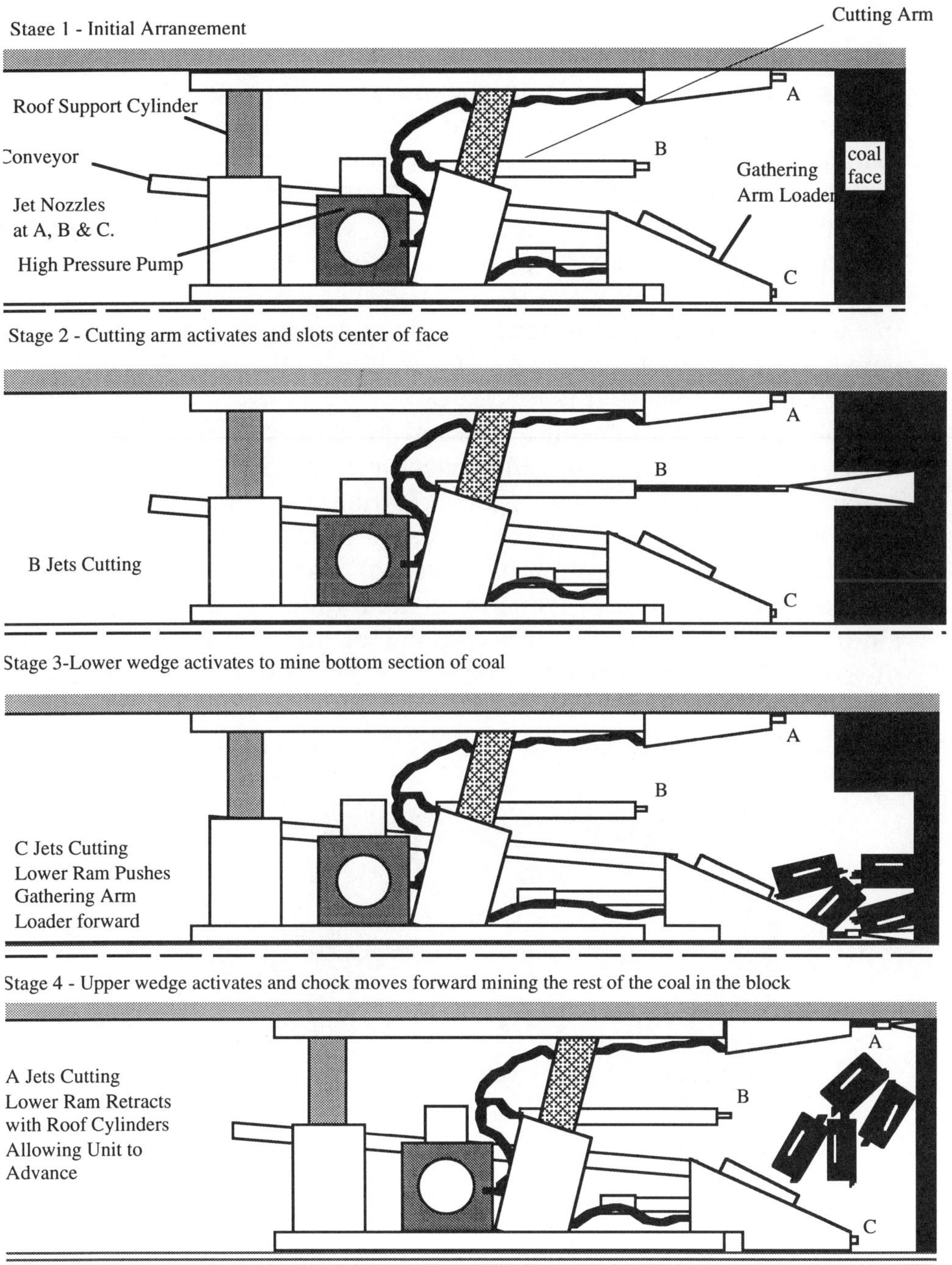

Figure 6.30 Sequence of mining for the RAPIERS system.

The face is first cut by a pair of oscillating waterjet lances, which cut a horizontal slot in the center of the face, and a second vertical cut along each rib of the tunnel. With the machine held in place by two sets of hydraulic cylinders (the recovered halves of a roof supporting chock) the loading apron is then pushed into the coal, which an oscillating jet system under the arm cuts a slot 5 cm wide into which the apron wedge will advance. This lifts and breaks the lower wedge of coal, pushing it up onto the apron, where it is steered to a conveyor section which carries it away from the face. An upper wedge, is then activated and moves forward with the jets cutting along the leading edge of the wedge. As it advances it enters the slot which has been cut and breaks the coal down from the roof onto the apron, where it is loaded out.

The body of the machine then advances in two separate movements of the cylinders, in turn, and the machine is ready to recycle. Because the machine uses only one pair of jets at a time, the amount of horsepower tied up in the jet system has been reduced to less than 100 kW. Advance rates from Hydrominer operation suggest that a cycle time for this unit should be less than 4 minutes. Thus, in a seam which is 2 m high, with the jets cutting a 3 m wide path, and taking a 1 m bite at a time, a production rate of over 1.5 tons/min can be achieved. While this was the rate suggested by the JPL report, it is possible that this could be considerably exceeded. A machine has recently been completed (Fig. 6.31), and is currently awaiting field trial.

Figure 6.31 Initial RAPIERS prototype.

6.2.4 JAPANESE OPERATIONS

Mitsui Mining first started hydraulic mining in Japan in 1963, at the Noborikawa mine. By 1979 [6.34] the technique was being used to mine five steep seams which were also quite gassy. Entries were driven some 100 m apart with two cross cuts, through all five seams, being used to give access to the coal. Of the 1.1 million metric tons of coal produced by the mine in 1978, 70% was won by hydraulic mining.

The mine operated 3 panels at a depth of 600 m. Face advance of over 12 m/day were achieved on 18 m wide panels, giving an output of nearly 800 tonnes/face/day of clean coal (Fig. 6.32). In order to mine the coal access drifts were driven through the seam, at an angle to allow the coal to be flumed away under gravity, once it had been mined. The drifts were some 15 to 18 m apart and located in the center of the coal seam. Thus, the monitor could be set in the lower drift and inclined to point upward and forward, mining out the overlying coal, and that forward of the machine from the last location. Once the coal had been removed, the machine was moved back down the drift, repeating the exercise. The extraction pattern would be that the upper levels were worked first, and the mine then moved to a lower drift to mine the coal underlying the first layer extracted.

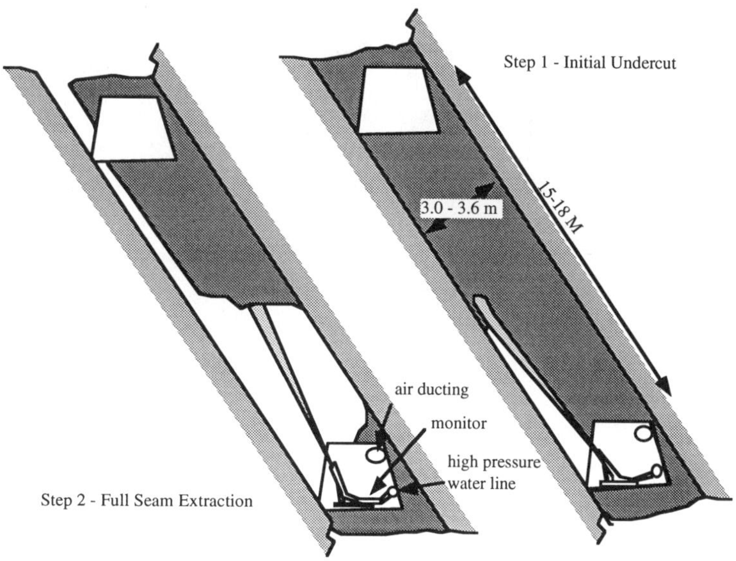

Figure 6.32 View along the seam of the layout in a Japanese hydraulic mine [6.5].

The coal seam was between 1.5 and 2.4 m thick, and employed 9 people in each section each day, for a face productivity of 87 tonnes/day/person, more than four times that of conventional mining, while overall recovery was increased by 30%. The monitor used either a 20.3 or 25.4 mm diameter orifice, at a pressure from 100 to 140 bar, using approximately 2730 lpm of water. All Japanese hydraulic mines are, reportedly, now closed, due to the difficult working conditions.

6.2.5 CHINESE HYDRAULIC MINING

Hydraulic mining was introduced into the People's Republic of China in 1956 [6.35]. Following the success of early experiments underground mines were established and, prior to the devastating 1977 earthquake, some seventeen million tonnes of coal were mined by this means each year. Operating mines have been established in mines with a variety of different coal conditions. Hydraulic operations can be found in seams varying from 1 - 8 m thick in gently inclined seams (i.e., in the range from 10 to 30 degree slope) where production rates of 600,000 tonnes/year can be achieved for an operating unit; in a 10 - 14 m seam dipping at 30 degrees, producing 450,000 tonnes/year; and in seams ranging from 0.6 to 4 m and steeply inclined (30 - 75 degrees) which produce around 250,000 tonnes/year [6.36].

Until fairly recently the monitors, in contrast to those in Japan and Russia, were still operated by hand and productivities on the order of 200 tons/hr were reported. Nevertheless, the Lazato Mine in the Lailan Coal District has been able to produce between 2 and 3 million tonnes/year. Typically in China, jet output pressures are on the order of 200 bar with flow rates being up to 250 - 300 cu m/hr. Giving a total power consumption on the order of 7 to 10 kWhr/tonne.

Over the last 20 years, coal mining technology in China has gone through several stages of development. The first stage, from 1958 - 1963, when waterjet pressure used for coal mining was on the order of 20 - 60 bar and, as in the earlier days in Russia, the pressure was only sufficient to wash away the coal after blasting, without being able to break the coal from the solid. In the period from 1964 - 1974 water pressure was increased to 80 - 125 bar and this allowed the elimination of the blasting process. Since 1975 waterjet pressures have been increased up to 200 bar. The advantages found elsewhere, high productivity, low manpower requirements and a safe dustless method have also been found in China.

Higher flow rates are used in seams which range from 4 - 8 m thick in order to increase productivity from 30 tons/hr to the current 200 tons/hr, and with the annual output of each monitor increasing from about 1/2 million to 3/4 million tonnes of coal. This gives an increase in output/person shift of 50%, while reducing the overall costs of mining by some 30%. The practice of working the coal out from unsupported areas, while lowering costs, also raises the risk of premature roof failure, with the resulting loss of coal that could have been mined. Thus, a careful planning of the excavation process is required and it can only be applied under special conditions [6.37]. In areas of high ground pressure the mining method is often to drive raises forward up the dip of the seam, and then to extract the rib coal on each side of the tunnel as the face retreats. However this is restricted in applicability and cannot be used, as coal angle increases, in seams of decreasing thickness. Ventilation of such areas requires special precautions and negative pressure is recommended. Not all the coal can be extracted from the section, some 70 - 80% is lost because areas of the block become inaccessible, some 10 - 25% is lost because the jets only mine part of the seam thickness, while 5% is lost through loose coal which escapes the flumes and collectors.

Roof control is particularly important in hydraulic mining, where large expanses of the roof may need to be self-supporting while the coal underneath, and in the immediate vicinity is mined. As the seam angle increases it has been reported that the risk of roof collapse reduces, but the problem of coal dilution from dirt falling from the overlying waste increases [6.38]. Both problems increase with seam thickness. Because mine roof stability is time dependent it has been found that the increase in monitor pressure to increase the speed of mining, and thus lower the time the roof must be stable, can be economically justified. However, if the jet power is too high then the operator may bring dirt down from the overlying waste area, and create too much dilution of the product. Where roof pressures become too great then the mining plan must be changed to ensure that the overlying strata pressure is transferred to the solid coal lying outside the mining area, rather than being concentrated around the point of mining or the entry where the monitor is located. Again a more rapid mining process is required to ensure maximizing the production. Where these modifications were introduced at the Yanchung Colliery a monthly maximum output of up to 90,000 tons was achieved (Fig. 6.33).

Where relatively friable coal lies under weak roof rock, the practice has evolved of working alternate sub-panels such as at the Hsiehchiachi #3 colliery of the Huainen Coal Mines (Fig. 6.34). Here the 5.6 m seam, which dips at 23 degrees is overlain by a mudstone. The monitors are operated at 60 bar and the 30 - 32 mm diameter jets mine an average of 200 tonne/hr and have mined over 5.3 million tonnes from this one mine.

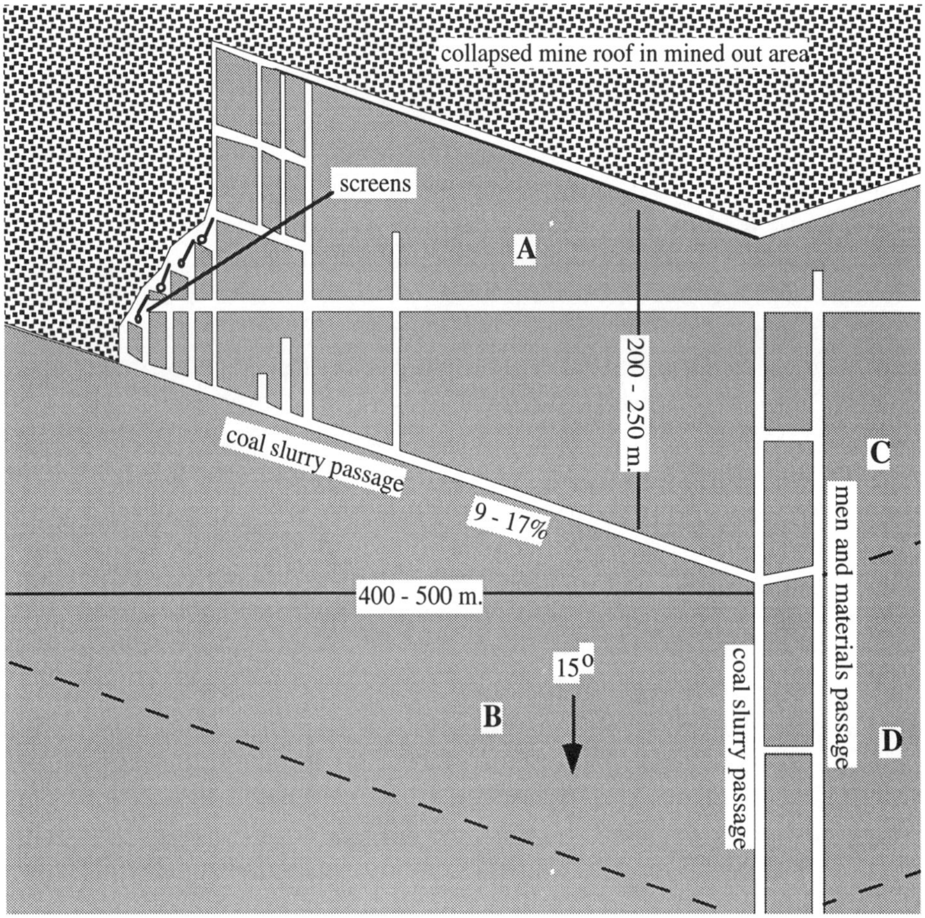

Figure 6.33 Plan view of the mining pattern at Yanchung colliery [6.38].

The use of flow straighteners is now widespread, particularly in those conditions where a lack of space does not allow the long entrance sections into the nozzle orifice, which produce the best jets. Various different geometries have been developed, for example, the nozzle designs used in China differ in shape, as do the flow straightener shapes. In that country recent studies have also led to increases in operational pressure, in order to increase productivity [6.39].

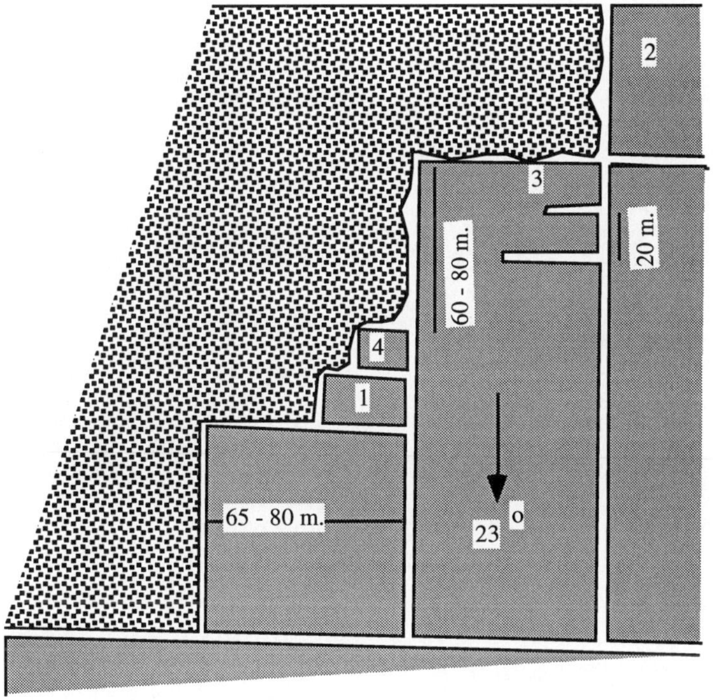

Figure 6.34 Plan view of the extraction pattern at the Huainen mines [6.38].

The need for rapid development of entries to provide working sites for the monitors has led the Chinese to develop a different approach to the driving of entries. While other international efforts have looked toward mechanical means of driving these entries in coal, the Chinese have developed, at their Xuzhou Institute of Mining an oscillating cutting head which excavates the face using a series of passes of the cutting lance (Fig. 6.35, [6.40]). The technique recognizes that, in different coal seams, the coal may be undercut, but adjacent cuts of the jet across the face must be tailored to that coal, to ensure that the intervening ribs between passes also come down. For maximum effectiveness the slots cut into the coal should have a finite width. The lance on the Xuzhou machine is therefore set to oscillate. This has several advantages. It is well known that waterjets cut more efficiently at higher traverse speeds and yet, with most mechanical systems, the higher, more efficient cutting speeds are difficult to attain. Ingeniously the oscillating miner allows this by combining the lateral sweeping motion with a high (5 - 20 cycles/sec) lateral oscillation of the lance. The result is a more efficient use of the jet, which cuts a wider, and in repeated passes, deeper cut into the face.

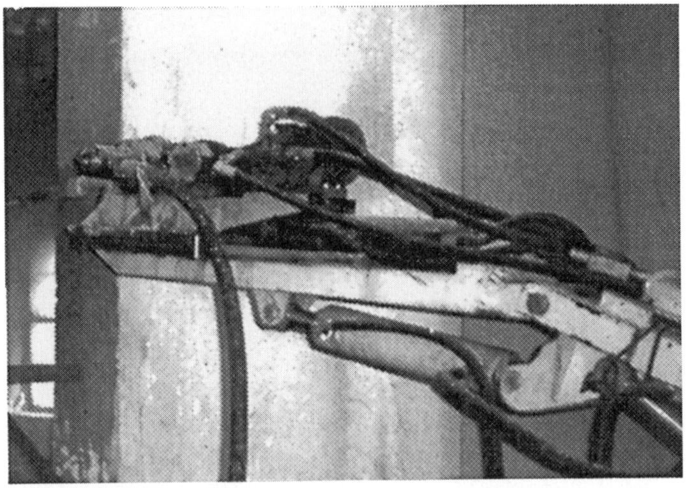

Figure 6.35 Oscillating head waterjet miner from Zuxhou.

The equipment was typically operated at a pressure of 250 bar with an amplitude of vibration of 44 mm [6.41]. In this mode, and at the correct stand-off distance, the oscillating jet would remove four times the amount of coal that would be removed by a single non-oscillating jet, traversed at the same speed. In addition, by careful planning of the traverse path up through the seam (Fig. 6.36) the intervening ribs would also break off, and the coal, being infused with water, would not generate significant dust as it collapsed. Through this technique it was possible to reduce the amount of coal actually cut by the waterjets to 7.5 - 24% of the seam, with the rest being broken out by indirect cutting and coal collapse under gravity [6.42].

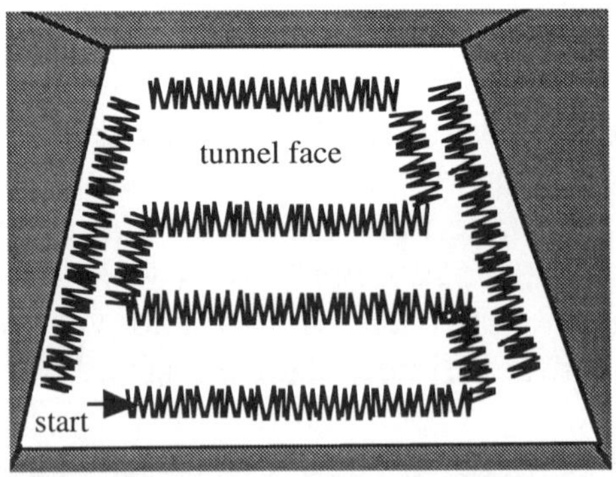

Figure 6.36 Cutting pattern of the Zuxhou waterjet mining machine [6.40].

In order to extend the cutting ability of the jets a flow straightener "with a star like cross section and a fish shaped in length" [6.43] was fitted behind the nozzle. The improvement in jet pressure, at a distance of 200 nozzle diameters, was from approximately 40% of nozzle pressure to roughly 70%. By 1983 the machine had been used to drive a tunnel 500 m long at the Mentougou Mine, with a dust make of 1.6 mg/cu m. With an operating pressure between 350 and 500 bar it was possible to mine coal at the rate of 0.8 - 1.3 tonnes/min with the correct extraction pattern. The machine was then tested in the high methane seams of the Matian and Xiandewang mines, where conventional mining practice is limited by the high volume of gas given out by the coal seam.

The procedure developed was to cut a slot ranging from 1 - 1.5 m deep at the bottom of the seam. To achieve as great a depth as possible several passes might be needed at a traverse speed of some 100 -150 mm/sec [6.40]. The arm was then moved up the face in a zig-zag pattern with some 350 - 450 mm between the centers of adjacent passes. Once the central block of coal in the drive had been removed in this manner, the edges of the tunnel were trimmed to shape by the cutting arm. Typical operating parameters for the machine were:

Table 6.3 Operating parameters for the oscillating waterjet miner [6.40]

working pressure	300 - 500 bar
nozzle diameter	1.9 - 2.4 mm
horizontal speed	100 - 300 mm/sec
amplitude of nozzle oscillation	100 mm
frequency of oscillation	about 14 Hz
stand-off distance	0 - 150 mm

It is interesting to contrast conventional mining practice in China with that achieved with the oscillating water jet miner (Table 6.4).

It is interesting to note, in addition, that the stability of the roof was significantly improved with this technique. While rapid roof convergence, with potential risk, was found with conventional drilling and blasting, roof convergence was much slower, and more stable when hydraulic mining was used (Fig. 6.37). This provides an intangible improvement in operating cost over conventional mining, in addition to the 30% reduction in actual mining cost at the face.

Table 6.4 Comparison of operational time between drill and blast and oscillating waterjet coal mining [6.40]

Procedure	*Drill and blast time (min)*	*Waterjet time (min)*
face preparation	3	5
drilling	20	-
loading explosive	4	-
workers leave face area	3	-
measuring gas levels	2	-
workers return to face	4	-
cutting holes for supports	-	15
loading coal	13	10
supporting roof	20	15
transporting coal	12	12
total time required	83 min	57 min
max dust level	274 mg/m^3	27 mg/m^3
person power required	11	9
relative drivage costs	100%	67%

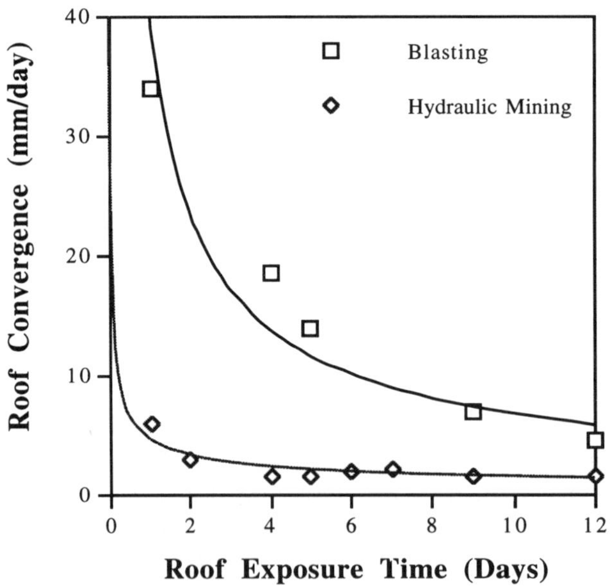

Figure 6.37 Ground convergence when the Zuxhou mining machine is used, relative to blasting [6.40].

6.2.6 CANADIAN HYDRAULIC MINING EXPERIENCE

The most successful of the hydraulic mining programs to date has been carried out at the Sparwood Mine of what was first Kaiser Resources and then BC Coal. Although the original technology was acquired from the Soviet Union, considerable improvement in performance was achieved during the Canadian operations. This mine operated from 1969 until the early 1980s and refined the technique of hydraulic mining to the point where 2 people operating a monitor, with a third person on hand as a supervisor, are able to achieve in excess of 4,000 tonnes of coal per 6 hour shift with a single monitor. The monitor typically was operated at a pressure of about 120 bar at the pump, putting 5250 lpm through the 31.7 mm diameter nozzle [6.44]. The water stream was collimated by a flow straightener before leaving the nozzle, and was able to cut and mine coal at distances of up to 30 m from the nozzle [6.45].

The extraction pattern was designed so that initially 3.6 m high entries were driven up through the coal seam and then retreated in 12 m lifts (Fig. 6.38). The coal washed from the seam fell from the 15 m thick layer, which dips at an average of 45 degrees, into the entry whence it was picked up in the water from the mining process. This water transported the coal into a crusher device from which it passed into a plastic lined flume which carried the coal out of the mine. By the use of this plastic it was possible to reduce the angle of the drift, and thus increase the amount of reserves which could be extracted from a drift.

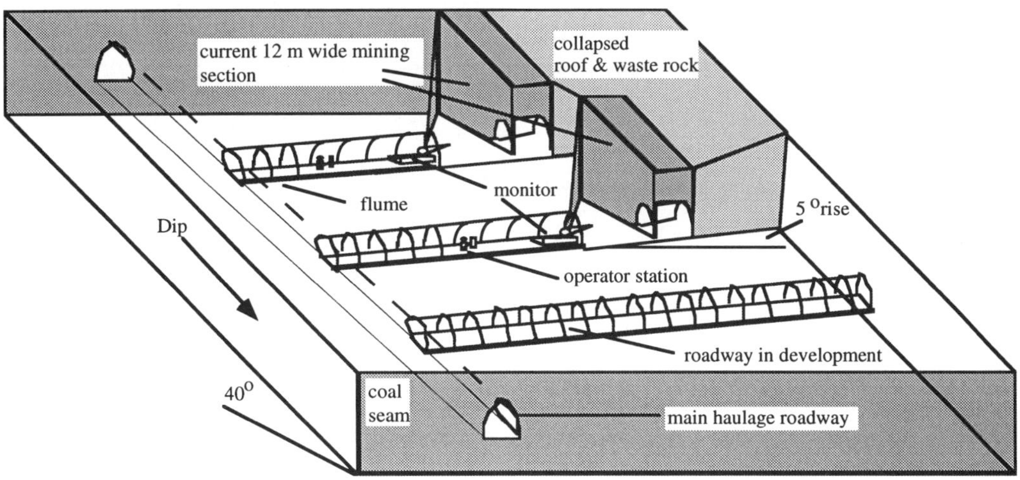

Figure 6.38 Mining pattern at the Sparwood mine in Canada.

This technology has been shown to be a very highly successful method of coal mining. It should also be borne in mind that, because no supports were used in the working area, the lifts which were 15 m apart and 15 m long, were washed out by the waterjet alone and when the roof fell down the mining process was over. Thus, should the roof collapse prematurely, then much of the coal can be lost and in fact although figures are not available from Sparwood, the Chinese report losses of up to 50% in some of their operations from premature failure to the mine roof.

Despite these conditions, the process is much safer than conventional mining as well as being more productive. The Sparwood Mine, for example, won the Canadian Safety Award for the safest mine in Canada for most of the years since that it was open. That does not mean that there were no problems. Because water and coal can be trapped in the waste, where the coal may oxidize and heat up in the presence of air, spontaneous combustion can occur. This fire in the waste has happened in several mines including the coal at Sparwood. This of course, leads to abandonment of these areas of the mine and can cause troubles in others.

The results from the few mines still in operation show that hydraulic mining has a future beyond the 15 million tons currently mined. However, although a paper was presented by Mr. Arentz of Storm King, describing successful 1983 experiments carried out in Colorado on hydraulic mining [6.20]; and although additional experiments have been carried out under Bureau of Mines funding by Kaiser Resources and by a team from Gulf Research Corporation who have examined hydraulic mining possibilities in Colorado and Washington state, no current hydraulic mines are in operation on the American continent. Capital costs for a hydraulic mine have been quoted (ibid.) as being half of those for a conventional operation, and in Canada hydraulic mining competed favorably with surface mining at a 5.5:1 stripping ratio. It has been shown to be a safer operation, and its high potential productivity hold out promise for the future.

6.2.7 GERMAN HYDRAULIC MINING EXPERIENCE

In Western Europe the major developmental effort in hydraulic mining has come from Germany. The work began in 1957 at SkBV (Steinkohlberg-bauverein) and used a 4.0 mm diameter nozzle at 3,000 psi to push coal from a steeply dipping face in place of a pneumatic pick. The monitor was attached to a pit prop and manually operated, but did improve productivity, although its cutting range was less than a yard.

Following a series of in-house tests, hydraulic mining operations were tried at Carl Funke Mine between 1962 and 1974, at the Robert Muser mine from 1965 - 1971, and at the Gneisenau mine from 1971 - 1977. These trials were followed by the creation of a complete hydraulic mining section at the Hansa Mine in October of 1977. The mine was set up to extract coal from a seam at a depth of 850 m at a pressure of 100 bar. With an investment of $56 million the mine was designed to extract 15,000,000 tons/yr of coal for 17 years, after a three year development [6.46]. By October, 1978 [6.47] the mine had produced up to 3,544 raw tons in a 10.2 hour operating time, but of this some 28% was waste rock. The initial plan called for an output of 250 tons/hr, but had reached 330 tons/hr by 1979. Transportation costs of 0.84 - 1.32 DM/tons compared similarly to conventional transportation costs.

The program did not finally prove successful, reportedly because:

a) the seam was not thick enough;
b) the coal was harder than expected;
c) the development roads could not be driven fast enough;
d) the production in the retreat mining of pillars was too low;
e) the initial development costs were too high;
f) there was insufficient central control because of the scattering of the workings;
g) too high a strata pressure;
h) ventilation problems.

Siebert has reported in Gluckauf that the production was well below that of local mines, and costs were much higher [6.48]. The mine was, accordingly, closed on November 30, 1980. As mentioned earlier, the German industry and government also cooperated in the development of longwall waterjet assisted plows. one with oscillating cutting arms similar to the UMR machine, and one [6.49] where the jets were directed behind individual picks on a fixed head plow. Both of the concepts were tested at the Lohberg colliery. Unfortunately, although the trials showed some success, the limited market for the machine led to the abandonment of this program.

6.2.8 POLISH EXPERIMENTS

A similar type of waterjet longwall waterjet mining machine had earlier been developed in Poland [6.27]. This unit, named the GIG-AH3, used a jet pressure of 300 bar, at a flow rate of 360 lpm, was tested on a 65 m long, 1.6 - 2.0 m high, longwall panel in the Rhymer mine in Poland in 1965. The 4 - 5 mm diameter waterjet was oscillated along the back of a rock of picks on the plow, but only in the vertical plane. This did not, thus, break the pillar of coal, which the jet would develop. The machine cut a web of 300 - 350 mm but had some difficulty in bringing down hanging coal when the web was increased to 500 mm, and at that thickness the plow was too weak to move the pillar. Problems were also encountered with dirt bands and rock in the coal seam (as found in the later German experience). The machine was nevertheless able to advance at speeds of up to 5 m/min mining up to 250 tons of coal/hour. The machine was subsequently taken for testing in the Soviet Union. Since the Russian development of longwall machines initially was more directed at rotating a waterjet on a coring path, as the machine cut along the face [6.50] which led to the development of a trepanner type of mining machine [6.51] and then into the development of waterjet assisted shearers, which will be discussed in Chapter 9, little further has been heard of this device.

It is interesting to note that in the 1976 Annual Report by the Director of Mining Research and Development of the British National Coal Board, hydraulic mining is cited as the only alternative to established methods of mining under development [6.52]. Recent studies have, however, been more directed at the use of high pressure waterjets as a means of assisting the use of mechanical means of cutting, rather than being used by themselves alone as cutting and material removal tools. One additional area, in which low pressure, high volume jet mining has been applied will first be discussed.

6.3 BOREHOLE MINING

During the earlier trials on the various different methods of mining in the Soviet Union, attempts were made to mine relatively vertical seams of coal by first boring a hole down through the seam to the next adjacent sublevel, putting a reaming waterjet on at the drill head at that bottom end and then slowly raising the drilling head while the waterjet was rotated. This was anticipated would mine and flush the coal down into the underlying entry where it could be transported away. This technology was not very successful in the Soviet Union, however it supplied an answer to a problem that was at that time bedeviling engineers at the American Gilsonite operation in the Green River Basin, Utah.

The development has been described earlier in section 6.2.3 (Fig. 6.14). Gilsonite was first mined hydraulically in 1957, using two car-mounted monitors which each mined an average of 500 tons/shift [6.53]. 3.6 m high benches were driven at a 2-1/2 degree grade, with a flume at one side acting as a transport system between the two shafts (located 300 m apart). A more modern system was developed in 1961, wherein a waterjet head was mounted on the bottom of a raise boring drill. The drill was first used to drill a 15 cm hole down 180 m to a collection drift and then, with the jet head mounted in place, slowly raised back up the hole. The holes were set 6 m apart and the drill rotated at 1 rpm. The drill was raised either automatically or manually, but usually manually under guidance from men in the mine. Average production was about 600 tons/shift.

The success of this technique led to trials of a similar method in a coal seam in Canada. The seam, however, was of relatively shallow dip, and the water dispersed without effectively carrying the coal into a collection flume [6.54]. Without being able to effectively collect the coal, the method did not work well, due, in part to the larger volumes of water required (as reviewed in section 6.2.1) and so it was abandoned. This result would also indicate that the JPL Coal Worm (Fig. 6.28) would likely have similar problems in mining relatively flat seams unless a method was found for confining the coal and the water together after mining, until the slurry could be transported to the surface.

In the course of trials of conventional hydraulic mining at the German Carl Funke Mine, a series of tests was carried out along the same line as those of the Russian and Vickary Creek experiments. In these tests waterjets were initially used to drill holes down through steeply dipping coal seams from one tunnel to the next, a distance of some 15 - 25 m [6.55]. The jets operated at pressures below 100 bar and with nozzle diameters of 7 - 10 mm. Once the initial passage way had been developed, then a reaming jet head was fitted to the drill, which was retracted, washing the coal in the panel down to the lower tunnel for collection (Fig. 6.39), in much the same way as in the gilsonite mining operation. The technique was tried from both directions, i.e., with reaming of the holes from the top down, and from the bottom up. It was found that the large pieces of coal which were broken by the jet would block the delivery borehole when the drill was retracted downwards, and so the upward reaming was the more effective method. Predictions from the initial experiments suggested that an output of 8 - 10 tons/person shift could be anticipated, with a cost savings of 15 - 20 DM/tons over conventional mining prices.

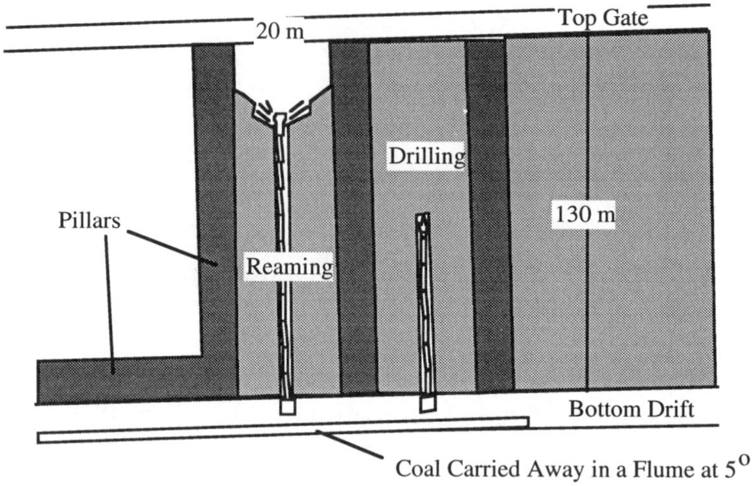

Figure 6.39 Experiments in borehole coal mining in Germany [6.55].

This technique showed that it was possible to ream a cavity up to 15 m in diameter around the borehole and provided a means of effectively extracting the coal. This technique was also found useful in extracting pillars in retreat. There were problems of aligning the boreholes in the centers of the pillars, however, and, concurrently the mine had to be closed due to a lack of sales into the domestic market. The experiments were thus brought to an end.

One result of this research was, however, the German conclusion, that in order to throw a jet a good distance and still achieve an adequate cutting pressure, it is necessary to start with a lower nozzle pressure and higher flow rate. For example, where a 20 and a 25 mm diameter nozzle are compared, at pressures of 92 and 72 bar respectively, then within 3.6 m of the nozzle a greater force is measured with the smaller nozzle, but at 15 m the larger force now comes from the larger nozzle, even though the jet starts out at a lower pressure (Fig. 6.40).

This borehole reaming research, although moribund in Germany, and in Russia, became of interest in the United States, largely as a result of an effort by the Bureau of Mines under the direction of Dr. Savanick. He saw, in this procedure, a potential method for excavating small lenses of minerals which would otherwise be uneconomic to extract because of their location; being too close to the surface for underground mining and too deep to economically surface mine. The concept had been developed from ideas which had been patented as early as 1932 [6.56].

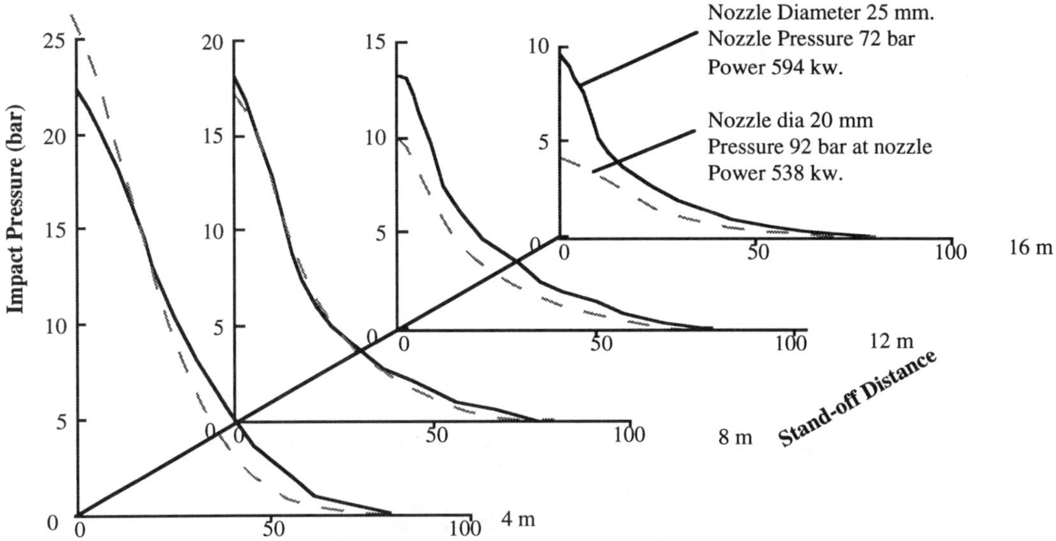

Figure 6.40 Comparative performance of high pressure, small jets and lower pressure large jets as a function of distance (after [6.55]).

Three systems had been developed, external to Government funding, one by FMC for the mining of phosphate ore in North Carolina, one by Marconoflow [6.57] and one by the A.B. Fly Company [6.58]. The latter system has been shown capable of mining sand and other material to depths of some 120 m at a production rate of up to 1 cu m/min. Further experiments in this technique, now known as **borehole mining**, were carried out under government funding, starting in 1975 [6.59]. Initial success led to further field tests, carried out initially internally at the Bureau [6.60] and then under funding to Flow Industries, Inc. Advantages foreseen for the technique were as a means of improving safety in the extraction of coal and other minerals, reducing both the time and the manpower required to develop mining sites.

The method which evolved was to drill a relatively large hole (approximately 50 cm in diameter) down from the surface to and through the mineral deposit. Into this borehole, a composite drill stem was lowered made up of three adjacent flow passages within the body of the stem [6.61]. Through one of the pipes high pressure water was pumped down to two nozzles located on opposite sides of the lower end of the drill stem. As the stem was rotated, using a Kelly of the type common in oil well operations, the ensuing jets cut a circular cavity out into the material on the walls of the borehole. By slowly raising or lowering the string the initial slot could be enlarged both vertically and horizontally until a chamber up to 7 m in radius could be created. The jets washing the broken rock down to a sump at the bottom of the drilled hole, where a small crushing device would break it into small fragments. At this point the slurry containing the material and the used water was picked up by a jet pump fed by water passed down through the second of the three passages in the drill steel, and this directed the water and slurry combination up through the third flow channel and thus out of the drill hole to a collection pond (Fig. 6.41).

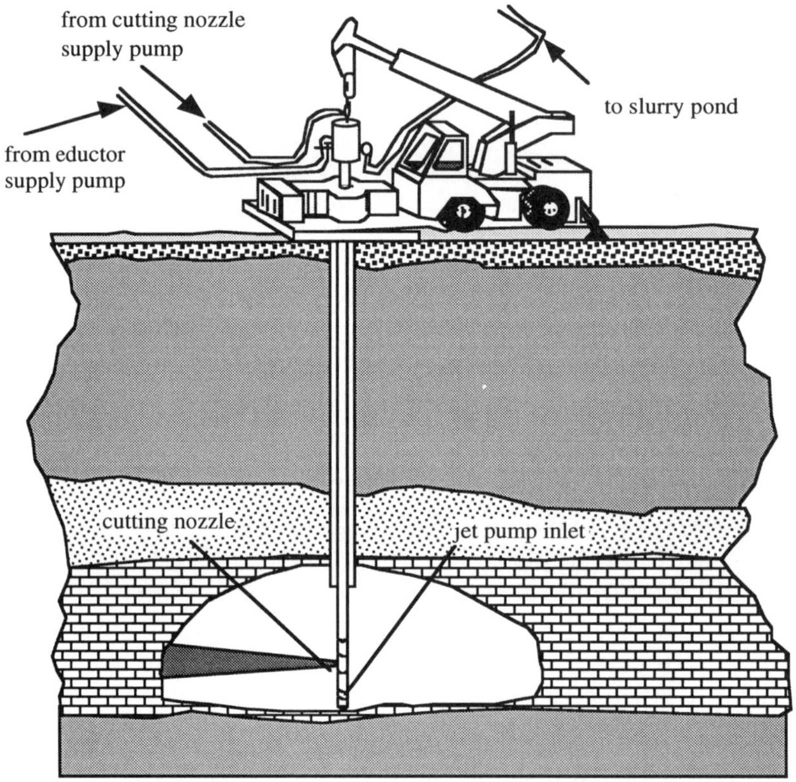

Figure 6.41 Early borehole mining layout [6.59].

The tool was first tried in a metallurgical coal seam outcropping near Wilkeson, Washington in 1975 and 1976. In that trial the unit mined at a rate of up to 7.3 tons/hour from a depth of 26.8 m and could cut the coal to a distance of up to 4.6 m from the original drill hole axis [6.56]. The equipment was operated at a pressure of 310 bar, at a flow rate of 378 lpm. The tests allowed some understanding of the equipment performance, with optimum results coming where the jet was traversed across the coal at a speed of 10 - 15 cm/sec cutting slots some 30 cm apart, and working from the bottom to top of the seam. Performance was somewhat restricted since shale from the roof blocked the jet pump, and the unit mined coal more rapidly than the pump could extract it. Thus, the pumping problem required solution.

At the same time the jets issuing from the unit were of limited range, given the need to sharply turn a corner from the borehole axis to the radial direction of the nozzle. In order to overcome insufficiencies in this area the Bureau of Mines contracted with TRW to develop an improved design [6.62]. The resulting design included flow channeling vanes within the nozzle (Fig. 6.42), and a quartic-straight design with a 20 degree angle, required since there was insufficient length available to use a more conventional nozzle. Testing of this nozzle indicated that it performed nearly as well as the conventional straight nozzle would, were space available for it, and that coherent jet streams up to 15 m long could be obtained [6.63].

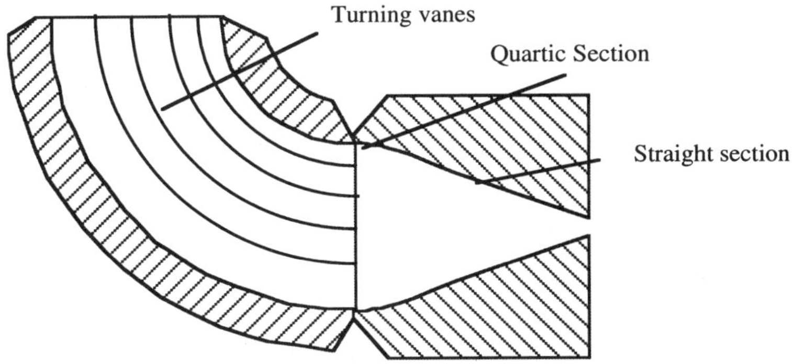

Figure 6.42 Turning vane assembly for a borehole mining nozzle (after [6.52]).

The economics of the coal mining aspects of the tool have undergone several reviews. It was pointed out [6.64] that during the initial trials the jet was cutting only a quarter of the time. Thus, while the overall system production was 7.7 tons/hr, the maximum production rate was 15.4 tons/hr and the jet cutting rate reached 34.5 tons/hr. When the problems of controlling the crushing of the coal and particularly of dealing with the shale were addressed then production rates of 18.1 tons/hr over the short term, and 10 tons/hr for sustained runs of over an hour were obtained. When the system was fielded in the Black Diamond coalfield in WA, intermittent operational production rates of over 45 tons/hr were obtained, with an average production of 18 tons/hr.

Based upon a review of the coal mining experiments, and subsequent experiments in mining uranium Evers and Knoke [6.64] drew the general conclusions that:

- the mineral to be mined had to be worth at least $20 a ton, as mined.
- the ore should be located below the economic stripping limit.
- the ore must not be too deep for pumping to be too expensive.
- the material being mined must require a pressure of less than 700 bar.

Prior to the Black Diamond trials the staff at the Jet Propulsion Laboratory working on the Advanced Coal Extraction Project (see section 6.2.3.5) had reviewed the potential for this method becoming economic in mining deposits of coal in Wyoming and Colorado [6.65]. A computer program was written to evaluate the relative economics, and was operated under the assumption that the unit would be capable of achieving a mining rate of 36 tons/hr. (As the subsequent trials showed this was not an unreal assumption.) The program included cavity backfilling and environmental restoration costs. It assumed a cavity diameter of 7.5 m in a 9 m seam, each of which would produce some 640 tons of coal. This would give 50% recovery with adjacent holes located on 10 m centers. A system designed to produce 6.64 million tons/year was, at 1977 prices, estimated to be able to produce coal for a direct cost of $15.33/tons, and an overall selling price of $27.11/tons in a thick horizontal seam, while a selling price of half that could be achieved in mining steeply dipping seams. The analysis was very sensitive to local geological conditions and system operating parameters.

This review was subsequently analyzed by the Engineering Societies Commission on Energy [6.66] and this report concluded that the technique did not hold out much promise for the United States, given that operating costs in thick coal seams were roughly three times that of existing methods, while steeply dipping seams, were sufficiently rare as to not justify an effort to exploit them. It must be considered unfortunate that this report very largely relied on the work by the JPL group and did not examine the results, for example, of the Gilsonite operation, or the Russian or German experiments. Further in assessing the design of the mine plans in thick seams, the author did not discuss the possibility of allowing the land to subside over the extracted coal - an event which realistically occurs even when the coal is strip mined.

Interestingly also, the limitation on the diameter of the chamber, set at 3.75 m would appear to be artificial given recent developments in Poland [6.67]. In seeking means of mining brown coal, the technique of excavating the coal under water, to provide cavity stability has been developed. Since the presence of the water reduces the cutting range of the jet, the two cutting nozzles are located on extensible booms which can take the nozzles out some 3 m from the initial bore (Fig. 6.43). This procedure has previously been suggested for reaming geothermal wells [6.57] on a smaller scale in harder rock. In the Polish design the design was fabricated and tested in the field. In the initial tests it proved possible to mine 1 cu m of brown coal in 3 minutes using 1.3 cu m of water. The coal was lifted to the surface using an air lift.

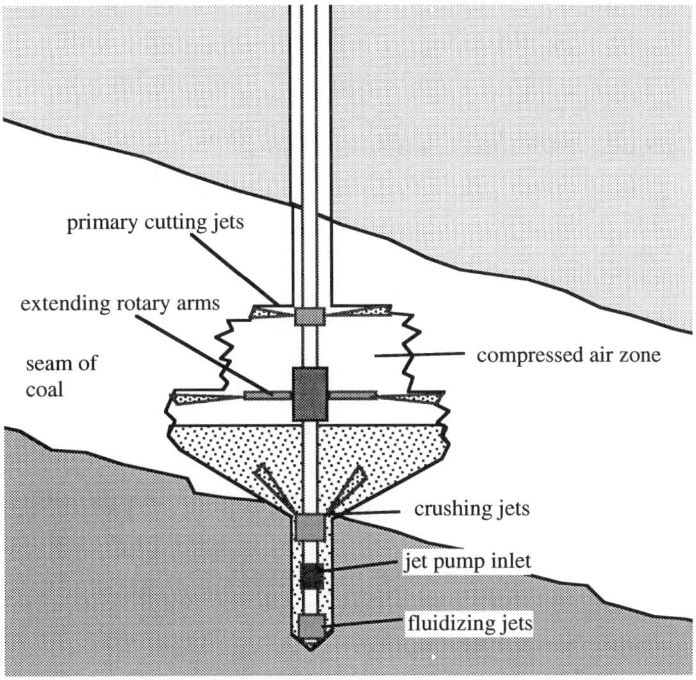

Figure 6.43 Polish method for increasing borehole chamber radius [6.69].

The experiments continued in Poland beyond the timeframe of this effort, [6.69] and jets have been successfully used to cut out to a distance of 5.5 to 6.5 m from the borehole axis. In addition it was possible to achieve in-hole crushing of the coal, although it was necessary to use compressed air, at a pressure of 3 to 5 bar, in order to provide an air atmosphere in which the jets could cut to the required radius, without it they were reduced to cutting between 0.7 and 1.0 m.

Field experiments have subsequently been carried out in borehole mining for non-coal materials. Flow Industries, under Bureau of Mines funding, mined uranium bearing sand, tar sand, and potash. In the second case, which was carried out in a uranium lens near Casper, Wyoming in the summer of 1977, some 900 tons of sand were mined from the deposit which was between 22 and 30 m below the surface. The unit averaged 8 tons/hr with a maximum rate of over 16 tons/hr The jets cut up to 10 m from the bore, and were operated at 138 bar pressure and a flow of 1135 lpm [6.56]. Once the cavity had been created it was subsequently backfilled, and up to 90% of the sand removed from the cavity could be replaced, using a slurry jetting technique, at a rate of around 27 tons/hr. Because of the friable nature of this rock, a secondary breaker at the bottom of the sump was not necessary. A method of operating the jet so that it could move up and down the hole without disturbing the inlet to the ejector pump would, however, have been of advantage since this movement affected the ability of the system to pump the slurry at optimum concentration from the hole.

Typical deposits of this nature are the uranium bearing sands near Casper, WY and the tar sands in and around the Bakersfield, CA. Mining in these deposits the waterjets, at a pressure of approximately 300 bar and 140 lpm, mined an underground chamber up to 6 m in diameter. A commercial system was sold to Rocky Mountain Energy for use in Uranium mining although due to the state of the market at the end of the 1980s, it is not currently in operation.

The system has one drawback in that when one is working in areas relatively close to the surface it is possible, if the device is rotated through a complete 360 degrees, to create a cavern large enough underground that the overlying burden will collapse into it taking with it, the reaming equipment. In such soft ground, therefore, it is recommended that the reaming operation take only 75% of the available mineral. One advantage of this technique is that the material which comes out can be treated so that, for example, the uranium or tar oil can be removed from the sand. This sand can then be pumped back down the cavity and used to restabilize the ground. The pipe can then be removed. the surface structures also taken away, and the net result is very little environmental impact. The process was demonstrated following the extraction of uranium ore at the Rocky Mountain Energy's Nine Mile Lake mine in Wyoming [6.70].

Following the experiments described above, more than 90% of the material originally mined was pumped back into the cavity. It was necessary to eject the slurry in jet form, under some pressure to drive the material to the back of the 15 m diameter chamber. The technique was also used successfully to backfill the chambers created in the potash mining experiments. Although the technology has demonstrated a number of commercial advantages, it has as yet had little commercial application, although the recent studies in Poland indicate a resurgence of interest.

The technique has more recently been tested in the phosphate near St. Augustine, where it was found to provide an advantageous means of reaching the deeper deposits beyond the depth where surface mining is practical [6.71]. More recently (1985) it has been tested by the Hawley Resource Group, under state funding in Alaska [6.72] as a means for mining gold under a layer of overburden. Jet pressures of up to 350 bar were used at a flow of 400 lpm in trials to cut up to 22 m below the surface, with a projected chamber diameter of 9 m. The state felt that the reduced water demand, some 10% of that required for surface placer mining, was a distinct advantage. Mining rates of some 15 - 20 cu m/hr were projected, for the tool which was estimated to cost over $100,000. It has subsequently been suggested that the heavy weight of the gold particles extracted by the miner caused them to easily sink to the bottom of the chamber cut by the jets, where they have proven very difficult to pick up and entrain into the slurry pump. In consequence the Bureau of Mines has developed a secondary system of mining in which two borehole mining units are used to develop the deposit. By using the jets from one to flow over the rock surface and flush the gold particles to the jet pump intake of the second unit, it is possible to extract the valuable material [6.73]. It is reported that, in relative terms, the gold recovered increased from 1.09 g to 3.65 g using the two borehole approach. In addition the authors noted the benefits of using hotter water to increase the system production rate.

Table 6.5 Summary of early borehole mining tests

Date	*Ore type (m)*	*Depth (bar)*	*Jet pressure rate (tph)*	*Av mining tonnage*	*Total*	*Location*
1976	coal	7.5 - 2.2	31.5	9	48	Wilkeson, WA
1977-78	uranium	22 - 30	17.5	6	940	Casper, WY
1979	oil sand	33 - 45	3.5	14	990	Taft, CA
1980	phosphate	39 - 75	2.1	35	830	St. Augustine, FL
1980	coal	0 - 6	31.5	20	?	Black Diamond, WA

6.4 NOTES AND REFERENCES TO CHAPTER 6

In the original development of this chapter its was anticipated that the work carried out in waterjet application to geotechnical materials could be covered in a single chapter. This idea was slowly modified, as the text was written, and the full extent of previous work in this area became apparent. It has been necessary, therefore, to separate the work into a number of different chapters, beginning with Chapter 5, which dealt with Civil engineering applications, and ending with Chapter 9, which deals with the use of waterjets in combination with mechanical tools.

The programs of development are, however, closely related, and many developments from one area can equally well be applied in others. The separation has therefore had to be an artificial one, and the reader is encouraged to read the other chapters in order to obtain a better perception of the activity and other tools which might be useful to their endeavors.

6.5 REFERENCES

6.1 W.A. Summers personal communication with D.A. Summers, 1990.

6.2. V.S. Muchnik, Candidate thesis, 1935.

6.3. Ofengenden, N.E., and Dzhvarsheishvili, A.G., Technology of Hydromining and Hydrotransportation of Coal. (initially published in Russian by Nedra Press, Moscow, 1980) translated by A.L. Peabody, Terraspace Inc., 304, N. Stonestreet Avenue, Rockville, MD., 20850.

6.4 Summers, D.A., "The Development of Waterjet Cutting," World Mining Equipment, Vol. 17, No. 11, November, 1993, pp. 24 - 28.

6.5 Cooley, W.C., Survey of Underground Hydraulic Coal Mining Technology, Final Report on Contract HO242031, Dept. of Interior, U.S. Bureau of Mines, October, 1975.

6.6. Yufin, A.P., Hydromechanisation, State Scientific Technical Press of Literature on Mining, Moscow, 1965 (Translation U.S. Army Foreign Science and Technology Center, Translation J-1061, 1967).

6.7 Mills, L.J., "Hydraulic Mining in the USSR." The Mining Engineer, June, 1978, pp. 655 - 663.

6.8 Ekber, B.J., and Kuzmich, I.A., "Coal-Cutting Technology with Water Jets in the USSR," paper E5, 5th International Symposium on Jet Cutting Technology, Hanover, Germany, June, 1980, pp. 297 - 302.

6.9 Summers, D.A., Disintegration of Rock by High Pressure Jets, Ph.D. thesis, Mining Engineering, University of Leeds, UK., 1968.

6.10 Okrimenko, V.A., Hydro-Monitor Operator in Coal Mines and Pits. State Scientific Technical Press of Literature on Mining, Moscow, 1962, 264 p. (Translation U.S. Army Foreign Science and Technology Center Document AD 820634, 1967).

6.11 The Coal Industry of the USSR - a Report by the Technical Mission of the National Coal Board, Part 2/6 Special Mining Techniques, 1957.

6.12 Jenkins, R.W., "Hydraulic Mining" The National Coal Board Experimental Installation at Trelewis Drift Mine in the No 3 Area of the South Western Division, M.Sc. thesis, University of Wales, 1961.

6.13 Moodie, K., "Coal Ploughing Assisted with High Pressure Waterjets," paper D6, 3rd International Symposium on Jet Cutting Technology, Chicago, IL., May, 1976, pp. D6 65 - D6 79.

6.14 Baker, J.H., "Mining by Hydraulic Jet," Mining Congress J., Vol. 45, No. 5, May 1959, pp. 45 - 46, 52.

6.15 Borden, P.E., "Latest Technological Developments in Mining Methods," 9th Annual Minerals Symposium, Moab, UT., May 22, 1964.

6.16 Wallace, J.J., Price, G.C., and Ackerman, M.S., Hydraulic Coal Mining Research: Equipment and Preliminary Tests, BuMines Report of Investigation 5915, 1961, 25 pp.

6.17 Malenka, W.T., Hydraulic Mining of Anthracite: Analysis of Operating Variables, U.S. Bureau of Mines, Report of Investigation 7120, 1969, 19 pp.

6.18 Palowitch, E.R., "Hydraulic Coal Mining in the United States," paper E5, Conference Internationale sur L'avancement rapide dans les chantiers d'exploitation des mines de houile, Liege, 1963.

6.19 Frank, J.N., Fogelson, D.E., and Chester, J.W., "Hydraulic Mining in the USA," paper E4, 1st International Symposium on Jet Cutting Technology, Coventry, UK, BHRA Fluid Engineering, Cranfield, Bedford, England, 1972.

6.20 Arentz, III, S.S., "Hydraulic Mining Studies at Storm King Mines," 2nd U.S. Conference on Water Jet Technology, Rolla, MO., 1983, pp. 243 - 253.

6.21 Read, R.G., Tarter, J.H., Smith, C.K., and Walter, R.P., Design of a Hydraulic-Jet Coal Miner, Final Report on Contract HO133052 by Bendix Research Labs, to the U.S. Bureau of Mines, April, 1974, PB241882, NTIS, 83 pp.

6.22 Singh, M.M, Labus, T.J, and Finlayson, L.A., Design of a Hydraulic-Jet Coal Miner, Final Report on Contract HO133119 by IITRI, to the U.S. Bureau of Mines, February, 1974, IITRI Report D6088, 104 p.

6.23 Labus, T.J., and Silks, W.M., "A Hydraulic Coal Mining Machine for Room and Pillar Operations," paper D7, 3rd International Symposium on Jet Cutting Technology, Chicago, IL., May, 1976, pp. X33 - X40.

6.24 Summers, D.A., Heincker, W., Eck, R.S., and Raghavan, S.H., Excavation of Coal Using a High Pressure Water Jet System, Final Report on Bureau of Mines, Contract HO232064, University of Missouri-Rolla, November, 1974.

6.25 Barker, C.R., and Summers, D.A., The Development of a Longwall Water Jet Mining Machine, Final Report on Department of Energy Contract DOE-AC01-75ET12542, University of Missouri-Rolla, July, 1981.

6.26 Summers, D.A., "Die Anwendung von Hockdruckwasserstrahlen im Bergbau," Gluckauf Forschungshefte, 41, (1), February, 1980.

6.27 Duczmal, M., Perek, J., and Wojtal, W., "Experience with the longwall face set for hydraulic cutting of coal at Rymer Colliery," Symposium on Hydraulic Transport of Coal Underground and at the Surface, Central Mining Institute, Katowice, Poland, October, 1966.

6.28 Schwarting, K.-H., Goris, H., Kramer, T., and Wille, G., "Development and Underground Testing of a Winning Method with High Pressure Jets," Gluckauf, 117, No. 23, December, 1981, pp. 642 - 644.

6.29 Du Toit, P.J.G., Black, S., Sakhuja, A., and Graham, T., Study of a Water Jet Continuous Coal Mining System, Final Report on Bureau of Mines, Contract JO155138, Ingersoll-Rand Research Inc., March, 1977.

6.30 Lavin, M.L., Estus, J.M., et al., Design and Evaluation of a Continuous, In-Place, Hydromechanical System of Underground Coal Mining, Final Report on Bureau of Mines Contract JO134054, Jet Propulsion Laboratory, July, 1984.

6.31 Miller, C.G., and Stephens, J.B., Coal Worm: A Remote Coal Extraction Concept, JPL Report 5010-7, December, 22, 1976, Jet Propulsion Laboratory, Pasadena, CA.

6.32 Summers, D.A., Barker, C.R., and Keith, H.D., Preliminary Evaluation of a Water Jet Drilling System for Application in In-Situ Gasification, Final Report to Sandia Laboratories, on Contract 130214, University of Missouri-Rolla, Rolla, MO., April, 1979, 66 pp.

6.33 Summers, D.A., and Yao, J., "Room And Pillar In-seam Excavator and Roof Supporter (RAPIERS)," 11th International Conference on Jet Cutting Technology, St. Andrews, Scotland, September, 1992, pp. 185 - 204.

6.34 Wakabayashi, J., "Slurry pumped 3,000 ft vertically," Coal Age, June, 1979, pp. 84 - 87.

6.35 Wang, F-D., "Status of Hydraulic Coal Mining in the People's Republic of China," 2nd U.S. Water Jet Conference, Rolla, MO., 1983, pp. 263 - 268.

6.36 Lin, Yilin, "A Brief Survey of Hydraulic Coal Mining in China," Proc. Hydraulic Coal Mining Seminar-Hydraulic Coal Mining in China, Colorado School of Mines, November 1-2, 1979.

6.37 Zhao, Yongcai, "Non-Supported Hydraulic Coal Mining System," Proc. Hydraulic Coal Mining Seminar-Hydraulic Coal Mining in China, Colorado School of Mines, November 1-2, 1979.

6.38 Lin, Yi Lin, Yan, Peng, and Wong, Zu-Lung, "Roof Control at the Hydraulic Working Face," Proc. Hydraulic Coal Mining Seminar-Hydraulic Coal Mining in China, Colorado School of Mines, November 1-2, 1979.

6.39 Li Hai, Jou, "Test of High Pressure Water Jet and Hydraulic Coal Winning," Proc. Hydraulic Coal Mining Seminar-Hydraulic Coal Mining in China, Colorado School of Mines, November 1-2, 1979.

6.40 Cheng, D., Zou, Z., Guo, C., Zhang, X., Li, G., Jia, Y., and Xiao, Z., "Major Advantages of Entry Drivage with Swing-Oscillating Jet in a High Methane Concentration Coal Seam," 5th American Water Jet Conference, August, 1989, Toronto, Canada.

6.41 Tang, Xuenan, "An investigation of the use of a swing-oscillation jet in underground in-seam drivage," paper III-6, 1st U.S. Water Jet Symposium, Golden, CO., April, 1981.

6.42 Cheng, D., Zhou, C., Zhou, Z., Jiang, S., Wu, B., Zhao, W., Wang, S., Jin, D., Mine, T.S., Cheng, B., Mine, M.T., Hu, N.Y.J., Xiao, Z., Jia, Y., Qian, F., Mine, B.Y., and Zhao, Z., "Preliminary Development of High Pressure Water Jet "Less Cutting Drivage" Study," International Water Jet Symposium, Beijing, China, September, 1987, pp. 5-12 - 5-17.

6.43 Cheng, Dazhong, Sou, Changsheng, Tian, Benzhao, Liang, Guolin, Li, Yiming, and Xing, Chengliang, "Preliminary Practice in the use of a Swing-Oscillation Water Jet in Coal Mine Drivage," 2nd U.S. Water Jet Conference, Rolla, MO., May, 1983, pp. 269 - 278.

6.44 Parkes, D.M., and Fisher, M., "Underground Coal Mining in Western Canada," Mining Engineering, Vol. 35, No. 4, April, 1983, pp. 313 - 316.

6.45 Lewis, R., "Underground Hydraulic Mine can point to Safety Advantages," Mine Safety & Health, Vol. 4, No. 2, pp. 2 - 6, April/May, 1979, U.S. Dept. of Labor, Mine Safety and Health Administration.

6.46 Anon, "Extraction by water cannon begins in Hansa coalfield," World Coal, May, 1978, pp. 5 - 6.

6.47 Jordan, D., "Hydromechanical Winning and hydraulic conveying," Coal International Supplement, April, 1979, pp. 42 - 47.

6.48 Siebert "Hydraulic mining at Hansa mine," Gluckauf.

6.49 Henkel, E.H., and Kramer, Th., "In-seam trials with the Hydro-Hobel," Proceedings 1st U.S. Water Jet Conference, Colorado School of Mines, April, 1981, pp. III 5-5 - 5-7.

6.50 Nikonov, G.P., "Research into the Cutting of Coal by small diameter, high pressure water jets," Chapter 34, Dynamic Rock Mechanics, ed. G.B. Clark, Proc. 12th Symp Rock Mechanics, University of Missouri-Rolla, November, 1970, pp. 667 - 680.

6.51 Cooley, W.C., Workshop on the Application of High Pressure Water Jet Cutting Technology, University of Missouri-Rolla, November, 1975, pp. 18 - 35.

6.52. 1976 Annual Report by the Director of Mining Research and Development of the British National Coal Board.

6.53 Anon, "Tunnel borer and shaft drill teamed at AGC's hydraulic mining operation," Engineering and Mining J, No. 7, July, 1964, pp. 69 - 70.

6.54 Disken, J.J., and Heiner, C.P., "Hydraulic Pitch Mining At Vickary Creek," Proc. 60th Rocky Mountain Coal Inst., June, 1964, pp. 63 - 65.

6.55 Benedum, W., Harzer, H., and Maurer, H., "The Development and Performance of two Hydromechanical Large Scale workings in the West German Coal Mining Industry," paper J2, Proc. 2nd Int. Symp. Jet Cutting Tech., BHRA.

6.56 Savanick, G.A., "Borehole (Slurry) Mining of Coal and Uraniferous Sandstone," SME Preprint 79-53, AIME Annual Meeting, New Orleans, February, 1979.

6.57 Anderson, A.K., "Hydraulic Jet Mining," First Conference on Uranium Mining Technology, University of Nevada-Reno, April, 1977, p. 36.

6.58 Fly, A.B., "Hydro-Blast Mining Shoots Ahead," Mining Engineering Vol. 21, No. 3, March, 1969, pp. 56 - 58.

6.59 Cheung, J., "Hydraulic Borehole Mining," Workshop on the Application of High Pressure Water Jet Cutting Technology, University of Missouri-Rolla, 1975, pp. 36 - 43.

6.60 Savanick, G.A., Ricketts, T.E., Lohn, P.D., and Frank, J.N., Cutting Experiments using a Rotating Water Jet in a Borehole, U.S. Bureau of Mines, Report of Investigation 8095 (1975).

6.61 Cheung, J.B., "Hydraulic Borehole Mining of Coal," paper D3, 3rd International Symposium on Jet Cutting Technology, Chicago, IL., May, 1976.

6.62 Lohn, P.D., and Brent, D.A., Improved Mineral Excavation Nozzle Design, Interim Report on USBM Contract JO255024, April, 1976, PB 264138.

6.63 Lohn, P.D., and Brent, D.A., "Design and Test of an Inlet Nozzle Device," paper D1, 4th International Symposium on Jet Cutting Technology, Canterbury, UK., April, 1978.

6.64 Evers, J.L., and Knoke, G.S., "Hydraulic Borehole Mining," ASME paper 81-Pet-6, Annual Meeting, New York, January 18 - 22, 1981.

6.65 Floyd, E.L., Borehole Hydraulic Coal Mining System Analysis, JPL Publication 77 - 19, April 21, 1977, Jet Propulsion Laboratory Contract NAS 7-100.

6.66 Boyce, T., Review of the Borehole Hydraulic Coal Mining System, ESCOE Report FE-2468-38, Final Report on DOE Contract EF-77-C-01-2468.

6.67 Kownacki, R., and Marek, A., "Hydraulic Borehole Mining of Brown Coal with High Pressure Waterjet Technology," 5th American Waterjet Conference, August, 1989, Toronto, Canada, pp. 315 - 320.

6.68 Summers, D.A., and Sebastian, Z., The Reaming out of Geothermal Excavations, Final Report to Sandia National Laboratories on Contract 13-3246, UMR, April, 1980, 155 pages.

6.69 Klich, A., Jura, W., and Mazurkiewicz, M., "Application of Water Jet Energy in the Borehole Mining," paper 33, 7th American Water Jet Conference, August, 1993, Seattle, WA., pp. 473 - 483.

6.70 Anon, Backfilling of Cavities Resulting from Borehole Mining, Technology News, U.S. Bureau of Mines, No. 95, March, 1981.

6.71 Savanick, G., Borehole Mining of Deep Phosphate Ore in St. Johns County, FL, SME Preprint 83-104, Annual Meeting, Atlanta, GA., March, 1983.

6.72 Joling, D., "New Mining Tool may cut pollution," Daily News-Miner, Fairbanks, AL., Friday, October 4, 1985 pp. 1, 8.

6.73 Miller, A.L., and Savanick, G. A., "Borehole Mining of Gold from Frozen Placers," paper 34, 7th American Water Jet Conference, August, 1993, Seattle, WA., pp. 485 - 499.

7 WATERJET USE IN ROCK CUTTING

7.1 ROCK DISAGGREGATION

Waterjet mining at lower pressure, i.e., at pressures below 700 bar, has been used in a number of rock cutting operations. Many of these were outgrowths of the work carried out in California during the development of the original technique for hydraulic mining (Chapter 5). One such development led into the extraction of clay and alluvial surface deposits, described above, another has been in mining high quality sand particles, particularly those of larger size [7.1]. There is some advantage to waterjet use for this, under certain conditions. The benefit comes from the other side of the coin which, on its negative side, led to the problems of river silting which arose during the early hydraulic mining in California.

In granular and crystalline material, the waterjet removes material from the solid by a sequence of events. First the water will penetrate into any cracks, crevices or grain boundaries of the solid. These small "fluid wedges", are then pressurized by the impact of subsequent segments of the waterjet. This has the effect of growing the crack, to the point of crack coalescence and particle liberation. In many larger particulate solids the largest cracks are along the existing boundaries of the individual grains. This means that the jet will break the rock down into its individual grain constituents, if the process is correctly applied. The process has been called **rock disaggregation**.

This method of rock removal can be of considerable benefit when it is recognized that most metallic ores contain only a small fraction of valuable material, and are mainly composed of waste rock. Yet in the extraction of the valuable fraction by conventional means, the entire volume of rock must be fragmented, transported to the surface, crushed and processed in order to remove the small proportion which is of value. Often this whole process not only must include the valuable vein of ore, but also a significant proportion of the surrounding rock, which must be removed in order to allow the miners to access the valuable material.

The advantage of the waterjet method of extraction is that the separation of the valuable mineral can occur as the mineral and rock is mined. If this separation can be made at the mining machine, then most of the waste material may be left in the mine, while only the valuable material is transported and processed. UMR has demonstrated this process in the mining of lead ore [7.2], but it can also be effective in other situations.

In the UMR demonstration samples of lead ore with both dolomite and sandstone as the host rock, each of which contained crystals of galena, were eroded by sequential passes of a waterjet over the surface. By tailoring the jet pressure to the ore it was possible to remove material from the solid, as individual particles. When the particles were then passed through a simple screen prior to analysis, the different grains segregated into different size ranges, (Fig. 7.1 [7.3]), so that a relatively inexpensive method of both extraction and concentration could be achieved in a single process.

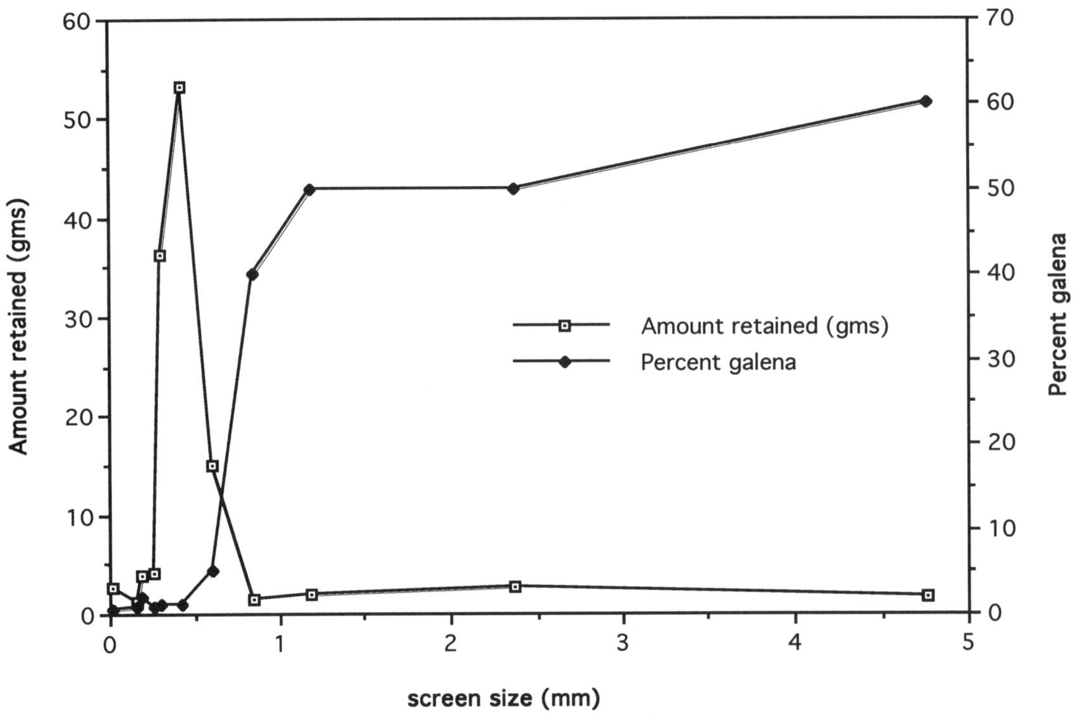

Figure 7.1 Size distribution of mineral particles, waterjet mined from a sandstone ore [7.3].

In order to demonstrate that high pressure waterjets could be advantageously used to mine sand in an underground operation, Dr. Savanick carried out tests in an underground sand mine in Iowa [7.1]. In the normal mining process the sand was blasted from the face, and carried to a surface processing plant. This plant was required to grade the sand and to remove a contaminating layer of clay found in the deposit. The clean sand was then sent to refineries for making into glass. By using the waterjet to mine the sand, the clay particles could be separated from the sand, during mining, and were carried away with the water, while the heavier sand particles sank to the bottom of the channel. This reduced processing costs.

An additional advantage to the mining of the sand also became evident during the mining. The layers of sand in the mine were made up of differing sizes of sand grains. With normal mining the layers were, however, mixed during the mining process, and could not be economically separated later. The deep cuts made by the waterjet nozzle allowed the individual layers to be extracted individually, however. This meant that the larger particulate beds could be economically separated. This would allow it to be collected and sold at a higher price as fracking sand into the oil industry. During the trials the mineral was mined at rates of up to 50 tons/hr, and could be pumped from the face at rates of up to 40 tons/hr.

7.2 DEVELOPMENT OF WATER CANNONS

The major research on cutting rock began, however, at the high end of the pressure spectrum. Early investigations into waterjet cutting of hard rock, whether in Russia, Western Europe, or the United States, began by considering that it would be necessary to use relatively high pressures in order to effectively cut rock materials. This could, in part, be traced to the same sort of logic inherent in the theory proposed by Powell and Simpson [7.4]. Based on that theory, which will be discussed in more detail in Chapter 11, it was deduced that the lowest specific energy of rock cutting would require that the waterjet pressure was somewhere on the order of 30 times the rock uniaxial compressive strength. **Specific energy** has been previously defined as the amount of energy which must be applied to the rock in order to extract a given volume of rock. Given that the uniaxial compressive strength of a typical granite, for example, will be approximately 2,000 bar, this would require a waterjet pressure of up to 60,000 bar for most effective cutting. Such a pressure cannot be sustained by a normal high pressure pumping systems and, therefore, specialized pieces of equipment, known as **water cannons** were developed. These were single shot devices designed to achieve the high pressures required by the theory.

The initial approach to developing this technology took place in Russia [7.5], and was followed, among others, by developments in the United Kingdom [7.6] and [7.7] and the United States [7.8], [7.9], and [7.10]. The basic premise upon which most water cannons were constructed, was to develop a sudden gas pressure behind a slug of water, accelerating it into a specially shaped nozzle. The design of this nozzle shape was based on observations of water acceleration as the tide enters a fjord which gradually narrows along their length. This accumulation effect was the initial basis for a nozzle design (Fig. 7.2) which has since been designated as the Voithsekhovskii set of nozzles [7.11]. With the use of this accumulating nozzle which typically has a converging section which follows a defined arithmetic equation, it is possible to input a water slug at a velocity of 550 m/sec and at the nozzle orifice, achieve an exit speed of 3.5 km/sec. Such a velocity will impact a surface and generate pressures on the order of 70,000 bar or above.

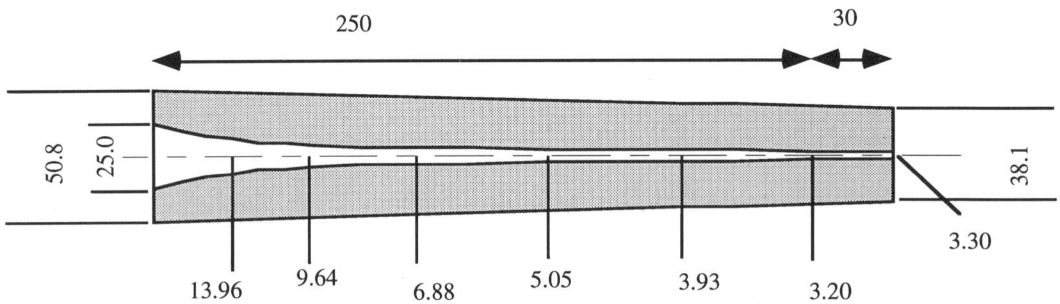

All dimensions in mm. Internal Diameters of the Nozzle are shown at intervals.

Figure 7.2 Configuration for a Voithshekovskii design of nozzle.

The first cannons of this nature were built in Russia for impact testing on rock and were introduced into the United States mainly through the efforts of Dr. William Cooley who began this work at Exotech and then headed Terraspace Inc., where he built a water cannon under funding from the U.S. Department of Transportation [7.12]. The initial studies carried out at Exotech [7.13] dealt with the construction of an air driven cannon, which could fire slugs of water at pressures to 7,000 bar, and contained energies of 1,100 joules. Specific energy values of the order of 320 joules/cc were reported for cutting sandstone, and it was, in part, these results which led to the recommendation that jet pressures should be ten times the uniaxial compressive strength of the rock. Concurrent testing by Singh [7.14] of a device which fired a much longer pulse, of up to 1.5 sec duration, at pressures to 8,010 bar, indicated that the energy levels required for rock cutting were perhaps two orders of magnitude higher than those reported by Cooley. For a 290 bar Indiana limestone, the specific energies were 25 - 35,000 joules/cc, while for Milford Pink granite, with a strength of 2,240 bar, a specific energy of 475,000 joules/cc was calculated. Singh recognized that this may have been due to the larger volume of water fired at the target.

Farmer and Attewell [7.15] had already shown that firing a continuous jet into a rock, without relative movement was a process of increasing inefficiency. (A point which will be further addressed in sections 7.3 and 7.5). Thus, the longer duration of the IIT shots were expending energy into the cavity cut into the rock which was no longer returned in additional rock volume removed.

The groups increasingly diverged in the processes which they developed. Exotech licensed Cooley's small extrusion devices in which a shaped glycerine gel (Fig. 7.3) was used to raise the jet pressure from the 10,000 bar attainable with a steel device, to a pressure of 20,000 bar [7.16]. A small, 5 kW system was demonstrated capable of generating pulses of 35,000 bar, as a result of which the company began to develop a tunneling machine with the Caldwell division of Smith International to drill in harder rock. Little public information on subsequent activity in this device is available after 1970.

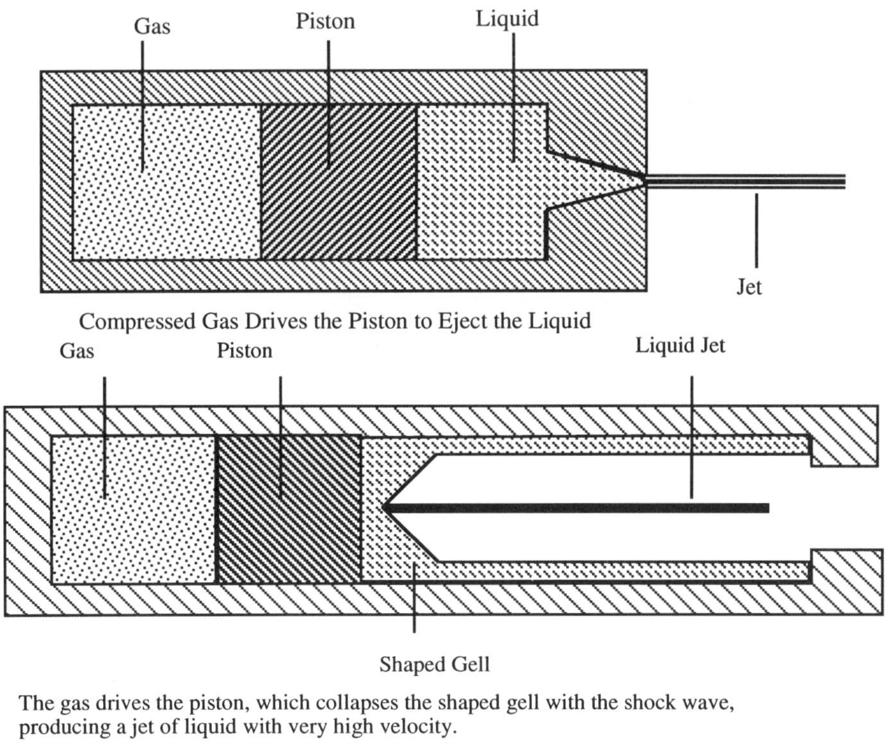

Figure 7.3 Conventional and shaped annulus jet drivers developed by Cooley [7.17].

Dr. Cooley continued his work, at Terraspace, obtaining a Russian nozzle from the Institute for Hydromechanics in Novosibirsk, and powering this, initially with a piston driven by a gas gun [7.18]. Initial results with the 7.16 mm diameter nozzle, showed that the jet would excavate granite with a specific energy of 500 joules/cc at a jet pressure of 16,200 bar. Continued experiments [7.19] indicated that a jet length to diameter ratio of less than 1,000 gave most efficient cutting (Fig. 7.4). This was then developed by Cooley to suggest that the jet velocity should be less than 1,000 times the lateral traverse velocity of the nozzle. Alternately this can be used to suggest that, with a lower velocity jet, moving at say 400 m/sec, that the lateral traverse velocity of the nozzle should be as high as 0.4 m/sec. Interestingly subsequent work (see section 7.4) has shown that most efficient cutting occurs at traverse velocities of up to 2 m/sec. Thus, this early finding by Cooley has been validated.

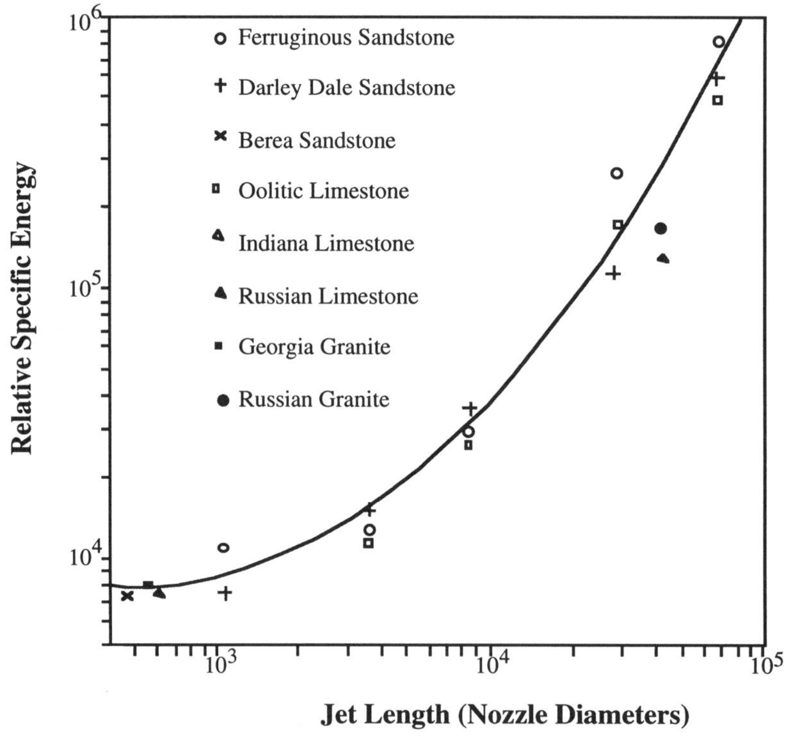

Figure 7.4 Specific energy as a function of jet length (after [7.19]).

A second, 6 mm diameter nozzle was built by the Speco division of Kelsey-Hayes for Terraspace, and proved capable of generating pressures of up to 21,00 bar when the device was fired with air in the nozzle [7.20]. When a vacuum was pulled on the nozzle, however, the pressure could be almost doubled, reaching 44,000 bar. Interestingly, however, the data where the vacuum was pulled appeared to be show a lower efficiency for the American nozzle, than when the larger diameter Russian nozzle was fired with air inside the nozzle [7.21]. This point will be discussed further, in the next section. The water cannon proved to be quite noisy in operation. Levels as loud as 134.5 db were reported for its use. However, a relatively simple baffle could be built around the end of the unit and this reduced the noise to an average of around 126 db [7.22]. The device was tried in an underground mine, and was capable of breaking off large pieces of rock measuring several cm in diameter. The driving device had, by this time, been modified and used a pneumatic-hydraulic actuator. The specific energy values reported at the higher pressures were equivalent to those of existing tunneling machines, and the data indicated (Fig. 7.5) that higher pressures were most efficient [7.22].

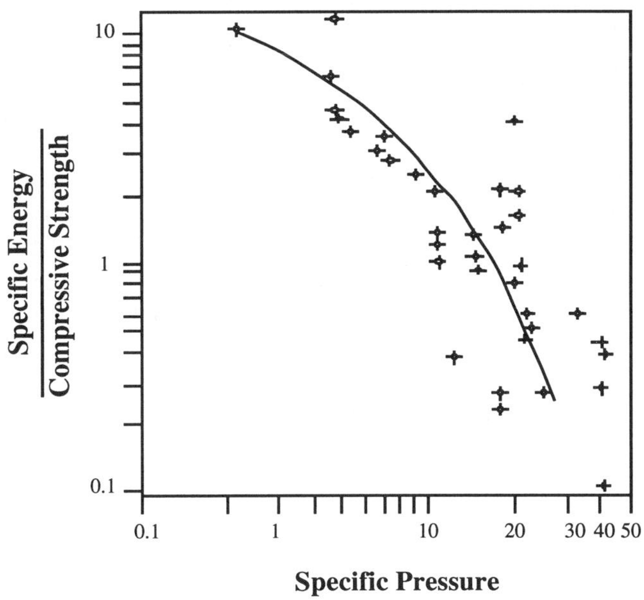

Figure 7.5 Specific energy as a function of specific jet pressure (as evaluated by Cooley [7.22]).

During the test program one of the nozzles suffered local failure after 80 shots. In discussing the price of nozzles at the 1972 BHRA conference, Dr. Cooley quoted a price of $2,450 for each Russian nozzle, with a fatigue life of several thousand cycles at 10,000 bar, but only several hundred cycles at a pressure of 15,000 bar.

To drive a tunnel 30 cm over a 6 m diameter would require 3,000 shots [7.23]. Thus, one could anticipate replacing the nozzle at around that frequency, given that the optimum jet pressures found lay between 12,600 and 42,000 bar (ibid.). This would give a cost, in 1972 dollars of $7.9 million/km for nozzles alone - a price too high to pay. (Dr. Andrew Conn first evaluated this problem in public).

Similar concerns had arisen in developing the Russian cannons. Early signs of wear at very high pressures had led to a decision to lower the effective operating pressure of the units to 6,000 bar. At such a pressure it was possible to develop a unit which would fire at up to 60 shots/min and with a nozzle which survived, albeit with significant internal wear, through 298,166 shots [7.24]. This pressure has since become something of an upper bound in water cannon design for this reason. Even small amounts of wear inside the nozzle can create problems since, with waterjet velocities of over to 1 km/sec any deviation of the leading edge of the jet can cause impact wear on the walls and make the wear rate much more rapid. One way the Russians have overcome this is to fill the nozzle with water prior to firing. This has the additional advantage of reducing the problems of cycling the system (ibid.).

A new machine, the KUB heading machine was introduced into the D.S. Korotchenko mine in the Soviet Union and used to remove an area of 12 sq m of rock from the face of the heading. The unit required 41,000 shots to drive the heading 30 m averaging 0.009 cu m of rock per shot. Dust generated in cutting the sandstone was below 0.1 mg/cu m. Optimum spacing between shots was 120 - 150 mm unless working to a free face, where the distance could be increased by 150 - 200%. Nevertheless, the barrel of the device had to be replaced after between 8,000 and 10,000 shots, while the piston was replaced after 1,500 - 2,000 shots [7.24].

A comparison of the results of water cannon impact with that of conventional continuous jets [7.25] in cutting Indiana limestone indicated that the continuous jets cut at one-third or less of the specific energy required when using the impulsive water cannon. This study used a water cannon which fired pulses at pressures of up to 3,500 bar, at diameters to 25 mm (Fig. 7.6). Comparative tests with the continuous jet were at pressures to 1,720 bar at a nozzle diameter of 1 mm. In contrast with the work by Cooley, different nozzle diameters were tested for the cannon, but the significant difference in efficiency arose because of the higher traverse speeds with which the continuous jet could be moved.

Figure 7.6 UMR water cannon firing 45 liters of water, using a powder charge.

One problem with the idea of the single pulse water cannon (whose future development then moved into the fragmentation of concrete - discussed in Chapter 5) lies in the shape of the pressure pulse generated (Fig. 7.7). Because only the initial leading edge of the jet cuts at full pressure, and develops fractures in the surface, it is often the case that the jet pressure has fallen below that required to exploit these cracks even as they develop. The water cannon device also suffers from the fact that only a small portion of its water is at maximum power, and that the rebound of this water from the hole will also reduce the effectiveness of the following, lower pressure water. This is because it is almost impossible to move the very heavy units during the very short time (below 0.005 seconds) of the effective pulse. This apparent disadvantage can be overcome, in some instances, by a more rapid pulsation of the jet, an approach which will be discussed below.

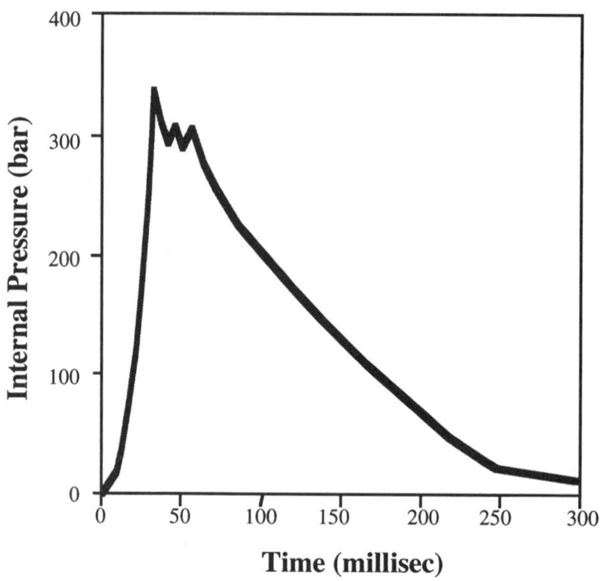

Figure 7.7 Shape of the pressure pulse from the UMR water cannon [7.25].

7.3 VERY HIGH PRESSURE CONTINUOUS JETS

Concurrent with the Cooley effort, a similar effort was underway at the IIT Research Institute under the direction of Dr. Madan Singh to develop systems capable of generating very high pressures [7.14]. Their device worked on the same principal as that developed in the United Kingdom, at the Safety and Mines Research Establishment, a team led by Dr. Graham Artingstall [7.26]. These two groups each developed water cannons powered by large volumes of gas. Nitrogen, or air, was compressed and stored in a reservoir, often made up from a series of accumulator bottles. When a sufficient volume had been brought up to pressure, a high speed release valve was opened instantaneously, pressurizing the volume behind the water slug and accelerating it through the nozzle (Fig. 7.8).

Waterjet pressures of up to 10,000 bar were achieved, with the advantage in this latter design, that the large volume of air stored in the gas reservoir was sufficient to sustain the pressure for as long as a second, in contrast to the Voitshekovskii/Cooley device which had a velocity peak with a duration on the order of several microseconds followed by a relatively rapidly decay in pressure because of the relatively low volume of water ejected by the cannon. It is interesting to note, in light of the earlier discussion on optimum traverse velocities that Moodie and Artingstall (ibid.) found that the optimum traverse velocity of the jet was in the range from 0.25 to 0.75 m/sec, increasing with jet pressure to a value of 5,500 bar, at which their experiments stopped.

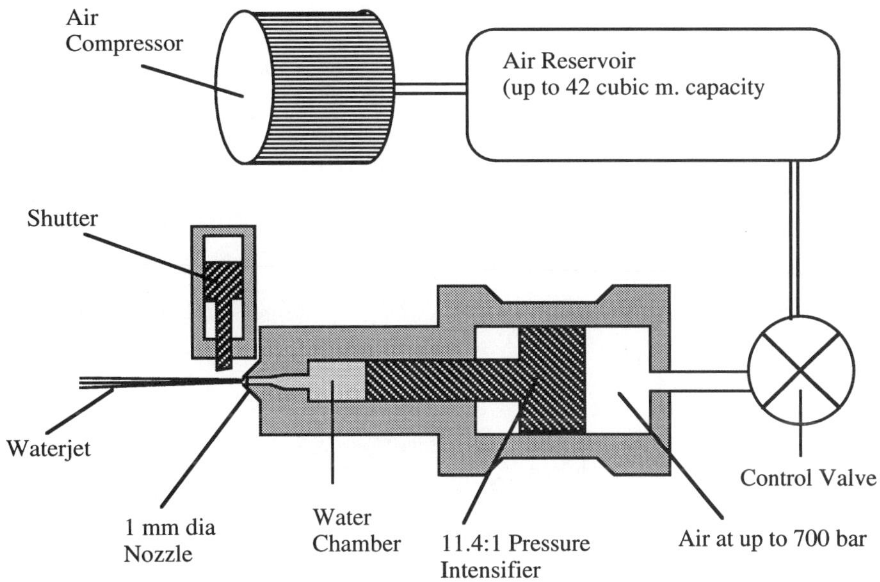

Figure 7.8 Schematic of an air driven intensifier [7.26].

Early work with water cannons seemed to indicate that Powell and Simpson were correct in estimating the pressures required for most effective waterjet cutting. As pressure increased, the resistivity declined as did the specific energy required for excavation. A study using water cannons cannot, however, address all the pertinent aspects of the waterjet cutting process. This is, in part, because a water cannon type of device works best, and most efficiently, when it is cratering the rock, or breaking blocks of rock to existing free surfaces. Moodie and Taylor [7.27] illustrated this by plotting values of specific energy obtained by pulsed, as opposed to traversing waterjets (Fig. 7.9).

The results are apparently more favorable for pulsed jets, since the existence of a free passage of the water away from the cut in a traversing jet allows the water to escape without pressurizing the cavity and inducing larger rock fracture. Such rock fracture is, however, not always successfully achieved. The presence of cracks in the rock, or the firing of a shot into a very large block can block the fracturing effect, dramatically reducing the effectiveness of the pulsed technique, as discussed earlier in regard to the operation of water cannons.

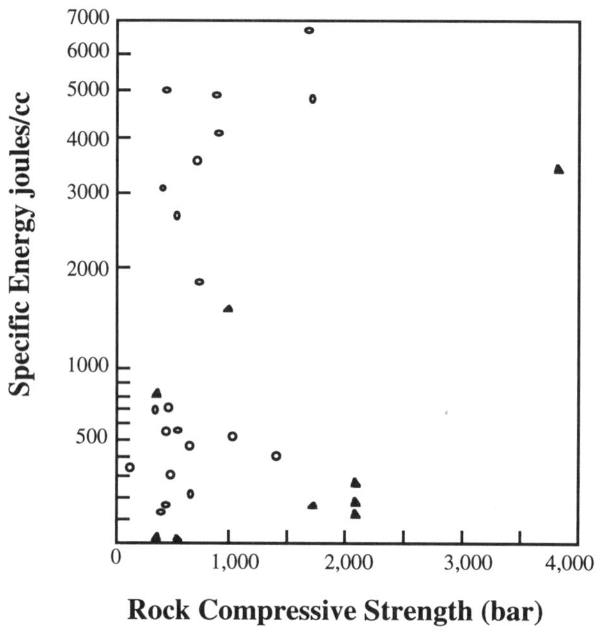

Figure 7.9 Variation in specific energy of cutting with rock strength and manner of attack - triangles show pulsed jet results, open points are from traversing tests (after [7.27]).

It has also since been found that waterjet cutting efficiency is very sensitive to nozzle diameter values and concurrently, as described elsewhere, waterjet ability to cut is very sensitive to the ability of the waterjet to penetrate into the microstructures of the rock surface. This can be illustrated by an anecdote.

Research in waterjet rock cutting began at the University of Leeds in 1965 following visits to the Safety in Mines Research Establishment to observe the action of the water cannon built by Dr. Leach [7.6]. He had shown that when one fired a water cannon onto a rock surface, that the pressure within the hole very rapidly decayed with hole depth (Fig. 7.10). This is partially because of the friction which develops along the sidewalls as the jet moves into the hole. More importantly, however, it is due to the return flow of the water which, once fired into the hole, blocks the entry of the next slug of water as it then tries to get out.

This, incidentally, indicates a major inefficiency of the system, if not corrected, and explains why many reported specific energies of cutting with jets are very high. Nevertheless, in the initial Leeds work (restricted to a maximum operating pressure on the order of 700 bar) results showed that as the waterjet pressure increased, specific energy decreased. The jets were also unable to cut any particularly hard rock such as granite [7.27].

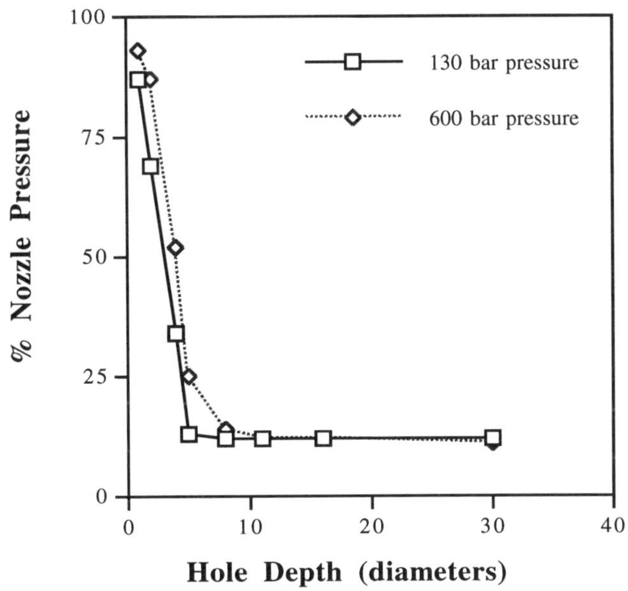

Figure 7.10 Pressure at the bottom of a crater as a function of depth [7.6].

However, it was obvious that one of the reasons for the inability to achieve great penetration was the masking effect of the water once it had created and entered a hole in the rock surface. Two alternative ways to get around this problem were examined. One was to insert an interrupter disc between the nozzle and the rock surface, spin this disc at a high rate of speed, and thereby chop the water length into slugs so that the water from one impact would be able to escape from the hole before the second impact occurred (Fig. 7.11). Where this was tried, it was possible to increase both the overall depth and the rate of penetration. However, this wasted at least one-third of the energy of the water, that being the length of jet which was impacting on the disc surface and never reached the rock.

An alternative approach was to rotate the waterjet, or in this case, the rock, slightly off-center from the jet axis. Where this was done, the waterjet always had a free surface to one side of the cut, into which the water could flow and thereby escape from the impact point. When this was tried, an improved penetration even over that achieved by the waterjet that was interrupted was achieved (Fig. 7.12).

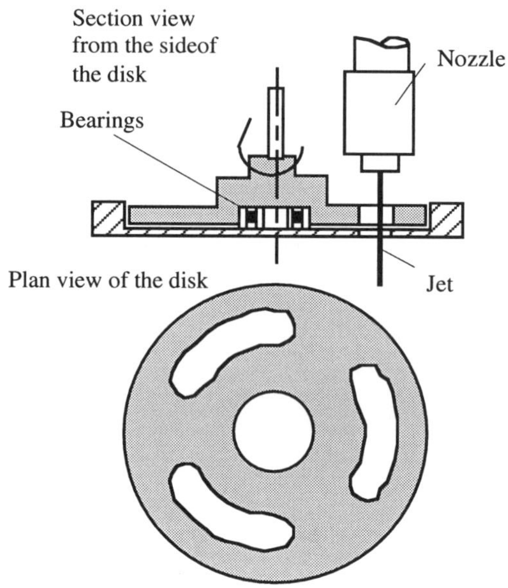

Figure 7.11 Location of an interrupter disc used to pulse the waterjet [7.28].

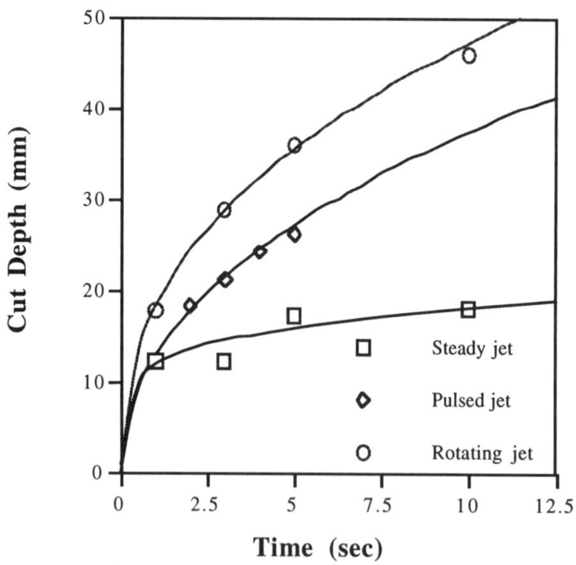

Figure 7.12 Relative penetration of continuous, pulsed and rotating waterjets into rock [7.28].

In one of those serendipitous events which are so often an encouragement to progress in research, it was during this phase of the test program that a block of granite was erroneously placed in this test rig. (Erroneously since this rock "could not be cut" at this jet pressure). Since it was easier to test the rock than to take it back out, a test was run at 700 bar to determine whether rotating a waterjet across the surface would give any penetration. Within a relatively short space of time a hole had been drilled at 700 bar through a 22 cm thick block of granite with a compressive strength of around 2,000 bar (Fig. 7.13). Thus, in one short afternoon, the premise that it was necessary to go to 30 times the rock compressive strength in order to achieve penetration was disproved, since a pressure, only 10% of that considered optimum, had been found to be more than adequate for the purpose required.

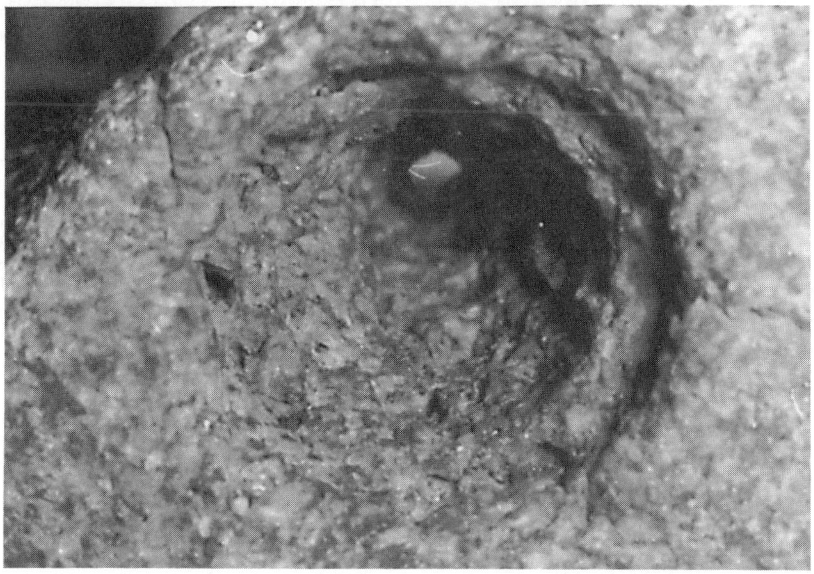

Figure 7.13 First block of granite drilled at 700 bar at the University of Leeds.

7.4 MECHANISMS OF MATERIAL REMOVAL

7.4.1 JET IMPACT ON A ROCK

It is pertinent to emphasize this point, and to explain why, under static testing the waterjet could not cut the granite, while when the rock was rotated, full penetration occurred. It may also serve to explain why, for example, it is easier for a waterjet to penetrate granite than it is for it to penetrate a "weaker" marble. The reason is that the jet penetrates a rock by firstly flowing into the existing cracks, weakness planes and grain or crystal boundaries in the rock surface. This fluid "wedge" is then pressurized by the arrival of the subsequent slug of water, creating a higher pressure wedge which acts to lengthen the crack in which it has been located. When hitting on a single point, the jet soon penetrates to reach a surface which an unaided jet of that pressure cannot cut, and penetration stops.

When the jet is rotated, however, there are two cumulative improvements. Firstly as the jet passes over the edges of crystals of different materials these will compress at a slightly different rate opening the crack between them. Secondly the jet will continually meet new crack surfaces at the edge of the blocking surface, and thus can penetrate around the resistant grain and remove it (Fig. 7.14). This is a major advantage for this new tool. It can be emphasized by reviewing the cutting of concrete, discussed earlier (and which with a more detailed discussion will also be reviewed in Chapter 11).

Figure 7.14 Section of a cut in granite showing attack on the grain boundaries, the sample has been magnified to illustrate the cut surface.

Where a conventional mechanical cutting tool is used to cut concrete, the tool must physically cut a passage through all the constituents of the concrete as it advances. This includes cutting though whatever material the aggregate is made of, which may include quartzite pebbles with a compressive strength of up to 4,000 bar. Mechanical tools must be manufactured to cut through this material. On the other hand the waterjet tool can be designed so that it just cuts out the softer matrix of cement. If all the holding material is cut out then the aggregate will fall out without needing to be cut. Since the cement strength is rarely as high as 400 bar, it can be seen that, with proper design, a much lighter and more effective concrete removal system can be designed based upon waterjet cutting than by mechanical cutting means. This was demonstrated by the development of the Bandit, a waterjet tool developed in Dakota [7.29]. This machine could remove the damaged concrete from a bridge deck over the width of a traffic lane, at a speed of 20 m/hr advance. The topic has been discussed in more detail in the previous chapter.

Where waterjets are used to cut rock the jet will cut a slot approximately 3 times its own width. It has been consistently reported that the waterjet will cut to a depth which is linearly related to the jet pressure, at constant nozzle diameter. Where, however, one considers the effects of pressure and diameter on the relative efficiency of rock cutting, then a different result is obtained.

The inter-relationships between jet parameters and cutting efficiency can be exemplified by reference back to the work carried out in finding out how effective a waterjet could be in the cutting of coal. The conclusions have since been confirmed by tests on a number of other different rocks. In the work, reviewed in section 6.2.3.4, a full factorial experiment was carried out at 5 levels of pressure, 5 levels of nozzle diameter, and 5 levels of stand-off distance to establish how these variables affected the cuttability of coal [7.30]. An increase in cutting pressure from 350 to 1,750 bar linearly increased the depth of cut at an almost constant cut width so that the specific energy of cutting decreased only slightly as pressure was increased (Fig. 6.22). However, when in contrast, the horsepower in the jet was increased a like amount by increasing the diameter of the flow, then not only did the width increase, but the depth also. It can be seen that as the diameter is increased so the energy in the jet increases but at a much slower rate than the increase in the volume of material removed. Thus, the specific energy was reduced quite dramatically (Fig. 6.23).

To demonstrate this difference in a more illustrative manner, it can be said that if the horsepower in the jet was doubled by increasing the pressure, then approximately twice the amount of material was removed. In contrast, if the amount of power was increased by increasing the diameter of the jet then the volume removed was quadrupled. Thus, it proved much more efficient to increase diameter than to increase pressure. This is a relatively important general finding, and applies in many situations. However, there are important restrictions on this rule which must also be borne in mind. Firstly it will only hold true when the jet pressure exceeds the threshold penetration pressure for that particular material.

The **threshold pressure** of the target is defined as the pressure at which the jet first starts to cut the material. It is controlled by the mode in which the surface is attacked. A simple way which has previously been used to establish this pressure is to direct a waterjet at a flat surface of the target material. As long as there is no penetration of the surface the jet will splay out along the surface after impact. Once penetration begins to occur then the jet will rebound at an angle to the surface, a quite visible change. This is because the jet is now entering a spraying back out of a hole in the target surface. However, the jet must exceed this limit by some measure to achieve effective penetration, and that is controlled by the mechanism of material removal.

This is illustrated, in part, by the description of the tests which were carried out on cutting the granite at 700 bar in the "serendipitous" test described above. Without rotation it was not possible to cut the material, whereas, once the jet began to rotate over the sample surface, then a 22 cm penetration could be achieved.

Interestingly, there is a second threshold to the cutting of many granites. At the lower pressure the jet cuts its way, as it did in the above case, by growing the cracks around the boundaries of the crystals which make up the rock. As the pressure increases a point is reached where the jet will cut through the individual crystals, rather than removing them as a unit. In this case the slot cut becomes much narrower, since the internal cracks in the crystals are now being exploited, rather than the larger external perimeter cracks and the material being removed is from within the crystal, rather than comprising the entire grains. This has become an area of investigation at the University of Cagliari [7.31].

There is, however, a need for caution in evaluating the results of rock property evaluations and their correlation with jet cutting ability. While this is not intended as a criticism of the Italian work it should be taken into consideration, particularly in evaluating some of the results of earlier experiments in this field. The problem relates to the way in which the jet was applied to the rock, and requires a discussion, therefore, of the jet delivery or equipment parameters.

7.4.2 THE EFFECT OF EQUIPMENT PARAMETERS

One of the problems in evaluating the effect of jet or target properties on the effectiveness of jet erosion is that there are inter-related effects which may change the interpretation of the results if a second parameter is changed as well as the first. The simplest example of this lies in the interpretation of jet pressure and nozzle diameter effects as discussed earlier. To take this one step further, the pumps normally used for waterjet cutting can be run at different flow volumes, but the same horsepower, by changing the plungers in the pump. Otherwise if pump pressure is lowered by increasing the diameter of the nozzle, and the total flow remains the same, then the horsepower being used is reduced.

This is a relatively important fact to bear in mind, since many of the conclusions which have been drawn on the performance of waterjets have extrapolated from individual results, without a full understanding of the real meaning of the numbers being quoted. Consider, for example, a slug of water which impacts on a piece of sandstone, for a period of 14 m/sec at a pressure of 700 bar with a diameter of 1 mm. The jet creates a quasi-cylindrical cavity 1.5 cm deep and 3 mm in diameter.

The energy in the jet is given as a product of flow and pressure by the equation:

$$Energy = const. flow. pressure. time$$

This can be rewritten, in terms of the nozzle orifice diameter, and the jet pressure as:

$$Energy = \pi . \frac{diameter^2}{4} . \sqrt{2 . \frac{pressure}{density}} . pressure . time$$

From which, inserting the appropriate values for the density of water and converting Energy to Megajoules (a Megajoule [MJ] is equivalent to a kg-m), produces the relationship (with diameter in mm and jet pressure in bar):

$$Energy = diameter \cdot \sqrt[3/2]{pressure} \cdot time$$

Substituting for the values cited above shows that the jet slug which created the cavity has an energy of 287 MJ. Given that the cavity has a volume of

$$1.5 \text{ x } \pi \text{ x } (0.15)^2 = 0.106 \text{ cc}$$

This gives a specific energy of cutting of approximately 2,870,000,000 joules/cc. This might then be used as a parameter against which to compare, for example, the rock uniaxial compressive strength to determine any significant relationship between the values. Such a step has been carried out in the past (and curves have been plotted throughout this book in which specific energy is compared with other properties). Before much credence can be placed on this result, however, one must consider how reliable the above value is, and how much variation there may be in the value, and whether, in fact, that value was correctly calculated.

A full factorial experiment was run varying jet pressure, nozzle diameter, traverse speed and number of passes at various levels of each on a suite of rocks, to observe the independent and inter-related effects [7.32]. In a factorial experiment [7.33] the experiment is designed so that each level of the different values of each parameter are used in combination once. Thus, a test which uses five values for each of four variables will require a total of 5 x 5 x 5 x 5 = 625 tests. This allows a more sensitive analysis of the data and permits an evaluation of interactions between parameters which control the experimental results.

In this particular case when a pink sandstone was cut, it was found that the depth of cut appeared linearly related to nozzle diameter at a low pressure (350 bar) but as the pressure was increased to 560 bar a greater effect of diameter could be observed (Fig. 7.15). If, in contrast, the full pump output was run through the three nozzle sizes, then the intermediate nozzle gave optimum cutting depth. This is because the largest nozzle was being operated at the lowest pressure.

This would indicate that, for a given flow rate, neither the highest pressure achievable (the result with the smallest diameter) nor the largest diameter of flow will give best results, but rather some intermediate combination of the two. This becomes even more evident if the specific energy of the process is considered, since the energy in the jet will be reduced as the diameter is increased (since the flow rate is the same and the jet pressure drops). Since the width of the slot is related to the nozzle diameter, the largest nozzle diameter now becomes the most efficient cutting jet (Fig. 7.16).

In this process, however, a single value for the specific energy of cutting the rock was not evident. Rather both the pressure at which the test was carried out, and the diameter of the jet, controlled the value for specific energy. The analysis is not, unfortunately even this simple. Consider that these tests were carried out with the rock surface at a fixed distance, 5 cm, from the nozzle. The problem of loss in jet energy with distance has previously been discussed, and will also be referred to again in following chapters. As an illustration of the effect which this has on the current case consider the reduction in cutting depth with a relatively poor nozzle even relatively close to the nozzle (Fig. 7.17).

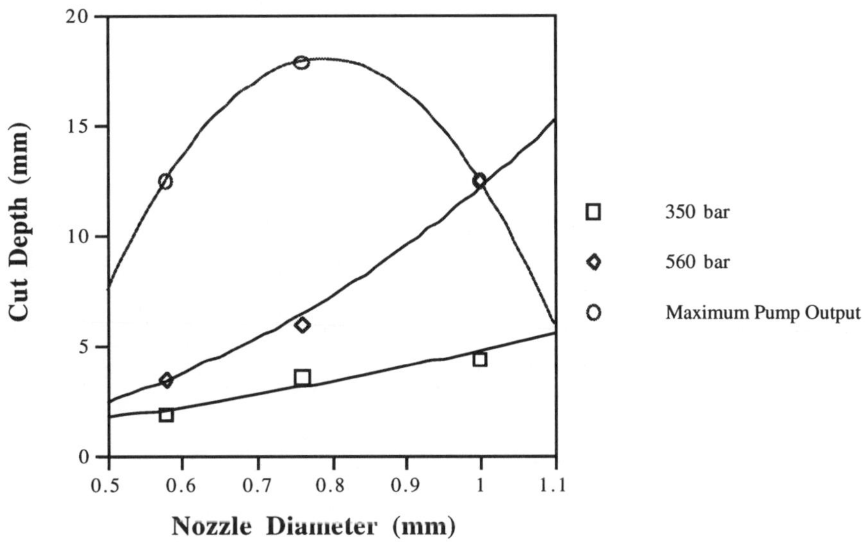

Figure 7.15 Depth of cut as a function of nozzle diameter at different pressures [7.32].

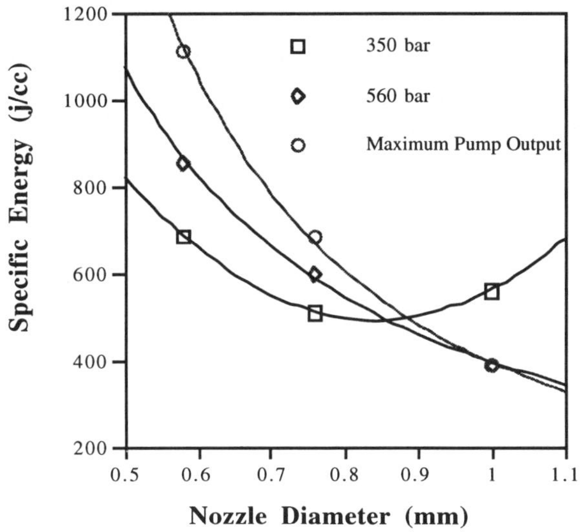

Figure 7.16 Specific energy as a function of nozzle diameter at different pressures [7.32].

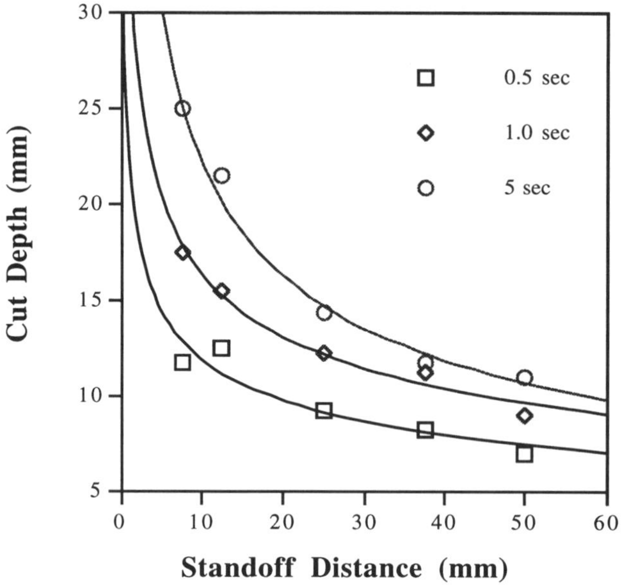

Figure 7.17 Effect of stand-off distance on depth of cut [7.28].

It can be seen that had the tests been carried out a different distance from the nozzle, then, although the jet energy used would have been the same, the volume of rock excavated would have been significantly changed. One should, therefore, in calculating the energy of the cutting process use the impact energy of the jet, rather than that delivered to the water at the pump. Unfortunately even this cannot be simply quantified. The effect of nozzle manufacturing condition, entrance flow condition and edge quality will all influence the quality of the jet produced, and each may change during and between experiments. (Replacing nozzles during the tests to check for the effect of nozzle diameter may alter inlet flow conditions in each case, if not well guarded against in the procedure). At greater distances, as for example demonstrated in a German experiment (Fig. 6.34) a larger diameter, lower pressure jet may deliver a greater pressure, than a higher pressure, smaller diameter jet [7.34].

Considerable caution must therefore be used in assessing the value of the energy which reaches the target, particularly where data from different investigators or different systems are being compared.

Further cautions must now be raised once the jet reaches the target. As Leach and Walker [7.6] and the above curves (Fig. 7.10) have shown, as a jet penetrates into the rock the pressure is attenuated by the flow of water in the hole trying to get back out of the cavity. This situation gets worse as the hole gets deeper, until the point is reached that the pressure at the bottom of the hole no longer is sufficient to allow further penetration.

Some relief from this can be obtained if the jet is traversed over the surface, since the jet, as discussed above, can escape through the previously cut section. But is it better to make a single slow pass over the surface or a number of faster passes to achieve the same depth of cut? To evaluate which option is best, the usual criterion is to refer back to the relative specific energy. In terms of cutting speed, providing the jet is penetrating the target, and within a normal range of values, the faster the jet traverses the target, the more efficiently it cuts. This can be appreciated in part given the fact that under normal cutting conditions a jet at a pressure of 700 bar jet is moving at a speed of around 360 m/sec. The actual speed with which the waterjet achieves its maximum penetration is rarely understood by investigators. Some work in this area was carried out by Summers [7.28] and then by Dr. Page [7.35] who ran tests of short jet exposure, to examine penetration rates where the nozzle and target were fixed. It was shown in this work that the waterjet achieved a majority of its penetration in the first 5 milliseconds of impact (Fig. 7.18). If one translates this into an effective traverse velocity then if the effective width of a waterjet is 3.5 times its diameter, that the most efficient traverse speed for a waterjet would be on the order of 700 times the nozzle diameter/second. (For a 1 mm diameter nozzle this is a speed of 70 cm/sec or 42 m/min.)

In the factorial experiment discussed above traverse velocities up to 2 m/sec were evaluated, and specific energy was found to continuously decline over that full range (Fig. 7.19). Subsequent experimental evidence from tests at UMR, at the Skotchinsky Institute [7.37], and at the University of Newcastle [7.38] have indicated that an optimum value to traverse speed, in terms of jet cutting efficiency, is reached at speeds in the range of 1 - 2 m/sec for a good, round waterjet at a jet pressure in the range of 700 bar, and a diameter on the order of 1 mm. Subsequent work at UMR has indicated that, at larger nozzle diameters, this minimum will occur at higher speeds.

This does not necessarily mean that it is better to make a series of superincumbent passes over the same path at high speed in order to best cut into a rock. The results from the factorial showed that successive passes of a single jet over the same line were increasingly inefficient (Fig. 7.20). The reason for this was that as the waterjet cut into the material the effective width of the slot reduced. Part of the energy in the jet in successive passes was therefore used to widen the channel to gain access to the back of the hole, in order to cut deeper, this reduces overall effectiveness.

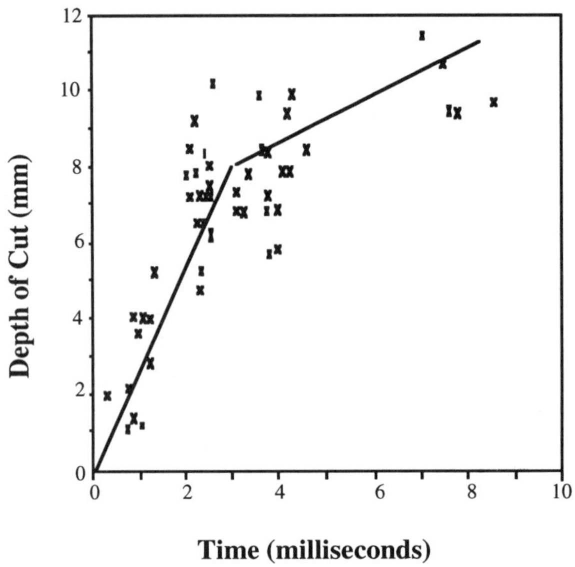

Figure 7.18 Penetration as a function of short impact time [7.36].

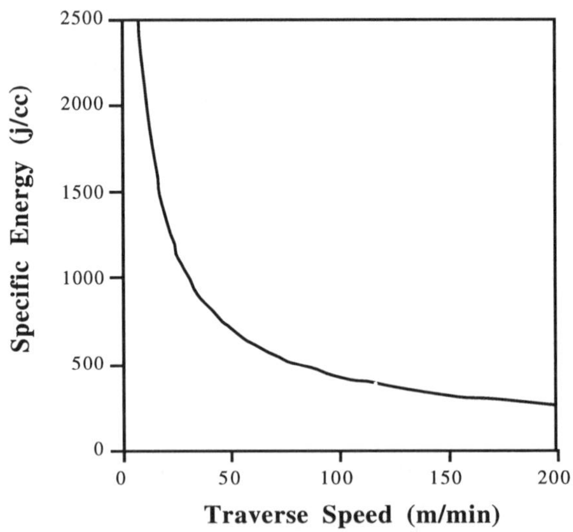

Figure 7.19 Effect of traverse speed on specific energy of cutting [7.32].

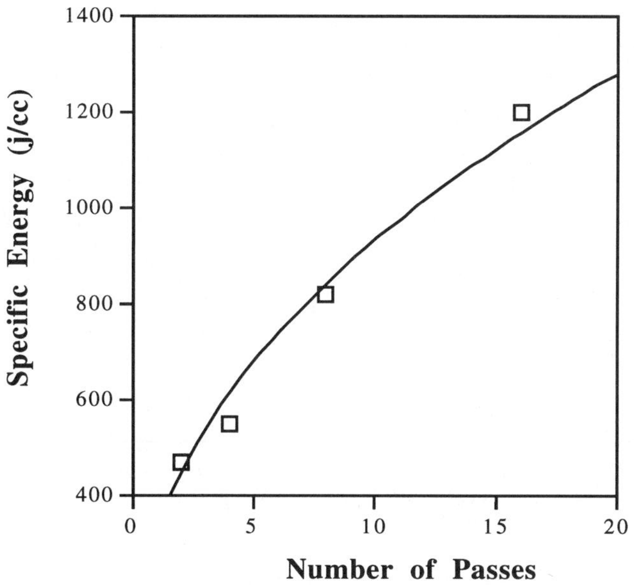

Figure 7.20 Specific energy as a function of number of passes [7.39].

This problem can be overcome, if a dual orifice nozzle is used, or if a single nozzle is oscillated [7.40] or rotated at a slight angle to the major hole axis, and the jet is moved forward a little between successive passes, then this impediment is no longer the case. With a slot cut by a dual jet, the jets do not encounter rock until they reach the back of the existing slot, and in this case the subsequent slots are cut at the same efficiency as the initial cut (Fig. 7.21).

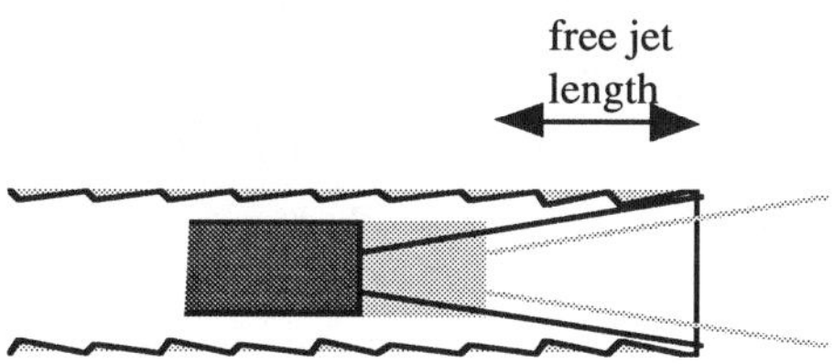

Figure 7.21 Dual jet cut with the effect of nozzle movement [7.41].

This discussion has been provided, not for the sake of confusing the issue, even though it might well have that effect. Rather it is provided to highlight a problem in the analysis of jet cutting experiments. Too often the results from single experiments have been used to derive a value for the specific energy of jet cutting, which has then to be used to show the inefficiency of the process. The experimental evaluation of the specific energy has often, however, been flawed since the value for the input energy to the target has been overestimated, based upon the loss in energy from the pump to the nozzle (which can be improved by better system design); in energy losses from the nozzle to the target (which may be improved by better placement of the nozzle or a change in jet parameters or nozzle design considerations); or in energy losses due to interactions of the jet with the outflowing jet or the sides of the previously made cut (each of which may be improved by a change in cutting technique). Since many of these considerations have only been fully understood since the early work on jet performance correlation with rock properties was finished, the data in the following segment should be accordingly treated with some caution.

7.4.3 THE EFFECT OF ROCK PROPERTIES

The problem of evaluating the correct physical properties of a target rock to correlate with the cutting performance of an incoming waterjet has several aspects. The first, as described above, relates to the correct evaluation of the form of the impacting energy. This problem is quite significantly more complex than described and this will be explained more fully in the Chapter 11. But even when we can define the size of the arriving jet energy pulse and can also describe its structure, one must still ask what property of the target will govern the volume of rock which the jet will remove? Of similar relevance is a subsidiary question, namely, can we easily measure that property, and if we cannot, can we correlate another property which is more easily measured with the property we need?

The earlier Russian work described in section 6.2.1 indicated a correlation between the energy required to remove unit volume of material and a combination of the Protod'yakonov number and the jointing index for Soviet coal. The Protod'yakonov number is a measure of the crushing strength of the material, and is obtained by dropping a 2.4 kg weight from a height of 60 cm onto five 10 mm or larger particles of rock, a known number of times between 5 and 20. The fragments are screened and the volume of powder generated below 0.5 mm is measured using a standard cylinder [7.42]. The Protod'yakonov number is then calculated as being equal to 20 times the number of blows divided by the powder column height.

This procedure gives a poor repeatability when used for evaluating rock [7.28] and a more rigorous test, based on this was developed at the University of Leeds by Brook and his students [7.28], [7.35], and [7.43]. This was called the Rock Impact Hardness number (RIHN) and is defined as the number of blows required to produce 25% fines smaller than 0.5 mm from a core 25 mm in diameter and 50 mm long. The value is obtained by running several tests at different numbers of impacts and plotting the best curve through them [7.43].

As a part of the correlation of jet performance with rock properties carried at UMR [7.32] this value was calculated for the seven rocks tested in the program. The values obtained for the jet performance values were obtained bearing in mind the problems described in the section above, and with a significant effort at minimizing their effect. A total of twelve properties was evaluated for each of the seven rocks used in this experiment, and a Pearson Correlation Coefficient calculated relating jet performance to these properties (Table 7.1) using the statistical program SPSS [7.44].

Table 7.1 Pearson correlation coefficients between waterjet cutting parameters and rock properties (after [7.32])

	Waterjet performance criteria		
Rock property	*Depth of cut*	*Specific energy*	*Energy ratio*
uniaxial compressive strength	-.249	.470	.030
Young's modulus of elasticity	-.066	.479	.519
Shore hardness	-.013	.286	-.060
Schmidt hardness	-.079	.434	.081
RIHN	-.279	.597	.278
fracture toughness	-.312	.494	.242
specific fracture energy	-.223	.379	-.099
inverse compressive strength	.352	-.547	-.223
inverse Young's modulus	-.004	-.496	-.495
inverse Shore hardness	-.078	-.193	.063
inverse Schmidt hardness	-.028	-.412	-.221
inverse RIHN	.052	-.505	-.446

When the various values were combined with jet properties to try and develop a predictive equation, the equation for the depth of cut contained two jet parameters jet velocity ratio (jet velocity/traverse speed) and jet pressure and two rock properties, the inverse compressive strength and the

Shore Hardness. The regression coefficient was 0.72 with an F value of 413, and a standard error of estimate of approximately 1 cm. The equation for the specific energy contained one jet parameter, jet pressure, and three rock properties, RIHN, the inverse compressive strength and the inverse Specific Fracture Energy. The regression coefficient was 0.89 with an F value of 1531, and a standard error of estimate of approximately 5,000 joules/cc.

To put this is perspective, when a correlation was similarly sought for a tunnel boring machine, when cutting through 72 rocks, the predictive equation for specific energy included cone indenter hardness (squared), uniaxial compressive strength (one third power), Shore Hardness (cubed) and cementation coefficient (cubed). The correlation coefficient was 0.866 and the standard error of estimate was 4.2 MJ/cu m [7.45].

Cooley had previously [7.46] used the process of dimensional analysis for both pulsed and steady jets. To characterize the rock he initially used a calculated value of rock specific energy at fracture (the square of rock uniaxial compressive strength divided by twice the Young's modulus value) which correlated with Protod'yakonov number [7.47]. However, this did not prove to give as good a correlation to the data as rock compressive strength alone, which he has used thereafter. A significant feature of the paper was the correlation of the relative specific energy of rock removal with the length of the impacting slug, minimum values being found at relative impact lengths below 1,000. Further the ability of the jet to spall or chip rock was found to reduce energy levels required by a factor of 10 - 30 where it could be sustained. Much of the data which Cooley analyzed for the paper came from earlier Russian work, for which little in the way of additional rock properties were available.

Singh and Huck [7.48] used data they obtained in tests on rock from the United States to examine correlations with the depth of crater created. Depth was chosen, rather than eroded volume to overcome the problems of dealing with variable spalling of the rock around the original impact point. Of the nine rock properties they measured and sought to correlate with depth of penetration, only two were found to fit best into the regression equation of the data. These were the inverse of uniaxial compressive strength and the inverse of the Schmidt hammer reading of the rock. The first gave a partial correlation coefficient of 0.89 and the second 0.79. Tests were carried out at jet pressures of up to 12,000 bar with a contained jet energy of up to 0.95 MJ. These pressures were significantly above the threshold pressures at which the rocks begin to cut.

Contemporary work in Japan with jets at 1,790 bar [7.49] also found that, at these higher pressures a correlation could be made with the inverse of compressive strength. In this case the equation derived for traversing jets, was of the form:

$$\frac{h}{d} = C_1 \cdot \left(\frac{W_w^{\alpha}}{W_t^{\beta}}\right) \cdot \left(\frac{\ell}{d}\right)^{\gamma} \cdot \phi^{\delta}$$

where W_w is the Weber number of the jet stream
W_t is the "Weber number" of the target (found by using traverse velocity rather than jet velocity in the jet Weber number)
d is jet diameter
ϕ is the Protod'yakonov number
C_1, α, β, γ, δ are constants

Average values for the coefficients established were $C_1 = 0.225 \times 10^{-10}$; $\alpha = 2.1$; $\beta = 0.62$; $\gamma = 0.369$; $\delta = 1.66$. However, it should again be stressed that this derivation was as a result of experiments at relatively high jet impact pressure.

The value of the jet pressure is rather significant since, as has been discussed earlier, it is relatively easy to demonstrate, at lower pressures, that compressive strength of the rock is not actually the correct parameter to use. For example, this can be established by any experimentalist who places two similar sized slabs of rock, one of granite, and one of marble adjacent on the test bed. If a waterjet at a pressure of between 700 and 1,400 bar is then traversed, at the same speed and condition, over both blocks in succession, the depth of cut in the granite will, normally, be significantly greater than that of the marble. The uniaxial compressive strength of the marble is, however, roughly 50 - 75% of that of the granite.

The explanation for this is not, as yet, fully developed, but relates to the mechanisms by which waterjets penetrate material. While this will be discussed in greater detail in Chapter 11, it is relevant to briefly discuss this at this point, since it is helps understand how best to use waterjets in removing material. Waterjets penetrate material, and induce material removal, by exploiting existing fractures and cracks in the material. These fractures in rock are frequently found at the boundaries of the crystals and grains of material within the rock [7.50].

The material property of most relevance to the rock cutting process thus becomes a structural property of the rock, rather than a physical one. Obviously within some degree of correlation it is possible to relate some of the physical properties of the rock to the structure, and thus, physical measurements can provide a guide. However the underlying relationships are to structure, and this will, in certain conditions, such as, for example the granite:marble relationship defy the corresponding match to the physical properties.

This can be illustrated by examining the waterjet cutting of granite in more detail. It has been well known to sculptors and quarrymen for centuries that granite, although apparently homogeneous to the layman, has, in fact, exploitable bedding planes. Thus, it is easier to split granite along, generally two directions, at right angles, than it is in any other. One such bed, generally close to horizontal, is sometimes referred to as the "lift," while the perpendicular vertical direction is known as the "rift." The orthogonal vertical direction, which is harder to split, is then known as the "hardway." A series of experiments was carried out at UMR to determine the effect of these orientations on the volume of rock removed by waterjet impact [7.51].

The experiments were carried out in two parts. In the first, single jet pulses were fired into small samples of rock cut at different inclinations into and along the two microfissure planes (Fig. 7.22). In the second series of tests, jet traverses were made to a common intersection point on slabs cut at different orientations. The results of the single impact tests (Fig. 7.23) showed a significant effect of orientation. Where the jets were fired along the orientation of the microfissures (the lift and rift directions) the craters cut into the rock were smaller, and confined laterally. In contrast, as the jets were fired perpendicular to these weakness planes, the spalls resulting would be larger. In relative dimension, volumes of up to ten times greater in size could be removed when the jet was cutting perpendicular to one of the weakness planes (Fig. 7.23).

The intentions in running the traversing tests were two-fold. Firstly it was to determine if the results from the single impacts would transfer to conditions where the jet was traversing. This was questionable since the spallation was due to the confinement of the jet pressure in the hole, as it was cut, and this would cause a pressurization of the microfissures out from the cavity created. Such was less likely to occur where the jet was traversing, and thus cutting a free path for the water to exit from under the following jet slug. Secondly, it was to determine the effect of the jet orientation on the relative distance at which two adjacent jet cuts would interact. The results can be clearly seen by examining the surface profile traces of the traverses made on a block oriented at 15° to the hardway (with the jet cutting almost along the rift) and that at 60° to the hardway (with the jet directed more perpendicular to the orientation of the rift) (Fig. 7.25). At greater angles to the rift the spallation was such that the test had to be carried out on two blocks due to cut interference. The results (Fig. 7.26) clearly showed the effect of jet orientation on the volume of rock removed and on the interaction distance. The traverses were carried out at a nozzle traverse speed of 30 cm/min with a 0.76 mm diameter jet at a pressure of 1,240 bar.

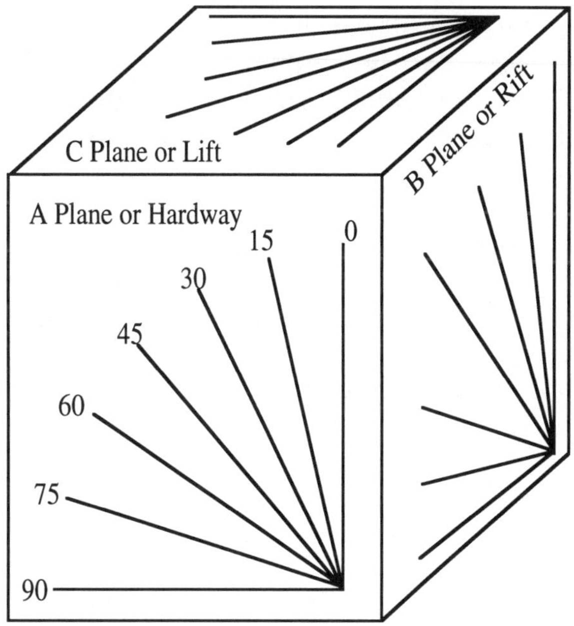

Figure 7.22 Sample angles relative to the grain of a granite block [7.51].

The relative differences in rock direction are extremely difficult to discern to the untrained eye. They do not appear to give significant differences within the statistical scatter of normal uniaxial compressive strength evaluations. They are, however, identifiable by measuring relative wave velocities through the rock [7.52]. It is the wide scatter in values of rock removed, relative to this almost indistinguishable change in rock structure, which thus suggests the difficulty in writing a controlling equation for jet performance in terms of target values.

The problem is further illustrated by considering the case of marble. Marbles can be tested which have grain sizes of about the same size as a granite. And yet the granite is easier to cut. Both investigators at UMR and at the University of Cagliari [7.53] have concluded that this might be due to differences in relative compressibility of the constituent materials in the surface. For marble, all crystals are of the same material, and in granite they are of different materials. In the latter case the difference would be sufficient to open the gap between crystals to a sufficient distance as to allow the jet easier passage in penetrating the rock. As yet, however, this is only a hypothesis.

One of the driving forces for this study, and subsequent efforts, has been toward the application of waterjets for cutting rock. While the actual uses will be discussed in the following section and chapters, their development required some initial parameter studies. The relevance, for example, of the distance at which two jet cut paths interact is critical to several uses of this new tool. Consider that a waterjet stream will typically cut a slot in a rock surface to a width of around 3 nozzle diameters, which will give a slot of some 2.5 mm, in width or less. The jet will issue from a nozzle held in some form of a holder, with the holder having a width typically on the order of 25 mm or more. Thus, if the jet is being used for some form of slotting or drilling operation, in which the nozzle body must advance into the opening being cut, in order to reach a required depth, then the jet path must be sufficient to sweep out a much wider width than that of a single jet.

This point has been partially addressed in earlier discussion, and will be returned to in Chapter 9 which follows on mechanically assisted jet cutting. What is required is some way of creating a wider path for the jet. Normally this is done by either using a series of jets set up to cut a wide slot, or a smaller number of jets are oscillated or rotated at an angle, in order to sweep out the wider path required for nozzle entry.

In either case, optimum efficiency in system design suggests that adjacent cuts of either the same or alternate jets are sufficiently close together that the material between them is totally removed. Tests in granular material have shown that two adjacent cuts may almost need to overlap before the intervening rib of material is removed [7.54]. In crystalline material such as granite, however, where rock spalling can be anticipated between cuts, the distance can be significantly greater. It can also be greater in coal, where the structure of the target material lends itself to jet exploitation with adjacent cuts as far apart as 50 mm removing all the intervening material, as shown earlier.

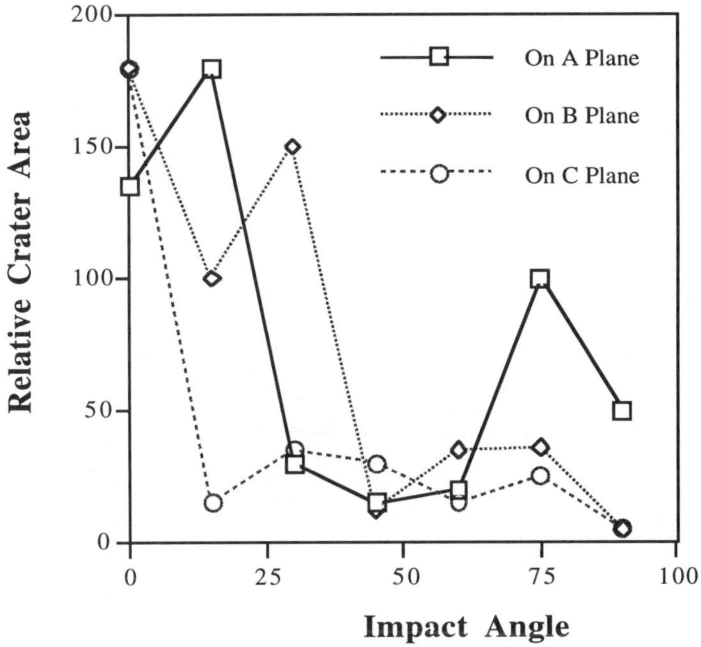

Figure 7.23 Variation in rock crater size with angle of jet impact on granite.

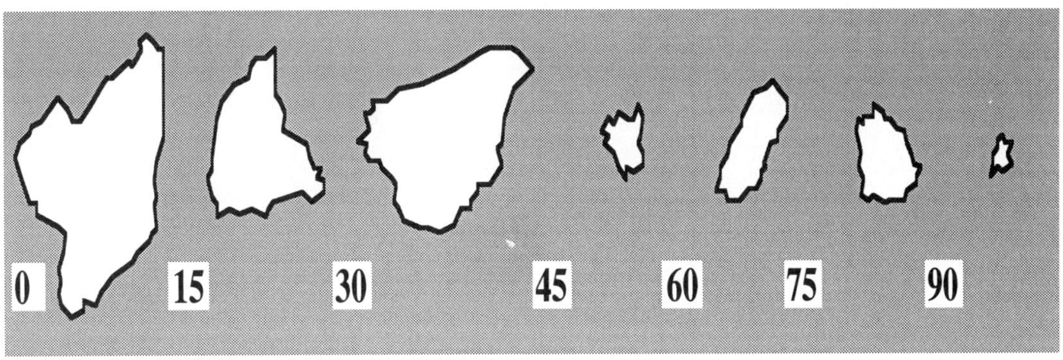

Figure 7.24 Surface profiles of typical craters at orientations to the rift [7.51].

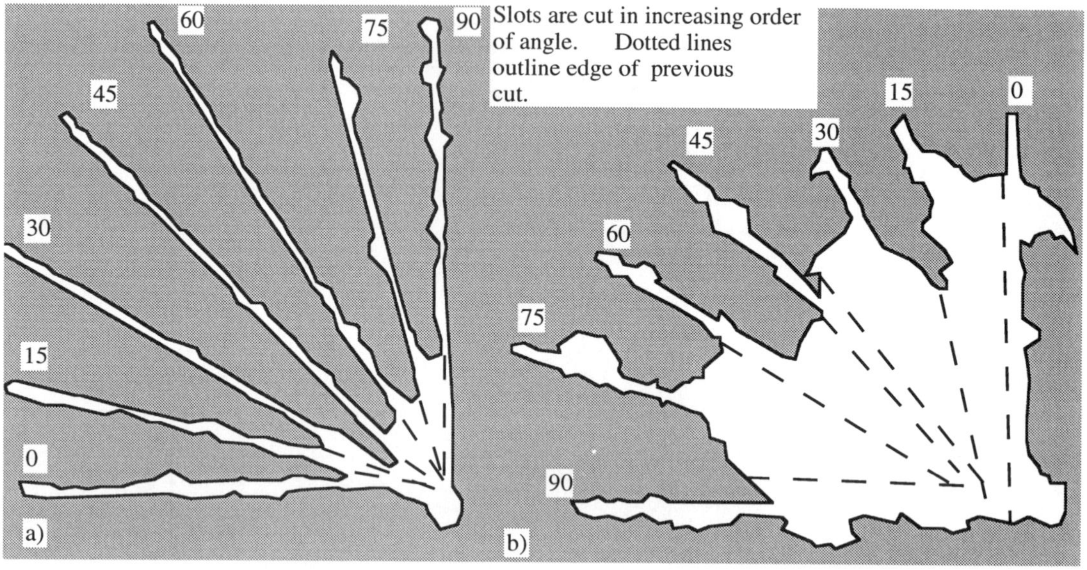

Figure 7.25 Effect of jet impact direction on the traverse profile in granite a) at 15° to the rift and b) with the block oriented at 60° to the rift (after [7.51]).

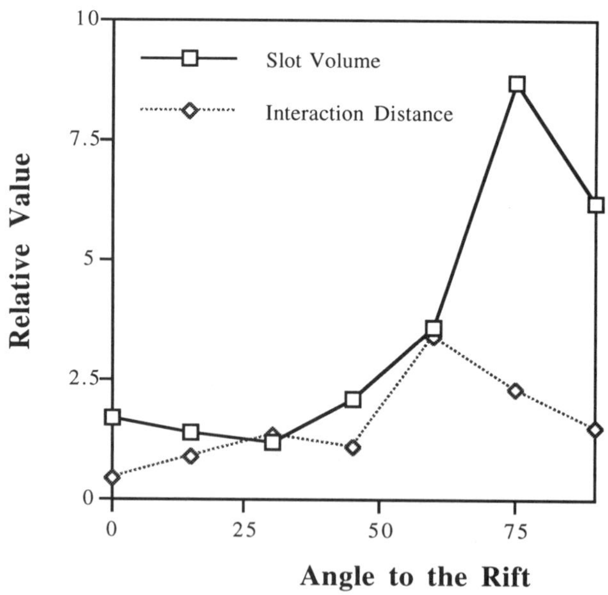

Figure 7.26 Effect of jet orientation on the volume of rock removed and the interaction distance (after [7.51]).

Experiments in plain jet cutting have shown that the jets cut more efficiently if the water flow is diverted to as few a number of nozzles as possible, since this allows the jet diameters to be larger. Further, as shown above, the jets cut better at higher traverse velocities. This led to the design philosophy of using a dual orifice nozzle, with the jets equally inclined to the axis (Fig. 7.27). The nozzle assembly is rotated to sweep a wide enough path ahead of the tool as it advances to cut a slot in the rock face. For most efficient cutting two parameters should be established, the optimum angle of the jets to the axis of the slot, and the advance distance between adjacent cuts of a jet over the slot width to remove all the remaining material.

Figure 7.27 Dual orifice nozzle design used for slot cutting in granite.

A series of tests evaluating these criteria was carried out by Bortolussi, while at the University of Missouri-Rolla [7.55]. To simplify the experiments, and to more easily collect data which showed the effect of rock anisotropy, the tests were carried out on rock cores which were rotated during the test process. A single orifice jet was then traversed at varying impact angles to the core at various advance rates (Fig. 7.28). In this manner the incremental distance between adjacent passes could be studied as well as the effect of impact angle. It had been established in field tests that if the angle of impact was too close to the angle of the wall of the slot (in this case the core axis) then the cut would not be of constant thickness and, in slot cutting, the slot would taper with depth.

The range of angles tested were at 2.5^o increments from 15° - 37.5^o, and increments 2.3 and 3.174 mm were tested on granites with a coarse grain and a fine grain. The tests were carried out at a jet pressure of 900 bar with a 0.45 mm diameter nozzle and with the core rotating at 18 rpm to give a linear traverse speed of 1.4 m/min. At angles below 20 degrees the jet was deflected and would not consistently remove all the crystals of the rock to give a straight edged cut. As the angle was increased, however, the depth of cut achieved in the direction of potential slot advance was reduced. For these reasons an optimum angle for the jets would appear to lie between 20° and 22.5° from the axis of the slot.

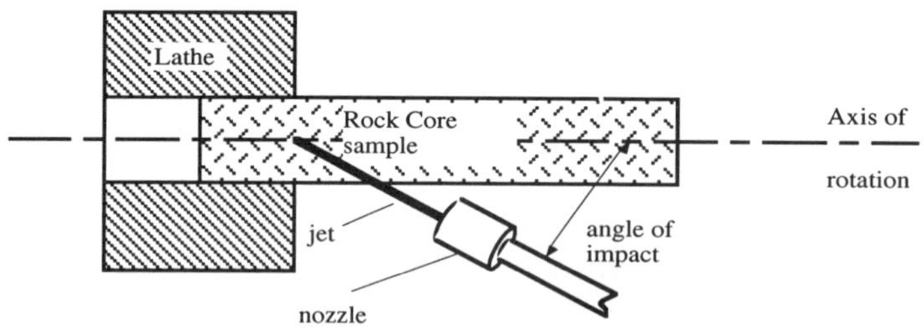

Figure 7.28 Experimental arrangement to observe the effects of angle on interaction distance and cut depth [7.55].

One additional point might be mentioned. In a demonstration of the effect of impact angle on the volume of material removed, carried out at Cagliari, it was shown that the direction of attack is also important. In that test an oscillating (as opposed to a rotating) waterjet was first directed against the direction of motion of the nozzle and then with the jet pointing in the direction of motion. In both cases the jet angle to the grain of the rock (actually concrete) was the same, but in the first case the nozzle was moving into the cut being made, while in the second it was leaving it behind. It was found (Fig. 7.29) that the jet removed more material when it was moving with the cut ahead of the nozzle, rather than behind it.

Figure 7.29 Slots cut by an oscillating waterjet with the jet moving a) with the direction of motion, and b) against it.

The results of the study of incremental distance were less clearly defined since the optimum value appeared to be a function of not only the grain size of the target material, but also the pressure and diameter of the cutting jet. For most conditions the optimum value appeared, however, to lie between 3 and 6 mm. Thus, by setting the rotational speed of the nozzle assembly relative to the advance rate of the nozzle down the slot to achieve such a result optimum cutting should result. While this appeared theoretically logical it did not give consistent results with earlier field data, predicting a cutting efficiency significantly below that achieved. This can partially be explained by the fact that the field work involved sequential passes down a slot, rather than a single pass as in the laboratory. Jets appear to cut better on rough rather than smooth surfaces.

Although the subject of how jets cut will be discussed in Chapter 11, it is appropriate to conclude this section with results recently reported from Italy [7.56]. The efficiency of a jet as it cuts into a rock is measured by the specific energy, which, until now has been suggested to be a process driven parameter. However, there is some minimum specific energy which any rock, or other material being cut, will require at the most efficient level of cutting. Agus et al., have correlated the specific energy of cutting with two measurable parameters of a rock. These are the average point load strength of the rock (Fig. 7.30) and the sonic wave velocity of P-waves transmitted through the rock (Fig. 7.31). The authors note that these values include a considerable scatter and that the mineralogy and porosity of the rock is also important.

In light of the discussion earlier in this chapter concerning correlation coefficients and their relevance, the data for the above curves has been recalculated by the current author to show the correlation coefficients for the two properties. It is interesting to note that the best correlation is with sonic velocity, given that the earlier work at UMR with Bur [7.52] indicated that this measurement was sufficiently sensitive to detect and correlate with relatively small changes in rock structure.

There are, also some other factors in play, not as yet fully understood. But to discuss these factors one should look at the ways in which the waterjets are now being used.

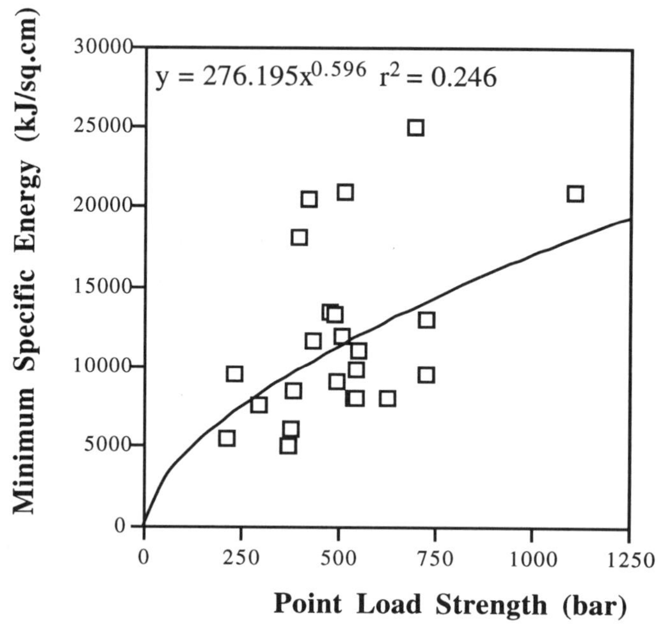

Figure 7.30 Variation in the specific energy of cutting granite with point load strength [7.56].

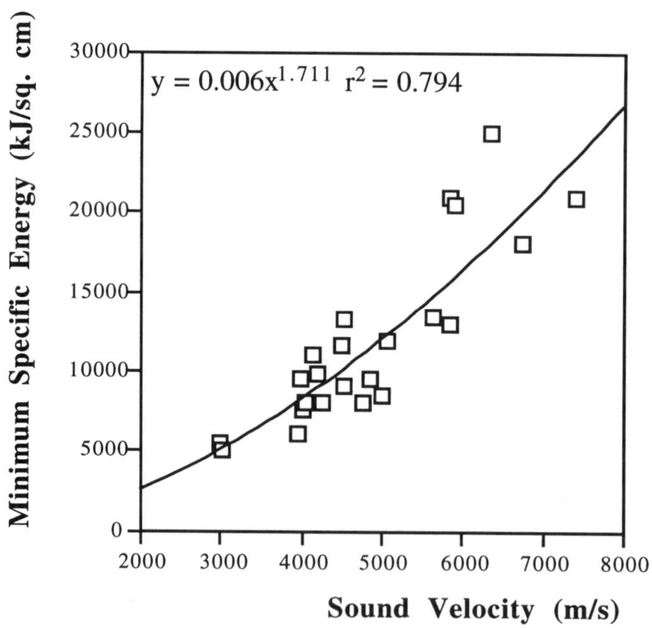

Figure 7.31 Variation in the specific energy of cutting granite with P-wave velocity [7.56].

7.5 APPLICATIONS OF WATERJETS IN CUTTING ROCK

There are many advantages to the application of waterjets in mining. However, the requirements that the jets must be able to cut through all the material to which they are exposed has limited their application. In underground mining applications the major significant change in technology has been with the addition of waterjets to conventional mechanical tools, which will be addressed in Chapter 9. Apart from such applications plain waterjet use has not, as yet, achieved a significant market penetration.

An early apparent market for waterjets appeared to lie in their use in quarrying. In particular the requirements for primary block removal in granite operations suggested a possible market. Although most marketed granite appears to be of consistent quality and pattern, this is not how much of the rock occurs. Near the surface the rock is often weathered, and the pattern of crystallization often yields undesirable "flaws" and marks in the stone. Thus, where a block of consistent pattern and quality is found it must be carefully extracted to yield the greatest volume possible.

The classic, and some of the more recent technologies of granite quarrying for dimension stone have been described by Bortolussi et al. [7.57] and relate somewhat to the manner in which the rock is being removed from the solid. In essence large blocks are first cut from the solid and then these are successively split into smaller pieces. This continues until the size is small enough for the blocks to be moved to a production plant where they can be cut to final shape and polished where necessary.

A simple method of making the first cut is to drill a series of holes along the line of the required fracture and then to fill these with a form of gunpowder or equivalent slow burning explosive. The resulting blast is designed to break the rock from the solid, and move it out some distance. In some cases the round is also designed to topple the block over onto a cushioning pile of dirt at the same time. Bortolussi reports [7.57] that from 10% - 20% of the block volume could be lost by excess breakage, uneven fracture lines and weakening of the block at this point in the process.

To break the large blocks thus formed into smaller, handleable size pieces, the rock can be split, often along the weakness planes discussed above. A series of short holes is drilled along the line of projected fracture. Small wedges are set in the holes between two "feathers", hemi-cylindrical sections of steel which grow larger with depth. Thus, as the quarry worker sequentially taps each wedge along the line deeper into the hole, the stress applied through each feather increases, until a crack is generated separating the block. The split face is usually somewhat irregular with this technique, requiring that some of the rock be removed to provide a flat working surface.

The block can be separated with better control where the holes are drilled through the block using a line drilling technique. With the advent of hydraulic powered rock drills, higher penetration rates can be achieved, making this technique economic for small block splitting. The technique requires accurate drill alignment however, and is not as viable where the holes must be of significant length.

A more recent development for creating the "first cut" on a block has been with the development of the "flame torch." This comprises a long lance through which a combination of air and kerosene is blown to a combustion chamber which directs an intense flame from the free end. This technique was first applied to drilling holes in hard rock by the Linde Company in 1947 [7.59]. The high temperature directed at the rock surface will induce spalling "as a result of decrepitation resulting from differential expansion of rock crystals due to the thermally induced stresses" (ibid.). The small particles of rock removed are carried away by the force of the combustion gases. In this way a slot is created around the end of the burner as it is manually advanced down the slot. Typical production rates for such a unit might be on the order of 1.1 sq m/hr. The technique was widely used in the United States into the late 1970's. However, it produces large quantities of dangerously small rock particles, and noise levels of up to 140 db well above those set by Federal Regulation. Fuel costs make up some 60% of production costs and the wide slot which the lance cuts, together with the zone of thermally weakened rock around the block may generate losses of up to 40% of the block volume [7.58]. A typical burner may use between 40 and 60 l/hr of fuel, and, depending on operator skill, may excavate some 0.75 sq m/hr [7.60].

Diamond sawing was a practice introduced in the late 1970s to the United States from Europe. Saws measuring over 1 m in radius were mounted on a traversing carriage and used to cut thin slots in the rock surface. While the tool minimized the amount of rock wasted in the cut, the blade could become trapped when cutting through rock which was still geopressured and which would then stress relieve, by expanding into the cut. Blade costs were similarly high, and the difficulty in setting and aligning the traverse mechanism has reduced its impact. Because of the presence of an arbor in the center of the blade, required to provide a drive to the wheel, cutting depths for these wheels are usually limited to less than 1 m.

On the other hand, diamond wire cutting has proven more successful. In its original conception, wire sawing used silicon carbide particles which were drawn into the cut by the wire, and thus cut into the rock. Production rates in Minnesota ranged from 1.5 - 2.7 sq m/hr at a cost, in 1974, of $4.50/sq m [7.61]. The system has since become economically unattractive, with an increased price for the abrasive, and has been replaced by diamond wire cutting. In this latest development small plastic beads are set at regular intervals along the wire length. They are set with small industrial diamonds and driven in an endless loop by a small motor, usually set on a set of rails to provide a constant tension to the wire as the cut deepens. According to Bortolussi, et al., [7.58] diamond wear provides 70% of the cost of sawing, but the smooth surface of the cut means that almost all the rock extracted from the quarry becomes available for use. Equipment and operating costs are significant limitations for the use of this technique in cutting granite.

These practical limitations on existing technologies have led to several investigations of the use of waterjets as a means of granite cutting. As a warning to the layman, it should be known that the word "granite" has acquired a generic broadness of use quite beyond the limited geologically restricted rock type. This is important to bear in mind since the rock which can be effectively cut by the waterjet at different pressures is a function of the rock structure and content, (discussed above). Thus, "black granites" and other igneous rock which does not have the relatively large, multi-component crystalline structure of the main granite family are less likely to be as easily cut with the waterjet systems discussed at this time.

Early experiments in Russia had shown that waterjets could cut granite at relatively low pressures, and this was confirmed by experiments at the University of Leeds [7.28] in the 1960s. Thus, when the American granite industry began to examine methods for modernizing granite operations in 1977, under funding by the National Science Foundation [7.62], waterjet technology was one candidate under consideration. Because there was considerable discussion at the time as to the most effective pressure for cutting the rock, two tests were subsequently funded by the Elberton Granite Association (EGA), in co-operation with three other major American quarry owners. These were carried out in quarries at Elberton in Georgia, and were trials to establish if, and under what conditions, waterjets could make the primary cuts to delineate the blocks in the rock.

For the waterjets to be effective they must be able to cut slots to a depth of up to 10 m. Since this depth of cut is not achievable from the surface, the nozzle must advance into the slot, as it is being developed. This, in turn, requires that the slot be wide enough to allow the nozzle body entrance. Such a cut width is typically on the order of 5 cm, relative to the narrow width of the jet itself. Two methods to cut a slot of this width were investigated relatively early in the development of the technology, one by the NRC in Canada [7.63], used a dual-orifice rotating waterjet head to cut the slot, the other, developed at Flow Research, used an oscillating nozzle technique [7.64].

The first series of tests carried out at Elberton was by personnel from UMR [7.65]. Based on a series of experiments at the University, a simple rotating nozzle assembly, on the same principal as the Canadian concept, had been built which directed two outwardly pointing jets at the rock. This device was mounted on the traversing frame from a large diamond wheel cutting system, and used to cut a slot into the rock under the device (Fig. 7.32). Experimental pressures were around 1,000 bar, with a flow rate of 60 lpm. Excavation rates achieved were on the order of 1.6 sq m/hr and a slot was cut into the rock some 5.4 m long and 1 m deep [7.66]. The equipment used was largely already commercially available, although the traversing mechanism was a modification of specialized equipment and the swivel in use to allow lance rotation had, at that time, a relatively limited life. The primary power generator for the system used some 20 l/hr [7.60].

The second experiment was carried out by IIT Research Institute, again comprising an initial laboratory test followed by a field demonstration at Elberton [7.67]. This used a higher pressure delivery system (2,000 to 2,750 bar) at a lower flow rate, but due to the use of a very small feed line there was a significant pressure drop before the water reached the cutting orifices. Slots up to 3.5 m deep were cut using this system, at a cutting rate of 1.17 sq m/hr. As a result of these tests it was possible to make some comparisons with flame jet cutting (at the time the most common method of cutting rock in Elberton). Based on 1980 costs it was estimated that waterjet cutting of granite could be achieved at a cost of $7.12/sq m as opposed to a cost of $18.3/sq m for the flame torch method. The waterjet method was anticipated to reduce cutting time by up to three times.

The results of these trials showed that it was feasible to use waterjets in quarries and, as a result, three different programs were undertaken. The EGA undertook a program with the Georgia Institute of Technology, at a cost of $170,000 to develop a prototype machine for use in quarries [7.68].

The system was built and tested in Elberton. Unfortunately, the equipment included the design of a new pump, as well as related equipment, and it proved difficult to satisfactorily achieve the performance anticipated for the unit. As a result the major program was discontinued.

Figure 7.32 Early quarry cutting with waterjets at Elberton.

A second set of equipment, operated at a pressure of up to 1,700 bar, at a quarry in South Dakota [7.60]. The equipment proved capable of cutting slots up to 52.5 m long and 4.8 m deep. Typically, since the nozzle must advance into the slot as it is being cut, a slot width of approximately 5 cm was maintained. One problem became evident in the use of this system. Although the waterjet system was providing a faster cutting rate than other techniques, and could be automated to cut without continuous operator presence, the unit built was robust. This meant also that it was heavy and required a significant effort to move and align. The effect of this was that while a faster excavation rate was achieved while the unit was operating, it took longer to set up, and thus, the overall daily production was not significantly increased [7.69]. This system used water flow rates of up to 400 lpm.

Such a flow rate was not practical in the Rocky Mountains of Colorado, where the shortage of water has proven to be a continuing problem. A quarry owner has solved this by increasing the operating pressure of the jet to 3,100 bar [7.57]. At this pressure the flow rate could be reduced to as little as 5 lpm through a single orifice, some 0.36 mm in diameter and still achieve an acceptable performance. By oscillating this jet over the surface of the slot while traversing the nozzle at 4.2 m/min it has proved possible to achieve slot cutting rates of 0.6 sq m/hr.

An alternate to the use of a single oscillating jet has been developed in France [7.70]. Now known as the Loegel system, after the inventor and quarry owner, the system was initially used to quarry sandstone in the Vosges de Nord of France. Initial pressures of 650 bar were used to cut horizontal slots 8 to 10 cm wide over lengths of 10 m, and to height of up to 10 m in the quarry face. By interspersing vertical and horizontal cuts it was thus, possible to outline blocks of rock, with regular size, significantly increasing the volume of saleable product. The system is reportedly being modified at the beginning of the 1990s for use, at higher pressure, in quarrying granite.

Not all cutting work on rock is completed at the time that the block leaves the quarry, however. In the working of dimension stone, significant additional work, and expense, is incurred in bringing the rock to the final required shape and surface finish. Waterjets can be applied in two ways to this process. One is to reduce the volume of rock which must be finally carved to shape. The second, where a slightly rougher cut is an acceptable finish, is to cut the piece to the final contour. (Additional discussion of this will take place in Chapter 10 on the use of abrasive laden waterjets.) The use of waterjets in this latter case can be illustrated by example.

In 1983 it was decided to erect a half-scale model of the megalith Stonehenge on the UMR campus. The rock chosen was granite from the Elberton area (from the same quarry as some of the earlier test work). Roughly outlined blocks of granite were shipped to Missouri, where they were carved to final shape using a rotating pair of waterjets, similar in design to those used in the initial tests in Elberton. Jet pressures during the cutting of the 53 blocks was at pressures between 850 and 1,000 bar, at a flow rate of some 40 lpm. It was found that the jets would cut a straight edge to the rock, even where the amount of rock to be removed was less than half the cut width of the traversing jets (Fig. 7.33). The rig which was used to traverse the nozzle was made deliberately light, and could be easily moved.

Figure 7.33 Edge cut on a granite block for the UMR Stonehenge.

Thus, it proved possible to move the cutting mechanism more easily than the blocks, and a relatively rapid pace was finally obtained in cutting the blocks to the required dimensions. The simplicity and ease of moving the equipment can be illustrated by the use of bicycle chain to pull the nozzle assembly along the cutting path. Fractional horsepower electric motors were used (Fig. 7.34) for nozzle rotation, traverse advance, and to feed the nozzle forward after each pass over the rock surface [7.71]. As a practical note the same rotary swivel was used for the entire cutting program, several hundred hours of use, in contrast with the poor performance achieved with the state-of-the-art swivels some seven years earlier. Also the nozzles lasted 20 - 30 hours each, with failure being due to erosion of the holder by the abrasive impact of liberated particles of rock. The monument (Fig. 7.35) was carved, with the exception of the face of the analemma used to tell the date, entirely to shape using the high pressure waterjet system at UMR.

The monument was dedicated on July 20, 1984 with the assistance of John Bevan, a white-robed Druid of the Geffodd of Druids of the Isle of Britain. It was awarded one of the Ten Outstanding Engineering Achievement Awards in the United States in February, 1985 by the National Society of Professional Engineers.

Figure 7.34 Equipment used to carve UMR Stonehenge.

Figure 7.35 Completed UMR Stonehenge monument.

Research on the application of waterjets to the quarry industry is continuing. An investigation of the use of such jets in carving Sardinian granite and marble being carried out at the University of Cagliari, has been referred to above. The study [7.58] has already developed predictions for the required penetration rates and recovery levels which must be achieved if the technique, either alone or in combination, is to become an economic as well as a technical success. This work has sequentially examined the various factors involved in the use of waterjets for granite cutting. The lateral movement of the jets across the slot being cut to allow nozzle advance, can be achieved by either rotating the nozzle head, or causing the head to oscillate. Parameter effects of the oscillation technique were reported in a paper [7.72] in Czechoslovakia, as a function of traverse velocity and oscillation frequency on the depth of cut. The results were derived from a series of sequential passes over a sample surface and indicated that, for a 2,000 bar jet issuing through a 1.25 mm diameter orifice, that there was a correlation between the two parameters and their effect on the depth of cut (Fig. 7.36).

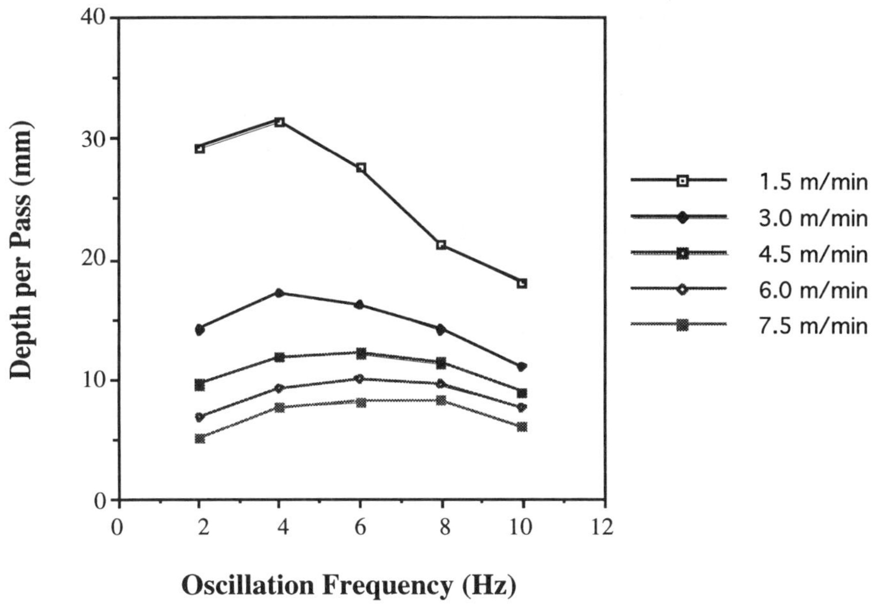

Figure 7.36 Depth of cut, as a function of oscillation frequency, as controlled by traverse velocity [7.72].

The trend in data can, perhaps be more readily observed if the data is recalculated in terms of the areal cutting rate (Fig. 7.37). Where the original graph was plotted the investigators also derived the pitch or feed of the rock forward between adjacent oscillations. For the combination of conditions it appeared that, as the jet velocity increased the optimum pitch gradually increased from 1 - 1.8 cm. It is interesting to note that the predictions for quarry performance with these tests, reported in late 1991, were that the system would cut at a rate of 4 sq m/hr (Table 7.2). In other granites, however, and under different jet parameters the shape and trends of the data shown below were not duplicated.

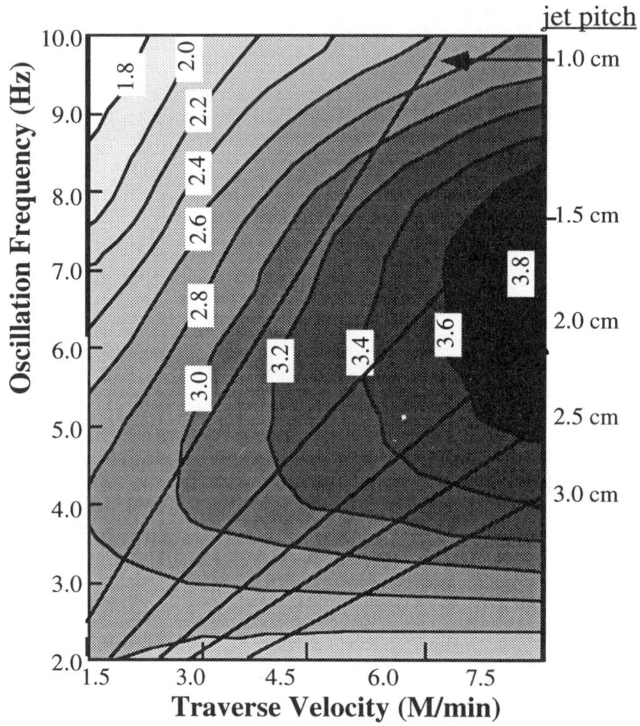

Figure 7.37 Area of rock removed, as a contour plot controlled by traverse velocity, and oscillation speed. The transverse lines are those of equal jet pitch [7.72].

Agus and his colleagues (ibid.) concluded that the optimum pitch was dependent on the diameter of the nozzle, but apparently independent of the grain size of the rock. Further they concluded that it was the correct selection of the combination of jet pressure and nozzle diameter which was the most effective choice, rather than simply designing for the largest hydraulic horsepower. The latter, as Table 7.2 shows, is not, in itself a guarantee of optimum performance.

Table 7.2 Effect of jet conditions on performance in granite (after [7.72])

Jet pressure (bar)	Nozzle diameter (mm)	Specific energy (KJ/cm^3)	Areal cutting rate (m^2/hr)	Power (kW)
2,000	1.25	3.9	4.19	193
2,000	0.70	5.3	0.98	61
1,250	1.00	2.6	1.99	61

Studies are also continuing in the United States into the use of waterjets to reduce the volume of rock which must be removed from a block before it can be finish carved and polished. Work has also been undertaken in Canada to examine the use of such systems in Canadian quarries [7.73]. The cutting rate achieved by the system, which operated at a pressure of 1,380 bar and achieved an effective exposure rate of 1.17 sq m/hr, with a maximum value achieved of 1.7 sq m/hr. Slots were cut to 5 m long and 3.4 m deep. It was found that the nozzles used were lasting on the order of 13 hours. When the equipment was used as a drill it achieved penetration rates of 1.5 m/min, some 3 to 4 times that of the existing equipment at the site.

Additional applications of waterjets are becoming evident in the use of waterjets which are assisted by abrasives, or with the use of mechanical tools. Such applications will be discussed in the relevant sections of subsequent chapters. This will include the effective cutting of marble which has, as yet, proved easier to cut with abrasive jets than with plain waterjets alone [7.74].

Perhaps one of the more dramatic uses of waterjets in recent years was the rescue of Jessica McClure from a well in Texas. The 18-month old young girl had fallen 7 m down a disused well shaft outside her home and become trapped [7.75]. A rescue well was dug alongside the original well, and rescuers had then to cut a path through to reach the child (Fig. 7.38). Conventional mechanical drills had significant problems in penetrating the ground, and a 2,800 bar waterjet system was brought in. The high pressure waterjets cut through the rock in an estimated 10 hours less than would have been taken with the drills [7.76], [7.77].

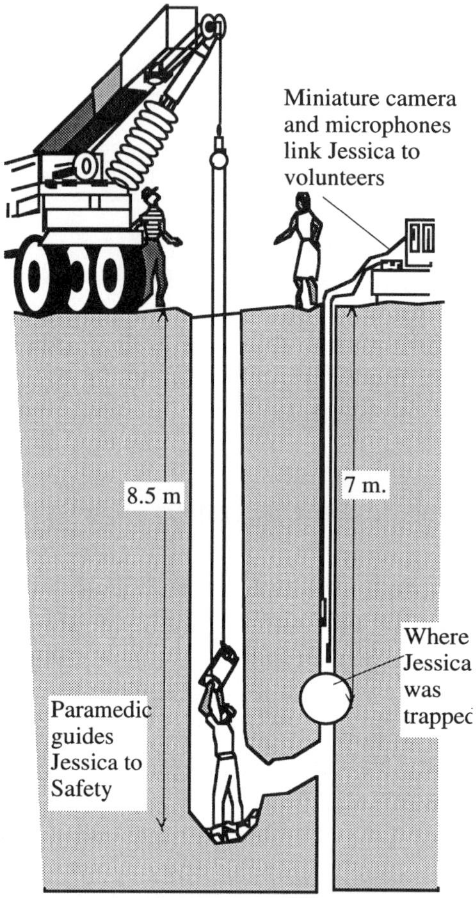

Figure 7.38 Layout of the wells in the Jessica McClure rescue.

7.6 NOTES - CHAPTER 7

The use of waterjets to cut rock has allowed an easier path to understanding jet cutting behavior than with other materials. This is because of relatively large grain structure of many rocks, and more obvious development of failure process. The mechanisms of material removal have only qualitatively been described, without an adequate theoretical basis for predicting jet performance. This is because of the relative complexity of the process. The subject will be returned to, in more detail, in Chapter 11.

Development of waterjets for cutting rock included several specialized applications and techniques. Rock drilling and waterjet assisted cutting are dealt with in the following two chapters. These should be read with this one for a more complete understanding of this subject.

7.7 REFERENCES

7.1 Savanick, G.A., "Hydraulic Mining Experiments in an underground mine in the St. Peter Sandstone, Clayton, Iowa," 2nd U. S. Waterjet Conference, Rolla, MO., 1983, pp. 187 - 202.

7.2 Summers, D.A., and Mazurkiewicz, M., The Use of Cavitation in Rock Disintegration, Final Report to the U.S. Bureau of Mines on a UMR MMRI Contract, 1982.

7.3 Mazurkiewicz, M., and Summers, D.A., "The Enhancement of Cavitation Damage and its use in Rock Disintegration," paper A2, 6th International Symposium on Jet Cutting Technology, Surrey, U.K., April, 1982, pp. 27 - 38.

7.4. Powell, J.H., and Simpson, S.P., "Theoretical Study of the Mechanical Effects of Water Jets Impinging on a Semi-Infinite Elastic Solid," International Journal of Rock Mechanics and Mining Science, Vol. 6, No. 4, 1969, pp. 353 - 364.

7.5 Voitsekhovsky, B.V., Solovkin, E.B., Grebennik, O.I., Kuvshinov, V.A., Shoikhet, G.Ya., Nikolaev, V.P., and Lesic, N.P., "On Destruction of Rocks and Metals by High Pressure Jets of Water," paper G8, 1st International Symposium on Jet Cutting Technology, Coventry, U.K., April, 1972, pp. G8 93 - G8 112.

7.6 Leach, S.J., and Walker, G.L., "The Application of High Speed Liquid Jets to Cutting," Philosophical Transactions, Royal Society (London), Vol. 260 A, 1966, pp. 295 - 308.

7.7 Farmer, I.W., Penetration of Rocks by Water Jet Impact, Ph.D. Thesis, Univ. of Sheffield, U.K., 1965.

7.8 Cooley, W.C., "Rock Breakage by Pulsed High Pressure Water Jets," paper B7, 1st International Symposium on Jet Cutting Technology, Coventry U.K., April, 1972, pp. B7 101 - B7 112.

7.9 Cooley, W.C., and Brockert, P.E., Optimizing the Efficiency of Rock Disintegration by Liquid Jets, Final Report on Contract HO 230005, U.S. Bureau of Mines, by Terraspace Inc., Rockville, MD., Report No. TR-4051.

7.10 Singh, M.M., Finlayson, L.A., Huck, P.J., "Rock Breakage by High Pressure Water Jets," paper B8, 1st International Symposium on Jet Cutting Technology, Coventry, U.K., April, 1972, pp. B8 113 - B8 124.

7.11 Voitshekovskii, B.V., Jet Nozzle for Obtaining High Pulse Dynamic Pressure Heads, U.S. Patent No. 3,343,794 September 26, 1967.

7.12 Cooley, W.C., Beck, F.L., and Jaffe, D.L., Design of a Water Cannon for Rock Tunnelling Experiments, Final Report on Department of Transportation Contract DOT-FR-00017, Terraspace Inc., Bethesda, MD., February, 1971, Report No. FRA-RT-71-70.

7.13 Clipp, L.L., and Cooley, W.C., Development, Test and Evaluation of an Advanced Design Experimental Pneumatic Powered Water Cannon, Exotech Inc., Report No. TR-RD-040 to the U.S. Dept. of Transporta tion, March, 1969, 62 pp.

7.14 Singh, M.M., "Novel Methods of Rock Breakage," 2nd Symposium on Rapid Excavation, Sacramento State College, CA., October 16-17, 1969, paper 4.

7.15 Farmer, I.W., and Attewell, P.B., "Rock Penetration by High Velocity Water Jet," International Journal of Rock Mechanics & Mining Sciences, Vol. 2, No. 2, 1964, pp. 135 - 153.

7.16 Anon, "Jet, mole team seeks tunnelling breakthrough," Engineering News Record, January 22, 1970, p. 48.

7.17 Cooley, W.C., Hypervelocity Jet and Projectile Velocity Augmenter, U.S. Patent 3,465,693, September 9, 1969.

7.18 Cooley, W.C., and Brockert, P.E., Rock Disintegration by Pulsed Liquid Jets, Final Report on ARPA Order No. 1579, Bureau of Mines Contract HO210012, Terraspace Inc., Bethesda MD., January 1972, Report No. TR-4032.

7.19 Cooley, W.C., "Correlation of Data on Erosion and Breakage of Rock by High Pressure Water Jets," Chapter 33, Dynamic Rock Mechanics, ed., G.B. Clark, 12th Symposium on Rock Mechanics, University of Missouri-Rolla, November, 1970, pp. 653 - 665.

7.20 Cooley, W.C., Workshop on the Application of High Pressure Water Jet Cutting Technology, UMR, Rolla, MO., November, 1975, pp. 18 - 35.

7.21 Olson, J.J., and Olson, K.S., ARPA-Bureau of Mines Rock Mechanics and Rapid Excavation Program, A Research Project Summary, U.S. Bureau of Mines Information Circular 8674, 1975. 191 pp.

7.22 Cooley, W.C., Performance and Noise Suppression Tests of a Water Cannon, Terraspace Inc., Report No. FRA-ORD and D-75-9 to the U.S. Dept. of Transportation, September, 1974.

7.23 Lucke, W.N., and Cooley, W.C., "Development and Testing of a Water Cannon for Tunnelling," paper J3, 2nd International Symposium on Jet Cutting Technology, Cambridge, U.K., April, 1974, pp. J3 27 - J3 44.

7.24 Chermensky, G.P., "Experimental Investigation of the Reliability of Impulsive Water Cannons," Paper H1, 3rd International Symposium on Jet Cutting Technology, Chicago, IL., May, 1976, pp. H1 1 - H1 14.

7.25 Summers, D. A., and Henry, R.L., "The Effect of Change in Energy and Momentum Rates on the Rock Removal Rate in Indiana Limestone," paper B5, 1st International Symposium on Jet Cutting Technology, Coventry U.K., April, 1972, pp. B5 77 - B7 88.

7.26 Moodie, K., and Artingstall, G., "Some Experiments in the Application of High Pressure Water Jets for Mineral Excavation," paper E3, 1st International Symposium on Jet Cutting Technology, Coventry U.K., April, 1972, pp. E3 25 - E3 44.

7.27 Moodie, K., and Taylor, G., "A Review of Current Work on the Cutting and Fracturing of Rocks by High Pressure Water Jets," Conference on Fluid Power Equipment in Mining, Quarrying, and Tunnelling, I. Mech E., U.K., February, 1974, pp. 41 - 48.

7.28 Summers, D.A., Disintegration of Rock by High Pressure Jets, Ph.D. thesis, Mining Engineering, University of Leeds, U.K., 1968.

7.29 Summers, D.A., Raether, R.J., "Comparative use of Intermediate Pressure Water Jets for Slotting and Removing Concrete," paper J1, 6th International Symposium on Jet Cutting Technology, Guildford, U.K., April, 1982, pp. 387 - 396.

7.30 Summers, D.A., Peters, J.F., "Preliminary Experimentation on Coal Cutting in the Pressure Range 35 to 200 MN/m^2," paper H2, 2nd International Symposium on Jet Cutting Technology, Cambridge, U.K., April, 1974, pp. H2 17 - H2 28.

7.31 Agus, M., Bortolussi, A., Ciccu, R., Manca, P.P., Massacci, G., and Bosu, M., "Jet Impingement Tests on Mineral Crystals," First Asian Conference on Recent Advances in Jetting Technology, Singapore, May, 1991, Ci-Premiere Conference Organization.

7.32 Summers, D.A., "Water Jet Cutting Related to Jet and Rock Properties," 14th Rock Mechanics Symposium, June, 1972, ASCE, pp. 569 - 588.

7.33 Box, G.E.P., Hunter, W.G., and Hunter, J.S., Statistics for Experimenters, John Wiley and Sons, 1978, 653 pp.

7.34 Benedum, W., Harzer, H., and Maurer, H., "The Development and Performance of two Hydromechanical Large Scale Workings in the West German Coal Mining Industry," paper J2, 2nd International Symposium on Jet Cutting Technology, Cambridge, U.K., April, 1972, pp. J2 19 - J2 26.

7.35 Page, C.H., Penetration of Rocks with High Pressure Water Jets, Ph.D. thesis, University of Leeds, U.K., 1972.

7.36 Brook, N., and Page, C.H., "Energy Requirements for Rock Cutting by High Speed Water Jets," paper B1, 1st International Symposium on Jet Cutting Technology, Coventry, U.K., April, 1972, pp. B1 1 - B1 12.

7.37 Kuzmich, I.A., "Some Relationships in the Coal Penetration by High Pressure Thin Waterjets," paper E1, 1st International Symposium on Jet Cutting Technology, Coventry, U.K., April, 1972, pp. B1 1 - B1 12.

7.38 Fowell, R.J., Johnson, S.T., and Tecen, O., "Studies in Water Jet Assisted Drag Bit Cutting," paper F2, 7th International Symposium on Jet Cutting Technology, Ottawa, Canada, June, 1984, pp. 315 - 329.

7.39 Summers, D.A., and Henry, R.L., Water Jet cutting of Rock with and without mechanical assistance, SPE Preprint 3533, Fall Meeting, 1971.

7.40 Cheng, D., Zou, Z., Guo, C., Zhang, X., Li, G., Jia ,Y., and Xiao, Z., "Major Advantages of Entry Drivage with Swing-Oscillating Jet in a High Methane Concentration Coal Seam," 5th American Water Jet Conference, Toronto, Canada, August, 1989, pp. 307 - 314.

7.41 Summers, D.A., Heincker, W., Eck, R.S., and Raghavan, S.H., Excavation of Coal using a High Pressure Water Jet System, Final Report on Bureau of Mines, Contract HO232064, University of Missouri-Rolla, November, 1974.

7.42 Protod'yakonov, Jr., M.M., "Mechanical Properties and Drillability of Rocks," 5th Symposium on Rock Mechanics, University of Minnesota, C. Fairhurst, ed., 1963, pp. 103 - 118.

7.43 Brook, N., and Misra, B., "A Critical Analysis of the Stamp Mill method of Determining Protod'yakonov Rock strength and the Development of a Method of Determining a Rock Impact Hardness Number," Chapter 8, 12th Symposium on Rock Mechanics, G.B. Clark, ed., UMR, September, 1970, pp. 151 - 165.

7.44 Nie, N.H., Brent, D.H., and Hull, C.H., Statistical Package for the Social Sciences, McGraw Hill, 1970, 343 pp.

7.45 McFeat-Smith, I., and Fowell, R.J., "Correlation of Rock Properties and the Cutting Performance of Tunneling Machines," in Rock Engineering, A Conference at the University of Newcastle, U.K., 1977, pp. 581 - 602.

7.46 Cooley, W.C., "Correlation of Data on Erosion and Breakage of Rock by High Pressure Water Jets," in Dynamic Rock Mechanics, 12th Symposium on Rock Mechanics, G.B. Clark, ed., UMR, November, 1970, pp. 653 - 665.

7.47 Clark, G.B., Haas, C.J., Brown, J.W., and Summers, D.A., Rock Properties Related to Rapid Excavation, Final Report on Contract 3-0143 for the U.S. Dept. of Transportation, UMR, March, 1969.

7.48 Singh, M.M., and Huck, P.J., "Correlation of Rock Properties to Damage Effected by Water Jet," in Dynamic Rock Mechanics, 12th Symposium on Rock Mechanics, G.B. Clark, ed., UMR, November, 1970, pp. 681 - 695.

7.49 Kinoshita, T., Hoshino, K., and Takagi, K., "Rock Breaking with Continuous High Speed Water Jet Stream," paper B2, Proc. 1st ISJCT, Coventry, U.K., April, 1972, BHRA.

7.50 Brace, W.F., "Dependence of Fracture Strength of Rocks on Grain Size," 4th Symposium on Rock Mechanics, 1961, Pennsylvania State University, pp. 99 - 103.

7.51 Summers, D.A., and Peters, J.F., "The Effect of Rock Anisotropy on the Excavation Rate in Barre Granite," paper H5, 2nd International Symposium on Jet Cutting Technology, Cambridge, U.K., April, 1974, pp. H5 49 - H5 62.

7.52 Bur, T., Anisotropic Wave Propagation, Engineering Mechanics Seminar, University of Missouri-Rolla, April, 1971 (and thesis to the Mining Department thereof).

7.53 Agus, M., and Ciccu, R., private discussion with D.A. Summers, Cagliari, July, 1991.

7.54 Summers, D.A., and Henry, R.L., "Waterjet Cutting of Sedimentary Rock," Journal of Petroleum Technology, Vol. 24, No. 1, July, 1972, pp. 797 - 802.

7.55 Bortolussi, A., Yazici, S., and Summers, D.A., "The Use of Waterjets in Cutting Granite," paper E3, 9th International Symposium on Jet Cutting Technology, Sendai, Japan, October, 1988, pp. 239 - 254.

7.56 Agus, M., Bortolussi, A., Ciccu, R., Kim, W.M., and Manca, P.P., "The Influence of Rock Properties on Waterjet Performance," paper 30, 7th American Water Jet Conference, August, 1993, Seattle, WA., pp. 427 - 442.

7.57 Bortolussi, A., Ciccu, R., Mance, P.P., and Massacci, G., "Prospects of Water Jet Technology in Granite Quarrying," 13th World Mining Congress, Stockholm , Sweden, June, 1987, pp. 149 - 159.

7.58 Bortolussi, A., Ciccu, R., Manca, P.P., and Massacci, G., "Granite Quarrying with Water Jets: A Viable Technique?" 5th American Water Jet Conference ,Toronto, Canada, August 29-31, 1989, pp. 49 - 58.

7.59 Calaman, J.J., and Rolseth, H.C., "Technical Advances Expand Use of Jet Piercing Process in Taconite Industry," 5th Symposium on Rock Mechanics, University of Minnesota, C. Fairhurst, ed., 1963, pp. 473 - 498..

7.60 Raether, R.J., Robison, R.G., and Summers, D.A., "Use of High Pressure Water Jets for Cutting Granite," 2nd US Water Jet Conference, Rolla, MO., April, 1983, pp. 203 - 209.

7.61 Zink, G., "The Application of Wire Saws to Underground Excavation" 1974 Rapid Excavation and Tunnelling Conference, AIME, p. 848.

7.62 Anon, "NSF Overview Committee Meets in Elberton," Elberton Graniteer, Spring, 1978, Vol. 22, No. 1, Elberton Granite Association, p. 8.

7.63 Harris, H.D., and Brierley, W.H., "A Rotating Device and Data on its Use for Slotting Sandstone," International Journal of Rock Mechanics and Mining Science, Vol. 11, 1974 pp. 359 - 366.

7.64 Reichman, J.M., and Cheung, J.B., "An Oscillating Waterjet Deep-Kerfing Technique, "International Journal of Rock Mechanics and Mining Science, Vol. 15, 1978 pp. 135 - 144.

7.65 Cann, C., "Machines Tested at Quarries," The Elberton Beacon, December 6, 1978, p. 1.

7.66 Anon, "Water Jet Machine Tested in Elberton," The Elberton Graniteer, Spring, 1979, Vol. 23, No. 1, Elberton Granite Association, pp. 4 - 5.

7.67 Hilaris, J.A., and Bortz, S.A., "Quarrying Granite and Marble using High Pressure Water Jet," paper D3, 5th International Symposium on Jet Cutting Technology, Hanover, FRG, June, 1980, pp. 229 - 236.

7.68 Anon, "Prototype Water Jet Machine Built in Elberton," The Elberton Graniteer, Fall, 1980, Vol. 24, No. 3, Elberton Granite Association, pp. 8 - 9.

7.69 Raether, R.J., conversation with D.A. Summers, August, 1983.

7.70 Roy, G., "Decoupe au jet d'eau en carriere de l'experimentation a la production dans la gres," Le Mausolee, October, 1985, pp. 1596 - 1599.

7.71 Summers, D.A., and Mazurkiewicz, M., "Technical and Technological Considerations in the Carving of Granite Prisms by High Pressure Water Jets," 3rd U.S. Water Jet Conference, Pittsburgh, PA., 1985, pp. 272 - 290.

7.72 Agus, M., Bortolussi, A., Ciccu, R., Manca, P.P., and Massaci, G., "Jet Energy Requirements for driving Deep Kerfs into Eruptive Rocks," Geomechanics 91, Hradec A.M., Czechoslovakia, September, 1991.

7.73 Hawrylewicz, B.M., Remisz, J., Paqueste, N., and Vijay, M.M., "Design and Testing of a Rock Slotter for Mining and Quarrying Applications," 9th International Symposium on Jet Cutting Technology, Sendai, Japan, October, 1988, pp. 377 - 386.

7.74 Miranda, R.M., Lousa, P., Mouraz, Miranda, A.J., and Kim, T.J., "Abrasive Water Jet Cutting of Portuguese Marbles," 7th American Water Jet Conference, Seattle, WA., August, 1993, pp. 443 - 458.

7.75 Anon, "Baby in Well Miracle," Brisbane Sunday Telegraph, October 18, 1987, p. 1.

7.76 Archibald, J.J., "He was at the Cutting Edge of Drill that rescued child," St. Louis Post Dispatch, April, 1988.

7.77 Olmert, P.M., "High Tech to the Rescue," NSF Directions, May - June, 1988.

8 WATERJET DRILLING SYSTEMS

8.1 EARLY DRILLING DEVELOPMENT

In evaluating the ways to improve waterjet performance, it became clear, early in the study of the process, that as the jet cut a deeper and deeper hole, that its efficiency was increasingly reduced by the energy required to overcome the power of the water leaving the hole. In the last chapter two ways were examined for overcoming this problem, one of which was to pulse the waterjet, the other to rotate the jet slightly off-axis [8.1]. However, if when the jet was rotated it was also turned at an angle to the axis of rotation, then the jet would cut a path wide enough for the nozzle to feed into the rock. As the nozzle was moved forward, the waterjet would continue to cut a groove ahead of the pipe and this would continue until the jet had drilled a hole through the rock (Fig. 8.1).

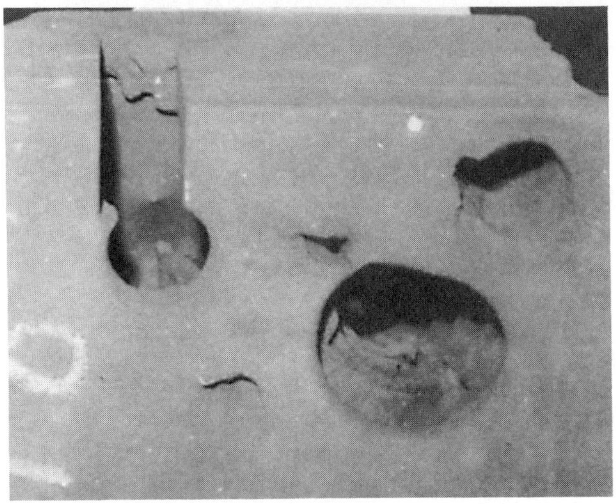

Figure 8.1 First holes "drilled" with a waterjet at the University of Leeds, note that the edge of one hole has been removed to show the central cone left as the jet rotates [8.1].

This initial concept led to the development of a waterjet drilling system. Where a single inclined jet is used, then the jet must be oriented in the nozzle holder so that the jet cuts over the center of the bore. It must also then cut out all the way to the gage of the hole. This hole diameter must exceed the size of the nozzle holder, which should be shaped to allow the spent water and rock debris space to flow out of the hole (Fig. 8.2).

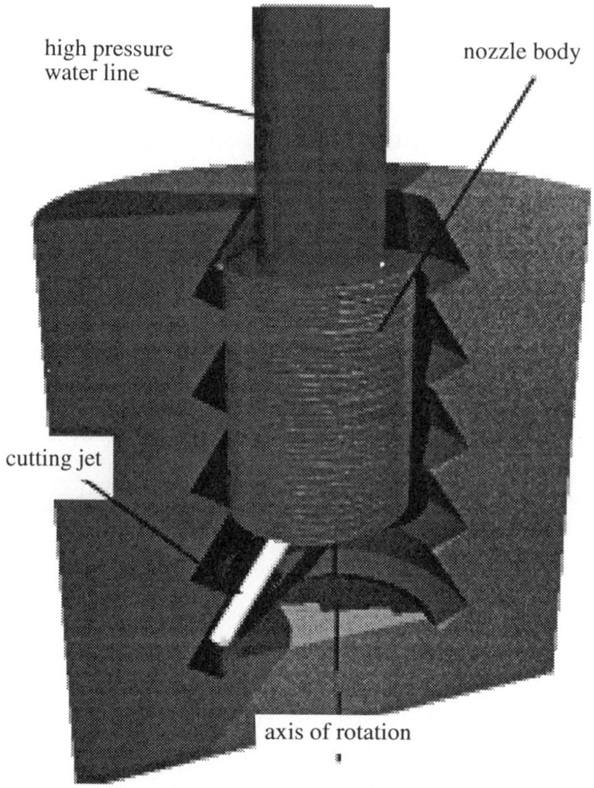

Figure 8.2 Conceptual scheme for the development of hole drilling.

The use of a single jet, however, requires that the jet must be able to cut more than half the diameter of the hole and, as vertical holes deepen and the gap between the nozzle holder and the hole walls reduces, back pressure in the hole may reduce this jet cutting range. The result is to slow the effective speed at which the drill can advance. As an example, drilling rates of around 40 cm/min were achieved in drilling relatively soft material, such as Berea sandstone [8.2]. However, it was discovered early in the 1970's [8.3] that the performance of the jet could be significantly improved by adding a second, axially aligned jet, through the center of the bit (Fig. 8.3). This jet, which could be significantly smaller in size than the sideways pointing jet, had only to erode the rock directly in front of the center of the bit to be effective. By removing this central core, however, it allowed the second orifice to be stepped over toward the side of the hole, reducing the burden which the inclined jet needed to cut.

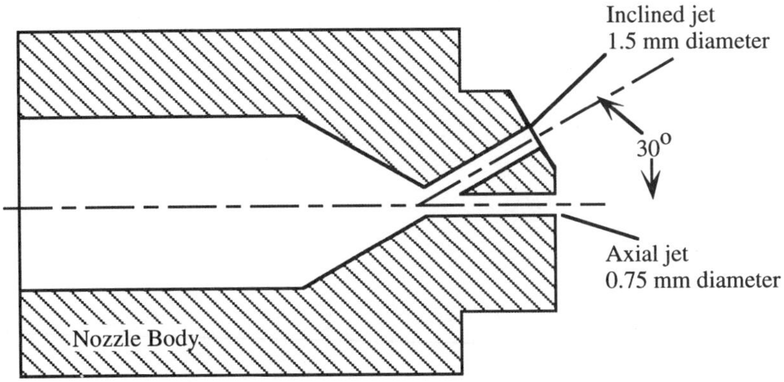

Figure 8.3 Improved nozzle design [8.3].

Following this original design studies were carried out in which the central jet was offset slightly from the axis of the pipe and hole [8.4]. The results reported were not as good as for the original design, and the offset central jet was, therefore, not as effective a drill. This has led to the conclusion that the forward pointing jet has an additional effect, and it is believed there is a pressurizing effect of the jet directed forward into a sealing hole, which improves jet penetration into the rock ahead of the drill and helps induce failure of the cone of rock ahead of the face [8.5].

This combined drill design increased the speed at which soft rock could be drilled by two orders of magnitude. The speed at which the hole was drilled was, however, a function of the diameter of the hole required. As the advance rate was reduced, for equivalent other conditions, so the hole diameter was increased, and vice versa (Fig. 8.4).

Where the hole diameters are small, on the order of 5 cm. diameter in most rocks, but in soft materials such as coal or sandstone perhaps as large as 20 cm. in diameter, then waterjets alone or abrasive-laden waterjets can be very effective in drilling holes through material.

This is particularly true where the waterjet hole must be drilled in a straight line over a very long distance or at some angle to the initial target surface. Because there is not cutting contact between the mechanical parts of the drill and the rock it has been possible for example, to use brass nozzles to drill a 6 m hole in extremely abrasive sandstone with little wear [8.6]. Because there is no thrust through the drill steel to deflect it, the holes are drilled straight, particularly if a guide vane is used directly behind the jet contact plane (Fig. 8.5). Use of such a gage, tied to the feed circuit to control advance, can ensure that the jets will drill a straight hole of the required diameter. It should be stressed that it is important to gage the hole which is being drilled. If this is not done, and the hole is drilled in other than a vertical position, then the hole created may be significantly larger than the nozzle assembly at the head of the drill. In this case, as the drill advances, gravity will pull the head down to rest against the bottom of the hole. The result will be that the hole slowly inclines downward, but at an accelerating rate. This needs to be compensated for in the design of the drill head.

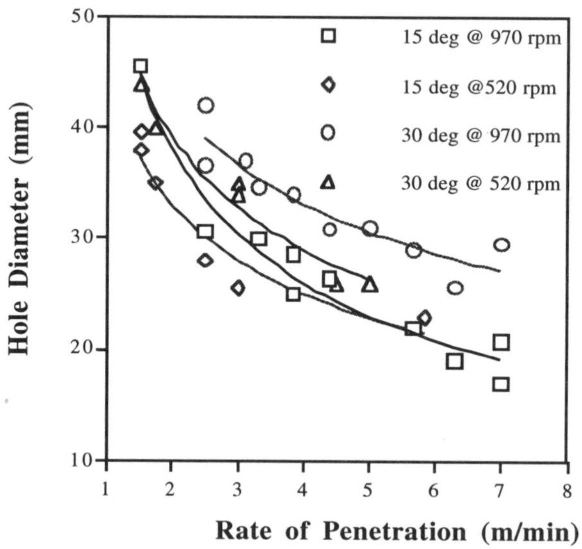

Figure 8.4 Rate of penetration (ROP) as a function of hole diameter.

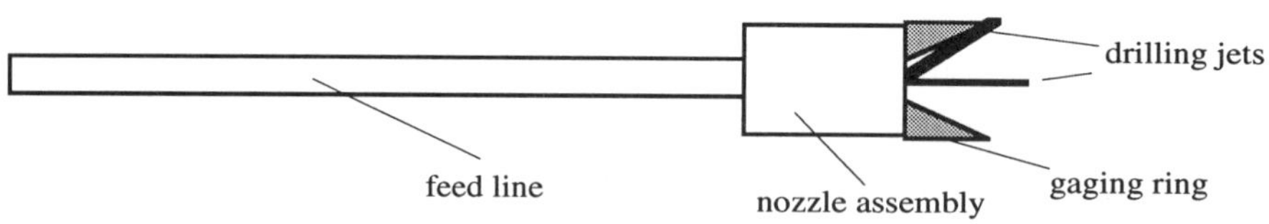

Figure 8.5 Use of a gaging ring to hold hole diameter in waterjet drilling.

It was in pursuit of a simple, waterjet based drill, as a part of contract for ERDA, that drilling tests were carried out in the early 70s [8.7]. During these trials it was possible to drill holes of approximately 25 mm. diameter in an abrasive sandstone at rates of up to 6 m/min (Fig. 8.4).

It is somewhat difficult to isolate the threads of the developing technology in rock drilling at that particular time in its development since several different ideas, many building from earlier work, were evolving at the same time. Certain of the ideas developed can, however, be grouped by concept, the material being drilled or by the purpose of the activity. Such arbitrary divisions have been used in the following sections to provide some organization to developments which were often contemporaneous and to some extent inter-related.

8.2 WATERJET DRILLING OF ROCK BOLT HOLES

The need for a high speed, small diameter drill arises, in part, from the requirement, in underground mining and tunneling operations, for a system which will hold the rock above the working area from falling on those working and traveling within that space. To ensure that the rock is held in place, long metal rods, known as **rock bolts**, are inserted into holes which are specially drilled into the rock to hold them. One effective form of these steel rock bolts is thrust into the hole through a series of plastic bags holding a mixture of either a plastic resin or concrete. This is mixed and pushed against the walls of the hole and the bolts, as the rod passes through the bag. As this binder then sets hard it acts to grip both the rod and the surrounding rock wall, holding them together and restraining the rock from moving. The effectiveness of these bolts is controlled, in part by the diameter of the hole, since the larger the rod inserted, then the stronger the anchorage will be. However, the larger the hole diameter the more expensive the device becomes, since more steel and binder must be inserted into the hole.

Resin anchored bolts are often much stronger than those set in concrete, and can be used in smaller holes. They would, for example, be economically effective at hole diameters of between 20 and 25 mm. It is, however, quite difficult to drill a straight hole at high speeds, where the hole must be more than 2 m long and 25 mm or less in diameter. This is because conventional mechanical drills require a high level of thrust to penetrate the rock, and the small size of the drill rod which carries the load to the bit, has a tendency to buckle under the high thrust loads required in order to drill at such a speed. Bonge and Ozdemir [8.8], for example, have suggested that the maximum penetration rate for a 28 mm diameter bit on a 25 mm steel is 2.5 times faster than for a 22 mm diameter bit on a 19 mm steel.

Concurrently, the borehole wall which is produced by a conventional mechanical rotary or percussive drilling bit is relatively smooth, reducing the ability of the resin to grip the wall and thus lowering the effectiveness of the bolt. In contrast it has been shown that if a rough hole could be created, then the wall anchorage would increase the effective bolt strength so that it would be equivalent to that of a bolt in a hole approximately 50% larger [8.9]. The results of early tests showed that, with care, a programmed roughness could be created along the wall of the borehole drilled by high pressure waterjets (Fig. 8.5). This indicated that there was an overall performance improvement to be achieved from using a waterjet drilled hole. In addition, since there is a little thrust from the jets back onto the steel it is easy to move the drill forward and keep it going straight regardless of the speed of penetration since there is no contact with the rock. Given also (Fig. 8.4) that the faster the rock was drilled the smaller the hole, and thus, the less steel and resin would be required, there were considerable economic benefits to be achieved, were a high speed waterjet drill to be developed.

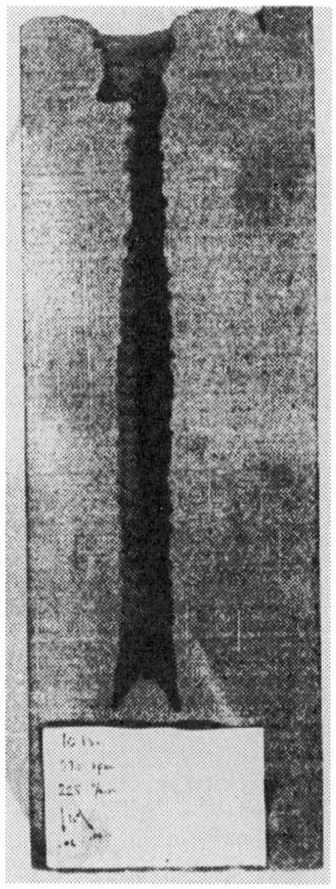

Figure 8.6 Section through a waterjet hole drilled in Berea sandstone.

The programmed roughness of the hole is created due to an additional factor, which should be considered at this time. This is the **incremental distance** advanced by the drill as it moves forward during each revolution of the head. It is often inadequately understood that the waterjet cuts sufficiently far ahead of the nozzle, that there is no mechanical contact between the drill and the rock until the cutting zone is passed. Therefore, it is possible for a single, inclined, rotating waterjet to cut a single slot ahead of the bit but with a width only three times that of the jet. The hole drilled may be cut wide enough where the slot has been created to allow the nozzle body to move forward, however, ribs of material may be left between the adjacent slots cut by the jet as the nozzle advances, which are of sufficient size that the nozzle body cannot pass through them (Fig. 8.7).

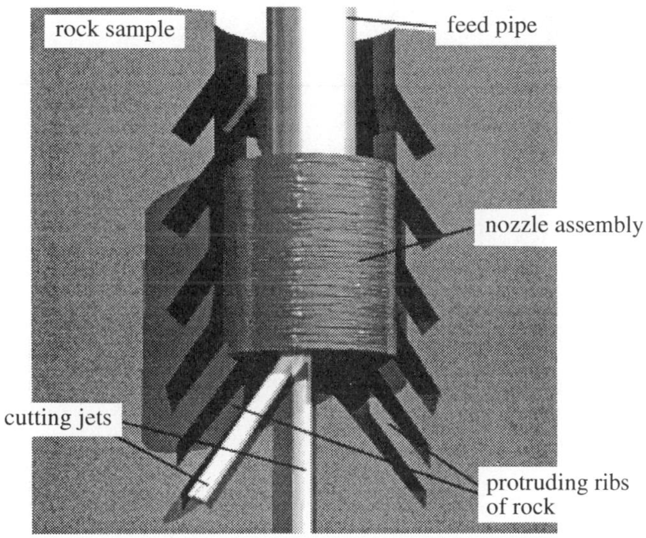

Figure 8.7 Ribs left behind the jet cutting plane interfering with the penetration.

Further, since the waterjet is aimed forward from the nozzle so that it is cutting ahead of the edges of the holder, should this come into contact with a protruding rock ledge, such as the two ribs shown, the waterjets are already cutting beyond this blockage and cannot cut the material. In order for the drill to cut the rib and clear the passage, it must be retracted back up the hole. This stop and start, back and forth motion can create a very poor impression of the capabilities of a waterjet drill. To guard against this it is considered advisable that the rate of penetration (ROP) of the waterjet drill be not very much greater than the width of the waterjet cutting zone (3 diameters) on each rotation. Thus, a drill with a 1.5 mm inclined jet rotated at 1,000 rpm should not be advanced at much more than 4.5 x 1,000 = 4,500 mm/min or 4.5 m/min. (Note that such a drill could be have an ROP of up to 6 m/min in a soft sandstone as shown in Fig. 8.4. A slower ROP would likely be required in a stronger rock). The benefits of this control of the incremental distance include that the jets can produce a greater cutting depth as successive jet paths move closer together (Fig. 8.8). The data was obtained from use of a 0.58 mm diameter jet cutting into the rock at a pressure of 700 bar.

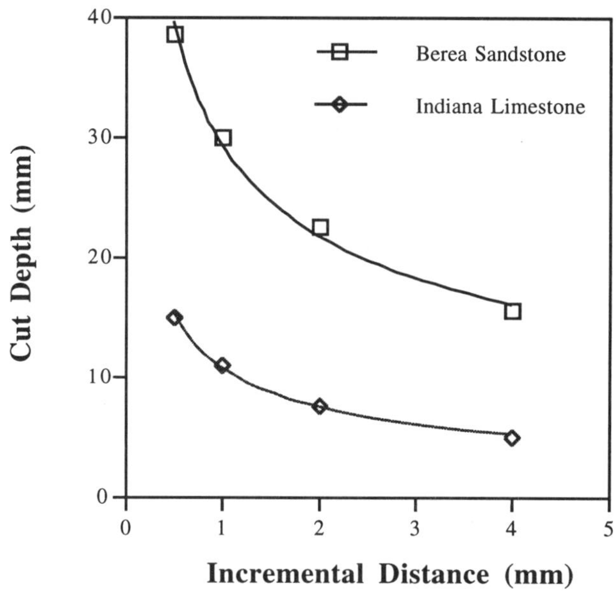

Figure 8.8 The effect of incremental distance on depth of cut.

The development and subsequent demonstration, at UMR, that waterjets could be used to drill roof bolt holes, led to two contracts being issued by the U.S. Bureau of Mines [8.10, 8.11]. Both developments made reference to the work in Missouri, but as the technology was evolved by the investigators at Colorado School of Mines (CSM) and at Flow Research (FRI), different solutions to the problems to be faced in developing a practical rock drill were evolved. The major problem which had to be overcome lay in the need to be able to drill as wide a range of rock types as possible. The original development had been at jet pressures of 700 bar and could effectively drill only a limited range of rock. Two different ways of extending this range were developed by the two organizations.

At CSM the direction taken addressed both the problem of removing the intervening ribs of rock between the jet passes, and that of drilling harder rocks [8.8]. The solution proposed was to create a hybrid bit which combined both the benefits of high pressure waterjets and the cutting ability of mechanical bits. Considerable effort was made in optimizing the design to locate the jets relative to the ribs of rock which would be carved ahead of the mechanical face of the bit. However, interestingly, the jet action was not combined with that of the bit in a synergistic manner, as had, by that time, been developed by Dr. Hood (see Chapter 9). Rather the jets were used to create slots in the rock, which were then sequentially removed by the mechanical breaking action of the carbide insert on the bit (Fig. 8.9).

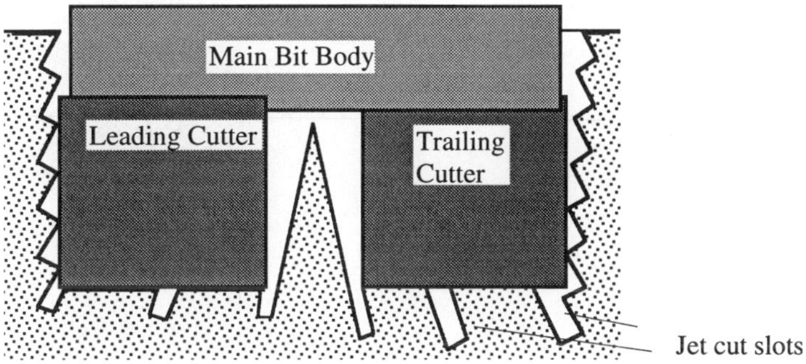

Figure 8.9 Pattern of cutting with the CSM rock drill [8.8].

A drill was built, based on this concept, capable of drilling holes some 1.8 m long. The total power required for the unit was some 40 kW with three jets operating at pressures of up to 4,200 bar. In the comparative tests carried out to prove the equipment a test series of three rocks was assembled, with the rocks stacked one above the other. The suite was made up of a medium strength, fine to medium grained sandstone, a weak, poorly cemented sandstone, and a mudstone. Typically the waterjet drill penetrated the two sandstones at twice the rate of a conventional drill, and penetrated the mudstone four times as fast as a conventional drill. It was, in addition, found that hole diameter remained constant over the life of the bit, and that holes were drilled within an accuracy of 1.6%.

Testing of the initial system developed were thus very encouraging and an improved, and mobile unit was manufactured with assistance from the U.S. Bureau of Mines. This was tested in some of the hardest rocks found in American coal mines, Geneva sandstone and shale from a U.S. Steel mine near East Carbon, Utah. During typical mining operations in that installation a mechanical bit would require a thrust of between 1,300 and 2,300 kg and would wear out after drilling only between 0.5 and 1.0 m of hole length. At a pressure of 2,800 bar, with a thrust of 360 kg the CSM drill penetrated some 10 m between bit regrinds and achieved average penetration rates of 0.3 m/min in the sandstone and 0.96 m/min in the shale. In the final design of the bit the jets issued from a central well in the bit, being directed out to cut ahead of the carbide inserts, which were set some distance ahead of the nozzle orifices (Fig. 8.10).

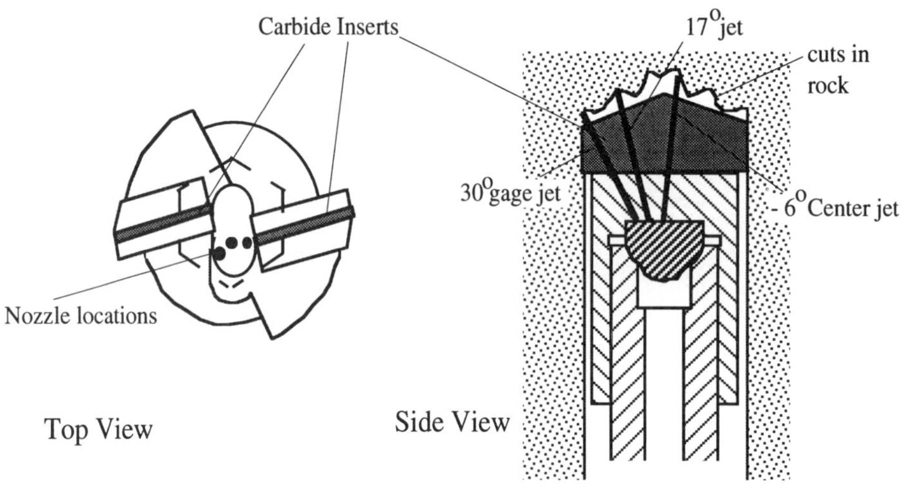

Figure 8.10 Optimized CSM bit design [8.8].

It is interesting to note, in reviewing the work done at CSM, that the study ended with tests of the diamond Stratapax bits with waterjet assist. These bits lasted over 90 m before needing replacement when assisted by the 2,800 bar waterjets. As mentioned earlier, conventional mechanical bits are replaced after approximately 1 m of drilling. An economic analysis as part of the contract final report suggested that the use of conventional carbide assisted waterjet drilling would lower bolting costs by 13 - 18%. When the jets were combined with a Stratapax system this was anticipated to lower costs by up to 30 - 44%. While an underground fieldable system was then constructed, the program fell into abeyance shortly after that time due to the lack of an available underground test site [8.8].

Some of those who worked on the development of the CSM drill went on to form the waterjet equipment company Stone Age, working out of Durango, CO. This company developed and has used small hand-held waterjet drills for both drilling relatively soft rock, and for use with borehole modification techniques for enhanced excavation [8.12] (see Section 8.8 below).

It is pertinent to include a comment at this time of subsequent work on this type of design. As part of a combined effort under funding from NSF through MIT, a project was carried out at UMR in 1993, to revisit the benefits of adding high pressure waterjets to Stratapax cutters [8.13]. The results of this study were quite promising and a small subsidiary study carried out for the U.S. Bureau of Mines indicated that with proper jet orientation penetration rates of the diamond tool could be increased by over 300% when waterjets, at a pressure of 420 bar were added to the cutting head [8.14]. That study was still in progress at the time this text was being prepared. Additional discussions on the combination of waterjets and mechanical bits will occur later in this chapter and in the following one.

The concurrent development of an alternate drilling strategy to that of the CSM drill took place at Flow Research, Inc. [8.15]. The evolution of this concept has been less thoroughly described in the available literature, due to its rapid development into a commercial system. The early work at lower pressures was, as discussed above, insufficient to achieve acceptable drilling rates in harder rock and, as a result the system was modified to operate at pressures of up to 4,200 bar. The initial tests of the system took place in a coal mine in Colorado. As with the testing of the CSM drill, costs were compared with conventional rock drilling for placing resin grouted roof bolts in the overlying rock over the tunnels. It was found possible to achieve bolting rates of up to 30 1.5 m long bolts in the roof each hour, which was some 50% greater productivity than that achieved using a conventional mechanical drilling system. In addition, the smaller holes drilled meant that only 60% of the resin was required to fill the holes, providing an additional saving [8.16].

A full scale bolting machine was ultimately developed and was tested by Jarvis-Clark in a mine in Kentucky to assess equipment reliability and system maintainability. One of the early problems with the system was in the requirement for very clean water to operate pumping equipment at this high pressure. The system was subsequently marketed by Jarvis Clark, after this initial debugging trial [8.17] and [8.18] but with changes in corporations dealing with the tool it is not currently available.

8.3 CANADIAN WATERJET DRILL BIT DESIGN

Staff at the National Research Council of Canada began an examination of water jet drills in 1979 [8.19] and their work was reported in a series of papers [8.20, 8.21, 8.22 for example]. The nozzle design chosen was somewhat different from that of the earlier Rolla work (Fig. 8.3) in that, where a central axial jet was used in the UMR design, with the Canadian design this central jet was inclined to the axis of the hole (Fig. 8.11). Given that there has been some discussion which has suggested that these two designs were similar, it is important to make this rather critical distinction.

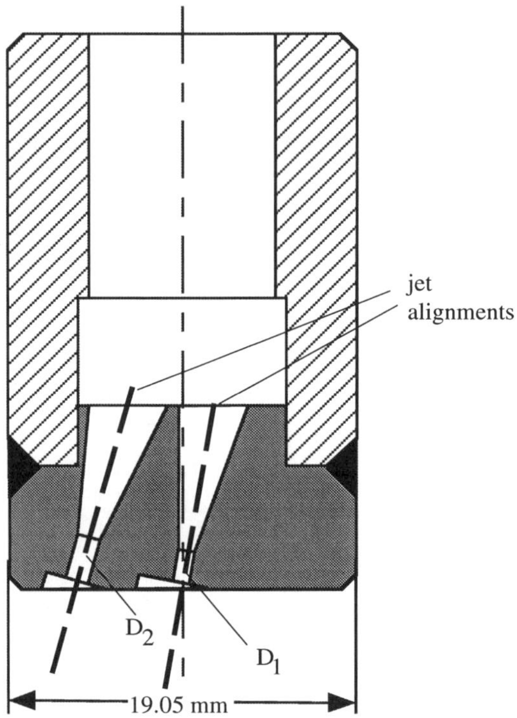

Figure 8.11 Canadian drilling nozzle design [8.21].

The use of inclined nozzles will produce the same type of erosion or rock removal process under each jet, rather than including the internal pressurization which occurs with the UMR design. This becomes critical to the consideration of location and performance of the outer jet orifice. For example, where, as in the above figure, the outer orifice is placed approximately half-way from the center to the perimeter of the nozzle holder, the volume which must be removed by the outer jet is at least three times that which must be removed by the center jet. The actual amount is somewhat greater, since it is necessary to cut some additional distance to allow clearance for the nozzle passage.

Given that the jet which is traversed around the outer perimeter of the hole is traversed much faster over the rock surface at this diameter, this argues further for a much greater flow and larger diameter for the outer jet. From that point of view it is unfortunate that in the test data published only one set of nozzles was tested in which the outer jet was as large as approximately twice the size of the inner jet. In this case the jets were only inclined outward at an angle of 10 degrees, a shallower inclination than that recommended from the UMR study.

This particular design feature is to be regretted since experiments at UMR have shown that, at this shallow an angle the jets tend to incline inwards and do not always drill a straight hole forward [8.23]. As a result the conclusions drawn from the first Canadian study on nozzle design, that low rotational speeds (around 200 rpm) are optimum; that the optimum nozzle ratio was where the outer jet carried 60% of the flow; and that the outer jet should be inclined at 20 degrees to the hole axis are all subject to some question.

The problems generated in deriving a simple analysis from the data can be overcome by treating some of the data as being derived from a factorial design and re-analyzing it in a simplified format. Dr. Vijay considered two dependent variables, the speed of the drill advance and the diameter of the hole created. A third can be derived as the product of the two in the form of the volume of rock removed per minute. While three factors, diameter ratio of the two nozzles, rpm, and outer jet angle were considered, only the latter two were tested in sufficient quantity to evaluate in this manner.

Simple plots can be made showing the effect of change in nozzle angle on speed of advance (Fig. 8.12), hole diameter (Fig. 8.13), and volume rate of rock removal (Fig. 8.14). Examination of these curves suggests that the optimum angle for highest rate of advance lies at around 19 degrees, that there is no optimum in terms of hole diameter, while the optimum angle for volume rate of rock removal lies at around 28 degrees.

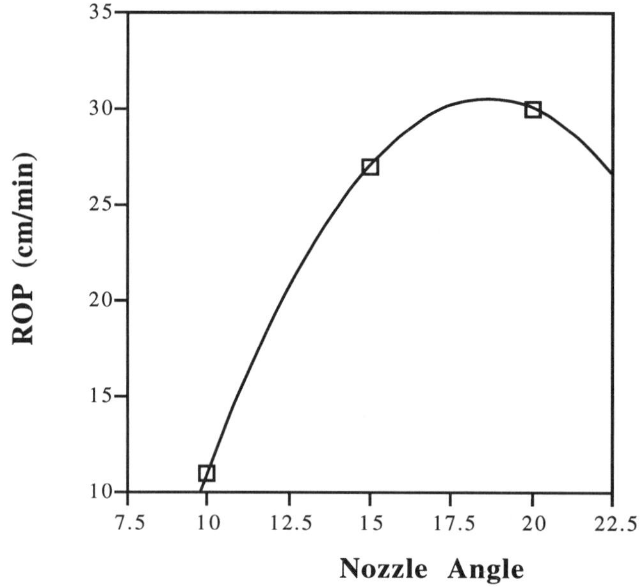

Figure 8.12 Effect of outer nozzle angle on ROP (after [8.19]).

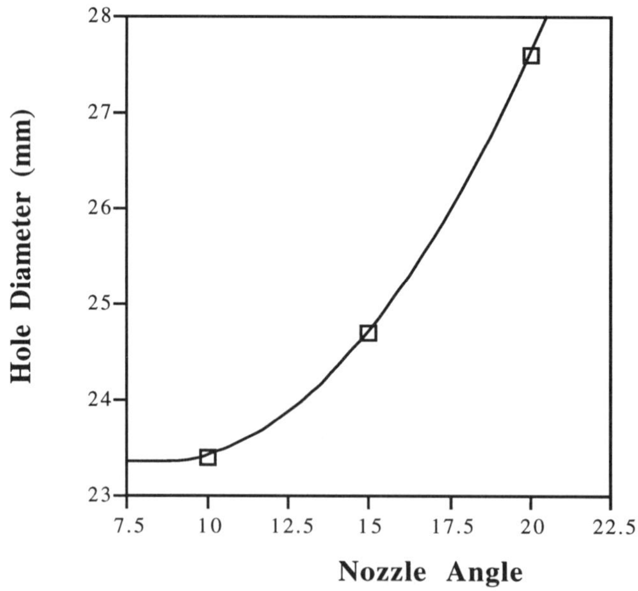

Figure 8.13 Effect of outer nozzle angle on hole diameter (after [8.19]).

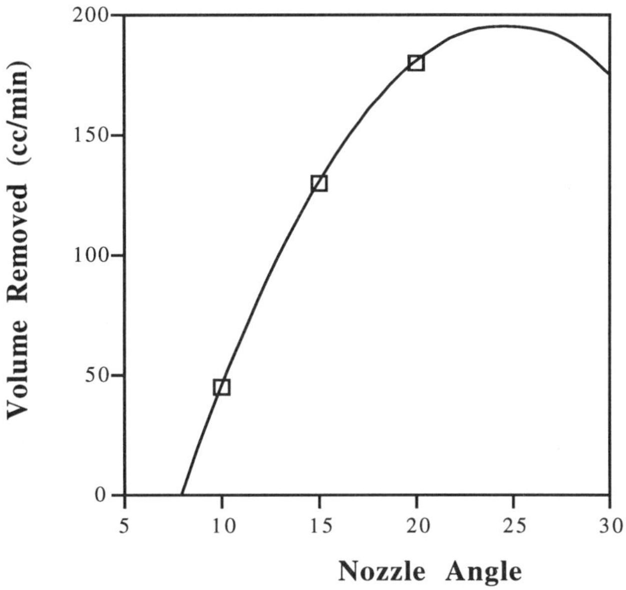

Figure 8.14 Effect of outer nozzle angle on volume removal rate (after [8.19]).

The problem with this simplified analysis is that it fails to recognize that there are two different actions going on when the jets drill the hole. The first of these is that, as a control on the advance rate both the jets must cut sufficient clearance for the head to advance. Thus, not only must the outer jet cut beyond the required diameter of the nozzle body, but the inner jet (which is at a fixed average diameter and a constant angle for these tests) must also cut clearance not only for itself but also for the inner part of the outer half of the outer jet. Increasing the angle of the outer jet thus has less effect on advance rate, while the control on advance comes from the inner nozzle. However, changing the angle of the outer jet will have some effect on the performance of the inner jet. This is because the effort to turn the outer jet to a greater angle will force more water through the shallower angle central jet and thus allow it to cut more effectively.

Once the central jet is sized correctly to cut the central cone of material out to the path of the outer jet, then the outer jet effectiveness is gaged by the hole diameter being cut. Thus, where the outer diameter is larger than the minimum required, this suggests that, for optimum advance rate, less energy need be directed at this outer jet, and more to the central jet. The two jets must, therefore, be tuned to work together, depending on the hole diameter required, to achieve optimum drilling performance.

In contrast, where the effect of rotational speed is concerned this is less difficult to optimize. For although there is still some disagreement between the curves showing the dependence of ROP (Fig. 8.15), hole diameter (Fig. 8.16), and volume removal rate (Fig. 8.17) on rate of rotation (rpm) the variation in the optimum value predicted from the data varies only from 240 to 280 rpm.

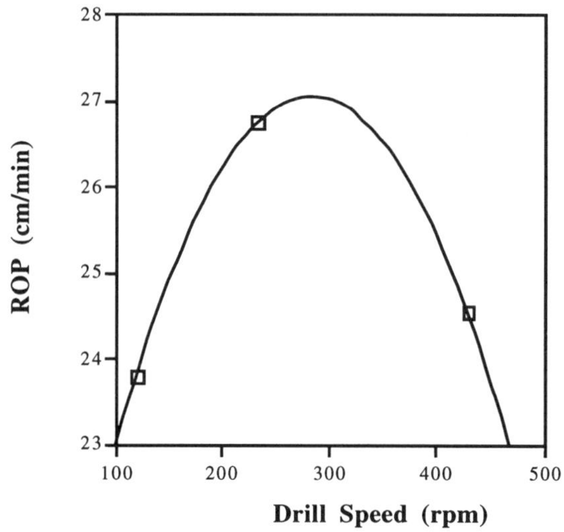

Figure 8.15 Effect of rotation speed on ROP (after [8.19]).

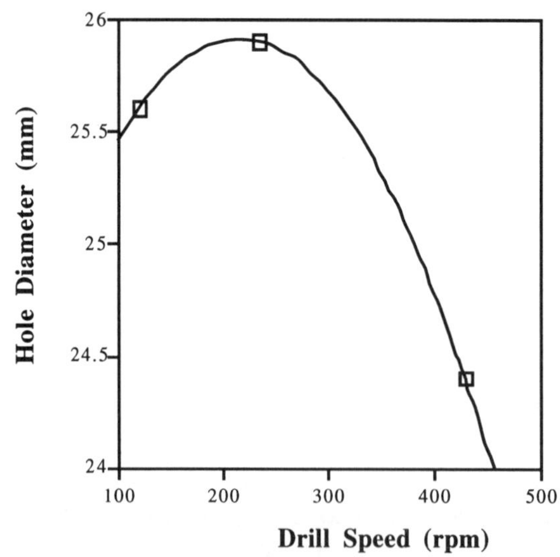

Figure 8.16 Effect of rotation speed on hole diameter (after [8.19]).

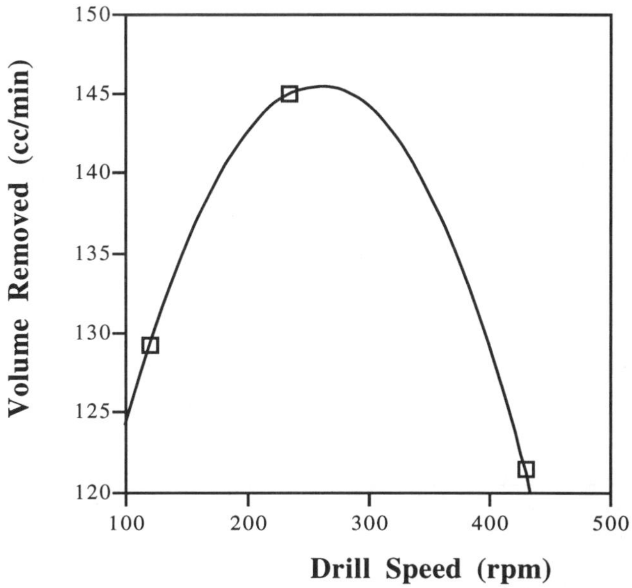

Figure 8.17 Effect of rotation speed on volume removal rate (after [8.19]).

Understanding the inter-relationship between the actions of the two jets, yet recognizing their different roles in the drilling and rock removal process is critical to understanding the meaning of the data. It serves to explain, for example, the difficulty in analysis of the data on the effect of change in the relative diameter of the two jets on drill performance. After the initial paper had been presented Vijay et al. continued the study at greater angles for the outer jet [8.21]. This study included tests which increased the angle of the outer jet to 30 degrees, while continuing to hold the angle of the inner jet at 10 degrees, with the same relative position of the two orifices. The results indicated an optimum performance, for the rock drilled in the earlier work, at an outer jet angle of 24 degrees.

There is, however, some change in the optimum conditions for the nozzle design where rocks which fail in a different mode were drilled (8.21). In particular where the rock drilled was an Ottawa limestone the results reported are somewhat less convergent. When based on the effective diameter of the hole the optimum results obtained by factorial analysis of the data indicate an optimum value for the outer angle of 23 degrees, very similar to that found for drilling the granite. In contrast, however, the optimum value increases to between 27.5 and 30 degrees when one considers the effect on volume of rock removed, and advance rate. This variation is strongly controlled by varying the flow of the water to the outer jets from a nominal 60% to a nominal 70% of the flow. At a 60% flow of water to the outer jet there is no optimum found for the data, which indicates that better performance could be achieved at an even greater nozzle angle than those tried in the program.

Interestingly, however, the data obtained uniformly shows that, in drilling this rock, a rotational speed below 120 rpm would be more effective, in contrast with the more than double that value which was the optimum speed for drilling granite. It should again be pointed out that this data has been re-analyzed by the current author, using a factorial summation of data, and this has led to the slightly different conclusions drawn from the data than those reported by the original authors.

In light of the comments referred to in Chapter 7 by Ciccu et al. on the effect of grain size on jet performance [8.24], it should be noted that the grain size of the granite was 1.98 mm on average, while that of the limestone was 1.5 mm, and the jet diameters varied from 0.91 to 1.4 mm. While the Canadian team continued to develop this tool and reported on further work in 1984 [8.22], the emphasis of the work was changed to examining the slotting of rock, rather than drilling, and this has been discussed both above, and, in its further development, in the section of Chapter 7 dealing with abrasive injection.

8.4 WATERJET ASSISTED PERCUSSIVE DRILLING

Japanese investigations of the use of high pressure waterjets predated the first international waterjet symposium, at which a paper [8.25] was presented on water jet assistance to tunnel boring machines. This subject will be dealt with in more detail in a subsequent chapter, but because the jets improved mechanical cutter performance from 2 to 5 times, it led the investigative team to examine waterjet use in drilling. A series of three papers [8.26, 8.27, and 8.28] in the three successive BHRA symposia describe the development of this tool. In contrast with other developers who had investigated the addition of waterjets to rotary drilling tools, the Japanese team chose to integrate high pressure water into the operation of a pneumatic drill, largely for use in developing railroad tunnels.

A series of laboratory experiments was first carried out [8.26] in which the relative volume broken from the face of a drill hole by the impact of a mechanical bit was established (Fig. 8.18). In order to simulate the action of the pneumatic drill, the bits were dropped with a known energy onto rock which was either un-notched or pre-cut by a waterjet with a 6.7 mm deep slot, initially notched into the rock at the perimeter of the hole.

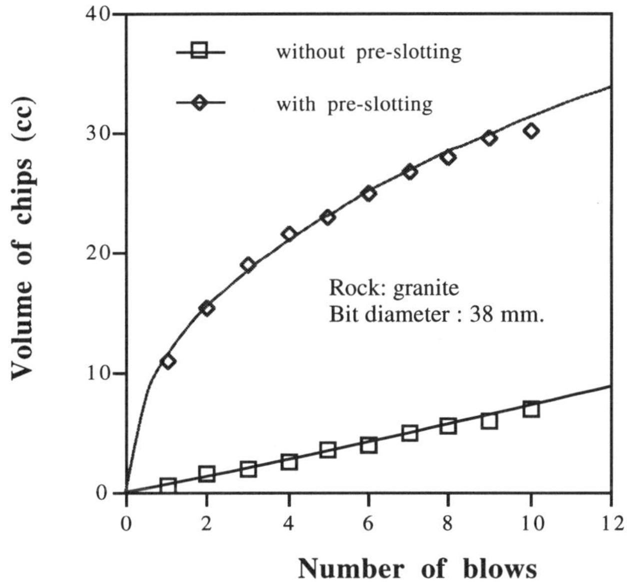

Figure 8.18 Effect of pre-slotting a drill hole on the rock removed by an impacting bit [8.26].

The test data clearly showed that slotting the rock face improved performance, and when the area of the cut (i.e., the product of slot length and depth) was varied, a linear relationship was found between the improved volume of rock removed, and this area of pre-slotting (Fig. 8.19). These cuts were made in holes with the slot diameter varying between 30 and 40 mm and with the depth of the pre-cut varying from 3.8 to 13 mm. Again the energy input per blow was 6 kg/m and the tests were carried out in a borehole in granite.

The experiments were again carried out using a 38 mm diameter drill bit, and such was their success that a pneumatic drill was modified to accept a high pressure waterjet feed. Jet pressures of up to 4,000 bar were used with two orifices on the carbide bit (Fig. 8.20). Three factors were examined in the first series of tests of this new design, the pressure of the jet, the rotational speed of the bit, and the blow frequency of the impact hammer. Rates of penetration between 2 and 5 times that obtained without jet assistance were reached. The results indicated an increasing improvement in drilling performance with jet pressure

(Fig. 8.21), that there was some modest improvement in ROP with increased rotation speed of the bit, with the best value found at 250 rpm which was the highest rpm tried. At this speed the outer edge of the nozzle was traversing over the rock at a speed of 50 cm/sec. It was concluded that varying the frequency of the hammer cycle did not significantly influence the drilling rate of the tool.

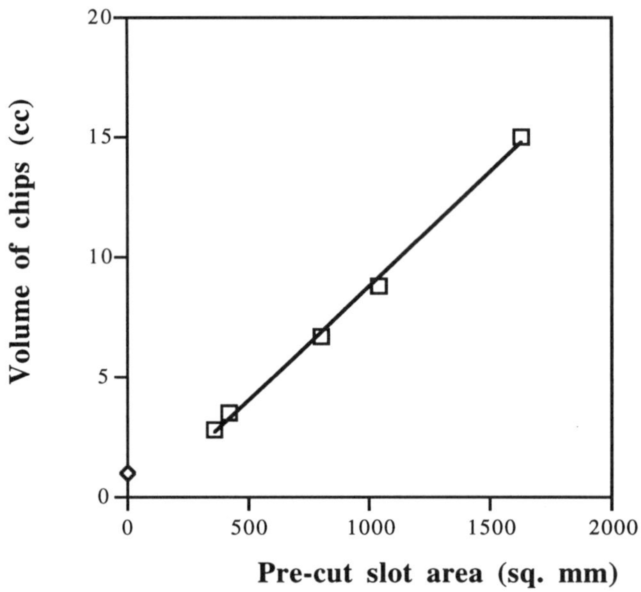

Figure 8.19 Volume of rock removed by a single blow of an impacting bit as a function of slot area [8.26].

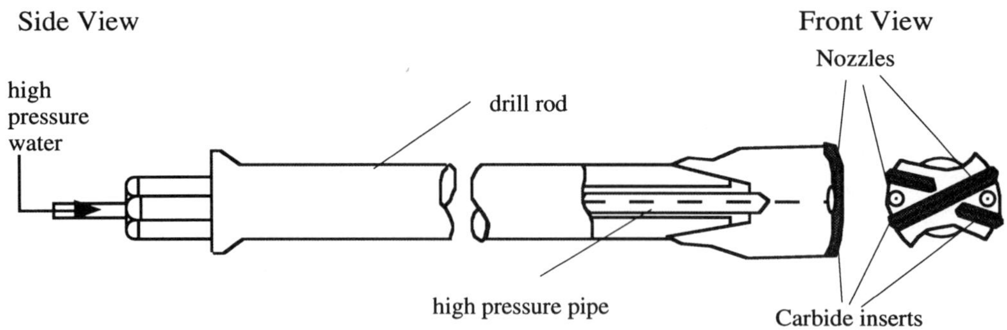

Figure 8.20 Location of the jet nozzles on the Japanese waterjet assisted drill bit [8.26].

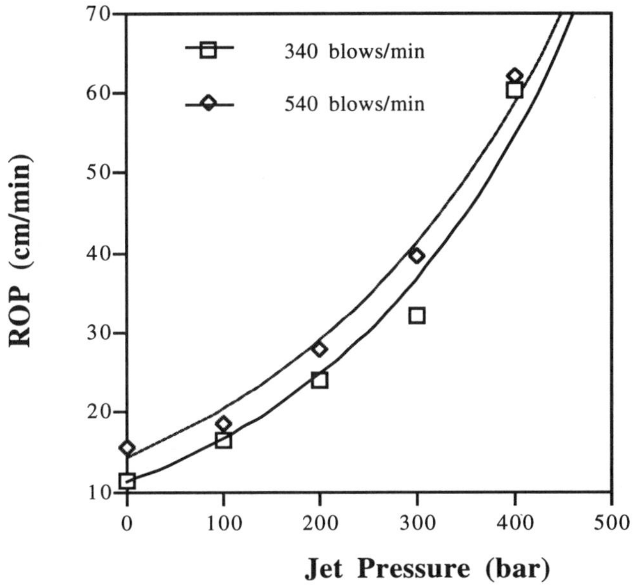

Figure 8.21 Effect of jet pressure on percussive drill ROP, at 160 rpm [8.26].

The results of the trials of the first machine were sufficiently promising that a second generation machine was constructed, and its development described at the third symposium [8.27]. Such a development was necessary since the initial equipment used in the earlier experiments had been too complex for practical use in a geotechnical environment.

An immediate problem arose in that such a drill would need a swivel to allow the high pressure lance to rotate within the drill rod. However, the swivel available required increasing torque as jet pressure increased. Higher torque would reduce the power available through the drill to rotate the drill rod and drive it into the rock. As a compromise, therefore, a lower jet pressure was set when developing the technique for use with a pneumatic jack-leg unit (Fig. 8.22). It is interesting to note that two alternative designs were considered for the location of the swivel. The first of these, and that ultimately chosen, was to feed the water into a swivel attached to the drill rod itself. An alternative, in which the swivel was spring loaded to the back of the drill was not pursued. A second unit, operated with a separate drive for rotation did not have such a limitation on pressure and rotary power, and thus was also built. This unit also fed the high pressure water through a swivel on the drill rod, and was designed to operate at up to four times the jet pressure of the first unit (4,000 bar) while retaining the same 2,500 blows/min performance of the jack leg drill at a rotational speed of up to 250 rpm.

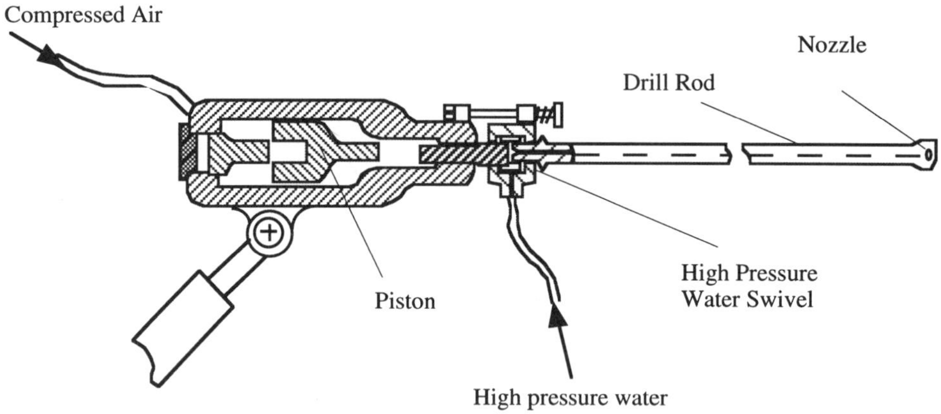

Figure 8.22 Location of the high pressure water swivel on a jack leg drill with the first Japanese unit [8.27].

When the drills were taken to the field site at the entrance to the Kinshoji tunnel there was some concern with the novelty of the tool and the lack of experience of the workers. Thus, the jet pressure was lowered to 600 bar at the start of the program, with a flow rate of some 150 lpm. It was, however, subsequently raised back to the 4,000 bar with a flow rate of up to 180 lpm. Early test results with the jack leg drill showed that the tool required a change in the swivel design to allow freer rotation. Even at low pressure the penetration rates for the drill were increased by between 10% and 50% and the improvement was seen to increase when the jet pressure was raised higher. Although there was considerable scatter in the data it also appeared that using two jets directed at the perimeter of the hole gave better results than directing a single jet through the center of the bit.

In the final paper reporting on this work in 1978 [8.28] a small single stage intensifier had been developed and attached to the back of the drill, in order to conserve space within the limited confines of a tunnel. The swivel was also moved to the face of the intensifier, and before the rod fed through the drifter section of the drill (Fig. 8.23). As the pressure of the jets issuing through the single 0.35 mm nozzle was increased from 1,000 to 3,500 bar the ROP of the unit, when drilling through granite, was increased from just over 30 cm/min to nearly 70 cm/min. Although this was a promising development, its future, since that time, has not been reported outside of Japan.

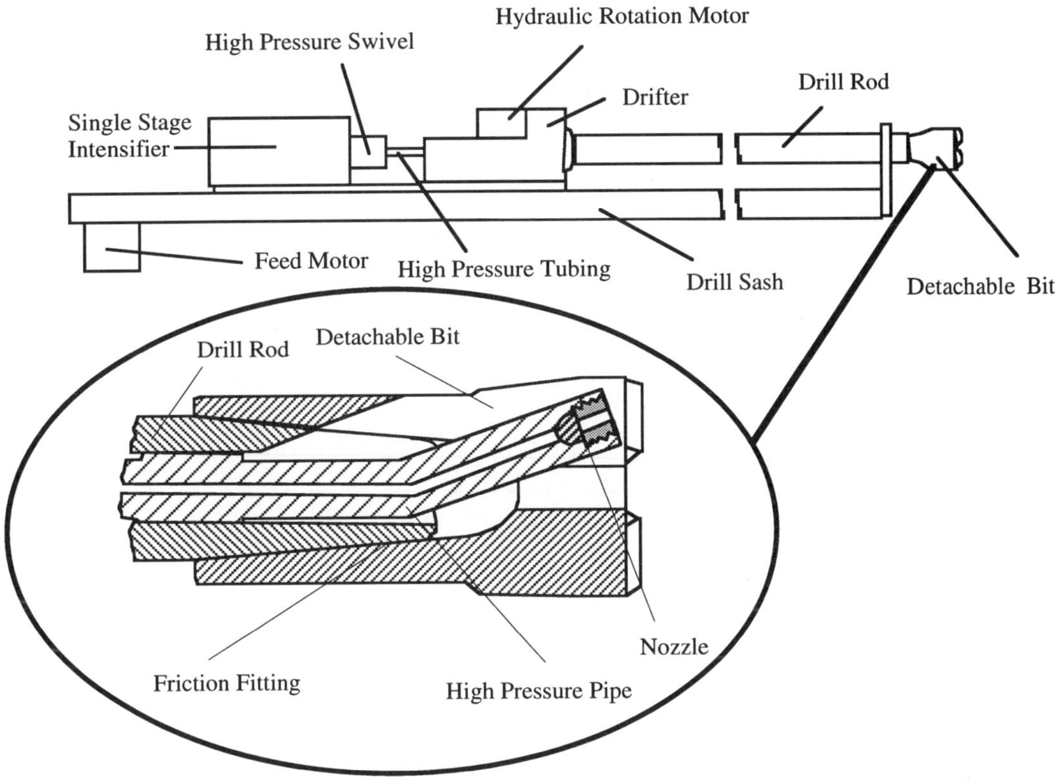

Figure 8.23 Schematic of the final Japanese waterjet assisted drifter design [8.28].

A similar study was, however, undertaken by CERCHAR in France, and reported in 1984 [8.29]. The intention of the study was not only to increase the performance of the rotary drill in rock, but also to extend its range of application into harder and more abrasive rock than the drill would otherwise be able to economically penetrate. The work resulted in a new patent for an improved bit design. This arose from a series of comparative tests examining the effect of different nozzle positions on the performance of the drill. These can be exemplified from the simplified curves provided in the paper showing the effect of jet position on penetration rate (Fig. 8.24) and on the relative life of the bits (Fig. 8.25). It should be noted that the first curve was obtained by using the drill to create horizontal blast holes at a diameter of 37 mm, while the second curve was derived when the drill was used to drill vertically upward holes for installing rock bolts, at a diameter of 22 mm.

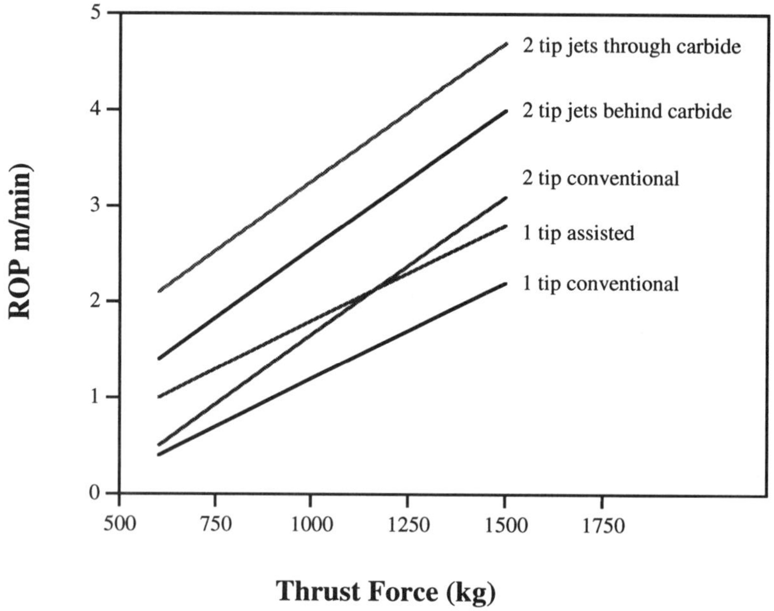

Figure 8.24 Relative performance of differently designed waterjet assisted mechanical bits. Bit diameter 37 mm, at 600 rpm and a jet pressure of 1,500 bar (after [8.29]).

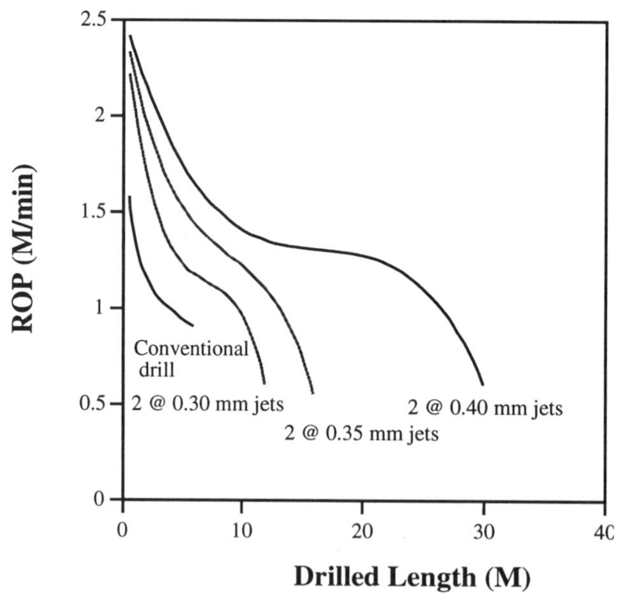

Figure 8.25 Effect of differently designed waterjet assisted mechanical bits on bit life (after [8.29]).

The results of the study indicated that the addition of the waterjets to the bit allowed either an increase in speed at the same thrust or a reduction in thrust at the same speed to prolong bit life. At the same time the optimum rotational speed of the bit could be increased, thereby reducing the bite per revolution of the bit while maintaining a constant rate of penetration. This again had the benefit of increasing operational lifetime of the equipment.

A qualitative judgment was noted in that, for harder rocks, no gain in penetration was achieved once a certain critical waterjet pressure was reached. Penetration rate also increased with the diameter of the nozzles used, although this was only tested over a small range (0.3 to 0.4 mm). In a comment which will be revisited in a later chapter it was noted that penetration was optimized when the jet was directed, through the carbide of the bit, onto the rock directly ahead of it. As a results of tests in different rocks, a suite of recommendations for optimum nozzle designs for these different conditions was also established (Table 8.1).

It was believed that the increased bit life was due to the additional cooling which the bit received, but it was necessary to modify the design by increasing the distance between the tips and allowing a larger clearance around the bits for them to work most efficiently. This was, in part because of the larger chips which this assisted bit would produce. Unfortunately the work at CERCHAR in the application of high pressure waterjets has since been discontinued.

The theoretical aspects of combining waterjet assistance with the advantages to be gained by changing to polycrystalline diamond compact (PDC) inserts, rather than the more conventional tungsten carbide, was first discussed, together with some preliminary results, at the first U.S. Water Jet Symposium [8.30]. This study used square diamond insets and found that, while the orientation of the inserts was of little concern in softer rocks a negative rake of some 10 degrees was optimum in harder rock (Fig. 8.9). As opposed to the CERCHAR study, the CSM design used the center well of the bit as a location to place the jet nozzles, and distributed them along the face of the bit. The investigation showed that, because the jets were cutting from the center of the bit outwards, that there was a critical speed of rotation, beyond which the jets were not cutting to the gage of the hole and that, as a result, bit ROP reduced for the same thrust force (Fig. 8.26).

Table 8.1 Comparative performance of jet assisted drills in various rocks [8.29]

Type of Rock

	a	*b*	*c*	*d*	*e*	*f*
rock hardness	low	medium	medium	high	high	very high
rock abrasivity	high	medium	high	low	high	low
comp strength (bar)	<400	400-800	400-800	800-1600	800-1600	>160
22 mm hole dia						
conventional drilling rate (m/min)	2	3.5	0	1.7	0	0
specific energy (j/cc)	202	105	0	153	0	0
jet assisted drilling rate (m/min)	10	6.7	3.5	4.5	2.7	1.6
specific energy	37	54	61	133	92	139
optimum design						
no. of nozzles	2	2	1	2	1	1
nozzle dia (mm)	0.3	0.4	0.4	0.4	0.4	0.4
thrust reqd (kg)	600	800	600	1000	700	700
rotation speed	700	700	700	700	700	700
37 mm hole dia						
conventional drilling rate (m/min)	n/a	3.5	0.9	0.7	n/a	0
specific energy (j/cc)	n/a	133	291	229	n/a	0
jet assisted drilling rate (m/min)	n/a	5.0	2.4	2.8	n/a	1.0
specific energy (j/cc)	n/a	86	99	107	n/a	163
optimum design						
no. of nozzles	n/a	2	2	2	n/a	2
nozzle diameter (mm)	n/a	0.3	0.3	0.3	n/a	0.3
thrust reqd (kg)	n/a	1600	1200	1600	n/a	1600
rotation speed (rpm)	n/a	600	500	500	n/a	500

The combination of the diamond cutting surface and the jet assistance have given a dramatic improvement in drilling performance in hard rock, reducing both the thrust required to penetrate the sandstone, but also increasing the life of the bit by up to seventy-five times (Fig. 8.27). Although there was, at the time, little commercial interest in this development, interest in the technology has recently been rekindled. This is, in part, because changes in technology can now increase the life of the PDC bit to between 200 and 600 times that of a conventional carbide [8.31]. Because the diamond surfaces are still, however, sensitive to temperature the benefits of clearing and cooling the bit insert using high pressure water will not only allow the bit to be operated at higher thrust, increasing ROP, but it will also allow it to be used to drill in harder rock [8.14]. This subject will continue to be discussed later in this work.

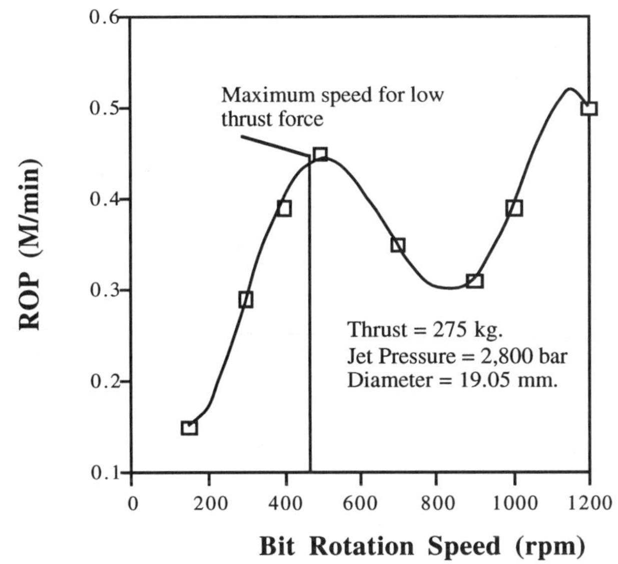

Figure 8.26 The effect of rotational speed of a PDC assisted bit on ROP [8.30].

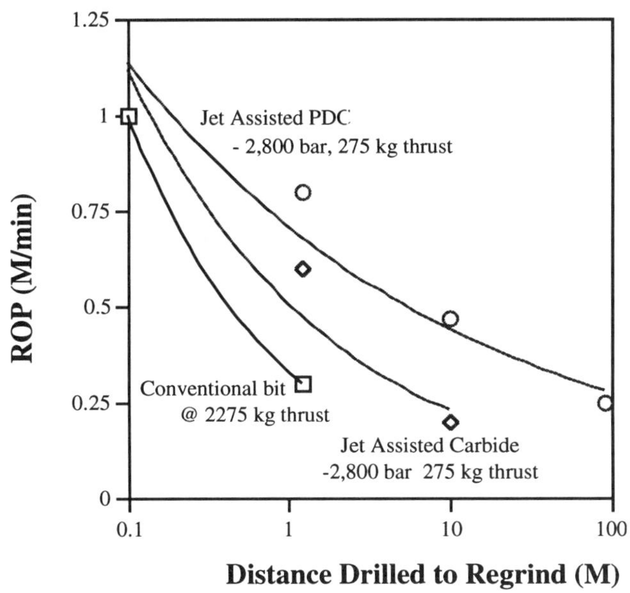

Figure 8.27 Relative ROP with time for early PDC and other bits [8.30].

The potential benefits of the system might, however, be envisioned from a comparative table of costs developed [8.32] for using resin bolts to pin a mine roof, using 1978 costs.

Table 8.2 Approximate comparison of conventional and waterjet assisted drilling costs, using resin bolting [8.32]

Item	*Conventional*	*Jet Assisted*
Machine Cost	$70,000	$110,000
Drilling Rate (M/min)	0.75	2.4
Bolting Cycle (min)	5.0	3.2
Bit Reconditioning Cost	$6.00	$40.00
Bit Life (M. drilled)	6	112.5
Drill Rod cost	$28.00	$110.00
Drill Rod Life (M. drilled)	180	1,500
DRILLING COST ($/M.)	$1.17	$0.43
Bolting Materials		
Bolt/Hole Diameter (mm)	19.3/26.92	16.0/18.8
Bolt Cost	$1.70	$1.20
Resin Cost	$1.60	$0.40
Plate Cost	$0.20	$0.20
TOTAL MATERIAL COST	$3.50	$1.80
TOTAL COST/INSTALLED BOLT	$5.61	$2.57

The basis for the comparison is fully described in the reference, but can be summarized as: installing bolts through a combined sandstone and shale roof, with the bolts being installed to a depth of 1.8 m. With conventional drilling the drill is plugged approximately once every third hole, and bits are changed after each hole. The data is taken from observations at the Somerset Mine of U.S. Steel.

One of the advantages of the waterjet assisted drills is that they have significantly lower thrust force requirements for drilling below those found with conventional drills. For this reason they can be readily hand-held, as was early demonstrated by the Stone Age equipment. This advantage led the U.S. Bureau of Mines to evaluate the use of hand-held jet-assisted drills for use in more conventional rock drilling [8.33]. This application is a significant change from the major direction of development of the drills just described, since these were primarily developed for use for upwardly vertical drilling. Hand-held drills, on the other hand, are more commonly used in drilling horizontally, often for blastholes.

The test program carried out by the Bureau of Mines used a jet pressure of 2,450 bar at a flow rate of up to 10 lpm. feeding water through 4 orifices, two of which (at 0.23 mm dia) slotted the central section of the face of the hole, while the remaining two (at 0.31 mm dia) were directed to cut a slot at the gage of the hole. A central, cross bit carbide insert was used to break out the remaining ribs of material in these trials (Fig 8.28).

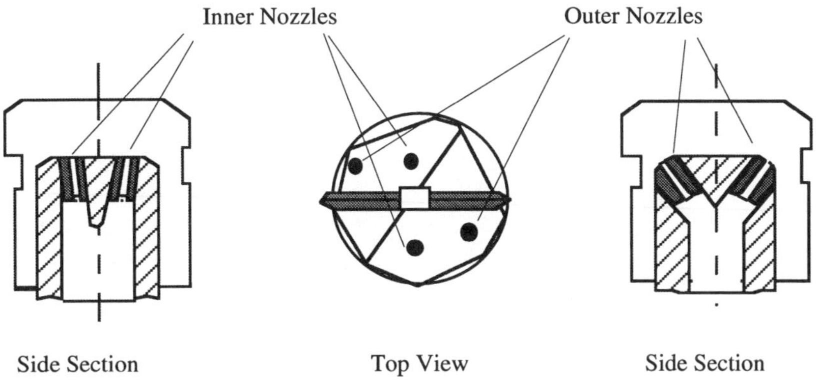

Figure 8.28 Design of the jet-assisted bit tried by the U.S. Bureau of Mines [8.33].

Tests were carried out at three jet pressures and several levels of forward thrust, drilling into a suite of five materials, including three sandstones, coalcrete and trona. It is interesting to note that the gain in drilling rate with thrust was significantly affected by the jet pressure at which the test was carried out (a sample result as projected by the investigators is shown in Fig. 8.29).

Drilling rate increased with increase in jet pressure, and at the same time, specific energy of drilling was reduced as the thrust on the drill was increased. It should be borne in mind, however, that in the softer rocks the jets were drilling a hole slightly larger in size than the bit itself, and it was only in the harder sandstone that the hole diameter was maintained at that of the bit. It should be also remembered that while this test was to determine the improvement in penetration rate when using jet-augmented drills, a sometime practical advantage with hand-held equipment is that it would allow the operator to thrust at a lower level and still drill the rock at an acceptable rate. The results of these tests did indicate, however, that such a practice would not be as efficient as the alternative.

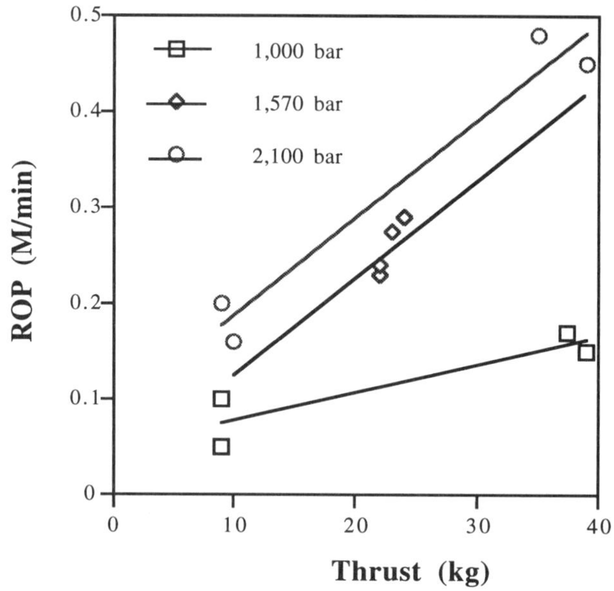

Figure 8.29 Drilling ROP as a function of thrust in Greenwich Sandstone [8.33].

8.5 WATERJET DRILLING IN COAL AND ADJACENT ROCK

The major emphasis for using waterjets by themselves for drilling has concentrated on its use in drilling into coal. This is because of the relative ease with which water can drill this material, in contrast to other, harder rocks. In coal the ability of the waterjet to cut at a considerable distance from the nozzle and to drill holes of relatively large diameter, is enhanced because of the unique structure of the coal itself. Specifically because of horizontal and vertical planes of weakness which are naturally formed within the coal, coal has very little strength in tension. Where waterjets can intersect these planes of weakness, then the jets can penetrate along them, pressurizing the surrounding material and moving into any adjacent free space. Since this is usually an adjacent part of the hole the result is that the jets can remove large volumes of coal with little effort, and under these circumstances, it is possible to drill holes up to about 20 cm in diameter without additional assistance to the waterjets.

The first study of coal drilling at UMR was funded by the Department of Energy, through Sandia Laboratories [8.34]. The first tests which were carried out used the same design for the drilling nozzle that had been used in the earlier rock drilling work (Fig. 8.3). Such a nozzle was attached to a drill rod and tested in an exposed coal outcrop in June 1978. The first assigned task for this drill was in horizontal drilling and to reach hole depths of 15 m, with initial ROP targeted at around 1 m/min. The first hole drilled on the test site, within minutes of setting up the equipment, exceeded this goal.

It was noted that the holes drilled at this rate were roughly square and measured some 17 cm on a side (Fig. 8.30). Although this larger hole size indicated that the waterjets were drilling the coal much more effectively than had been anticipated, and that greater performance levels could be anticipated, there were some problems with creating holes of this size.

Figure 8.30 Initial "square" holes drilled by waterjets in coal.

One problem in drilling this larger hole size using waterjets was that it increased the annulus or spacing between the drill steel and the surrounding coal. This, in turn, reduced the speed at which the water was forced to flow out of the hole. At a slower speed the water does not carry as much or as large a size of coal particles. The result was that the cuttings from the drilling which were being carried out of the hole by the water tended to settle out along the length of the hole. Once settling started, particles mounded around each other creating "dunes" and eventually formed a dam across the hole, so the debris became increasingly trapped, until the hole became instantaneously sealed. This blockage trapped the spent water in the hole, pressurized it as more water entered from the nozzles, and reduced the cutting ability of the jets issuing from the nozzle. To keep the cuttings from building up and to encourage flow, the drill assembly was frequently pulled back and partially out of the hole before being fed forward again, in this way breaking the "dam" and allowing water and coal particles to escape. This problem has been, and remains, a significant problem in many horizontal drilling applications, particularly, as will be referred to below, where the drill is used to "drill around a corner" from an initially vertical well into a horizontal section. The problem is not restricted to drilling in coal and has been and continues to be addressed in drilling for oilfield development [8.35].

Continued field tests of the horizontal waterjet drill in coal showed that where the rotational speed of the drill was reduced from 200 rpm to 60 rpm that the hole diameter would reduce from 17 cm to 10 cm. Similarly an increase in the ROP of the drill reduced the diameter of the hole being drilled. The major purpose in carrying out this research program was to see if a tool could be developed which could drill a horizontal hole out from the bottom of an existing vertical well. As the technology developed it became known as the RTC (for Round The Corner) concept while it was being developed at UMR, and then became known as the Cornering Drill as it was subsequently developed by Sandia. This aspect of the development will be described in more detail in Section 8.7.

One of the requirements of such a system, which would make it easier to bend the drill around the "corner" more easily, was a drilling system where the drill rod supplying the head would not need to rotate. Since the drilling head itself had to rotate then some way of turning just the head had to be provided. (Sections 3.4 and 8.7 discuss variations where this rotation <u>may</u> not be necessary.) Not only the head, but also the high pressure waterjet connections must turn, however, and thus, the head would need to contain not only a drive motor but also a high pressure water swivel. While it was not difficult to find a small motor to fit within the 15 cm diameter envelope of the drill developed it was much more difficult to find a reliable swivel in 1979 [8.36]. Samples of commercially available swivels capable of carrying water at 700 bar were obtained by UMR from most suppliers and tested in the laboratory. A number of these original models failed within a test time of ten hours and only one model was identified which could operate for this length of time. Subsequently several companies have produced swivels which can operate for much longer times without significance failure, but one offshoot of the design was an orbiting nozzle (Fig. 3.33) which does not require a swivel in the drilling head [8.37]. Field trials of this new tool (Figs. 8.31, and 8.32) were carried out and holes were drilled which were up to 30 m long at an average ROP of 0.6 m/min with this equipment [8.36].

The use of waterjets to drill holes in coal has several potential advantages. Not least of these is the ability of the operator, by tailoring the jet pressure to the geological conditions as described below, to hold the drill so that it only drills into and thus moves forward inside the coal seam. This was seen, by Australian investigators, as a possibly significant advantage in developing an exploratory and methane drainage network of holes in coal. A program was therefore set up between the University of Queensland and Capricorn Coal to develop such a drill, and test it as a means for drilling holes to a depth of 1 km along the seam [8.38].

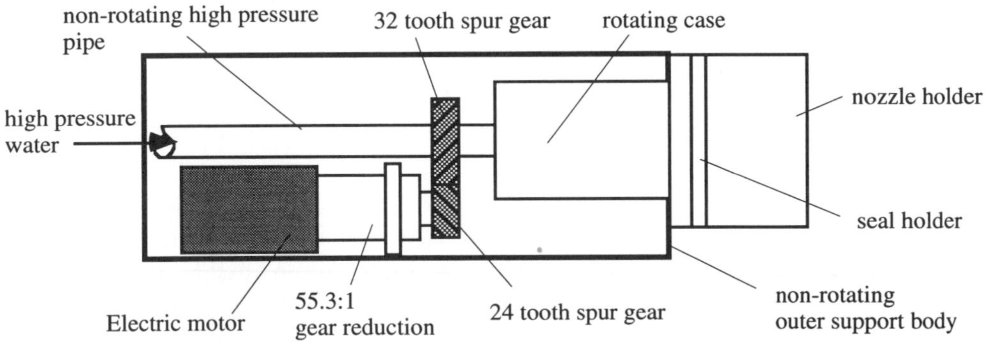

Figure 8.31 Schematic representation of an in-hole drilling head layout [8.36].

Figure 8.32 Internal components of the in-hole drilling head outside the cover and with a simpler nozzle mounting.

Results from preliminary experiments at the University emphasized the need to drill holes large enough to allow the cuttings free passage out of the hole, and were sufficiently successful to justify field testing of the design. Initial tests in an outcropping coal seam re-emphasized the need for proper control of cuttings. In waterjet coal drilling the particles are not removed in a consistently small size, as is the case in conventional drilling. As a result larger particles can be anticipated, and these will settle out, or become trapped along the hole more easily than smaller particles. As a result it becomes easier for the hole to block, with the unfortunate results discussed above. Changes in nozzle design to the use of self-rotating nozzles overcame this problem, but introduced the problem of a residual cone in the center of the hole when a conventional self-rotating nozzle was used. This problem, in turn was overcome, using a second self-rotating nozzle (the Woma FR 22).

This nozzle initially drilled holes at a large enough size (60 mm) to overcome the problems of tailings blocking the hole by keeping the outlet flow of water fast enough through the narrow annular channel [8.39]. When this design was taken to a second site, however, it was found that the volume of water supplied to the nozzle was no longer sufficient to clear the cuttings from the hole as it was drilled deeper. To overcome this problem a set of retro-jets was added to the assembly just behind the cutting nozzle assembly. It was interesting to note that the diameter of the retro-jets could be adjusted so that they provided a greater forward thrust to the drill than the reaction force from the cutting nozzles. In this case the drill became "self-advancing" as well as "self-rotating." Given however, that the diameter of the hole is a function of advance rate (see for example Fig. 8.4) at the faster advance rate the drill was no longer cutting a hole large enough for the larger diameter retro-jet section to move through. As a result it was necessary for the operators to manually pull back on the drilling rod assembly to slow down the advance of the head so that it would cut a large enough hole diameter. This need to slow a drill down is a relatively uncommon practice in drilling.

Tests using hose as part of the supply line allowed holes of up to 38 m in length to be drilled, but these tended to drift towards the floor, as the drill advanced into the hole. This was, in part, because the jet would tend to overdrill the hole, cutting it to a larger diameter than was expected. As the unit advanced, therefore, the weight of the drill head and supporting lance would cause it to droop until it again rested on the floor of the hole drilled. The result was a gradually increasing downward curve for the hole until it intercepted the floor of the seam. To overcome this trend a bent rod was located directly behind the drilling nozzle (Fig. 8.33) and this **bent sub** (as it is referred to in drilling) raised the location of the drilling nozzle, initially at an upward angle of one degree, to the center of the hole, minimizing the droop [8.40]. Concurrently the self-rotating nozzle was changed for the larger FR47 model.

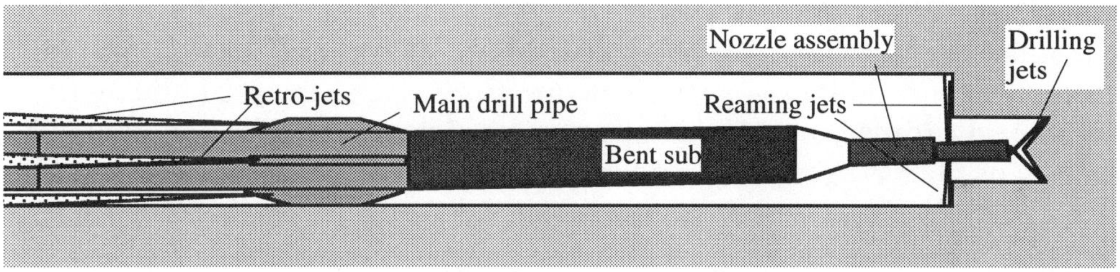

Figure 8.33 Australian long hole drilling head, showing method of steering [8.40].

This drilling combination was able to drill holes out to a depth of 52 m, although it was found, when the holes were subsequently surveyed, that the direction of the hole had turned toward the direction of the primary cleat. Subsequently the drill string was modified, by including a non-magnetic austenitic stainless steel drill rod section behind the bent sub. This allowed the hole to be surveyed while the tool was in the hole, and thus drilling directions could be changed, by turning the bent sub. At a pressure of 650 bar and a flow rate of 160 lpm this drill penetrated 123 m into the seam, at a rate of 1 m/min on average, maintaining position within the seam as the drill advanced. A second hole was drilled to a depth of 225 m, penetrating through a 2 m wide intrusive dyke at a depth of 46 m into the hole.

It is interesting to note that the change in surveying technique lowered the cost of the in-hole components by 80% (1991 costs from A$50,000 to A$10,000). In addition the tool showed very little sign of wear after drilling the two holes. At the end of 1991 preparations were underway to drill to a depth of 500 m.

Concurrently an alternative design to achieve the required clearance of the hole, while using an alternative assembly of self-rotating nozzles has been pursued by Smet [8.41]. In the design,which has been fitted to the end of a 200 m long, 12.4 cm diameter hose, two self-rotating nozzles are attached to a supporting swivel, which itself rotates driven by the reaction forces from the jets (Fig. 8.34). The unit is designed to operate at a pressure of 420 bar, at which pressure the head rotates at up to 1,000 rpm, and has been reported to drill at an ROP of up to 5 m/min.

Tests both by UMR and the University of Queensland were carried out from the surface, with the waterjets drilling into outcrops of coal. While, in the latter case the length of the hole and the increasing depth of overburden subjected the coal to some overburden pressure, this is not as high as that encountered in the majority of underground drilling. In underground coal mining operations it is occasionally advantageous to pump water, under pressure, into the coal ahead of a mining operation. Such holes do not have to be deeper than 75 m and need only be of sufficient size to deliver a sufficient pressure to the back of the hole that the water will infuse into the cracks and weakness planes in the coal. This system is used to weaken coal and reduce dust levels when coal is mined.

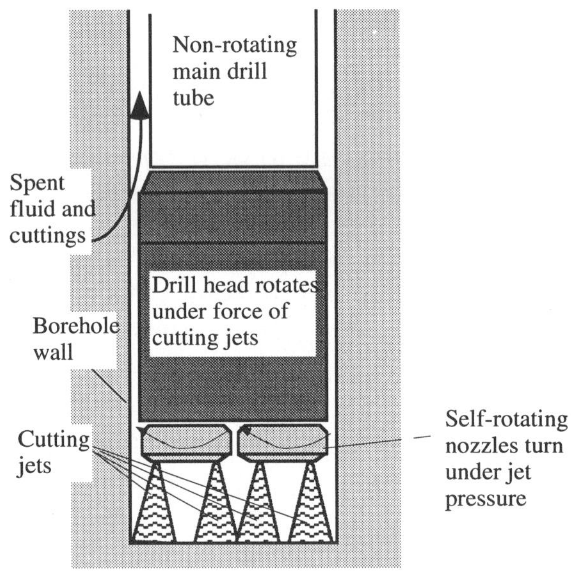

Figure 8.34 Smet dual self-rotating coupling and nozzle drilling head [8.41].

Given that the water must be under pressure if it is to infuse into the coal, an experiment was carried out in Germany to determine if a high pressure waterjet system could also be used to drill the holes [8.42]. This used a similar design to that initially used in the Australian experiments, except that a shield was placed around the cutting nozzle, and a stabilizing rod was placed directly behind the nozzle (Fig. 8.35). At a pressure of 420 bar the drill was able to penetrate 30 m into the coal, at an ROP between 4 and 5.2 m/min in an underground mine.

It is interesting to note that a test of this drill design in the subsequent Australian tests did not prove successful. One possible reason for this was that the tests in Germany were with the drill inclined to drill at an upward angle, while in Australia the holes were drilled at a slight downward angle into the coal. There have been no subsequent reports of any continuation of these trials in Germany.

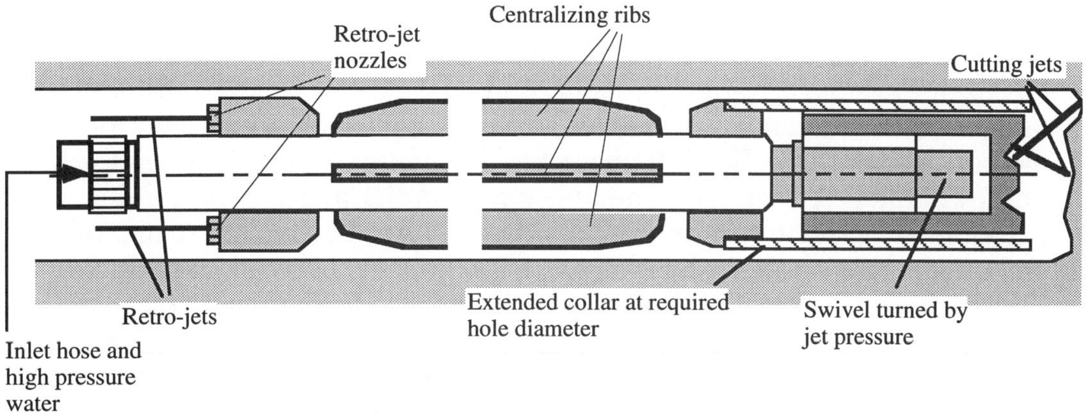

Figure 8.35 German design for an underground waterjet drill [8.42].

One of the reasons for infusing water into the coal ahead of mining is to weaken the coal, which will reduce its tendency to burst out into the workings under high pressures. An alternative use of waterjets to resolve the same problem has been tried in Japan [8.43]. The technique for alleviating the problem differed from that described above in that it required that much larger holes be drilled into the coal, in order to create stress relief. In Japan the coal being mined lies at a depth of around 800 m and holes 250 mm in diameter have been found to be large enough to successfully relieve the stress. However, because the machines required to conventionally drill such holes are large and difficult to maneuver in the mines, tests were carried outs using waterjet drilling at the Akabira and Taiheiyo coal mines.

In order to allow the drill to function a special swivel was developed, capable of rotating at 700 bar, and this was found to have an operational lifetime of about 200 hours underground. The drill was set up largely to ream boreholes out to a radius over 65 mm, and successfully did this at a penetration speed some 3.5 times that of conventional mechanical drills, with a need for only half the labor force. What is worthy of additional note is that the Japanese used a scroll feed on the drill pipe to assist in getting the cuttings out of the hole. This hole clearing solution was of greater significance, since the combination of gas content and high pressure in the coal around the hole, would cause the coal walls to burst into the hole as it was being drilled. This would generate more cuttings (up to 1.98 cu m/m of hole length) than would be created in drilling a more conventional hole, and these would need to be cleared so that the drill could move forward.

The use of such a scroll is a common feature of some earth drills, and has been developed for larger earth and coal augers, and these have been used to excavate coal from, holes up to 70 cm and larger, as a primary method of mining. Because these are relatively simple tools and can be made of relatively low cost they have a potential market for mining coal, particularly in relatively thin seams [8.44]. As the thicker seams of coal are mined out, so there has developed a greater need to exploit thinner seams of coal, which increasingly become a significant part of the remaining reserves. However, the cost of mining these thinner seams conventionally is one which society finds itself increasingly unwilling to pay [8.45] and, thus, more remote methods of extraction are required.

Auger mining has some considerable advantages for this form of mining, in that the tool (much like a large drill bit) can be fed into the coal seam and serves to mine out the entire layer, and move all the coal back out of the seam, in one operation. The range of the tool is limited, by the friction between the scroll and the borehole wall, and the high levels of thrust required in a conventional head restrict its range of operation to less than 30 m because of problems of steering to retain the drill-head in the coal beyond that distance. By lowering the force required to cut through the coal, and tailoring the jet pressure to selectively cut the coal, rather than the surrounding rock, a waterjet assisted auger can successfully drill to much greater depths than conventional systems (Figs. 8.36 and 8.37).

There are a number of significant additional advantages which can come from refining the design of the auger drive and the supply and transportation network, associated with the auger. These can be made because of the lower forces required to push the jet-assisted cutting head into the coal. An illustration of this comes from an early experiment at UMR where a waterjet assisted auger was being used to penetrate into an artificial coal seam (Fig. 8.38). It was found that the auger could quite easily mine the "coal" out to a diameter in excess of 70 cm (Fig. 8.39), and that it could be advanced into the hole, at a steady speed, by an undergraduate student using a manually operated "Sylvester" or "come-along". What made the demonstration particularly memorable was that the student had suffered an injury to one arm, and thus was pulling the auger into the hole and thus, mining coal at a steady rate, using only one hand

(Fig. 8.40). While this technique has been licensed, current development at a commercial level is still restricted.

Figure 8.36 Front view of an early design of waterjet assisted auger head, showing the nozzle pattern and the scrolls behind the head to move the coal [8.44].

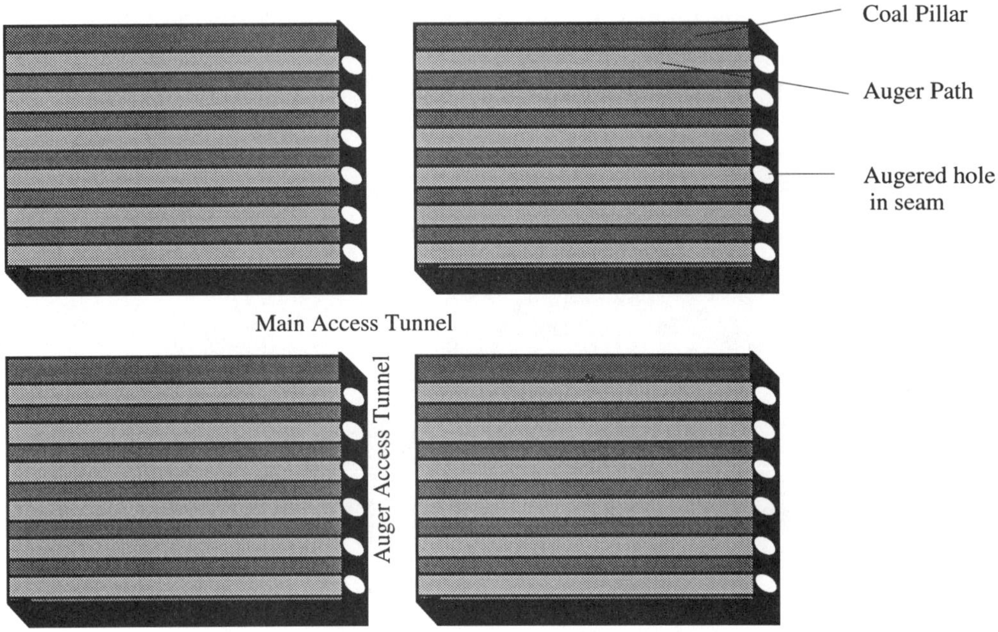

Figure 8.37 Possible mining layout using a waterjet assisted auger [8.44].

Figure 8.38 UMR coal auger starting a cut.

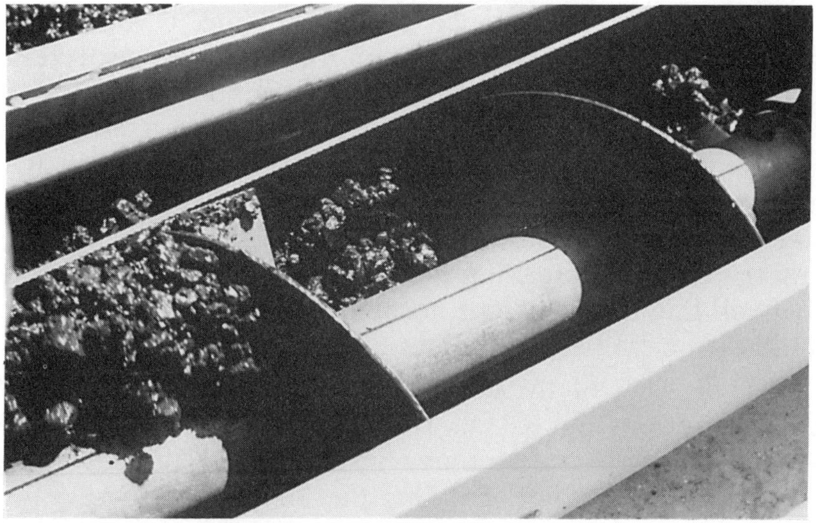

Figure 8.39 Coal moving along the scrolls of the auger.

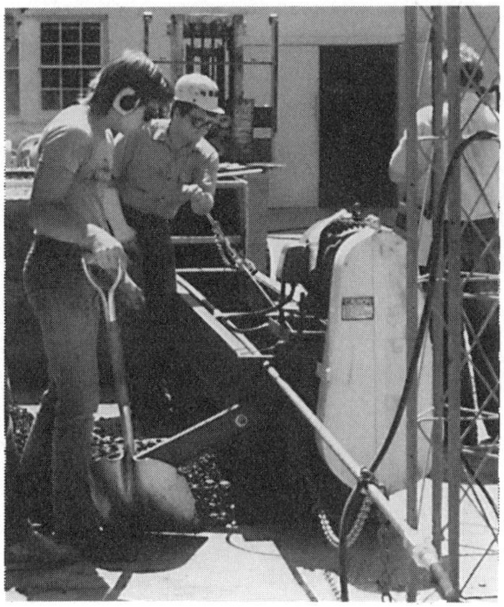

Figure 8.40 Student (Mr. C. Canon) moving the auger into the artificial coal, note the cast on his right arm.

The use of the scroll auger in the United States was created to solve a peculiar problem which arose while trying to address an apparently relatively simple task [8.46]. UMR had been asked, by the Department of

State Lands in Montana for assistance in extinguishing a coal seam which was burning at its outcrop on the banks of the Yellowstone River in that State. Because of the difficulty in getting conventional equipment to the site the use of high pressure waterjets was investigated as a means of extracting and extinguishing the burning coal. The lightweight nature of the waterjet rig made it an ideal candidate for this task.

The waterjets, at a pressure of 700 bar, had little difficulty in extinguishing the coal, and breaking down both it and the superheated clay immediately overlying the seam. In normal operations, the water entering the hole would then have been sufficient to carry this debris back out of the hole. However, the heat of the combustion zone was over 1,500 ^{o}C and this vaporized the water (Fig. 8.41). The cuttings were left in the hole and made it difficult for the lance to move in and out of the cavity, and for other devices to be inserted to put out the fire. This was of critical importance since the coal was only extinguished along the line of the nozzle advance, although the jet reamed a cavity out to over 30 cm in radius. Within 2 cm behind the wall of the cavity the coal was still glowing red, and could easily re-ignite the coal in the hole, were it not removed. In addition, the plan required that the hole be backfilled with clay. This also required that the hole be cleared of existing material. An auger scroll on the drill steel would have solved this problem, but was not then available.

Figure 8.41 Extinguishing a burning coal seam in Montana - showing lack of anything coming out of the hole. The drill pipe is hanging under the triangular section.

This loss of water while drilling within the hole is not unique to drilling in burning seams. Kennerley has reported [8.39] that in-seam blockage of the hole can create enough pressure inside the hole to open fissures along which water can escape. Thus, a way must be provided of making sure that the cuttings are positively removed from the hole. This removal can be achieved either with retro-fitting jets, such as has worked well in Australia, or through a mechanical scroll on the drill rods (which requires that the drill steel rotate).

At the beginning of the above discussion it was pointed out that the waterjet pressure could be adjusted so that a waterjet would cut into coal but not into the surrounding rock and that this could be used to help steer a drill. It is pertinent to digress to explain a method by which this same idea could be used to help detect the interface between coal and overlying shale. This formed part of a project carried out at UMR for the National Aeronautics and Space Administration [8.47].

The problem which was to be solved was the need to locate the interface between a coal seam and either the overlying shale rock, or the rock which lay under the coal seam. This is of growing importance in the drive to automate mining equipment, since automatic sensing of the machine position relative to the roof and floor will allow mining machines to be able to steer forward and mine coal instead of the surrounding rock.

As previously mentioned, a major feature of coal response to waterjet attack lies in the presence of weakness planes within the material. Thus, in the use of the Hydrominer machine [8.48], a dual orifice nozzle cut two separate slots which diverged to more than 5 cm apart. Yet, because of the weakness of the coal, all the intervening material was removed by the jet action, thus, creating a single wide slot (Fig. 6.25). Such a characteristic response does not happen when the jets cut other rocks which have a significant tensile strength (Fig. 8.42). Thus, there is a property which allows a way of differentiating between coal and other materials.

Simply put, if a waterjet nozzle were rotated around the perimeter of a 5 cm diameter circle in coal, all the material within the hole will be removed, and a cylindrical hole will be created. In contrast if that waterjet cuts into another rock, such as a sandstone or shale, then either the pressure of the water will be too low to cut that material, or the jet will cut a slot around the perimeter of the hole leaving a central core intact (Fig. 8.43). Thus, when drilling upward into the roof through a coal seam. the jet would encounter the shale and perimeter slot, but remove all the coal lying below the interface. To a sensing probe looking up the hole it would appear that the hole ended at the coal:shale interface.

Figure 8.42 Spiral cut on a 15 cm wide sandstone block showing the lack of rib removal.

Figure 8.43 Jet cut through coal to an overlying rock strata [8.47].

By incorporating a measuring device, such as a ruler or the range finding device found on current automatic focus cameras such as the Polaroid or the Honeywell system for the Konica camera, is it possible to measure the distance to the back of the hole. A combined drilling and measuring system could relatively, and in real time, give the distance from the cutting drum of the machine to the overlying rock, by measuring the relative height to the coal, and then the depth of the successive holes drilled to the interface [8.49]. To date, however, this concept has seen little further development, even though it, by inversion, might provide a way of identifying the point, in a changing lithography, at which the admixture reaches the point where it is no longer coal and more shale.

8.6 WATERJETS IN OIL WELL DRILLING

The discussions on drilling up to this point have dealt mainly with the drilling of relatively small holes, close to the surface. Even in this conditions it has been necessary to address the problem of effectively removing the cuttings from the area around the bit, and out of the borehole. Both of these problems occur with a much greater impact when drilling oil wells. While practically true for virtually all oil well applications, in the past wells have largely been drilled within a narrow range of being inclined only vertically downward. In the recent past, however, the benefits of drilling out horizontally from the bottom of an accessing vertical well have been found to be significant [8.50]. This development will, however, be discussed in the Section 8.7.

Oil wells must frequently penetrate through several thousand meters of rock to reach the pool of oil which is their target. In penetrating down through this rock the drilling tool has most commonly been a set of two and three conic bits which are inset with teeth, and which rotate over the rock surface. The two-cone bit was first patented by Howard Hughes Snr. in 1908, while the three cone bit was introduced in 1933 [8.51]. As the cones rotate over the face of the hole, the teeth penetrate into the rock causing it to chip away, under the applied pressure, and thus, incrementally, the hole is deepened.

In order to stabilize the sides of the hole, and to carry the rock chips up out of the hole, the drilling team will circulate a specially mixed mud down through the drill pipe to the bit. At the bit this drilling mud is forced out through specially designed nozzles, and directed forward against the rock surface. The intent of the resulting jets is to have sufficient force to lift the chip away from the rock so that the bit can continue to drill in fresh rock. At the same time the mud is supplied at sufficient pressure, density and volume that as it then moves back up through the annulus between the drill pipe and the rock wall, it will carry the chips and other debris back up out of the hole.

Even where the jet power is insufficient to penetrate the rock, and where the streams do not interact with the bit action, there can be significant benefits to increased jet pressure and flow rate. This can be achieved in several ways. One of the more common is to increase the delivered power down through the feed pipe to the bit. However, considerable gains can be made, for the same delivered power, if the system losses are reduced by bringing the nozzle closer to the cutting surface. While some of the reasons for this will be discussed later in this section, the results of moving the nozzle closer to the rock surface have been documented by Baker, [8.51] among others. The greater effective power of the jets leads to a more efficient cleaning of the chips from ahead of the bit, but since the jets on the bit must effectively clean the entire surface of the rock around the hole, the proper design of the jet nozzles, and their correct sizing has been critical to good bit performance. Improved cleaning of the chips from the bottom of the hole, can lead to an increase in ROP of between 15% and 40% in softer rock formations. Where such good design is not practiced then cavitation may also occur. Cavitation can be a means of improving the performance of the drilling bit, as will be discussed in Chapter 10. When the cavitation occurs in the bit itself, it will attack the bit instead of the rock and leads to premature bit failure, which is not desirable.

This improvement in performance with improved hydraulic power being delivered to the rock is not restricted to drilling in soft formations. For example [8.52], a series of 36 wells was drilled in the Virginia Hills area of Central Alberta in Canada. Seventeen of the wells were drilled with conventional hydraulic units, while the remainder were drilled with improved hydraulic power. The wells were drilled to an average depth of 2,850 m through what is considered by the oil industry to be hard rock. In order to improve the performance of the drill fluid the hole diameter was reduced and the applied hydraulic power to the rock was increased by 128%. It took an average of 19 days to drill the wells with this new program, as opposed to an average of 29 days per well for the conventionally drilled set. Problems with hole cleaning and other causes was reduced by 85% and the bit performance was improved so that, in the same rocks, the amount of time spent actually drilling each hole dropped from 256 hours to 175 hours with increased ROP ranging from 15 - 80%.

An additional benefit to the increased performance of these bits is that, because of the improved cutting conditions for the bits, they did not wear out as quickly. Thus, each hole was drilled with only four bits, or less, while the conventional holes had all required that at least five bits be used. Since the entire drill string must be pulled from the hole to replace a bit, and this can take, for deeper holes, over 24 hours for each change, the benefits of this improvement can be financially quite significant. The study did not show that there was much gain by increasing jet velocity, in comparison with the overall gain from increasing total hydraulic power (i.e., the combination of volume and velocity). There was, however, some additional hole enlargement by using the high fluid flows, but this was more than offset by the reduction in the original hole diameter drilled.

There is a minimum velocity required in the annulus to move the chips and clear the holes, which Maurer [8.53] has reported as lying around 24 to 36 m/sec. The flow velocities in the above example were more than twice that high, and did give rise to some erosion of the sides of the hole. If such jets could erode the sides of the hole, it appeared logical to see if they could not similarly be employed to remove rock ahead of the bit.

Maurer had carried out a survey of some twenty-five novel methods suggested for improving the drilling of oil wells [8.54]. The majority of these were not likely to improve penetration rates or costs over conventional drilling, with the potential exception of the use of high pressure jets on the drill bits. This provided the impetus for Esso Production Research to begin their study of high pressure jet cutting. Initial trials with a pulsed water cannon indicated that, in contrast to the results reported above with smaller nozzles, that nozzle diameter had little effect on the specific energy of cutting [8.53]. The first jet bit tested contained two jets each 3.81 mm in diameter and was tested at pressures to 700 bar. A second bit contained five nozzles and could be operated at pressures of up to 1,000 bar. ROP between 54 and 90 m/hr were obtained in laboratory tests of this second design, and this led to the testing of the concept in four shallow wells near Ft. Worth in Texas.

The bits used were designed so that all the cutting would be carried out by the fluid jets, which were operated at pressures of up to 950 bar in the field. While conventional drilling of offset wells was, at that time, usually carried out at an ROP of 3 to 6 m/hr, the jet drills were able to achieve rates of 32 to 85 m/hr in three of the wells and 6.9 m/hr in the fourth, (in which there were some hole gaging problems).

The success of that effort led to the addition of high pressure jetting nozzles to the suite of conventional drilling bits. Some problems with blocking nozzles reduced the number of effective tests which could be carried out, yet the results were promising. ROP were more than doubled in some holes achieving rates of up to 60 m/hr and drilling to depths of up to 3,900 m. Bits not only drilled faster they required less thrust or weight on the bit in order to drill and they lasted up to 75% longer than conventional bits. An economic evaluation, by Maurer, at the time, suggested that the increased cost of the equipment would require that the bits at least double and most advantageously should treble drilling rates before they would be sufficiently advantageous to justify their use.

Because of the high cost of equipment and operations Maurer suggested [8.55] that it would be better if the diameter of the holes being drilled were somewhat smaller, since this would concentrate the effect of the jets and improve the ROP (Fig. 8.44).

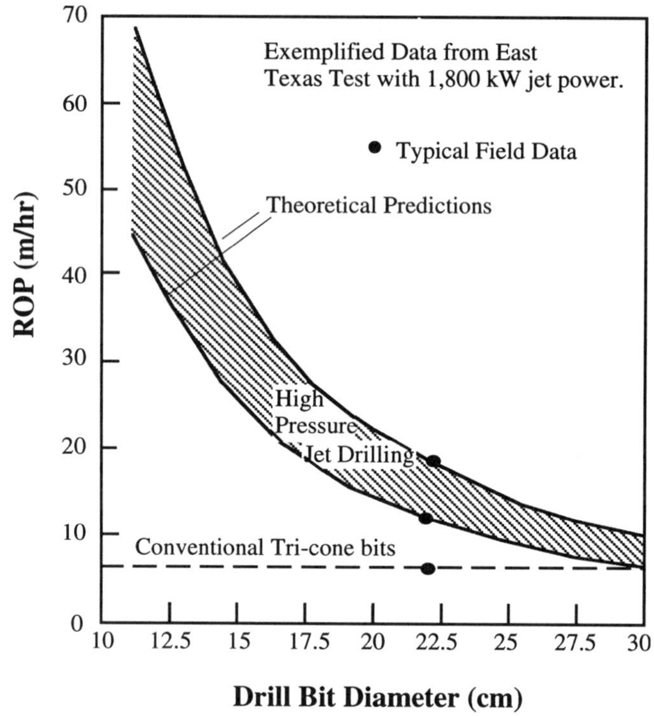

Figure 8.44 Effect of hole diameter on jet-assisted ROP (after [8.55]).

This ability of high pressure fluids to deliver high levels of power to small areas of a rock is much greater than that of conventional tools. For example, a mechanical roller bit can only transmit from 15 to 75 kW of power to the rock, even though there may be over 1,000 kW available on the drilling platform. In contrast the use of high pressure jets can allow power levels of over 4,000 kW to be transmitted down through the bit to the rock. Such bits could drill at speeds of five to eight times the ROP of conventional tools. In developing those tools Exxon had to solve many of the problems associated with running high pressure jet bits at depth. Out of that study they developed, and patented means for replacing operating nozzles on the bit, by changing the flow to spare nozzles already fitted to the bit, as well as other useful techniques.

A co-operative research program between eight oil companies then funded jet-assisted drilling of an additional five wells. Penetration rates were double those of conventional methods but there appeared to be a reduction in the improvement in performance as the holes drilled deeper, with little gain being observed at hole depths of 3,700 m [8.56]. This was particularly true when comparisons were made between the gain which could be achieved by working at lower jet pressures (420 bar) but with an extended nozzle array, in contrast with the results at the higher pressures but with the nozzles in their conventional position (Fig. 8.45). The program was, therefore, terminated in the 1976-1977 year [8.57].

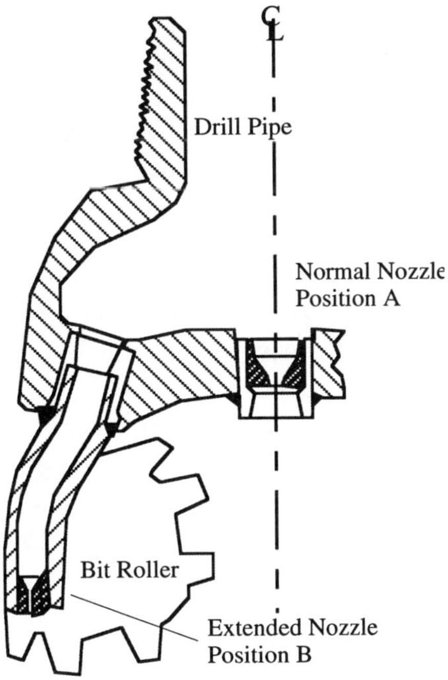

Figure 8.45 Location of high pressure and extended nozzle orifices on a schematic section of a tricone bit [8.55].

The work has subsequently been revisited by Maurer [8.58] who developed a combination of high pressure (700 bar) jets to kerf rock, and then to remove the intervening ribs using PDC inserts set in a drill bit, rotating at speeds of 800 to 1000 rpm. The drill has been reported to drill medium strength rocks at rates of up to 300 m/hr.

Beginning in the early 1970s, Shell Exploration and Production Laboratory began tests of jet-assisted oil well drilling in Holland [8.59]. The Dutch study placed a greater emphasis on the effects of cutting fluid on the performance of the jet drills, and the role of threshold pressure on the jet drill performance. It had earlier been established by several investigators that there is a threshold level of pressure below which rocks are not cut by waterjets. This had been verified in drilling, but the study had not previously shown whether this pressure was controlled by borehole pressure.

The investigation found that at borehole pressures below 100 bar there was no discernible effect on threshold pressure for cutting rock. Concurrently it established that when the drilling fluid was changed from water to conventional drilling mud that there was similarly no effect on threshold pressure. However, the depth of cut achieved with drilling mud containing bentonite, Limburgia clay, barytes and Calgon which together increased the specific gravity to 1.42 reduced the depth of cut achieved by a jet under otherwise equivalent conditions. This is in contrast to the conditions where the back pressure was atmospheric and the mud jets cut a deeper groove.

It was reported, however, that when the pore pressure in the sample was equilibrated to that in the borehole then the reduction in cutting depth due to the change in the properties of the cutting fluid was overcome in some cases. The different additives were each found to have a different effect on the ROP, in some cases, as with barytes, the penetration rate dropped by 86%. This was considered to be because the barytes bridges the pores in the rock, reducing the ability of the water to penetrate into the rock face. Under some conditions bentonite improved drilling rates, but this was considered to be because the particles were separated and able to penetrate into the rock, rather than flocculating and blocking the entry to the pores. The laboratory study revealed that erosion nozzles should be placed no further than 3 orifice diameters apart to totally erode the rock ahead of the bit. While a greater spacing can be achieved under atmospheric conditions, rocks under confining pressure are "tougher" and the intervening ridges left at greater spacings become more difficult to remove.

The combination of high pressure jets and diamond bits was found to be effective particularly in drilling harder rocks, where the greater penetration rates and longer bit life has a greater effect on drilling economics than is the case of softer rock formations. The study was extended into a greater range of rock types [8.60, 8.61] and it was concluded that it was important, particularly in hard laminated rock, to combine the jets with a mechanical tool both to control the hole geometry and to achieve a satisfactory ROP. Given that it is always possible to encounter harder layers of rock in a drilling operation, Shell began to concentrate only on studies where jets were combined with conventional mechanical bits.

Field tests over a drilling interval of 1,267 m in late 1976 showed that it was possible to lower the cost of the well (by the then sum of $134,000) by increasing the fluid jet pressure to 500 bar from the 350 bar that would normally have been used. Greater cost savings were achieved in the upper levels of the hole relative to those obtained as the hole went deeper [8.62].

The effect of increasing borehole pressure is quite significant in reducing the effective cutting distance of submerged jets after they leave a nozzle. This effect is separate from that which may be identified as a change in the threshold pressure level, such as has been identified by Melaugh [8.63] and correlated with changes in the failure mechanisms of the rock. Rather, this effect is due to a reduction in the cutting distance of the jet, and is more particularly visible with the smaller nozzle diameters associated with higher pressure jet cutting.

This can be illustrated by the results from experiments carried out at UMR [8.64]. A rock sample was cut into a wedge shape and placed in a pressure chamber so that it would lie under an offset nozzle which could rotated over the rock surface (Fig. 8.46). By setting the pressure in the chamber at different values, and then pressurizing the jet and rotating the nozzle over the inclined rock surface, the distance over which the jet retained its cutting power could be measured (Table 8.3). Under ambient conditions the 700 bar jet would have cut completely through the sample of Berea sandstone used for the test.

This result which may help to explain the reduction in the improved ROP achieved with jet-assisted bits at higher pressure with increasing depth of the well, as reported above, and why extended bit nozzles give better drilling performance than bits with the nozzles set in the conventional position at greater depths [8.65]. The results suggest that increasing the diameter of the jet nozzles to compensate for the increase in the depth of hole would allow the drill to maintain the same cutting improvements from jet assistance which are currently only reported at shallower depths. Such an increase would require either a higher flow rate through the delivery line, or that the size of the hole be reduced, so that a smaller number of nozzles would be required to cover the area of the rock ahead of the bit. Given that the hole diameter of an oil well normally reduces with depth, this latter requirement might not be difficult to achieve. This explanation also underlines the difficulty in analyzing the causes of reduced drill bit performance, since the reduced ROP in this case is due to the inability of the fluid jets to reach the rock at an effective cutting pressure, rather than that the rock has become harder to cut due to confinement.

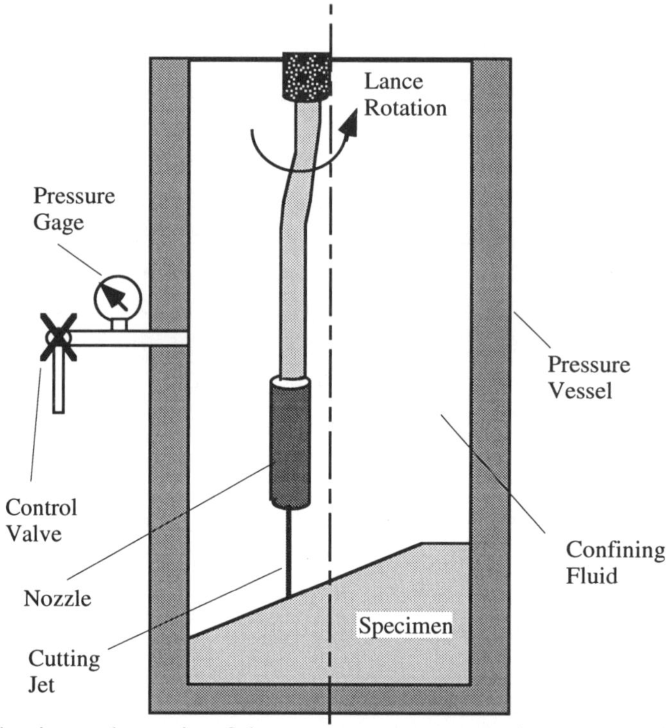

Figure 8.46 Side view schematic of the test arrangement to demonstrate jet cutting range under hole back pressure.

It is more difficult to establish the exact role of rock confining pressure on drilling performance. Two examples might illustrate factors that should be considered in analyzing this effect. In the first example a waterjet drill was taken into a mine in Missouri [8.66] and used to drill small holes into the pillars supporting the overlying roof. The combined pressure on the rock, due to the mine geometry and the depth, were on the order of 280 bar on the rock at the point where the 700 bar jet pressure waterjet drill cut into the wall. The rock was a coarse sandstone and contained galena particles. It was difficult and expensive to drill using a conventional carbide drill due to the abrasive nature of the sandstone.

Figure 8.47 Top view of a typical result showing the range of the jet cutting path.

Table 8.3 Effective cutting distance of a high pressure fluid jet under simulated borehole conditions [8.64]

Jet Range is given in mm

Nozzle diameter	*Hole fluid pressure (bar)*					
(mm)	*0*	*35*	*70*	*105*	*140*	*175*
.75	28.7	16	16	7.9	--	--
1.0	47.5	22.4	22.4	11.2	--	--
1.62	69.9		31.8	23.9	20.6	17.5

Where the jets drilled a hole in the pillar the holes which were drilled were oval in shape (Fig. 8.48(a)) reflecting the increased stress on the sides of the hole, compressing the voids in the rock and reducing the ability of the jet to penetrate. The rock at the top and bottom of the hole undergo some stress relief (Fig. 8.48(b)) and thus the cracks are opened and the waterjets can penetrate more easily and more deeply into the rock. In contrast where a slot was cut into the rock around the drill hole, relieving the ground stress before the hole was drilled, then the resulting hole was round, and slightly larger (Fig. 8.48(c)).

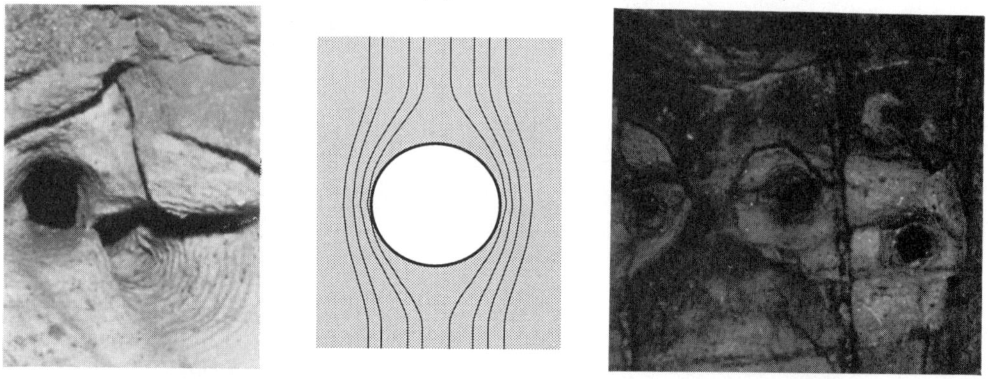

Figure 8.48 Holes drilled at 700 bar into an underground pillar a) in ground at a stress of approximately 280 bar b) the stress contours around the hole c) hole drilled under the same conditions as (a) but without ground stress.

This clearly isolates the effect of ground pressure in closing the pores and small fracture planes through which the jet can enter the rock. The result corroborates the effects seen by Shell in their study on drilling muds and their effects on jet performance [6.59]. However, there is often a combination of the two effects between the reduced jet cutting ability with increased distance from the nozzle, and more greater difficulty which the jet has in penetrating through the more tightly packed rock at depth which makes it difficult to isolate the role of each (for example Table 8.4).

Table 8.4 The effect of borehole confinement and borehole back pressure on fluid cutting ability [8.64]

Jet Pressure 700 bar; Nozzle Diameter 1.62 mm; Target Rock Berea Sandstone; Stand-off Distance 6.35 mm.

Rock confining pressure (bar)	*Hole pressure (bar)*	*Depth of cut (mm)*
0	0	62.5*
140	175	11.0
280	175	8.74
420	175	6.75

*jet pressure 616 bar

The interaction between the effects of jet pressure, jet diameter and the effective stand-off distance from the nozzle to the rock, as they are controlled by borehole pressure, rock type and ground pressure have yet to be thoroughly studied. The strong effects identified do illustrate, however, the advantages of using the initial cutting jet to cut a sufficiently deep kerf around the opening to distress the rock within that perimeter, and thus making it easier to excavate the rock.

It is useful to explain that point a little further. Rock within the ground is subject to pressures both from the weight of the overlying material, and from the changes which have occurred in the Earth since the rock was deposited. These pressures are often referred to as **in-situ stresses**. When a hole is made in the rock the pressure which the removed material carried must be redistributed. This usually occurs by increasing the load on the rock immediately around the hole, with the greatest increase being in the rock at the edge of the hole, and the increase reducing with distance away from that edge [8.67]. However, when a drill penetrates into the rock it will often break the central zone of the rock face first, and only sequentially remove the rock from that center out to the edge or gage of the hole (Fig. 8.49, 8.51).

The result is that the rock which is being cut by the different segments of the bit is always under a high compressive (or hoop) stress which keeps the particles of the rock pressed together and makes them harder to cut and remove. If, however, in contrast (Fig. 8.50), the rock is first cut on the outer edge of the hole, then the central rock within the hole is relieved of the surrounding stresses and their multiplication and becomes easier to cut. At the same time the creation of this outer cut gives the rock an outer free surface to break to which will, in turn, make it potentially easier to break the rock than if it were still being broken into a central, smaller cavity.

The need to reduce the stand-off distance from the nozzle in order to maintain ROP with increased hole depth helps to explain why waterjets may be a more effective tool when combined with diamond cutting bits than they are with conventional carbide bits. With a conventional bit, unless the nozzle is extended, which makes it more vulnerable to damage, there is a considerable distance between the nozzle and the rock surface, leading to a considerable power loss in the jet stream before it reaches the rock. In contrast, with diamond bits the "cutting assembly" is set in a much shorter distance above the rock, allowing the jet nozzles to be mounted much closer to the rock surface (Fig. 8.51). In addition, the lack of large roller cones, which block easy location of the nozzle assemblies near the rock, will permit a greater number of nozzles to be set on the face of the bit. This development has been commercially pursued by Hydronautics [8.68] with some commercial success. The high rates of penetration which can be achieved with this combination have also been referred to above in the work by Maurer et al. [8.58].

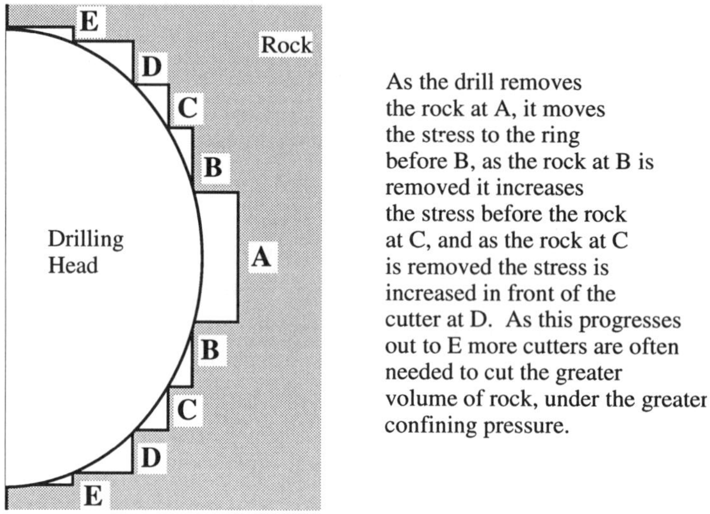

Figure 8.49 Sequence of cutting with a more conventional bit.

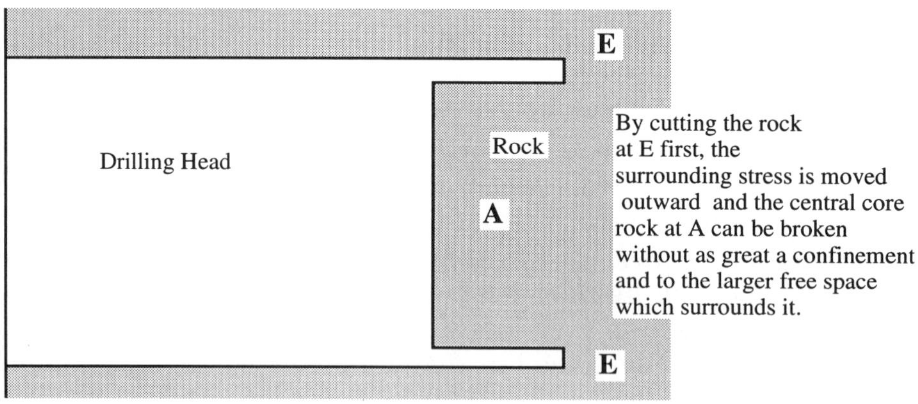

Figure 8.50 Sequence of cutting with the edge, or gage cut made first.

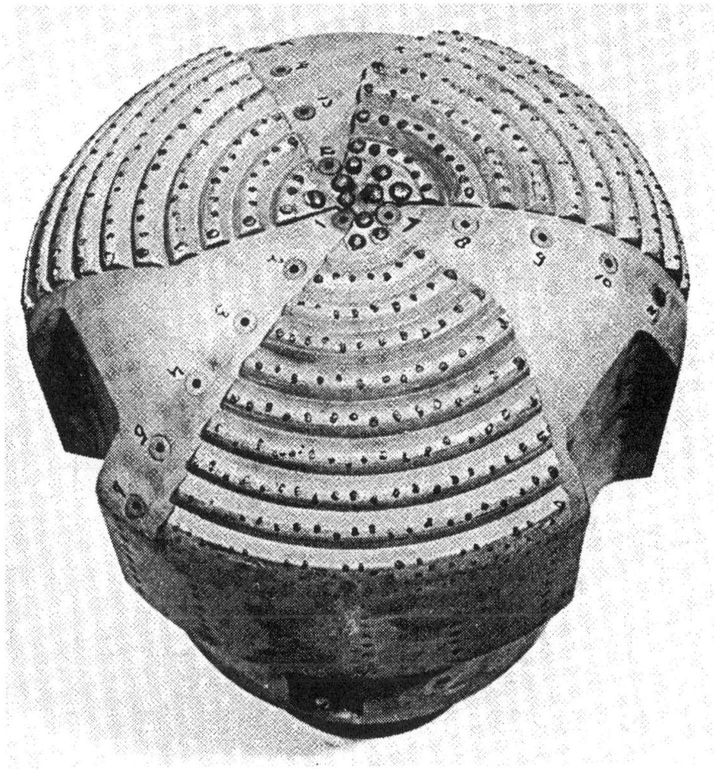

Figure 8.51 Jet-assisted diamond drilling bit, the numbers mark the nozzle locations [8.61].

The work at Hydronautics grew out of an initial program at the U.S. Department of Energy to improve the economics of geothermal energy production in the United States [8.69]. One way of improving the economic efficiency of a geothermal well was to reduce the cost of drilling the hole, which is, in turn, controlled in part by the drill ROP. The concept of using cavitation, which was known to occur in deep hole drilling, as a means to enhance the drilling jet performance, had earlier been proposed by Angona [8.70], however, using the results of earlier studies for other applications, Hydronautics worked with NL Hycalog to develop a combined cavitating jet:roller cone drilling bit. Because the impact pressure generated by collapsing cavitating bubbles can exceed 7,500 bar [8.71] these bubbles can significantly weaken the rock ahead of the tricone bit. Concurrently the turbulence generated around a cavitating jet can enhance the cleaning of the chips and mud from the rock on the bottom of the hole. Again this will enhance drilling performance. The phenomenon of cavitation and its more general role in waterjet applications will be discussed in more detail in Chapter 10.

Because the damage caused by the impact of a cavitating jet is most significant when the target lies relatively close to the nozzle, performance of the Hydronautics nozzles improved fourfold when this stand-off distance was reduced from 89 mm to 12.5 mm. A laboratory comparison of a conventional extended nozzle bit with one fitted with cavitating nozzles [8.69] was carried out with a pressure drop of 168 bar across each nozzle, and showed that the cavitating design improved the specific ROP (used to account for slightly different flow rates of the two designs and defined in terms of mm/kW-hr). This was sufficient encouragement to lead to full field trials of the concept [8.72] which, in turn gave very promising results (Table 8.5). The technology has since been commercialized.

Table 8.5 Initial comparative field trials of cavijet-augmented roller cone drill bits [8.72]

Bit No	*Bit Type/Nozzle Type*	*Distance Drilled (m.)*	*Rotating Time (hrs)*	*ROP (m/hr)*
a) In Shale, at 50 - 70 rpm, 85 - 107 kN weight on bit, Trial in Canada				
7	Standard/Standard	101	28.5	3.5
8	Two cone/Cavijet	37	3.5	10.5
9	Two cone/Cavijet	115	11.0	10.5
10	Standard/Standard	210	36.0	5.8
b) In Shale, at 70 - 90 rpm, 36 - 72 kN weight on bit, Trial in Canada				
19	Standard/Standard	89	34.8	2.56
20	Standard/Standard	95	35.8	2.66
21	Two cone/Cavijet	164	41.0	4.00
22	Two cone/Standard	118	41.0	2.87
23	Standard/Standard	180	91.5	1.96
24	Two cone/Cavijet	124	35.0	3.54
25	Standard/Standard	89	33.8	2.63
26	Standard/Standard	90	30.5	2.95
c) Trial in Texas with only limited Data reported				
i	Standard/Standard			2.1
ii	Two cone/Cavijet			3.0

The use of cavitation for drilling was explored at the Institut Francais du Petrole, beginning in 1984 [8.73]. The initial concept was to include a second jet, or decompression sub which would reduce the pressure around the bit in order to create the cavitation. Where a second jet was used it would be pointed back up the borehole, rather than downwards, while the decompression sub contained a small pump to pump fluid back up the hole, thereby lowering the pressure around the bit. Laboratory experiments [8.74] validated that it was possible to induce cavitation down to depths of 4,000 m where the hole pressure would be on the order of 500 bar, and where, therefore, a driving pressure of 700 bar would be required to create the cavitating clouds of bubbles.

An alternative approach to enhancing the ROP by adding waterjets to conventional bits has been taken by Flowdril and Grace Drilling Co. In this approach much higher pressures are supplied to the drilling bit through an inner high pressure, beryllium-copper tube passing down through the conventional drill pipe [8.75]. As a further validation of the efficacy of the Geothermal program funded by the U.S. Department of Energy this concept appears also to have been initially developed under that program [8.76], although in the early analysis of three competing systems for delivering high pressure to the bit, this was considered a less potentially successful technique than developing a downhole pump.

At a pressure of nearly 2,100 bar, initial field tests of the Flowdril system were carried out at a well in West Texas in February 1989. The augmented bits drilled through dolomite at an ROP of around 27 m/hr in contrast to a conventional bit rate of 9 m/hr [8.75]. Four 225 kW intensifiers delivered a total of around 160 lpm of mud to the drilling bit. By the time of the initial test some $17 million had been spent, over four years, in getting to the first field trial. By 1991 four operational prototype systems had been assembled [8.77], capable of drilling depths of 3,750 m.

The system uses either pure water or brine as the cutting fluid which is delivered to the bit through the high pressure circuit. A separate circuit circulates the lower pressure drilling mud through the annulus between the two pipes in order to conventionally remove the cuttings generated (Fig. 8.52). Tests of an improved version of the system in East Texas showed that the rate of penetration could be increased up to three times (which it might be pointed out was the target predicted by Maurer for the technology to become economically practical [8.53]). ROP for the new drill are related to the weight on the bit; at reduced weight the gains from the jets are less significant, about 1.4 times conventional at one-third the normal thrust, but three times the rate when the full weight was applied to the bit. In order to ensure the full benefit from the jetting action the nozzle is located close to the face of the bit, with the jet directed to cut close to the gage of the hole. The final design of the bit, especially for larger and deeper hole drilling was still being developed at the time that this work was written.

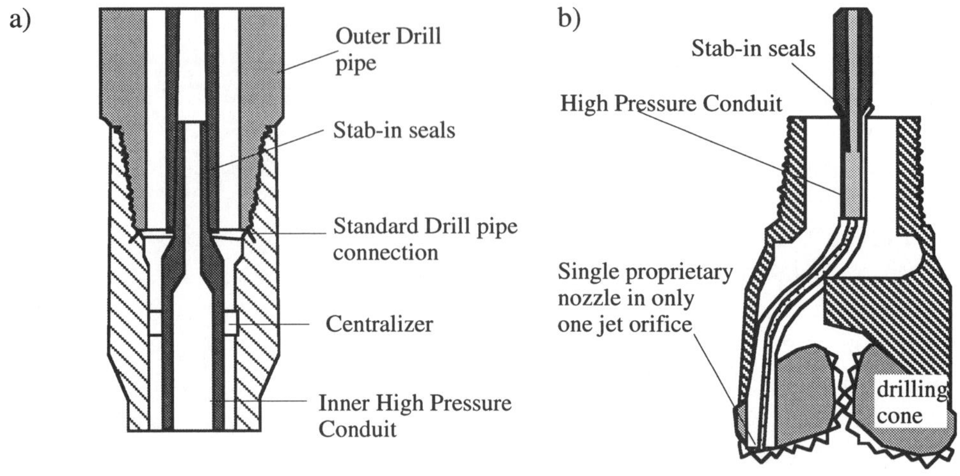

Figure 8.52 a) Inner pipe connection of Flowdril system and b) drilling head configuration [8.77].

The development of drill heads which combine mechanical and waterjet cutting elements has considerable potential. At the present time a number of investigations has been spread over a range of different techniques and for a number of different applications. Thus, it has been rare for the potential benefits of one aspect of the technology or the results from one investigation are integrated into approaches by other investigators in the field. This is particularly the case when there are ideas which did not work. For the sake of going against this trend, and also therefore possibly providing otherwise unavailable information, the following experiment at UMR has been included at this point.

A drilling bit design was developed at UMR, at the beginning of the work on geothermal bit design, which was along the lines of the design described above. Although the initial tests with this nozzle were with the jets directed vertically, this was changed so that the jets were directed across the face of the cutting bit teeth (Fig. 8.53).

When this rock was tested in rock of a consistent geology a significant improvement in ROP was achieved over the unassisted bit at the same thrust. When the bit was used to drill through several rocks, where the rocks were relatively easy for the jets to cut, a significant improvement in ROP occurred. However, when the bit was used to drill through rock of increasing hardness, a point was reached where the rock under the mechanical teeth failed, before the jets had adequately removed the material in the path of the nozzles. In this case, the full weight of the bit came on the nozzle body (Fig. 8.54), which failed.

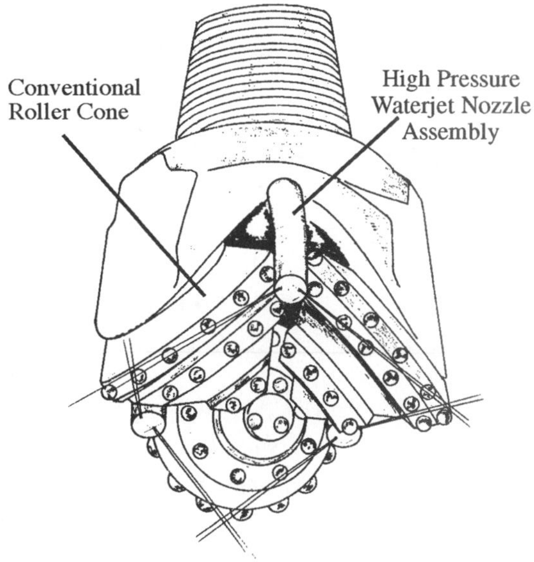

Figure 8.53 Schematic showing the location of the jets to assist roller cone bits in a UMR design.

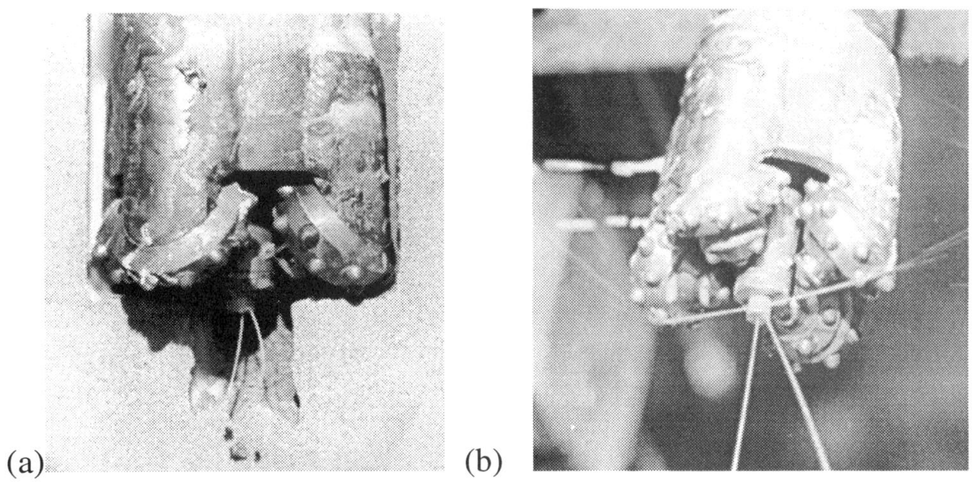

Figure 8.54 UMR design for jet-assisted mechanical drilling bit, which did not always work a) showing pre-drilling in a sectioned hole in rock b) outside the hole showing the lateral "cleaning" jets.

The geothermal program itself did not continue to actively fund the further development of waterjet assisted drilling, although it had been the cultivator of several of these concepts. One somewhat different approach, as yet only tested in a laboratory concept, also came out of this program. The initial requirements from the Department of Energy had been that the program should reduce costs of the geothermal operation by 25%. While the recent achievement of a threefold improvement in drilling rate by Flowdril brings that goal to within practical reach, the alternative concept related to an alteration of the borehole geometry. Flow into and out of a geothermal well is controlled by the diameter of the well. Thus, by reaming a well out from an original radius of 115 mm to a final radius of 1 m, flow would be increased by 36%. This would, in turn, require that the number of wells that must be drilled would be reduced, providing the necessary cost savings [8.78].

In order to ream the cavity out to the required size, at a depth of up to 2,000 m it was necessary to design a tool which could be fed down through the existing well bore. Once at the require depth the head had to be expandable to cover the full face of the rock to be removed over the required change in diameter. This was determined based on the reduced range of a jet cutting device at that great depth [8.66]. While such a concept was developed, the demise of the emphasis on the geothermal program has not led to a full scale or field test of the concept. This approach to the unconventional abilities of waterjets, relative to conventional drilling systems, however, led to their consideration in a different venue, and an investigation as to their capability to solve other problems.

8.7 ROUND THE CORNER DRILLING

Because the pressure exerted through the high pressure jet will be directed along the direction in which the orifice of the nozzle is pointing as the jet leaves it, this not only allows high levels of horsepower to be transmitted through relatively narrow pipes, it also allows the force at the delivery end to be directed other than along the line of the pipe. This principle was exploited in conceiving what is now known as the round the corner (RTC) or cornering drill which was conceived, designed, and field tested under sub-contract from Sandia Laboratories to UMR under a contract from the United States Department of Energy [8.79]. This drill, and its subsequent developments, are described in some detail so as to show some of the advantages which waterjet drilling can provide.

The problem initially to be addressed was to resolve an ongoing problem in the in-situ combustion of coal. To control and extract energy from a burning coal seam in the ground, one must first have two vertical shafts, or wells, which have been sunk down to the coal seam, which is normally too deeply buried to be economically strip-mined from the surface. Once the coal seam has been set on fire, then air is blown down one of the shafts to the seam so as to sustain combustion, while, the gases and other valuable by-products of the burning are recovered through the second shaft (Fig. 8.55).

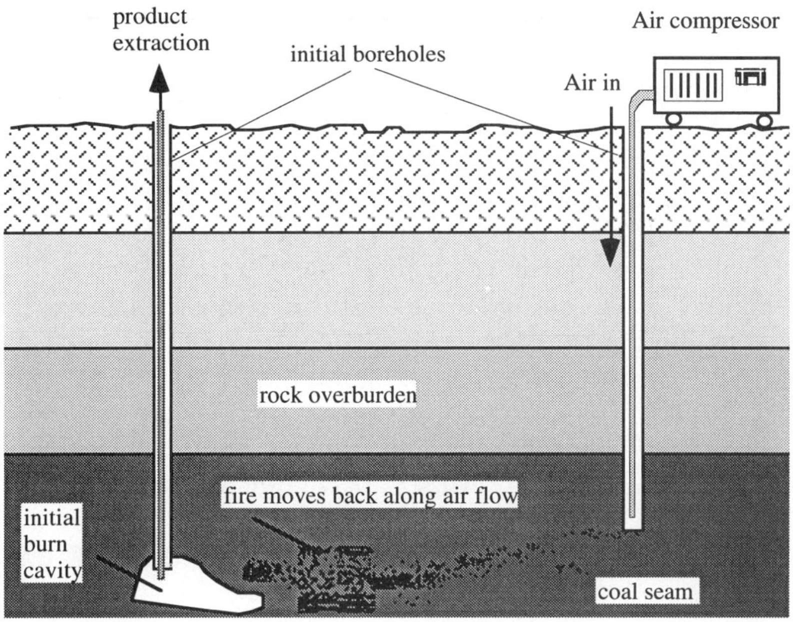

Figure 8.55 Reverse burn system for preliminary well connection and subsequent in-situ combustion of the coal [8.79].

A second requirement for a successful operation is that there should be a path between the two shafts so as to control the direction of the burn, and to collect the gases from the fire at the second shaft. Historically this has been created by starting an initial burn at the second shaft (as shown above) and having this work its way back along the air flow back to the first shaft. Hopefully this occurs with the fire migrating along the bottom of the coal seam, and with the fire confined to a narrow path. The problem is that this does not always work, and if the flame burns up to the top of the coal, it is harder then to go back and burn the material under the passage. At the same time the speed of the flame front is usually less than 1.0 m/day, and so it might take more than a month to connect two wells which are only 30 m apart.

Conventional mechanical means of connecting the two wells are limited, since the minimum turning radius of a mechanical drilling tool is on the order of 6 m, while the depth of the wells is likely to be less than 50 m. It was suggested that, because of the directional freedom of waterjet drills, that they might be able to make the turn from the original vertical well into a horizontal direction within a shorter turning radius. A program was started to develop a drill which could be lowered down a vertical well, and would drill out horizontally from this well, into the coal seam. The preliminary experiments to find out how rapidly and under what conditions that the waterjets could effectively drill these connection passages through the coal have been described above [8.34], [8.36]. Once these had been carried out, the next step was to develop the method for turning the corner and advancing the drill horizontally from the vertical bore.

The RTC drill was designed around the very low thrust forces that would be required since the waterjets require almost no thrust across the drilling head to direct them forward into coal. Further, once the drill is aligned along a straight line, then it should be possible to maintain the alignment of the hole, as the head advances, since there are no deviatory forces on the drill stem from high thrusts such as are found with conventional mechanical drills. This is particularly important in long hole drilling in coal mines where any deviation of the hole from the horizontal can cause it to leave the coal seam and penetrate into the rock either above or below the seam.

In order to demonstrate that waterjets could not only drill the horizontal section, but also make the turn from the vertical, a machine was built and several versions field tested over the period from 1978 to 1981. The basic idea on which the design developed, was that if it was possible to only rotate the head and drive mechanism of the drill, then all the other parts of the drill stem leading to the head need not rotate, and their ability to turn could be simplified. The drilling unit was accordingly built around this principle, and the tests showed that it was possible to turn the drill through 90° in a small space. Field trials of this device were carried out in 1978 and verified that the concept worked (Fig. 8.56).

Figure 8.56 Field layout of equipment for drilling around corners.

The drill consisted of a nozzle assembly containing a rotating nozzle assembly, a small high pressure swivel and a small drive motor. This was connected to the supply tubing. This tube contained the three hoses, one of high pressure to the cutting nozzles, the other two supplying hydraulic fluid to and from the drive motor. The drive tube was made up in segments each 45 cm long, and hinged on their top surface. A drive chain was welded along the length of the tubes. As the tubes came down the vertical well, they were disconnected on their lower side, and thus, were able to rotate around the drive pulley at the bottom of the vertical well (Fig. 8.57). As they returned to the horizontal, they could be reconnected, and the chain was engaged by the toothed sprockets of a drive gear. This was, in turn, driven by a motor in the vertical section of the well. Thus, the pipe would feed down the vertical well as a rigid unit, and would also feed into the horizontal as a solid unit.

The nozzle design was initially similar to that used for hard rock, in that one waterjet was directed forward into the coal, while a second jet reamed the hole out to the required diameter. This was changed since this design tended to drill a square hole (Fig. 8.30), and Sandia requested that the drill cut a round hole. This necessitated putting a waterjet to cut along the line of the required perimeter of the hole, so as to make a free face for the reaming jet to cut toward. In retrospect this may have been a mistake. The reason that cutting a square hole might have been better relates to the need to keep the drill within the coal seam.

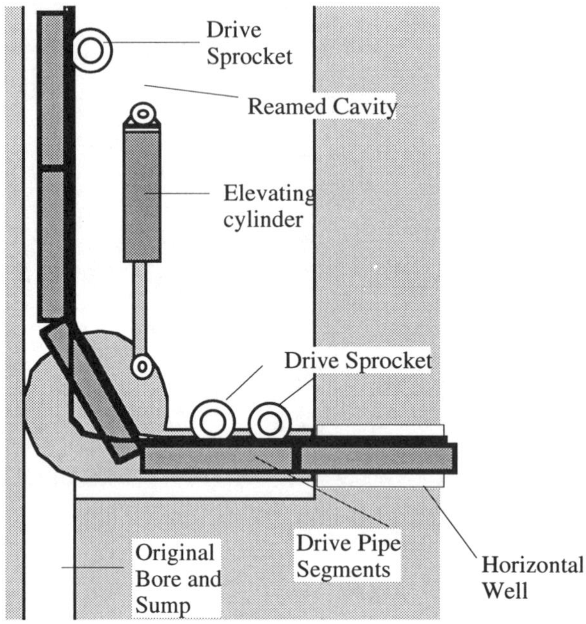

Figure 8.57 Schematic showing the drive components of the RTC drill [8.79].

One cannot guarantee, due to geological changes that the coal seam itself will remain horizontal. Thus, while it is important that the waterjet retain its relative position within the coal section, the coal section may be moving up or down, as the coal gets further from the drill insertion point. If the drill is to remain in the coal over long distances, the hole drilled must be able to align itself within the seam. To do this it would help if the drill could control the exact position within the seam in which it lies.

This principle can be used in the RTC drill to keep the head drilling in the coal seam, since the drill can be patterned to only drill a full volume hole in coal. This however, requires no perimeter slot be cut. Other advantages also accrue to a square hole being drilled. Square holes are drilled, because the cutting out from their central position to find and exploit weakness planes within the coal structure. The water will penetrate and infuse along this plane of weakness and force the coal off into the free space which is the hole. In this manner it has been observed that, provided the two original horizons are relatively clearly defined, the waterjet will continue to cut to these planes as it drives forward into the hole. This device could then be more accurately controlled since it would be inherently self-steering in the vertical plane, rather than relying on electronic metering and control, which has a short life in this difficult environment.

A second advantage of this system is that the drill was able to turn a 22.5 cm turning radius, the minimum bend radius of the pressure hose used in this equipment. The whole device could thus be lowered down a vertical well which has been reamed out by 30 cm, over a vertical interval of around 1 m. This additional space allows the turning mechanism to be raised into the horizontal position. The drill can then be activated and will drill forward into the coal.

This device was built and was demonstrated through use of a simulated borehole from which the drill was fed forward into an outcrop exposure of the coal. These tests showed the validity of the concept, but also highlighted the problems of clearing the debris from the hole. To overcome that problem additional jets were put on the assembly in order to increase the flow of fluid carrying fragments back to the vertical well. Advance rates in these trials were at rates greater than 15 m/hr [8.80], and proved that the drill was capable of meeting the assigned tasks. An improved version of the drilling head was developed, [8.81] and development of the unit was then turned over to Sandia Laboratories who had designed the instrumentation for the further progress of the technology [8.82, 8.83]. This instrumentation was largely concerned with indicating the orientation and location of the drill head, and adjusting this to driving the drill in the required direction and azimuth (Fig. 8.45).

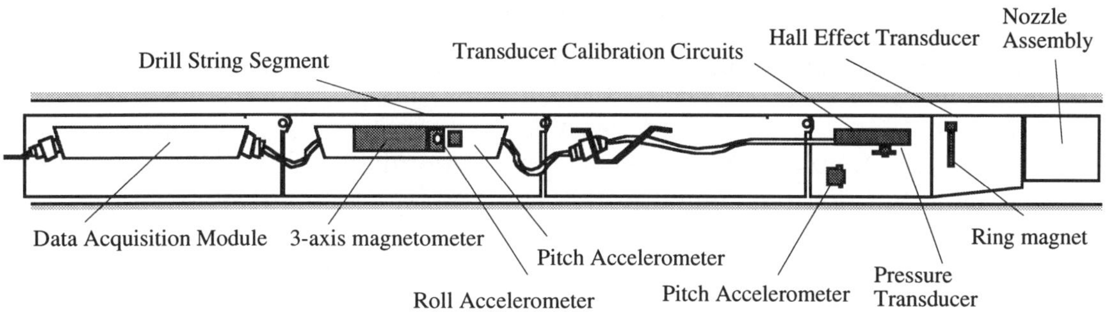

Figure 8.58 Schematic of the RTC drill head, showing head instrumentation (after [8.83]).

The device was tested in bench scale at the Sandia Laboratories in Albuquerque, and then taken to an abandoned mine area in New Mexico for a full field test [8.84]. The drill was lowered down a well to a coal seam some 30 m below the surface and the drill raised into the horizontal position. Some equipment problems limited the completion of the full field program, however holes were easily drilled out for a distance of 12 m, and one of the test holes extended some 22 m into the coal (Figs. 8.59 and 8.60). The program has since been placed in abeyance, although there has been recent interest in revitalizing it for methane drainage applications and for use in remedial work around underground waste storage facilities.

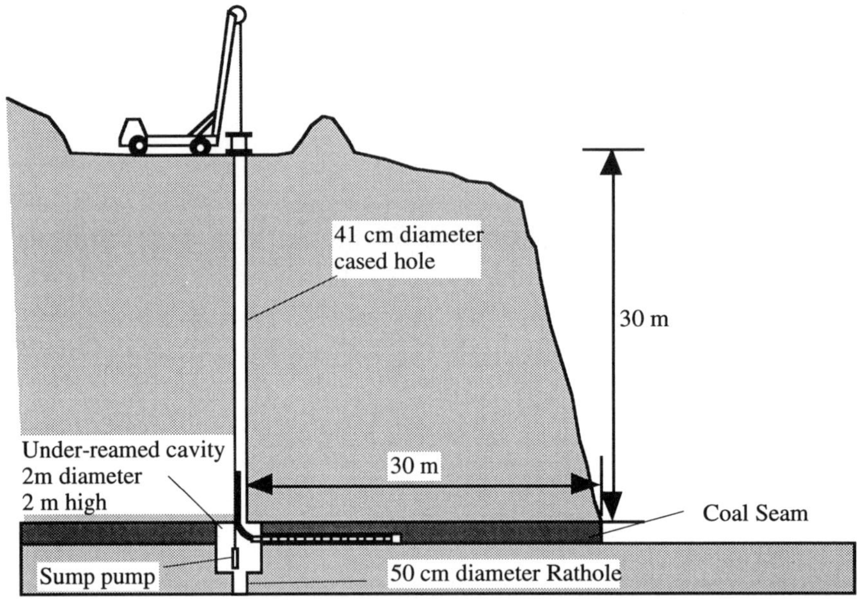

Figure 8.59 Schematic equipment layout for the test to verify the "round the corner" drilling ability carried out by Sandia Laboratories [8.84].

This development has opened up a number of different possible applications for the technology. For example, recent demonstrations that horizontal drilling can significantly improve the production from oil reservoirs was first sparked by work at the Rospo Mare field in Italy. The first horizontal well was sunk into this oilfield in 1982 [8.50]. By 1988 the field was producing 22,000 b/d of crude, 19,000 of which came from six horizontal wells, with the remainder coming from three vertical wells. The wells were drilled with conventional equipment [8.85], which meant that the drill had to be inclined slowly around until it reached the horizontal (Fig. 8.61).

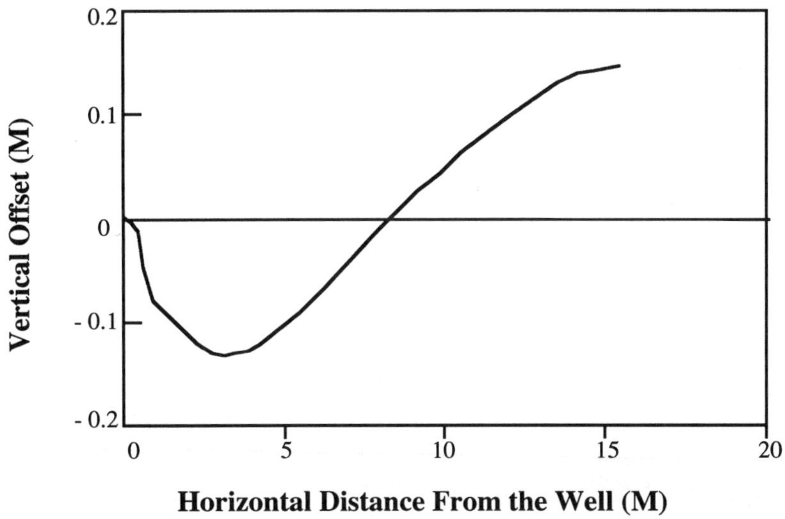

Figure 8.60 Plot of relative alignment of drill with horizontal penetration [8.84].

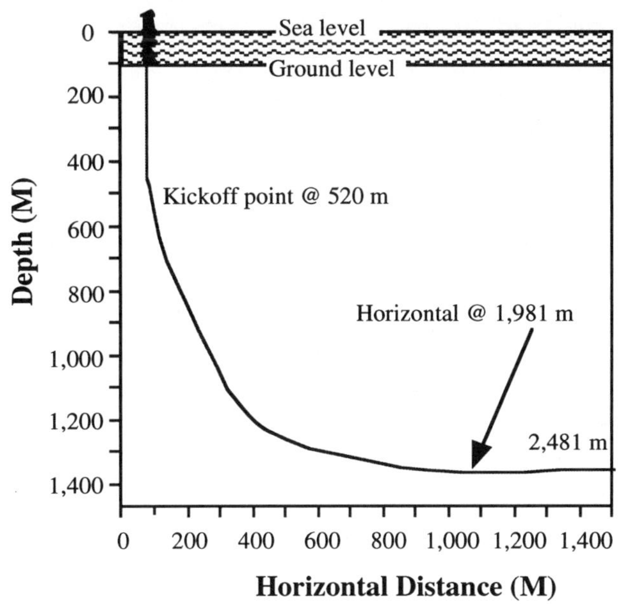

Figure 8.61 Horizontal well profile from the Rospo Mare field [8.85].

This success has lead to a major development of this technique into the oil industry, particularly in the United States, although the technique has also been implemented in Canada, Europe and the Soviet Union. The technique has been investigated for possible use in oil mining [8.86], and for use in methane drainage of coal gas (methane) from underground coal seams.

In 1981 Bechtel began a program to develop waterjet drilling techniques for use in horizontal well development [8.87], now known as the BecWell Horizontal Drilling System. The concept followed along the lines of that originally developed at UMR in that high pressure waterjets were to be used as the method of cutting through the rock. In addition, the design makes use of an access vertical well, and a reamed cavity from which to drill out into the formation (Fig. 8.62). However the drill design, the drive mechanism and the configuration of the drill feeding into the horizontal are all different. The drill head is mounted on a continuous coil of carbon steel tubing 31.75 mm O.D. and 27.7 mm I.D., operates at pressures of 420 - 700 bar, and does not rotate. After the vertical well has been drilled and under-reamed to a diameter of 1.2 m, over a vertical height of 2.4 m, the drilling system is lowered into the well and raised to the horizontal position. The jets cut a hole varying from 10 - 15 cm in diameter as the drill advances, which it does under the driving pressure developed from the operating pressure of the jets.

Relatively high advance rates have been claimed for the tool, particularly in the softer formations of the Kern River. In one well at that site, the tool was used to drill four radial wells out some 30 m into the reservoir, and these were then used as a means of injection steam into the formation to stimulate oil production. Before the well had the radials drilled it was producing 5 b/d of oil. After the treatment it produced up to 700 b/d of crude. In order to penetrate the steel pipe and allow the steam access to the formation (the pipe is difficult to retract and is usually cut off and left to lie in the bottom of the hole drilled), the pipe was perforated electrochemically. The technique was developed by Petrolphysics, who have continued to refine the technology. This further development was, in part, designed to extend the range of cutting capability beyond that of the relatively unconsolidated formations of the California reservoirs [8.88]. A paper in 1987 [8.89] described experiments with a conical nozzle (Fig. 8.63), which would appear to use cavitation to improve the penetration of the drill. This topic will be dealt with in Chapter 10.

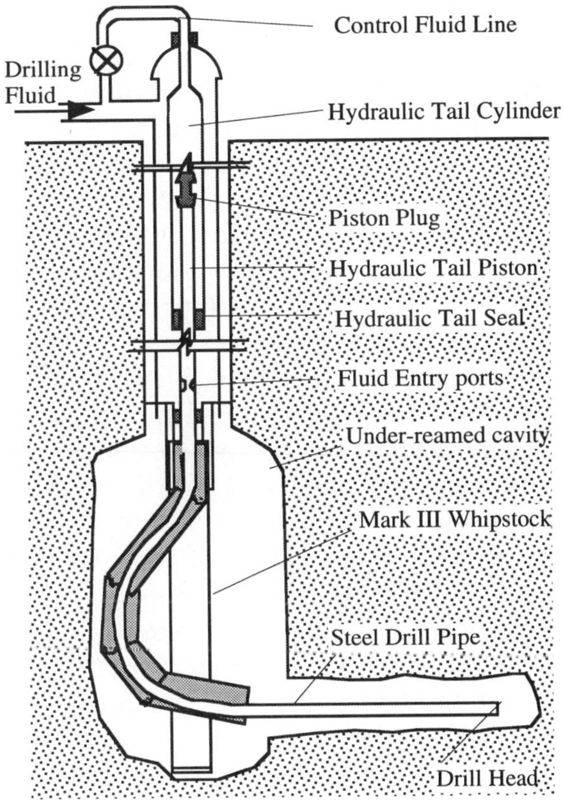

Figure 8.62 Drive mechanism for the Petrolphysics drilling system [8.88].

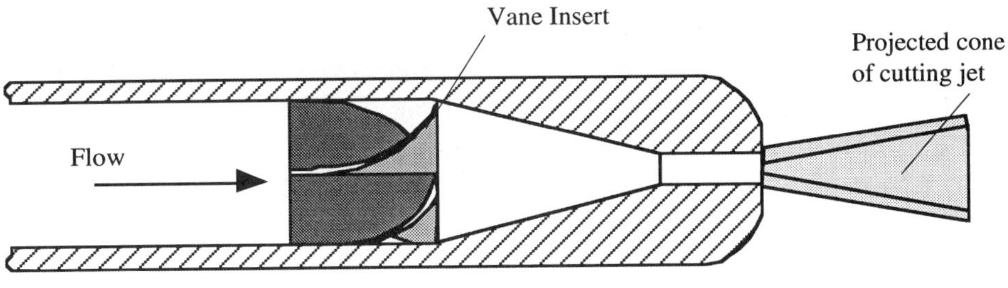

Figure 8.63 Conical nozzle design investigated by Petrolphysics for drilling horizontal holes [8.89].

The rig is designed to make the turn into the horizontal with a radius of 25 - 30 cm. Some 750 kW of hydraulic power are utilized to power both the drive mechanism and the cutting jets. By using the driving pressure to push the front of the drill into the formation a force of some 2,700 kg is developed at 560 bar, and since this is exerted on the front of the drill, it tends to pull the drill string into the hole in a straight line. By 1985 some three hundred radials had been drilled into wells in the California basins of the San Joaquin Valley and the Los Angeles basin.

A second generation tool [8.90] has subsequently been developed which requires that the reaming diameter can be cut in half. However, at that time the technique had still only been developed for unconsolidated deposits, with the work ongoing to develop the different design for the cutting of the harder rocks. It is reported that, using this technique with the new nozzle design, that they have been able to send out as many as 500 radials from a single well, working at four different levels [8.91].

There are other circumstances, however, where drilling "round-the-corner" in unconsolidated materials may be of advantage. For example, Kaback [8.92] has described the problems of containing the wastes from nuclear operating facilities. Solvent spills, for example, have penetrated into the ground and contaminated large volumes of weakly consolidated and unconsolidated ground. By running horizontal wells above and below this contamination it is possible to bubble air through the solvent, stripping it from the ground and capturing it in the overlying well. The technique requires that very shallow wells be driven, to depths of less than 50 m, an extremely expensive operation with conventional drilling equipment, necessitating that the rig be tilted over on its side.

An alternate use of the "round-the-corner" drill is in mining. There are many seams where the coal is quite thin. At the same time the rock overlying the coal requires, for best support, that rock bolts be inserted which are longer than seam height. The Bureau of Mines has spent many years investigating this problem [8.93] and has developed special drills and equipment for bending the bolts into the hole and then straightening them out. The inherent simplicity of a flexible waterjet drill, and its proven ability to drill through even the harder rocks [8.94] suggests that this is an area for future development with considerable potential.

There is one additional benefit which can be applied with the use of high pressure waterjets. In the holes drilled using the BecWell system the horizontal well is left with the carbon steel pipe lying in the bottom of a 10 cm diameter hole. In order to either inject or withdraw fluid from the surrounding rock along the length of the hole some means of perforating the hole is required. While electro-chemical means have been used by the operators of the BecWell system, there are simpler waterjet techniques which do not carry with them the potential for contaminating the ground.

Where the pipe is of a softer material, such as, polyvinyl chloride, fiberglass, or cement a plain waterjet can be used which may penetrate up to 1 m into the surrounding rock. In its original application [8.95] the technique was developed for use in stimulating uranium leaching wells.

Where the pipe is of steel, then it is necessary to add abrasive to the fluid. While this has been achieved by mixing the abrasive and the water in the pipe, the small diameter of the BecWell system makes it more advantageous to use a single pipe. This requires an entrained slurry abrasive system of the DIAjet type be used. Such a system has been demonstrated and shown to be effective [8.96]. The abrasive laden jet, at a flow of 40 lpm at 350 bar was able to penetrate the pipe in approximately six seconds. It should be mentioned that the use of abrasive jets for cutting pipes is not, in itself new (for example [8.97]). However, earlier tests of the system used lower pressures, around 175 bar, and larger jet diameters, around 4.75 mm diameter. With such a system it would take ten to twenty minutes to give the required penetration, which must be through the casing and into the rock. It has been found advantageous, once the jet has penetrated the casing and the cement to minimize the amount of rock removed.

This ability of waterjets to cut out from the borehole and modify the shape of the surrounding ground has additional benefits.

8.8 BOREHOLE MODIFICATION

The discussion at the beginning of this section dealing with the use of waterjets for drilling rock showed that, by varying the advance rate of the drill, it was possible to change the diameter of the hole drilled (Fig. 8.4). This feature of waterjet drilling can have several advantages. For example, even when drilling small holes for blasting purposes it is possible to "chamber" the hole over the interval where the explosive must be placed (Fig. 8.64). This has two advantages, firstly it means that the explosive will be more effectively confined than would normally be the case. This is because it is conventional practice to fill the hole above the charge with waste rock or stemming, in order to confine the explosive products and thus force them to assist in fracturing the surrounding rock. By using a smaller diameter hole to access the chambering section, the amount of stemming required will be smaller, and the confinement achieved by the rock over the explosive will be greater. This may well have the potential for requiring less explosive to achieve the same amount of ground fragmentation. This becomes more likely if it is realized that the explosive works more efficiently if it is packed as a spherical charge, rather than the conventional long thin sausage in the blast hole [8.98]. To date, no experiments have been carried out, to this author's knowledge, to validate and quantify this likely economic improvement in blasting performance.

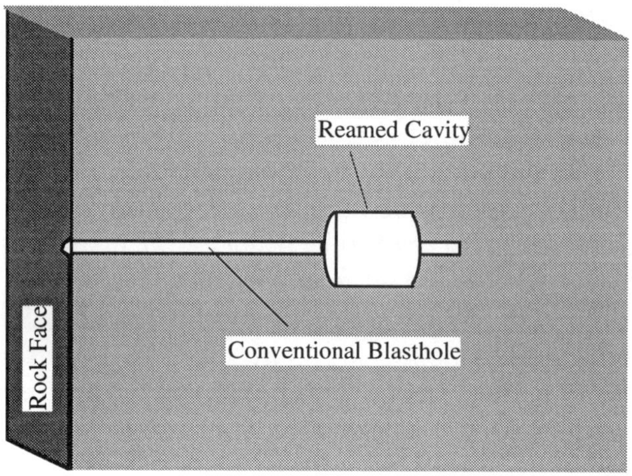

Figure 8.64 Chambering a blast hole to allow more efficient use of explosive.

An alternate benefit to blast-hole chambering has already been demonstrated in two different applications. The first of these, and one of the pioneering demonstrations of the technology, was carried out by Stone Age [8.99].

While the first demonstration was in the mining of uranium ore, a more novel development came in the cutting of a grout trench for a dam and in enlarging the excavation of a tunnel in Utah [8.100]. In normal rock excavation where the rock to be broken is fragmented, it is often necessary to protect the remaining wall which will be left after the blast. One way of doing this is to drill perimeter holes around the required edge of the opening and to lightly charge these with explosive. By firing these together a crack may be grown around the edge of the projected opening. This crack then acts as a barrier to the growth of the cracks generated when the main blasting round is then fired. The procedure is known as **pre-splitting** the rock around the opening.

What Stone Age did in Utah was to go one stage further than this, and to use waterjets not only to drill the perimeter holes, but by then using a lance with two jets pointing out at right angles to the borehole (Fig. 8.65), it was possible to cut deep enough notches into the side of these holes that they connected. This created an even stronger barrier to crack growth into the surrounding wall and produced a stronger rock wall as a result. The 25 mm diameter holes were typically drilled to a depth of 3.6 m in 40 - 45 seconds. The drill used 23 lpm of water at 620 bar. With two 0.8 mm diameter nozzles the jets cut to a depth of 10 cm into the rock on average, and frequently intersected one another. Where the connection was not made, then 100 grain detonating cord was strung through every second or third hole. This was then detonated to make the required fracture. The resulting accuracy of the wall was within 12 mm of the required contour. The project took two men four days, during which they drilled and notched 1,200 m of hole.

Figure 8.65 Notches cut out from a drill hole wall to ease fracture and help pre-splitting.

The improvement in quality of the wall created using this technique was demonstrated by the more conventional use of perimeter blasting by McGroarty [8.101]. In a set of experiments carried out in the development drift for a new portal at the UMR Experimental mine a series of blasts was carried out comparing conventional pre-split blasting with the results achieved when the holes were first notched, using high pressure waterjets. Jet notching of the local dolomite was carried out, after the holes had been conventionally drilled, using a pressure of 1,100 bar with the lance being withdrawn at the rate of 70 cm/min. It was established that a higher degree of rock wall quality could be obtained and that hole spacing could be increased by almost 50% for equivalent results.

The improvement which could be obtained in the quality of the rock wall could be observed both visually (Fig. 8.66) and by examining the quality of the rock, by passing a sound wave through it. The velocity of the wave is a function of the degree of fragmentation or cracking of the rock. Such a study was carried out in the blasted area by Worsey [8.102] and this showed (Fig. 8.67) the improved rock quality where the holes had been pre-notched by the jets.

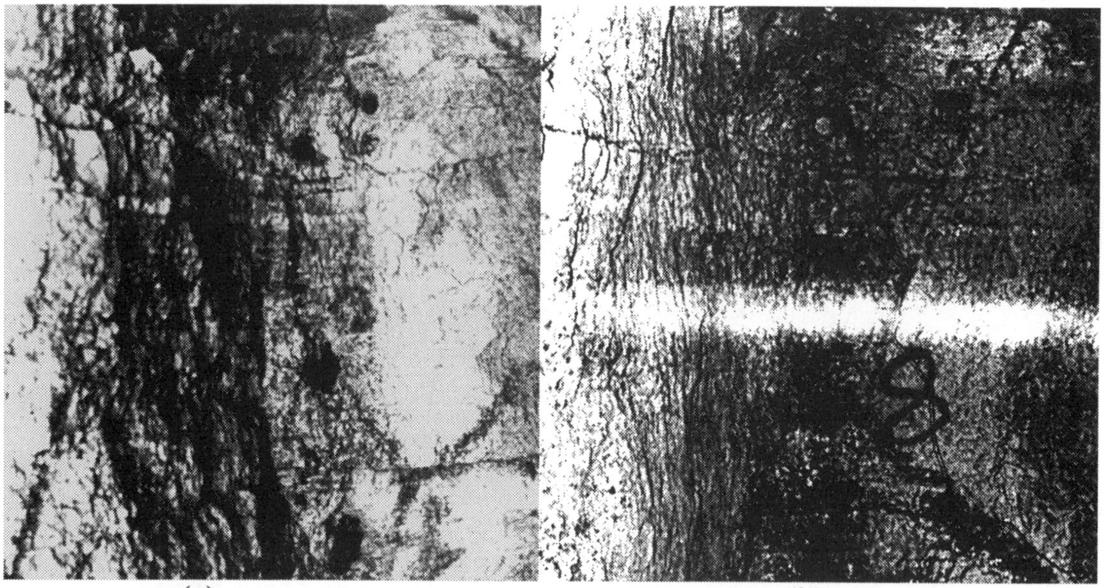

(a) (b)

Figure 8.66 Wall condition with a) conventional blasting b) waterjet pre-splitting of the perimeter holes (note the half-collars of the blast holes).

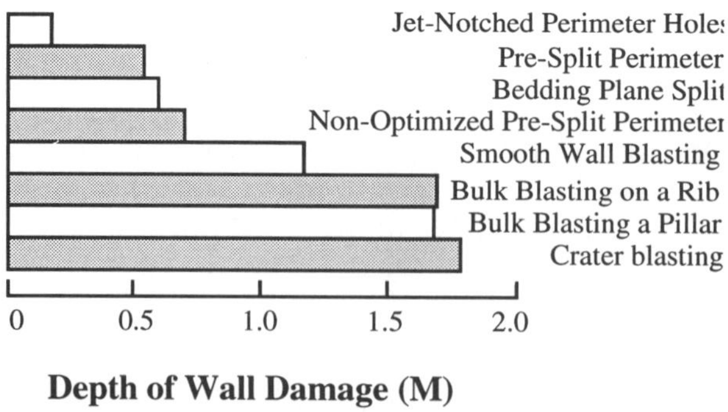

Figure 8.67 Average disturbance depths for different wall protection techniques [8.102].

The results were significantly better than when a mechanical tool was used to create the notches in the borehole wall, as had previously been tried in a tunnel in Boston [8.103]. In that operation, although the notching was shown to give an improvement in blasting performance, the problems of removing the mechanical tool from the hole, after it had been used were often quite large. In contrast the waterjet lance, with no contact against the wall, was easily held and manipulated by two people, and could be advanced and retracted along the length of the holes with almost no effort. The benefits from even the shallow notching achieved with the mechanical tool can be seen from the data obtained from the tunnel test in Boston.

The table below is presented, even though the notching in this case was not carried out by waterjets, to illustrate that even though the local cost of an operation - notching boreholes, may be higher than the conventional, the overall savings from the practice can be very significant.

A similar technique was developed in Japan [8.104] where the technique was first demonstrated for cutting boulders along preferred lines of fracture. When slots were cut in boulders of granite two passes of the 3,000 bar jets cut slots 10 - 15 mm deep. When these were fired and compared with holes without the notching, it was seen that the crack grown had preferentially started where the notches were, giving a much smoother fracture of the rock. The cracks even grew against the preferred direction of fracture of the block, and with lower charges than would be required without the notching.

A field trial was carried out at the Kamioka mine in a 2,200 kg/sq cm uniaxial compressive strength gneiss. Slits ranging from 7 - 12 mm in depth were cut into the holes on the perimeter of the excavation. It was found that when the perimeter holes were notched and fired after the main round (opposite blasting sequence to that used in McGroarty's work) the post-split holes gave a rock face smoother than normal, required less explosive for same results, reduced fragmentation beyond perimeter line.

This process was carried one stage further in France [8.29]. The valuable minerals which one finds in the Earth can occur in thin veins. To extract this ore it is often necessary to blast this narrow vein and the adjacent rock. This is needed for a passage wide enough for workers to go forward into the rock, and to blast out the next segment of the vein. If the vein could be removed before the surrounding rock were blasted, this would reduce the amount of rock hauled out of the mine and processed, improving mine economics. The French experiment was to outline such a vein with a series of holes pre-notched along the edge of the vein (Fig. 8.68). The first trial of the concept was in a lead and zinc mine, and was not successful, because of poor notch quality. A second trial in a uranium mine proved the concept was practically viable, but because the holes were drilled at only half the speed of a conventional drill the economics were not considered favorable at that time.

Table 8.6 Summary of perimeter blasting experimental data [8.103]. Costs are in 1980 dollars

	Contractors normal smoothwall	*Modified smoothwall*	*Notched borehole*
Roof average overbreak (cu m)	3.9	2.8	4.0
Wall average overbreak (cu m)	3.7	3.4	2.3
Average overbreak/round (cu m)	11.3	9.6	8.6
Average advance as %age of hole length	85	89	88
Powder factor (kg./cu m)	1.6	1.44	1.26
Hole factor (number/cu m)	1.28	1.26	1.11
Cycle times (hours)			
Drilling		2.2	2.05
Notching		n/a	1.05
Load and shoot		1.2	1.3
Muck out		2.3	2.2
Total Cycle time		5.7	6.6
Drilling cost			
Labor - 2 men @ $0.25/min x 6 min	$3.00	$3.00	$3.00
Notch -2 men @ $0.25/min x 6 min	-	-	$3.00
Bits and Steel 3.6 m @ $0.66	$2.40	$2.40	$2.40
Notching tool 3.6 m @ $1.00	-	-	$3.60
Equipment 3.6 m @ $1.32	$4.80	$4.80	$7.20
Total Drilling Cost	$10.20	$10.20	$19.20
Loading Cost			
Labor - 1 man @ $0.25/min	$1.25	$1.25	$1.25
Blasting cap	$0.75	$0.75	$0.75
Primer Stick (0.24 kg.)	$0.25	$0.25	$0.25
Smooth Blasting Powder (3 m.)	$2.25	$1.20	-
Primacord (3 m.)	-	$0.60	$3.75
Spacer/spider tube	-	$1.20	$1.00
Stemming	$0.20	$0.20	$0.30
Miscellaneous	$0.30	$0.30	$0.30
Total Loading Cost	$5.00	$6.25	$7.50
Total Cost/hole	$15.20	$16.45	$26.70
Hole spacing (cm)	53	61	84
Number of holes	28	24	17
Total cost of perimeter holes	$426	$395	$454
Estimated overbreak volume (cu m.)	26	22.1	19.8
Cost of extra mucking	$35.50	$16.00	$4.50
Cost of extra shotcrete @ $75/cu m.	$532.50	$240.00	$67.50
Total Cost for Lined tunnel	$994	$651	$525

It might be noted, that the speed of waterjet cutting is often not well understood. It was observed, when new students were introduced to waterjets for notching blast holes at UMR, that they usually moved the lance out of the hole much slower than necessary to achieve the required depth of cut. Such performance would have a significantly negative effect on the overall economic evaluation of the economics of the process, especially since the waterjets can, as easily, notch a conventionally drilled hole [8.101] as they can a waterjet drilled one [8.29].

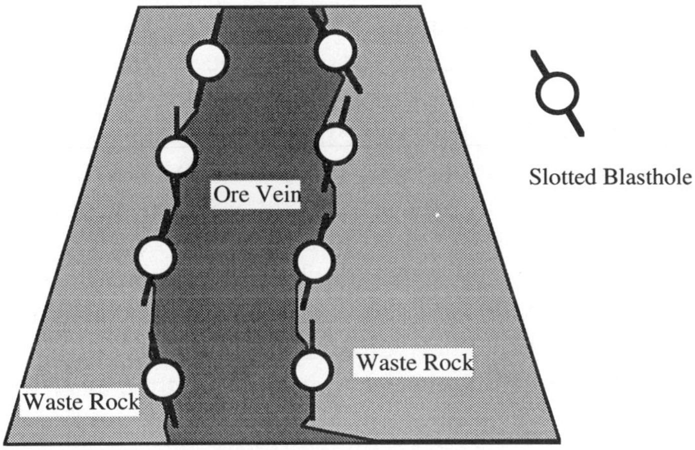

Figure 8.68 Pre-notched holes used to outline a vein of ore [8.29].

The energy required to notch the holes using a waterjet is not always necessary for the cracked hole concept to work. In a development known as CUSP mining (for CUt and SPlit) Summers [8.105] had proposed using notched boreholes and splitters as a means of breaking the rock in the center of a tunnel round, after the perimeter had been pre-slotted using an abrasive waterjet. Experiments were tried at UMR which showed that the rock could be split using relatively little pressure within the boreholes. Experiments in the laboratory with rock samples suggested that this pressure could be achieved with small rock splitters.

Field experiments during the excavation of the underground large-screen movie theater at the St. Louis Arch showed that this pressure was too small when the blocks of rock to be broken were over 0.5 m in size. When a more powerful splitter, a Darda hydraulic feather and wedge unit, was used in the boreholes the pre-notching by the waterjets was found to be unnecessary and satisfactory excavation rates could be achieved without the jetting. This was particularly the case where the cracks emanating from the holes were extended by using an impact breaker in the holes [8.106].

Experiments at the site had shown that the depth of crack required to orient the primary fracture out from the borehole did not need to be more than 2.5 mm long when using the pressurized bladder splitter. Not only was that splitter not powerful enough to break out the large size of rocks required, it also required a very smooth hole in which to work. This was not always possible to achieve, although more recent work at UMR has shown that it is possible to drill rocks, even with abrasive laden waterjets and achieve a relatively smooth wall to the hole [8.107].

At the present time the benefits from notched boreholes are not fully appreciated. They have particular benefit where there are constraints on conventional blasting. For example Japanese workers have used the notched borehole technique with an expansive grout to fracture rock and concrete. The expansive grout is inserted into the pre-notched hole while wet. When it dries it expands and exerts sufficient pressure on the walls of the blast hole as to simulate a "slow" blast. The result is equivalent to an explosive in terms of the damage, but there is a much reduced level of damage outside the crack in and there are no shock waves driven into the ground as is the case with the conventional explosive use.

As with other forms of water jet applications the range of applications where borehole modification is beneficial is a function of the rock and the cutting conditions. For example, in drilling deep holes into rock for blasting purposes it is possible to only extract a part of the length of the drilled hole when the explosive is fired. The remaining rock, or **bootleg**, must then be redrilled and fired. This can be an expensive waste of effort, since where the last third of the hole does not pull, this will require that half as many again holes must be drilled as a real necessity.

One way of possibly overcoming this is to rotate a waterjet drill at the bottom of the hole for a short period of time cutting a perimeter slot at the base of the borehole (Fig. 8.69). This will then act as a crack initiator when the blast is fired, and give a break in the rock along the required back of the cut. In this way it is possible to increase the likelihood of pulling the full length of the borehole when the charge is fired.

The range of rocks which must be drilled in a mine can vary daily. For any tool to gain widespread acceptance it should be able to drill through any rock it is likely to encounter. For many years this claim could not be made. The addition of abrasive to the waterjet system has removed that limitation.

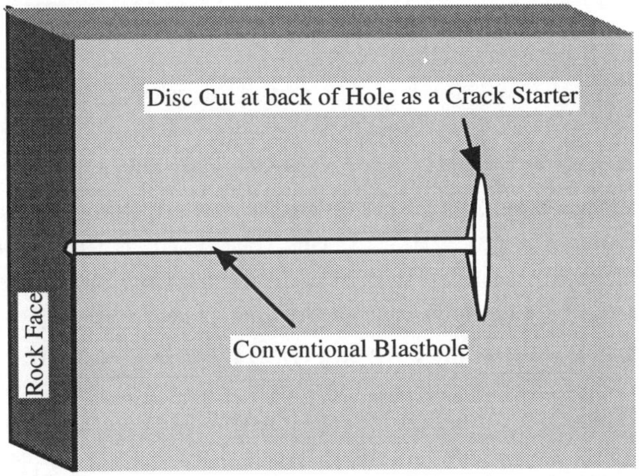

Figure 8.69 Bottom hole notching to improve blasting performance.

8.9 ABRASIVE WATERJET DRILLING

In the light of recent developments, it is interesting to note that it was as early as 1955 that the first papers on what was then known as Pellet Impact Drilling were presented [8.108]. By that time some 37 years had been invested in the project, which was directed at improving the drilling rate of oil field bits. The particles used in those early tests were significantly larger than those used subsequently, ranging in diameter from 9.5 to 19 mm. The principle behind their use was also original. Particles would be drawn into the jet stream, using a suction effect similar to that of conventional abrasive nozzle systems, and would flow down with the jet to the target (Fig. 8.70). After impact the particles would be carried up around the outside of the secondary nozzle by the recirculating fluid, and would move into a "cloud" position behind the inlets to the suction ports of the secondary nozzle, where they would be balanced in place by the upward force of the surrounding fluid. From the cloud they would be drawn back through the inlet ports and re-introduced back into the jet stream, and thus would work in a continuous closed loop until the particles became too small to be stabilized near the inlet ports. At that point the recirculating fluid would move them back up the drill hole and out of the well.

The bit was referred to as a GA (for Gravity-Aspirator) bit. The initial bit did not rotate, which had practical advantages in reducing drilling costs. However, maximum impact efficiency of the pellets required that there be a set of "feet" around the base of the secondary nozzle to keep a stand-off distance of 2.8 to 3.4 nozzle diameters. This ranged from 12.5 to 15 cm for the nozzle required to drill a 12 cm diameter hole. These feet left hidden shadows on the rock ahead of the bit, which had to be rotated to uncover and erode them. Early experiments showed that the bit was more effective when the rock was removed under an impact, rather than abrasive mode (a distinction which will be discussed in more detail under the brittle/ductile modes in a later chapter). It should be noted that this bit drilled an elliptical hole and the feet did not rest on the floor of the hole, but rather on a ledge of rock, roughly halfway between the bottom of the hole and the bit. This location, among other advantages, ensured that the hole was drilled to the required gage while the pellets were still effectively cutting the rock in the plane of any obstruction to the drill passage. To allow the drill to advance, and to continue to circulate the pellets back to the aspirating nozzle inlets mounted in the bridging structure, this gage had to be at least one pellet diameter larger than the outer diameter of the secondary nozzle and following assemblies.

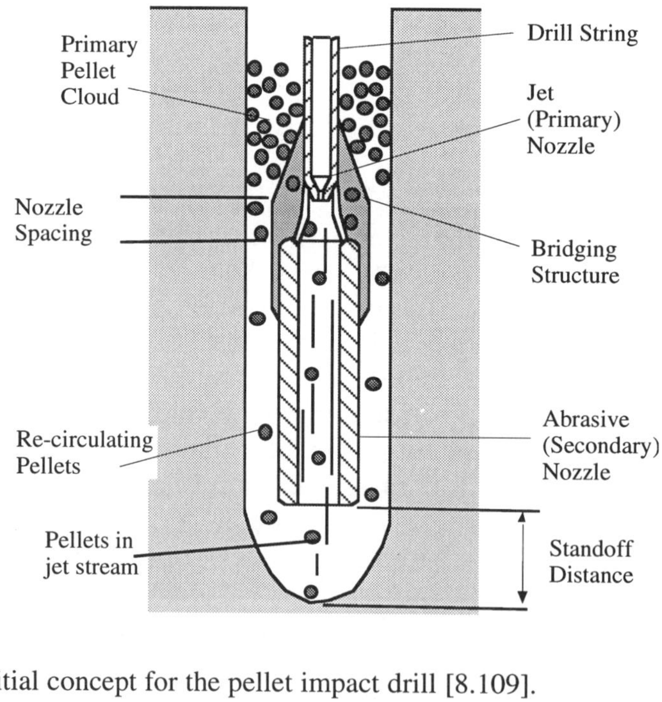

Figure 8.70 Initial concept for the pellet impact drill [8.109].

From the results of the early experiments it was deduced that it would require some 180 kW to drill a 23 cm. diameter hole through an Indiana limestone at a rate of 9 m/hr. Steel shot appeared to be the best candidate material for the pellets, not only did it last longer (a 64 kg charge lost 1.14 kg of steel over a 3.5 hr drilling test in pink quartzite), but it had a more consistent shape. As discussed earlier, in the section on airborne abrasive in Chapter 3, the larger pellets were more effective than smaller ones, although the size was restricted by nozzle diameter. In order to ensure that the nozzle did not block, the particle size was recommended to be less than 0.382 of the orifice diameter.

The length of the secondary nozzle was also studied, and the results (Fig. 8.71) suggested that a length of between 9 and 15 internal diameters would be most effective. In most of the tests this value was kept at 8.6 diameters to allow different nozzle orientations to be studied.

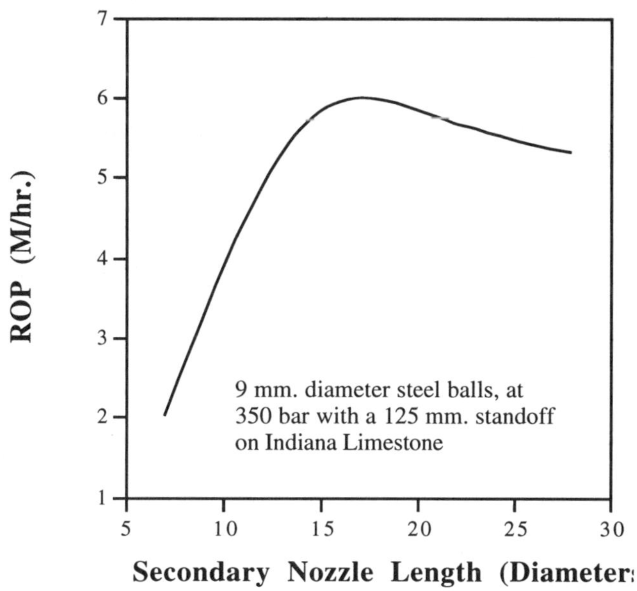

Figure 8.71 ROP as a function of secondary nozzle length (after [8.108]).

Particle velocities were, relative to current systems, quite low. The first tests were carried out at velocities of 60 m/sec, and indicated that this technique would not be as influenced by down-hole pressure as conventional drills, although an increase in fluid density from water to 5.45 kg mud did reduce drilling rates by some 25%. Tests were carried out both in the laboratory and in drilling down to 120 m from the surface.

Increasing the fluid velocity, by reducing the primary nozzle diameter at constant power, increased the penetration rate of the drill (Fig. 8.72).

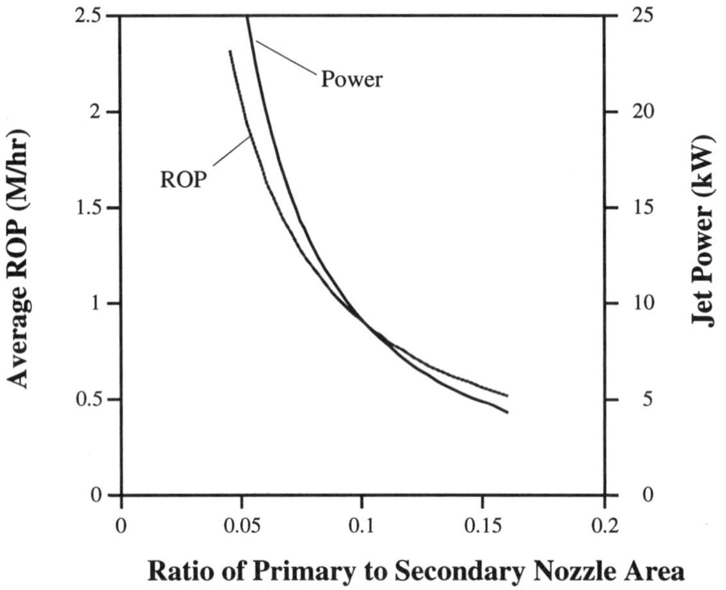

Figure 8.72 ROP as a function of power and relative nozzle size (after [8.108]).

It was found best to have the secondary nozzle inlet at a minimum distance of 0.2 diameters below the primary nozzle, but otherwise its shape had little effect on drilling performance. While inclined nozzles drilled faster, they also wore at a greater rate, a more than compensating disadvantage in the view of the research team.

The use of recycled particles meant that it was only necessary to feed a certain quantity of pellets into the drill. Drilling rates increased with charge size (which for a 12 cm diameter hole would be on the order of 3.6 kg of pellets) and then stabilized to a value of 2.5 times this "saturation" charge. Greater charge volumes caused a reduction in drilling rate.

Energy transfer from the surface pump to the bit and back was considered to use half the available power. The pumping efficiency of the jet pump was between 25% and 38%, but the transfer of energy to the pellets was only on the order of 2 - 3%. This limitation appeared due to the limited ability of the jet fluid to accelerate the pellets, and the energy dissipation from the pellets back into the fluid after they leave the secondary nozzle. Approximately 80% of the kinetic energy in the particles was calculated to go into rock fragmentation.

Tests were carried out using a 23 cm diameter bit and this was able to drill Oklahoma marble at 2.25 m/hr; Virginia limestone at 1.2 m/hr, and quartzite at 0.15 m/hr. For these tests some 140 of the 31.75 mm diameter pellets hit the rock each second, at a velocity averaging 22.5 m/sec. The driving pressure was approximately 4 bar over the 22 mm primary nozzle before the particles entered the 89 mm diameter secondary nozzle.

As a result of this study no further development of this program was then continued. The evaluation of the results [8.109] achieved suggested that it was not economical relative to the costs of conventional rotary drilling in part because of the inefficient acceleration of the particles in the secondary nozzle and partly because the particles broke the rock into fine powder. This required considerably more energy than that required to break the rock into chips, and thus the program became inactive.

The major disadvantage foreseen for the drill, that of pulverizing all the rock, could be overcome by combining the abrasive jet cutting with the mechanical breaking action of the conventional roller cones. Several concepts were suggested for this [8.110, 8.111, 8.112 for example] and have been described by Maurer [8.55]. The major development was, however, undertaken by Gulf Research and Development Co. [8.113].

The study was carried out at much higher velocities than the earlier one, and accelerated the particles with the drilling fluid, rather than using a secondary nozzle. In addition, by using the abrasive laden jets to cut grooves in the rock ahead of a conventional bit, significantly better drilling performance could be achieved. Early field tests of the system showed that the abrasive-jet assisted bits could drill at speeds of 4 - 20 times faster than conventional bits, while the drilling bit assembly lasted from 3 - 7 times longer down-hole. This latter meant that down-time on the rig while the bit was replaced could be substantially reduced.

Laboratory tests before the program went to the field had shown that drilling rate increased with jet pressure (Fig. 8.73), the square root of particle diameter, linearly with nozzle diameter and with the square root of abrasive concentration (Fig. 8.74). Although this suggests that faster drilling is achieved with larger nozzles, because the power increases as the square of the diameter, smaller nozzles were the most efficient.

Penetration rates were linearly related to pressure, however the abrasive jets cut grey granite about twice as fast as they cut gabbro or dense limestone. In the latter case a 350 bar jet would cut to a depth of 3 mm, while at 700 bar jet pressure would cut the dense limestone to a depth of 10 mm.

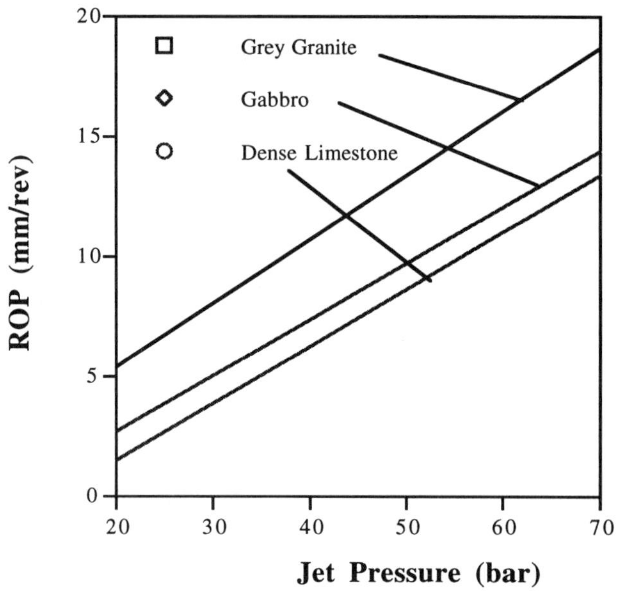

Figure 8.73 Effect of jet pressure on abrasive jetting ROP (after [8.113]).

It was recognized that these results only held within a range of investigation, and that, as with other work, too much abrasive, or too large an abrasive size, would adversely affect the drilling rate, for logistical reasons if nothing else. Nozzle diameters were on the order of 3.175 mm using 6% of a 20 - 40 mesh steel shot, with the nozzle located 1.25 cm above the target rock. The jet was moved over the surface at a speed of 240 mm/sec.

The use of the steel shot required that a special, relatively inexpensive mud be developed to suspend the steel. This is important since the shot must now be pumped all the way back up the hole (which might be as far as 4.5 km) after it had hit the rock one time, and before it could be reprocessed for recycling. One mud that was suggested was a 3% by weight hardwood fiber; 3% by weight attapulgite clay; 10% by volume fuel oil; and, where necessary, 1 - 2% by weight carboxy-methylcellulose.

As with the earlier program a significant part of the cuttings from the drilling was made up of fine particles. Unless these could be separated from the mud, this had a significantly deleterious effect on the drilling rate. This effect has been found in subsequent work, including the use of the DIAjet type of rock drilling which will be described at the end of this chapter.

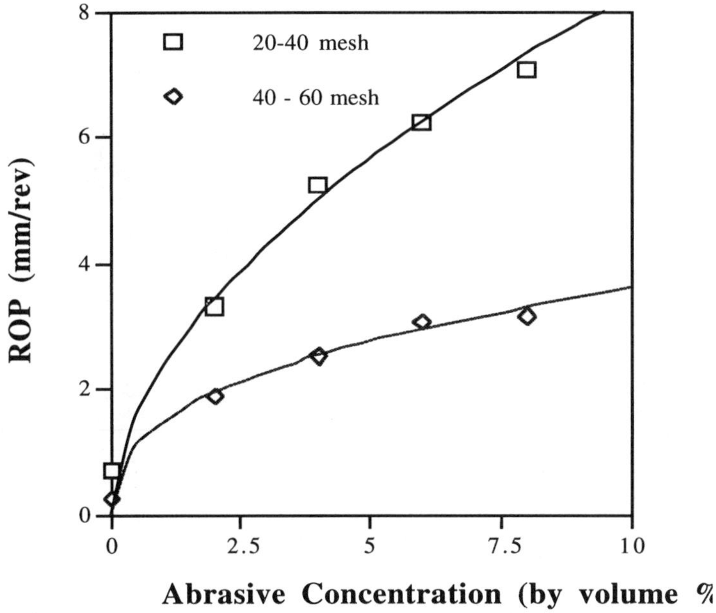

Figure 8.74 Effect of abrasive size and concentration on drilling rate (after [8.113]).

An early problem with nozzle location was identified when a two-cone roller was modified to accept two abrasive jet nozzles. While the jets improved drilling performance in granite four-fold, the rebounding sand completely eroded the teeth from the cones within thirty minutes. Subsequent designs used a rigid bar on the bit (similar to the carbide insert on a pneumatic drill except that it did not vibrate forwards) to remove the ridges between adjacent cuts. In an alternative embodiment the ribs were broken by hemispherical wedges set into the face of the bit (Fig. 8.75). This design kept the required stand-off distance between the nozzles and the rock.

Several features of this design are worth noting. The bit was chambered out behind the nozzles to slow the velocity of the mud and particles before they entered the nozzle, so as to reduce erosion. Each nozzle had its entrance section set 12.5 mm. above the inner base of the bit, so as to induce turbulence around the entrance and reduce the likelihood of bits plugging. The investigators pointed out that the erosion of the nozzles did not stop the bit from drilling, it merely required that more horsepower be supplied to maintain the penetration rate.

The first field trial of the technology was in 1964 in Texas, and a well was drilled from 2,225 to 2,347 m at a diameter of 15 cm with 6 - 10% by volume sand in the fluid, at a differential pressure across the nozzle of approximately 380 bar. Drilling rates ranged from 1.5 - 4 times faster than that of a conventional bit.

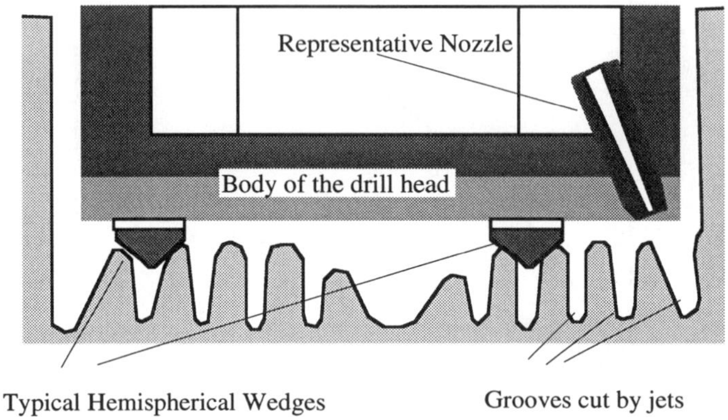

Figure 8.75 Gulf abrasive nozzle bit design (after [8.55]).

A second test in 1967 drilled from 701 to 2,627 m using steel shot, in the same volume concentration. The jet pressure drop across the nozzles was raised to 420 bar approximately, some portions of the hole were drilled at half the normal drilling rate, but other sections were drilled up to 10 times faster than conventionally could be achieved. On average the bits drilled three times more hole than conventional bits. Optimum rotation speed for the bit, which contained twenty 3.1 mm nozzles, was 60 rpm.

The third test was in a deeper well in Texas, in 1969. The bit diameter was 23 cm, and the technique successfully drilled the well from 3,226 - 5,364 m. Problems were with surface logistics, rather than down-hole drilling. A comparison of the then costs for this system (Fig. 8.76) showed the potential benefits which could be anticipated for this new method of drilling. Where the conventional drill could achieve ROP on the order of 0.3 - 3.6 m/hr, the abrasive jet bit averaged 8.25 m/hr. For the last 300 m of hole the abrasive jet drill penetrated at 11.1 m/hr in contrast with a conventional rate of 2.75 m/hr. Speeds in excess of 24 m/hr were monitored for the abrasive jet drill over short intervals of the hole, when pressure and abrasive concentrations could be optimized at the bit. Eleven bits were used to drill this interval and drilled almost twice the distance of conventional bits. The holes drilled were significantly straighter than for conventional roller drilling. Wear rate of the drill pipe and casing was about three times that of conventional drilling, but this was still within acceptable limits.

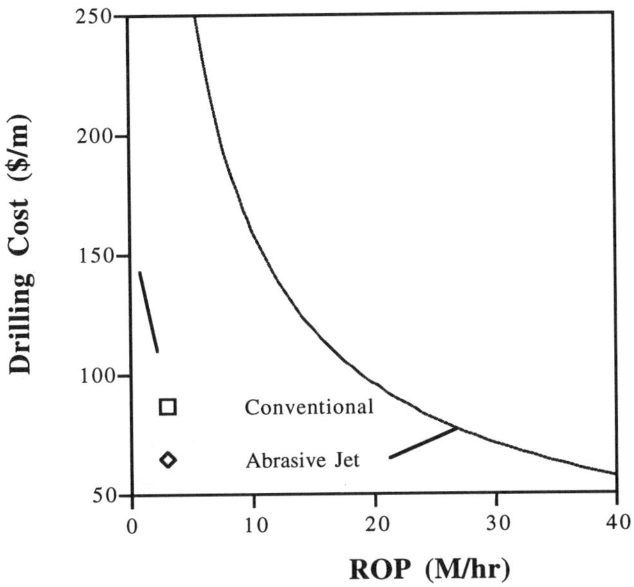

Figure 8.76 Relative costs for drilling with an abrasive jet drill (after [8.113]).

A significant problem with this system is that, at the longer lengths of pipe, if there is a significant change in the pressure (due for example to a temporary blockage of a nozzle or a change in the fluid density) this will cause the length of pipe to change under the pressure differential. This will either overload the bit, driving it hard into the rock and damaging it, or lift it off the rock, so that it cannot drill effectively.

One of the advantages of high pressure drilling fluid is that it allows much greater horsepower to be transmitted to the rock than can be achieved mechanically through a roller cone bit. For example, in the test just described the power issuing through each one of the twenty-three jets on the bit was equal to the power which could be delivered to the rock by all the cones of a conventional roller cone drill of the same size [8.114]. The test showed the potential for an abrasive-laden jet assisted drill bit to penetrate at speeds of up to 30 m/hr, where penetration rates of 15 m/hr would be cost competitive with existing conventional rotary drilling.

Continued development of the technology suggested that a magnetic separation of the steel shot was a much better system than trying to use a cyclone separator. The use of wedge bit designs was found to give better performance than a flat bladed bit (Fig. 8.77) during an extended program of development into 1973. Jet orifice diameters were reduced to 2.29 mm in order to improve penetration. Abrasive costs were $10.96 /m. in 1973, when the steel shot cost $235 a tonne. Improved performance of the drill could reduce this by at least half. Fair [8.114] has described many of the improved techniques which Gulf developed.

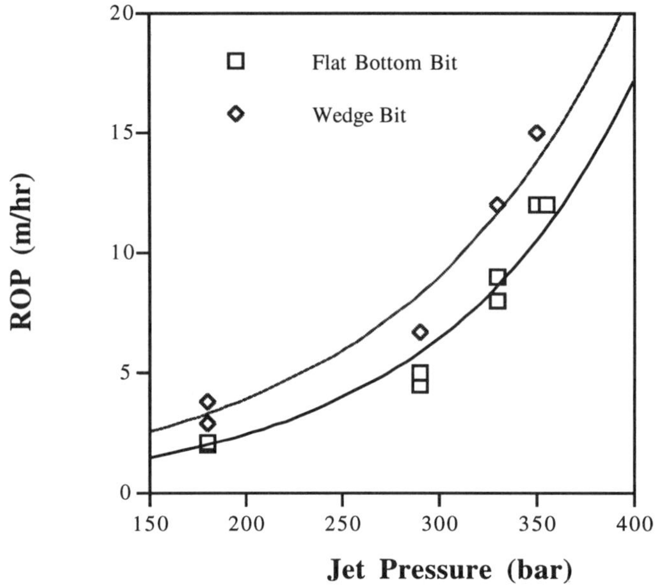

Figure 8.77 Comparative drilling rates of blade and wedge faced abrasive jet drills [8.114].

By 1973 equipment reliability had been significantly improved, rig operations had become routine, drilling rates from 7 - 12 times those of conventional bits had been achieved, and bit life had been increased 50%. However, there remained significant problems with some components of the equipment, excessive fines still were a significant problem, and lost circulation had not been successfully addressed. Unfortunately, a fire at the test laboratory in March, 1973, destroyed much of the equipment and the laboratory, and the program was never resumed.

Single impact studies of the depths of penetration as a function of impact velocity by Ripkin and Wetzel [8.115] had shown that a correlation could be made between the diameter of the projectile and the impact velocity on the resulting penetration (Fig. 8.78). A significant tool for enhancing drilling ROP could therefore be conceived. However, in order for this to be most productive a method had to be found for effectively introducing the abrasive into the fluid stream.

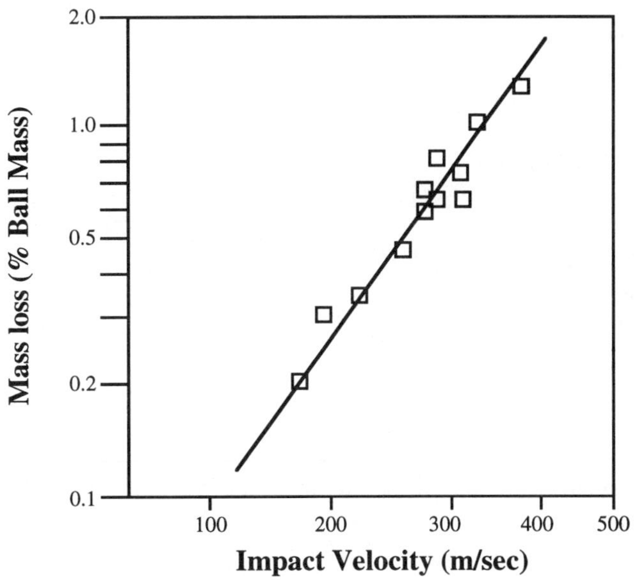

Figure 8.78 Rock mass loss under the impact of steel balls (after [8.115]).

One of the earliest attempts was carried out by Teledyne Brown Engineering under contract to the U.S. Department of the Interior, Bureau of Mines [8.116]. They examined two methods of injecting the abrasive into the jet stream (Figs. 8.79, 8.80) one in which the jet surrounds the abrasive flow, and one in which the abrasive is fed around the straight forward flow of the jet.

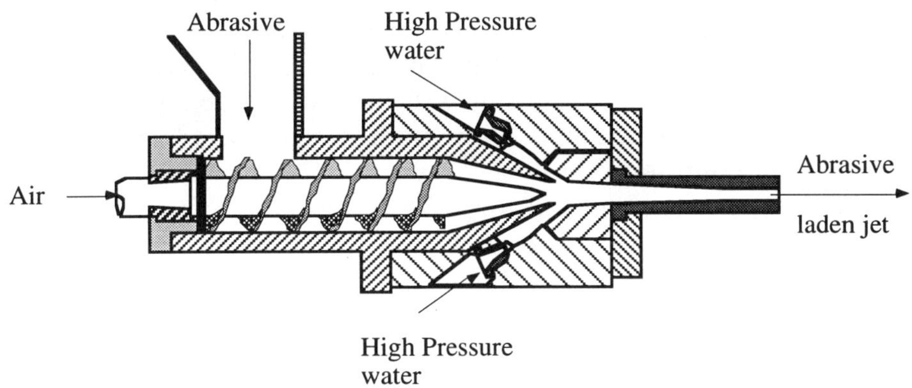

Figure 8.79 Nozzle assembly to add waterjets around an abrasive feed (after [8.116]).

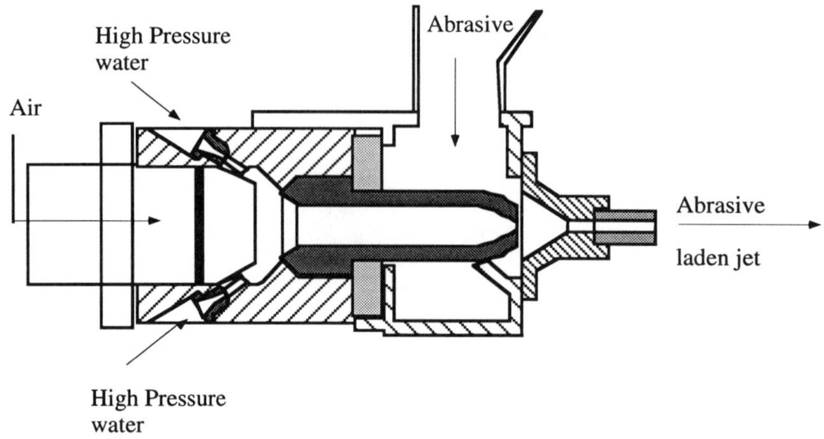

Figure 8.80 Nozzle assembly to add abrasive around a waterjet (after [8.116]).

When the nozzle design was perfected it was found that, for the centrally fed abrasive system, the screw feed served only to reduce the feed rate at which the abrasive entered the nozzle, and to reduce the chance of blockage. The aspirating effect of the jet was, itself, sufficient to draw abrasive into the jet stream and force feeding the abrasive was not necessary. The optimum design achieved (Fig. 8.81) was achieved by reaming out the center of the secondary nozzle until aspiration of the abrasive occurred. Some 1,700 particles each 3.8 mm in size could be drawn through the nozzle. Where the waterjet was run axially with a peripheral feed of abrasive (Fig. 8.79 version) this appeared to give a higher resultant jet velocity. When comparative cutting tests were made, even though the peripherally fed abrasive jet had a higher velocity it did not appear to cut any more effectively. Chatterton's study indicated that the excavation rate of the jet appeared to increase with particle diameter, that specific energy levels of cutting with abrasive-laden jets could be an order of magnitude lower than with plain waterjets. Nevertheless that program was not continued.

A study of alternative ways to improve waterjet cutting performance was carried out by El-Saie in 1977 as part of a study to accelerate the ROP of tunnels in hard rock [8.117]. In this design he used two holes drilled into the sides of a feed chamber to aspirate abrasive into the jet, and then recollimated the flow out of this chamber through a secondary nozzle (Fig. 8.82). By opening and closing these side ports it was possible not only to induct abrasive into the jet, or run the jet as a plain waterjet stream, but to induce cavitation into the jet stream. The results from those studies will be addressed in more detail, particularly regarding cavitation, in Chapter 10.

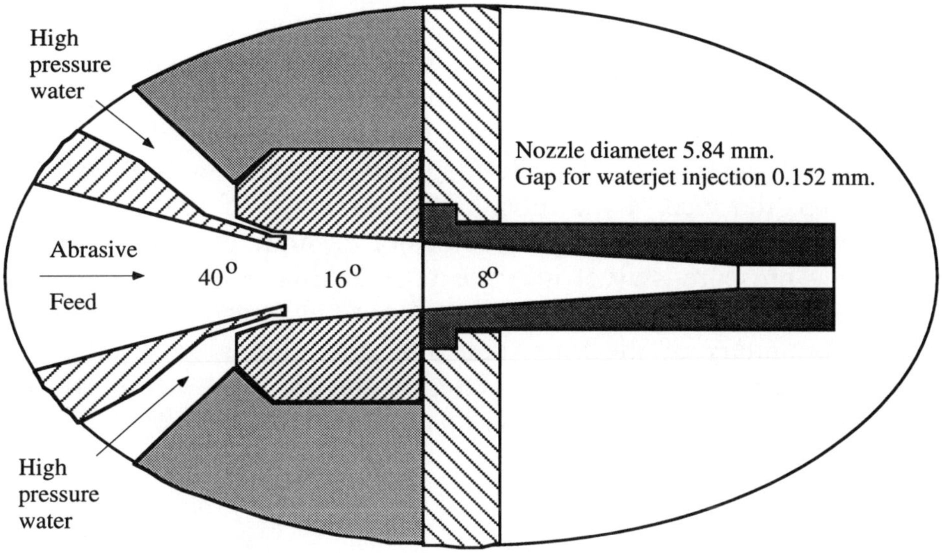

Figure 8.81 Simplified sketch of the optimum nozzle design for peripheral water feed around an aspirated axial abrasive flow (after [8.116]).

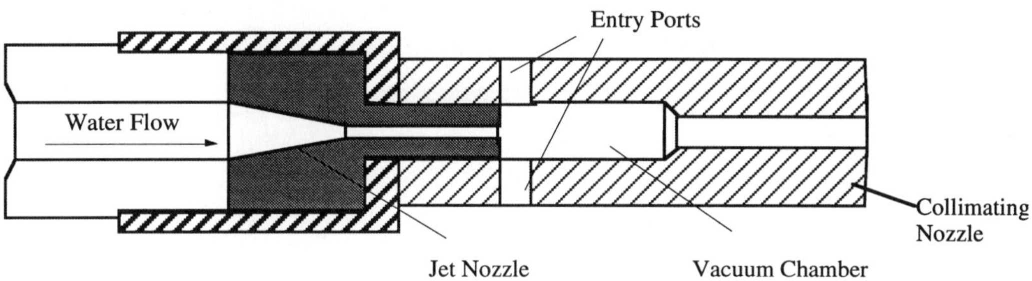

Figure 8.82 Dr. El-Saie's aspirating nozzle design [8.117].

Tests with this design indicated that it could be used both for cutting metal and for cutting rock. However the program of investigation was not continued beyond the dissertation, along the same direction. One of the problems which arose in trying to continue the study was that while the abrasive jet could be oscillated to cut a slot, it was difficult to design an effective swivel to allow the device to be used in deeper cutting.

An ingenious method of overcoming this problem was developed by the Bureau of Mines in Minneapolis [8.118]. Originally conceived as a simple method for rock cutting, high pressure water was accelerated through a nozzle and then directed into a focusing tube. However, in contrast to the development of abrasive injection systems for higher pressure, which had by then been developed (see, for example Fig. 10.58) this design did not use a second collimating nozzle to bring the jet back together. By using the open tube it was possible to disperse the jet over a wide enough area that it almost covered the face of the tube. One advantage of this system is that the jet, downstream of the jet nozzle is not at high pressure. Thus, by putting a low pressure swivel into the pipe at this point, it is possible to create a rotating device (Fig. 8.83). In order to ensure that the jet cuts out beyond the boundary of the pipe, and thus cuts clearance for the pipe to advance, two carbide inserts are inclined to the jet flow at the end of the pipe. In this manner it was possible to turn the slotting tool into a simple drill [8.119].

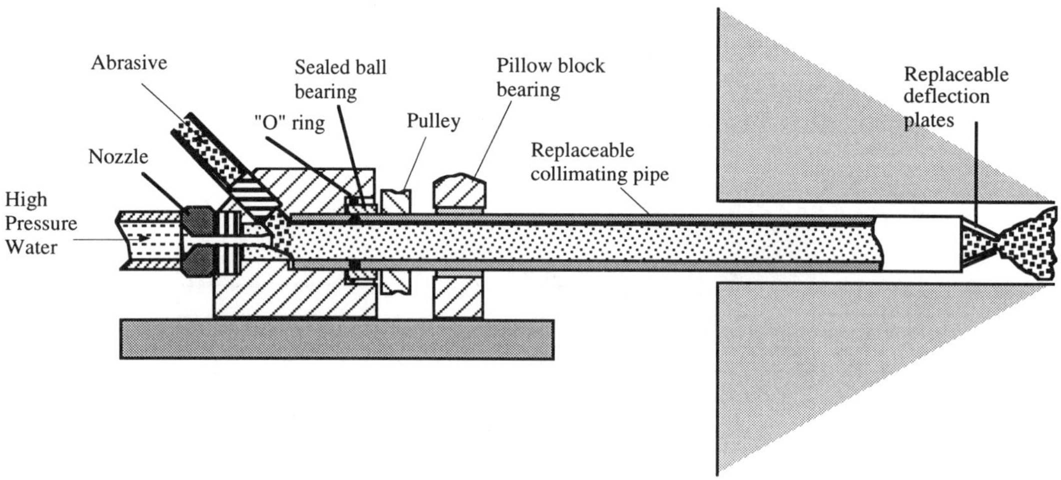

Figure 8.83 Schematic of an abrasive waterjet drill with guide plates (after [8.119]).

The deflection plates are made up of tool carbide and are set at an angle of 15° to the axis of the pipe, this shallower angle gives a narrower, but deeper cut than a wider angle would. Both the deflection plate and the pipe wear out, needing replacement after 2 hours of cutting at 700 bar, with a flow of 80 lpm of water containing roughly 10 kg of sand. Because there is no pressure on the swivel it could be made, in 1986, for $8.00. The pipe is rotated at 600 rpm and penetrated rock at rates varying with rock type.

Quartzite and granite were drilled at an ROP of 10 cm/min while dolomite was drilled at 15 cm/min. and limestone at 75 cm/min. The drill could also be used to drill through loose piles of rock, since there was little total force applied to the rock only localized points of high pressure. The drill has since been commercially licensed, and has been reported as capable of drilling reinforced concrete at speeds of up to 0.9 m/min.

Conventional methods of abrasive injection into high pressure waterjet streams have previously been discussed for their application in industrial cutting (Chapter 4). Some thought has been given to applying a similar technology to cutting rock. As with the work carried out by Dr. Savanick, the equipment was first designed to cut deep slots in concrete [8.120] and has since been modified for use in drilling holes. The advantage of this technology is, that by adding the abrasive to the high pressure water after it has left the acceleration nozzle, the swivels for the high pressure water, and the low pressure abrasive can be kept separate (Fig. 8.84), [8.121]. The low pressure swivel originally tested showed little signs of wear after 20 hours of use.

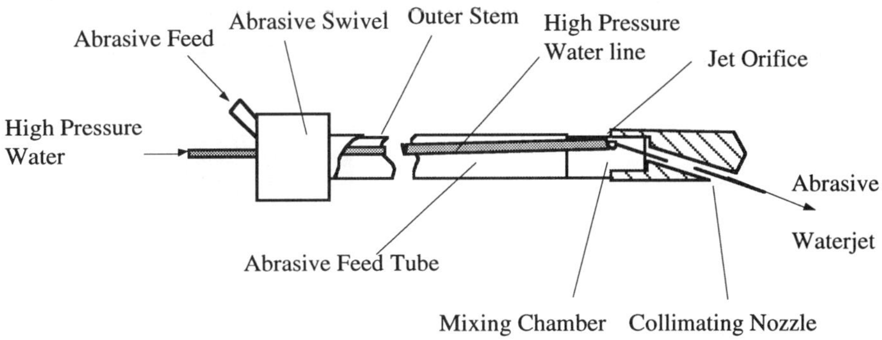

Figure 8.84 Components of the Flow rotating nozzle system [8.121].

At a pressure of 2,410 bar, through a water nozzle diameter of 0.762 mm and at a traverse speed of 1.7 mm/sec the system appeared optimized with an acceleration nozzle diameter of 2.3 mm, when using 30 mesh abrasive, at a flow rate of 45 gm/sec. In drilling concrete, garnet was found to perform slightly better than steel shot and steel grit, which both cut approximately four times as well as silica sand. The nozzle was rotated at 50 rpm. Interestingly when steel grit was recycled a deeper kerf was achieved than with unused grit. Hashish considers that this is because the grit is more optimally sized after an initial run through the screening process. Smaller grit (50 mesh) was reported to improve performance 17% over the use of 25 mesh grit.

Because the single jet does not completely and efficiently remove all the rock from ahead of the drill a variety of different designs were evaluated in order to better apply the tool as a drill [8.122]. Both non-rotary and rotating nozzle designs were evaluated in order to overcome these problems (Fig. 8.85).

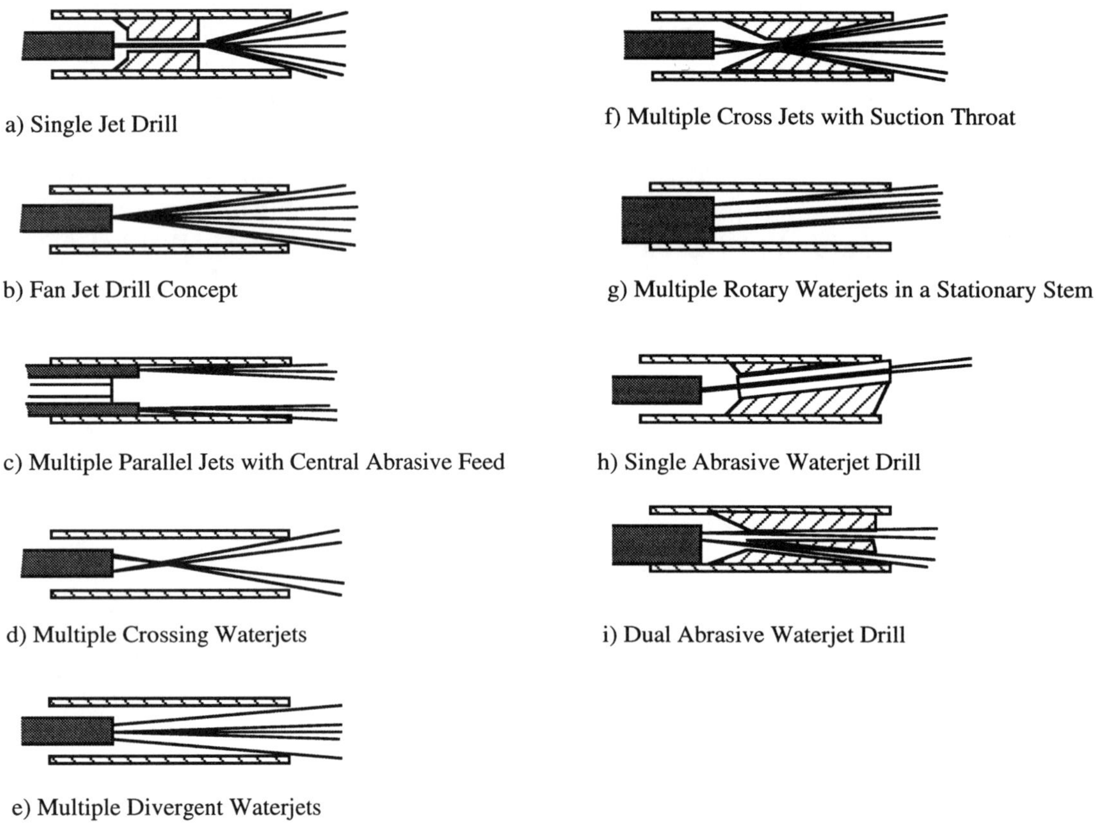

Figure 8.85 Designs of abrasive drill tested by Hashish [8.122].

The study was initially directed at determining the more favorable operating conditions based on different jet arrangements. A summary of the results obtained from nozzles comprising a number of slightly divergent jets surrounding an internal abrasive feed line gave a table of results which were calculated to include the specific energy of cutting (Table 8.7).

Table 8.7 Performance of various jet abrasive drills [8.122]

Jet Arrangement	*Abrasive Flow (gm./sec)*	*RPM*	*ROP (mm/min)*	*Specific Energy (KJ/cu mm)*
3 @ 0.559 mm	30	0	55	0.243
1 deg divergent	30	20	111	0.121
	41	0	79	0.170
	41	20	142	0.094
	41	1000	126	0.106
4 @ 0.457 mm	41	0	32	0.424
3 deg divergent	41	20	95	0.141
	41	1000	79	0.170
	51	0	39	0.303
	51	20	111	0.108
	51	1000	103	0.116
	69	0	55	0.216
	69	20	158	0.076
	69	1000	142	0.084
4 @ 0.457 mm	41	0	63	0.189
2 at 2 deg	41	20	111	0.108
2 at 3 deg	41	1000	111	0.108
5 @ 0.406 mm	41	0	47	0.249
4 at 2 deg	41	50	126	0.093
1 central	41	50	142	0.083

Tests were carried out on granite at a pressure of 2,410 bar using 60 mesh garnet, and the ROP was scaled from the volume removed during the 10 second test run. The bit diameter was 22.4 mm.

The data from the above table, should be read with some caution. One of the changes sought for an optimal design was one which would remove the central cone of rock otherwise left in the center of the bore by an outwardly inclined jet. The design had also to address a problem with the inward taper of the hole sides. This is a continuing problem with waterjet drilling, since the hole will tend to cone inwards with depth. This problem, and that of the central cone may not be immediately apparent when the hole is started, but becomes significant once the body of the drill enters the hole. At that time the annulus between the body of the drill and the rock wall provides some confinement to the escape of the drilling fluid, and a normally small back pressure is generated. This will reduce the effective cutting length of the jet, as discussed above. Thus, in the above table, where the rate of penetration is calculated, based upon penetration of a rock without significant drill advance, the numbers provided may be somewhat optimistic.

A second trial was conducted with a dual orifice nozzle, (similar to type (i) in Fig. 8.85). As with the data for Table 8.6, this data was obtained as a result of a ten second run without nozzle advance. Thus, problems which arise when the nozzle must advance into the hole are not addressed. The data (Table 8.7) should, therefore, be treated with some caution in regard to effective nozzle designs.

Table 8.8 Performance of dual orifice abrasive drills [8.122]

Inner Nozzle Dia (mm)	*Outer Nozzle Dia (mm)*	*Abrasive Flow (gm./sec)*	*Abrasive Size mesh*	*Drill RPM*	*Calculated ROP (mm/min)*	*Specific Energy (KJ/cu mm)*
1.78	3.18	28	60	325	166	0.046
1.78	3.18	42	60	325	213	0.036
1.78	3.18	59	60	325	237	0.032
1.78	3.18	29	40	325	201	0.038
1.78	3.18	43	40	325	261	0.029
1.78	3.18	59	40	325	332	0.023
1.78	3.18	31	36	325	261	0.029
1.78	3.18	47	36	325	373	0.021
1.78	3.18	60	36	325	426	0.018
1.78	3.18	31	36	325	190	0.024@ 1,720 bar
1.78	3.18	47	36	325	213	0.022@ 1,720 bar
1.78	3.18	60	36	325	308	0.015@1,720 bar
1.78	3.18	31	36	325	450	0.025@3,100 bar
1.78	3.18	47	36	325	509	0.022@ 3,100 bar
1.78	3.18	60	36	325	616	0.018@3,100 bar
1.78	3.18	31	36	600	225	0.034
1.78	3.18	47	36	600	367	0.021
1.78	3.18	60	36	600	403	0.019
1.78	3.18	31	36	180	213	0.036
1.78	3.18	47	36	180	355	0.022
1.78	3.18	60	36	180	379	0.020
1.78	2.34	31	36	325	308	0.025
1.78	2.34	47	36	325	249	0.031
1.78	2.34	60	36	325	367	0.021
1.78	3.80	31	36	325	261	0.029
1.78	3.80	47	36	325	332	0.023
1.78	3.80	60	36	325	320	0.024

Unless otherwise stated the jet pressure was 2,410 bar, the inner waterjet orifice was 0.457 mm (as opposed to the abrasive jet orifice given above); the outer waterjet orifice was 0.711 mm, the abrasive tube length was 76 mm and garnet was the abrasive. The target was granite.

Some trends can be discerned from these results. Penetration rate increased with abrasive size, and jet pressure although there was little change in specific energy with increase in jet pressure. Penetration rate increased, and specific energy decreased as abrasive feed rate was increased. There appeared to be an optimum for the rotational speed within the data evaluated, between 325 and 600 rpm. Increasing the size of the outer abrasive nozzle only seemed to improve cutting at lower feed rates and an optimum value did appear evident from the data. It was noted from observation of photographs provided with the text that the holes which achieved the greatest penetration also evidenced a fairly prominent ridge of rock left in the hole. This would prevent the drill from advancing. By changing the configuration of the nozzles to deal with this, a probable change in optimum choice of parameters would also be required. This is particularly true if the offset position of the jets is changed.

The problems associated with integrating an abrasive line feed to the end of the drill bit, where the flow must run parallel to the high pressure water can be overcome where the abrasive and the drilling fluid are pre-mixed, as was the case with the earlier oil well drilling nozzles. This approach has recently been revisited, but from the point of view of drilling small diameter, holes in rock for mining applications. While the intent has been to develop a long hole drill [8.123] the length and direction of the drill differ from the oil well application, in that they are normally at most only a few hundred meters long, and are usually inclined at an angle close to the horizontal.

In the development of the nozzle design for an abrasive-laden waterjet drilling head, Yazici has completed the most comprehensive evaluation of different designs [8.124], although that work has since been further developed by Wright [8.125]. The difficulties that arise in drilling head design are related to a number of conflicting problems.

The space available to install nozzles within the body of a drill traveling down a hole which is between 25 and 45 mm in diameter is quite restricted. Further, because an abrasive nozzle must have a certain length to properly accelerate the abrasive, this restricts the maximum angle at which the outer streams can be directed. This, in turn, has an effect on the outer shape of the hole drilled. If the angle of intersection of the abrasive jet with the wall is too shallow, then the abrasive will "bounce" along this wall on second and subsequent passes rather than cutting a deeper and thus wider hole. While the exact minimum angle is not well defined yet for abrasive drilling in hole, with pure waterjet drilling this minimum angle is around 12°. Related problems are that the steeper the angle of the accelerating nozzle to the central hole axis, the more rapid is the erosion of the rear entrance to the nozzle.

An additional problem arises where the rock breaks out ahead of the bit in large pieces. In contrast with mechanical breakage, there is a significant space ahead of the body of the drill, and fragments of rock in this space are not confined. These can be larger than the gap between the edge of the drill and the rock wall and thus cannot float out of the hole until they are further broken. When they are finally broken by the bit to a small enough size, not an insignificant problem in itself, they can then flow into the annulus between the drill head and the hole wall. Because the drill head is usually circular this gap is small, and the particles can become lodged, stopping the drill and jamming it in the hole. Since the normal torque to drive these drills is very small (fractions of a kW) this may stall the motor. Mechanical assistance to the jet bit, as Gulf achieved, is not currently contemplated since it removes the major advantage foreseen for the equipment, that of creating an extremely light and maneuverable drill.

As part of the program it was necessary to develop a rotary swivel for the feed of the abrasive fluid to the drill. This device, having been modified and rebuilt twice, was successfully used for two successive Ph.D. research programs [8.124 and 8.106], although it has since been redesigned for more effective use in new equipment [8.126].

Yazici evaluated the performance of a series of nozzles (Figs. 8.86, 8.87, and 8.88), and was able to compare their performance in a review which has been summarized (Table 8.8). In contrast to the results reported by others, this experiment was carried out at fixed advance rates. The results showed whether it was possible to sustain drilling at that speed (drilled OK) or if the bit jammed. Where a bit drilled successfully at one speed it was often tried again at a faster drilling rate. Jamming generally indicates that the hole was not being cut to a large enough diameter a fault of the outer jet, hitting bottom meant that there was a cone in the center of the hole. This, in turn, suggested that the inner jet was not properly sized or aligned.

Where the bit jammed, the hole was usually coning in so that the actual hole diameter at which the drill stopped equals the drill bit diameter. Where the drill continued to advance the hole diameter remained constant.

It can be seen that an outer angle of 30° was needed to prevent the hole from tapering inwards as the drill advanced. For the inner jet angle, a zero or 5° jet was able to remove the material ahead of the bit, at larger angles the data is not as clear, at 10° for example a cone was left which the bit hit. Part of this problem, however, relates to the position of the two orifices. The inner jet should cut a hole large enough to clear the inner half of the body of the outer nozzle, so that it does not bind. From the inner edge of the outer jet to the hole bore the rock should be removed by the outer jet or jets.

An additional problem which must be addressed lies in the balance of the nozzles. Later experience with larger diameter drills has shown that, particularly while drilling the first few cm of the hole, an approximate balance in the lateral forces imposed by the jets is critical. Large diameter jets, used here at the steeper angles at the outside of the bit, impose a much greater force on the bit, and at slower rotational speeds these out-of-balance forces can de-stabilize the drilling platform.

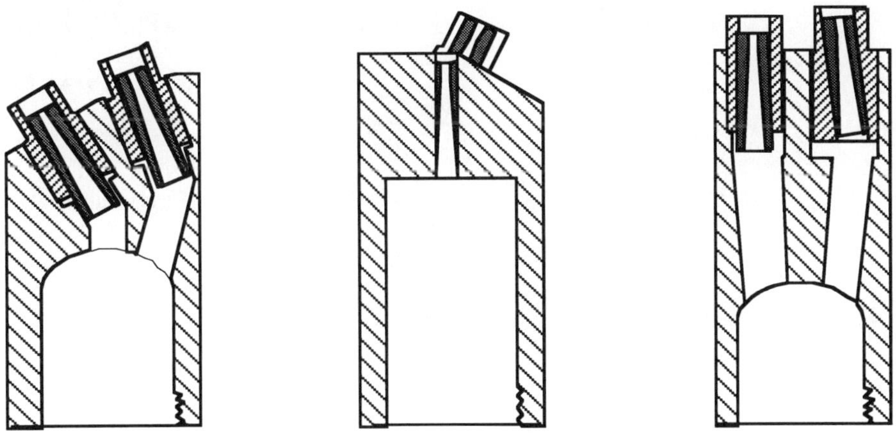

Style #3 20° and 30°

Style #5 straight and 27° the rear orifice is 8 mm offset

Style #6 with two inclined nozzles, the apparently straight one is inclined toward the viewer at 25°

Figure 8.86 Dual orifice nozzle designs evaluated by Yazici [8.124].

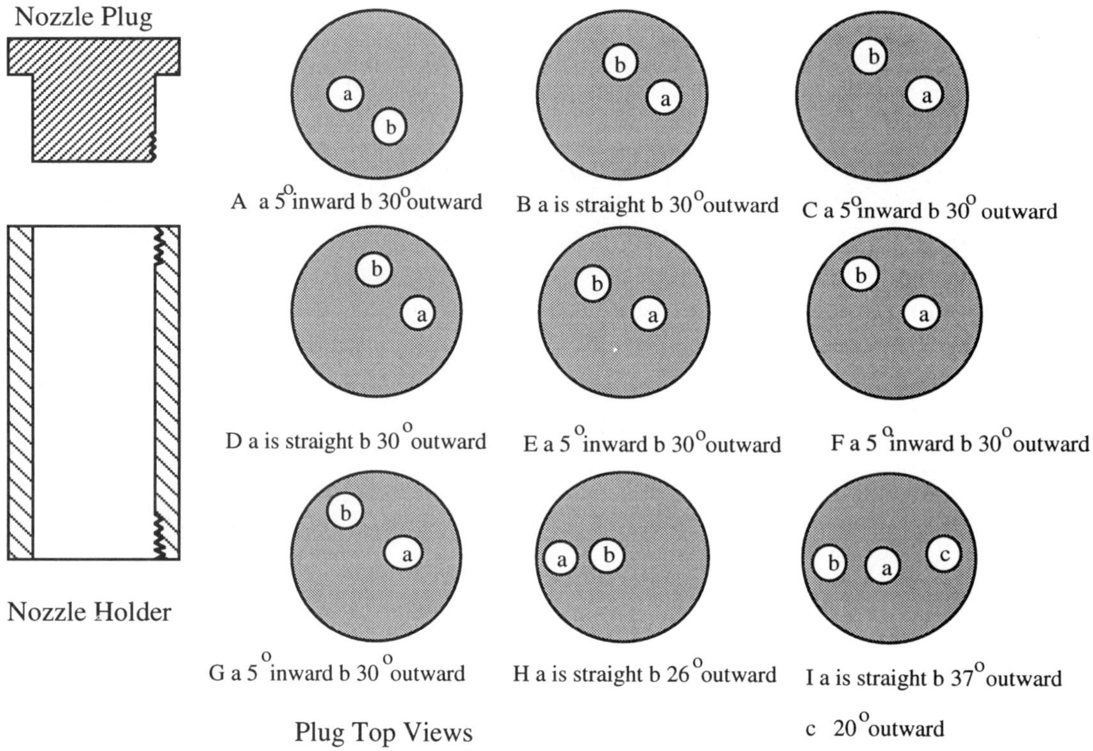

Figure 8.87 Nozzle inserts attached to a common holder, tested by Yazici [8.124].

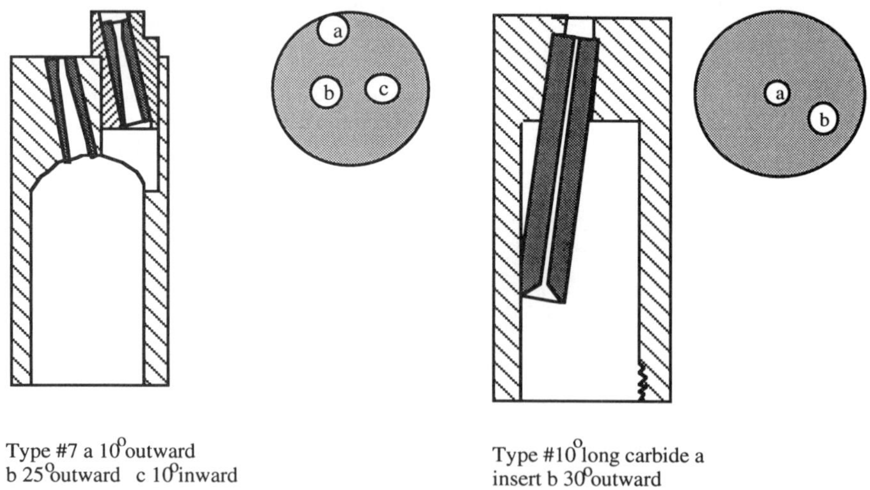

Figure 8.88 Alternate nozzle design combinations tested by Yazici [8.124].

Table 8.9 Drilling nozzle configuration test results (after [8.124])

Design #	Inner Angle	Outer Angle	Pressure (bar)	RPM	Drilled Rate mm/min	Specific Energy (KJ/mm^3)	Hole Dia (mm)	Erosion (cc/gm.)	Result
6	5	30	250	475	76	0.016	50.8	0.015	drilled OK
6	5	30	250	475	102	0.015	45.7	0.016	drilled OK
6	5	30	250	475	127	0.013	44.5	0.020	jammed
6	5	30	250	475	152	0.011	44.4	0.024	jammed
3	20	30	250	475	102	0.016	44.4	0.016	jammed
6	5	30	250	475	127	0.013	44.4	0.020	hit bottom
6	5	30	250	120	102	0.014	47.5	0.018	wobbling
7	3 @ 10 deg		350	475	127	0.005	46.7	0.063	jammed
5	0	30	350	475	127	0.021	40.6	0.016	drilled OK
6	5	30	350	475	94	0.022	46.0	0.016	drilled OK
10	7	30	370	150	127	0.012	43.7	0.027	jammed
10	7	30	370	220	127	0.011	49.0	0.028	jammed
10	7	30	370	300	127	0.013	42.4	0.023	jammed
10	7	30	370	475	127	0.015	40.9	0.021	jammed
3	20	30	370	475	138	0.013	44.5	0.028	jammed
3	20	30	370	475	162	0.013	40.0	0.027	jammed
8B	0	30	350	200	149	0.013	49.0	0.026	drilled OK
8C	-5	30	350	200	171	0.011	45.5	0.031	hit bottom
8D	0	30	350	200	164	0.012	42.9	0.028	hit bottom
8B	0	30	350	200	172	0.013	44.5	0.027	jammed
8B	0	30	350	200	166	0.012	45.2	0.028	jammed
8B	0	30	370	200	182	0.010	45.7	0.029	jammed
8B	0	30	370	200	155	0.022	45.2	0.027	drilled OK
8E	5	30	370	200	144	0.014	46.0	0.025	jammed
8F	5	30	350	200	150	0.014	41.4	0.025	jammed
8F	5	30	350	100	143	0.013	41.1	0.026	jammed
8G	10	30	350	100	155	0.010	50.8	0.033	hit bottom
8H	single 26 deg		350	100	126	0.012	38.6	0.027	jammed
8H	25	26	350	100	161	0.014	41.1	0.025	jammed
8H	25	26	350	100	158	0.014	39.5	0.025	jammed
8A	5	30	350	50	170	0.014	39.4	0.028	jammed
8A	5	30	350	100	162	0.014	39.0	0.024	jammed
8A	5	30	350	100	143	0.014	38.6	0.024	jammed

The need for a long section in which to accelerate the abrasive was examined by using both long and short carbide inserts in the body of the nozzle. The shorter nozzles gave a less focused jet, which drilled a wider but shallower cut in the rock, they also wore out quite rapidly when using the garnet abrasive. Longer inserts lasted longer, and gave a more focused jet. This cut deeper, but not as wide a cut. Wider shallower cuts were found more useful, and allowed a better spacing of the nozzle inserts.

Once the basic design for the drill had been established, it was used to drill 4.5 m long holes in dolomite as locating points for a rock reinforcement system [8.127]. The holes to be drilled were in the upper layers of a bedded dolomite which had been previously blast shattered, in places, as part of the initial excavation process at the site. The rock was additionally layered with beds of chert and clay. The holes to be drilled had to be rough in order to provide satisfactory anchorage for the resin encapsulated bolts to be inserted in the hole. Once installed and after the resin at the back of the hole was set, these bolts would then be subject to loads of up to 40,000 kg.

In preparation for that work a test was carried out to determine the most effective abrasive for use on the project [8.128]. Steel shot was found to give the best cutting result (Fig. 8.89). However, during the cutting program on site the difficulties in collecting the abrasive made this too expensive. Quartz sand was used instead, and appeared to undergo little reduction in size in a single pass through the system. Earlier tests using garnet indicated that this could be recycled several times through the unit before the problem of fines build up severely affected the drilling rate.

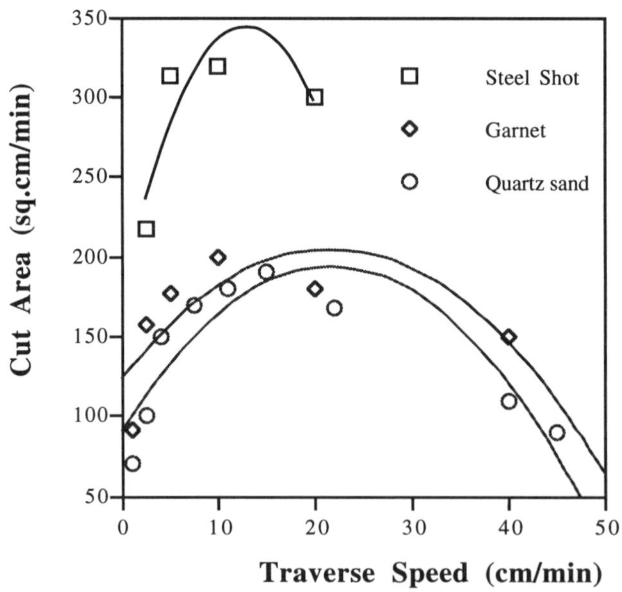

Figure 8.89 Comparison of abrasives for drilling dolomite (after [8.106]).

Using steel shot as the abrasive has several potential advantages, it gives a deeper cut, it can be recycled, and it does not damage the nozzle and holder as severely as does the quartz or garnet. In contrast it is a significantly more expensive abrasive, and has proven more difficult to mix into the DIAjet system. However, because of its increased effectiveness it is able to drill dolomite, basalt and other harder rocks with greater ease than some of the other abrasives tried [8.125].

While the performance achieved in drilling through dolomite and chert with sand as the abrasive at 350 bar was slow, on the order of 10 cm/min in the fractured rock, the machine which evolved was relatively light and maneuverable and the drilling rate was achievable through ground of varying geology, including chert. This could be contrasted with the performance of a conventional mechanical drill which was able to drill competent dolomite much faster than the abrasive drill, but had great difficulty in drilling though loose and broken ground, or when there were thick pockets (over one m thick) of clay along the line of the drilled hole.

Pull tests on the 25.4 mm bolts inserted into the waterjet drilled holes indicated (Fig. 8.90) that the hole diameter at the back of the hole was still small and rough enough to sustain the required maximum pull on the bolt and to validate the anchorage between the bolt and the resin and the resin and the rock.

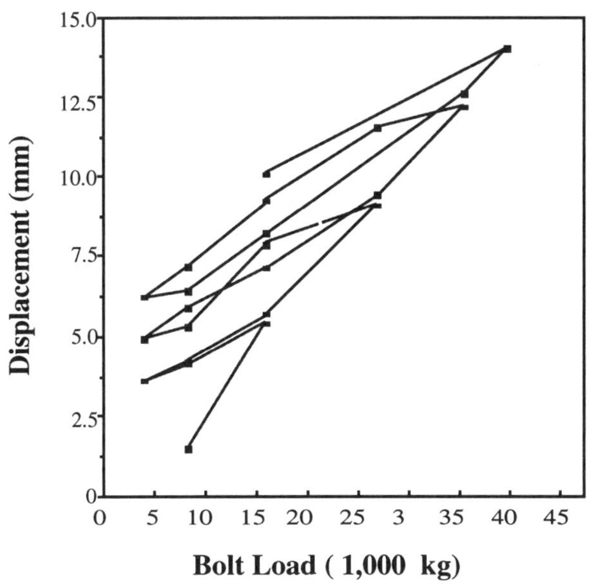

Figure 8.90 Load displacement curve for the first rock bolt installed in a DIAjet drilled hole [8.128].

The development of an alternative use of the DIAjet system for hole drilling has recently been advanced in Norway [8.129]. The underlying concept for the work was that while abrasive waterjets required up to ten times the specific energy to drill as conventional tools, yet the power available to them was one hundred times that available conventionally. This would give a drill capable of ten times the penetration rate of conventional equipment.

As with the work carried out by Gulf in the 1970s [8.114] the work by Vestavik concluded that using abrasive waterjets to slot the rock and then use a mechanical tool to fragment the ribs was a more efficient method of rock removal. In contrast with most designs, the study therefore moved away from the use of concentric slot cutting and, instead sought to derive a pattern which would have the mechanical cutters moving perpendicular to the slots in the rock, rather than along them (Fig. 8.91).

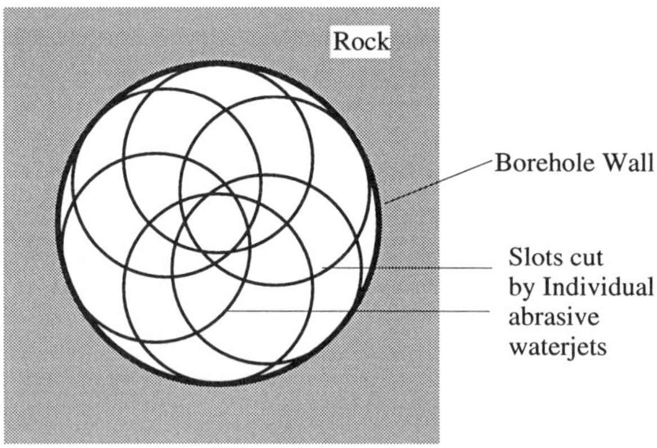

Figure 8.91 Slotting pattern developed in the face of a drill-hole by the jets in the drill developed by Rogaland Research [8.129].

In order to cut that pattern in the rock face a DIAjet abrasive injection system was connected though flexible hose to the back of the drilling head. Within the head the flow of fluid is split into six nozzles mounted on an assembly within the bit head. The bit is rotated by the rotation of the drill pipe, but as it moves the nozzle assembly is swept across the rock surface, without rotation. This is allowed by the bearings set between the assembly and the main bit body (Fig. 8.92). The bit body also contains small mechanical cutting bits set into the perimeter of the tool, which act to break off the ribs of rock left by the abrasive jet action.

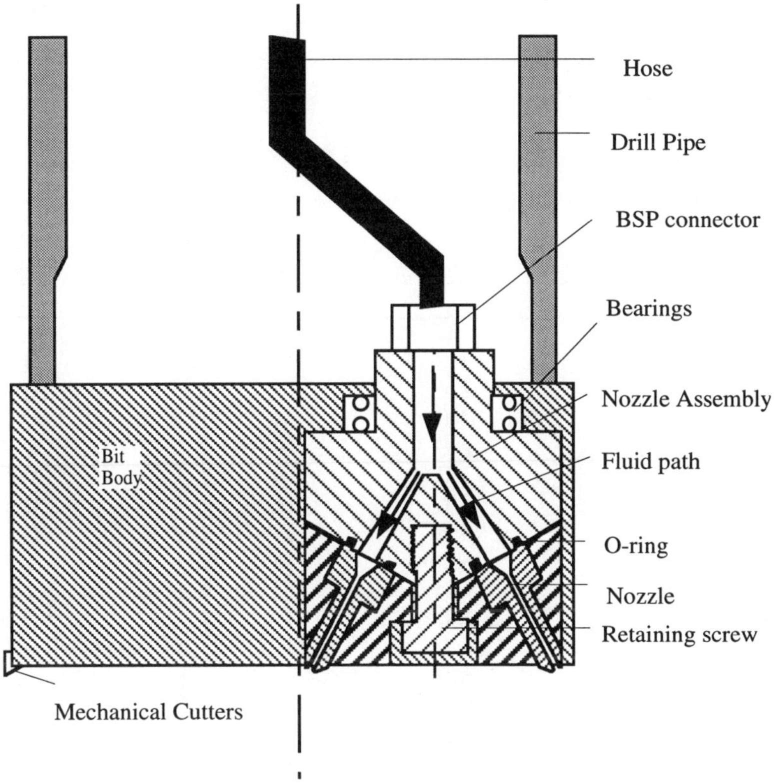

Figure 8.92 Drilling bit developed by Rogaland Research [8.129].

The nozzle holder was made of steel, while the nozzles themselves were made of tungsten carbide. Each nozzle is mounted with an O-ring seal at an angle of approximately 30° to the central axis of the nozzle assembly. Jet diameters of 1.0 mm. were used in the preliminary tests of the device. The sand used had an average grain size of 250 mm. Data from the first test series, in which a 145 mm. diameter bit was used is given in Table 8.9.

The experimental program was continuing at the end of 1991.

Because field experiments have shown the risk in trying to extrapolate short cutting runs in the laboratory with resulting field performance of drilling configurations it is dangerous to predict the future of this aspect of the technology. While further conclusions, therefore, depend on the results of ongoing field experiments, it is worthwhile to recollect some benchmarks set by Maurer in his review of drilling techniques [8.55]. He concluded that high pressure waterjet drills would likely succeed in the commercial market when they could penetrate at rates of around three times those of conventional equipment. The advent and continued success of the Flowdril development has been because it has finally been able to reach and sustain that type of drilling penetration.

Table 8.10 Data from the first orbiting DIAjet drilling test (after [9.129])

Rock type	*Jet pressure (bar)*	*Flow (L/min)*	*Abrasive conc.(vol %)*	*RPM*	*Bit force (kN)*	*Penetration rate (m/hr)*
concrete	195	52	2.0	14	1.1	0.34
concrete	255	56	2.0	14	1.1	0.54
concrete	295	64	2.0	14	1.1	0.60
concrete	335	72	2.0	14	1.1	0.79
concrete	300	65	2.0	14	1.1	0.60
concrete	300	65	3.6	14	1.1	0.88
concrete	300	65	5.8	14	1.1	0.97
concrete	300	65	8.0	14	1.1	1.13
concrete	300	70	6.0	14	1.1	1.02
concrete	300	70	6.0	8.2	1.1	0.99
concrete	300	70	6.0	5.4	1.1	0.97
concrete	300	70	6.0	5.4	1.8	1.15
concrete	300	70	9.0	5.4	1.8	1.29
concrete	300	75	12.0	5.4	1.8	1.42
granite	300	75	6.0	5.4	1.8	0.15
granite	280	95*	6.0	5.4	1.8	0.14
sandstone	280	95*	6.0	5.4	1.8	0.31

* one nozzle was at 1.6 mm diameter.

Two additional criteria were developed by the current author during the earlier development of the program at UMR. It became clear that the technology could not anticipate becoming widely accepted until such time as it could drill any rock that the drill was likely to encounter as it advanced. It was also clear that without the ability to drill a relatively smooth hole that abrasive waterjetting might well be restricted in its application to the drilling industry. Both these developments have now occurred and holes can now be drilled, at relatively widely available pressures (350 bar) through virtually all rock types, ranging in strength up to basalt and quartzite (Fig. 8.93). In addition the holes drilled in the latest generation of bits have been round and smooth, removing one of the final technical barriers to their introduction [8.125].

The experiments with abrasive jet drilling have also shown that penetration rates can be significantly increased beyond the threshold level set by Maurer. To some degree this is because the drill allows use of much higher horsepowers at the bit than are available with conventional equipment. That said, the development of reliable equipment over the last twenty years is likely to make some form of abrasive jet drill a significant factor in drilling at the turn of the century. This is particularly likely given the move toward drilling more and larger horizontal holes of increasing length. As the work carried out by Gulf, and more recently by Rogaland indicates, however, it may be most efficient to marry waterjet cutting with the assistance of mechanical tools to get the most effective system.

Figure 8.93 Hole drilled in basalt at 15 cm/min by a 350 bar waterjet bit, using steel shot as the abrasive [8.125].

8.10 REFERENCES

8.1 Summers, D.A., Disintegration of Rock by High Pressure Jets, Ph.D. thesis, Mining Engineering, University of Leeds, UK, 1968.

8.2 Summers, D.A., and Bushnell, D.J., "Preliminary Experimentation of the Design of the Water Jet Drilling Device," paper E2, 3rd International Symposium on Jet Cutting Technology, Chicago, IL, May, 1976, pp. E2 21 - E2 28.

8.3 Summers, D.A., Mazurkiewicz, M., Bushell, D.J., and Blaine, J., Method and Apparatus for Water Jet Drilling of Rock, U.S. Patent No. 4,119,160, disclosed March 5, 1976.

8.4 Vijay, M.M., and Brierley, W.H., "Drilling of Rock with Rotating High Pressure Waterjets: An Assessment of Nozzles," paper G1, 5th International Symposium on Jet Cutting Technology, Hanover, FRG, June, 1980, pp. 327 - 338.

8.5 Summers, D.A., "The Development of High Pressure Waterjets for Drilling Purposes," paper II-3, International Symposium on Water Jet Technology, Tokyo, Japan, December 6, 1984.

8.6 Summers, D.A., Lehnhoff, T.F., and Weakly, L.A., "Development of a Waterjet Drilling System and Preliminary Evaluations of its Performance in a Stress Situation Underground," paper C4, 4th International Symposium on Jet Cutting Technology, Canterbury, UK, April, 1978, pp. C4 41 - C4 50.

8.7 Summers, D.A., and Lehnhoff, T.F., The Design of a Waterjet Drill for Development of Geothermal Resources, Final Report on Contract DOE EY 76 S 02 2677.M003 to the Department of Energy, Sept., 1978.

8.8 Bonge, N., and Ozdemir, L., Development of a System for High Speed Drilling of Small Diameter Roof Bolt Holes, Final Report on Contract DE-AC01-76ET-12462, CSM, April, 1982, 236 pages.

8.9 Karabin, G.J., and Debrevec, W.J., Preprint No. 76-F-32, AIME Annual Meeting, Las Vegas, NV, 1976.

8.10 U.S. Bureau of Mines Contract No. HO262054 - work continued as U.S.D.O.E. contract DE-AC01-76ET-12462 - awarded to the Earth Mechanics Institute at Colorado School of Mines, June 21, 1976.

8.11 U.S. Bureau of Mines Contract No. HO262036 - awarded to Flow Research, Inc., Kent, WA.

8.12 Zink, E.A., Wolgamott, J.W., and Robertson, J.W., "Waterjets Used in Sandstone Excavation," 3rd Rapid Excavation and Tunneling Conference, Chicago, IL, June, 1983, Vol. 2, Chapter 40, pp. 685 - 700.

8.13 Summers, D.A., Wright, D.E., and Xu, J., "The Use of High Pressure Waterjets to Enhance Tunneling Machine Performance," Final Report on MIT Purchase Order CE-R-386168, issued 5/27/93.

8.14 Wright, D., Summers, D.A., and Sundae, L., "A Comparison of Abrasive Waterjet and Waterjet Assisted PDC Drill Bits," 12th International Conference on Jet Cutting Technology, Rouen, France, October, 1994, (in press).

8.15 Veenhuizen, S.D., Kirby, M.J., Cheung, J.B., and Summers, D.A., Development of a System for High Speed Drilling of Small Diameter Roof Bolt Holes - Phase 1 Report, November, 1976, under U.S.B.M. contract HO262036.

8.16 O'Hanlon, T.A., An Overview of Current and Future Applications for Ultra High-Pressure Waterjets in Mining, Canadian Institute of Mining, Annual Meeting, Ottawa, Canada, April, 1984.

8.17 Jarvis Clark, Full Trackless Service, a product brochure from Jarvis Clark Co., 4445 Fairview Street, Burlington, Ontario, Canada L7R 3YB.

8.18 Jarvis Clark, JB2-H 2 boom Waterjet Roofbolter, product brochure 5M-4- 84, Jarvis Clark Co., 4445 Fairview Street, Burlington, Ontario, Canada L7R 3YB.

8.19 Vijay, M.M., and Brierley, W.H., "Drilling of Rocks by High Pressure Liquid Jets: An Assessment of Nozzles," paper G1, 5th International Symposium on Jet Cutting Technology, Hanover, FRG, June, 1980, pp. 327 - 338.

8.20 Vijay, M.M., and Brierley, W.H., "Drilling of Rocks by High Pressure Liquid Jets: A Review," ASME Preprint 80-Pet-94, Energy Technology Conference, New Orleans, LA, February, 1980, 11 pages.

8.21 Vijay, M.M., Brierley, W.H., and Grattan-Bellew, P.E., "Drilling of Rocks with Rotating High Pressure Water Jets: Influence of Rock Properties," paper E1, 6th International Symposium on Jet Cutting Technology, Guildford, UK, April, 1982, pp. 179 - 198.

8.22 Vijay, M.M., Grattan-Bellew, P.E., and Brierley, W.H., "An Experimental Investigation of Drilling and Deep Slotting of Hard Rocks with Rotating High Pressure Water Jets," paper H2, 7th International Symposium on Jet Cutting Technology, Ottawa, Canada, June, 1984, pp. 419 - 438.

8.23 Bortolussi, A., Yazici, S., and Summers, D.A., "The Use of Waterjets in Cutting Granite," paper E3, 9th International Symposium on Jet Cutting Technology, Sendai, Japan, October, 1988, pp. 239 - 254.

8.24 Agus, M., Bortolussi, A., Ciccu, R., Manca, P.P., and Massaci, G., "Jet Energy Requirements for Driving Deep Kerfs into Eruptive Rocks," Geomechanics 91, Hradec n.m., Czechoslovakia, September, 1991, Czech Academy of Sciences.

8.25 Hoshino, K., Nagano, T. and Tsuchishima, H., "Rock cutting and breaking using high speed water jets together with "TBM" cutter," paper B6, 1st International Symposium on Jet Cutting Technology, Coventry, UK, April, 1972, pp. B2 13 - B2 28.

8.26 Nagano, T., Hoshino, K., and Narita, Y., "The Development of a Water Jet Drilling Machine," paper E1, 2nd International Symposium on Jet Cutting Technology, Cambridge, UK, April, 1974, pp. E1 1 - E1 10.

8.27 Hoshino, K., Nagano, T., Takagi, K., Narita, Y., and Sato, M., "The Development and the Experiment of the Water Jet Drill for Tunnel Construction," paper E4, 3rd International Symposium on Jet Cutting Technology, Chicago, IL, May, 1976, pp. E4 41 - E4 48.

8.28 Nagano, T., Takkagi, K., Narita, Y., and Sato, M., "Development of Water Jet Drifter," paper C6, 4th International Symposium on Jet Cutting Technology, Canterbury, UK, April, 1978, Vol. 2, pp. C6 67 - C6 75.

8.29 Lefin, Y., and Hurel, A., "Rotary drilling assisted by waterjets and other recent developments of high pressure waterjets for the cutting of rocks in France," paper H3, 7th International Symposium on Jet Cutting Technology, June, 1984, Ottawa, Canada, pp. 439 - 454.

8.30 Bonge, M.J., and Wang, F.D., "Advanced Applications of Water Jets to Small Hole Drilling," 1st U.S. Water Jet Symposium, Golden, CO, April, 1981, pp. III-3.1 - III-3.14.

8.31 Sundae, L., and Cantrell, B.K., "Breakthrough in Roof Bolt Drill Technology Provides 200 to 600 times greater Bit Life," 4th Conference on Ground Control in Midwestern U.S. Coal Mines, Mt. Vernon, IL, November, 1992, pp. 291 - 313.

8.32 Zink, G., Wang, F.D., and Wolgamott, J., Development of a Water Jet Augmented System for High Speed Drilling of Small Diameter Roof Bolt Holes, Phase III report on U.S. Bureau of Mines Contract HO 262054, CSM, May, 1978.

8.33 Kovscek, P.D., Taylor, C.D., and Thimons, E.D., Evaluation of Water-Jet-Assisted Drilling with Handheld Drills, Bureau of Mines RI 9174, 1988,15 pp.

8.34 Summers, D.A., Barker, C.R., and Keith, H.D., Preliminary Evaluation of a Water Jet Drilling System for Application in In-Situ Gasification, Final Report on Contract 130214, UMR to Sandia Laboratories, April, 1979, 66 pp.

8.35 Tomren, P.H., Iyoho, A.W., and Azar J.J., "Experimental Study of Cuttings Transport in Deviated Wells," SPE Drilling Engineering Journal, February, 1986, p. 285.

8.36 Barker, C.R., and Timmerman, K.M., "Water Jet Drilling of Long Horizontal Holes in Coal Beds," 1st U.S. Water Jet Symposium, Golden, CO, April, 1981, pp. III-2.1 - III-2.10.

8.37 Barker, C.R., "Orbiting Nozzle," U.S. Patent.

8.38 Kennerley, P., An Interim Report on the Development of a High Pressure Water Jet Drill for Drilling Long Holes in Coal Seams, University of Queensland, Dept. of Mining and Metallurgical Engineering, July, 1989, 10 pages.

8.39 Kennerley, P., Development of a High Pressure Waterjet Drilling System for Coal Seams, Thesis for the degree of Master of Engineering Science, Dept. of Mining and Metallurgical Engineering, University of Queensland, January, 1990, 212 pages.

8.40 Kennerley, P., Phillips, R., Just, G.D., and Summers, D.A., "A High Pressure Waterjet System for Inseam Longhole Drilling of Coal," 6th American Water Jet Conference, Houston, TX, August, 1991, pp. 213 - 222.

8.41 Hix, G.L., "Something New under the Sun: a New Drill Rig," Water Well Journal, December, 1990, pp. 32 - 34.

8.42 Becker, H., and Schmidt, B.H., "Drilling Bore Holes in Coal Mines using High Pressure Water," 2nd U.S. Water Jet Conference, Rolla, MO, May, 1983, pp. 179 - 183.

8.43 Ohga, K., and Higuchi, K., "Application of Water Jet Technology at Coal Mines in Japan: Water Jet Drilling for Large Diameter Stress Relief Borehole in Coal Seam," 8th International Symposium on Jet Cutting Technology, Durham, UK, September, 1986, pp. 9 - 19.

8.44 Summers, W.A., The Oldest Miner of them All, Presidential Address, North of England Institute of Mining and Mechanical Engineers, Neville Hall, 1984.

8.45 Gup, T., "The Curse of Coal," TIME magazine, Vol. 138, No. 18, November 4, 1991, pp. 54 - 64.

8.46 Summers, D.A., Mazurkiewicz, M., and Galecki, G., "Horizontal Waterjet Exploration of Burning Coal Seams," 8th International Symposium on Jet Cutting Technology, Durham, UK, September, 1986, pp. 21 - 32.

8.47 Summers, D.A., and Barker, C.R., The Detection of the Coal Roof Interface by Use of High Pressure Water, Final report to the National Aeronautics and Space Administration, Marshall Space Flight Center, Alabama, 1979, 50 pages.

8.48 Barker, C.R., and Summers, D.A., The Development of a Longwall Water Jet Mining Machine, Final Report on Department of Energy Contract DOE-AC01-75ET12542, University of Missouri-Rolla, July, 1981.

8.49 Summers, D.A., Barker, C.R., and Tyler, J.T., "The Use of Water Jets to Detect the Coal:Shale Interface in Automatically Steering Mining Equipment," paper E4, 5th International Symposium on Jet Cutting Technology, Hanover, FDR, June, 1980, pp. 287 - 296.

8.50 Dussert, P., Santoro, G., and Soudet, H., "A decade of drilling development pays off in offshore Italian Oil field," Oil & Gas Journal, February 29,1988, pp. 33 - 39.

8.51 Baker, W., "Extended Nozzle two-cone bits require precise nozzle sizing for optimum effect," 1979 Drilling Technology Conference, International Association of Drilling Contractors, Houston, TX, pp. 153 - 169.

8.52 Sutherland, J.A., "Increasing Water-Drilling Hydraulics helps Improve Drilling Performance," Oil and Gas Journal, Vol. 89, No. 29, July 22, 1991, pp. 91 - 94.

8.53 Maurer, W.C., Heilhecker, J.K., and Love, W.W., "High Pressure Jet Drilling," SPE 3988, 47th Annual Fall Meeting of SPE, San Antonio, TX, October 8 - 11, 1972.

8.54 Maurer, W.C., Novel Drilling Techniques, Pergamon Press, (1968), 114 pp.

8.55 Maurer, W.C., Advanced Drilling Techniques, Petroleum Publishing Co., 698 pp.

8.56 Deily, F.H., Heilhecker, J.K., Maurer, W.C., and Love, W.W., "A Study of High Pressure Drilling," Drilling Technology Conference, 1977, New Orleans, LA.

8.57 Maurer, W.C., Drilling Research Report, No. 7, p. 87.

8.58 Maurer, W.C., McDonald, W.J., Cohen, J.H., Neudecker, J.W., and Carroll, D., Laboratory Testing of High Pressure High Speed PDC bits, LA/UR-86-2269, Los Alamos National Laboratory, 1986, 17 pp.

8.59 Feenstra, R., Pols, A.C., and Van Steveninck, J., "Tests show jet drilling has promise," Oil & Gas Journal, July 1, 1974, pp. 45 - 57.

8.60 Feenstra, R., Pols, A.C., and Van Steveninck, J., "Rock Cutting by Jets," Mining Engineering, June, 1974, pp. 41 - 47.

8.61 Pols, A.C., High Pressure Jet-Drilling Experiments in some Hard Rocks, Publication 479, Shell Exploration and Production Laboratory, Rijswijk, Netherlands, October, 1975.

8.62 Van Strijp, A.J.R., and Feenstra, R., 1977 ASME Energy Technology Conference, Houston, TX.

8.63 Melaugh, J.F., The Effects of Confining Pressure on the Jet Erosion Cutting of Rock, M.S. thesis, The University of Tulsa, 1972.

8.64 Summers, D.A., and Sebastian, Z., The Reaming out of Geothermal Excavations, Final Report to Sandia Laboratories on DOE Contract 13-3246, April, 1980.

8.65 Pratt, C.A., "Increased Penetration Rates Achieved with New Extended Nozzle Bits," Journal of Petroleum Technology, August, 1978, pp. 1191 - 1198.

8.66 Summers, D.A., and Lehnhoff, T.F., The Design of a Water Jet Drill for Development of Geothermal Resources, Final Report to the Department of Energy on Contract DOE-EY-76-S-02-2677.M003, September, 1978.

8.67 Obert, L., and Duvall, W., Rock Mechanics, Pergamon Press.

8.68 Conn, A.F., and Radtke, R.P., paper 77 -Pet 54, 1977 ASME Energy Technology Conference, Houston, TX, 1977.

8.69 Conn, A.F., and Radtke, R.P., "Cavijet Cavitating Jets for Deep-Hole Rock Cutting," 1980 ASME Energy Technology Conference, New Orleans, LA, February, 1980.

8.70 Angona, F.A., "Cavitation, A Novel Drilling Concept," International Journal of Rock Mechanics & Mining Science, Pergamon Press, Vol. 11, 1974, pp. 115 - 119.

8.71 Robinson, M.J., and Hammitt, F.G., "Detailed Damage Characteristics in a Cavitating Venturi," J. Basic Engineering Trans. ASME Series D, Vol. 89, 1967, pp. 161 - 173 (quoted in [8.70]).

8.72 Conn, A.F., Johnson, V.E., Lindemuth, W.T., and Frederick, G.S., "Some Industrial Applications of Cavijet Cavitating Fluid Nozzles," paper V.2, 1st U.S. Waterjet Symposium, Golden, CO, April, 1981.

8.73 Bardin, C., and Cholet, H., "Jet Assisted Oil Drilling," paper A3, 7th International Symposium on Jet Cutting Technology, Ottawa, Canada, June, 1984, pp. 33 - 50.

8.74 Bardin, C., Cholet, H., and Lecoffre, Y., "Assistance for deep drilling by cavitation damage," 9th International Symposium on Jet Cutting Technology, Sendai, Japan, October, 1988, pp. 611 - 628.

8.75 Killalea, M., "High Pressure Drilling System Triples ROPs, Stymies Bit Wear," Drilling, March/April, 1989.

8.76 McDonald, M.C., Reichman, J.M., and Theimer, K.J., Evaluation of High Pressure Drilling Fluid Supply Systems, Flow Technology Co., Contractor Report SAND81-7142, under DOE contract DE-,AC04-76DP00789.

8.77 Cure, M., and Fontana, P., "Jet-Assisted Drilling nears Commercial use," Oil & Gas Journal, March 11, 1991.

8.78 Summers, D.A., "The Potential Advantages of a Water Jet Reaming Device," paper IV.2, 1st U.S. Waterjet Symposium, Golden, CO, April, 1981.

8.79 Summers, D.A., Barker, C.R., and Keith, H.D., CYGNET: Preliminary Evaluation of a Water Jet Drilling System for Application in In-Situ Gasification, Final Report to Sandia Laboratories on Contract 130214, University of Missouri-Rolla, Rolla, MO, April, 1979, 66 pp.

8.80 Summers, D.A., "Potential Advantages to Water Jet Drilling for Use in In-Situ Coal Gasification Processes," paper 80-Pet-76, 1980 ASME Energy Technology Conference, New Orleans, LA, February, 1980.

8.81 Barker, C.R., Water Jet Drilling Head Development, Phase Report to Sandia Laboratories on Contract 13-8747, University of Missouri-Rolla, Rolla, MO, December, 1980, 58 pages.

8.82 Anon, "Water-jet drill simplifies horizontal boring task, Machine Design, December 9, 1982, pp. 92 - 93.

8.83 Engler, B.P., "Instrumentation for the Cornering Water Jet Drill," 8th Underground Coal Symposium, Keystone, CO, August, 1982.

8.84 Shirey, D.C., and Engler, B.P., "Development of a Cornering Water Jet Drill," paper H4, 7th International Symposium on Jet Cutting Technology, Ottawa, Canada, June, 1984, pp. 455 - 468.

8.85 Anon, "Elf completes Adriatic horizontal boring," Oil & Gas Journal, May 18, 1987, p. 24.

8.86 Anon, "Field test recommended for oil mining method," Oil and Gas Journal, June 5, 1989, p. 73.

8.87 Pendleton, L.E., and Ramesh, A.B., "Bechtel develops innovative method for horizontal drilling," Oil and Gas Journal, May 27, 1985, pp. 95 - 99.

8.88 Dickinson, W., and Dickinson, R.W., "Horizontal Radial Drilling System," SPE 1985 California Regional Meeting, SPE 13949, March, 1985.

8.89 Dickinson, W., Wilkes, R.D., and Dickinson, R.W., "Conical Water Jet Drilling," <u>4th U.S. Water Jet Conference</u>, Berkeley. CA, August. 1987, pp. 89 - 96.

8.90 Dickinson, W., Anderson, R.R., and Dickinson, R.W., "A Second-Generation Horizontal Drilling System," <u>1986 IASC/SPE Drilling Conference</u>, Dallas, TX, February, 1986, IADC/SPE 14804.

8.91 McNally, R., "Making New Drilling Technology work for you," <u>Petroleum Engineer International</u>, January, 1988, pp. 15 - 18.

8.92 Kaback, D.S., "Savannah River Site Environmental Restoration and Waste Management Needs," <u>Mining Workshop for Nuclear Waste Cleanup</u>, ASME, Colorado Springs, January, 1991, pp. 39 - 44.

8.93 Derby, G.K., and Bevan, J.E., <u>Longer than Seam Height Drill Development Program</u>, U.S. Bureau of Mines Report of Investigations 8273, 1978, 16 pp.

8.94 Summers, D.A., and Lehnhoff, T.F., "Waterjet Drilling in Sandstone and Granite," <u>18th U.S. Symposium Rock Mechanics,</u> Keystone, Colorado, May, 1977.

8.95 Savanick, G.A., and Krawza, W.G., <u>Water jet perforation: a new method of completing and stimulating in-situ leaching wells</u>, U.S. Bureau of Mines PB 82-135583, October, 1981, 37 pp.

8.96 Summers, D.A., Fossey, R.D., and Blaine, J.G., "DIAjet use in an Environ- mental Application," <u>11th International Symposium on Jet Cutting Technology</u>, St. Andrews, Scotland, September, 1992, pp. 281 - 291.

8.97 McCauley, T.V., "Backsurging and Abrasive Perforating to Improve Perforation Performance," <u>Journal of Petroleum Technology</u>, October, 1972, pp. 1207 - 1212.

8.98 Livingston, C.W., U.S. Patent 3,762,771 October, 2, 1973.

8.99 Zink, G.P., Wolgamott, J.E., and Zink, E.A., "New Uses of High Pressure Water in Uranium Mining," 5th Annual Uranium Seminar, Albuquerque, NM, September, 1981, AIME, Chapter 12, pp. 77 - 79.

8.100 Zink, E.A., Wolgamott, J.W., and Robertson, J.W., "Water jet used in sandstone excavation," 1983 Rapid Excavation and Tunneling Conference, Section 10, Chapter 40, pp. 685 - 700.

8.101 McGroarty, S.J., An Evaluation of the Fracture Control Blasting Technique for Drift Round Blasts in Dolomitic Rock, M.S. thesis, Mining Engineering, University of Missouri-Rolla, 1984.

8.102 Worsey, P.N., "In-situ measurement of blast damage underground by seismic refraction surveys," 26th U.S. Symposium on Rock Mechanics, Rapid City, SD, June, 1985, pp. 1133 - 1140.

8.103 McKnown, A.F., and Thompson, D.E., "Experiments with Fracture Control in Tunnel Blasting," 22nd U.S. Symposium on Rock Mechanics, MIT, 1981, pp. 223 - 230.

8.104 Hoshino, K., and Shikata, S., "Application of Water Jet Cutting on the Smooth Blasting," paper X1, 5th International Symposium on Jet Cutting Technology, Hanover, Germany, June, 1980, pp. 165 - 180.

8.105 Summers, D.A., "High Speed Quarrying Techniques applied to the Tunneling Industry," Third International Conference on Innovative Mining Systems, UMR, Rolla, MO, November, 1987, pp. 253 - 259.

8.106 Yao, Jianchi, High Pressure Waterjets in Rock Excavation, Ph.D. Dissertation, Department of Mining Engineering, University of Missouri-Rolla, June, 1991.

8.107 Wright, D.E., and Summers, D.A., "Performance Enhancement of DIAdrill Operations," paper 39, 7th American Water Jet Conference, Seattle, WA, August, 1993, pp. 549 - 559.

8.108 Eckel, J.E., Deily, F.E., and Ledgerwood, L.W. Jr., "Development and Testing of Jet Pump Pellet Impact Drill Bits," 30th Annual Fall Meeting Petroleum Branch of AIMME, New Orleans, October, 1955, paper 540.

8.109 Ledgerwood, L.W. Jr., "Efforts to Develop Improved Oilwell Drilling Methods," Journal of Petroleum Technology, April, 1960.

8.110 Bobo, R.A., Method of Drilling with High Velocity Jet Cutter Roller Bit, U.S. Patent 3,112,800, December 3, 1969.

8.111 Cleary, J.M., Apparatus and Method for Earth Drilling, U.S. Patent 3,231,031, January 25, 1966.

8.112 Mori, E.A., and Schaub, P.W., Continuous Coring Jet Bit, U.S. Patent 3,424,255, January 28, 1969.

8.113 Anon, "New Gulf method of Jetted-particle drilling promises speed and economy," Oil & Gas Journal, June 21, 1971, pp. 109 - 114.

8.114 Fair, J.C., "Development of High-Pressure Abrasive-jet Drilling," Journal of Petroleum Technology, August, 1981, pp. 1379 - 1388.

8.115 Ripkin, J.F., and Wetzel, J.M., A Study of the fragmentation of rock by Impingement of water and solid impactors, ARPA Contract HO210021, Report No. 131, University of Minnesota, February, 1972.

8.116 Chatterton, N.E., Development of a Safer, More Efficient Hydraulic-based Technique for Rapid Excavation of Coal, Rock and Other Minerals, Final Report on U.S. Bureau of Mines Contract HO232062, Teledyne Brown Engineering, PB 245 343, April, 1975.

8.117 El-Saie, A.A., Investigation of Rock Slotting by High Pressure Water Jet for Use in Tunneling, Ph.D. Dissertation, Department of Mining Engineering, University of Missouri-Rolla, 1977, 144 pages.

8.118 Savanick, G.A., Krawza, W.G., and Swanson, D.E., "An abrasive Jet Device for Cutting Deep Kerfs in Hard Rock," 3rd U.S. Waterjet Conference, Pittsburgh, PA, 1985, pp. 101-122.

8.119 Savanick, G.A., and Krawza, W.G., "An Abrasive Waterjet Rock Drill," 4th U.S. Waterjet Conference, Berkeley, CA, Aug., 1987, pp. 129 - 132.

8.120 Hashish, M., Halter, M., and McDonald, M., "Abrasive Waterjet DeepKerfing in Concrete for Nuclear Facility Decommissioning", 3rd U.S. Waterjet Conference, University of Pittsburgh, PA, 1985, pp. 123 - 144.

8.121 Echert, D.C., Hashish, M., and Marvin, M., "Abrasive Waterjet and Waterjet Techniques for Decontaminating and Decommissioning Nuclear Facilities," 4th U.S. Waterjet Conf., Berkeley, CA, Aug., 1987, pp. 73 - 81.

8.122 Hashish, M., "The Potential of an Ultrahigh Pressure Abrasive Waterjet Rock Drill," paper 32, 5th U.S. Water Jet Conference, Toronto, Canada, August, 1989, pp. 321 - 332.

8.123 Summers, D.A., and Yao, J., "First Steps in Developing an Abrasive Jet Drill," 8th Annual Workshop, Generic Mineral Technology Center for Mine Systems Design and Ground Control, Reno, NV, November, 1990, pp. 49 - 60.

8.124 Yazici, S., Abrasive Jet Cutting and Drilling of Rocks, Ph.D. Dissertation, Department of Mining Engineering, University of Missouri-Rolla, June, 1989, 191 pp.

8.125 Wright, D.E., and Summers, D.A., "Performance Enhancement of DIAdrill Operations," paper 39, 7th American Water Jet Conference, Seattle, WA, August, 1993, pp. 549 - 561.

8.126 Wright, D.E., M.S. thesis, Department of Mining Engineering, University of Missouri-Rolla, (in preparation).

8.127 Yao, J., Summers, D.A., Galecki, G., Blaine, J.G., and Tyler, L.J., "Field Trials and Developments of the DIAdrill Concept," 6th U.S. Water Jet Conference, Houston, TX, August, 1991, pp. 545 - 560.

8.128 Yao, J., Summers, D.A., and Galecki, G., "DIAjet cutting of Dolomite and Chert - A Case Study at the St. Louis Arch," 6th U.S. Water Jet Conference, Houston, TX, August, 1991, pp. 529 - 543.

8.129 Vestavik, O.M., Abrasive Water-Jet Drilling Experiments, Progress Report, Rogaland Research, Stavanger, Norway, May, 1991.

9 JET ASSISTED MECHANICAL CUTTING

9.1 INTRODUCTION AND DEFINITIONS

Between 1972 and 1975 the U.S. Department of the Interior, Bureau of Mines planned a three-year, $6.6 million dollar program seeking to find more effective ways for improving rock excavation. This sum was, at the time, some 18% of the entire Federal budget for research into rapid excavation. The program was sponsored by the Advanced Research Projects Agency of the Department of Defense and in may ways underwrote the initial development of many of the programs which have been described in this chapter. A review of those programs [9.1] describes some of the earliest experiments and concepts developed for the use of high pressure waterjets, and also describes some of the initial negative results which led to the termination of some of those creative ideas. Regrettably the U.S. Congress deleted funds for this program after the first year of funding, thus, many of the potentially promising new programs were not initiated, and many of those funded did not reach the levels of field experimentation initially planned to demonstrate their potential commercial value. This may be one reason it has taken this technology so long to reach its current level of maturity, and why, as yet, some of the ideas first developed many years ago have not yet become a commercial reality.

It is difficult to compare the potential benefits from different methods of excavation based on the promises and evaluations made when the ideas were first put forward, and before they could be verified in field trials. Possible advantages and disadvantages of the techniques are rarely fully considered when the concept is first developed. At the same time there have been a number of cynics who have taken part in various reviews of new technology who reject novel techniques for superficial reasons. It has been necessary to create some form of screening of these ideas, to give some idea of their relative worth. This evaluation can then be used, both to compare the benefits of the new ideas, one to another, but also to contrast them with existing technology as a means of indicating what particular benefits would be obtained from investing in their development.

Some of the new ideas proposed have other problems (for example, the presence of dangerous levels of radiation or the generation of noxious fumes) which led to their rejection as the techniques were advanced. One of the major points of discussion was based on the use of the relative specific energies with which different methods could be calculated to remove rock. As defined earlier in this text, specific energy is taken to mean the amount of energy required to remove a unit volume of rock.

This term was first proposed by Teale [9.2] in examining the processes which control the mechanical drilling of rock. That paper ends with the statement "It is interesting to speculate whether, on physical grounds, any process of breaking rock from the face of a semi-infinite solid can be conceived which would require a lower specific energy than do the mechanical methods now available."

The use of specific energy provides one comparative means of evaluation of the relative effectiveness of different methods for breaking rock (Table 9.1). It should be recognized, however, that the values given in the table below do not necessarily reflect the levels of energy which have been found to prevail, as the techniques have been refined with additional use and investigation. Nor does it indicate any other potential benefits of any one technique which might have been found during an investigation which would encourage the use of the technique, over existing methods. (As an example the use of high pressure waterjets for cutting will usually eliminate any problems generated with dust - since there very rarely is any with this method). It is interesting to note, some 20 odd years after that program was initiated that high pressure continuous jet kerfing and mechanically assisted hydraulic jet kerfing are the only two techniques which have made significant market impact in rock cutting. It is the purpose of this chapter to discuss some of the side benefits which have come from the development of this idea and its implementation.

It is interesting to note that the study did not include the development of polycrystalline diamond composites, the development of abrasive jet systems, cavitation cutting or other developments of the last two decades. The fact that a number of "more efficient" rock cutting methods have not been further developed also speaks to the point that it is not just the relative efficiency of rock fracture which is the deciding factor in developing new technologies into the commercial market place.

An additional consideration which is not always clear must also be understood. There is a considerable evaluation of novel technology in terms of the specific energy of the excavation process. But the specific energy is, itself, related to the size of the particles which the process generates. Thus, it must be a part of the evaluation to assess the size of such particles, and whether, by changing the process slightly to create larger particles, a lower specific energy can be generated. The initial aspect of this point was illustrated by Cook and Joughin in 1970 [9.15] both in terms of a number of different excavation techniques, and in regard to the typical size of the rock particles which they might generate (Fig. 9.1).

Table 9.1 Specific energy of novel rock breaking methods evaluated by ARPA [9.1]

	Normalized typical specific energy values
THERMAL METHODS	
high power laser [9.3]	450
high power plasma [9.4]	120
electron beam gun [9.5]	8
HYDRAULIC METHODS	
low-velocity water slugs [9.6]	85
high pressure continuous jet kerfing [9.7]	45
high pressure pulsed jets [9.8]	1
COMBINATION METHODS	
mechanically assisted hydraulic jet kerfing [9.9]	40
mechanically assisted thermal fracture [9.10]	0.1
SOLID, NON-EXPLOSIVE PROJECTILE IMPACT	
medium sized, moderate velocity projectiles [9.6]	0.2
large high velocity projectiles (the REAM gun) [9.11]	0.07
MECHANICAL METHODS	
sonic Energy Rock kerfing [9.12]	1.5
conical borer [9.13]	0.7
high energy mechanical impact [9.14]	0.5

The reference numbers refer to project reports available in 1975, and cited in the reference list at the end of this chapter.

The above data assumes a characteristic product size, and because of this, a given specific energy range for that technique. These numbers are not necessarily accurate (consider for example, earlier discussions of how much of a jet energy content is actually used efficiently in cutting). Further, they make fixed assumptions about product size which have only a limited accuracy. This point can be illustrated by two examples. Among the projects listed above (Table 9.1) is that of mechanically assisted hydraulic jet kerfing. This in part arose because of a study which had been separately funded by the Department of Defense [9.16]. In that study a series of passes had been made over a rock surface at different spacings (Fig. 9.2). These had initially been directed at establishing how closely together two adjacent waterjets would need to be when cutting rock before the intervening rib was removed. It was noted that even when this rib was relatively narrow, say 1 mm thick, and the adjacent slots quite deep, perhaps 5 cm, that the intervening rib would not be broken by the jet action. Thus, to completely excavate the rock by water jet would have required that the rib be removed by application of a 60 kW waterjet. However the rib was sufficiently weak that it could be easily broken by the pressure from one finger.

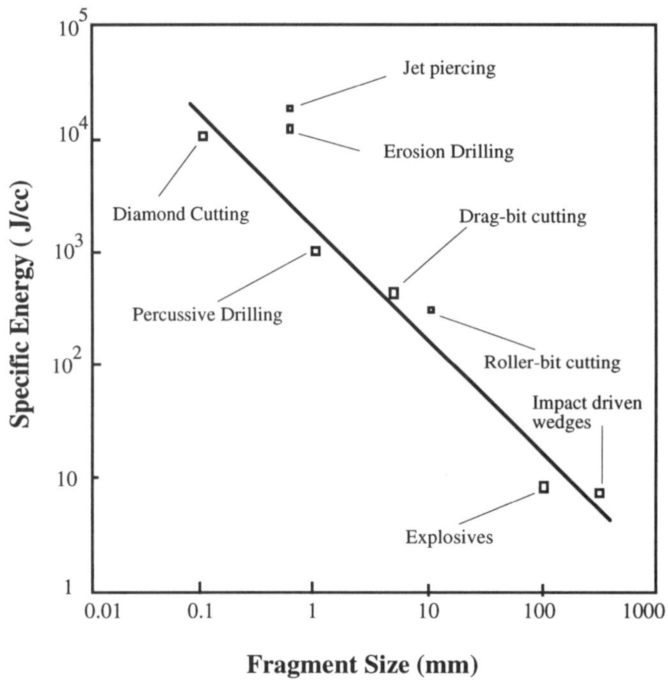

Figure 9.1 Specific energy values suggested by Cook and Joughin for some novel excavation techniques as a function of particle size produced [9.15].

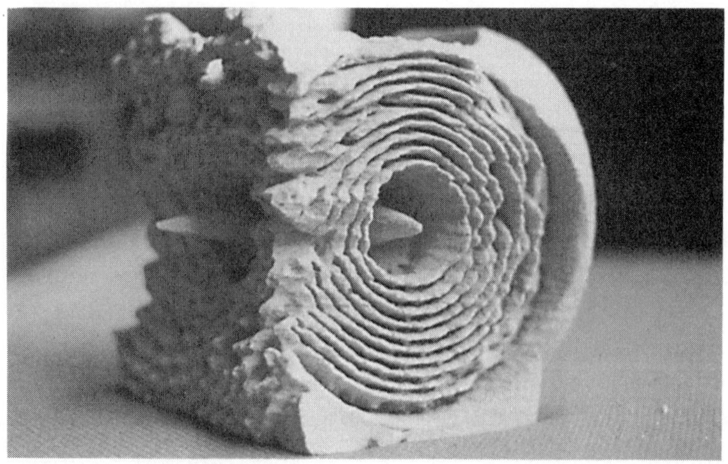

Figure 9.2 Ribs of rock left by cutting adjacent passes with a waterjet.

This study suggested that it would be more effective to combine mechanical cutting with the jet cutting of slots to more effectively remove the intervening material (Table 9.2). This concept has been developed, with abrasive jet cutting both by the Gulf Research effort [9.17] and the Rogaland studies [9.18] discussed at the end of the previous chapter. It should be noted that the results of the combined method of fragmentation (Table 9.2) bring the specific energy of rock cutting considerably below that of any technique identified by the Cook and Jachim graph. The second example combined abrasive waterjet rock drilling with the use of hydraulically driven wedges in the CUSP mining method. In this technique the rock "particles" produced were in the metric range rather than the millimetric [9.18} and the rock was largely liberated by the growth of a single crack around the fragment. The combined excavation method produced, again, a much lower specific energy than that suggested by the earlier reviews of potential energy levels available with new techniques. It should be noted, further, that both these examples use combined techniques to achieve the overall reduction in specific energy.

The initial study in which waterjet slotting was combined with mechanical breakage led on into the subsequent studies, both with linear cutters and mechanical cutter discs, in which a mechanical tool was used to further break up a rock which had been scored by the action of a waterjet. This will be referred to as **Mechanically Assisted Waterjet Cutting.**

Table 9.2 Specific energy of breakage for the mechanical removal of ribs left by waterjet slotting of Berea sandstone [9.16]

Rib Width (mm)	*25*	*12.5*	*6.25*
Slot Depth (mm)	*Specific Energy Values in joules/cubic cm.*		
6.6	-	6.12	0.51
11.2	0.44	1.52	0.39
15.0	0.30	0.46	0.17
15.75	0.40	0.50	0.05
17.0	0.23	0.21	0.05
23.9	0.19	0.20	-

This process must be separated from the action of the mechanical cutting tool, where a waterjet is directed to hit the rock right under the cutting tool and the waterjet then works with that tool to synergistically remove the rock. That process, developed and demonstrated first by Dr. Hood, will be referred to as **Waterjet Assisted Mechanical Cutting**.

9.2 MECHANICALLY ASSISTED WATERJET CUTTING

The concept of using a mechanical tool to break out the rock had also been recommended by Bendix, who had been working, under ARPA funding, on studies of rock cutting by waterjets at pressures between 2,500 and 5,600 bar [9.20]. Their conclusion, from cutting harder rocks including basalt and quartzite, was that the system would be more efficient if the intervening ribs of rock between the jet cuts could be removed by mechanical means, and a disc cutter was suggested as a fracture wheel. The concept proposed was to roll the wheel down a central cut, concurrently breaking off the rock out to the next adjacent cuts on each side (Fig. 9.3).

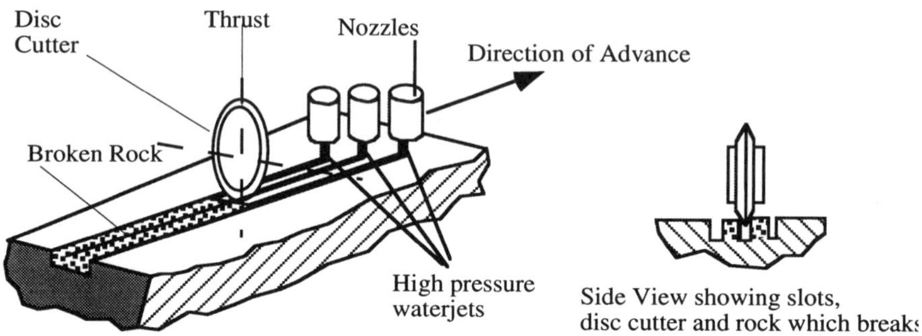

Figure 9.3 Bendix concept for more efficient rock fragmentation using a mechanical assistance to the jet cutting [9.9].

A second contract was let to Bendix [9.9], to try out this process in a field test. A series of waterjet cuts was made into the face of a granite quarry in Vermont, leaving intervening ribs of material. A mechanical cutting wheel was then rolled down the slots breaking out the material to either one or both sides of the cutter. The slots were cut by a waterjet with a diameter of either 0.2 or 0.3 mm, with waterjet pressures of either 2,100 or 4,550 bar. The typical slots resulting were on the order of 6.35 - 9.5 mm deep and were spaced from 2.5 mm - 8.9 mm apart. Only a slight reduction in cutting efficiency was achieved with this design, in contrast with the more dramatic predictions made with lower pressure jets on softer rocks [9.16].

A contemporary study in Japan had concluded [9.21] that the optimum spacing of the jet slots should be a function of the uniaxial compressive strength of the rock. Further, that where this was carried out, that the cutting and breaking out of the rock would be from twice to five times as great as for a TBM alone depending on the cut spacing and depth (Fig. 9.4).

One point which the Japanese were careful to address was that the cutting speed of the jet over the rock surface be in the same order of magnitude as that required on a tunneling machine head. Because of the large diameter of these drums (4 m is common) even at slow rotational speeds the nozzles may traverse at several m/sec. Accordingly, the test program ran tests at up to 1 m/sec. Tests were also carried out using a gear cutter between the slots, rather than a disc cutter. A doubling of the rock volume excavated was achieved by that change.

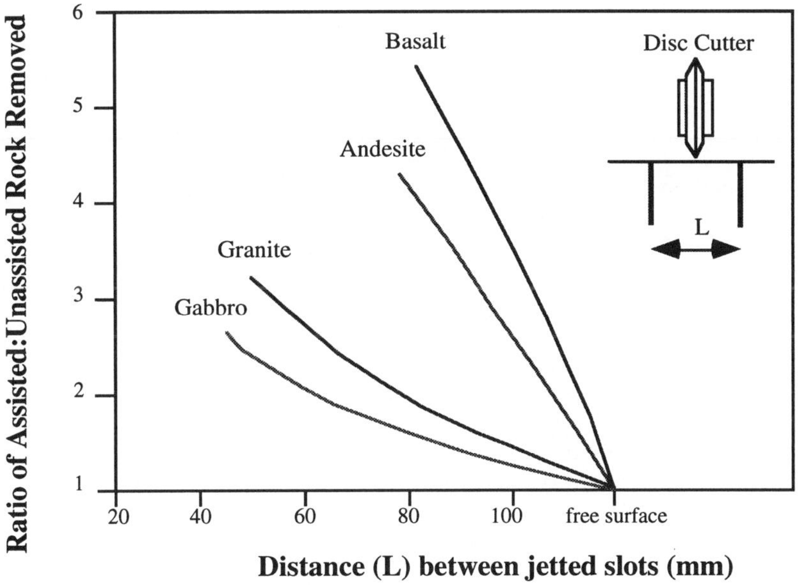

Figure 9.4 Effect of pre-slotting a face on the volume of rock removed by a TBM cutter (after [9.21]).

The work with the gear cutters was subsequently continued, examining the use of the hybrid system for drilling the 200+ mm diameter holes used as the relieving holes for blasting rounds and similar applications in tunnel drivage [9.22]. The results of the study indicated (Fig. 9.5) that it was possible to increase boring speed by up to four times, using toothed cutters and jets 0.3 - 0.4 mm in diameter, at pressures of 4,000 bar. The improvement was related to bit thrust, jet pressure and nozzle location. Where only a single jet was added, then it did the most good when located at the gage of the hole.

The spacing and depth of the relieving slots created by the jets plays a major part in determining how efficient the process might be, particularly in ground which is under rock pressure. This point, partially addressed in an earlier chapter (see discussion of Figs. 8.48 (a) and (c)), may have been neglected in later developments of this technology, and so it is important to have some grasp of the potential benefits which might accrue if this combined waterjet and mechanical cutting system were to be used most efficiently. For example in the initial study by Summers and Henry [9.16] it was possible to break the rib between adjacent jet cut slots with a specific energy of 0.5 joules/cc.

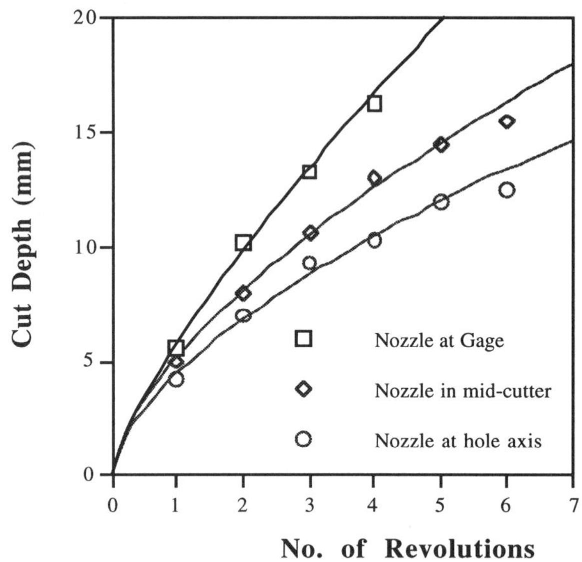

Figure 9.5 Increase in bit cutting depth as a function of jet location [9.22].

Chadwick [9.9] calculated that this value could theoretically be reduced to a level of 1.4 x 10^{-5} joules/cc if the confining conditions on the rock ribs could be further reduced (perhaps in the way being tried by Rogaland and shown in Fig. 8.90). Mechanical excavation specific energies of argillites drilled by a TBM in Boston averaged 93 joules/cc, and for a quartzite in Arizona averaged 464 joules/cc. Bendix calculated the jet specific energies to cut the slots at 3,963 joules/cu c for the argillite, and 6,611 joules/cu c for the quartzite. For an optimized jet cut (i.e., slot depth being equal to slot spacing) the mechanical energy to remove rock equivalent to the argillite was 2.39 joules/cu c and for the quartzite 2.71 joules/cu c. With the shallower kerfs and wider spacing for the Lawrence-HRT-12 discs, the study calculate that the jets, when combined with the existing cutter pattern, would excavate rock at 1,275 and 1,584 joules/cu c in each rock, which would still more than double excavation costs for a 1500 m tunnel. Costs and performance were based availability in 1972 when the study was made.

Although that study concluded that an optimum spacing between the slots cut into the rock, was at twice the depth of the cuts which the jet made [9.9] an improvement on the initial 1:1 ratio suggested by Summers, still the values for the required levels of energy appeared too high. This was because, under normal operating conditions, at the speeds of the TBM heads, the slot which the jets could produce would be too shallow, at the allowed spacing, to produce the larger energy savings suggested by Summers and Henry.

However, the argument about specific energy is not in itself, conclusive. This was shown in a subsequent study by Crow [9.23]. Although some of the data and particularly the theoretical analysis on jet cutting accompanying it has only limited validity, yet the basic argument which Crow applied, has been echoed by others. This is that although the hybrid system requires more power than a conventional system, and may not mine, in the aggregate, as efficiently, yet the combined system delivers more power to the rock and thus can achieve penetration rates which are not available with plain mechanical cutting. It is a similar argument to that which has finally led to the development of the waterjet assisted oil well drilling systems discussed above. Thus, Crow calculated that it would require a machine power of 1,400 kW to drive a 6 m diameter tunnel at 3 m/hr through red granite. This provides roughly three times the power of the Lawrence TBM cited above, and would require, by Crow's suggestion, (Fig. 9.6) a combination of 20 disc cutters, and 149 jets (of 0.25 mm diameter) at a pressure of 2,600 bar and a flow of some 350 lpm of water. This was not considered to be an impossible requirement. (It should be noted that the present author does not believe this to be an optimum configuration.)

The spacing between adjacent jet-cut slots, which also becomes a function of the spacing between the disc cutters on the face of the machine, has further significance. The size of this rib will control the height of material through which the jet must cut, and also the distance which the nozzle must be mounted back from the face of the rock, if it is not to be damaged while making its traverse. The necessity for considering this was made clear in tests carried out in Washington by a consortium under the overall direction of the Colorado School of Mines [9.24]. Funded by the National Science Foundation, and together with the Robbins Company and

Flow Industries, who provided the pump, a modified large scale tunnel boring machine was assembled (Fig. 9.7).

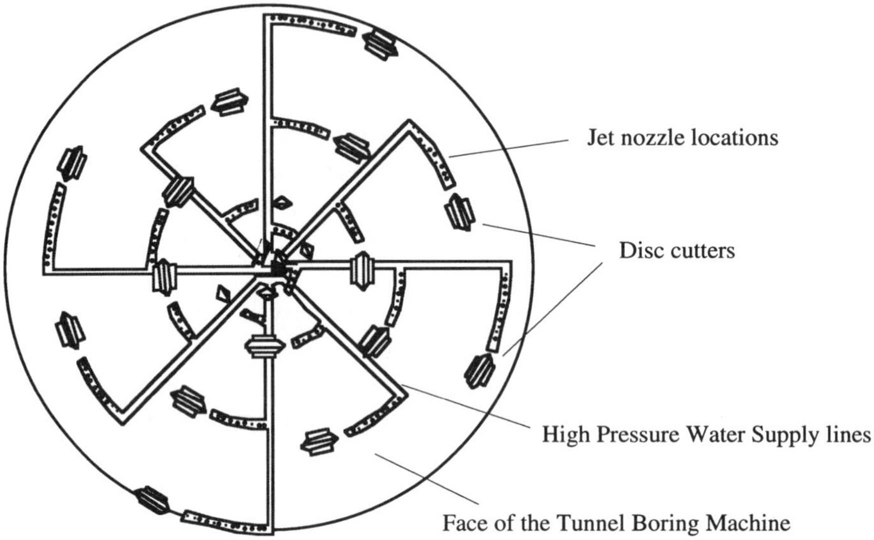

Figure 9.6 Suggested layout of nozzles and cutters for a 6 m diameter TBM face, (after Crow [9.23]).

Figure 9.7 Tunnel Boring Machine fitted with high pressure waterjet nozzles, by CSM.

This 2.1 m diameter machine incorporated high pressure waterjets in an array between the conventional mechanical cutters. In total, some 750 kW of hydraulic power was made available to these nozzles from four intensifiers, each operable at 3,850 bar. The TBM had 16 cutters and could apply a thrust of 18,000 kg. through each cutter. Three jet patterns were tested in the program. Jets were located between the cutters, under the cutters, and in combination. In one of the first tests of the system the head advanced sufficiently rapidly, under waterjet assist, that in 20 seconds of operation the ribs between the cutters reached the nozzle manifold and destroyed it. It therefore became necessary to move the nozzle manifold back from a starting location of 25 mm from the face to one where the nozzle block was 37.5 mm from the face. With an optimum nozzle diameter of 0.3 mm, this meant that the rock face was 125 orifice diameters from the nozzles. Jets placed under the cutter did not appear to improve the penetration rate, even though they cut up to 50% deeper when running in this broken ground (Table 9.3). Comparative tests of cutting with and without jet assist indicated a maximum improvement of 375% over mechanical cutting (Fig. 9.8) although the design was not completely optimized, and the relative improvement in performance was controlled by the achievable jet pressure. Because the pumps were the first developed at this size by Flow, there were some initial equipment problems, not helped by the need to carry out the experiments in the very early morning due to power limitations at other times [9.25].

Table 9.3 Effect of jet location on TBM ROP (after [9.25])

Jet location	*Slot depth (mm)*	*Thrust per cutter (kg)*	*Advance (mm)*	*Specific energy (j/cc)*
under cutters	7.87	12,320	4.57	560
between cutters	5.08	12,320	5.08	560
under & between cutters	5.08	12,320	3.30	843

Although the trials at Skykomish were only a limited success the data from them suggested that tunneling advance rates could be more than doubled, with the potential to lower the cost of driving tunnels by some 14 to 25% [9.26]. The study also concluded that water flow rates to the cutting head should be kept in the range of 40 - 120 lpm to maintain good working conditions. Advance rates were improved by between 50 and 60% on average, depending on the thrust levels through the cutters. There was an average reduction of 25% in the torque required to turn the cutting head when the waterjets were running.

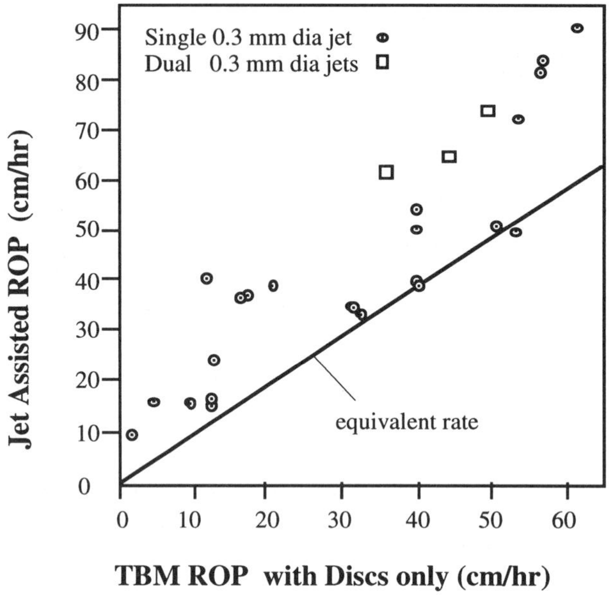

Figure 9.8 Conventional and jet assisted TBM performance with full face tests at Skykomish [9.24].

Results of the project were subsequently reviewed by Hustrulid [9.27]. One problem which he had in interpreting the information available, arose from the considerable scatter in the reported data on the exact way in which the jet parameters controlled the depth of the slot cut. Some fifteen years later that confusion has still not been completely eliminated. Part of the problem for example, arose in the relationship between nozzle diameter and depth of cut. Other work since has shown that this relationship is a power law with an exponent on the order of 1.5. However, because the larger nozzles were perhaps more sensitive to poor inlet flow conditions (as Wang has suggested) this relationship was not clear at the time of this analysis. This had some bearing on the analysis, as did the assumption that it would be better to use multiple jets to increase the depth of cut on the outer gage of the cutter, rather than use an equivalent flow through a single, larger diameter jet. The current author would favor this latter approach.

To give some measure of the performance parameters which were discussed, the TBM cutters were spaced an average of 75 mm apart. The maximum head ROP was 0.6 m/hr, at a rotation speed of 9 rpm. This gave an advance of 1.14 mm/rev. If the nozzles are equally spaced between the cutters, this gives a depth:spacing ratio of just under 1:70. This is far from the optimized use of the jets which can be achieved where the depth:spacing ratio is closer to unity. Tests on the Skykomish granite indicated that a 0.3 mm diameter jet would cut to a depth of 1 mm when traversed over the rock at a speed of 250 mm/sec at a jet pressure of 3,850 bar.

Hustrulid satisfactorily explained the reduction in advance where the jets cut ahead of the cutter path. Because the jet width is narrower than the cutter face it crushes the sides of the slot into the cut. This creates a zone of crushed rock under the cutter (Fig. 9.9), which no longer allows the development of the stresses in the adjacent rib of rock which cause the rock to chip, and lead to its easy removal.

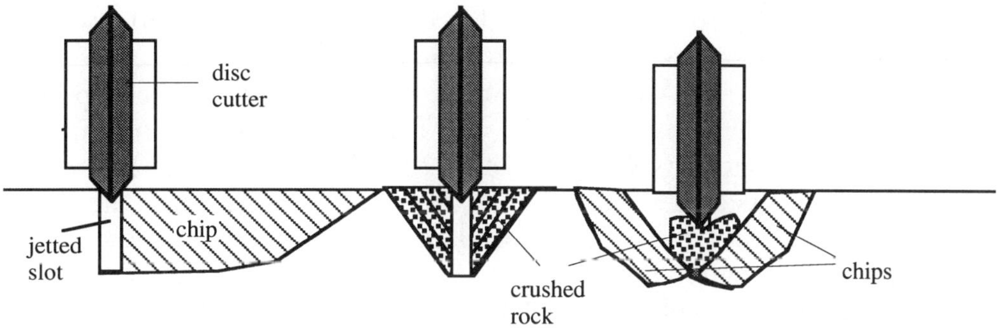

a) The concept of Disc Breakage   b) Rock Crushing   c) Reduced Fracturing due to the plastic zone created by the crushed rock.

Figure 9.9 Mechanisms of rock failure of a jet-weakened rock under a cutter [9.27].

Hustrulid concluded that while the data on performance held considerable promise, the question as to whether the technique could be economic might relate to the ability of the jets to cut deeper slots ahead of the cutters than had been achieved in this initial test. It was not clear that that was a commercially economic possibility. His subsequent analysis of cost and required nozzle conditions to drive a 7 m diameter tunnel at an ROP of 1.2 m/hr through granite suggested (in contrast to Crow's figures) that it would require 192 nozzles each 0.3 mm in diameter and operated at 3,500 bar in order to double the machine advance rate. Since this would require some 4,425 kW and give an additional required force of 1480 kg and a water flow rate of 740 lpm, this was considered to be of marginal benefit particularly in those conditions where the long term reliability of the system could not be guaranteed.

While the effects of this review have been to some degree overtaken by later developments, the prestige of the author cast a significant pall over the future of waterjet-assisted TBM research in the United States. This has been regrettable since there is now a growing body of data which might suggest that many of the assumptions used by Hustrulid were not valid. Because of a desire to keep the machine configuration viable in case the jets failed to work, and through lack of consideration of larger diameter, lower pressure jets the system may have seriously overestimated the number of jets required to significantly improve performance. Alternative design developments, discussed below, may have made this discussion, possibly, moot. The negative impact of this study did, however, cause a revision by some investigators of their basic approach to waterjet cutting, and some of those changed developments are still continuing.

Following the trials in America the high pressure pumping equipment was transferred to Germany where it was used for similar experiments carried out by Bergbau Forschung [9.28, 9.29]. The rock chosen for the first trials of the machine was a relatively softer sandstone than the Skykomish granite. Jet pressures of 3,000 bar were used in the tests, through 0.25 mm diameter nozzles. Nozzle and cutter spacing was set at 50 mm and the traverse velocity was 0.125 m/sec. The preliminary data from the tests was sufficient to justify adding water jets to a Wirth-TBM with a diameter of 2.65 m. This machine had an installed power of 320 kW with 14 disc cutters. It was fitted with a nozzle array which could include up to 100 nozzles in five manifolds. Of these some 65 could be simultaneously operated, at a pressure of 4,000 bar and 0.25 mm diameter. With a diameter of 0.5 mm it was only possible to operate 15 of the nozzles at one time.

The test program carried out in Germany lasted two years and documented significant advantages to the use of waterjets on the cutting head, as well as some restrictions on its use. For example, while the jet gave significant improvement in performance at lower thrust forces (Fig. 9.10) the benefit was diminished as thrust increased. This is not unreasonable given that the jet will continue to slot to the same depth in either case. If the thrust on the head is linearly related to the depth of advance of the cutter, this will approach and surpass the depth cut by the jets as the thrust goes up. The benefits of jet cutting will obviously go down.

The advantage that this yields is not necessarily in the increased drilling rate for a particular tunnel. Rather, it may be that the same rate can be achieved but with a machine which needs only exert some 50% of the applied force. Such a machine could be smaller, lighter and less expensive.

In turn, it would, therefore, pay for itself over a shorter tunnel length, and would increase the market for such machines. However, should a constant thrust be desired then the tests indicated that penetration rate for the machine could be doubled. When the size of the machine was extrapolated out to the 6 m diameter, which is the size of units typically used in mine tunneling, it was calculated that an additional 2500 kW of waterjet power would be required, based on the results obtained. The additional water and the heat from the rock would have a significant negative effect on the mine atmosphere. In contrast, the volume of rock dust generated was almost negligible.

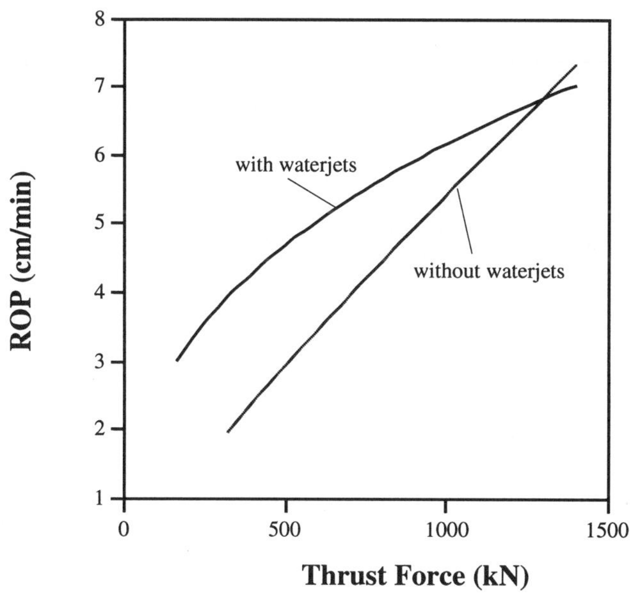

Figure 9.10 TBM ROP as a function of thrust, with and without jet assist at 3,200 bar [9.29].

A successful method of reducing the misting, and improving jet cutting power was to add a long chain polymer to the water (Fig. 9.11), which either allowed an increase in cutting depth of up to 70% or a reduction in the required pressure, for the same cutting depth of 40%. On this basis it was concluded that it would not be necessary to use jet pressures above 2,000 bar on the jet assisted TBM.

The trials in Germany indicated that waterjet assisted TBMs would have the following advantages:

- lower thrust forces required
- less bulky equipment design
- easier transportation of equipment
- more maneuverable equipment
- reduced wear on the cutting tools
- more efficient dust suppression
- elimination of sparks
- reduced amount of fine material in the muck pile.

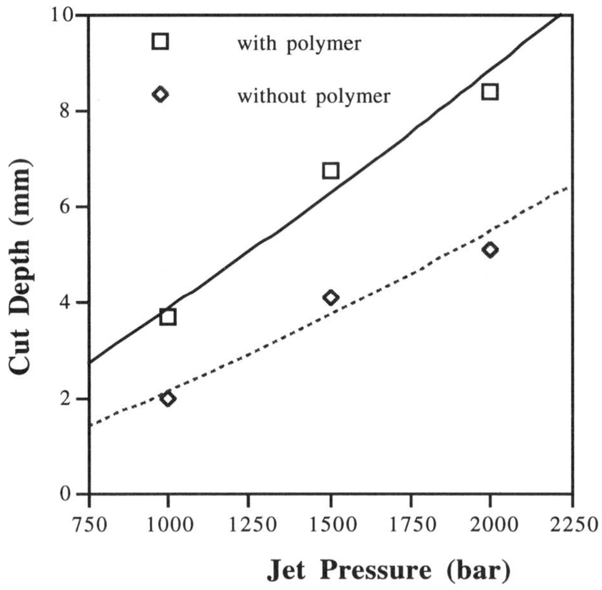

Figure 9.11 Depth of jet cut in rock, with and without the addition of polymers to the cutting water (after [9.29]).

A 6.0 m diameter machine was then constructed by Demag and operated in the Saar Berkwerke with jets operating at 4,000 bar jet pressure, using 60 lpm of water over nozzles distributed across the head of the machine. Performance improvements of 50% were reported from this use [9.30].

An investigation in the application of similar technology to the mining of coal has been carried out in Moscow [9.31] which has shown similar benefits. The shape of the disc cutter was found to have a significant effect on the way in which the coal would break. Further, as with the German work, the size product of the coal was significantly affected by the location and depth of the jet cut relative to that of the disc cutter. The data indicated that specific energy was minimized where the disc cut into the coal some 7 cm from the free face (defined in the article as the pitch of the cutter). An additional benefit arose in that the dust generated by the cutters was tied to the depth of cut and the spacing of the cut from the free surface (Fig. 9.12).

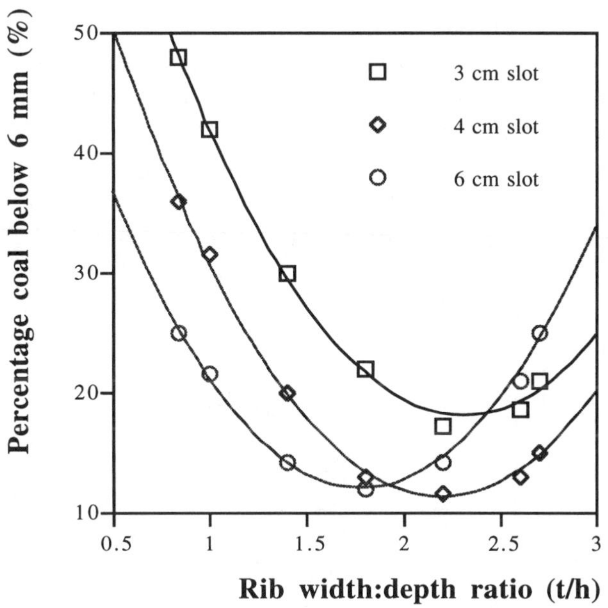

Figure 9.12 Dust generated as a function of rib geometry between the slot and the free surface [9.31].

Other conclusions drawn by the Russians from this study were:

- That if the depth of cut of the disc cutter was less than the slot depth, that the intervening pillar failed within its length, so that any increase in depth of cut had no effect on the reduction in cutting forces.
- That if the cutter penetrates deeper than the waterjet cut slot then the interslot pillar fails along its base. In this case the deeper the slot cut the lower the force exerted on the cutter at failure, and the smaller the area of contact between the cutter and the coal.
- By free-floating the cutter so that it can align with the slot, side forces on the cutter disk were eliminated and this also reduced both cutting and thrust forces, to make the cutting process much more efficient.

A subsequent investigation was carried out in South Africa, examining the improvements which could be made in cutting the hard quartzite rocks using jet assisted disc cutters [9.32]. This work was subsequent to the work carried out by Hood and discussed below. The emphasis on cutter type was changed since disc cutters are less prone to wear than drag bits. Four 1.2 mm nozzles were mounted in pairs at a distance of 35 mm from the rolling contact point between the cutter and the rock and on either side of the disc (Fig. 9.13). Cutting speed was maintained at 0.6 m/sec a typical speed for cutters in the mine. Tests at two cutting depths and four cut spacings (Figs. 9.14 and 9.15) were reported. On average, thrust forces were 53% higher without jet assist, and cutting forces were 66% higher without the jet assist at the shallower depth of cut, at the deeper depth the jet improvement was less pronounced. There was relatively little influence of jet pressure in these trials above a level of 400 bar.

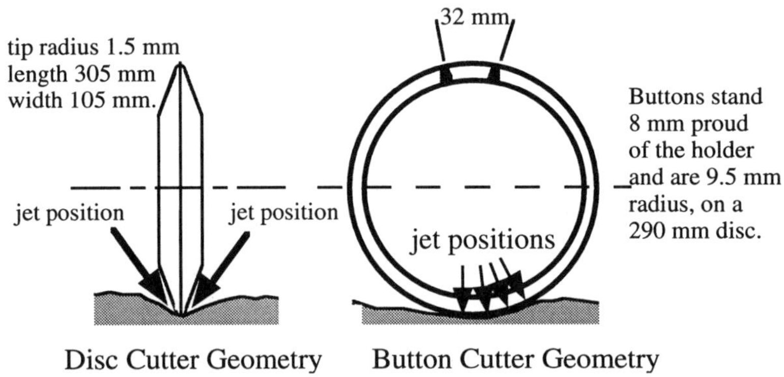

Figure 9.13 Location of jet nozzles in S. African study of jet assist to a disc cutter [9.32].

Where the sharp continuous roller cutter was replaced with a disc faced with button bits the results were quite different. While the unassisted cutter required a 28% greater thrust to achieve the same performance as the jet assisted cutter, the cutting forces were the same in both cases. However, a subsequent test showed that the role of the jet is best applied in cleaning out the crushed material under each button. This zone is more difficult to reach with the button cutter than with the disc cutter, where it forms a continuous line instead of pits. Where the jet was applied to this crushed zone then the unassisted cutter required 50% more thrust and 36% higher cutting forces than the jet assisted cutter in similar applications.

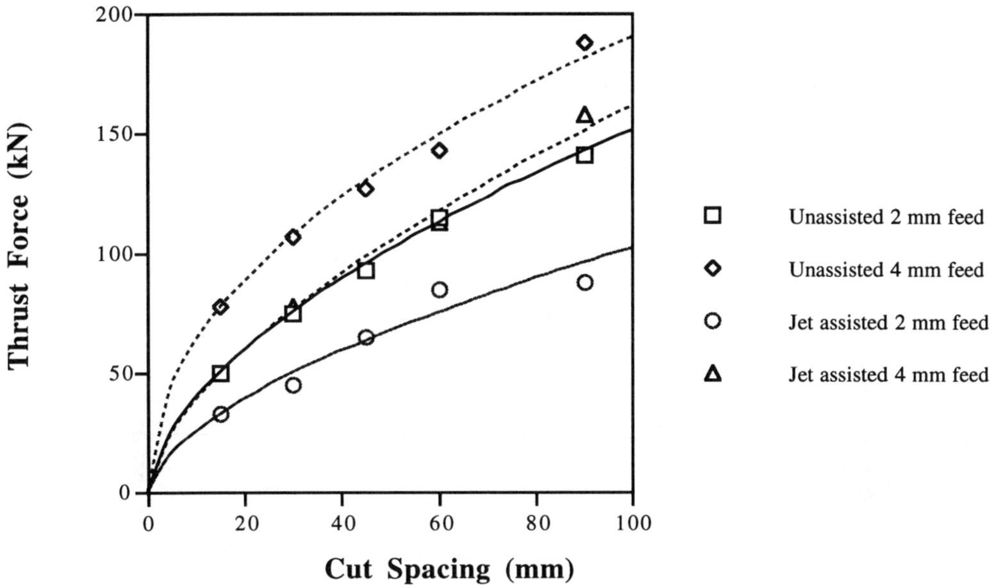

Figure 9.14 Thrust forces as a function of cut depth and spacing, with and without jet assist to a disc cutter [9.32].

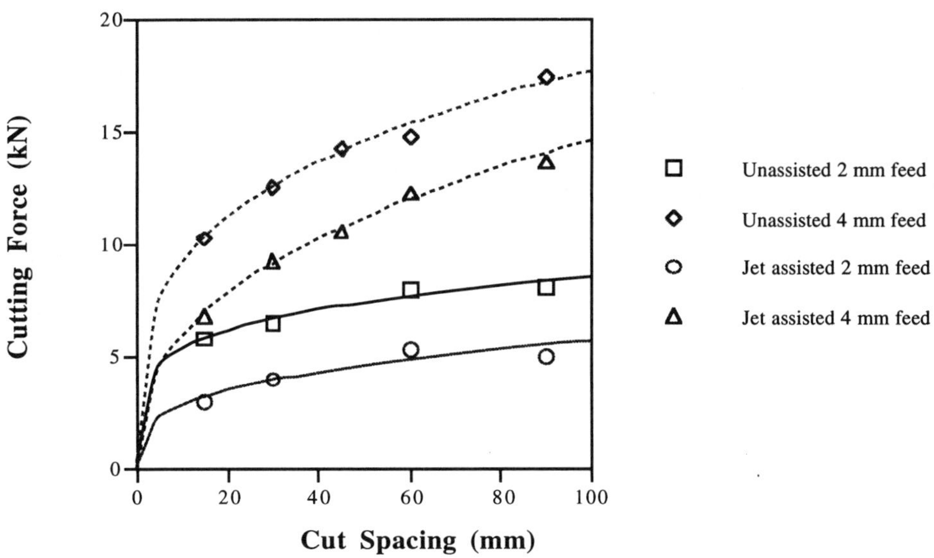

Figure 9.15 Cutter forces as a function of cut depth and spacing, with and without jet assist to a disc cutter [9.32].

This data was somewhat validated by subsequent work at UMR [9.33] in which the penetration of a button bit into dolomite under a fixed thrust was monitored, with and without the hole being washed out by waterjets at a pressure of 420 bar between loading cycles(Fig. 9.16). In this case the jets were not on the rock while the button was penetrating, and yet the clear benefit of the cleaning can be seen.

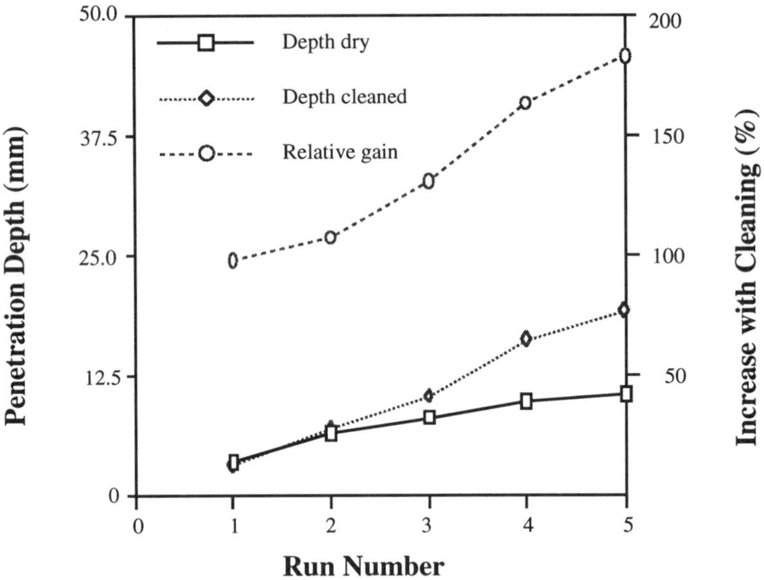

Figure 9.16 Improvement in the thrust:penetration curves of a button bit into dolomite at constant thrust, when the holes are cleaned with 420 bar waterjets [9.33].

The results of the S. African study program led to the development of a waterjet assisted model of the Robbins 52RE blind-hole boring machine [9.34]. The decision was based, in part, on the larger chips achieved by the combined system, and that the spacing between the cutter discs could be increased by 50% for the same cutting force. This would reduce the number of cutters on the cutting head, which would compensate for the slight increase in power costs for the machine due to the power required for the waterjets.

Laboratory tests of various combinations of jets with button cutters in the United States prior to the mine testing indicated that the best combination of nozzles was to direct two pairs of jets so that they combined their flow at the point of contact of the buttons with the rock. Four jets were used since the cutters consisted of two rings of buttons on the disc.

The jets were directed from the back of the disc into the cutting zone (Fig. 9.17). Regrettably, the initial tests of this system in the mine did not give the levels of improvement obtained by changing the mechanical design of the head, and the program of research was therefore not continued.

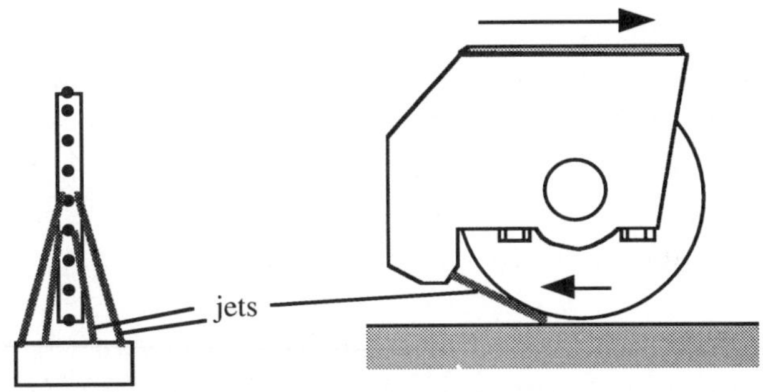

Under-view showing jet angles and relative impact locations

Side View of Jet Addition showing disc rotation and direction

Figure 9.17 Layout of jets on a button cutter disc to improve performance [9.34]

9.3 WATERJET ASSISTED MECHANICAL CUTTING EQUIPMENT

Perhaps the most dramatic development in waterjet application to mining, and certainly the one with the most rapid and greatest influence over the past two decades began as a doctoral thesis in S. Africa in 1974 [9.35]. In light of the developments since that time it is pertinent to review some of the highlights of that thesis, both for what was and what was not found.

The study arose because the economics of gold mining at depth required an improved method of mining. Typical drill and blast methods broke out the vein of gold, and the surrounding rock over a width sufficiently wide to allow the mining to proceed, but at the cost of mixing the gold with that surrounding waste rock. If the gold bearing rock layer or "reef" could be separately mined, and then the surrounding waste rock broken out, but packed into the abandoned hole behind the mining operation, a considerable savings would arise. The savings would not only arise from the lowered cost of transporting all the waste rock, but also from the reduced processing costs involved.

A promising method of mining the reef had been developed. In this method a drag bit was pulled across the rock, cutting a channel in the rock face, to which the surrounding rock could be broken. However, the S. African host rock is very hard, typical uniaxial compressive strengths run around 1,500 to 3,000 bar, and the rock may contain up to 98% quartz. Thus, pulling the bits across the rock surface required both a high **thrust force** (the force holding the bit into the rock) as well as a high **cutting force** (the force required to move the bit forwards). Under these forces the drag bit tooth, although made of tungsten carbide, would very rapidly heat up causing tool failure, generally after cutting approximately 9.2 sq mm of rock. Given that the depth of cut was from 4 to 8 mm, this meant that the bit would cut a linear distance of some 2,300 to 1,150 m before needing to be replaced. A program was undertaken to increase the life of this bit.

Preliminary study indicated that the bit did not fail the rock by chipping it out as the wedge head moved forward, but rather (Fig. 9.18) the rock failed under the compressive force applied through the bit, and its underlying wearflat to the rock beneath it. Hood showed that this was the case, in part, by pointing out that the shape of the deformed carbide included a flowing forward of the leading edge of the blade, an impossibility if this was providing the continual contact and high levels of force required to chip out the rock above and ahead of the bit.

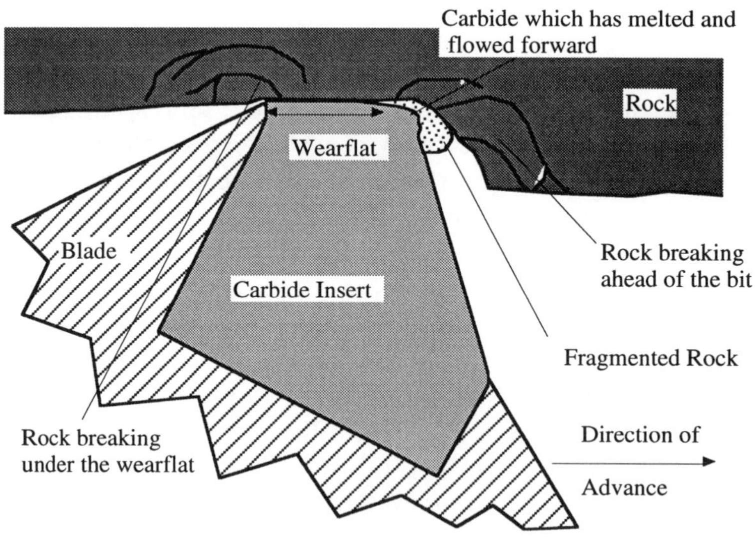

Figure 9.18 Outline of carbide bit and rock showing zone of rock failure [9.35].

A study of the thermal conditions of normal bit operation indicated that temperatures stabilized after the bit had moved approximately 1 m along the reef, and very high temperatures could then build up in both the tool and the holder (Fig. 9.19). Various metallurgical and tool geometry solutions were attempted in order to either reduce the temperatures in the tool, or to provide a greater resistance to the results on the tool structure. While some progress was made in this area, the gains were incremental and did not resolve the operating difficulties with the tool. An alternative approach was, therefore, examined, in which a spray of water, through a fan jet, was directed at the carbide to cool it.

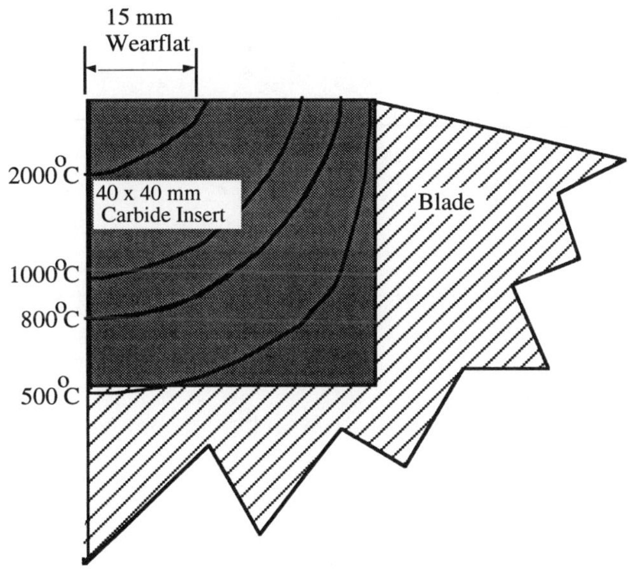

Figure 9.19 Temperature profiles in an uncooled drag bit on rock [9.35].

It was anticipated initially that the cooling would prolong the bit life, however, it also reduced the thrust force required to hold the bit in the rock, both in terms of the peak values measured and the average force both by approximately 10%. The reduction in the fluctuation of the thrust force and the lower temperatures on the bit were similar to those which could be found by using a low pressure flooding of the bit with 120 lpm of water. However, the reduction in the measured thrust force was significantly greater (Fig. 9.20). The fan jet pressure was 80 bar, with the nozzle mounted 50 mm from the rock:bit contact, and the jet consumed 40 lpm of water. This improvement led to tests with coherent cylindrical jets directed at the rock ahead of the corners of the bit instead of using fan jets with their more rapid dissipation of water. These jets were used to reduce the energy loss which occurs with the travel of a fan jet through air.

Water flow rates were maintained at 40 lpm since this was a logistics limit set for operations of equipment in the mining area, or stope.

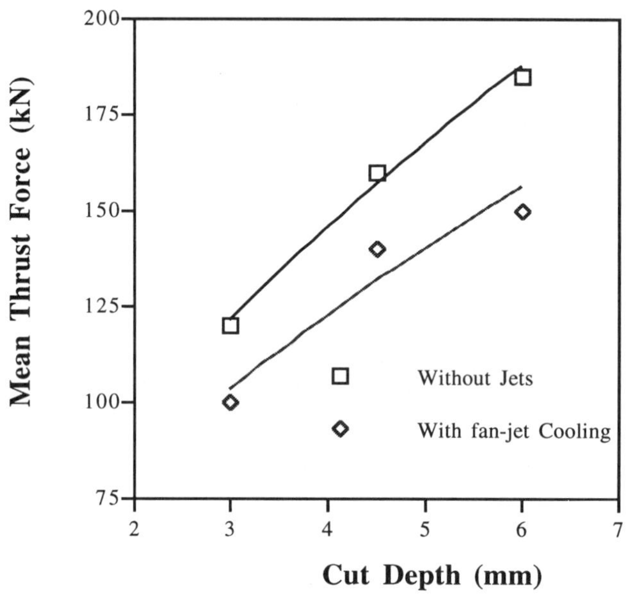

Figure 9.20 Initial reduction in thrust force with fan-jet cooling of the bit [9.35].

It is noteworthy that the nozzles were kept within 100 nozzle diameters of the rock, based on Leach and Walker's findings that jet pressure stayed relatively constant out to this distance [9.36]. Four jet positions were evaluated, with the jets impacting outside the corners of the inserts in the bit plane; with the jets impacting inside the corners and 2 mm ahead of the bit; with the jets impacting within the corners but 10 mm ahead of the bit; and with a single jet striking the rock 2 mm ahead of the bit, but in the center of its path (Fig. 9.21).

The results have been summarized in terms of the mean thrust and cutting forces acting on the rock as a function of depth of cut (Figs. 9.22, 9.23). Jet pressures of 100, 150 and 500 bar were used for the evaluation. The improvement in conditions was clearly evident. It can be illustrated by the increase in allowable cutting depth for the bit before the tool stalled. Under dry conditions the force would stall the bit at a depth of 4.5 mm, but with the jets the same force would not be reached until a depth of 10.5 mm was reached.

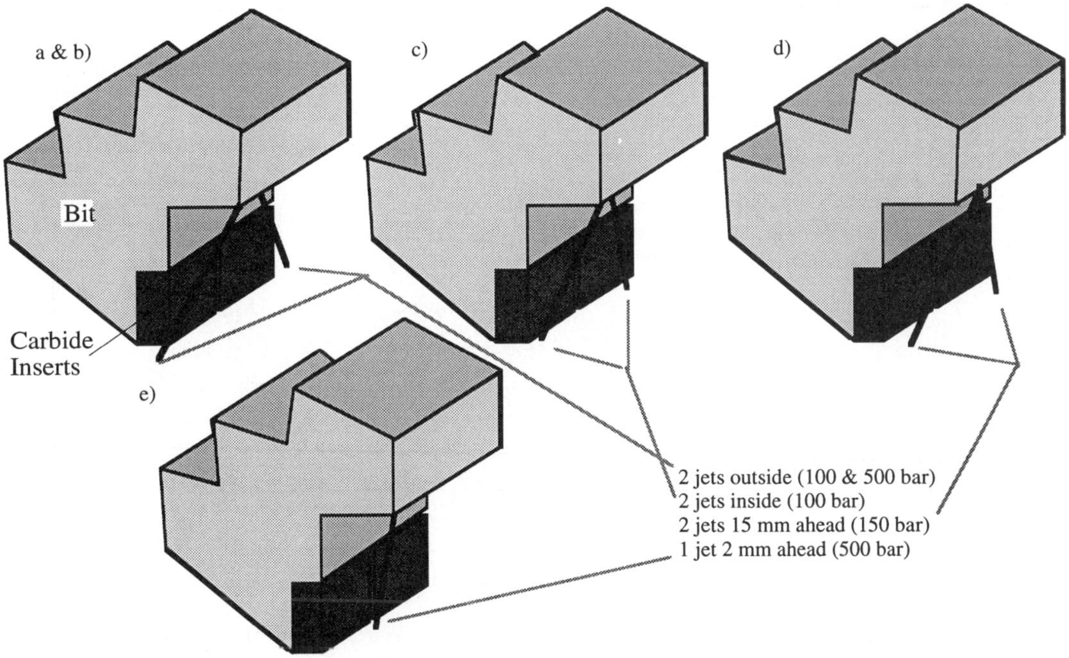

Figure 9.21 Location of the waterjets relative to the cutting bit positions for tests carried out by Hood [9.35].

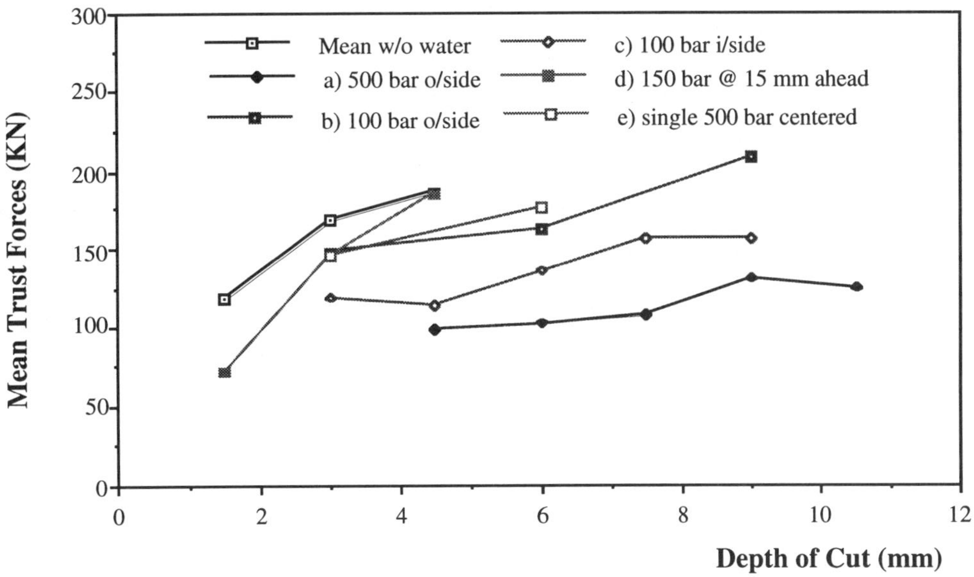

Figure 9.22 Thrust force as a function of depth of cut for various bit cooling geometries [9.35].

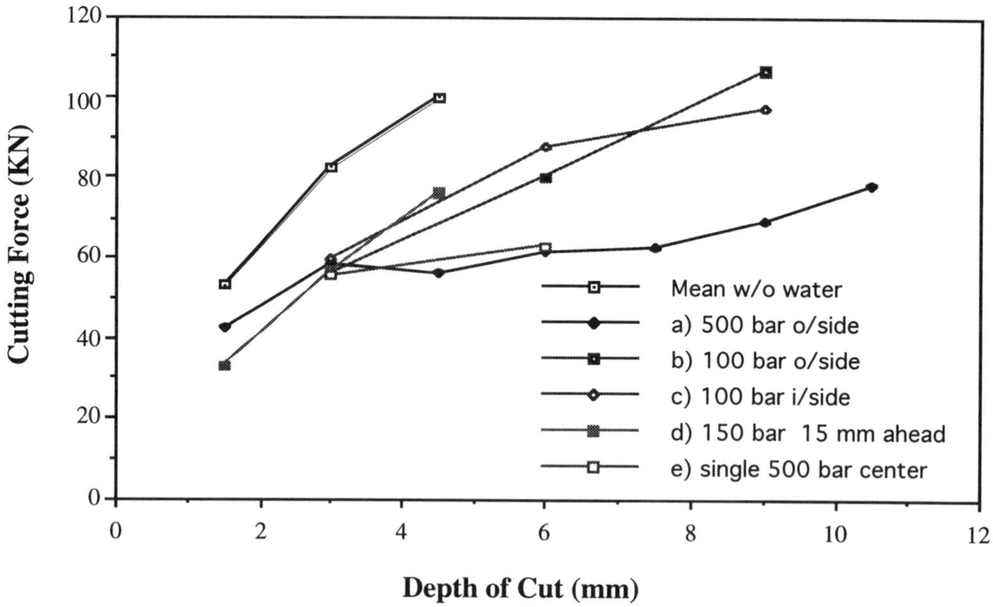

Figure 9.23 Cutting force as a function of depth of cut for various bit cooling geometries [9.35].

The reduction in the thrust force was even more significant with the force required to hold the bit at 10.5 mm depth being equivalent to that required to hold the bit at a 2 mm "bite" with the bit running dry. Tests with a variation in jet pressure indicated that while the reduction in the cutting force occurred at both pressures, the most significant reduction in thrust force required the higher pressure. While cutting force was not significantly reduced with a change in the jet position from outside to inside the bit corners, this change further reduced the thrust force required to hold the bit at the required depth. This change also influenced the side forces applied to the bit.

Following the success of the laboratory trials five of the rock cutting machines at the Doornfontein mine were modified to allow 400 bar jets to impact on the rock ahead of the carbide bits. The flow was restricted to 40 lpm, and the jets were directed at the rock immediately ahead of the bit, and at its corners. Where the uncooled bits were able to cut 2-3 mm into the rock, the waterjet assisted bits cut to a depth of 10 - 15 mm where the rock was unfractured. In fractured ground the normal 8 - 10 mm cut was increased to 40 mm. Deflections of the tool which made cutting at the top edge of the opening with a normal bit were reduced with the jet assist, and allowed the cut to be made along this edge. This had previously not been possible. It was found that the jet cutting just outside the bit corner increased the average life of the bit by 68%. The improved cutting at this position allowed a reduction of two men in working the stope. Hood also reported that the smoother cut on the roof made working conditions safer. Bit failure became commonly due to excessive wear, rather than the thermal failure which had previously been the case.

Hood concluded his initial study by examining the rock failure under the bit and seeking to understand how this was enhanced by the action of the jet. His conclusion was that the jets were entering the cracks ahead of the bit, causing them to grow, and assisting in the adjacent rock failure. The pressure in the crack is not, however, confined, and thus, the crack growth is relatively slow since the intensification of pressure at the tip is not high. The effect was controlled, however, by the pressure of the impinging jet. A secondary effect was considered to arise from the flushing of crushed rock from under the bit. That this is secondary is illustrated by the case where the bit was run, at maximum depth, until it stalled, without water. When water was then added to the bit it immediately began to advance, under the lower force required, without the jets having initially perceptively had the chance to remove the crushed rock under the bit. (This is, however, a debatable point given that the jets move at a speed of several hundred m/sec and with the crushed rock comprising only a layer perhaps 1 - 2 mm thick the time for the jets to clean out the crushed material may well be undetectably small.)

In another study, Wang has found that the jet pressure was unable to remove this crushed material without bit movement, given the high confining pressure on the powder, relative to the jet pressure. However, once the jet removed the powder ahead of the bit, then it does not have the chance to accumulate and build up the pack of crushed material. (To comment on that point, Wang was directing the jet through the center of the bit into the crushed zone, in most conditions the jet is working from the outside in, and the central pressure can only build up due to the increased confinement of the surrounding material. This point is discussed, with regard to crushed pillar strength by Wilson [9.37] and will be further discussed relative to pick advance later in this chapter.)

Hood similarly found that in static indentation tests the jet pressure did not remove the crushed rock under the bit, showing that the reduced force could be ascribed to the assistance in crack propagation. A tertiary effect was considered to be the role of the pressurized water in lubricating the passage of the bit over the rock surface.

Hood presented these results as an appendix to the Proceedings of the 3rd BHRA meeting in Chicago in 1976 [9.38]. They provided the basis for a significant research program carried out as a joint effort between the U.S. Department of Energy and the British National Coal Board [9.39].

Research in the U.S., however, began along a somewhat different direction. McNary, et al. [9.40], added a central high pressure jet to the cutting head of a conventional continuous mining machine to see if this would lower the amount of dust which the machine created. The nozzle was mounted between the two cutting heads of the machine (Fig. 9.24) and inclined into the cut being made by the cutter head on the right side of the unit. As the machine mined the coal the cutting head was advanced into the face and then cut on the right side of the machine to remove a swath of coal over the face. The jet was, therefore, cutting into this zone. The picks used were of the "point attack" type.

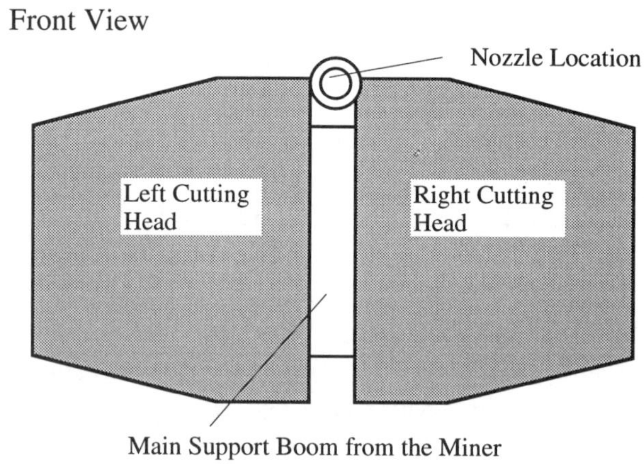

Figure 9.24 Location of the nozzle on the Bureau of Mines first jet-assisted machine (after [9.40]).

The major aspect of the test was the reduction in respirable dust. Such dust can come from several sources during mining. For example, some of it is caused by the crushing of the rock under the impact point (as observed by the Russians and discussed in [9.31]). Other dust is created as part of the production of coal from the original plant matter, and forms as thin layers of powder along the bedding planes and weakness planes in the coal. When the coal is broken along these planes this dust is liberated and becomes airborne. In order to reduce the amount of dust created in mining the coal it is thus necessary to catch both types of dust. The mechanism by which the jet enters the coal is an advantage in this, in that it can penetrate into the coal ahead of the cutting bit, and wet the dust which is already there. The Bureau study showed that this depth of penetration into the coal was some 5 cm in the tests they carried out. The jets were not aimed directly at any of the picks during this test so that the dust generated by coal crushing was not immediately effected by the jet impact. Nevertheless, the dust made by the machine was reduced by 70% when the high pressure waterjets were added to the head (Fig. 9.25).

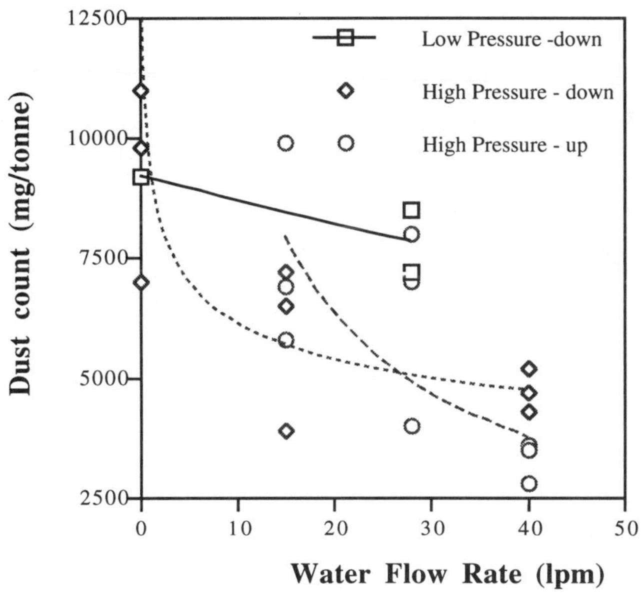

Figure 9.25 Effect of jet flow and pressure on dust created in continuously mining coal [9.40].

The previous two studies led on to work at the Colorado School of Mines [9.41] in which the potential of using high pressure waterjets to assist the cutting of coal measure rocks, much softer than the S. African quartzite was examined.

In understanding the results which came from this and subsequent studies one must appreciate that there are several different shapes for the cutting tools used in the mechanical excavation of rock, and several thoughts as to the best way in which these can be applied. The different shapes can be simplified into three basic types (Fig. 9.26). These are the pointed pick, sometimes referred to as the **point-attack pick**, the chisel blade, somewhat similar to that used by Hood. This is sometimes called the plow blade or the **radial pick**. The third pick, which will be used in later discussion is the **forward attack pick**. During the tests, because of scatter in normal cutting data from rock tests, averaged values for runs of up to 15 m were obtained. Tests were carried out in cutting both hard and soft sandstone, shale and limestone, as being typical of the types of rock to be found near coal seams in mines. Some 1,200 cuts were made in this initial test program, which largely examined the benefits of adding waterjets to the point attack pick, since these provided a more consistent result than coping with the wear of the radial picks.

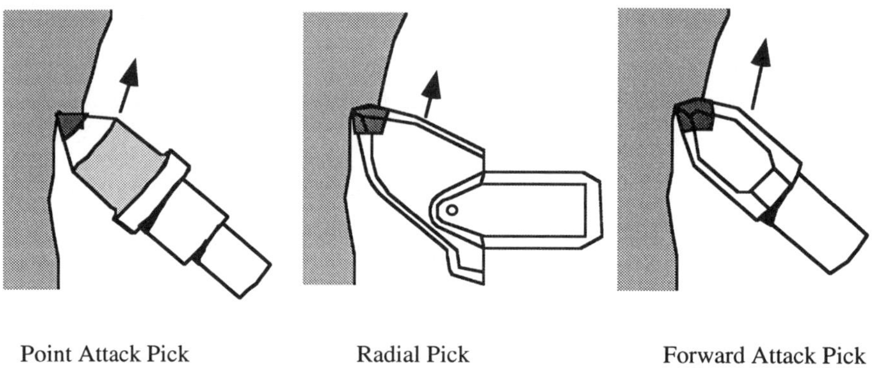

Figure 9.26 Point attack, radial and forward attack pick shapes [9.42].

Nozzle sizes of 0.3 and 0.6 mm diameters were used in the tests. It was reported that no significant change in the depth of cut of jets occurred when these jets were operated at 700 bar and at stand-off distances ranging from 5 - 25 cm. This seems a little odd given that normal jets show a significant stand-off distance effect (see for example the data from Leach and Walker [9.36]). However, this range is beyond the 100 nozzle diameter range that several reporters have suggested is the most effective for jet action, and thus, the tests may have been occurring in a range where jet performance had already been substantially degraded. This may explain the relatively poor jet performance, with cutting depths into the rock of 1.27 mm for the 0.3 mm diameter jet, and 2.5 to 3.7 mm for the larger jet. Because of this question on the effective use of the jets beyond their most effective stand-off distance, it may be that the results reported understate the benefits which could accrue from the use of waterjet assist on the bits. This point may be substantiated by examining the more effective results obtained where a larger (1.2 mm) jet was used in subsequent tests. Since this has a 100 diameter reach which includes the test stand-off distance it could be anticipated, as was found, that this gave a significantly better performance.

The two types of bit show different results as the bits wear, under normal use. For example, point-attack bit forces reduce after a new bit has cut some 1,500 m, while for a radial pick the forces significantly increase. It was noted, in the only tests carried out with the radial picks, that the cutting and thrust forces were reduced by approximately 60% when a 0.6 mm nozzle was used to direct a 350 bar jet into the cutting zone from behind the bit. The cuts were made to a depth of 12.5 mm and were some 37.5 mm apart. It was noted during these tests that the bit life was considerably prolonged when the waterjet was used.

The small point of the point-attack pick, and its rotation within the holder during use to keep itself sharpened, makes it a more difficult bit to which to direct a high pressure jet. Thus, to more thoroughly study the effect of the jet on cutting performance, the jet was directed both in front, to the side, and in from behind the bit. The results are exemplified by the data given for the softer sandstone tested in the program (Figs. 9.27 and 9.28). When the jet was directed at the side of the pick poorer cutting results were obtained than when it was directed either ahead of or behind the pick.

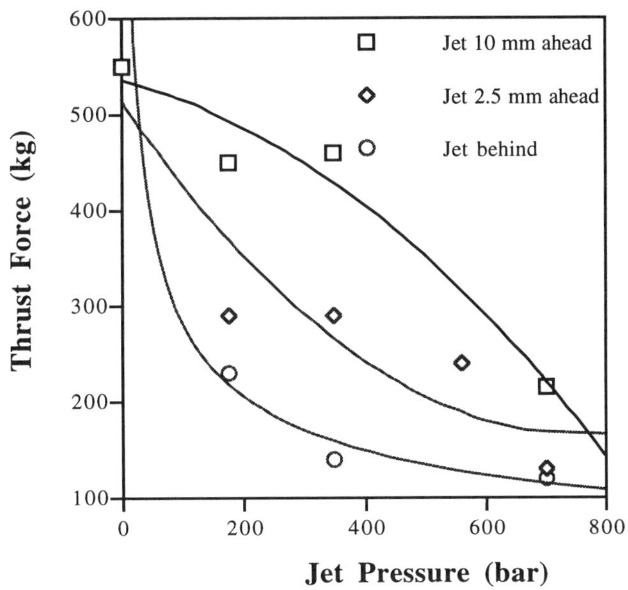

Figure 9.27 Thrust forces on a pick as a function of jet assist position [9.41].

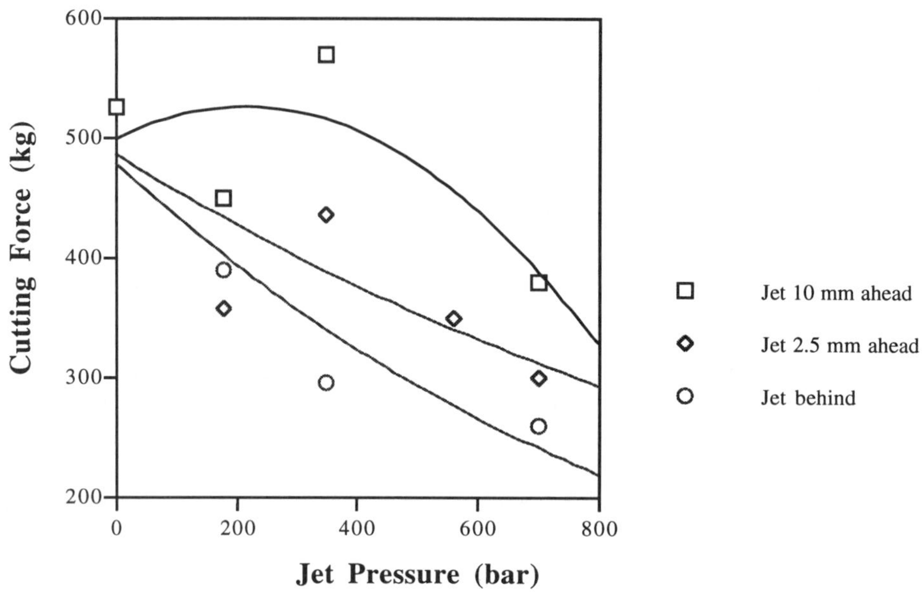

Figure 9.28 Cutting forces on a pick as a function of jet assist position [9.41].

Some conclusions can be drawn from this data, however, it is difficult to generalize these since the results were not replicated with the other rocks tested. For example, while a 350 bar jet improved the pick performance in the softer sandstone, it had less effect on the harder sandstone and virtually no effect on the cutting of the shale and the limestone. In the latter two rocks it was necessary to increase pressure to 700 bar to see an effect, and for the shale this required that the jet be directed into the bedding planes of the rock.

This data, and the results shown, confirm the concept which Hood had proposed that the jets were exploiting the cracks generated around the impact point to improve the cutting performance. The jet penetration would, however, be enhanced where the rock was granular and weaker where the passages for the jet to enter the rock would be enlarged. Thus, even at the greater stand-off distance, the jet could find pathways into the rock to weaken it and enhance performance. In a more resistant rock, or where the cracks generated by the pick do not penetrate into the zone where the jet impacts, then the effect of the jet on pick performance is much less.

The National Coal Board (NCB) had, meanwhile, been carrying on its own program of investigations. Beginning in 1977 [9.43] tests were carried out, as a precursor to the developing co-operative agreement with the Bureau of Mines. The tests were accordingly designed to conform to the likely conditions of a pick on a roadheader. Cutting speed was set at 1.14 m/sec, and the nozzle was set to impinge a jet within 2 mm of the pick tip. This was closer than the CSM study, but was consistent with a defined condition apparent from preliminary trials of the concept. A stand-off distance of 50 mm was required to keep the nozzle assembly from being damaged during equipment operation. This was located to direct a jet perpendicular to the rock face. Radial picks were studied in this program, of the type usually fitted to British roadheaders. The test was set up to cut a spiral path over a rock sample, with a length of 57 m. When the jet was run over the rock without a pick in place a slot was cut to a depth of 15 mm at a pressure of 350 bar.

It was found that the system worked better at higher jet pressures, although there was less impact due to a change in flow volume. The reduction in forces was more evident when cutting through harder rocks than through softer rock (a contra-indication to the CSM data, which was with another pick design). Typical results were a reduction of 15.5% in cutting force and 33% in the thrust forces on the bit.

These were promising data, and accordingly, under the joint program, a roadheader was fitted with a 700 bar jet system, designed to send jets to up to fifteen of the picks on the head. Because of the geometry of the head, stand-off distance from the nozzles had to be increased to 10 cm for the trials. Initial trials with the system were undertaken in an artificial rock face built underground at the Middleton mine in the UK. The face was built up of sandstone blocks, which could be excavated by the roadheader to demonstrate any reduction in sparking and dust, as well as to determine what force reductions might be seen on the machine.

Tests in the sandstone did not show any significant improvement in the performance of the machine, either by reduction in the cutting forces or increase in the cutting rate. However, frictional sparking and dust generated during cutting were both completely eliminated during the high pressure jet tests. The machine was able to cut this rock quite well, however, without the use of the waterjets. In contrast, with these results a significant change was made when the machine was tried in the harder host limestone at the mine.

To understand the significance of the result one should be aware of the typical performance of the Dosco Mk2A machine. Roadheaders work by the hydraulic pressure through the ram on the cutting boom forcing the picks into the rock face. In order to obtain a high enough thrust to achieve effective performance it is necessary to use higher thrust levels as the rock becomes harder to cut. As a very crude measure the difficulty in cutting is related to the uniaxial compressive strength of the rock. To cut stronger rocks a higher thrust is required, and thus, a larger, heavier, and more expensive machine (Fig. 9.29). Without jet assist the machine would normally be used to cut rock of less than 700 bar compressive strength. In order to cut rock of the 1,400 bar compressive strength typical of that at the mine would require a machine which weighed around 110 tons [9.44]. Such a machine would cost on the order of $750,000, in contrast to the $250,000 approximate cost of the 24 ton Dosco machine.

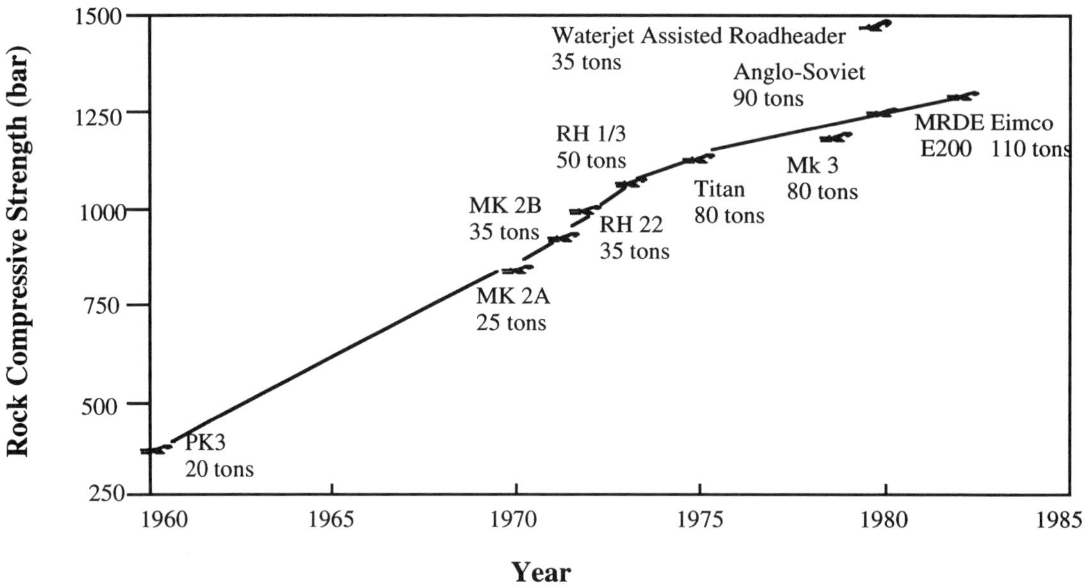

Figure 9.29 Machine size required to cut rock of varying compressive strength [9.44].

The trials of the unassisted machine in the host rock bore out this "rule of thumb". When point attack picks were used the head could not be kept consistently cutting, and the motor continuously stalled under overload. With jet assist the head would cut, albeit poorly, and a significant reduction in pick wear was seen. However, the reaction on the boom was still highly stressful and the trials continued using only the radial picks.

Improvement was immediately seen, whether the jets were operated at 420 or 700 bar and whether 6, 9, or 12 picks were assisted. Cutting rate improvements of up to 250% were obtained. In addition, pick life was considerably extended. The unit was then used to drive a heading some 5 m deep in order to obtain more consistent measures of performance.

These runs showed that it was best to operate the equipment with high pressure water delivered to the leading nine picks of the head, using a flow of 3.6 lpm to each 0.64 mm nozzle at 700 bar. The picks assisted formed the first three picks of the three helical pick lacings. These results do not agree with the data from CSM but it should be noted that while the CSM tests showed little improvement in cutting limestone, those tests were carried out with the point attack picks, which were similar to those used in the early trials at Middleton. That design did not work well in the Middleton trials either, and so there may not be as much disagreement in the results as might appear.

A considerable number of design changes in the equipment could be identified as a result of the first trials and these were incorporated in a second generation machine which was taken into a coal mine in 1983 [9.45]. While that machine was being prepared the prototype machine continued to be used at the Middleton mine, where it drove a 90 m conveyor roadway. The following results comparing the jet assisted and unassisted performance were obtained in cutting this 4.27 m wide drift.

Table 9.4 Relative performance of a Dosco Mk2A roadheader at the Middleton mine (after [9.46])

Performance parameter	*Without assistance*	*With 700 bar jet assist*
cutting time (min)	59	29
cutting rate (cu m/hr)	0.921	1.875
specific energy (kWh/cu m)	35.8	17.6
picks consumed/cu m of rock	2	1

Two British companies, Anderson Strathclyde and Dosco, found this new technology of sufficient interest to develop prototype machines of their own design, building on the experimental results obtained by the National Coal Board (Fig. 9.30). Anderson Strathclyde chose to modify an RH22, medium duty machine weighing 35 tons and fitted with a 112.5 kW cutter boom [9.47]. The machine was substantially modified to include the design changes required. It was first tested in driving a slightly dipping tunnel at Sutton Manor Colliery in July 1983. During the trial, the machine cut through a wide variety of rock, while driving the 3.6 m wide tunnel (Fig. 9.31). On average the waterjet assist improved cutting rates by 50% and reduced the specific energy of cutting by some 30%. Both dust and frictional sparks were almost eliminated with the water assistance. Pick costs averaged 0.7 picks/meter of advance. The experience provided sufficient evidence of benefit that the company proceeded to develop add-on packages for other models of roadheader it produced.

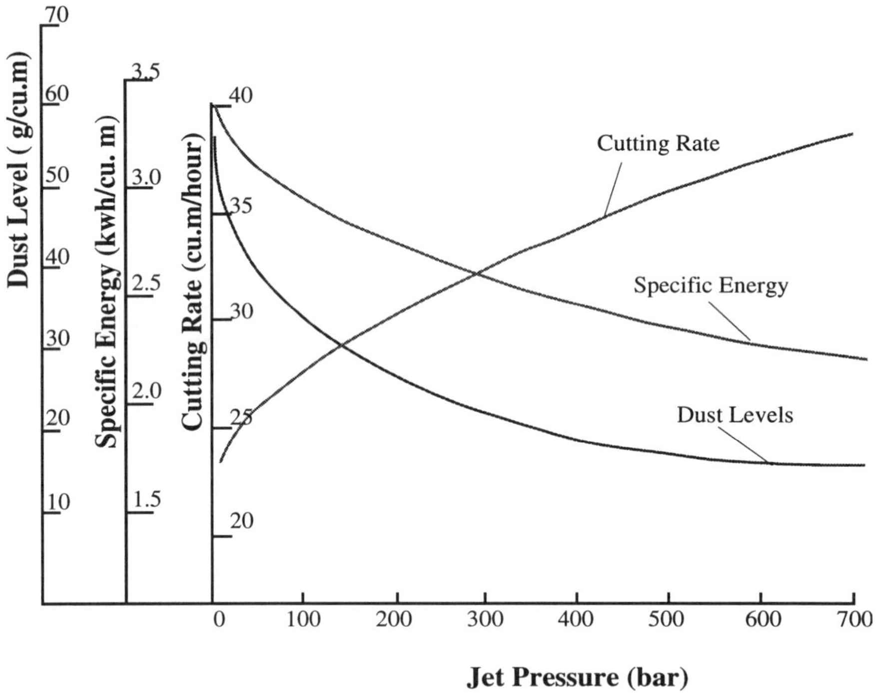

Figure 9.30 Summary of National Coal Board jet-assisted cutting results [9.48].

Further trials were carried out using a Dosco machine at Bentinck Colliery and with the NCB modified Dosco unit at Calverton Colliery. In the Calverton trials [9.50] the machine was used to drive a 73 m tunnel in sandstone with a compressive strength of 896 bar. Bit life was doubled, and dust and frictional ignitions were not a problem during the operation of this test.

The trials at Bentinck Colliery took place in a rising tunnel (angle 1 in 7) through relatively soft mudstone. Considerable problems with water mixing with rock on the floor gave transport problems and the test proved inconclusive. The design was modified and a unit was installed at Devco coal mines in Nova Scotia in 1984 [9.51]. Working on a 22% gradient the Dosco machine drove one tunnel while an Anderson Strathclyde machine drove the other. The success of both machines led the company to buy another four Dosco machines and one Anderson Strathclyde unit. Trials of a Dosco unit at Agecroft Colliery in 1985 [9.49] showed that in cutting a 360 - 440 bar shale the jet assist increased the cutting rate of the machine by 75%. At Kellingley Colliery in 1986 a 350 bar pressure assist to an RH22 roadheader reduced the amount of picks used/m of advance from 60 - 25, a 15% savings on the cost of driving the roadway. This was improved at Cresswell in cutting a sandstone with a strength of 1,640 bar. The pick consumption dropped from 80 picks/m to 15. From this point the technology rapidly moved into industry and by 1987 [9.46] there were over 60 units operating world-wide. A Dosco kit to modify existing roadheaders for waterjet assisted cutting was then selling for $75,000.

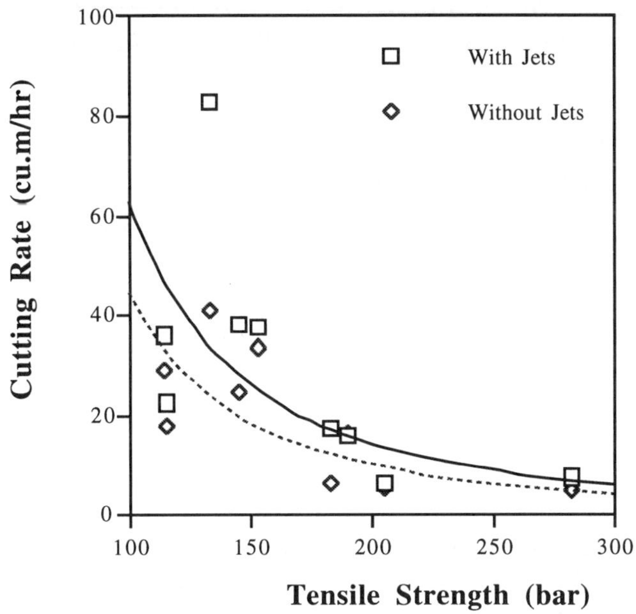

Figure 9.31 Cutting rate for a waterjet-assisted roadheader as a function of rock tensile strength, at Sutton Manor (after [9.47]).

The somewhat mixed results from the various field applications of the technique provided an impetus for additional laboratory research on ways of better understanding and improving the hybrid system which has become known as waterjet-assist.

Hood had begun his work by seeking ways to reduce the heating of the carbide tooth of a drag bit. Subsequent work by Powell and Billinge [9.52] examined this phenomenon as it applied to cutting coal measure rocks in the United Kingdom. The problem is quite serious in that the very hot rock left behind a pick may be at a temperature of up to 1,800° K. The zone of rock above 1,400° K can stretch some 100 mm when a pick is cutting at 3 m/sec. A zone of such hot rock 400 sq mm in size is capable of igniting dangerous concentrations of methane in the mine air, with potentially fatal results for miners. In a typical year some ten to twenty-five such ignitions occur on machines in the U.K.

The speed of a pick was found to be an important factor in causing frictional heating only at pick speeds below 1.25 m/sec, for new picks, and 0.7 m/sec for worn ones. Above those speeds the likelihood of ignition of a combustible atmosphere was more than 50% within five seconds. In order to reduce this possibility it was found that applying a conventional jet of water to the front of the bit, while effective in dust suppression, did not reduce the risk of ignition. Water applied behind the bit was, however, effective with a spray of water being more effective than a solid stream. The water was found to be most effectively applied as a fan jet just covering the path width of the tool. Of the conditions tested a 40° fan producing 90 μm droplets at a surface density of 8.7 l/sq m/sec was found to give the best results. If such a spray it confined to the zone immediately behind the pick then the water required per nozzle would be on the order of 0.82 lpm at a pick speed of 3 m/sec. If a broader solid cone spray is used then it may require 2.8 lpm. This figure may be compared to the 3.6 lpm at 700 bar which was applied to each jet on the NCB first trial machine at Middleton. The high pressure jet, applied ahead of the bit, was therefore able to penetrate into the zone under the rock to quench the rock, if the rock and bit are to be cooled to the levels reported.

The question of the best position for a high pressure jet was studied in more detail by Dubugnon at CERAC [9.53]. In a study of three locations for the jet, but in all cases directing the jet at the corners of a the tool, this study concentrated more on the slow speed drag bit cutting such as that examined by Hood. Jets were directed into the corners of a bit from outside the bit path to study the effect of rock type, jet pressure and jet flow rate. It was found that the type of rock had a significant effect on the importance of jet pressure (Fig. 9.32). In a sandstone the improvement in cutting ability appeared to reach an optimum value at just over 300 bar, while for a granite the improvement in cutting continued as the pressure increased up to 650 bar, at a constant flow rate.

It should be noted that while only the results for average forces have been plotted equivalent results were also obtained for the peak forces. Where the jet pressure was held constant at 700 bar a test showed that there was a significant effect on the results from a change in the volume of flow; achieved by altering the nozzle diameter (Fig. 9.33). Nozzle diameters studied ranged from 0.4 to 1.6 mm and it appears that the nozzle stand-off remained the same for all tests. This would affect the nature of the results since sketches of the equipment suggest that with the smallest nozzle the stand-off distance was more than 100 diameters from the rock.

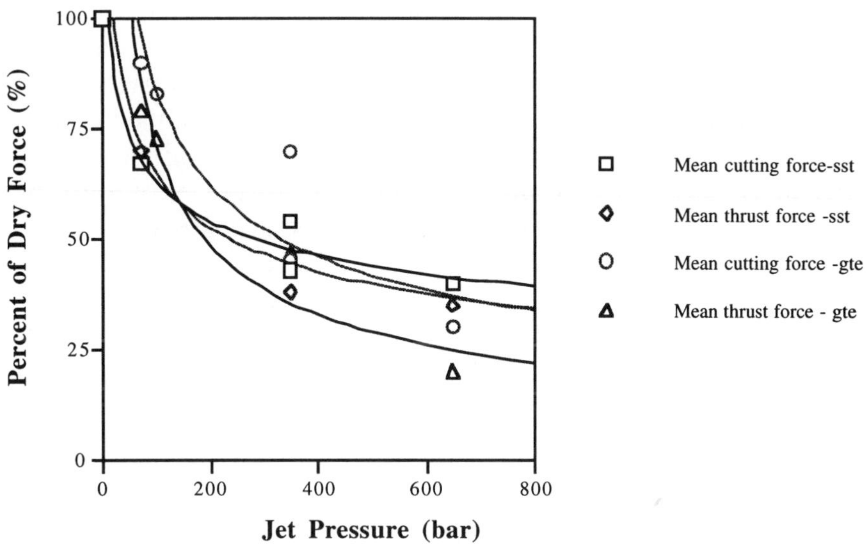

Figure 9.32 Effect of increase in jet pressure, as a function of rock type (after [9.53]).

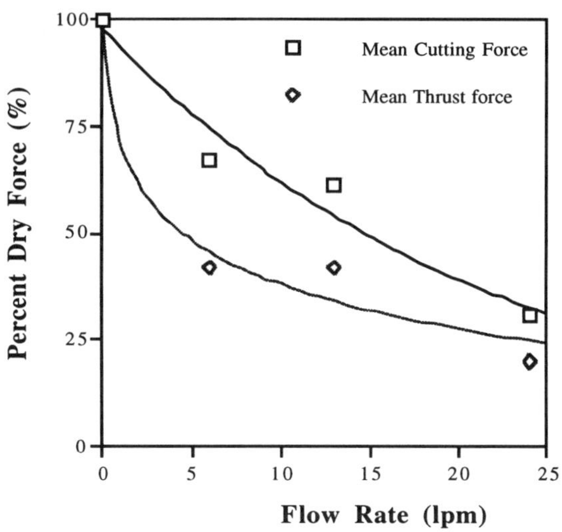

Figure 9.33 Bit force reduction as a function of jet-assist flow rate [9.53].

A comparison in which the jets were directed from the sides into the edges of the bit, from 2 jets in front pointing straight back at the bit and from a central feed line with the jets inclined outwards to the bit corners indicated that the results were all relatively similar, but given that the best configuration for deep slot cutting would require that the nozzle follow into the slot, a central nozzle with outward pointing jets was recommended.

The exact position of the jets ahead of the tool contact was also studied, in the range from 0 - 4 mm. While the results were sufficiently close to suggest that there was not a significant difference, over the range, it would appear that a 2 mm stand-off distance was most effective (Fig. 9.34). Dubugnon then theorized that the benefits from jet assistance came from a combination of removal of the crushed zone of material under the bit, a fluid penetration of the developing cracks and their exploitation, and a pressurizing of the pores in the rock around the cutting zone.

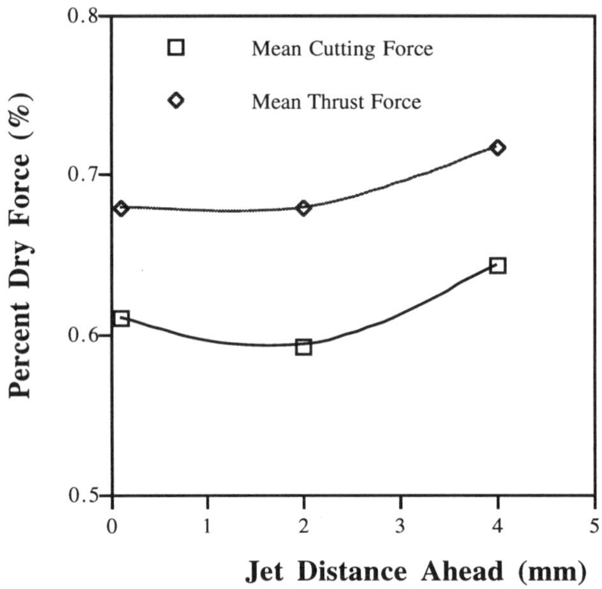

Figure 9.34 Bit force reduction as a function of nozzle lead distance ahead of the bit (after [9.53]).

The importance of the point of impact of the jet on the rock surrounding the pick point of contact was further reviewed by Fairhurst [9.54]. After reviewing the theories relating to the failure of rock under the action of the pick alone, he used a computer model known as MUDEC (Micro Universal Distinct Element Code) from Itasca Consulting Co. Fairhurst showed that the stress from the tool is not evenly distributed across the crushed rock zone, but rather as Cundall had earlier shown [9.55], the stress is transmitted through a series of discrete particles from the tool surface to the rock. He then used Rittinger's theories on material pulverization [9.56] to explain why up to 98% of the energy applied in cutting the rock goes in creating the crushed particles under the tool, while less than 2% is used to generate the large chips formed in the process.

Part of the reason for this relatively high energy level consumed outside of the chip formation process comes from the need to generate the required rock geometry for effective chip creation. It is this phase of the rock cutting process which can most significantly affect the overall efficiency of the mechanical cutting of rock. Fairhurst pointed out that the crushed material which passes under the pick is a small fraction of the total amount of rock crushed, but that this material is much finer in size than that ejected ahead of the pick, it absorbs significant amounts of energy.

Some of the crushed material, however, is created immediately after a large chip is formed. This is because the chip will dip forward into the rock ahead of the tool, before turning upwards towards the free surface (Fig. 9.35). Thus, there is a small ridge of material left between the face of the tool and the void left by the ejection of the chip. This material is usually immediately crushed by the tool advance.

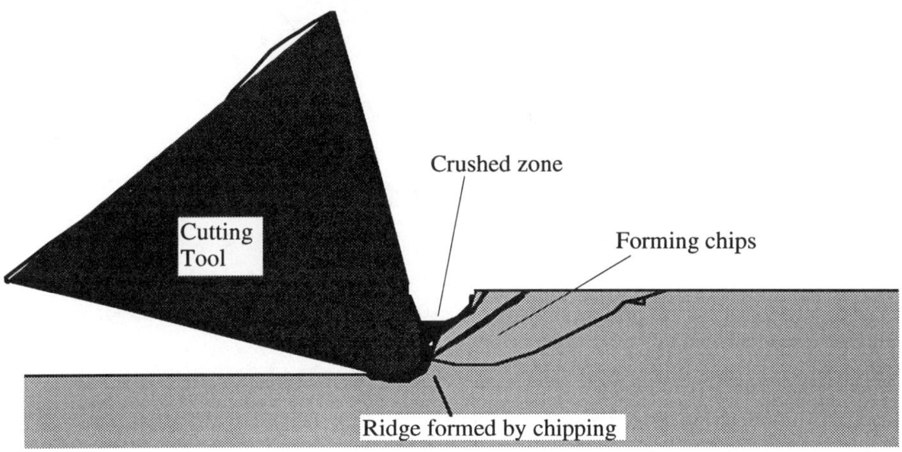

Figure 9.35 Direction of chip growth ahead of an advancing cutting tool (after [9.54]).

Where it is not, as Hood has separately pointed out [9.57], the small rib of material may block the path of the jet into the crushed rock zone reducing, at least temporarily, the benefits of the jet assisting.

The major emphasis of Fairhurst's study, however, was in the effects of the jet working in that crushed zone of rock which surrounds the tip of a working drag bit in rock.

This point has been missed by several investigators. For example, as Hood et al. have also pointed out [9.57] Tecen [9.58] in the U.K. and Ozdemeir in the U.S. [9.59] both failed to grasp the need to keep the jet in close proximity to the bit, in order for optimum assistance to be provided.

Fairhurst considered two main ways in which the jet could assist a pick. In the first he directed the flow of water through the pick body, but then, initially, had the orifice issuing 5 mm from the tip and inclined forward at 40°. This meant that the jet was striking the rock ahead of the bit but significantly above the point of the tool. With this design, (Fig. 9.36) the jet provided little assistance to the bit unless the depth of cut was on the order of 10 mm. At that point the jet was impacting on the crushed rock zone ahead of the bit and could remove at least part of it reducing the cutting force required by up to 25%.

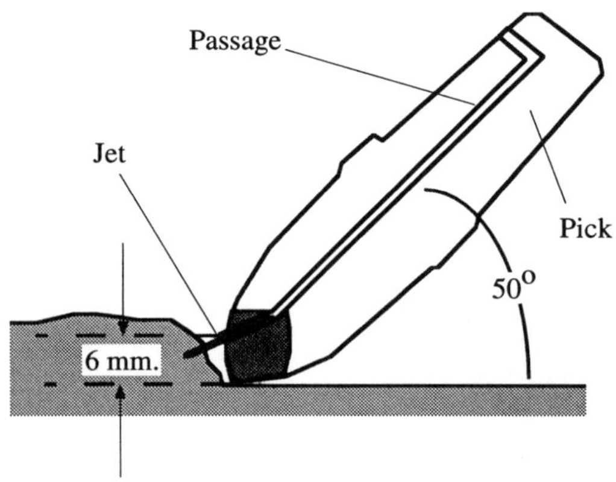

Figure 9.36 Location of the jet fed "through the pick" by Fairhurst [9.54].

When the jet was lowered until it was more directly fed into the crushed zone, then it was possible to lower both cutting and normal forces by up to 25%, however, the bit failed early in the evaluation of that design, which was considered a relatively impractical one for conventional use.

The more practical design is one in which the nozzle is aimed so that the jet hits the rock within 1 mm of the tool tip. At that point, for the jet to do most good, it must penetrate through the crushed rock ahead of the bit, and also reach and remove some of the crushed rock which would pass under the bit. Even where the rock is crushed this still requires some considerable effort, and some time. This becomes an important consideration at the normal speeds at which these picks cut into the rock. At a speed of 2 - 3 m/sec the depth of penetration of the jet can be quite small, depending on the strength of the material being removed and its degree of confinement. Thus, it is not surprising that as one moves into these velocities, the effect of the jet on improving cutting is reduced (Fig. 9.37), although there is some improvement at higher jet pressures.

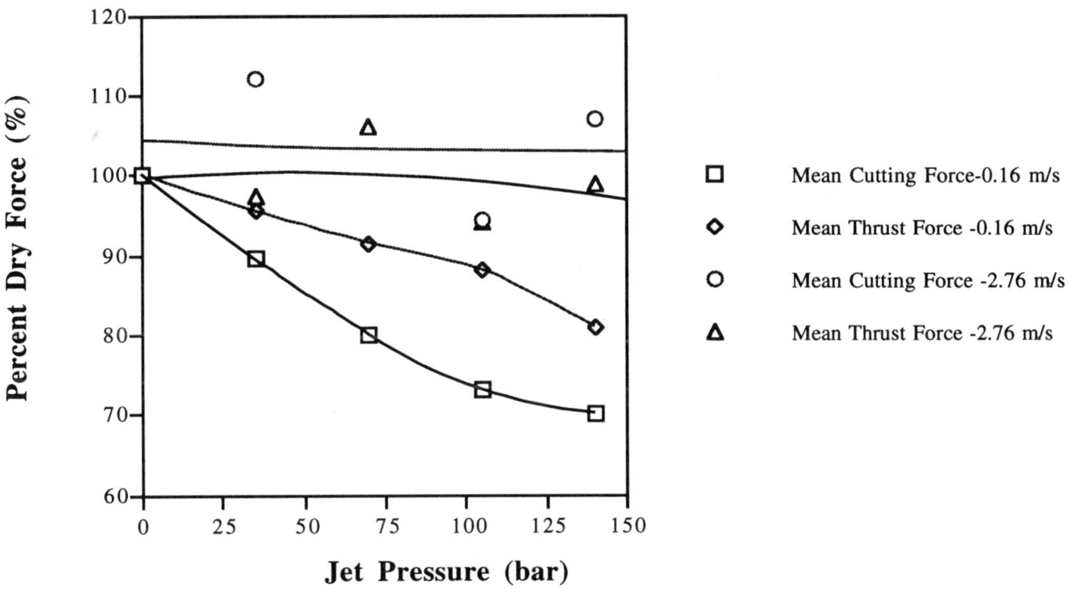

Figure 9.37 Effect of traverse speed on jet-assist benefit (after [9.54]).

The improvement in benefit also varies with the rock being cut. While a granular material may crush significantly ahead of the bit, providing material which the jet can easily remove, other crystalline rocks may break into larger pieces which "bridge" the crushed zone and which, individually, are still hard for the jet to cut through, and too large for it to go around. For the jet to exploit a crack it must not only be directed into that crack, but the crack must confine the jet so that it can rebuild pressure. Where the fragments have voids and free passages beyond them then the jet may not be able to pressurize the mass and erode it. Thus, Fairhurst concludes, even with jet assist the fabric of the rock remains an important parameter.

Comparison of data from different researchers is compounded by the changes made in the bit shapes which were tested. Such changes can significantly affect the results of the tests. For example, as discussed above, when Hood directed two jets at the corners of a drag bit he was able to see a significant improvement in performance whenever the jets were turned on, even if this was only after the bit had been cutting for some distance. In contrast Fairhurst reported a much lower gain when a similar test was carried out using his experimental arrangement, and a differently shaped tool.

The importance of getting sufficient water power at the right place to be effective cannot be overstressed. Tecen's work, for example [9.58], had the jet lead the point of a point attack pick by 5 mm and cut a slot ahead of the bit before it arrived. The jet had to cut through up to 11 mm of rock in order to reach the point of the tool. In contrast, had it been directed closer to the point it would have had to cut through significantly less material because of chipping ahead of the bit, and the rock would have been fractured making penetration easier. To cut through the solid rock to the required depth in the sample material required between 450 and 550 bar. However, the jet was sufficiently far ahead of the bit that no significant step in the results was obtained when the jet reached the depth of attack of the bit. Improved performance was achieved in part because the slotting of the bridge of rock over the bit path made it easier to break out the material than would have been the case without it.

There is a danger to this approach which has been recognized both by Morris and MacAndrew [9.60] and by Vasek [9.61]. If the jet is sufficiently far ahead of the tool that it does not play continuously on it, then the tool will still heat up under the frictional forces generated, particularly when the crushed rock is not flushed out from under the bit. Thus, without the cooling, the bit will build up the high temperatures noted earlier (see Fig. 9.19). Should the rock then fracture so that the leading jet can rebound along the cut back to the bit, then the tool will be instantaneously "quenched" and consequently thermally shocked. Morris and MacAndrew found this to be true even when the jet was 2 mm ahead of the bit, when the tests were carried out at speeds above 0.5 m/sec. Where the carbide could be continuously cooled then a hardened surface layer was maintained, at higher speeds and with the jet position too far ahead of the bit to continuously cool it, this layer was eroded by the rock abrasion. Interestingly with the pointed drag bit used a "nib" of material was formed on the bit at lower traverse speeds, as though the tool had melted and flowed forward (as Hood had shown) until the metal flowed sufficiently far forward that the point was exposed to the continuous impact of the jet which would cool and harden it. Vasek showed, in laboratory experiments, that the intermittent impact of a jet on a very hot cutting bit significantly reduced its operational life [9.61].

Interestingly, Fowell, Ip, and Johnson [9.62] followed the earlier study in which the jet was directed to cut the rock ahead of the bit, with a second stage in which the jet was brought closer to the bit tip. In evaluating the application of this jet, the authors felt that the best application of the jet was to relieve the forces on the tool tip but not to remove the crushed rock around the tip, since this helped to create the tensile stresses ahead of the bit. Those tensile stresses, in turn, create the chips which break the rock ahead of the bit. Where the jet pressure is increased to the point that the jet cuts through that material, then the mechanism of rock cutting is believed to change, and less efficient rock removal is claimed to occur. In order to optimize this cutting parameter it was suggested that the jet be run under such conditions that it would cut to a depth between 20% and 30% that of the mechanical tool. These conclusions were drawn, in part, from results in cutting a hard sandstone at a speed of 1.1 m/sec (Fig. 9.38). Larger jet diameters were also found to give somewhat better rock penetration than smaller jets, at equivalent pressure. The nozzle was located 63.5 mm above the rock surface, and the pick was set to cut to a depth of 10 mm. The increase in depth for each nozzle diameter was achieved by increasing the jet pressure and the collected data from four nozzle sizes 0.6, 0.9 1.2 and 1.5 mm diameter have been summarized in the figure (full comparative data is given in the paper).

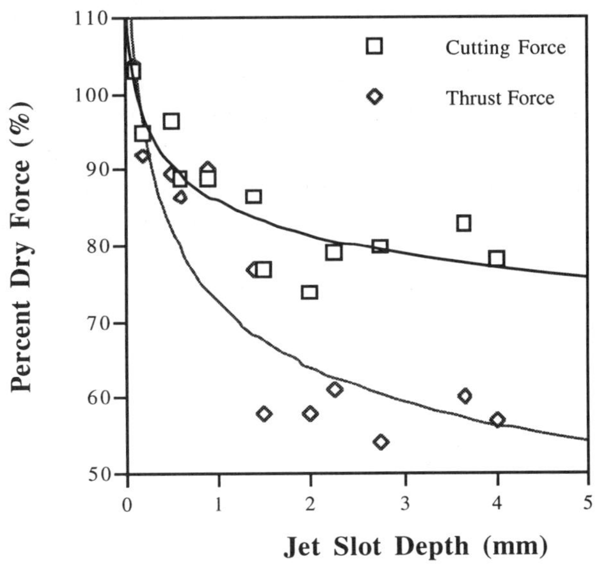

Figure 9.38 Mean force reduction as a function of jet penetration in Grindleford sandstone [9.62].

Jets located ahead of the bit gave improvement in performance of up to 75% when leading a blunt tool, but in contrast with other work, tests with jets directed from behind and under the bit did not show any improvement in performance. The reason for this is likely to be twofold. Firstly, the nozzle was held at a distance of 80 mm from the rock; and secondly, because of the more complex shape of the bit being used, it would appear to be more difficult to get the jet to the point where it would be beneficial. This is because, in contrast with the studies using the basic drag bit, these were carried out with a wedged faced bit. Thus, part of the cutting action becomes a lateral wedging in an area to which the jets have no access.

As with other investigators, Fowell [9.63] has, however, concluded that it is too simplistic to anticipate a single mechanism predominating in all rock failure. He agrees with Fairhurst that greater benefits accrue when the jet assisted system is used in rocks which easily crush, such as sandstone, rather than those, such as limestone, in which chipping failure predominates. This is in contrast with Dubugnon's conclusion [9.53] that it is the dilatancy of the rock which is more important. Rocks with high dilatancy are more easily cut with jet assistance than those which do not exhibit this property.

The study continued from the use of a linear cutting table, from which the above data was generated, to adding the jets to a full face roadheader assembly, albeit one used in a laboratory setting [9.64], [9.65]. These tests were carried out with the same limestone and sandstone as previously, but with the jet operated at higher pressure, up to 1,400 bar. The sandstone could be cut by the jet alone, the limestone could not. The work carried out by the NCB. had shown that the waterjets were effective in assisting the drag bit or radial cutter, but less effective in assisting the point attack picks. The tests in this series examined the use of the forward attack pick, which has a curved face and a wedge shaped contact surface. Although there was a clear improvement in the performance of the picks where the jet impacted on the cutting tool just above the contact point in cutting limestone, where the jet impacted 1 mm ahead of the bit and where the rock was sandstone, no significant improvement in performance occurred.

Tests were also carried out with a waterjet flowing through the pick, however the jet was directed at the back of the wear flat, and there was no clear benefit in using this design, in terms of gains in mechanical specific energy. Unfortunately, modifying the pick shape to allow the jet passage also weakened it, and the pocks failed prematurely. Fowell concluded [9.65] that for jet-assisted cutting to be effective:

- the jet needs to impact the rock directly at the contact between the tool and the surface, from the front of the tool.
- that the jet pressures need to be no more than 300 bar, since this allows the water to be supplied to more picks on the head, and reduces the problems of dealing with higher pressure fittings and seals.

It is of concern to note however, that Fowell reported that most nozzles used in the many waterjet assisted roadheaders which are in the field are of 0.6 mm diameter, while the stand-off distance lies in the range from 75 to 85 mm. This moves the rock beyond the point where the jet force on the rock is still a significant percentage of its original value. It is germane to note that in some of the test data which Fowell reported [9.62] there was, if anything, an increase in forces on the bit with this smallest of the nozzles tested, due to its inability to effectively transfer the jet energy to the rock at the cited stand-off distance.

The importance of this has been shown by a study by the U.S. Bureau of Mines [9.66] in which the measured pressures at such distances are reduced by 65 - 70% over those measured at the nozzle. The Bureau team showed [9.67] that, by adding a short length of straight pipe behind the nozzle, some 100 mm long, that a significant improvement in delivered pressure can be achieved (Fig. 9.39). While an additional 10% or so gain could be achieved by adding another 50 mm to this pipe, the logistics of arranging such a design on a cutting head may make this additional gain more difficult to achieve.

This point is stressed, as it has been throughout this section, because it is critical to an effective use of the hybrid system to ensure that the jets are used effectively, and their energy not just inefficiently dissipated. This approach has been most clearly stated by Hood. He has pointed out [9.68] that if the speed of the cutting tool is increased, then if the improvement in performance of the jet is to be sustained then it is logical to require a greater level of power be input to the rock. However, the two are related by the need to ensure that a consistent level of jet energy be applied <u>over the same length of the cut</u>. Thus, if the cutting speed of the head is increased from 0.5 m/sec to 3.0 m/sec then it is logical that the amount of jet power delivered through the assisting nozzle be increased by a factor of six accordingly. Such an increase in jet power has rarely been considered.

Hood has reported [9.69] that at speeds of up to 0.5 m/sec he was able to see an increase in jet effect on both cutting and normal forces which was clearly related to the jet energy delivered to unit length of cut, but which was not related to either the parameters of the jet or the traverse speed.

Unfortunately, the data was not continued into velocities above 1.0 m/sec where the controversy continues (in 1994) as to the size of gains to be achieved from using jet assistance. It is regrettable that many of the negative opinions come from mine operators who are reported to find the system inconsistent, particularly where the jets are improperly aligned a point not all of them are aware of.

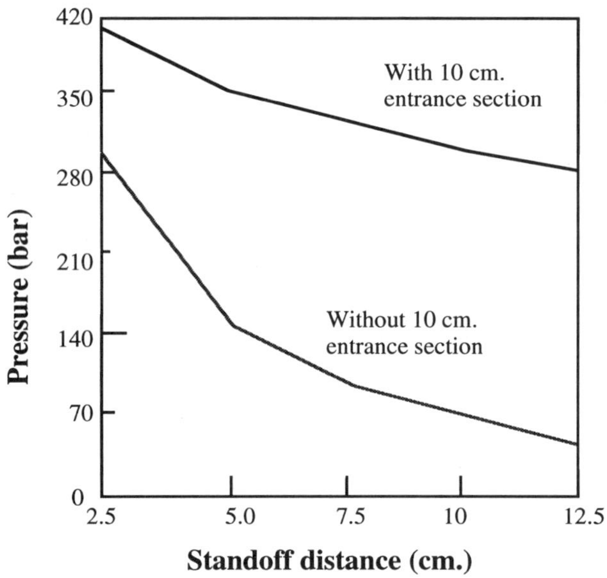

Figure 9.39 Jet pressure as a function of distance from the nozzle, with and without a 100 mm lead-in pipe [9.67].

Hood has argued, along the same lines as Fairhurst, that removal of the crushed rock is a major factor in the force reductions which are seen. Using data from Tutluouglu [9.70] he points out that, for Indiana limestone, the force required to generate chips ahead of a drag bit are on the order of 5 J/m. Given that the total energy required to make the cut is on the order of 4.5 kJ/m the relatively low importance of improving the chipping relative to the reducing the effort required in crushing becomes evident. This can, perhaps, be seen more clearly graphically (Fig. 9.40).

Hood et al. have pointed out [9.57] that because the jet must penetrate into the very narrow crushed zone ahead of the bit, then if the nozzle is to be at some considerable stand-off, required by many cutting head geometries, that the jet must parallel the front face of the bit. This recommendation is somewhat difficult to follow for several bit geometries, particularly, as is pointed out, for point and forward attack tools with a dual angle on the tool. For such tools a jet directed into the cutting zone from behind the bit may prove a more effective solution.

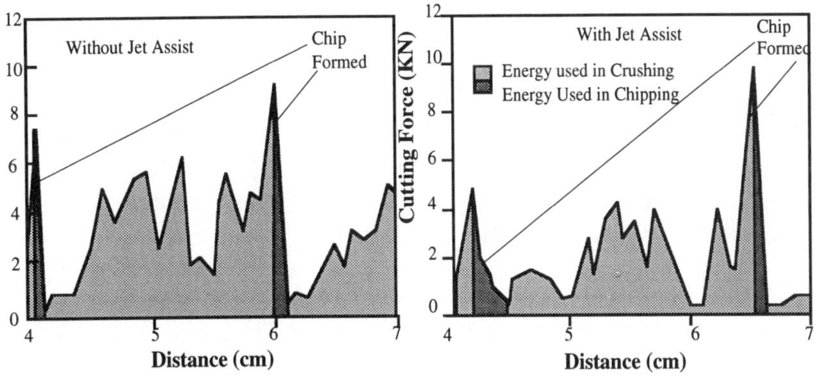

Figure 9.40 Cutting force on a bit, showing the relative levels of energy going into crushing and chipping the rock (after [9.69]).

Removing the crushed rock from around the bit is not easy. The bit width may vary from the narrow point of a point-attack tool to over 25 mm for wider drag bits. For many tools it is on the order of 12 mm. The jet cutting into the crushed zone has a width of around 0.6 mm. For it to be completely effective in removing the crushed rock it should be at a pressure which will penetrate through that material, but at insufficient pressure to cut a slot deeper into the rock. The reason is that when the jet hits a harder relatively flat layer of material it flows outwards along that surface, and will clear the rock of overlying crushed material. If the jet penetrates into the solid rock under the bit then it will flow back out of the hole created along the axis of the incoming jet. Its ability to clear crushed material outside the actual jet path thus becomes significantly less.

In the data which Hood obtained in evaluating the role of jet impact energy on the effectiveness of jet assist [9.69] the data showed only a slight improvement in cutting force reduction with increased jet energy (Fig. 9.41) relative to the consistent increase in the thrust force reduction as jet energy was increased (Fig. 9.42). Given that the jets were directed to the correct point just ahead of the cutting edge of the bit, this might suggest that as the jet energy was increased so it was more able to penetrate the crushed rock passing into the high levels of compression immediately ahead and under the bit. It is the removal of this tightly packed and excessively crushed material which will reduce the area of rock in contact with the wearflat on the underside of the cutting tool, and at equivalent pressure applied, through the tool to the rock, reduce the overall force applied. Further, by removing this material from the vicinity as it is first broken free from the surface, there is less chance of this material being re-crushed between the rock and the pick as it continues to advance. This reduction in contact area, both at the front and under the tool, is likely, however, to be greater under the tool, where it is also likely to have the greatest effect. This relative improvement is shown by the two curves.

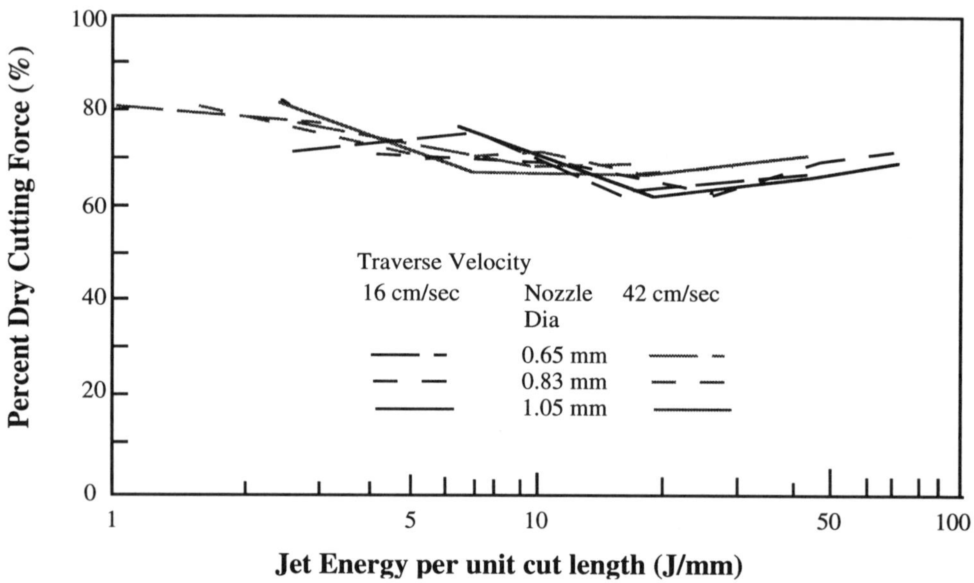

Figure 9.41 Reduced pick cutting force with increasing jet energy density on the cut (after [9.69]).

This seems a logical conclusion, given that Fairhurst [9.54] has correlated the normal forces applied to the bit with the area of the wearflat showing that part of the reduction in the normal forces which occur with jet assist occur because the bottom of the bit is not increasing in size as rapidly. Thus, for the same applied pressure, required to give the same amount of rock fragmentation, the force level is not as great since it is applied over this smaller area.

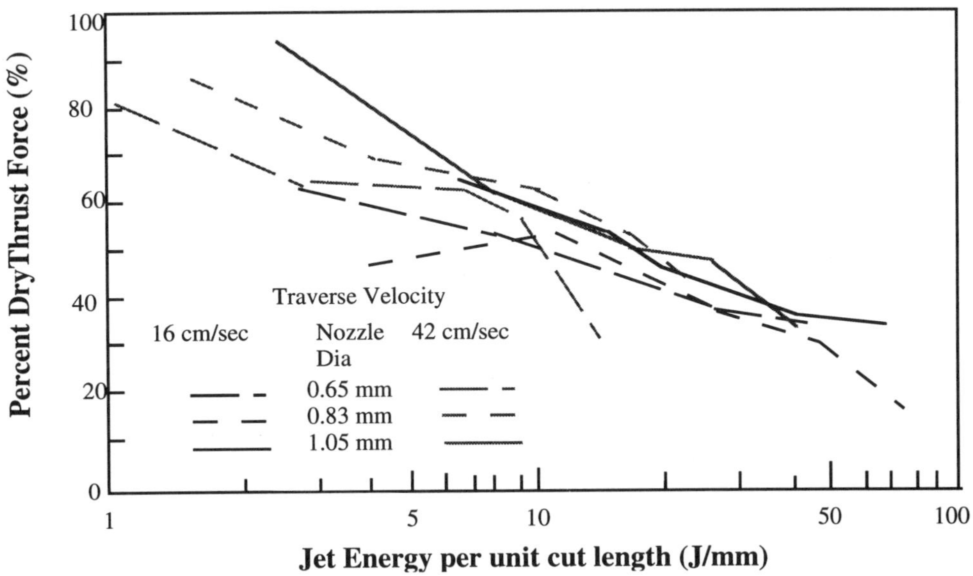

Figure 9.42 Reduce pick thrust force with increasing jet energy density on the cut (after [9.69]).

Keeping the wearflat area small can be achieved in two ways. Firstly, and the original impetus for this entire study, is by keeping the tool cool by continuously having the waterjet impact on the edge of the tool. This requires that the jet impact within 1 mm of the edge of the tool. The requirement is less critical if the jet is directed along the front face of the tool. It is much more difficult to control jet position and effectiveness with the jet directed in from the back of the tool and underneath. This is because of the very narrow gap that the jet must flow in along between the underlying solid rock and the overlying tool body to reach the crushed rock. Given the forces involved on the tool and its holder, this position may change during the course of the cutting stroke due to the compression of the tool and holder under the applied load. Secondly, this can be achieved by reducing the volume of fine material which must pack under the tool as it advances. This requires that the jet penetrate to the bottom of the crushed zone of rock, a zone under high compression, although small in size. Where this can be effectively achieved then the effective forces on the bit can be cut in half. In addition, since the chips become more isolated and not confined by the surrounding powder the peak forces on the head are reduced, and this in turn reduces the overall magnitude of the force variations (Fig. 9.43) increasing the operational life of the equipment. Performance data for such a machine in a Japanese mine [9.71] has shown the consistent gains where this technology is used, and follows to a large extent that previously reported for the results from field operations in the United Kingdom.

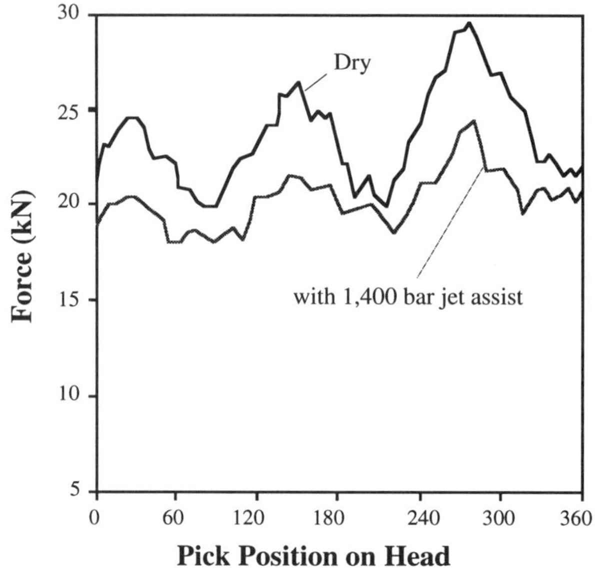

Figure 9.43 Reduction in magnitude of forces on a roadheader with waterjet assist to the picks (after [9.64]).

Many of the advantages which this may bring to the cutting machines of the mining and tunneling industry become more evident when the machine must cut harder rock. Work in Germany showed that a tunnel profiling machine using jet assisted picks could cut a slot around a tunnel, with force reductions on the tool of up to 50%, although the program was halted due to other greater priorities [9.72]. The dramatic improvements in performance when the NCB unit cut into the harder limestone at Middleton were a greater persuasion to move the technology forward than were the reduced improvements seen in the softer sandstone of the earlier trials. Recent Russian work [9.73] has shown that the addition of waterjets to smaller roadheaders allow their use in place of larger, heavier and more costly models in cutting harder rock. And Hood has continued his pioneering work [9.74] by developing a waterjet assisted rock cutting head demonstrated to be capable of driving a tunnel through rock of compressive strength up to and perhaps beyond 1,700 bar consisting of marbles and hornfels.

For the technology to reach a wider audience, however, the required precision in flow lines to the nozzle, in nozzle alignment and position, in stand-off distance and location of the jet impact point must all be understood and adhered to. This is sometimes difficult to achieve, given that those who build mining and tunneling equipment are often more prone to think of precision as meaning fractions of a centimeter, rather than fractions of a millimeter. Further, the hybrid system of jet assist may well only reach its full potential when machines are designed more to fit the optimum conditions for application, rather than, as is still the case, trying to modify existing machine designs to this new technology. Given those caveats, and the poor results reported from some investigative laboratories where these conditions were not fully followed, it is interesting to note that the technology is now being use, with apparent success, in industry on a world-wide basis.

9.4 JET ASSISTED CUTTING OF COAL

The collaborative work between European and American government research groups continued to grow, as the benefits of jet assisted cutting became more obvious. While there is a need for assistance in cutting rock with jets, the benefits from the technology appeared also possibly to be of benefit in the mining of coal. The initial work on jet assisted cutting in the U.S. had been to assist the actions of a machine in cutting coal, but with a small "roadheader" type of cutting head. In Europe, and increasingly in the United States, coal is won using a shearer-loader type of machine on a longwall face. (See section 6.2.3.4 and Fig. 6.19.)

Such machines have one or two drums, of relatively large diameter, with a series of helical vanes set with picks around their perimeter. As these rotate they grind the coal from the solid face, and the vanes carry it over onto an adjacent conveyor (Fig. 9.44). The picks undergo the same types of wear as those of the roadheader, and thus, studies began on the possibility that jet assist might improve shearer performance.

An early study in the Soviet Union developed a shearer with jet assist [9.75] tested underground at the Nagornaya mine in the Kuznetsk Coal basin. The seam was 2.3 m thick, dipped at 12°, and contained a mudstone layer 0.07 m thick. Nozzle diameters were varied from 1.9 - 3.7 mm, optimum configuration of 2.5 mm, jet pressure of 300 bar. Preliminary data indicated that the overall haulage force required on the machine, and the power demand of the motor, were reduced with the jet assist.

Figure 9.44 Waterjet-assisted shearer.

The major development of the technology in the West began in 1984. The U.S. Bureau of Mines awarded a contract to a subsidiary of Gebr. Eickhoff to develop a shearer-loader which was assisted by jets of 700 bar pressure [9.76]. The machine was first tested underground from June - November 1986 at the Gewerkschaft Auguste Victoria colliery in Marl, Germany. Cutting speeds of the picks on the machine were approximately 2.1 m/sec, a matter of some concern since the data at the time indicated that this speed was above that at which waterjet assist provided an improvement in cutting. Even this speed was a reduced one of 50% of that of a normal cutting head. Nozzle diameters were on the order of 0.6 mm.

Results from this test [9.77] indicated that there was no statistically significant reduction in shearer power, although there was a reduction of 80% in the respirable dust generated. Concurrently the product size produced by the machine increased with the percentage of material below 6.3 mm falling from 37 - 28%. As with other jet-assist systems, the picks lasted longer on the machine. The machine was fitted with a phasing system to feed water only to the 54% of the picks which were in contact with the coal at the time. Dust suppression occurred when the jets were operated at 125 bar. Although there was no significant effect on the power of the machine this depended on the method of interpretation, since there was an increasing reduction in the increase in normal force required on the picks as they wore, when jet-assist was provided (Fig. 9.45).

Concurrently, with the development of the Eickhoff machine the Bureau began its own work using a modified Joy shearer [9.77]. The machine was tested at the Bureau surface test facility in Bruceton, PA, and the data obtained in those tests largely correlated with the results of the German program. It should be noted, relative to the study which follows, that both

German and U.S. trials used point attack picks on the cutting drums. It has been reported [9.47] that jet impingement distance from the cutting point was not well controlled, with values ranging up to 7 mm ahead of the bit.

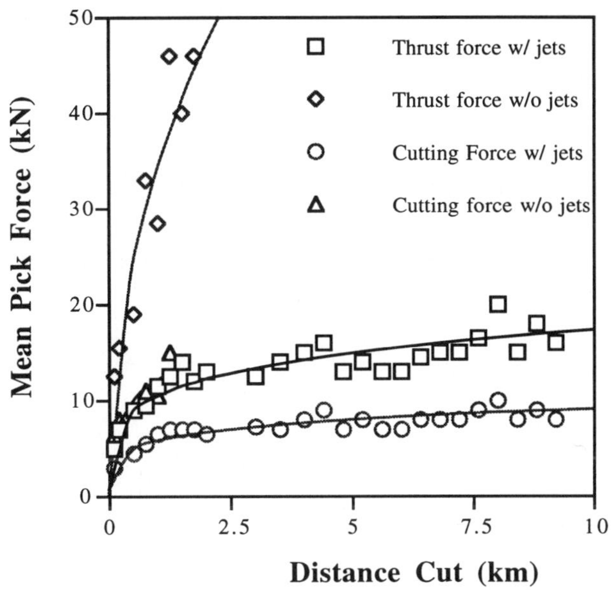

Figure 9.45 Relative increase in force on a pick as a function of cut length, with and without jet assist [9.78].

The reduction in the amount of small coal produced was a consideration in a concurrent development of jet assisted shearers in the U.K. At that time, 1984, it cost $4.50 to clean coal over 0.5 mm in size, as opposed to $10.50 for coal smaller than that [9.79]. In this case the machine modified was an Anderson Strathclyde AB16 which was fitted with a Harben pump to produce up to 55 lpm at 690 bar. The drum was a three vane unit and was modified so that the jet nozzles were mounted 50 mm from the pick points and so that the jet hit the coal less than 2 mm from the pick point. Nozzle diameters were adjusted between 0.5 and 1.0 mm, using the change in diameter to control the pressure of the jets. Radial picks were used on this machine. The machine was first tested at Golborne Colliery taking a 2.1 m cut from a 2.3 m high seam, containing a 0.45 m dirt band and dipping at 1 in 3.5. Jets were directed to assist both the vane picks and those cutting clearance for the machine (Fig. 9.46).

The machine ran at an advance rate between 0.047 and 0.076 m/sec with a drum speed of 62 rpm, which gave a pick speed of 3.8 m/sec. The depth of cut for each pick ranged up to 21 mm [9.47]. Interestingly, the British results did not confirm those reported in either Germany or the U.S. There was little change in product size, except when the jets were only directed at the picks cutting clearance. When only nine picks were used on this ring, but fifteen jets were retained, the size of coal produced which measured larger than 50 mm increased by 23%. This was taken to suggest that more work be done on jet placement. The reduction in fines by 20% (or 1% of total output) would give a payback time of 1 year for the equipment if priced at $24,000. (This calculation originally derived by Mort has been converted from 1983 English pounds at a rate of one pound to 1.5 American dollars.) The British tests also showed that there was a significant reduction in the haulage force required to move the machine down the face (Fig. 9.47).

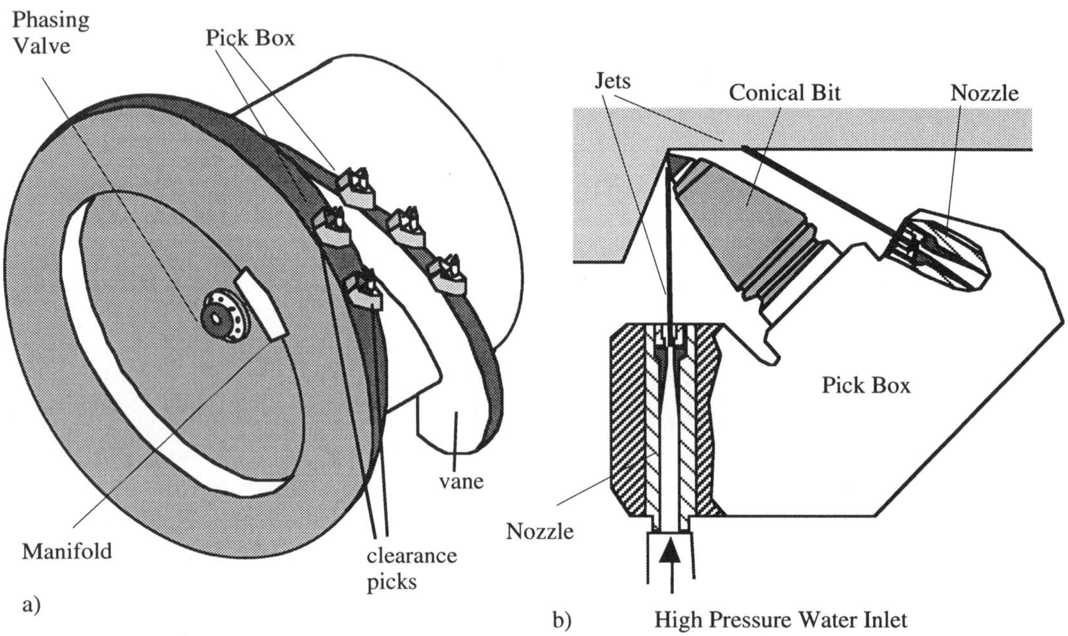

Figure 9.46 a) Arrangement of pick boxes, b) location of jets to assist vane and clearance picks [9.76].

Similarly, there was a reduction of 45% in the power required to cut and load the coal applied through the gearhead to the cutting drum. However, when the power required to power the waterjet pump was included there was no reduction in the total amount of energy required to mine a ton of coal. It was also found that when the machine was cutting at the highest speed, 0.076 m/sec, the haulage force required when cutting "dry" would stall the machine with a pull of 18.5 tons. When the machine was operating with the jets on at 655 bar the machine ran without stalling. This translated into an effective increase of 40% in the maximum cutting speed of the machine.

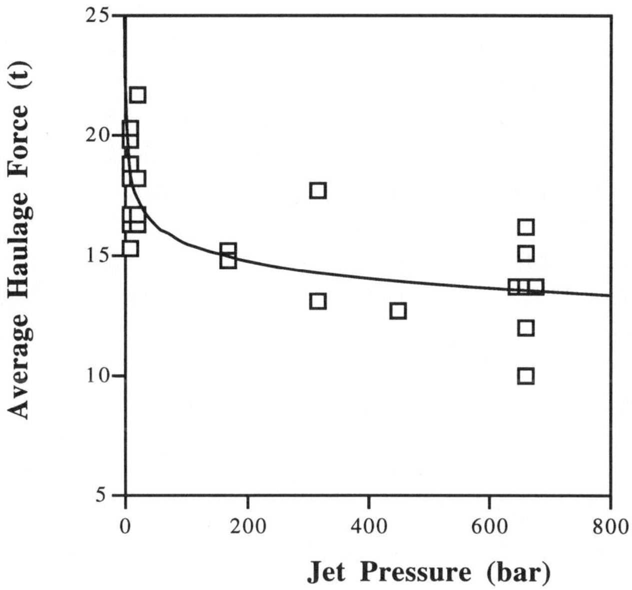

Figure 9.47 Reduction in haulage force on a shearer as a function of jet pressure (after [9.78]).

In the original trials the jets were not phased (i.e. controlled so that they were on only when the jet was in the coal). This wasted a considerable amount of energy and reduced the quality of the working environment. The trials did not show the large reduction in respirable dust values reported from the other trials. This may, however, have been a function of the coal being mined.

The results were sufficiently promising that a second trial was undertaken, using a different method of jet pressure generation. This system, called the "Epytensic" system, uses radially operated pistons within the cutting drum to generate the jet pressure [9.80]. Water is then supplied, sequentially to nine 0.6 mm diameter jets on the cutting drum as they cut into the coal. The machine was tested over a six month period at Bolsover Colliery in 1986. It was found that there was a 6.4% reduction in specific energy during the cutting of the coal , including the power for the jet pumping system. Haulage and drum forces (the normal and cutting forces for a single pick) were reduced by 18.9% with the difference made up by the power required for the pump. The web depth of the machine (i.e. the thickness of the slice of coal removed) was reported to have been increased by 25% [9.81], and there was such a reduction in shock loading and cyclic peak forces on the unit that the life of the cutting arm was doubled and the internal parts were all found to be still within the manufacturers original tolerances at the end of the trial [9.82].

Pick costs were significantly reduced, over a six week period only two picks were changed because of wear problems [9.47] in contrast with at least 200 picks being changed under normal conditions. Interestingly, in light of Dr. Hood's comment about the need for higher jet parameters to compensate for faster pick cutting speeds, the trials at Bolsover, and the subsequent units installed at 6 U.K. mines and 1 Australian mine all have operated at jet pressures of 1,570 bar. Models have now been shipped with 12 or 18 pistons.

It is interesting to note that the significant improvement in performance with the work in the U.K. may be due to the shape of the picks with which the jets are reacting. Should this be the main cause for the change in performance it would underline the comments made at the end of the preceding section on the need to rationally review the best combination of tools to achieve optimum cutting. As has been stated elsewhere [9.83] in order to produce the best jets consideration must be given to system losses in the delivery line, and to a proper location of the nozzle so that the resulting jet does some good. The Minnovation Epytensic system, with a straight flow from the pump piston to the nozzle with minimum bending and change in flow diameter illustrates the benefits which can be achieved with such a choice.

This technology has been developed for less than ten years but has already found a place in the arsenal of the excavation engineer. The mechanisms by which it works, and the optimum choice of tools are not yet totally evident for the different rocks which might be cut by such hybrid systems. There remains a considerable amount of research to be carried out, with the promise that the benefits to accrue from it will have a significant payback.

9.5 REFERENCES

9.1 Olson, J.J., and Olson, K.S., ARPA-Bureau of Mines Rock Mechanics and Rapid Excavation Program, A Research Project Summary, U.S. Bureau of Mines Information Circular 8674, 1975, 191 pages.

9.2 Teale, R., "The concept of specific energy in rock drilling," International Journal of Rock Mechanics & Mining Science, Vol. 2, No. 1, March, 1965, pp. 57 - 74.

9.3 Zar, J.L., The Use of a Laser for ARPA Military Geophysics Program (Rock Mechanics and Rapid Excavation), AVCO-Everett Research Laboratory, (Everett, Mass.), Final Technical Report on BuMines (ARPA) Contract HO210039, October, 1972, AD 749982, 83 pp.

9.4 Poole, J.W., and Thorpe, M.L., Study of High Powered Plasmas for In Situ Hard Rock Disintegration, Humphrey's Corp. TAFA Division (Bow, NH), Final Technical Report on BuMines (ARPA) Contract HO220051, July, 1973, AD 772506/2GI, 76 pp.

9.5 Schumacher, B.W., and Holdbrook, R.G., Use of Electron Beam Gun for Hard Rock Excavation, Westinghouse Electric Corp. (Sunnyvale CA) Final Technical Report on BuMines (ARPA) Contract HO110377, December, 1972, AD 761208, 283 pp.

9.6 Ripkin, J.F., and Wetzel, J.M., A Study of the Fragmentation of Rock by Impingement with Water and Solid Impactors, St. Anthony Falls Hydraulic Laboratory, University Minn. (Minneapolis Minn.) Final Technical Report on BuMines (ARPA) Contract HO210021, February, 1972, AD 746510, 39 pp.

9.7 Kurko, M.C., and Chadwick, R.F., Continuous High Velocity Jet Excavation - Phase 1 -Final Report, Contract HO210034, on ARPA Order 1579, Bendix Research Labs, Report 6241, May, 1972.

9.8 Cooley, W.C., and Brockert, P.E., Optimizing the Efficiency of Rock Disintegration by Liquid Jets, Terraspace Inc., (Rockville, MD) Final Technical Report on BuMines (ARPA) Contract HO210012, December, 1972, AD 757120, 27 pp.

9.9 Chadwick, R.F., Continuous High Velocity Jet Excavation - Phase II, Bendix Research Laboratories, (Southfield, MI) Final Technical Report on BuMines (ARPA) Contract HO220067, October, 1973, AD 773417/1GI, 33 pp.

9.10 Clark, G. B., Lehnhoff, T.F., Patel, M., and Allen, V., An Investigation of Thermal Mechanical Fragmentation of Hard Rock, University of Missouri-Rolla, (Rolla, MO) Final Technical Report on BuMines (ARPA) Contract HO210028, June, 1972, AD 746196, 267 pp.

9.11 Watson, J.D., Full Scale Field Test Results of the REAM Concept for Hard Rock Excavation. Physics International (San Leandro, CA) Final Technical Report on BuMines (ARPA) Contract HO220015, January, 1973, AD 757116, 70 pp.

9.12 Graff, K.F., Fundamental Studies in the Use of Sonic Power for Rock Cutting, Ohio State University Research Foundation (Columbus, OH), Final Technical Report on BuMines (ARPA) Contract HO220037, April, 1973, AD 766047/5, 200 pp.

9.13 Hug, H., and Peterson, C., Design, Fabricate and Test a Conical Borer, Foster-Miller Associates Inc. (Waltham, MS), Final Technical Report on BuMines (ARPA) Contract HO210044, March, 1972, AD 744989. 55 pp.

9.14 Gangal, M.D., The Design of an Experimental High Energy Impact Tunneler, Ingersoll-Rand Research Inc. (Princeton, NJ) Final Technical Report on BuMines (ARPA) Contract HO230006, June, 1973, AD 774983/1GI, 50 pp.

9.15 Cook, N.G.W., and Joughin, N.C., "Rock Fragmentation by Mechanical, Chemical and Thermal Means," paper 1c, 6th International Mining Congress, Madrid, Spain, 1970.

9.16 Summers, D.A., and Henry, R.L., Water Jet Cutting of Rock With and Without Mechanical Assistance, SPE 3533, New Orleans, LA, October, 1971.

9.17 Fair, J.C., "Development of High-Pressure Abrasive-jet Drilling," Journal of Petroleum Technology, August, 1981, pp. 1379 - 1388.

9.18 Vestavik, O.M., Abrasive Water-Jet Drilling Experiments, Progress Report, Rogaland Research, Stavanger, Norway, May, 1991.

9.19 Yao, J., Summers, D.A., and Galecki, G., "DIAjet cutting of Dolomite and Chert-A Case Study at the St. Louis Arch," 6th U.S. Water Jet Conference, Houston, TX, August, 1991, pp. 529 - 543.

9.20 Chadwick, R.F., and Kurko, M.C., Continuous High Velocity Jet Excavation - Phase I-Bendix Research Laboratories, (Southfield, MI) Final Technical Report on BuMines (ARPA) Contract HO210034, May, 1972, AD 744014, 85 pp.

9.21 Hoshino, K., Nagano, T., and Tsuchishima, H., "Rock Cutting and breaking using high speed water jets together with "TBM" cutter," paper B6, 1st International Symposium on Jet Cutting Technology, Coventry, U.K., April, 1972, pp. B6 89 - B6 100.

9.22 Hoshino, K., Nagano, T., and Takagi, K., "The Development of a Water Jet-boring Machine for large diameter holes," paper E3, 2nd International Symposium on Jet Cutting Technology, Cambridge, U.K., April, 1974, pp. E3-19 - E3-29.

9.23 Crow, S.C., Design Principles for Jet Tunnelling Machines, Final Report, NSF Grant GI 37193, UCLA Report UCLA-ENG-7471, October, 1974.

9.24 Wang, F-D., "Remarks," Workshop on the Application of High Pressure Water Jet Cutting Technology, Rolla, MO., November, 1975, pp. 149 - 170.

9.25 Wang, F-D., Robbins, R., and Olsen, J., "Water Jet Assisted Tunnel Boring," paper E6, 3rd International Symposium on Jet Cutting Technology, May, 1976, Chicago, IL., pp. X63 - X71.

9.26 Miller, R.J., and Wang, F-D., Development of High-Pressure Water Jet Equipment for Underground Application of Excavating Energy Materials, Final Report on U.S. Bureau of Mines Contract JO155133, July, 1976, 64 pp.

9.27 Hustrulid, W., A Technical and Economic Evaluation of Water Jet Assisted Tunnel Boring, Report NSF/RA-760174, University of Utah, July, 1976, 152 pp.

9.28 Henneke, J., and Baumann, L., "Jet Assisted Tunnel Boring in Coal-Measure Strata," paper J1, 4th International Symposium on Jet Cutting Technology, Canterbury, UK., April, 1978, pp. J1-1 - J1-12.

9.29 Baumann, L., and Henneke, J., "Attempt of Technical-Economical Optimization of High Pressure Jet Assistance for Tunnelling Machines," paper C4, 5th International Symposium on Jet Cutting Technology, Hanover, FDG, June, 1980, pp. 119 - 140.

9.30 Reich, P., and Reese, H., "Remarks in Discussion," Seminar on Water Jet Assisted Roadheaders in Rock Excavation, Pittsburgh, PA., May, 1982.

9.31 Kouzmich, I.A., and Merzliakov, V.G., "Coal Massif Breakage with Disc Cutter and High Speed Water Jet," paper I4, 8th International Symposium on Jet Cutting Technology, Durham, UK, September, 1986, pp. 147 - 155.

9.32 Fenn, O., Protheroe, B., and Joughin, N.C., "Enhancement of Roller Cutting by Means of Water Jets," 1985 Rapid Excavation and Tunnelling Conference, AIME, New York, 1985, Chapter 21, pp. 341 - 356.

9.33 Summers, D.A., Wright, D.E., and Xu, J., "The Use of High Pressure Waterjets to Enhance Tunneling Machine Performance," Final Report on MIT Purchase Order CE-R-386168, issued 5/27/93.

9.34 Marlowe, A.C., and Page, J.L., "Water Jet Assisted Boring in South African Quartzites," Chapter 22, 1985 Rapid Excavation and Tunnelling Conference, AIME, New York, 1985, pp. 357 - 378.

9.35 Hood, M., A Study of Methods to Improve the Performance of Drag Bits used to cut Hard Rock, Chamber of Mines of South Africa Research Organization, Project No. GT2 NO2, Research Report No. 35/77, August, 1977.

9.36 Leach, S.J., and Walker, G.L., "The Application of High Speed Liquid Jets to Cutting", Philosophical Transactions, Royal Society, London, 260A, 1966, pp. 295 - 308.

9.37 Wilson, A.H., "Stress and Stability in Coal Ribsides and Pillars," 1st Annual Conference on Ground Control in Mining, West Virginia University, 1981, pp. 1 - 12.

9.38 Hood, M., "Contribution to Paper E6," 3rd International Symposium on Jet Cutting Technology, Chicago, IL., May, 1976, pages X72 - X74.

9.39 Tomlin, M.G., MRDE/Department of Energy Collaboration Agreement, Contract No. ET-78-C-01-3126, Field Trials with a Roadheader equipped with a 10.000 psi Water Jet Assist System, MRDE, NCB, Tunneling Branch, Technical Memorandum TU(81)10, July, 1981.

9.40 McNary, R.O., Blair, J.R., Novask, D.D., and Johson, D.I., "Augmentation of a Mining Machine with a High Pressure Jet," 3rd International Symposium on Jet Cutting Technology, Chicago, 1976, paper D2, pp. D2-13 - D2-20.

9.41 Ropchan, D., Wang, F-D., and Wolgamott, J., Application of Water Jet Assisted Drag Bit and Pick Cutter for the Cutting of Coal Measure Rocks, Final Technical Report on Department of Energy Contract ET-77-G-01-9082, Colorado School of Mines, April, 1980, DOE/FE/0982-1, 84 pp.

9.42 Anon, "Pick Ignitions," Colliery Guardian, January, 1992, p. 23.

9.43 Tomlin, M.G., Trials held at SMRE using a Water Jet to Assist Rock Cutting with a Drag Bit, MRDE, NCB, Tunnelling Branch, Report No. TU(79)1, 1978.

9.44 Morris, A.H., "The Development of Boom-Type Roadheaders," Seminar on Water Jet Assisted Roadheaders in Rock Excavation, Pittsburgh, PA., May, 1982.

9.45 National Coal Board Seminar on Water Jet Assisted Roadheaders in Rock Excavation, Pittsburgh, PA., May, 1982.

9.46 Marham, D.K., and Tomlin, M.G., "High Pressure Water Jet Assisted Rock and Coal Cutting with Boom Type Roadheaders and Shearers", paper 6, <u>8th International Symposium on Jet Cutting Technology</u>, Durham, UK., September, 1986, pp. 57 - 70.

9.47 Barham, D.K., and Buchanan, D.J., "A Review of Water Jet Assisted Cutting Techniques for Rock and Coal Cutting Machines," <u>The Mining Engineer</u>, July, 1987, pp. 6 - 14.

9.48 Morris, A.H., "Practical Results of Cutting Harder Rocks with Picks in United Kingdom Coal Mine Tunnels," <u>Tunneling '85</u>, 4th International Symposium on Tunneling, Brighton, UK, 1985, pp. 173 - 177.

9.49 Morris, A.H., and Harrison, W., "Significant Advance in Cutting Ability - Roadheaders," <u>1985 Rapid Excavation and Tunnelling Conference</u>, AIME, New York, 1985, Chapter 20, pp. 317 - 340.

9.50 Timko, R.J., Johnson, B.V., and Thimons, E.D., "Water-Jet-Assisted Roadheaders," <u>1987 Rapid Excavation and Tunnelling Conference</u>, AIME, New Orleans, LA., June, 1987, Chapter 49, pp. 769 - 782.

9.51 Straughan, J., "High Pressure Waterjet Application to Roadheaders," <u>3rd U.S. Water Jet Conference</u>, Pittsburgh, PA., May, 1985, pp. 194 - 213.

9.52 Powell, F., and Billinge, K., "The Use of Water in the Prevention of Ignitions caused by Machine Picks," <u>The Mining Engineer</u>, August, 1981, pp. 81 - 85.

9.53 Dubugnon, O., "An experimental study of water jet assisted drag bit cutting of rocks," Paper II-4, <u>First U.S. Water Jet Conference</u>, Golden, CO., April, 1981, pp. II-4.1 - II-4.11.

9.54 Fairhurst, C.E., <u>Contribution A L'amelioration De L'abbatage Mecanique De Roches Agressives: Le Pic Assiste Et Le Pic Vibrant</u>, Doctoral Thesis, L'Ecole Superieure des Mines de Paris, October, 1987, 221 pages (in French).

9.55 Cundall, P.A., "Distinct Element Models of rock and soil structure," in Analytical and Computational Methods in Engineering Rock Mechanics, Allen and Unwin, 1987.

9.56 Bond, F.C., "Crushing and Grinding Calculations," British Chemical Engineering, 1960, pp. 378 - 385, 543 - 548.

9.57 Hood, M., Knight, G.C., and Thimons, E.D., A Review of Water-Jet-Assisted Rock Cutting, U.S. Bureau of Mines Information Circular 9273, 1990, 17 pages.

9.58 Tecen, O., and Fowell, R.J., "Hybrid Rock Cutting: Fundamental Investigations and Practical Applications," 2nd U.S. Water Jet Conference, Rolla, MO., May, 1983, pp. 347 - 357.

9.59 Ozdemeir, L., and Evans, R.J., "Development of Waterjet Assisted Drag Bit cutting head for Coal Measure Rock," 1983 Rapid Excavation and Tunnelling Conference, AIME, Vol. 2, pp. 701 - 718.

9.60 Morris, C.J., and MacAndrew, K.M., "A Laboratory Study of High Pressure Water Jet Assisted Cutting," paper 1, 8th International Symposium on Jet Cutting Technology, Durham, UK., September, 1986, pp. 1 - 7.

9.61 Vasek, J., "Why Research on water jet hard rock disintegration in Czechoslovakia," Session 3, International Conference on Geomechanics, Hradec n. M, Czechoslovakia, September, 1991.

9.62 Fowell, R.J., Ip, C.P., and Johnson, S.T., "Water Jet Assisted Drag Bit Cutting: Parameters for success," paper 3, 8th International Symposium on Jet Cutting Technology, Durham, UK, September, 1986, pp. 21 - 32.

9.63 Fowell, R.J., "Mechanical Rock Excavation with Water Jet Assistance," Session 3, International Conference on Geomechanics, Hradec n. M, Czechoslovakia, September, 1991.

9.64 Fowell, R.J., Wagott, A., and Anderson, I., "Full Scale Boom Tunnelling Machine Water Jet Trials at Pressures up to 140 MPa." 9th International Symposium on Jet Cutting Technology, Sendai, Japan, October, 1988, pp. 323 - 339.

9.65 Fowell, R.J., Gillani, S.T., and Waggott, A., "Water Jet Assisted Rock Cutting - The Importance of Jet Position," 11th International Conference on Jet Cutting Technology, St. Andrews, Scotland, September, 1992, pp. 217 - 231.

9.66 Thompson, J.L., Thimons, E.D., and Timko, R.J., Evaluation of Moderately High Pressure Water-Jet Assist Applied to Single Drag Bit Tools, U.S. Bureau of Mines Report of Investigations RI 9239, 1989.

9.67 Kovscek, P.D., Taylor, C.D., and Thimons, E.D., Techniques to Increase Water Pressure for improved Water-Jet-Assisted Cutting, U.S. Bureau of Mines Report of Investigations RI 9201, 1988.

9.68 Geier, J.E., Hood, M., and Thimons, E.D., Water-Jet-Assisted Drag Bit Cutting in Medium Strength Rock, U.S. Bureau of Mines Report of Investigations RI 9164, 1987.

9.69 Hood, M., Geier, J.E., and Xu, J., "The Influence of Water Jets on the Cutting Behaviour of Drag Bits," 6th International Congress on Rock Mechanics, Balkema, 1987, pp. 649 - 654.

9.70 Tutluoglu, L., Mechanical Rock Cutting with and Without High Pressure Water Jets, Ph.D. thesis, University of CA-Berkeley, 1984, 165 pages.

9.71 Sato, K., Fujisawa, A., Sato, N., Ito, K., and Takahashi, K., "Development of Roadheader for Harder Rock Assisted by High Pressure Water Jets," paper G2, 9th International Symposium on Jet Cutting Technology, Sendai, Japan, October, 1988, pp. 341 - 356.

9.72 Knickmeyer, W., and Baumann, L., "High Pressure Water Jet Assisted Tunnelling Techniques," 2nd U.S. Water Jet Conference, Rolla, MO, May, 1983, pp. 325 - 334.

9.73 Brenner, V.A., Zhabin, A.B., Miller, M.M., and Katagarov, N.N., "Results of Experimental Research of Destruction of Mountain Rocks by Mechanical Hydraulic Cutters," Session 3, International Conference on Geomechanics, Hradec n. M, Czechoslovakia, September, 1991.

9.74 Hood, M., Li, X., Salditt, P, and Knight, G., "An Advanced System for Rock Tunnelling - Results from a Field Experiment," 6th American Water Jet Conference, Houston, TX., August, 1991, pp. 63 - 70.

9.75 Ekber, B.J., and Kuzmich, I.A., "Coal Cutting Technology with Water Jets in the USSR," paper E5, 5th International Symposium on Jet Cutting Technology, Hanover, Germany, pp. 297 - 302.

9.76 Neinhaus, K., Weigelt, H., and Thimons, E.D., "The Development of a Water-Jet-Assisted Shearer loader," paper 8, 8th International Symposium on Jet Cutting Technology, Durham, UK., September, 1986, pp. 79 - 92.

9.77 Thimons, E.D., Hauer, K.F., and Neinhaus, K., "Waterjet Assisted Longwall Shearer: Development and Underground Test," 4th U.S. Water Jet Conference, Berkeley, CA., September 1987, pp. 113 - 120.

9.78 Kovscek, P.D., Taylor, C.D., Handewith, H., and Thimons, E.D., Long-wall Shearer Performance using Water-Jet-Assisted Cutting, U.S. Bureau of Mines Report of Investigations RI 9046, 1986.

9.79 Mort, D., "The Application of High Pressure Water Jets to Longwall Mining," The Mining Engineer, January, 1988, Vol. 147, No. 316, pp. 344 - 350.

9.80 Parrott, G., "Design of cutting drums using high pressure water," Mintech '89, pp. 166 - 168.

9.81 Tuke, A.W., "The Application of Technology to Production," The Mining Engineer, September, 1988, Vol. 148, No. 324, pp. 137 - 146.

9.82 Anon, "The Cutting Edge," World Mining Equipment, Vol. 11, No. 11, November, 1987, pp. 24 - 25.

9.83 Summers, D.A., "Hydraulic Mining: Jet-Assisted Cutting, Mining Engineering Handbook, SME of AIME, Sec. 22, Chapter 22.3, 1992, pp. 1 - 25.

10 ENHANCING WATERJET PERFORMANCE

10.1 INTRODUCTION

The application of waterjets to a variety of industrial processes has led to the development of a new industry. With the growth in application, however, the limitations of conventional waterjets have become apparent. For while the jets can be very effective in cutting, their range of application has, in many cases, been somewhat limited. Thus, investigators, and engineers, have constantly sought to find means to improve jet performance. Performance improvements have been sought in two different ways. The first, by seeking to understand the mechanisms of waterjet cutting and cleaning, has aimed at optimizing the delivery of power to the target surface, and then to best use that power in order to remove material. The second approach has been to improve the performance of the system by adding different components to a plain waterjet operation.

10.2 ENHANCING THE DELIVERED JET POWER

10.2.1 THE USE OF A SHEATH AROUND A JET

One of the earliest attempts to this improvement was to seek, by sheathing the jet in an external stream, to improve the distance over which the jet would retain an effective pressure. This practice has been found particularly effective in improving the performance of waterjets which have been used in a submerged condition. One of the first examples of this was carried out in Japan, in improving jet performance in creating a grout curtain [10.1]. The addition of an air shroud around the jet significantly improved the throw of the stream in the submerged cavity.

The air sheath was created by setting the original, 2 mm diameter waterjet nozzle within a second nozzle assembly (Fig. 10.1). This outer holder was designed to leave a 1 mm gap between the inner jet nozzle and the outer air nozzle. The air delivered through this gap would thus form a high speed annulus around the waterjet (operated at a pressure of 700 bar) and help it remain coherent as it passed through the surrounding water.

Where the range over which a waterjet is effective in air is normally on the order of 100 nozzle diameters (see for example the discussion around Fig. 2.25), this drops significantly where the jet is placed under water, and more markedly as the pressure of the surrounding water increases. Thus, a selection of the results from the tests which Yahiro and Yoshida carried out to measure the range of the cutting jet, as a function of surrounding fluid back pressure, emphasizes the gain from the use of the air shroud (Fig. 10.2).

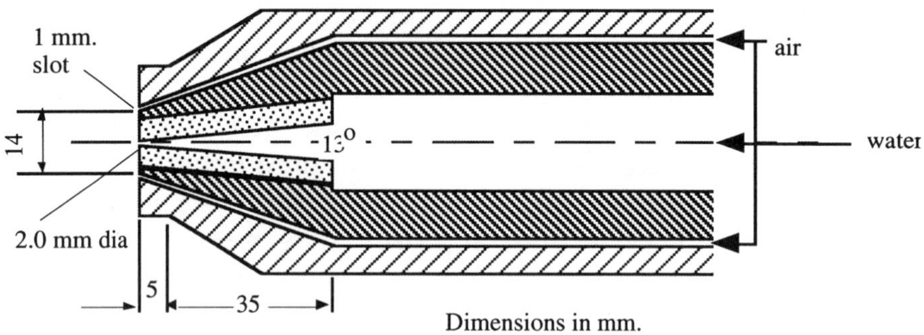

Figure 10.1 Air sheathing nozzle design (after [10.1]).

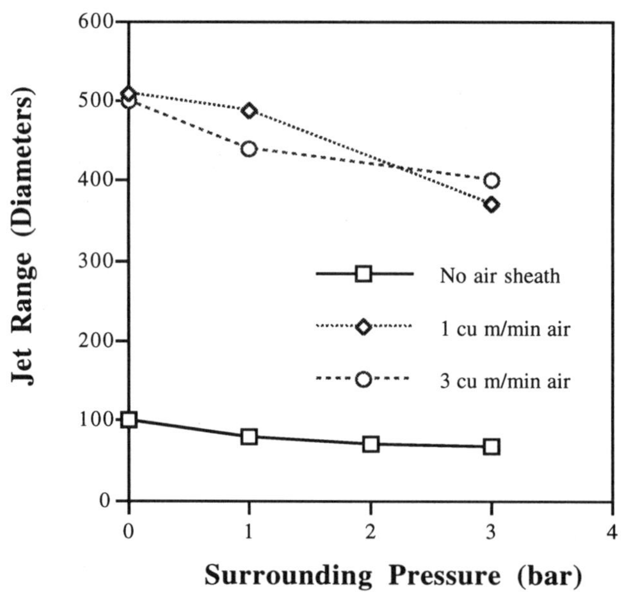

Figure 10.2 Effect of fluid pressure on the range of a waterjet under water, with and without an air shroud. Jet pressure 200 - 250 bar, diameter 2.0 mm (after [10.1]).

The reported results showed sufficient promise that the technique has been adopted by the Japanese in their development of curtain and column grouting. It has also been adopted and used successfully in several countries in Europe [10.2]. Baumann has noted that the air sheath provides better penetration and control of the jet grouting process, allowing a greater spacing between adjacent injection points. In contrast, however, he also notes that the technique requires more expensive drill rods and injection equipment and that the effect of the air are less clear or perhaps needed in soft and loose soils where larger holes might also be less desirable since the ground is less stable.

It is important to note, however, that the air shroud generated should be properly formed and tuned to the waterjet being protected. Eddingfield and Albrecht [10.3] showed, in some of their early work in this field, that where the air velocity falls below that of the waterjet and where the shape is not effective, that the shroud can reduce the effective length of the waterjet, rather than enhancing it.

Although air has been the most common fluid used in forming the external sheath, water has been used in cases where the intent has been to carry a cavitating jet stream out to greater distances, when the stream is operating in air [10.4]. In this case, (Fig. 10.3), the inner, high speed waterjet will induce cavitation in the boundary layer between it and the surrounding slower speed jet.

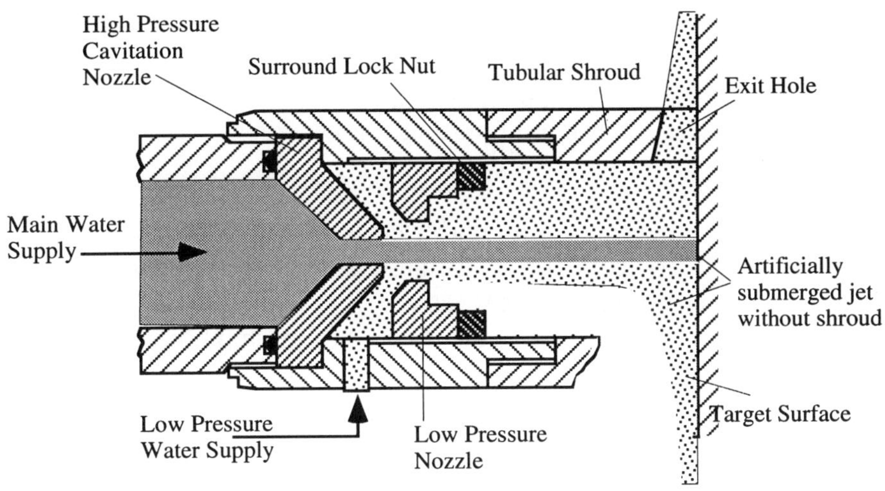

Figure 10.3 Water surrounding sheath on a waterjet for sustained cavitation [10.4].

The surrounding tube then ensures that the cavitation conditions are continued until the jet hits the target surface. The relative benefits of inducing this cavitation and sustaining it can be seen by comparing the effectiveness of a 200 bar, 3.2 mm diameter jet in cutting into aluminum where the jet is in air, under water, and then surrounded by a jet stream at a pressure of 2 bar (Fig. 10.4). A significant increase in cutting efficiency is achieved, and the distance over which the jet is capable of cutting is also increased.

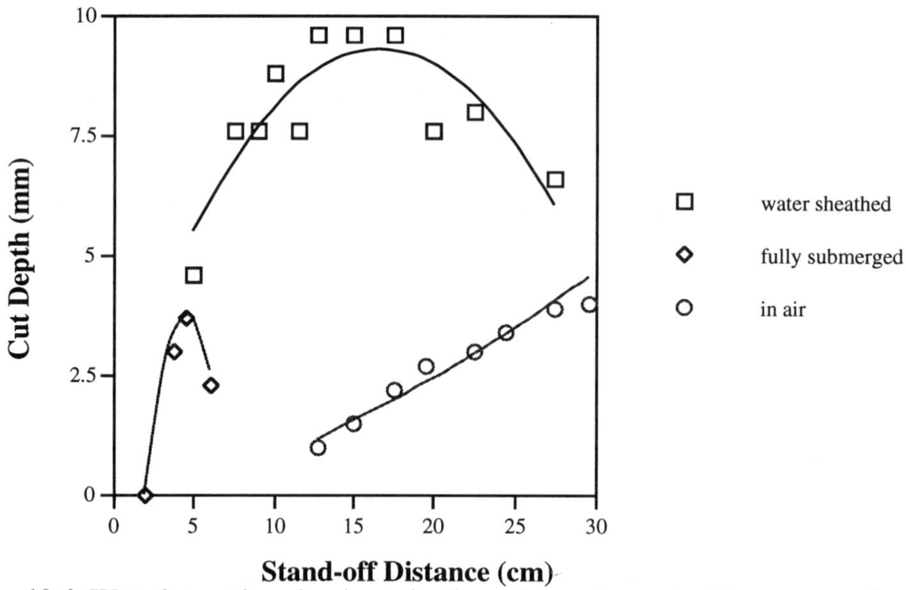

Figure 10.4 Waterjet cutting aluminum in air, under water, and with a surrounding water sheath. Jet pressure 200 bar, test time 5 minutes, surround pressure 2 bar, 3.2 mm diameter jet [10.4].

As with the air sheath of the waterjet it is critical to ensure that the resulting water flow is properly collimated and at the correct pressure and flow rate for the technique to be effective.

All these systems have, however, been somewhat complex to operate and maintain. Given that system simplicity has been one of the selling points of the technology, and that unrestrained currents of rapidly moving air are a potential hazard in surface operations, these techniques have not caught on, as yet.

10.2.2 WATER CHEMISTRY CHANGES

Investigators have also looked at changes in the chemistry of the water or a change in fluid entirely as a means of improving performance. Brunton and Rochester [10.5] reviewed several earlier studies in which it was separately reported that while water was found to be more erosive than either glycerol or paraffin [10.6] it was less so than brine [10.7]. They also noted that when Hancox and Brunton used carbon tetrachloride rather than water [10.8] erosion rates were more than doubled. This was considered to be due to the greater density (1.69 gm/cc) of the attacking fluid.

In these, and subsequent investigations [10.9, 10.10], jets made up from various fluids have been tested to find out if they would be practical or beneficial to use instead of plain water. The work generally has been directed both at determining the effect of change in fluid properties on stability of the fluid, and on the resulting jet ability to penetrate the target more effectively than water. Certain chemicals, for example, are known to be able to weaken the strength of rock, and thus, improve cuttability [10.11]. Of these studies, most have been carried out with relatively low volumes of material because of the potential difficulties in pumping chemicals of different types through high pressure systems into the atmosphere. Thus, for example, Rochester and Brunton [10.9] fired small slugs of different fluids at target surfaces, and evaluated jet performance on the basis of the craters generated.

The results showed, as a trend, (Fig. 10.5) that performance improvement could be achieved by increasing jet density, and would be reduced as fluid surface tension decreased. Although these results have some relevance to waterjet cutting, they should not be, in and of themselves, considered absolute criteria for evaluating the cutting capabilities of the jets.

To explain this caveat, one must consider, for example, the use of vegetable oils as a fluid for cutting confectionery and other foods containing a high concentration of sugar [10.12]. Oil is the preferred fluid in this application rather than water, since the presence of any humidity on the cut surface will dissolve the sugar and thus destroy the structure of the piece being cut. Given that restriction, confectionery manufacturers have switched to the use of an oil as the cutting fluid. While this provided a relatively high quality cutting ability and improved manufacturing performance, one unanticipated problem arose due to the increased viscosity of the fluid being pumped.

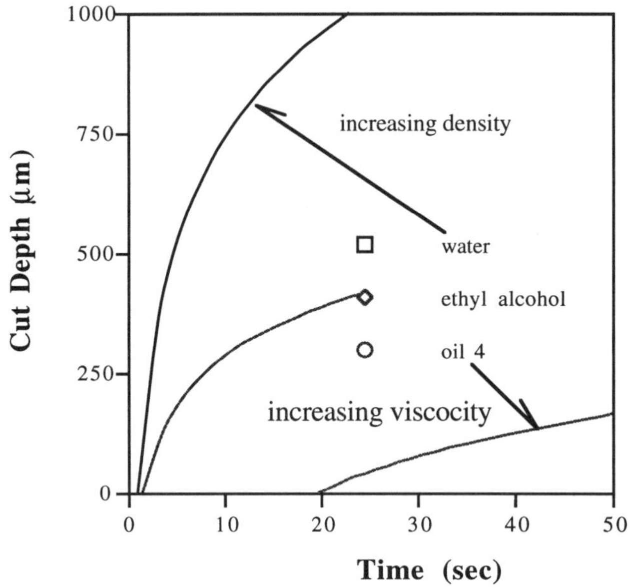

Figure 10.5 Variation in penetration with time for fluids of different properties (after [10.9]).

This change in fluid properties was not evident at the nozzle, in the jet structure or in the jet cutting ability, rather, it was a problem because of the greater resistance generated in pumping the fluid from the pump to the nozzle. This gave rise to a greater pressure drop between the pump and the nozzle, and caused much greater fluctuations in pressure at the pump when it cycled, over that normally encountered. The result was an enhanced fatigue load on the pump pistons and a much more rapid failure of the system components. After a series of tests to try and find a better cutting oil, the problem was resolved by moving the pump much closer to the cutting nozzles. However, the germane point in this case is that the change in fluid properties had their most dramatic consequence before the fluid got to the nozzle rather than after the jet left the nozzle and came down onto the target surface.

The change in fluid properties must also be considered in several different aspects. Firstly, as one improves the cohesion or energy transport condition of the jet then a greater amount of the jet energy available at the nozzle is delivered to the target. Secondly, as one increases the density of the fluid then, for an equivalent velocity, greater energy is delivered to the surface. Once at the surface the jet will remove material through its ability to penetrate the cracks in the surface and then, under subsequent pressurization, to grow those cracks. This can be achieved largely by mechanical action, but can be enhanced by chemical reactions also. Thus, for example Hancox and Brunton reported [10.8] that changing from plain water to a 3.5% NaCl brine solution enhanced the cutting of steel by 1.8 times but had no effect on cutting copper. Further raising the temperature of the water increased the shear damage on a surface, but not the number of impacts to induce initial damage. This would be consistent with the increased ability of the hotter water to penetrate into the cracked surface, and thus, exploit the damaged structure to a greater extent. It would also comply with Rochester and Brunton's data cited above, although they conclude that the result is due, not to an increased difficulty in penetrating the material, but rather because of the difficulty of the subsequent impacting jet in reaching the surface through the thicker layer of fluid lying on the surface [10.5]. The benefits of heating the water has also been discussed in Chapter 3 (see for example Fig. 3.22).

10.2.3 THE INTRODUCTION OF LONG CHAIN POLYMERS

Interestingly it was a consideration of the reduction in friction losses along a delivery line, that first led to the modern day introduction of long chain polymers into high pressure waterjet systems. In fighting fires it is necessary to direct as much water as possible to the area that is burning, in order to extinguish the fire. This water has historically been delivered to the fire through a nozzle, itself fed through a hose which can be of varying diameter. In the mid 1960s, a series of articles were published which indicated that the addition of small amounts of a long chain polymer to the water would significantly reduce the friction forces on the water traveling through small diameter hoses [10.13]. An early newspaper article [10.14] indicated that an equivalent amount of water could be delivered through a 2.5 cm diameter hose where a friction reducing agent was added to the water as would otherwise be delivered through a 5 cm line without the additive. Howells, [10.15], has since quantified the amount of drag reduction in regard to the concentration of one such long chain polymer, marketed under the trade name SUPER-WATER® (Table 10.1).

The change in the diameter, and thus, the weight of the hose required, would have a significant effect on the ability of firemen to fight a blaze and had other practical advantages. Given the delivery problems associated with the transfer of energy from the pump to the nozzle referred to, for example in Chapter 1, it was considered that the addition of a long chain polymer to the water could have a potential benefit.

Table 10.1 Drag reduction capabilities of a long chain polymer [10.15]

SUPER-WATER *concentration by weight (ppm)*	*Percentage* *drag reduction*
9	20
15	37
21	50
30	63
60	65

In commercial applications a 0.3% aqueous solution is usually used, which has a drag reduction of approximately 50%.

Initial experiments were carried out [10.16] in which a concentration of 0.1% of the long chain polymer polyethylene oxide, (marketed under the trade name Polyox) was added to the water before it went through the pump. The fluid performance was then compared with that of conventional water. It was found that the penetration of the jet was significantly improved (Fig. 10.6). Measurement of the flow through the system indicated that, in the pressure range from 150 to 700 bar, that the jet velocity appeared to increase by 10 - 15 m/sec with the addition of the polymer, but this was insufficient to explain the enhanced performance. Of perhaps more interest, it was noted that the improvement in performance appeared to get better with increased distance from the nozzle (Fig. 10.7).

The improved cutting performance was initially explained by the reduced pressure drop between the pump and the nozzle. This would thus deliver, for the same pump output, a greater jet pressure at the nozzle. And that improved jet pressure would allow a greater penetration of the target.

Interestingly, however, when a three dimensional plot is made of the data (Fig. 10.8) it can be seen that the polymer is of greater benefit to the jet cutting at the lower jet pressures, but at the greater stand-off distances from the nozzle. The polymer is less effective closer to the nozzle or where the jet is at higher pressure, under the conditions of test.

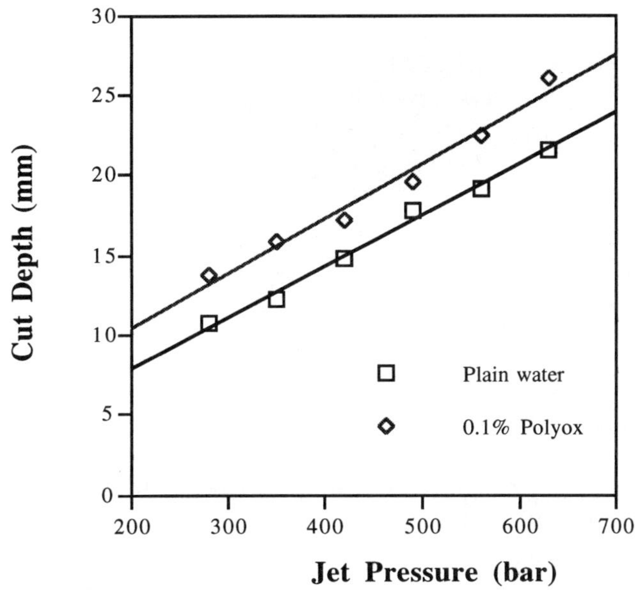

Figure 10.6 Improved cutting with the use of a long chain polymer (after [10.16]).

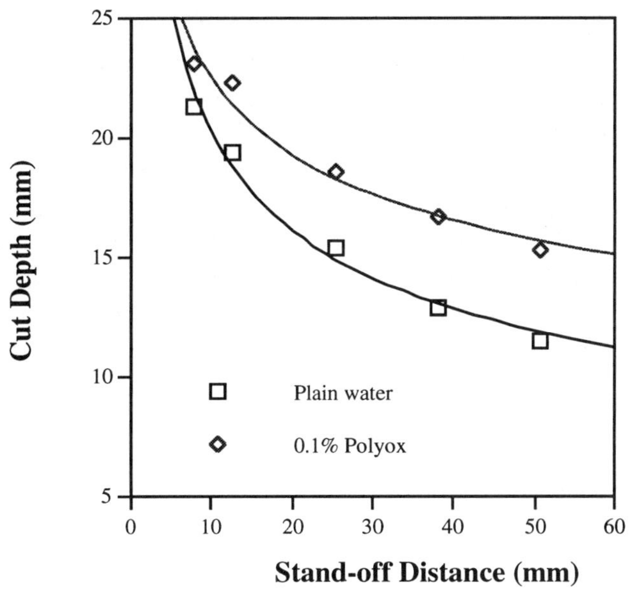

Figure 10.7 The effect of stand-off distance on cutting ability with plain and polymer enhanced water (after [10.16]).

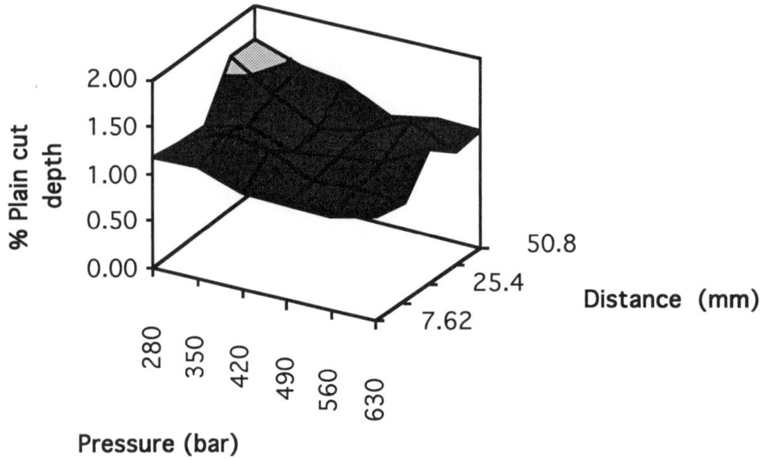

Figure 10.8 Improvement in jet cutting with polymer as a function of jet pressure and stand-off distance (after [10.16]).

The increase in jet pressure will reduce the effectiveness of a jet at increasing stand-off distance because of the more rapid jet attenuation which occurs at higher jet velocities. The improvement in jet performance with increasing stand-off distance seemed to indicate that the addition of the long chain polymer improved the cohesion of the jet as it moved away from the nozzle. This has proven to be the case and where jets are observed which contain sufficient concentrations of the long chain polymer a significant reduction in the dispersion of the jet has been demonstrated (Fig. 10.9). This does not explain the greater gain in performance at the lower pressure and greater distance. It is currently considered that this gain is due to the more significant enhancement of the jet power at a point where the jet is otherwise close to the critical pressure at which it starts cutting. Thus, any gain will be somewhat more dramatic than would be the case where the cutting regime is more established.

Franz was the first to apply this improved fluid in a commercial cutting evaluation, examining the gain to be achieved in jet performance when cutting wood products [10.10]. After examining the improvement in performance achieved by using gelatin and glycerin as different additives to water he found (Fig. 10.10) the addition of long chain polymer significantly improved the depth of cut over the alternate solutions, and gave depth of cutting improvements up to 300% over water alone. Where the polymer was then used to cut material a significant reduction in the wetting characteristic of the fluid when cutting corrugated board was identified by Szymani [10.18].

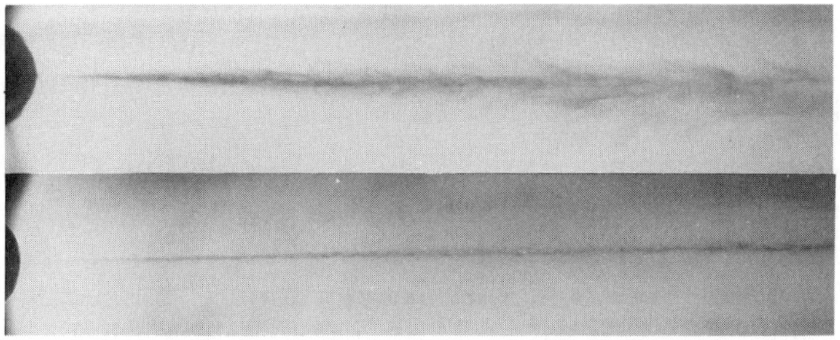

Figure 10.9 Jet diameter as a function of stand-off distance for a plain and polymer enhanced waterjet at 2,100 bar, through a 0.5 mm orifice, the length of the photograph is approximately 20 cm [10.17].

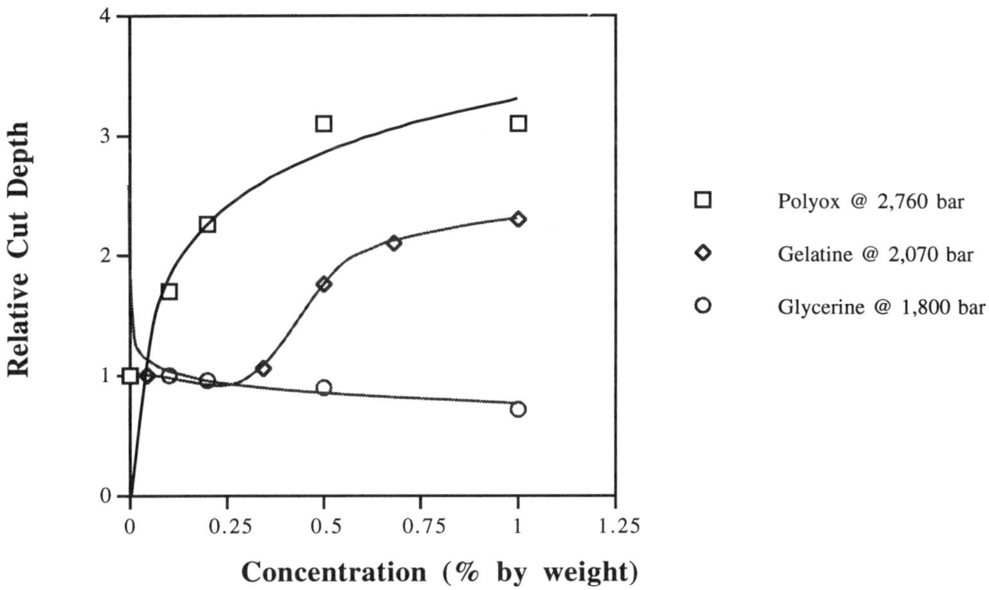

Figure 10.10 Ratio of cut depths with increasing concentrations of fluid additives. It should be noted that the concentration of gelatin is 3 times that of the Polyox, and that of the glycerin is 100 times that of the Polyox [10.10].

The improved cohesion of the polymer enhanced water not only holds the jet better moving from the nozzle to the target surface, but also reduces the amount of water that would be retained in the target surface once the cut has been made. Both of these factors were significant in the initial application of high pressure waterjets to the cutting industry, and it has been reported, particularly in Europe, that a significant number of installations continue to use polymers of different types and concentrations in order to achieve these benefits in high pressure industrial cutting.

The improved performance of the polymer in cohering the waterjet can be discerned by examining the pressure profile generated by jets containing polymer at different distances from the nozzle. Such a study [10.17] provides a better means of accurately determining the actual levels of improvement, rather than relying on photographs, although these can be of some benefit. (For photographs to be effective, however, they should be back-lit photographs taken with an exposure time in the microsecond range). When pressure profiles were taken through jets of varying pressure, and at different polymer concentrations, significant improvements in performance were found with polymer addition, particularly at greater distance.

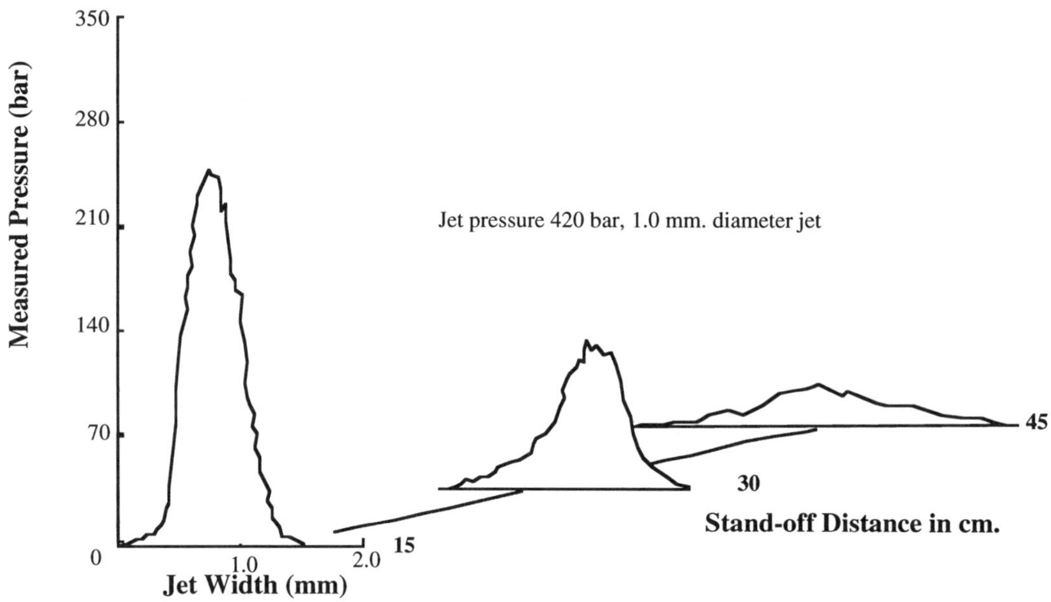

Figure 10.11 Pressure profiles through a 420 bar jet of 1 mm diameter (after [10.17]).

For example, a series of pressure profiles of a jet without polymer (Fig. 10.11) were taken along the jet issuing at 420 bar through a 1 mm diameter nozzle at distances of 15, 30, and 45 cm. By the 45 cm stand-off distance (450 nozzle diameters) the jet pressure had fallen to approximately 4% of that of the original jet. In contrast, where a 500 ppm concentration of polymer was added to the jet at the same jet pressure and nozzle diameter, the jet pressure at 45 cm has been improved by some 500% to approximately 20% of that at the nozzle. Significant jet pressure was now also detected at a distance of 60 cm from the nozzle (Fig. 10.12).

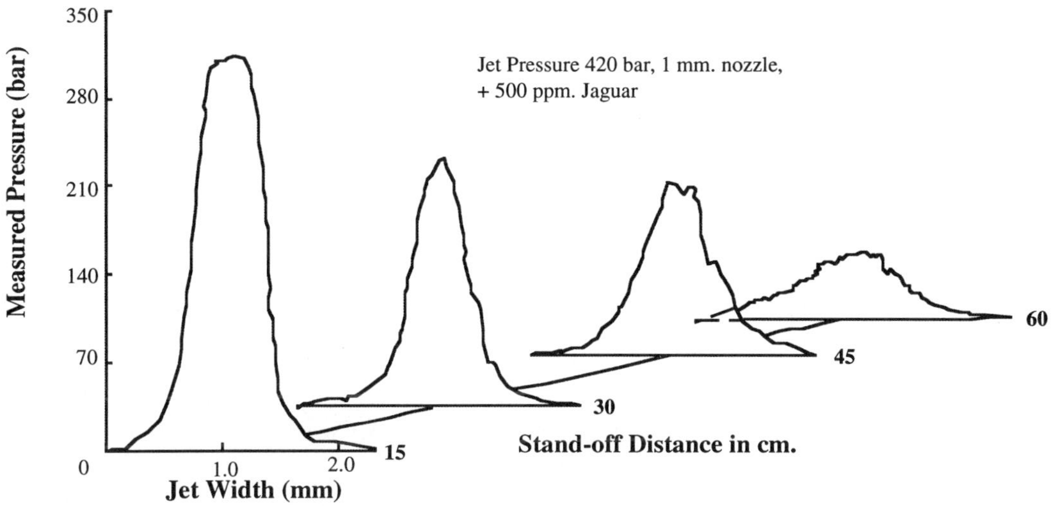

Figure 10.12 Pressure profiles along a jet similar to that of Fig. 10.11, but with the addition of 500 ppm of the polymer Jaguar [10.17].

Although polymeric additives have been found to be extremely effective in their application to waterjets of small diameter and very large pressure, where cutting occurs relatively close to the nozzle, similar improvements in performance have not been found by investigators working with larger nozzle diameters. One reason for this perhaps is the relatively smaller reduction in power loss which occurs with larger nozzles. For example, relative to the pressure profiles taken earlier (Fig. 10.9) consider the effect of a waterjet at the same pressure of 420 bar but with a 1.5 mm nozzle diameter (Fig. 10.13). Even though the jet is operated at a smaller delivery pressure, at a distance of 45 cm from the nozzle (300 nozzle diameters) the jet still retained approximately 75% of the delivery pressure. Relatively less gain in performance could be achievable by the polymer since there has been significantly less performance loss. Further, at a stand-off distance of 75 cm which is the equivalent to 500 diameters, the jet still retains 33% of its original pressure. The pressure decay with distance is obviously slower at the higher diameter.

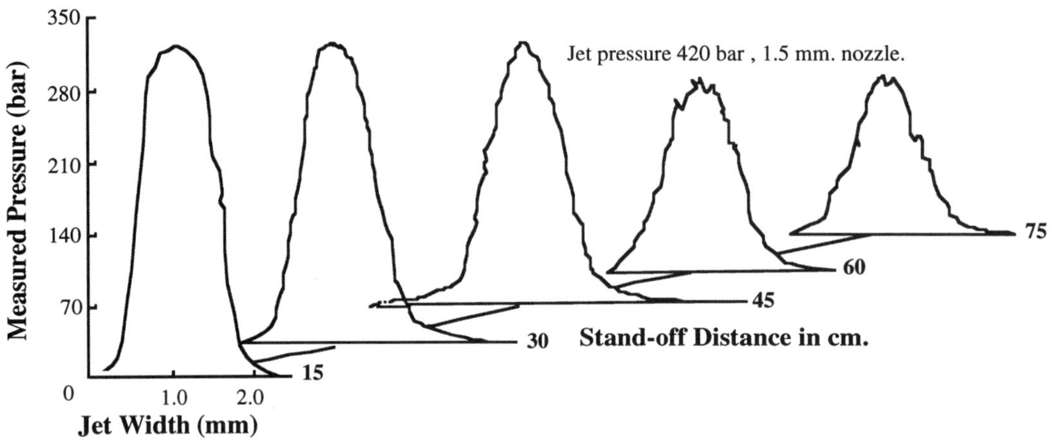

Figure 10.13 Pressure profiles through a 1.5 mm diameter jet, at a pressure of 420 bar [10.17].

The improvement in performance when polymers are used does, however, extend up to significantly larger nozzle sizes. For example, a Bureau of Mines study [10.19] examined the decay in jet pressure with distance for two nozzle sizes and then measured the increase in force, at various stand-off distances with several polymers, at different concentrations. It is interesting to note that while the two jets appear to decay at the same rate when the distance is plotted in absolute measure (Fig. 10.14), the larger jet decays significantly more rapidly when the data is plotted in terms of nozzle diameter (Fig. 10.15). The most improvement occurred with the product Separan-AP-273 (Fig. 10.16), with the optimum concentration occurring at between 750 and 1,000 ppm.

It should also be pointed out that polymers appear to work more effectively in those conditions where flow through the nozzle has not been optimized. This is particularly true with poor fluid paths in the line to the nozzle, where there are bends in the entrance to the nozzle, where surface finish has not been optimized, and under similar, often considered to be "normal", operational conditions. Under these circumstances, the improvement in jet cutting found by adding polymers is usually considerably greater than that found where optimum fluid flow conditions prevail.

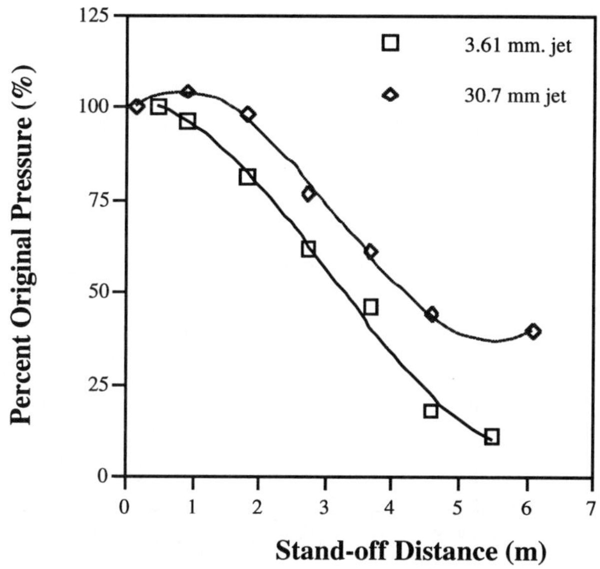

Figure 10.14 Jet pressure as a function of absolute stand-off distance for two nozzles [10.19].

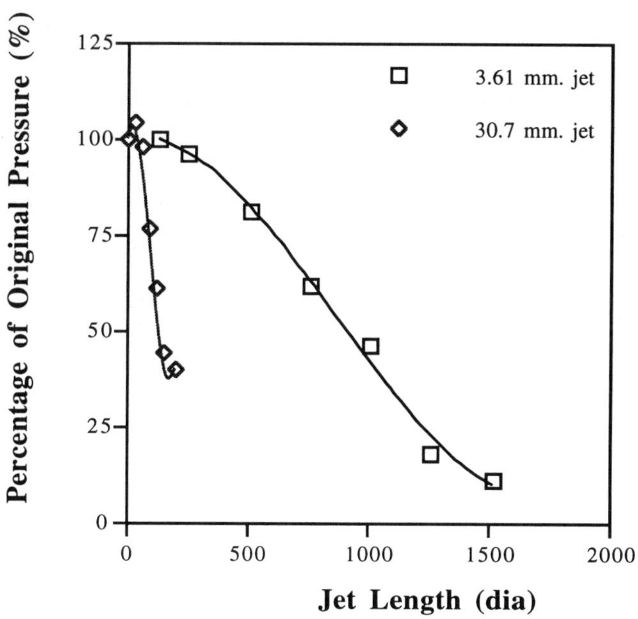

Figure 10.15 Jet pressure as a function of relative stand-off distance for two nozzles [10.19].

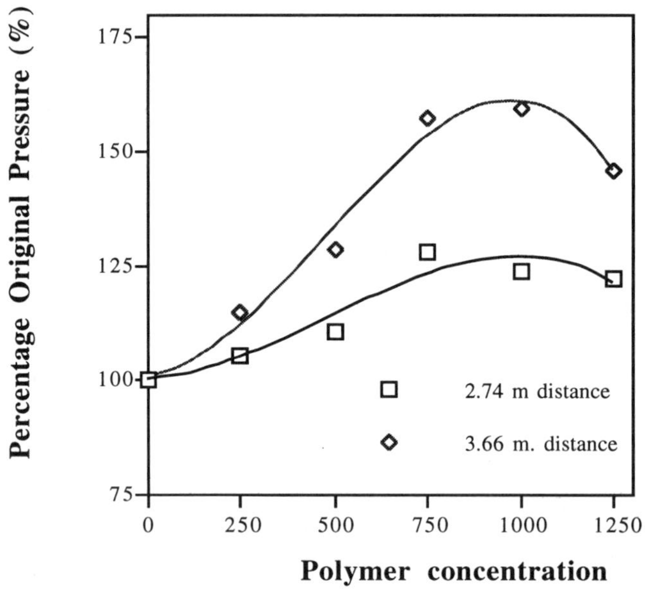

Figure 10.16 Improvement in jet pressure for the smaller jet as a function of polymer concentration at two distances [10.19].

It is for this reason that it is important in any potential application of waterjetting to consider the economic benefits which might arise from the use of long chain polymers in the application. A cost:benefit ratio and experiments to determine the relative effectiveness of the improvement the polymer can bring, should be carried out. Particular consideration should be given to potential changes in the environment. The polymer will improve the cohesion of the water, reducing splash back. It has been demonstrated to have some improved performance capabilities in the addition of abrasive to waterjets, albeit, this requires special consideration be given to the mixing of the abrasive into the water. This subject will be discussed further in the section on abrasive use later in this chapter.

The polymer may, under certain circumstances because of its structure, create more difficult operating conditions if it is allowed to fall on the floor in and around the working area since it induces some slipperiness to the water and can reduce the friction coefficient of the floor on which the operator must work. It may be necessary to install a temporary working surface (i.e., wooden pallet floor), or to find a way of overcoming this problem in conditions such as industrial cleaning operations where the distribution of polymerized water across the floor might increase the possibility of operators slipping and thereby having accidents.

The chemical mechanisms by which polymers enhance the performance of a waterjet system have been studied extensively over the years, with a summary of this knowledge published, inter alia, by Howells [10.20, 10.15]. However, the experimental application of this tool has not been as clear. The improvement in jet structure and performance when polymers have been added to the fluid, has been somewhat controversial since their initial introduction. Part of the problem with this, particularly at lower pressures, comes both with the nature of the polymer being used, the method by which high pressure waterjets are generated and the means by which the pressure is controlled.

Unpublished experiments, for example, at the University of Missouri-Rolla, showed that the initial polymer used for many experiments (polymerized ethylene oxide) was not only difficult to mix into the water but also exhibited an aging characteristic. This meant that while significant improvement in performance could be achieved if a solution was used fresh, if the solution was allowed to stand for a period of time then the effectiveness of the polymer began to decrease. Concurrently, in many waterjet operations, the pressure at the nozzle is controlled by adjusting a bypass valve between the pump and the nozzle. This bypass line returns part of the fluid from the pump to the supply reservoir. In this way, pressure can be controlled at the required level for different nozzle diameters. The problem that this gives, in evaluating the performance of polymers, is that some of the polymer in the water reservoir is sent through the pump and subjected to the high shearing actions generated as it passes through the pistons. It is then returned to the tank without issuing through the nozzle. Repeated cycling of this nature degrades the effectiveness of the polymer significantly. Thus, evaluation of polymer performance, which might be reported on quite negatively, may be related to the method of mixing and the treatment of the water prior to its use.

To properly mix the polymer polyethylene oxide, for example, the practice was established of suspending the small beads of polymer in a suspension of isopropyl alcohol. The container was then agitated to give a relatively even distribution of particles in the fluid, which was then dumped into the vortex created by a mixing paddle turning within the main water tank. Other mechanisms used have been to vibrate small controlled quantities of the polymer on a bead-by-bead basis into the water supply line, or into the water in the main supply tank, while a mixer was operating. Many of these methods do not guarantee an even distribution of polymer in the water. The thickness and cohesiveness of the partially hydrogenated polymer is such that it will agglomerate into large lumps if not rapidly dispersed, making it more difficult to achieve an even concentration of polymer in the resulting fluid as it is fed to the nozzle.

Because of these problems, the arrival of the liquid polymers, such as the products from Nalco Chemical, and SUPER-WATER[R] was welcome, since they made it much easier to evenly meter polymer into the water and to ensure an adequate dispersion and mixing. The use of specially designed mixing systems [10.21] to totally disperse the polymer within the water flow also made it easier to insure the quality of the product and to more effectively and consistently demonstrate its improved effectiveness.

It is important in evaluating the improvement in performance achieved with a polymer, to insure that a proper mixing of the polymer into the water has occurred and that the resulting fluid was not subject to high shear loading rates before it was fed to the pump and then to the cutting nozzle.

It is important to ensure that the polymer has been allowed sufficient time to properly hydrogenate in the supply tank, normally a process taking only a few minutes, before the fluid is delivered to the cutting nozzle. In this way, optimum performance can be achieved by the polymer, not only reducing friction from the pump to the nozzle, but also in cohering the water from that point in order to enhance jet cutting performance.

When the polymer is properly mixed in this way the improvement in cutting performance where polymers have been added to the operation can, in the right circumstances, be quite significant. Howells [10.15], has given several examples of significant performance improvement.

For example, in the cleaning of the vertical evaporators at the Fiber Board Company's Antioch Plant two vertical evaporators containing 700 stainless steel tubes each, were cleaned by use of a polymerized waterjet (for which the term **polymerblasting** has been evolved) over a period of four days and at a total cost of $13,500. This application took place in 1984 as did the cleaning of the heat exchanger tube bundles at the Hydroprocessing Reactor Exchanger at the Chevron Refinery in Richland, CA., which similarly had 1500 U-shaped tubes to be cleaned. The use of polymer blasting allowed the cleaning of these tubes within a 24 hour period. A time savings of between 30 - 35% can be anticipated where the use of polymer in the water is used in contrast to conventional high pressure cleaning. This is potentially due to the increased delivery pressure which is achievable with the system for the same pump operating pressure with the enhanced power being delivered not just at the nozzle, but also by reduction in transmission of power from the nozzle to the target surface.

Howells provided cost examples for the use of the SUPER-WATER[R] polymer which, at the time of the article cost $8.58/liter and of which approximately 76 liters were required to clean both shell and tube side of a heavily fouled 1,000 tube heat exchanger bundle. Where polymer was used at a 3,000 ppm or 0.3% by weight concentration in the water, this translated into a polymer cost of approximately $0.024/liter. The heat exchanger could be cleaned with a 3 - 4 hour savings in time, with an additional cost for the polymer of between $275 and $300. Given a contractor cost of $82.00/hour, the time saving alone would cover the cost of the polymer. However, this gain is not always achievable and currently it is reported that only approximately 10% of the industrial cleaning jobs at one company were being carried out utilizing the polymer.

Howells also points out one factor which is likely to become of increasing importance in future operations. Reference has already been made to the reduced wetting which a polymerized waterjet will inflict on the target surface. At the same time, because of the increased effectiveness of the jet, it is possible to achieve equivalent cutting conditions with a reduced volume of water over that required where no polymer is present. As environmental concerns become more prevalent, and particularly in the cleaning of potentially hazardous materials from target surfaces, the resulting fluid in and of itself becomes a disposal problem. Minimizing the volume of this fluid by increasing the concentration of polymer in the fluid may, therefore, have a subsidiary cost benefit over and above that achieved by the polymer in the direct cleaning operations.

The improvements achieved in cleaning can be significantly greater in an underwater application than in surface cutting. This is because of the higher energy dispersion rate which occurs when the waterjet seeks to deliver power in the submerged condition. Zublin has shown (Fig. 10.17) the significant improvement which can be achieved where the polymer SUPER-WATER[R] is used in underwater cleaning [10.22].

The reduction in wetting where the long chain polymer is added to the water, can sometimes also have unforeseen benefits. Howells reports on an application to remove adhesive attached cork from a concrete ceiling. Without the addition of the polymer the cork was absorbing all the water, and its structure was collapsing under the weight. With a polymerized jet, clean cuts could be made with little wetting of the substrate.

This reduction in the wetting problem is potentially of considerable significance in mining applications. There are many mines where the use of water must be minimized since it reacts with the rock in the floor to turn it into mud. By using a polymer in the water this can be mitigated. The potential advantages of this treatment of the water has already been established in oil well drilling. The addition of polymers to the drilling mud has inhibited the wetting and interaction with shales found in the Gulf region of the United States [10.23a]. Under normal conditions where a water-based mud is used in drilling through this rock it turns into a low grade mud and continuously flows into the cavity, reducing borehole stability dramatically. Where, however, a polymer is added to the water, this inhibits the wetting action of the drilling fluid, so that it is possible to drill through the rock without any damage to the walls of the hole.

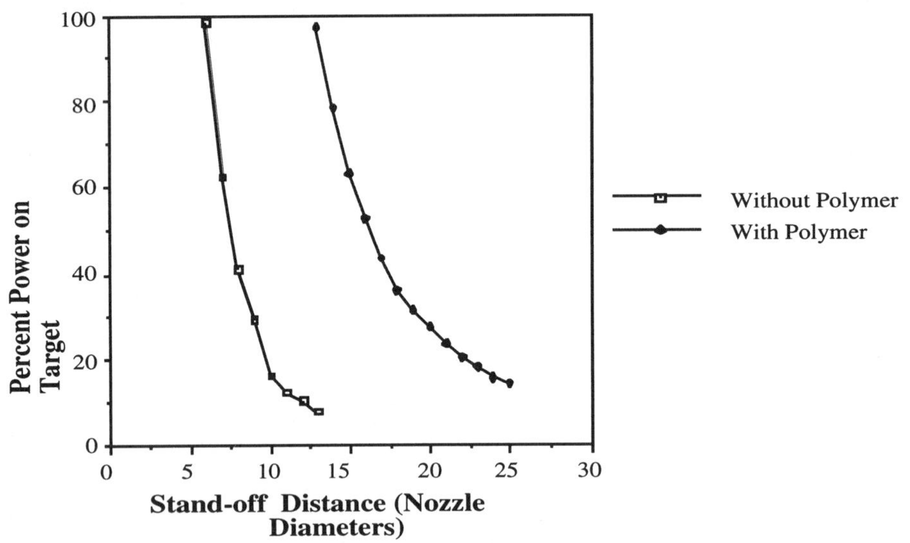

Figure 10.17 Benefits from the use of SUPER-WATER^{R} submerged (after [10.23]).

The potential advantages of polymer use in conventional waterjet cutting has led to an examination of their use in abrasive assisted jet cutting. The results which have been reported for this need to be properly understood to appreciate the value of this process.

In polymer use, one benefit from the addition is the water cohesion is enhanced. Conventional abrasive injection requires the abrasive must force its way into a coherent fast moving jet stream. While this is not normally difficult with plain water, the presence of polymer makes it more difficult to achieve. Results of tests where polymer is added to the water and used with conventional abrasive injection [10.24] have shown a poorer result than where no polymer was used.

However, where the polymer and the abrasive are mixed together before the water is concentrated into the narrow stream issuing through the first nozzle, and accelerated as in, for example the DIAjet system, then the results are significantly different (see Figs. 10.68 and 10.69). Hollinger has shown [10.25] that in this case the addition of the correct concentration of polymer can substantially increase the cutting ability of the jet, while maintaining the narrow kerf required for optimal cutting. Interestingly, however, his study went beyond the simple improvement in jet cohesion to the target in analyzing the benefits from the addition of the polymer.

Hollinger looked at the width of the cut made by an abrasive-laden jet traversing a 19 mm thick bar of aluminum. He deduced that, while the viscoelastic properties of the polymerized jet influenced the depth and width of the cut, the improved stability of the jet during impact held the jet together and led to the reduced kerf width. This he demonstrated by the narrower gage of the cut at lower pressures since, at higher pressures, the jet energy was sufficient to overcome the viscoelastic forces and cause the jet to spread, giving a wider cut. Greater polymer concentrations increased the viscoelastic forces and held the jet together, giving a narrower cut, to higher jet pressures. While a 2,070 bar methyl cellulose jet cut only half way through the sample at a speed of 159 mm/min a 1.3% solution of SUPER-WATER[R] cut through the bar at 2,070 bar, while at 3.9% solution the bar was cut through at 1,030 bar. This subject will be further discussed in the section dealing with abrasive use.

10.3 THE ACCUMULATION OF WATERJET PRESSURE

10.3.1 PULSATING WATERJETS

The use of long chain polymers acts largely to reduce the losses of energy the water undergoes as it moves from the pistons of the pump through the delivery line, is formed into a jet as it flows through the nozzle, and travels to the target surface. Investigators have also examined different ways of changing energy distribution within the jet in order to improve operating characteristics. Initial efforts to achieve this have mainly been directed along two different approaches. One has been to look at the arrival of the waterjet on the surface, where it can create much higher pressures than that of the steady state condition. The other is to bring two or more jets together in order to create a higher jet power from the combination.

The use of interrupted or pulsed jets as a means of improving jet cutting ability arose from initial studies of the early stages of water droplet impact. This work has been most thoroughly studied at the University of Cambridge, UK, with significant papers being presented by that group at successive Erosion by Liquid and Solid Impact Conferences and their predecessors the Rain Erosion Conferences [10.26].

Brunton and Rochester [10.5] identified Cook [10.27] as the first to include the compressibility of the water in the calculation of the impact pressure under a water droplet. He equated the impact pressure as being equal to the water hammer pressure as given by the equation

$$\text{water hammer pressure} = \rho c v$$

where c is the velocity of sound, a significantly higher value at lower jet stagnation pressures than that of jet impact velocity (v). One can see much higher pressures are created during impact than those of the jet stagnation pressure ($1/2\rho v^2$) which is developed during the later stages of impact.

The augmentation of pressure does not stop with that value, nor is it an even distribution of pressure over the surface. As the curved droplet surface hits the surface, the initial contact in the center, sends out a compression wave through the drop. This augments the impact velocity and hastens the downward movement of sequential rings of contact around the initial point. At the same time these rings have a shorter distance to travel to reach the surface than does the water at the initial point of contact, which is now trying to escape from the slug of water flowing down behind it (Figure 10.18). This confinement increases pressure on the contact surface and the fluid at the interface. This augmentation continues until the point where the curvature of the drop raises the travel distance of the next ring of fluid above that needed to allow the confined water to escape, which it does. The resulting lateral jet, presuming no significant target penetration, travels at a much higher velocity than the original impact. Fyall [10.28], reported measuring radial flow velocities some three times the initial 300 m/sec impact velocity of the droplets.

Brunton and Rochester [10.5] measured the pressure distribution under an impacting droplet and found that while the pressure under the center of the droplet was equal to that of the water hammer, a pressure value some 2.5 times this high was measured 0.1 diameters on either side of this point. The data reported was at an impact velocity of 100 m/sec. More recent work in this area has been reported by Lesser and Field [10.29].

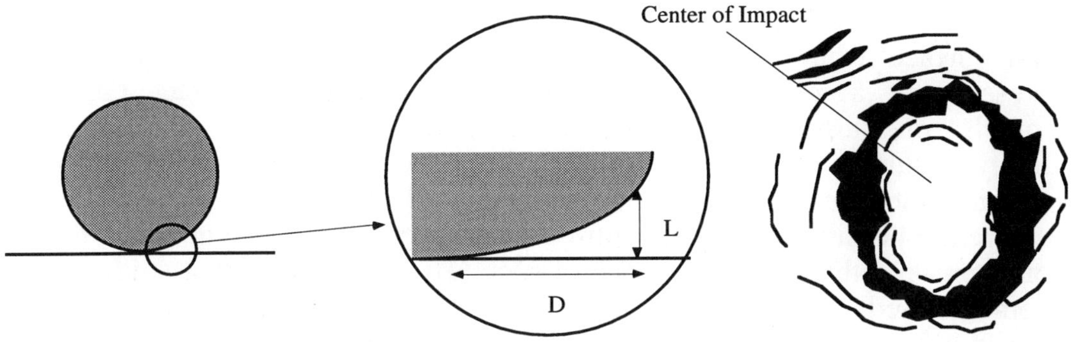

Figure 10.18 Impact of a droplet on a flat dry surface, including a sample of typical damage as a schematic.

This very high initial impact pressure can have a very significant effect, if it can be sustained, and it led to a series of investigations and developments of pulsed jet equipment. The concept on which they were based [10.30] relied only partially on this enhanced pressure, but also on the more rapid induction of failure from a rapid series of impacts, over a steady state condition. By pulsating a jet, using an internal rotating device within the nozzle, Nebeker was able to penetrate into a granite, with a uniaxial compressive strength of 3,800 bar, with a 1.5 mm diameter waterjet operated at 560 bar. The jet was interrupted at a frequency of around 10,000 Hz. For practicality, simple rotary devices were considered more effective in generating the pulsation [10.31] than alternate methods which have been suggested over the years. These alternative suggestions have included firing pulses of electricity across the nozzle to vaporize the jet at intervals. One practical consideration for those who might be tempted to try pulsating the nozzle by inserting a simple blocking device which alternately blocks and opens the passage from the supply line to the nozzle. While this technique induced a pulsation which can lead to water hammer impact pressures on the target, it can also send equal pressure pulses back up the supply line to the pump every time the nozzle blocks. This pulse will also generate a water hammer pressure and one such pulse, in this author's experience, was sufficient to blow both high pressure couplings on a hose and split the hose which intervened between the nozzle and the pump. In other cases it has caused significant damage to the cylinders on the pump.

Although there has been some subsequent development of this technique, [10.32] it has not been greatly successful. There are two possible reasons for this. Firstly, high initial impact pressures are generated on a relatively flat surface where there is relatively little fluid on that surface. The presence of a relatively thick layer of fluid will ameliorate the high impulse pressures and they may not be generated [10.33]. This will be likely as surface conditions change during penetration of any solid surface.

The other reason comes as a result of work at the University of Leeds, where the possibility of interrupting the waterjet was examined as a means of improving the jet performance of rock [10.16]. During that study it was demonstrated that pulsed waterjets were a more effective means of penetrating the rock than a steady continuous jet flow onto a single point on the surface (Fig. 10.19). The reason, however, for this occurring was not, as has generally been surmised, due to the water hammer effect of pressures induced by sequential impact on the surface.

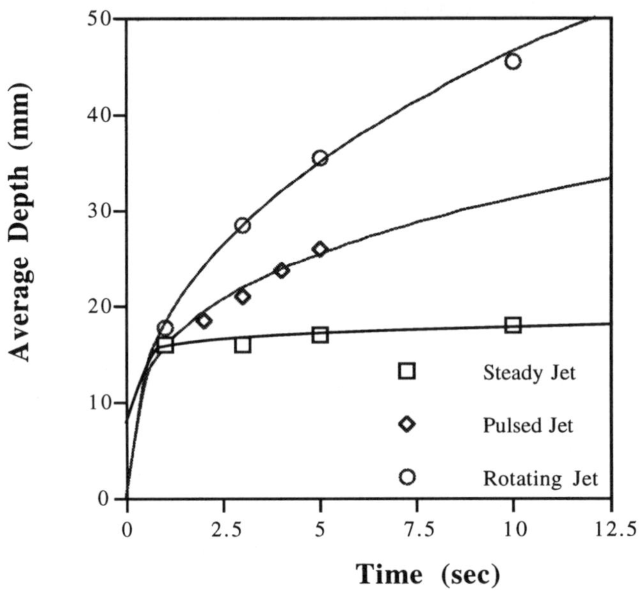

Figure 10.19 Penetration of sandstone by a steady, pulsed and rotating waterjet at 700 bar (after [10.34]).

The interruption of the jet in this case was designed initially for a different reason and worked for that same reason. When a waterjet comes down and impacts on a target material and penetrates into the surface, the cavity which is created is relatively small and does not exceed by any great amount the diameter of the incoming jet (Fig. 10.20). Thus, the fluid in the hole has relatively little option in the path which it must follow to exit the hole once its energy has been delivered. Unfortunately, in exiting the hole it must pass out along the same passageway as the incoming subsequent water slug. Interference is inevitable. Leach and Walker have shown [10.35] that the attenuation of the incoming water in a small hole can be such that the delivery pressure at the bottom of the hole will be completely attenuated within a penetration of only about five hole diameters. Under these circumstances, it was considered rational by the Leeds investigators to interrupt the jet flow before it struck the target surface and to allow the first slug of water to exit from the hole before the next slug was delivered.

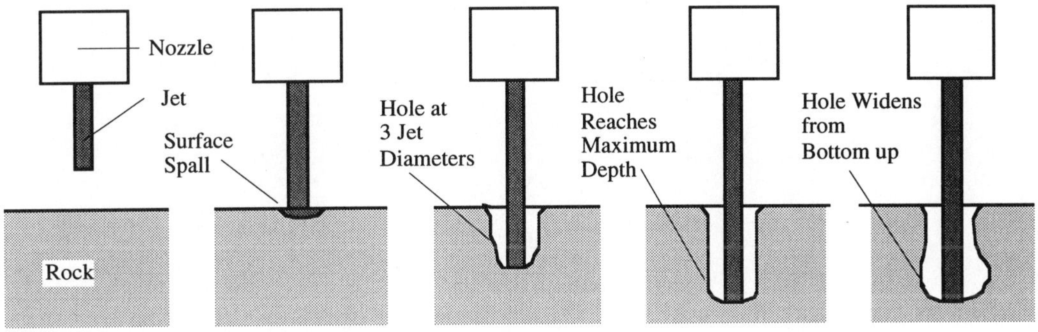

Figure 10.20 Developing geometry of a hole under steady waterjet impact.

Where this logic was applied by placing a simple interruption device between the nozzle and the target surface, (Fig. 10.21) significant and continuing improvement in performance of the jet could be achieved over the stagnation penetration which resulted where the jet was directed continuously into a flooded hole (Fig. 10. 19). Some care must be taken in selecting the length of slug, since the energy in the leading edge of the slug will be sufficient only to initiate or extend a crack into the target surface. If insufficient energy is contained in the subsequent length of water behind the leading edge, then the cracks will not fully be exploited and material will not be removed (the major reason for the reduced efficiency of the water canon type of system). It should, however, be remembered that the jet is moving at several hundred m/sec in most cases. Thus, even at relatively high interruption speeds the jet has a finite length. However, while plots of differing jet lengths show some trend toward an optimum value at around 0.8 m long (Fig. 10.22) this is not constant across the different jet velocities and becomes even more tenuous when the tests were repeated with polymer in the water (Fig. 10. 23).

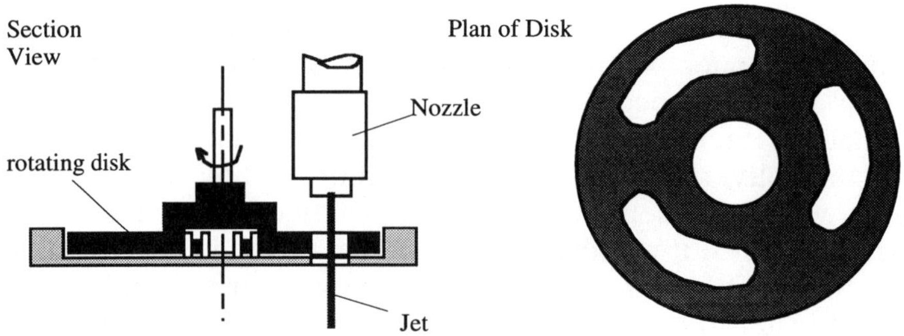

Figure 10.21 Early pulsating interruption device [10.16].

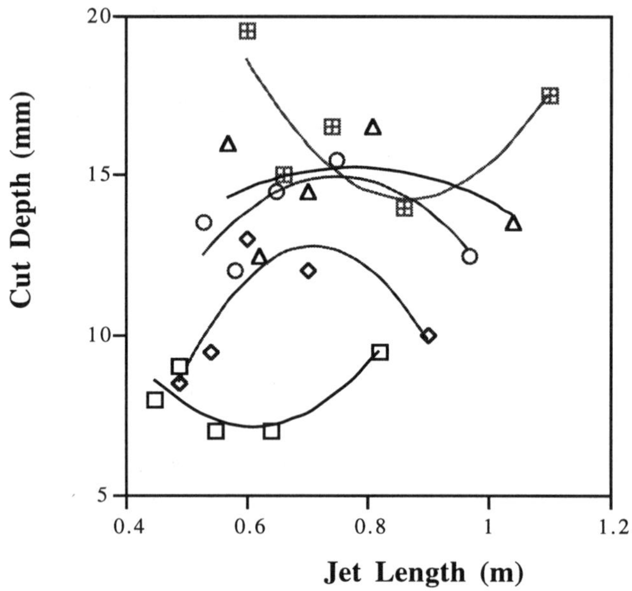

Figure 10.22 Effect of interrupted jet length on penetration depth for various pressures when impacting on sandstone [10.16].

For this reason it would appear that an interrupted waterjet system would be more effective than a single steady jet stream impact on a target surface. This only holds true where the nozzle is incapable of movement. Within the last 20 years operational systems have been developed which can allow the nozzle to move relatively freely, in continuation of the experiment described above, the nozzle was placed off-center to the axis of the required hole and rotated in this configuration (see Fig. 9.1). The water could escape along the area of the enlarged hole being drilled by the jet, rather than being restricted to the relatively narrow passage cut by the pulsed or steady jet. Under this circumstance, much faster material removal rate could be achieved and the loss in delivered energy dissipated by the interruption device would no longer occur (Fig. 10.19).

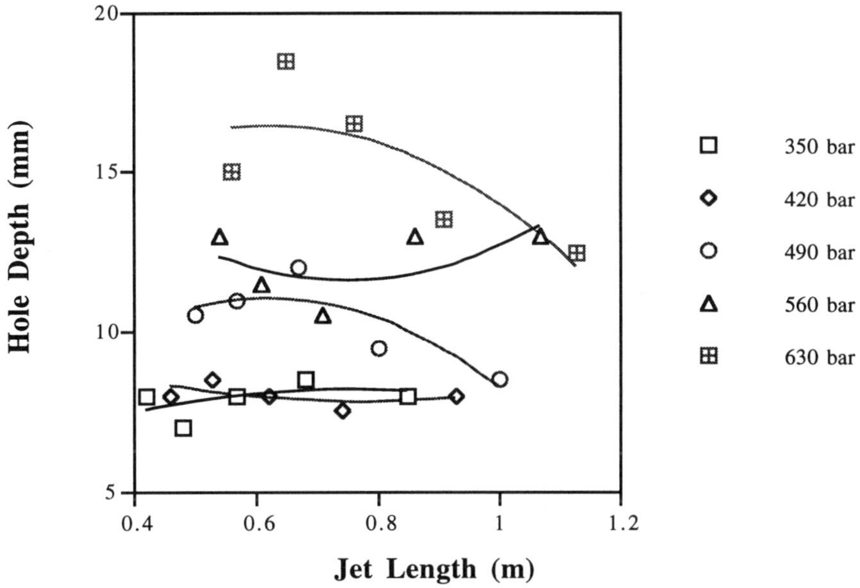

Figure 10.23 A similar test to that shown in Fig. 10.22, but with 0.1% Polyox added to the water in the jet [10.16].

10.3.2 SELF-RESONATING PULSATING JETS

An alternate use of the pulsating effect has been developed by Chahine and Conn [10.36] building on a tool which was first developed as an underwater means of generating cavitation. By running a jet flow through a specially designed nozzle (Fig. 10.24) it is possible to cause the resulting jet to pulsate as it leaves the nozzle. When this occurs underwater it creates ring vortices which can be used to accelerate the action of the jet, since these contain clouds of cavitating bubbles which subsequently collapse against the target surface. This has been used both for cleaning and for cutting applications [10.37, 10.38].

In a more recent application Conn [10.37] has adapted the technique for removing asbestos coatings from walls and pipes. Working with a system which might use up to 4 guns, each operating at 700 bar and a flow of 7.6 lpm through 0.79 mm diameter nozzles. The delivered pressure to the nozzle was designed to reach 560 bar, and this removed the asbestos with a favorable water ratio. That ratio is defined as the point where the asbestos absorbs the water used to remove it, and makes clean-up of the site a much simpler procedure. Cleaning rates of 23 to 28 sq m per person-hour have been reported for surfaces which, with manual cleaning, took up to seven times as long to clean.

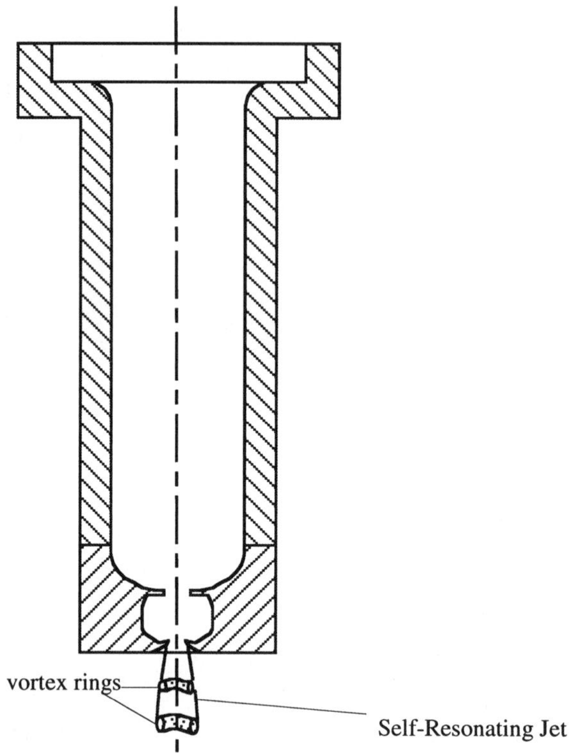

Figure 10.24 Self-resonating nozzle design (after [10.36]).

10.3.3 FOCUSING MULTIPLE JETS TO GENERATE HIGHER PRESSURES

Different applications of pulsed jets have been sought by investigators over the last 30 years. These not only seek to induce the water hammer phenomenon which can still be usefully applied where purely surface breakage is required such as, for example, in the removal of concrete by fracturing, but also because under certain circumstances the geometry of interrupted jets can be adjusted to give pulses of higher pressure.

The underlying idea that two converging jets will merge to form a slug of fluid moving at a higher velocity is not new. It underlies the design of the shaped charge military round, in which a cone of metal is collapsed on itself by a surrounding explosive charge. As the plane of the cone collapse moves forward, the metal liquefies and successive rings meet on the axis of the device (Fig. 10.25). This creates a high velocity slug of metal which leaves the cone at high velocity and has become a standard tool for metal cutting and for penetrating rock, under constraining circumstances [10.39]. The mathematics of the relationship have been well defined [10.40], but the method requires a high degree of precision in the manufacture of the cones for it to work well.

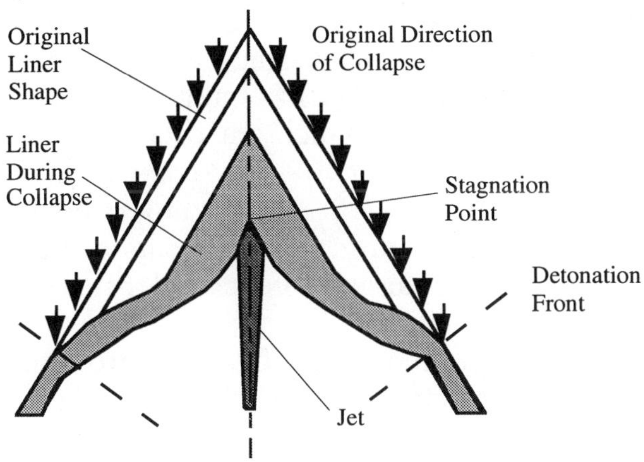

Figure 10.25 Generation of a high speed jet by cone collapse (after [10.39]).

The technique has been adopted for generating high pressure waterjets of ultra-high (above 6 km/sec) velocity. First developed in the UK [10.41], the technique has also been adopted at the University of Missouri-Rolla, for short term impact studies [10.42]. In order to form the cones with a water layer, rather than metal, a small amount of cerageenan gel is mixed with the water. This provides sufficient rigidity for the cone to be molded, yet has relatively little effect on the resulting jet.

For applications where a more continuous operation is required, continuous jets replace the intermittent jets formed by cone collapse. This approach was developed in 1968 [10.43] by Bowles and Stouffer, to augment jet velocity after both streams had left the accelerating nozzles. Where two high velocity fluid streams are caused to contact at a common angle to the major delivery access then during the impact a relatively small but much faster jet can be generated moving along the common axis. This fast jet is then followed by a slug of slower moving fluid which can be used to augment the damage made by the small additional delivery slug. This has the advantage, over a single pulse attack on the rock, in that there is sufficient subsequent flow of water to exploit the crack systems generated by the impact of the high speed jet.

In the Bowles study, high speed cameras were used to study a jet generated by the impact of two primary jets impacting at an angle of 10 degrees. This initial study showed an augmentation for the jet of approximately 11.5 times. This demonstrated that a system operating at a delivery pressure of 20 bar, producing a primary jet velocity of 60 m/sec, could be used to generate a jet moving at a velocity of 810 m/sec at a pressure of some 3,500 bar. Relatively little development of this interesting phenomenon has taken place since.

Mazurkiewicz has demonstrated [10.44] that the concept will also work with high velocity continuous jets which are oscillated in a vertical plane. The vertical oscillation of the two jets makes the jets sufficiently discontinuous as to allow the generation of the augmented central jet. Such a jet has been demonstrated in cutting experiments to be able to penetrate significantly further into material and to cut harder material than would be the case by the jet alone.

For example, in the operation of the UMR Hydrominer at a test site in Missouri in 1976, the operating pressure of the cutting jets was some 700 bar at the delivery pump. Subsequent analysis has indicated that there was a significant pressure loss between the pump and the nozzle. Nevertheless, the jets were sufficient to be able to cut through the coal at the test site to a depth of approximately 70 cm. Performance of the Hydrominer was significantly impeded by the presence of lenses of pyrite in extremely hard lenticular nodules measuring up to 10 cm in thickness and extending up to approximately 60 cm along the face. These proved a very strong impediment to the progress of the machine. After several experiments it was found that the lenses could quite easily be penetrated by changing the design of the cutting nozzles, normally a two orifice diverging pair of jets to a jet where the nozzles were manufactured to produce a converging pair of jets which met at a distance of approximately 15 cm ahead of the orifice plane. With this design of nozzles it proved relatively easy to cut through the pyrite and the machine could proceed down the face with relatively little difficulty [10.45]. Although demonstrated in both laboratory and in the field, it is only recently, and in somewhat different form, that this technique has been revisited [10.46].

Mazurkiewicz et al. had found it best to use relatively shallow angles of intersection for the two primary jets. This could be best achieved by creating a single nozzle, using a electroforming manufacture, with the alignment of the two orifices built into the design. The reason for this precision in manufacture has been shown by Selberg and Barker [10.47]. Converging angles of 2 degrees were found to give best results, in part because the need was for a jet which would cut at some distance from the nozzle. The Australian workers [10.46] however, found it more beneficial to converge the jets, closer to the nozzle, and within the rock body (Fig. 10.26). In this way stresses set up within the rock were found to invariably produce large single chips of rock roughly conical in shape with an angle similar to that of the impacting jets. Interestingly it was reported that there was little evidence of jet cutting action in these tests where a jet at a pressure of 2,800 bar was cutting into a 2,000 bar uniaxial compressive strength basalt. The results also were reported to show that the specific energy required for this technique was one to three orders of magnitude less than for conventional cutting of slots by jet action. Subsequent traversing tests on the rock were preliminarily reported to substantiate the results from the static testing.

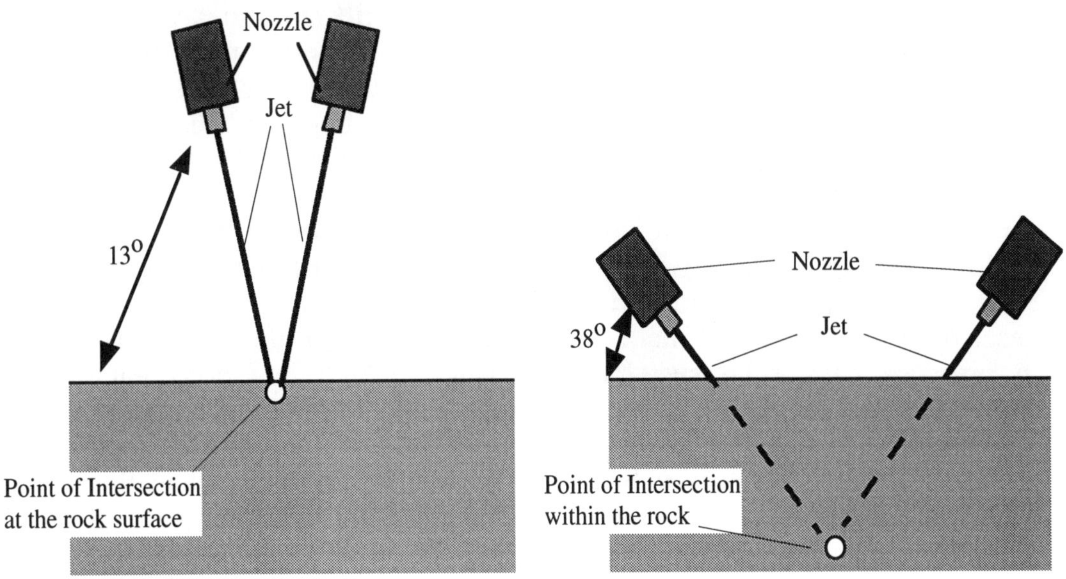

Figure 10.26 Comparison of techniques for generating accumulated jets either on the rock surface (after [10.44]) or within the rock (after [10.46]).

10.4 THE GENERATION OF CAVITATION WITH WATERJETS

10.4.1 THE MECHANISMS OF CAVITATION ATTACK

The high pressure pulses of water described above can also be generated in another way, but using the same basic concept. This is by taking advantage of the power of cavitation bubbles, deliberately created in a waterjet. The phenomenon of cavitation has been known for many years for its destructive ability in the operation of pumps and fluid delivery lines. However, it is only from the time of the first waterjet conference in Coventry in 1972 [10.48], that the potential that this tool has for a useful application in cutting and cleaning has been apparent.

Cavitation is a subject which can still generate considerable controversy. Following the presentation which Dr. Conn made at the first International Jet Cutting Conference in Coventry there was considerable discussion, at the second Symposium, as to whether the phenomenon which he had described was due to cavitation damage or to droplet impact. Subsequently, at the fifth Erosion by Liquid & Solid Impact Conference there was further discussion on the relative roles of individual bubble collapse, as opposed to the shocks generated by clusters of bubbles collapsing together [10.49]. This controversy still continues with a recent popular press article suggesting that it is the generation of the bubbles which causes the high level of damage, rather than their collapse. Given the nature of this discussion, it is appropriate to begin by discussing what cavitation is, why it occurs, where it occurs, and how useful it can be.

Cavitation in its simplest analogy has been described as small bubbles of vacuum traveling along in a waterjet stream. This is not a strictly correct analogy, but allows a simple picture to be drawn of what happens when cavitation occurs. Cavitation occurs where a waterjet or water flow of any description is so directed that a tensile force occurs in the water. This drop in pressure cannot be sustained by water, which has no great tensile strength. The net result is that the water, as it were, tears apart. As the bubbles of what might perhaps be initially considered as vacuum are created by the tearing apart of the water so the pressure inside the cavity drops below the vapor pressure of the fluid and the water on the surface of the bubble therefore will evaporate filling the void, or underwater bubble, with water vapor. This vapor pressure is, of course, below the ambient pressure of the surrounding fluid and as the water continues to travel then the bubble, which is at a low pressure, will be surrounded by a region of higher pressure which is sufficient to cause the bubble to collapse.

It has been shown through the use of very high speed photography, particularly by Dr. Ellis of the University of Southern California [10.50], that this collapse is not always symmetrical. In general the collapse of such bubbles is initially relatively uniform but, beyond a certain point, it becomes no longer symmetrical. This change may be brought about by a pressure differential, such as may be caused by jet turbulence, or by the presence of a nearby surface or flow disturbance. The result is to cause an area of the bubble to collapse, in on itself faster than the rest of the surface (Fig. 10.27). The stages of the bubble collapse have been charted both by Dr. Ellis and others [10.51].

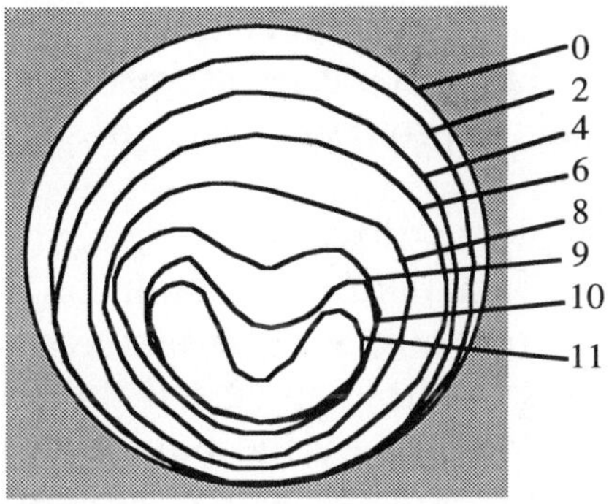

Figure 10.27 Representation of the stages of a cavitation bubble collapse (after [10.50]). The bubble was originally 4 mm in diameter and the stages shown are in microseconds.

The collapse of the bubble creates a high speed fluid jet which is forced inward from between the impacting bubble walls across the bubble and out the further side. This jet is known as a Monroe Jet, and is created in the same way as that used, in the collapse of the shaped charge cone, as described in the previous section. The size of the bubbles, and the jets thus created, are very small. Typical bubble diameters are on the order of 10 to 25 μm with the resulting jets being about one tenth of this size [10.52]. The jet range is also extremely limited with the photographic evidence from Lauterborn [10.51] suggesting that it will reach no more than the original diameter of the bubble before its energy is dissipated.

The advantage or disadvantage (depending on the application) of the very small jet created in this collapse of the cavitation bubble is that it moves at a considerably greater speed than the surrounding fluid. While numerous calculations have been made to relate this velocity to that of the surrounding fluid [10.53], the exact speed will depend on such factors as the surrounding fluid pressure, the size of the bubble, the amount of vapor present in the bubble, and the presence or absence of an adjacent surface. At the present time an accurate prediction of the impact force of this very small jet is not known. It has been predicted as being up to 70,000 bar in magnitude [10.54], a value confirmed by experiment [10.55]. What makes accurate measurement difficult, however, is the limited range over which this the small jet will travel into the surrounding fluid before its initially very high speed is slowed down by the water and it stops creating damage.

In most cases, the collapse of such a bubble on a very small surface will at best create only an extremely small impact crater. Brunton [10.56], has measured pits some 56 μm in diameter, but has also shown where the bubbles collapse over a crack, the stresses induced will cause the crack to propagate. It is only where one gets a very large number of bubbles collapsing, creating cumulative damage, the problem of cavitation becomes, for hydraulics personnel a problem, and those engaged in cutting, an advantage.

One of the first people to discuss the use of high pressure waterjet impact taking advantage of cavitation was Johnson who with Andrew Conn formed part of the cavitation team at Hydronautics Inc. [10.48].

In their pioneering paper they suggested that one could usefully take advantage of the damage induced by the bubble collapse to effectively cut through material. In their original work, however, they used only the cumulative effect of many bubble collapses to create the required hole in the target surface. Cavitation bubbles were formed in a high speed waterjet by placing a flat ended probe in the path of the nozzle, so arranged that the flat end lay beyond the acceleration section of the nozzle (Fig. 10.28). In this way as the jet reached maximum speed and moved out of the end of the nozzle it would pull a vacuum in the fluid directly over the tip of the probe and this would generate small cavitating bubbles which were then drawn into jet flow. The bubbles were carried by the main jet down to the target, where they collapsed onto that surface, where each would create a very small pit. The pits gradually coalesced, creating cracks and gradually tearing away the surface of the rock. In the original paper the penetration rate achieved was relatively slow, somewhere on the order of 75 mm/hr. But, it was the fact that the cavitation allowed the rock to be cut at a waterjet pressure much below that at which the rock is first eroded by a conventional high pressure waterjet which was impressive. It was followed by a preliminary investigation at Rolla in which it proved possible to erode a hole through an alumina plate roughly 6 mm thick within 30 seconds, using a jet pressure of 1,200 bar, a pressure at which the alumina is normally not damaged at all.

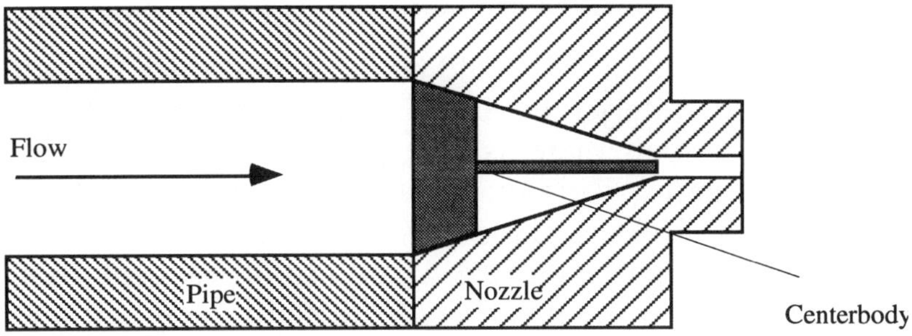

Figure 10.28 Design of a nozzle with center body to induce cavitation.

This initial paper was met with a certain degree of skepticism [10.57]. Much of the original argument centered around a debate on whether the probe in the nozzle disrupted the waterjet into droplets. If this were the case, then it would have been multiple droplet impacts which were causing the damage, a phenomenon discussed earlier. To a first order, similar levels of pressures may be obtained by the impact of the Monroe jet from the collapse of a cavitating bubble, and in the ring around the impact point of a single droplet. There is actually a difference in damage structure. For water droplet impact a circular crack around a central undamaged zone is characteristic of the initial damage to the surface (Fig. 10.18). In cavitated material attack, the damage surface is characterized by very small central pits caused by high impact of high velocity water. Although overall damage may be similar, the morphology of the surface is quite different.

There is one additional piece of evidence. If one examines the condition of centerbody probes which have been placed in a nozzle and run at higher jet pressure (say over 700 bar) it can be seen that significant cavitation damage occurs on these probes when they are not accurately aligned with the center of the nozzle (Fig. 10.29). In the case of the probe on the farthest right the entire titanium pin was destroyed in less than 2 minutes.

Figure 10.29 Titanium centerbodies and holder after use showing probe pitting and erosion.

It was this argument, taken with data obtained by running additional tests under water where droplets would not be created [10.58], which led to an acceptance, of the waterjet community, that cavitation damage was a valid phenomenon. Albeit there is still some controversy over what the exact phenomenon entails. This discussion still continues, although it has moved to a somewhat different plane than the initial discussion.

One design aspect which is critical to the use of the centerbody type of cavitation induction lies in the position of the centerbody within the throat of the nozzle. It is not always properly understood that if the probe length extends too far down the throat that the vacuum pulled on the tip will extend out to the outside air. At that point the vacuum bubbles will no longer form and instead the jet will be merely disrupted into droplets. This can be shown by data developed at Hanover [10.59] (Fig. 10.30), where it was also shown, as will be discussed later, that the range of damage of a cavitating jet is not necessarily short, where the cavitation is carried by the surrounding jet. The maximum damage with the centerbody located 3 mm back from the orifice of the jet shown below occurred some 300 mm from the nozzle. It is for this reason that experiments where data are reported from nozzles with the centerbody extruding beyond the orifice [10.60], should be reviewed with considerable caution since it is unlikely that cavitation was occurring in the jet when it was tested in air.

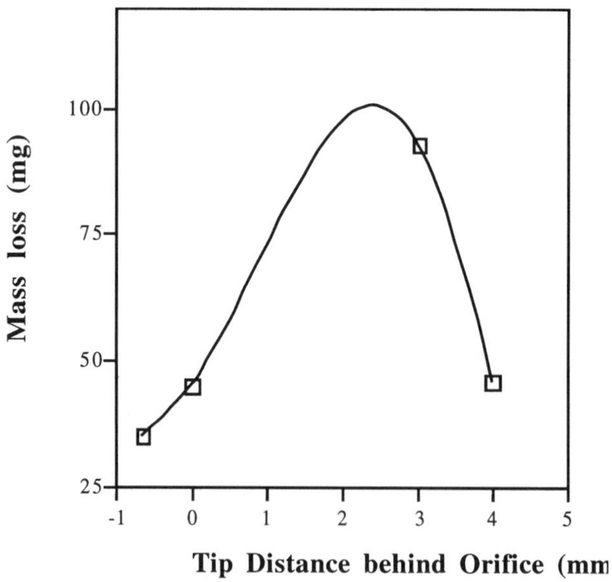

Figure 10.30 Mass loss from aluminum plates as a function of probe position, in a jet stream at 400 bar pressure (after [10.59]).

To further understand this developing argument it is necessary to first describe how a specimen's resistance to cavitation erosion is measured. This technique is described because, as may have been apparent from earlier chapters, as yet there exists no viable set of measurable parameters for a rock or other material which easily identifies its resistance to waterjet attack. Since waterjets are a part of the cavitation erosion phenomenon, it was initially hoped that the measure of a rock's cavitation resistance might be equated to its waterjet resistance. This is especially true since the cavitation resistance of a material has been measured and a standard test to determine this has been developed [10.61].

10.4.2 ASTM CAVITATION TEST

The most widely accepted cavitation test is that developed by ASTM [10.61]. In contrast to the above discussion, which dealt with flow induced cavitation, the existing standard, however, deals with cavitation induced in a normally stationary fluid. This is germane to the discussion, as will become clear, when the pressure levels generated are discussed.

In the case of this standard, cavitation is induced by ultrasonically vibrating the test sample within the test fluid. To achieve this, a small test sample of the test material is machined to follow a closely defined shape which can be threaded onto the tip of a specially shaped horn (Fig. 10.31).

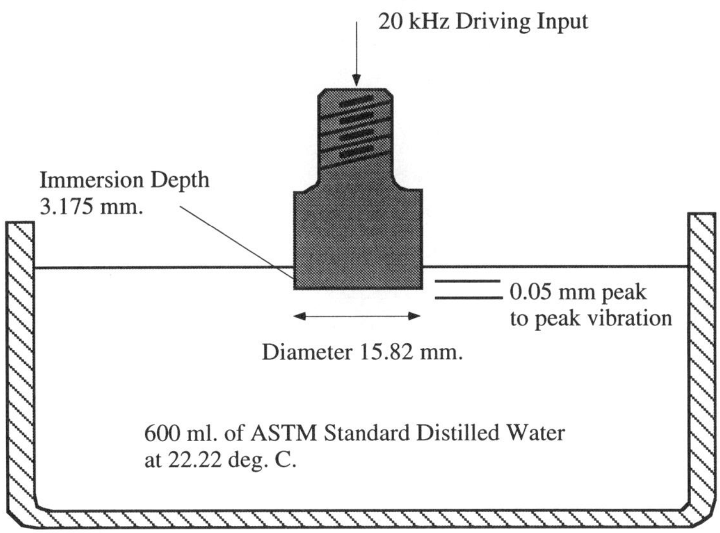

Figure 10.31 Required dimensions for a sample to be tested in the ASTM equipment (after [10.61]).

This long metal horn is then attached to the end of a magnetostrictive oscillation device so that, when the current is turned on and tuned properly, the specimen on the end of the horn will be caused to vibrate over a distance of approximately .05 mm (50 μm) at a frequency of 20,000 kHz, (20,000 times/sec), (Figure 10.32). By immersing this tip 6.35 mm under the surface in a specifically shaped water bath, it is possible, on the upstroke of the horn, to pull sufficiently on the water that small cavitation bubbles are created across the sample surface. Then, in the subsequent downstroke, these cavitation bubbles are driven back against the surface and collapse, causing erosion of the surface.

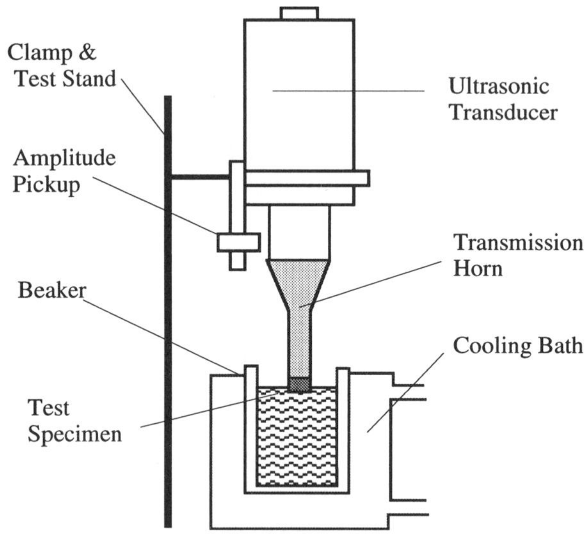

Figure 10.32 Test layout for the ASTM test (after [10.61]).

The very small amplitude of the movement generates a relatively small number of bubbles, and the resulting rate of erosion of the sample is typically very slow. Tests are normally run for periods of time that are measured in tens of hours. The sample is removed at regular intervals over this time frame to be weighed and the weight loss recorded. In a typical result (Fig. 10.33) the mass loss is, at first, zero, over an initial time, known as the incubation period. Once a certain deformation of the surface has occurred then mass loss will begin, and this accelerates to a constant level, which holds steady for a period of time, and then tails off to a second longer but slower erosion.

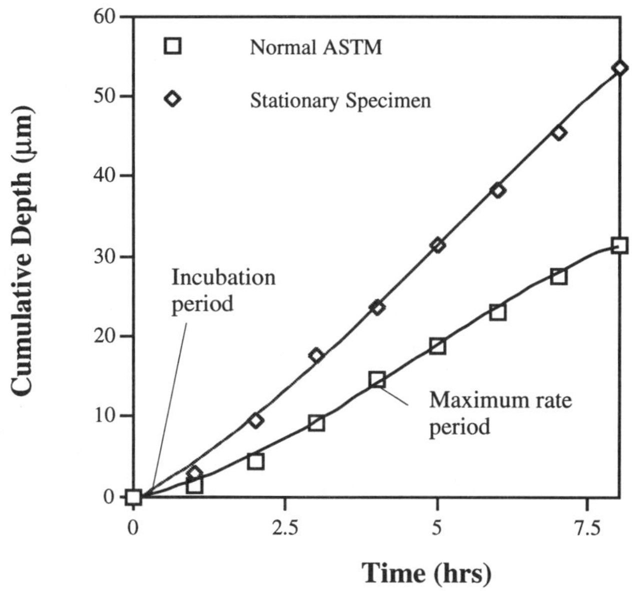

Figure 10.33 Typical result from a standard ASTM test [10.62].

Considerable thought has been given toward an explanation of these curves. One explanation is that the incubation period, prior to material removal, is brought about by the initial pitting of the surface and is a period that it takes for the pits to grow to a critical length, and then coalesce sufficiently to cause cracks. These cracks, in turn, must then join together and grow to the point that they isolate fragments of material which then become detached from the surface. Evidence for this opinion, which follows that of Brunton [10.56] will be described in a subsequent section.

This test has been evaluated in laboratories around the world. Unfortunately, even when specimens of the same material were tried in a particular round robin test, the results were not completely consistent [10.61]. While part of this confusion is related to the conditions of the test equipment, it is also related to the internal dynamics of the bubble, and the level of pressure generated in the bubble collapse.

For example, in calculating the pressures generated during bubble collapse Ellis and Starrett [10.54] did not consider that the internal vapor pressure, caused by water evaporation into the bubble, would play a great part. In larger single bubble experiments this is not the case. Either entrained air or vapor will generate an internal pressure within the bubble, which grows as the bubble shrinks. In time, this is sufficient to cause the contraction to reverse, and the bubble will begin to oscillate in size. Frequently, however, these bubbles have been generated by spark discharge into the water, rather than coming from a more conventional generator. Thus it is possible that the method of formation may have created some of the constraints in that situation.

Certainly there is now a significant body of evidence that high pressures, and associated events can be found during collapse. Summers and Fossey [10.55] have shown that the pressure is sufficient to ignite grains of an insensitive explosive, while blank [10.63] has been able to sustain a light within a fluid, by creating cavitation within it. This sonoluminescence is evidence that, within the body of the fluid, there is a highly localized energy intensification (of several orders of magnitude) to create this effect. Similarly it has been possible [10.64] to create chemical changes in fluids as a result of the high pressure (and thus temperature) conditions within the cavitation cloud.

One problem identified, was that it was found that size of the container played a part in the results that were measured. This phenomenon, among others, has led to a different theory as to the major cause of cavitation erosion of materials. To understand how this theory works it is pertinent to discuss a third method by which cavitation can be induced.

10.4.3 CAVITATION SHEATHING

Flow cavitation can be induced in the center of a flowing jet by passing that jet over a blunt ended probe centered in the jet stream. However, cavitation can also be induced by flowing that waterjet out into a lower pressure body of water. The shear stress generated along the interface between the waterjet and the surrounding fluid may be such that cavitation bubbles will be created around the interface. This technique has been developed by Daedalean Associates, of Woodbine, Maryland, who have used it in a number of commercial applications ranging from concrete cutting, through the demilitarization of missiles to the cleaning of the internal surfaces of geothermal wells [10.65, 10.66, 10.67].

In examining jet structure that develops with cavitation sheathing it has been observed by a number of observers, that the cavitation sheet does not occur as a continuous sheath around the jet, but rather, the bubbles seem to come down in a sequence of quite distinct clouds. In other words, there will be a period where there is an intense cloud of bubbles around the water and then a time with virtually no bubbles, before the next cloud forms. The frequency of this bunching of the cavitation cloud does not appear to correlate with pump performance. This bunching of cavitation bubbles is also not unique to this method of generating cavitation bubbles. Hydronautics staff [10.68] have been able to control frequency and bubble intensity generated by such a jet. From this they have developed a special nozzle to enhance and control the frequency, (Fig. 10.34).

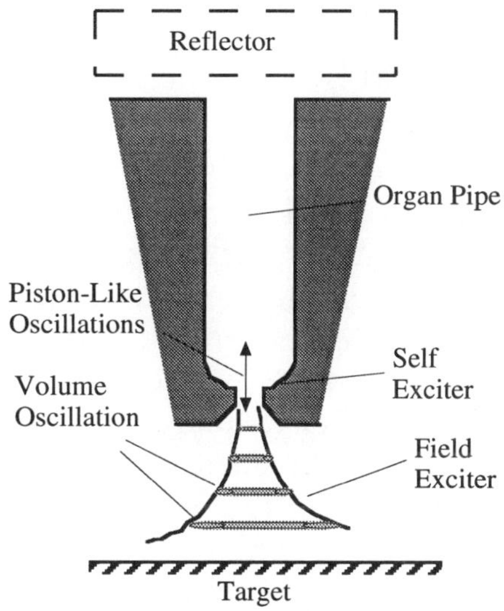

Figure 10.34 Design of nozzle to generate vortex cavitation [10.68].

When the bubbles reach the surface in clouds, they will tend to collapse on the surface simultaneously. This is similar to the events which occur when the cavitation bubble cloud formed in the ASTM test, is impacted by the downstroke of the specimen. This sequence has led to a second theory on a major cause of cavitation erosion [10.56]. The theory considers that as the initial bubbles in the cloud start to collapse that the shock generated will simultaneously induce additional bubble collapse and control its direction so that ultimately and, in a very short interval, the vast majority of the bubbles within the cloud simultaneously collapse. This will, in turn, induce a shock wave, traveling into the target surface, which will weaken and fatigue, that surface. The stress waves will of course also interact with existing flaw structure within the rock to induce further damage. The model was initially proposed by Vyas and Preece [10.69]. Brunton has produced some evidence that while there is some evidence of shock damage, that surface damage is most likely commonly related to single bubble behavior rather than that of a cloud collapse.

There has been further discussion on the validity of these theories of material failure, however the current author believes that the cause of the damage is most likely due to micro jetting though there is evidence of the types of damage that might well be due to shock impact. The contrasting evidence cannot of course be totally covered within the current text. But it can perhaps be illustrated, in part, from other procedures and results.

10.4.4 THE STATIONARY SPECIMEN TEST

Investigators studying the cavitation erosion of rock have been dissatisfied with the standard ASTM recommended method of test for some time. They are not alone in this, since others have also had problems in affixing materials other than metal to the end of the vibrating horn. This is partly because the specimens to be prepared must be machined with a thread which will attach the sample to the horn. This creates a problem in testing some ceramic materials and virtually all rocks, since it ranges from difficult to impossible to cut this threaded section and have it withstand the very high oscillatory speeds without fatigue in some of these materials.

A modification of the test procedure to overcome this problem, has been to change from locating the specimen on the end of the vibrating horn. In the modification the sample is replaced with a dummy specimen on the horn, and the true specimen is placed at a short distance, 6.35 mm, beneath the horn, and within the fluid bath (Fig. 10.35) This procedure has become known as the Stationary Specimen Test [10.70]. Results obtained compared favorably with those achieved with the conventional Vibrating Horn ASTM test when tested using metal samples (Fig. 10.33).

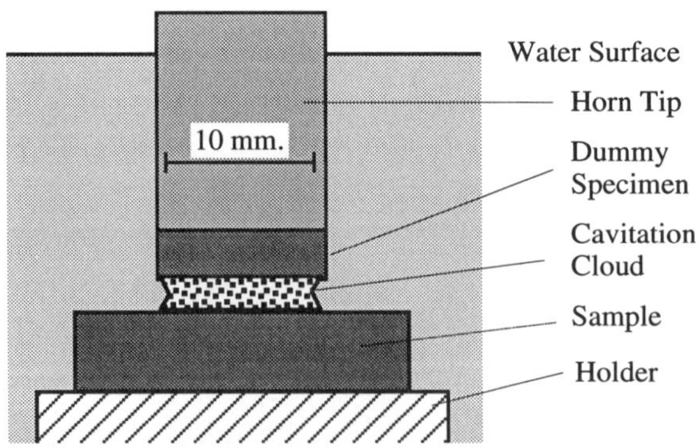

Figure 10.35 Layout of equipment for the Stationary Specimen test.

Unfortunately, this does not resolve the problem of evaluating large grain materials, such as rock. For example, one unanticipated problem arose in samples where relatively large pieces of material broke out from the test specimen. These became trapped in the narrow gap between the sample and the end of the horn, and were vibrated backward and forward between the horn and the specimen, doing damage to both [10.62]. A more profound problem is that the rock will often fragment in large pieces. The slow rate of erosion during most of the cavitation period thus tends to be overridden by the step increase when a large grain of material is removed [10.71]. In addition, the narrow gap between the specimen and the horn means that more energy is transferred from the horn into the surrounding fluid as heat. Since temperature has an effect on cavitation effectiveness [10.72] it was necessary to install a cooling circuit in the flow. This could resolve several problems if the water flowed across the specimen surface, and carried away the debris.

One method of supplying the flow was proposed by Hobbs [10.70]. He suggested drawing off some of the water in the test chamber and circulating this through a temperature controlled circuit, and then returning it to the test vessel. On its return, he suggested delivering the cooling flow upward through a small hole drilled in the center of the specimen. This raised a subsidiary problem, in that the erosion rate from the cavitation was found to be extremely sensitive to the fluid velocity [10.71]. Continued efforts to validate the method demonstrated that the technique required continuous control of an increasing number of parameters, and made the entire procedure less than effective as a simple method of material testing and evaluation. Another approach was therefore desirable.

10.4.5 CAVITATION IN A FLOW CHANNEL

There are a number of different approaches which have been tried, but, increasingly there has been a desire to shorten the time of test by increasing the erosion rate on the sample. This can be achieved by changing from what has been referred to as stationary cavitation, where the main body of the fluid does not move significantly, to flowing cavitation, where the fluid is moving as some form of stream across the specimen. For example, one approach was developed at the Technical University of Hanover initially under the direction of Dr. Erdmann-Jesnitzer and now being carried out by Dr. Louis [10.73]. In this system (Fig. 10.36) water is flowed past a restriction, which creates a contraction and then an expansion in the line, except that in this case the path change is sensibly two dimensional rather than the three dimensional configuration that is used with most jet formation in the applications described in this text. Cavitation is induced where the flow expands outward. By directing this flow against a specimen set in the side of the flow channel, cavitation can be caused which will impact on the specimen, causing erosion. This device has been used quite extensively to evaluate the cavitation erosion resistance of metals. It is normally operated with large volume flow rates at low velocity. This does not give rapid rates of material loss. For that reason this method was not further investigated during the UMR work, and instead a faster method of cavitation erosion was sought.

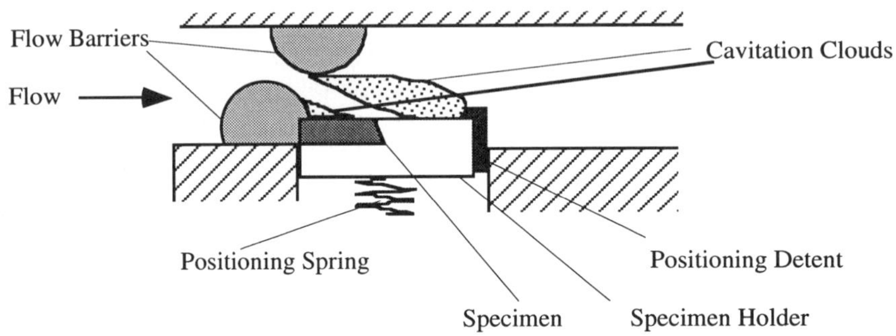

Figure 10.36 Sample and flow configuration for sample testing under channel cavitation [10.73].

10.4.6 THE LICHTAROWICZ CELL

Conventional methods of cavitation erosion testing operate under virtually ambient fluid pressures and with no intensification of the cavitation action. As a result the process is a relatively slow one, and the test periods required are extended. One method which overcomes this problem, is a device and procedure developed by Dr. Lichtarowicz at the University of Nottingham [10.74].

Unlike conventional methods, this technique can intensify the action of the cavitation erosion, by increasing the fluid pressure under which it occurs. To achieve this it was necessary to develop a specially designed test apparatus (Fig. 10.37). With this design a small, high pressure fluid jet is injected into the fluid-filled test chamber and directed at the specimen set on the opposite side of the chamber (Fig. 10.38). By adjusting the flow of fluid leaving of the chamber, the pressure of the fluid in the chamber can be accurately controlled to a set level. This controls the cavitation number of the experiment. This is a relationship between the pressure of the fluid in the chamber (P_d) and the pressure of the jet (P_u) which can be simplified to the ratio of chamber pressure to jet pressure. The equation should also include the fluid vapor pressure (P_v) in a form which generates a term known as the **cavitation number** (σ) which is given by the relationship:

$$\sigma = \frac{P_d - P_v}{P_u - P_v}$$

Thus, by increasing the chamber pressure the cavitation number can be increased, and, within a very narrow range, this has the advantage of increasing the intensity of the erosion.

In the original work, Dr. Lichtarowicz used lubricating oil as the test fluid in his equipment, in part because, at that time, it was difficult to find inexpensive high pressure pumps which would pump water. The pump supplied the oil through a small sapphire nozzle, typically 0.477 mm, and directed it onto a test sample of metal, approximately 2.5 cm in diameter. The test chamber was constructed with viewing ports at the front and back allowing both illumination and observation of the test in progress.

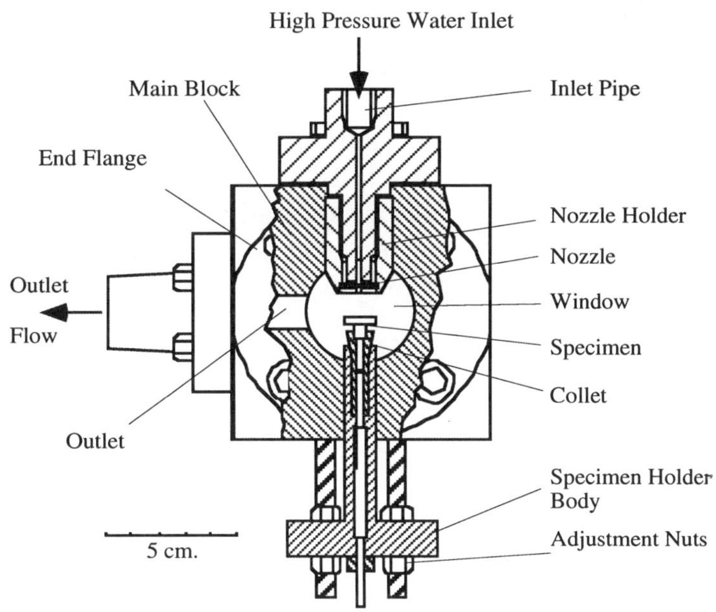

Figure 10.37 Schematic of the initial design of the Lichtarowicz Cell [10.74].

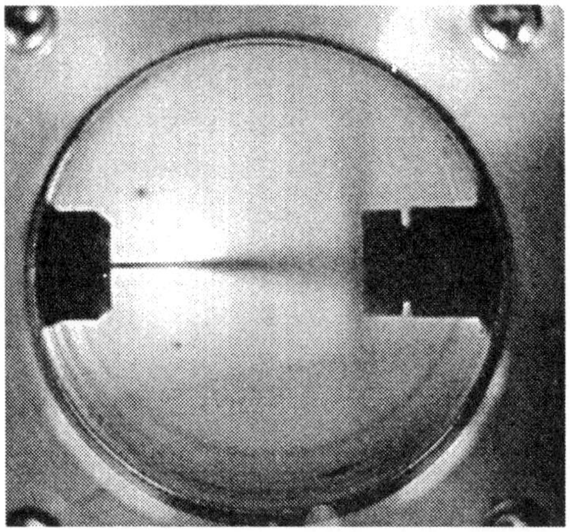

Figure 10.38 Lichtarowicz Cell in operation, showing the jet and surrounding cavitation cloud.

At the University of Missouri-Rolla initial testing of this concept showed that it was a much more effective device for the testing of brittle materials than previous devices [10.62]. Further, the erosion rates which could be achieved with the apparatus were significantly higher than those which could be achieved conventionally. At lower jet pressures with a jet pressure of 80 bar and a chamber pressure of 2 bar Lichtarowicz (ibid.) eroded 1 mg of aluminum within a period of 10 minutes. When, however, the test conditions were intensified, at UMR, by increasing both chamber and back pressure, then it was possible to remove grams of material in the same time frame.

The apparatus was, therefore tested, at UMR, as a means of evaluating the erosion resistance of rock. Initially the same equipment was used as that described by Lichtarowicz. It was rapidly determined, however, that the small size of the samples that the original equipment could hold were too small to contain the damage which was induced. During a test the impacting cavitating jet would create a cavity in the center of the sample and the network of cracks generated would lead to rapid sample disintegration. A second cell, built to twice the dimensions of the first, was therefore constructed to allow the testing of rock. These dimensions were chosen since it had been previously found [10.75] that where the rock samples were on the order of 5 cm or larger in diameter, that the samples would still retain their integrity under the stresses generated around the impact point, as the hole starts to grow.

Much of the work which has been carried out on the nature of cavitation erosion has tested metal samples (for example, [10.76]). However, when cavitation is applied to rock materials, which typically have much larger grain or crystalline sizes, the mechanisms of erosion, can be more easily studied and the rate of material removal significantly increased [10.71].

There is a very strong correlation between the cavitation number and the erosion intensity of the waterjet and, while the experiments have not been conclusive, there is some evidence that a cavitation number of .005 gives an intensified cavitation damage for the rocks tested. For example, a 700 bar fluid jet creates a jet with a high erosion intensity with a 3.5 bar chamber pressure. Secondly, it appears that there is a preferential attack by the cavitation bubbles on the crystal boundaries of the target sample. This can, perhaps be illustrated through a series of photographs taken sequentially as a sample of marble was eroded. During the course of the attack it can be seen that gradually the crystals which make up the marble become more exposed as the boundaries between them are emphasized (Fig. 10.39).

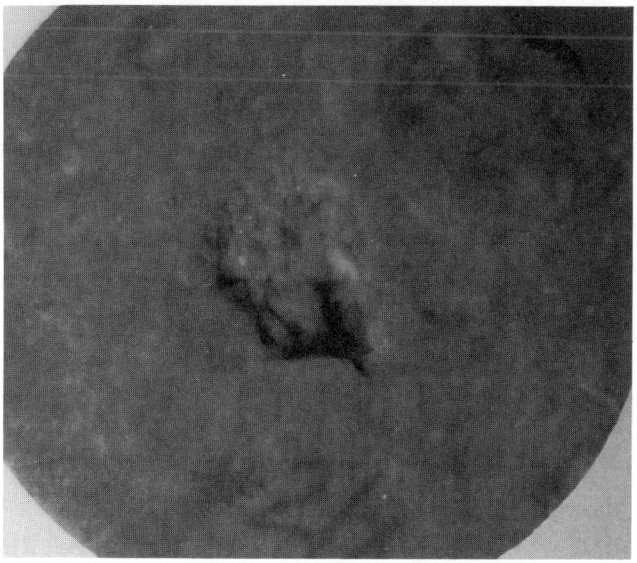

Figure 10.39 Erosion of a marble sample under cavitation attack [10.71].

Examination of a thin section through a typical cavity which has been created under cavitation attack (Fig. 10.40), shows that there is much greater erosion at the crystal boundaries. It was also of interest to note, early in the test program, that the zone of damage, generated by a cavitating waterjet is much larger than that created by a plain waterjet. The cavity shown below, for example, is approximately 12 times the width of the impacting waterjet, while a normal waterjet would cut a slot approximately three times the diameter of the jet. It is this ability of the cavitating waterjet to preferentially attack the existing flaws in the rock and to make them grow larger, which is perhaps of most interest for its use in waterjet cutting and related applications.

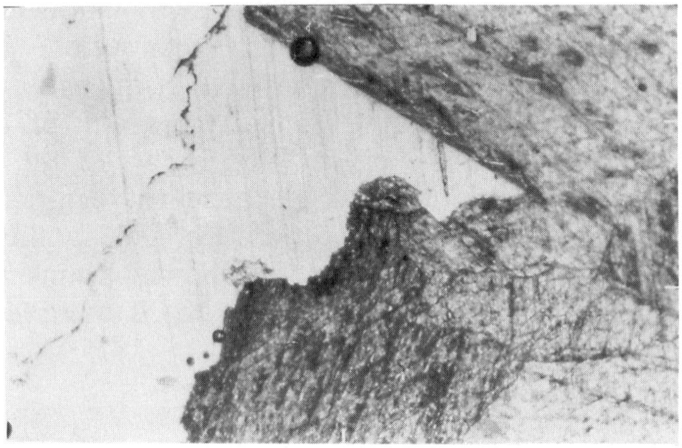

Figure 10.40 Thin section of a cavity in a rock sample made by cavitation attack [10.71].

Johnson [10.48] was only able to achieve cutting rates on the order of 7.5 cm/hr. Many drilling rigs which operate in the field make a hole at rates of penetration of up to 3 m/min. One therefore must find a faster way of operating a cavitating system if one is to find a practical use for it in mining and rock cutting operations.

The method which shows the most promise in this regard is to use cavitation in conjunction with the action of a continuous high pressure jet. One can combine the two mechanisms, so that the initial attack is by the cavitation bubbles which emphasize and extend the existing crack structure in the rock, or, in rocks where no previous cracks exist, create such cracks. Once these cracks have been established then they will be extended until the pressure of the main waterjet becomes sufficient to grow them further and cause large scale material removal. This can be illustrated with an example. However, because of the speed with which material could be removed under this combination, it was necessary to first develop a new test apparatus and procedure.

10.4.7 THE TRAVELING SPECIMEN TEST

The problem that exists with the conventional design of the Lichtarowicz cell is that both the specimen and the nozzle are held in a fixed position relative to one another. For most applications where waterjets are used, whether it be for cleaning or for rock cutting, either the target or the jet nozzle must move to effectively cover the surface of the target. This relative movement is not a feature of any standard cavitation cell. UMR investigators [10.71] therefore, built a test rig which would allow specimen movement (Fig. 10.41).

Figure 10.41 Equipment layout for the traversing sample cavitation test cell [10.71].

The design was based upon the enlarged version of the Lichtarowicz Cell, but incorporated a small carriage under the jet, which would allow the sample to move over a distance of 5 cm while the test was in progress. Under a normal test procedure the sample was located in the test chamber, and the distance from the nozzle to the top surface of the sample controlled, by adjusting the nozzle position. The front face plate of the chamber was then closed and locked in position. The supply pump was started and the relief valve which controlled flow from the chamber was closed. At the same time the exhaust valve from the chamber was opened. This opened a path out of the top of the chamber so that the chamber could be filled with water and all the air vented from it.

Once this had been achieved, then the relief valve was opened and the exhaust valve closed. Pressure in the cell could then be manually adjusted to give the required cavitation number for the test, as required. The test was initiated with the jet directed at a small cover plate held at one end of the sample by the holding fixture (Fig. 10.42). Once the test conditions had been stabilized, then the sample was moved under the jet at a controlled rate. This was controlled by using a small, variable speed motor to drive a traversing rod connected to the sample carriage. For the early test program advance rates on the order of 1 cm/min were used in the evaluation.

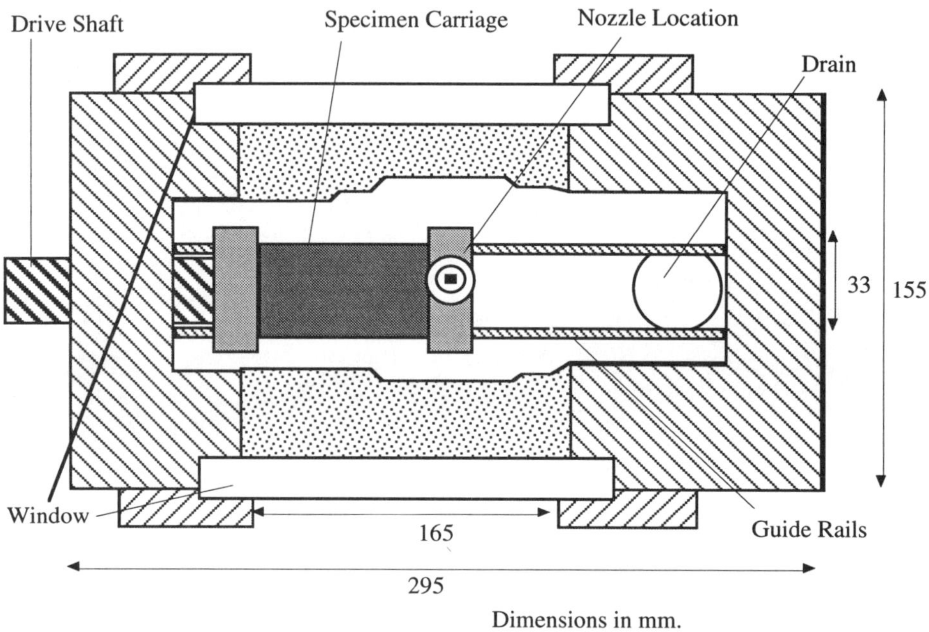

Figure 10.42 Schematic showing the nozzle position and the sample holder, from the top, for the traversing specimen cavitation cell.

The result of one of the first test series which were carried out with this rig illustrate quite clearly the development of erosion patterns of the type discussed at the end of the last section (Figs. 10.43, 10.44, 10.45). The rock sample was Missouri dolomite. The samples of dolomite were placed 25 mm below a 1 mm diameter nozzle operated at 500 bar with 5 bar back pressure in the cell. The first sample was traversed under the nozzle at a rate of 5 cm/min. It can be seen (Fig. 10.43) that a narrow groove was cut down the center of the specimen, showing the effect of the waterjet alone. On either side of this track there is a zone, approximately four times as wide, where the rock has been exposed to cavitation and which is pitted.

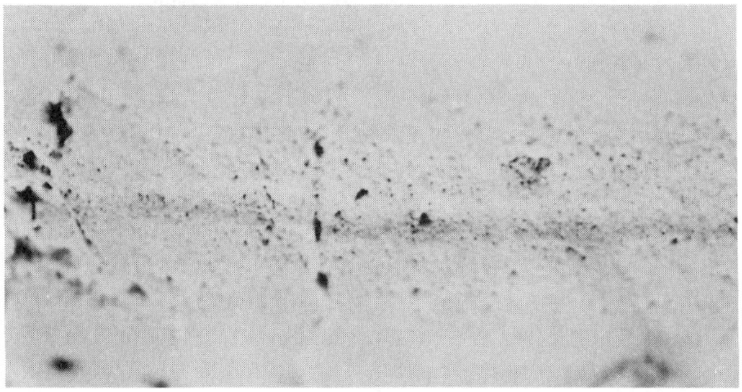

Figure 10.43 Dolomite sample traversed under a 500 bar cavitating jet at 5 cm/min.

As the traverse velocity of the sample was reduced in subsequent tests, the sample surface was no longer uniformly pitted, but became dominated by the preferential development of flaws and cracks (Fig. 10.44). As the traverse velocity was slowed even further from 12.5 mm/min to 10 mm/min (Fig. 10.45), the cracks created on the surface became large enough for the main jet force to exploit them and all of a sudden instead of having a rock with perhaps 6 mm of damage on the surface, the damage extended to a depth of about 18 mm with relatively large pieces of rock being removed in one piece.

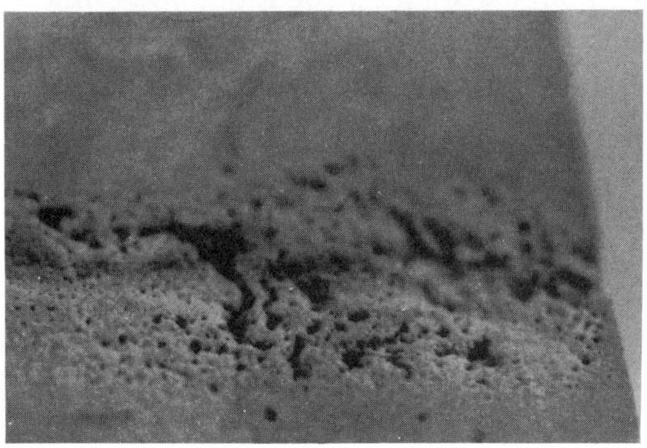

Figure 10.44 Dolomite sample traversed under a 500 bar cavitating jet at 12.5 mm/min.

Figure 10.45 Sample traversed under a 500 bar cavitating jet at 10 mm/min.

This combination of cavitation cracking and main jet crack exploitation is, possibly, the best way of using a cavitating waterjet for either cutting applications or for cleaning. This is because, when working with some marine growths one is working with material where there may be very few flaws within the surface that the waterjet can otherwise exploit. However, with a strong cavitation effect, then it is possible to create cracks sufficiently quickly, that the main waterjet force can exploit them and remove sufficient material from the surface, for an economic procedure to be developed. This combination, however, requires an accurate control of both the jet pressure and that in the surrounding fluid. This is not an easy requirement to be able to meet.

An additional advantage to combining cavitation into a high pressure jet is that, under normal conditions, the smaller crack size and density present in the target material requires that the waterjet pressure must be raised to a high level before the pressure inside the cracks is sufficient to cause them to grow sufficiently for large scale material failure. Where the cracks are extended by cavitation attack then as the cracks become longer they can be exploited at a lower jet pressure in order to create equivalent damage. To exemplify this advantage of the cavitating jet system, El-Saie showed [10.77], that by adding cavitation, that the pressure required to effectively cut granite can be lowered from 1,000 bar psi to around 700 bar. More interesting, to the long term application of high pressure waterjets to cutting and cleaning, preliminary experiments [10.78] indicated that, once the cavitating waterjet was operated at a higher pressure and under relatively effective cavitating conditions, that such a jet could out perform an equivalently powered abrasive laden jet (Figs. 10.46 and 10.47).

There are many other applications where a similar order of reduction can be achieved quite effectively, thereby saving horsepower, providing that the power of the cavitating jet could be properly applied.

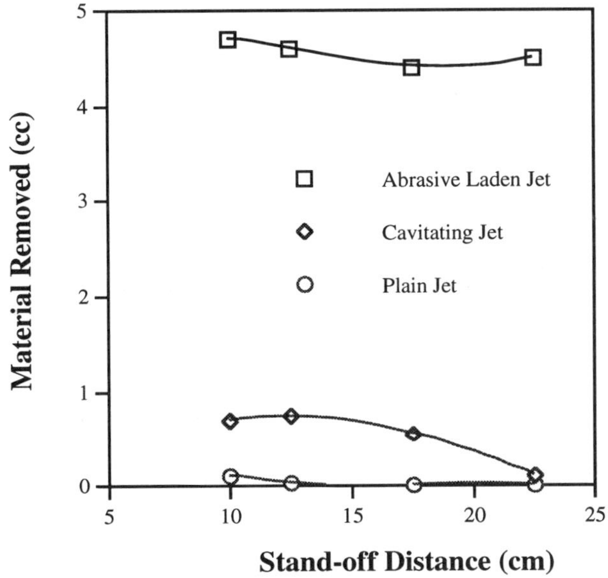

Figure 10.46 Comparison of plain, abrasive and cavitating waterjet cutting ability at 350 bar jet pressure on Missouri granite [10.78].

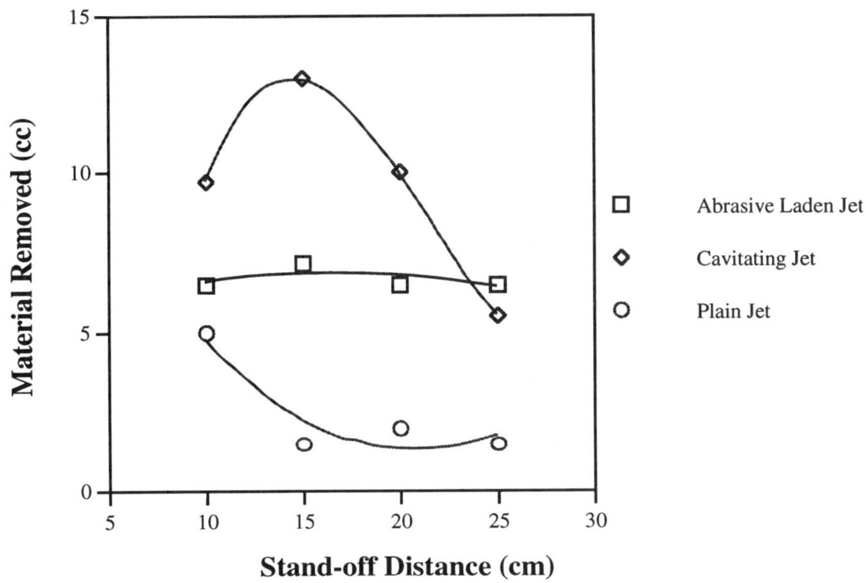

Figure 10.47 Comparison of plain, abrasive and cavitating waterjet cutting ability at 700 bar jet pressure on Missouri granite [10.78].

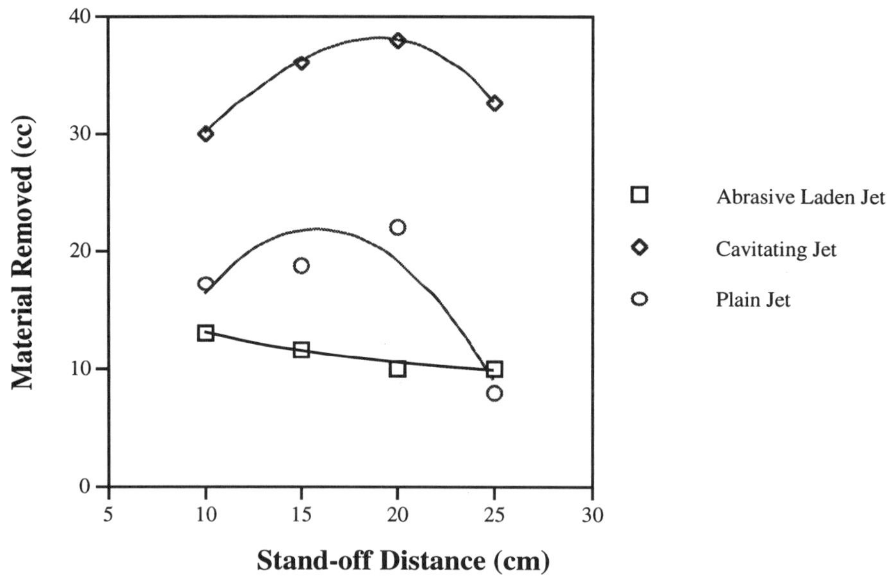

Figure 10.48 Comparison of plain, abrasive and cavitating waterjet cutting ability at 1,350 bar jet pressure on Missouri granite [10.78].

It should be pointed out however that there are many materials which do not need cavitation in order to be relatively effectively cut by a waterjet. There are also limitations to use of cavitation as a cutting or cleaning material and these include the following:

- The range of the cutting action is quite narrow (Fig. 10.49). While there have been some preliminary experiments in which cavitating jets have cut to depths in excess of 15 cm, these are quite rare, and under most conditions the range of a cavitating jet is on the order of 2 cm from the nozzle.

- At the present time the full range of variables which control the performance of cavitating bubbles in a flowing jet stream have not been well defined, or their control mechanisms well established. This is true, even though there have been significant texts which discuss and well describe conventional cavitation and its control (for example [10.79]). This makes it difficult to control and accurately predict the performance of cavitating jet systems.

- Even though the maximum intensity of cavitation attack is at the point of bubble collapse there are significant levels of erosion in the components of the system. This is particularly true where alignments are not precisely accurate, or materials of less than optimal resistance have been used. Thus it has not been uncommon to erode the nozzle generating the cavitation at a rate not much slower than that achieved in eroding the test sample. The use of the proper controls makes the procedure more expensive to apply.

- Higher levels of cavitation intensity generate higher frequencies of noise. This can be a serious threat to the health of operational personnel, particularly where these personnel are divers operating the equipment under water. Under those circumstances special precautions must be taken to protect the operators from harm.

Nevertheless there are several applications in which cavitating waterjets have already shown some useful application.

Figure 10.49 Damage to dolomite samples showing the range from the nozzle over which the cavitating jet was effective (the sample on the left was cut by a cavitating jet at 420 bar with 4.2 bar chamber pressure, that at the right at 500 bar with a chamber pressure of 2.5 bar. The samples are roughly 7.5 cm tall).

10.4.8 APPLICATIONS OF CAVITATING JETS

The initial application for which cavitation appeared to have potential benefit was in cleaning surfaces. The initial work was carried out in marine applications. Daedalean Associates [10.80], in their earliest work, reported being able to clean the fouling from ships' hulls at rates of just under 1.5 sq m/hr using a 16.5 kW system. This was carried out at about the same time as work at Hydronautics who adapted their CAVIJET system to removing the fouling from ships held in dry dock. They reported, by 1979, [10.81] being able to clean surfaces at rates of 500 sq m/hr using a six nozzle system at a pressure of 138 bar. This was compared with a rate of 90 sq m/hr which was achieved by a sandsweeping device. Interestingly they reported that using a sheath of water around the cavitating steam increased the width of the path being cleaned by about 50% [10.82].

When Houlston and Vickers compared cavitation with droplet impact it was found that cavitation could only be effectively used in certain combinations of nozzle systems [10.83]. They found that rates could be increased by surrounding the main jet with an external sheath of water [10.4]. It is also interesting to note, in light of the following discussion on abrasive use, that the Canadian investigators found erosion rate was enhanced when jets were inclined at 45° to the target surface (Fig. 10.50).

Cavitation has been proposed as a method for removing explosive components from loaded munitions, and one unit has been installed in Israel [10.84]. There has been work done on using cavitation to enhance performance of conventional jets in similar tasks in the U. S. [10.66].

In a comparison of performance between cavitating and plain jet results for this type of application it has been shown by UMR investigators [10.55] that the plain waterjets perform better than the cavitating jets (Fig. 10.51). This is explained as being because the range of intense cutting action of the cavitating jets occurs relatively close to the nozzle (see for example Fig. 10.49). In contrast the plain jets cut to a much greater distance and thus can more effectively remove material in this application.

There is a further caveat to the use of cavitating jets for work with explosives. As has been discussed above [10.54], [10.55] the implosion of cavitation bubbles can generate very high pressure pulses at the point of impact. These have been found sufficient to initiate a reaction in underlying explosive grains. Such a reaction has the potential to spread, and could lead to the reaction of the explosive mass. Since this is a dangerous result, the use of cavitating jets for this application cannot be recommended.

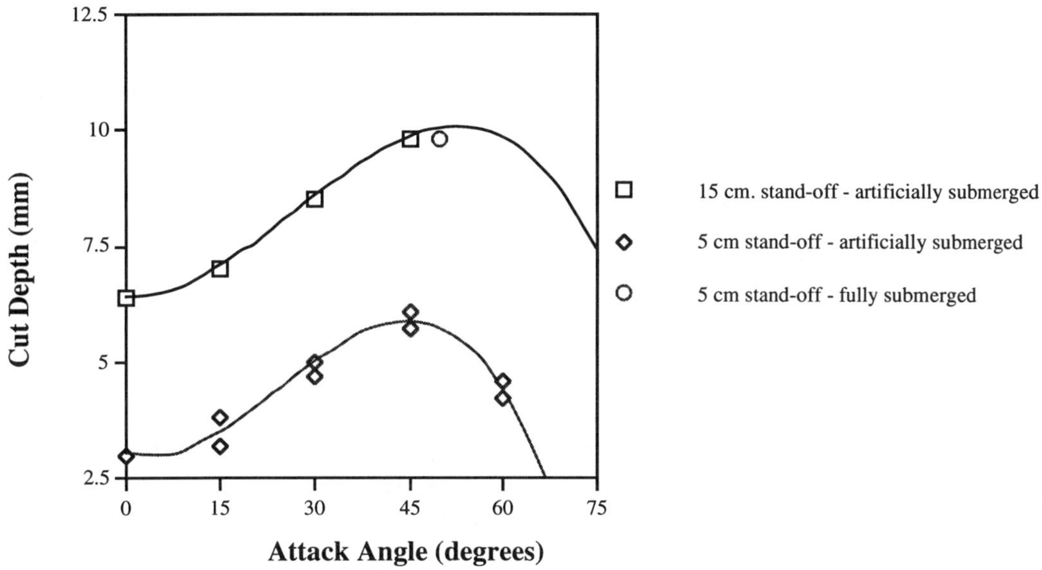

Figure 10.50 Effect of changing impact angle on cavitation effectiveness [10.4].

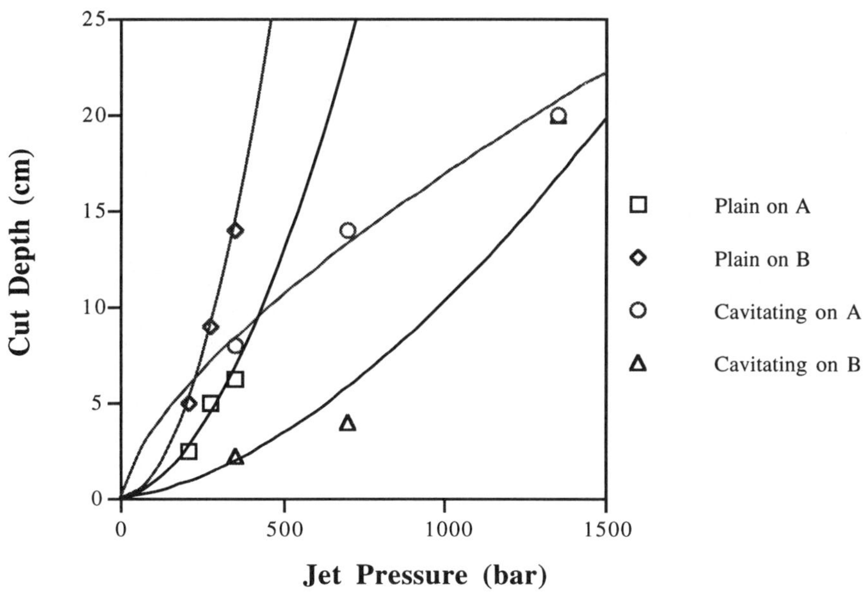

Figure 10.51 Cavitating vs plain waterjets in explosive removal, plain jet flow 20 lpm, cavitating flow 80 lpm [10.55].

A more interesting application, and one where, interestingly the word cavitation does not appear to be used, is in the drill developed by PetrolPhysics, and discussed in Chapter 8. The drill uses a swirling vane feed into the nozzle of the drill. While this is meant to generate a cone of cutting water out from the nozzle, and remove rock over the face of the drill, it will also (as discussed above) create a cavitation field. The result of this can be seen in the damage pattern generated on rock cut by the device [10.85]. These are similar to the cracked surface pattern shown (Fig. 10.45) above, which is a characteristic pattern for cavitation damage.

10.4.9 SHOCK LOADING

The discussion above may have given the impression that cavitation is very largely a micro jet impact phenomena. However, during several of the tests run at UMR, thin sections of the samples were prepared, following the test and one of these results was quite interesting. The rock being tested was calcite, which has the property that when it is stressed its ability to transmit light changes. Thus, if the material around the impact point is examined (Fig. 10.52) it can be seen that there is a zone, around the impact point, which has been sufficiently highly stressed for this change in light transmission to occur [10.71]. The level of intensity of an individual cavitation bubble is so small that it is unlikely that this would have had that range of action. This evidence, of itself, suggests some merit to the theory of damage suggested by Dr. Preece [10.69], that is that the damage on the target surface is due to shock waves from the cloud collapse of large number of bubbles at the same time.

Figure 10.52 Calcite sample after test showing the change in light transmissivity [10.71].

In this particular controversy, the available evidence suggests that there are several ways in which cavitation can induce damage into a surface. Some people will suggest that cavitation is one of the best methods of cutting material. This is no more true of cavitation than it is of anything else. But in its own particular area it has some major benefit though not as the universal tool. There is no such system available that can be discussed at this time. It is regrettable that research into this interesting area has currently declined. Given the increasing problems with the disposal of residual material after cleaning, induced by environmental regulation, the potential benefits from developing cavitation systems which do not use abrasive in the cleaning process would appear to justify further work.

10.5 SPARK GENERATED JETS

There has been some interest in the use of electrical sparks within waterjets as a means of enhancing the performance of the jet. The technique has been tested, both as a means of driving a waterjet out of a nozzle (Fig. 10.53) and as a means for developing a drilling device [10.87], [10.88].

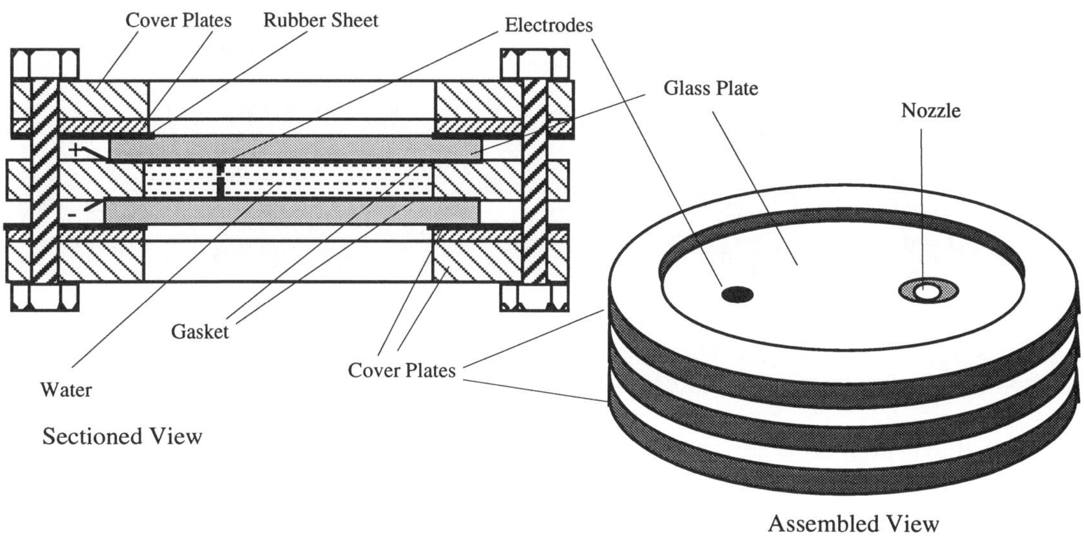

Figure 10.53 Method of developing a high pressure waterjet using a spark discharge [10.86].

The technique has considerable promise, and there have been some successes with the tests carried out. However, there have also been significant problems to overcome in any field application of the system, which has not, therefore, achieved any commercial success, and which is currently not being pursued.

10.6 THE INTRODUCTION OF ABRASIVES INTO WATERJETS

The ability of waterjets to cut a variety of industrial materials led to its adoption as part of the development of a new industrial process. With the growth in the number of applications, however, the limitations in the performance of conventional waterjets became apparent. Many of these have been documented in previous chapters. The inability, for example, of a waterjet to cut through any rock in its path, has restricted the introduction of waterjet drills into the mining industry. Similarly the problem of limiting crack growth around the cut has limited the usefulness of high pressure jets in cutting, for example, brittle plastic sheets (Fig. 10.54(a)), or in cutting through large billets of wood, and in cutting marble.

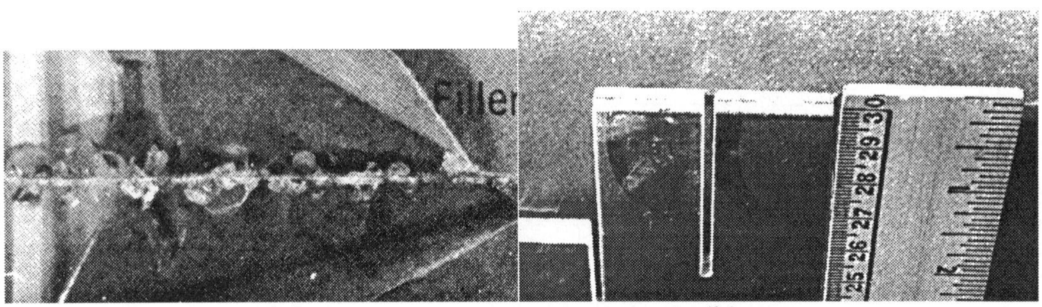

(a) (b)

Figure 10.54 Jet cuts in a) plastic sheet without the use of abrasive; b) a two-layer of glass with a plastic sheet interface with abrasive in the water.

A method was required which was capable of increasing the range of materials which could be cut by the jet. The idea of generating small impact fractures in the path of the jet has been explored above and found to improve the ability of lower pressure jets to cut harder materials, using cavitation to create the fractures. One alternative to using cavitation bubbles in order to induce the greater number of fractures in the surface is to entrain an abrasive in the jet stream.

The use of abrasive particles in a waterjet stream will create both positive and negative effects. The increased power of the jets makes it possible to extend the cutting range of the system so that the waterjets are able to cut through virtually all materials at commercially viable pressures. In addition, the edge of the cut created by the abrasive-laden jet can be significantly better in quality, and with a lower width of damage along the cut than that produced by other methods. In addition, significant other advantages were found. For example, in the cutting of safety glass. Safety glass is normally used in display cases and stores to limit injuries when a glass panel is broken. To retain fragments, the glass is made up of two separate sheets, with an interlayer of plastic. While glass is normally cut with a diamond scratch and a sharp blow, this becomes difficult with the two layers of material. As a result reject rates of up to 30% have been reported to this author when the material is cut. With abrasive laden waterjets, however, a cut can be made precisely through both the glass and the plastic on a repetitive basis (Figure 10.54(b)).

The technique has found applications in a wide variety of fields, many of which have been covered earlier in this text. Abrasive laden air streams have been usefully applied for many years for the cleaning of surfaces and the cutting of different materials. Air driven abrasive has, however, a number of different environmental disadvantages and its replacement by water borne abrasive would provide two advantages. The first is that the water can deliver a much greater energy level to the abrasive particle than can a normal air stream with considerably less potential risk to the applicant (since water is considerably less compressible and can be taken to considerably higher pressures without equivalent risk). The second advantage is that the waterjets potentially can exploit the flaws created in the surface of the material by the impact of the abrasive and thus the two components of the stream can synergistically interact to enhance the removal of material over that from plain abrasive impact.

As a result there has been a move afoot to develop predictive methods to allow modeling of the process [10.89]. Such a model would also make it possible to predict the performance of an abrasive jet system under a variety of conditions. These attempts should, however, be treated with some considerable caution. The reasons for this will be discussed below, as well as recognitions of some of the good features of the work in this area carried out to date.

10.6.1 THE MECHANISMS OF MATERIAL REMOVAL

To begin with it should be recognized that the removal of material under the impact of an abrasive laden waterjet is driven by a mechanism which varies as a function of the response of the material under attack. This knowledge is not new. The use of abrasives has been fairly widespread in the cleaning industry and the study of the effective abrasive particle impact upon surfaces has been examined by scientific investigators over a number of years, in order both to determine how to enhance the damage that induces and also to defend against it [10.90].

The differences in the failure mechanisms relate to the relative ductility or brittle nature of the surface under attack and while this can be explained, the contrast can also be simply illustrated by plotting the most effective impact angle of an abrasive flow in eroding a surface (Fig. 10.55). Those materials which are more brittle in their response such as glass, ceramic, or rock, will fail by crack growth and intersection and the best angle of attack is one close to perpendicular to the surface. In contrast, those materials which are more ductile such as most metals, fail under a regime where the material is either cut or caused to flow plastically under the impact, with more material removed when the abrasive attacks at a shallower angle [10.91].

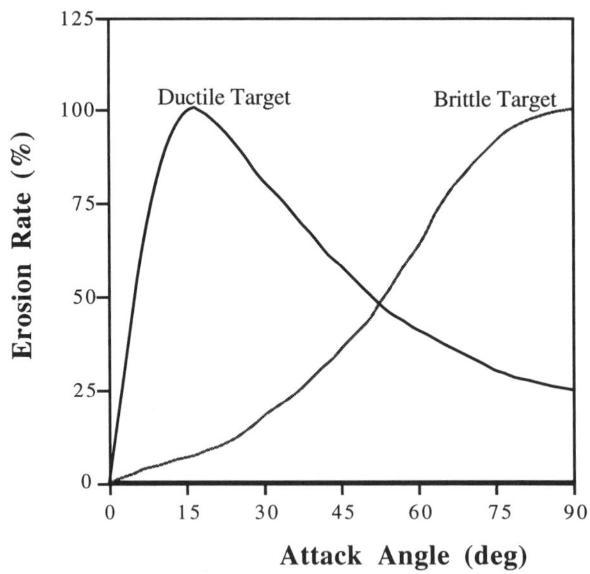

Figure 10.55 The effect of angle of impact on material removal rates (after [10.91]).

10.6.2 DUCTILE MATERIAL REMOVAL

To explain why this difference in mechanisms and behavior occurs it is easiest to look microscopically at surfaces damaged by abrasive particle impact. Consider the impact of a particle on a ductile metal surface. When the particle first hits the surface it will penetrate the material which tends to flow around the particle. At high angles of impact this creates a pit in which the particle may still remain, surrounded by a "wall" of material which has flowed out of the particle way, but not separated from the surface (Fig. 10.56). In some cases velocity of impact may be sufficient to fracture the particle into smaller pieces, causing them to break off and flow outward from the contact point, and these may remove this lip of material in a secondary erosive process. This most frequently happens when larger particles or particles of a friable nature, such as copper slag, are used.

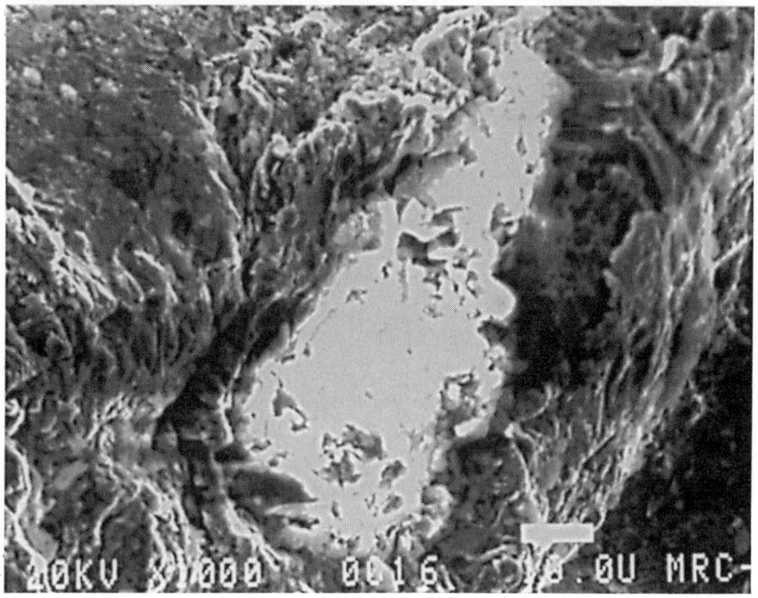

Figure 10.56 Particle of abrasive embedded in a metal surface (magnified).

Material will only be removed if it is significantly displaced, "plowed" or cut away from the surface by movement of a particle. A number of different ways in which this can occur have been suggested by Hutchings (Fig. 10.57). The physical evidence of this can be seen in the track of a limited number of particles which have hit a steel surface at an angle of 45° (Fig. 10.58).

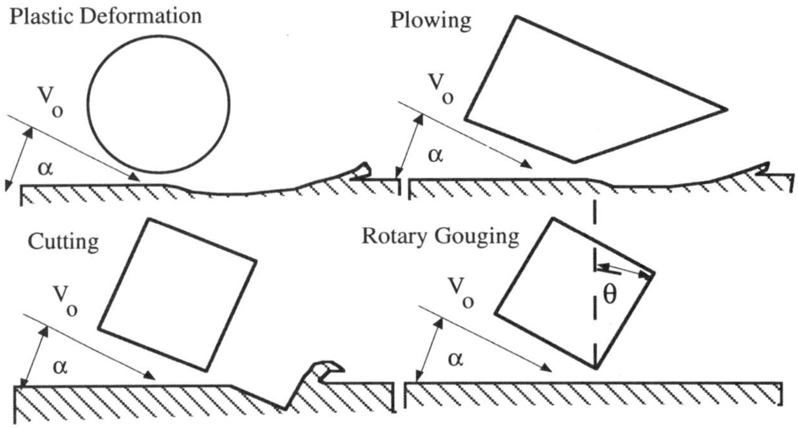

Figure 10.57 Mechanisms of ductile material erosion under abrasive attack [10.92].

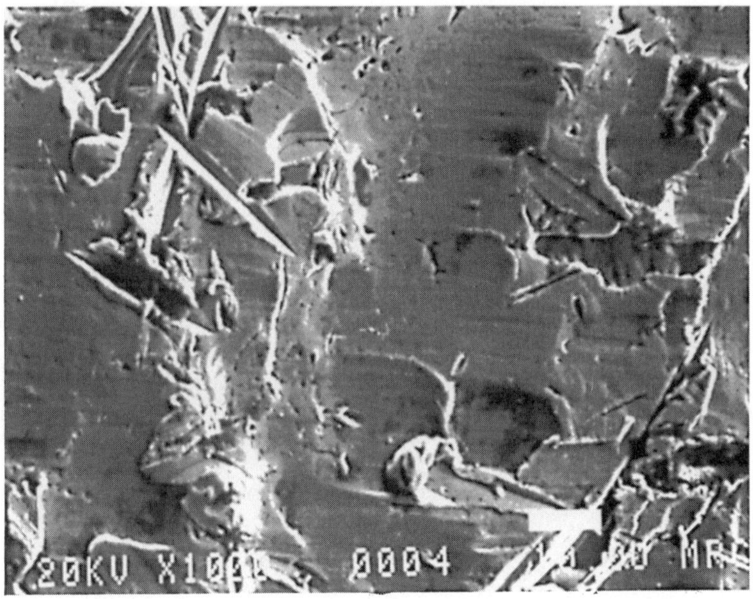

Figure 10.58 Microphotograph of a steel surface impacted by quartz particles at a 45° angle showing "plowing" and material removal.

Ductile material is removed through a process of either cutting or material flow, after a transition to a plastic state. As a side comment it should be noted this requires that the metal turn molten. This condition requires considerable heat, and may explain the white hot "sparks" which can sometimes be seen when metal is cut by an abrasive jet. Since the heating is an inherent part of the failure process it cannot be eliminated, but, as will be discussed in Chapter 12, it does not necessarily pose a hazard.

The failure of the material will also depend on the thermal properties of the target, as well its conventional material characteristics. The volume of material which is moved by the particle impact depends on how deep the surface has been stressed beyond its elastic limit, and how much force is then retained in the particle, as it continues to move, to deform the material. Material will only be removed if it is moved more than a critical distance, itself requiring a threshold input energy. That energy level will depend on the particle size, density, velocity and impact angle. Adler, however, has suggested that there is a lack of "a reliable experimental data base " [10.93], which makes an accurate assessment of the role of the individual parameters somewhat difficult.

The process of ductile material removal under abrasive attack can thus be grossly simplified into being viewed as one of getting the surface into a plastic state, and then pushing it until material is removed. This may occur either by inducing a large volume to undergo this process, or can be achieved, particularly at shallow angles of impact, by undercutting material with a broad based fragment of abrasive, which only needs to fail the bottom of a piece of the target in order to get it to move. The cutting of metal by a traversing abrasive laden waterjet can be quite effective, since along the cut, the abrasive-laden stream is flowing at a shallow angle across the surface to be removed (Fig. 10.59).

In this process, where the abrasive must "plow" along the target material, the angularity of the particle will control how deep the particle penetrates the surface, and the sharpness of the asperities will have an effect on the volume of material which is removed. Thus the shape factors of the individual particles will have a significant effect on the volume of material removed and the rate of cutting which is therefore achieved.

The relationships between the volume of material removed, and the impact properties of the abrasive are usually given in terms of the effective radius of the abrasive particle, and its velocity, by an equation of the form:

$$e = K \, . \, R^a . V^b . F(\alpha)$$

where e is the volume of target material removed/particle

K is a constant

R is the effective particle radius

V is the impact velocity
a and b are constants
α is the impact angle

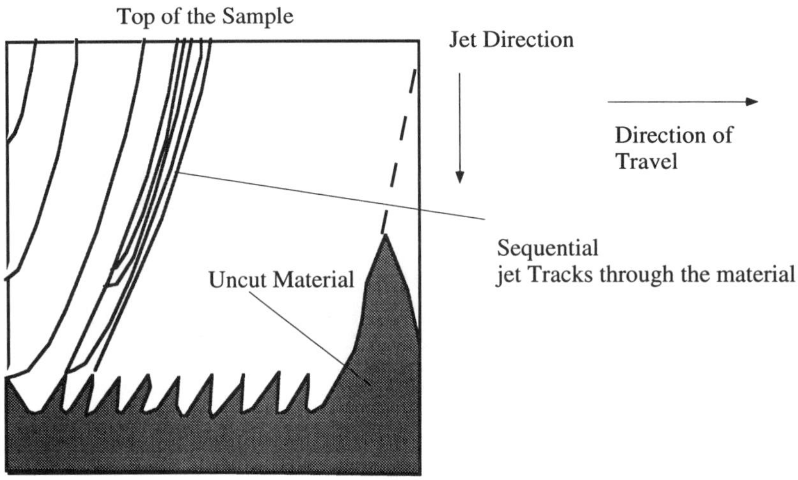

Figure 10.59 Cutting path of an abrasive laden jet through a ductile material (after [10.94]).

In ductile materials the values for the radial exponent have not been adequately evaluated, according to Adler [10.93]. There is some disagreement if the particle size has any effect above some critical value. The velocity exponent is also reported to have values ranging from 1 to 6.9 depending upon the target, the abrasive properties and the test conditions. Since the value of this exponent controls the relative volumes of material removed by an individual particle (Fig. 10.60), means by which one could increase the value of this exponent would be very valuable. There is however, an indication within the literature that the correct value lies in the range from 2.5 to 4.0.

The above discussion has dealt with the erosion as it has been induced by individual particle impacts. In a subsequent section, the effects of multiple particle impacts of the abrasive-laden water stream will be discussed further. The discussion will likely also become more controversial, given that many of the subsequent results will be reported for tests on rock, and there may be some disagreement as to whether rocks should be defined as ductile or brittle.

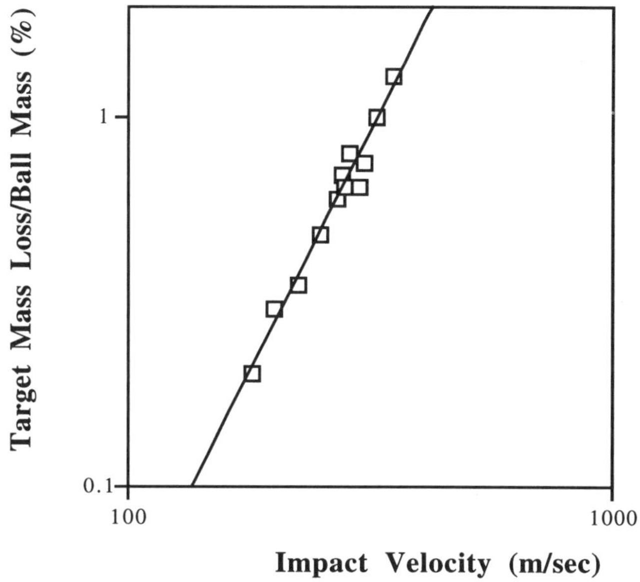

Figure 10.60 Relative mass loss when 9.5 mm steel balls strike a mild steel plate at an angle of 30° (after [10.92]) (exponent value 2.9).

10.6.3 BRITTLE MATERIAL REMOVAL

The failure of brittle materials is quite different in character to that of ductile material failure, and occurs by a process of crack creation and extension. This has been studied in some detail, perhaps an easier task, given that glass is a brittle material. Studies of impact on glass have, therefore, allowed visual observation of the extent and growth of damage as the velocity and frequency of impact increases. As a general rule rock is considered to be a brittle material, although there are some rock types within this classification (for example some shales) which could, more readily be described as ductile, both in overall behavior and in their response to particle impact.

The process of brittle material failure under impact can be equated to that which occurs when an indenter is pressed into the surface. At first the surface of the material under the indenter is compressed, as the face of the tool is moved into the material. At the same time the surface around the contact must stretch to maintain contact with the surface under the tool, which is, in relative terms, moving away. As load continues to increase, and the depression grows, radial cracks are generated in this stretched zone of material around perimeter of contact. Continued increase in pressure will lead to the formation of median cracks which develop in the material under the indenter, and the creation of lateral cracks, running almost parallel to the surface, but within the bounds of the radial cracks. As the force on the surface continues to build these, lateral cracks may continue to extend, and begin to join with other cracks, ultimately joining together to isolate a block of material leading to material removal. (Fig. 10.61).

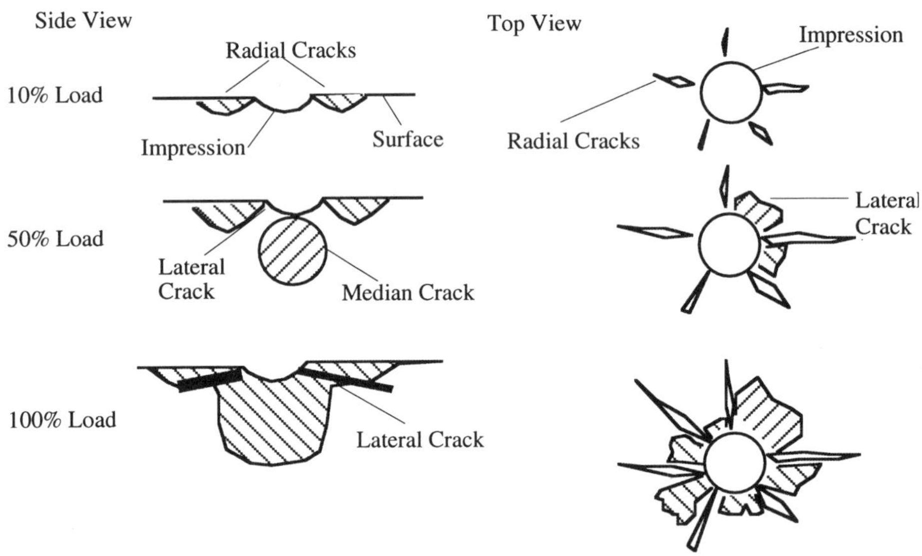

Figure 10.61 Schematic representation of the growth of damage in a brittle target under increasing impact pressures (after [10.95]).

Because the evaluation of the stress field generated around the impact is derived from an initial analysis by Hertz [10.96] this development of the circular fracture pattern around the impact point has been given the nomenclature of Hertzian.

As shown above, the most intense damage to brittle surfaces comes when particle impact is at close to right angles to the surface. With ductile materials, when particles strike brittle materials at shallower angles of impact, it is likely that the particles may glance off the surface without generating any significant effect. In order to generate the higher stresses in the material it is better if the abrasive presents a smaller, rather than larger footprint on the target. in this way the stress levels may be raised, and the resulting crack growth be more extensive.

The different aspects in the required aspects of the impacting particle and of the surface damage (for a relatively undamaged initial surface) play differing roles in the failure process. The ring and median fractures are reported (ibid.) to have, as their main effect, a reduction in material properties, and it is the lateral cracks most likely to cause material removal. Classical studies of these problems have been carried out by Evans [10.97] and Finnie [10.98]. Because damage is most intense at high angles of impact, formulation of the damage assessment has concentrated on perpendicular impact, with the generic equation given in the form:

$$e = K \, . \, R^{a} . V^{b}$$

where R is the particle radius and V the particle velocity.

The premise for this analysis is that the amount of material which will be removed as a result of the impact will be defined by the bounds of the damaged zone, notable its diameter and depth. Evans has shown that the average crack length is related, both in size and in location to the velocity of initial impact (Figs. 10.62 - 10.64).

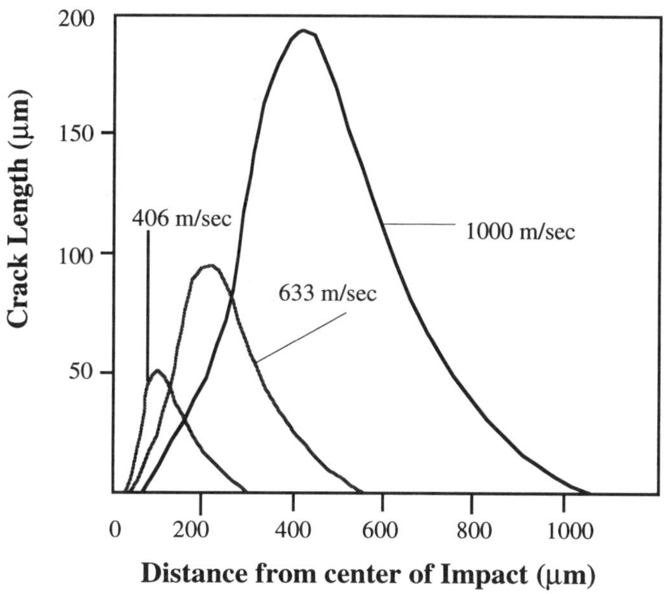

Figure 10.62 Effect of impact velocity on crack length, as a function of radial distance from the impact point (after [10.97]).

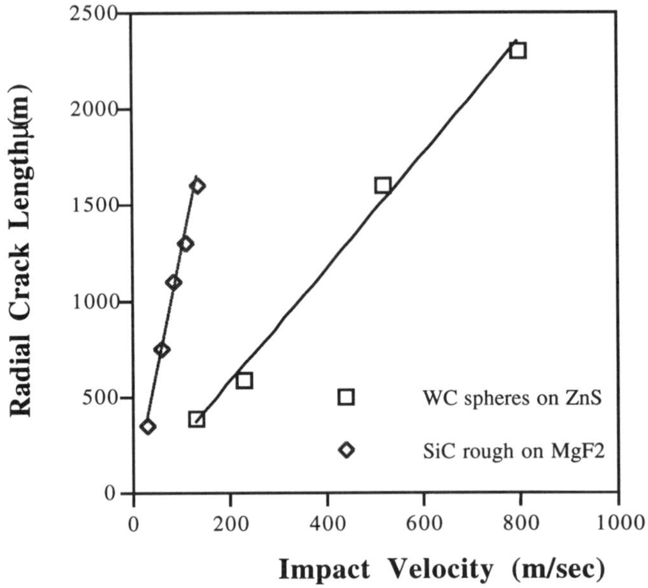

Figure 10.63 Effect of particle impact velocity on the radial length of cracks generated from a single impact (after [10.97]).

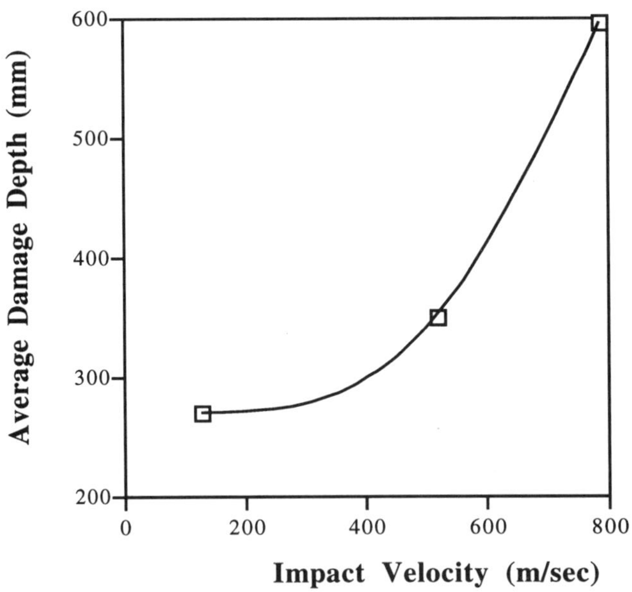

Figure 10.64 Effect of particle impact velocity on the depth of crack damage from a single impact (after [10.97]).

As mentioned in the case of rocks, the amount of material removed as a result of an impact will also on the number size and configuration of the cracks which exist in the surface before the impact. Many of the experiments dealing with the study of particle impact on a surface have been carried out on carefully prepared samples of glass or other ceramics, with the samples selected which have few large defects. Rock, in contrast, has a large number of defects, generally at the grain size of the material, and these will contribute significantly to the volume of material removed. The size, orientation and density of the flaws in a rock play a significant part in defining its strength. They have been described by Weibull [10.99] who has generated a statistical model to describe their effects on the material properties.

The effect of these pre-existing cracks has also been recognized in Finnie's analysis through the addition of the exponent (m) from the Weibull expression to describe the surface flaw effect.

$$e = K . R^{a} . V^{b}$$

where a = (3m-2)/(m-2)
b = 0.8(3m-2)/(m-2)
for spherical particles
and with "a" multiplied by 1.2 for angular particles.

This has been refined by Evans to give a more comprehensive equation for the erosion of material by a single particle. Given that there are few interactions between the cracks generated by each particle, due to the intersection of lateral fractures, Evans postulated that the material removed is a fraction (ε) of the volume through which these cracks grow. On that basis he defined the volume removed as:

$$V_i = \varepsilon \cdot \frac{(\rho_t \cdot \mu_t)^{0.67}}{K_c^{1.33} \cdot H^{0.25}} \cdot \left(v_p^{38} \cdot r_p^{44} \cdot \rho_p^{27} \cdot \mu_p^{8}\right)^{0.083} \left(\sqrt{\mu_t . \rho_t} + \sqrt{\mu_p \cdot \rho_p}\right)^{-2.67}$$

where the subscripts "p" and "t" refer to the particle and the target properties respectively (ρ is the density, μ is Lame's constant, r is the radius, K_c is the fracture toughness, and H is the target hardness, v_p is the particle wave velocity).

It should be noted that the underlying basis on which this analysis was built derived from quasi-static impact. Experiments reported by Adler [10.93] and Knight et al. [10.100] have shown that under the dynamic loading conditions which exist during abrasive impact, crack growth may exceed by a factor of 2 - 4 the lengths of the cracks generated in an equivalent static case. A semi-empirical estimate of the velocity exponent "b" by Adler, has suggested that the value should be 3.17. In limited special cases the values for both "a" and "b" may be approximated to 4. This suggests that both particle size and velocity may be more critical parameters in brittle material removal than for ductile materials. It should be borne in mind that this data is proven to be valid only for the impact of single particles on surfaces.

In contrast with the situation where there is the impact of a large number of small grains of abrasive impacting upon a target are the impacts of single, larger, typically steel projectiles into rock and soil. Such impacts can generate significantly different suggested relationships. For example, using steel projectiles on Berea sandstone Ripkin et al. [10.101] have found a linear relationship between the depth of cut achieved, and the impact velocity and the particle diameter (Fig. 10.65).

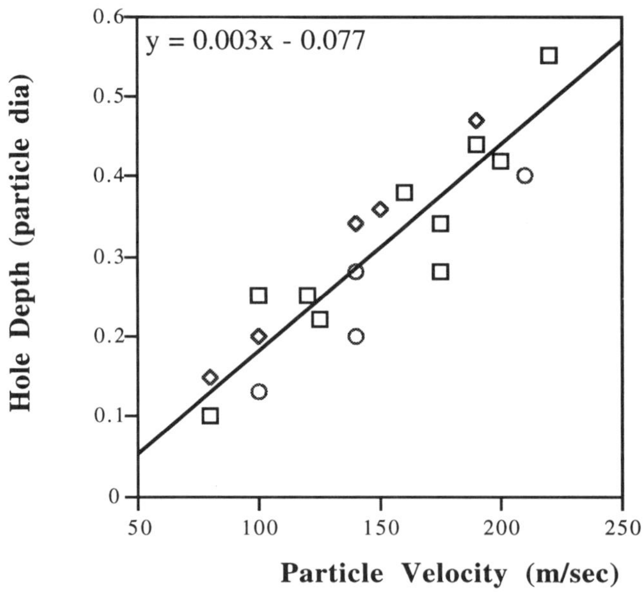

Figure 10.65 Relative amount of Berea sandstone removed by the impact of steel balls of varying size, compared with a predicted depth relationship (after [10.101]).

In this regard Chatterton [10.102] carried out a study which indicated that polished steel balls gave the best cutting performance in penetrating limestone and concrete, with specific energies of approximately half that of limestone particles. Specific energy was also found to decrease with pressure. It should be borne in mind that many of these tests were carried out a stand-off distances of 25 - 45 cm. This is considerably beyond the effective range of many waterjets and is at a point where the abrasive particles would be decelerating. Conclusions from this study were therefore considered to be only of limited reliability.

It is unfortunate that, by 1979, only one study had been reported [10.103] which had correlated data from individual particle impact with the mode of removal within the steady state condition which occurs with multiple particle impact. This is doubly unfortunate since not only does this limit the establishment of the conditions for continuous material removal under continued surface attack, but also that it neglects the increasing role which interactive crack growth will have on the material removal process. It also makes it somewhat difficult to correlate the conditions which prevail in the steady stream erosion of a surface under waterjet abrasive cutting with the predictions which have been made above for the effectiveness of those particles in removing material.

In the sections of this chapter which follow, the results and parameters which have been found to be effective in this new process will be reviewed. Following the introduction of that information it will be possible to revisit this discussion and to comment on the current state of material removal rates, relative to those predicted by the equations above, and what other factors might need to be considered in the analysis.

One should begin by recognizing that there is a major difference between erosion by solid particles and that brought about by liquid impact. One of the fundamental differences can be seen by the rates at which most effective cutting occurs. When a high pressure waterjet is used to cut a surface, experimental evidence is that up to a traverse velocity of well above 1 m/sec (depending on the target material) the faster that the jet is moved over the surface, the more efficient the cutting process is. Experiments have shown that the optimum traverse velocity with an abrasive laden jet lies at around 50 cm/min. This is much slower, and is likely related to the more isolated nature of the crack growth process with the solid impact.

With waterjet impact the fluid will penetrate and extend existing cracks within a zone of effect around the impact point. With solid impact the particle will induce fractures in a given pattern about the impact point, and only if these intersect existing fractures or reach the surface will material be removed. Thus, a critical number of impacts will be required within a defined area before the particles have induced a sufficient number of intersecting cracks to free material from the surface. This period is generally referred to as the incubation period in solid particle erosion testing (Fig. 10.66).

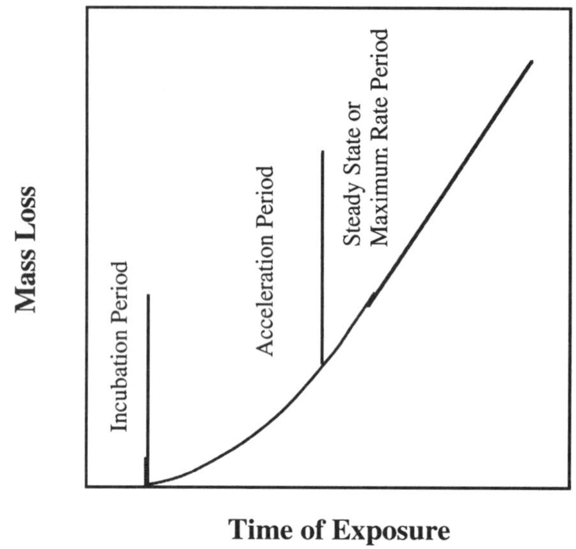

Figure 10.66 Stages in solid particle erosion from a surface [10.93].

The difference in cutting speed is also because of the change in the size of the fragments which are removed. With plain waterjet cutting the jets exploit the existing crack directions (see Chapter 7) and by growing the cracks which already exist can remove quite large fragments of material. With the use of the abrasive jets the particles generate an intense network of cracks within the target surface, and while these will work with existing fractures to remove material, as a general rule they are smaller than, for example, the normal grain boundaries which exist in the rock. The fragments are removed in much finer sizes (a problem in recirculation systems) than that of the grain size of the rock, and the grain boundaries play only a limited role in controlling the efficiency of the rock removal process. Crushing material requires energy, and breaking particles below the natural fracture size requires significantly more energy. In order to effectively remove the same volume of material in a smaller size fraction considerably more energy must be input to the target. Because the energy is absorbed more intensely in each layer being penetrated, the process will also therefore take more time. This much more localized effect of particle impact means that there is no significant effect of the orientation of flaws and cracks in the target surface on the volume of material removed. This was demonstrated by Bortolussi et al. [10.104] who ran tests at different orientations to the weakness planes in a series of granite samples, and found no statistical difference in the volume of material removed in any direction of attack to the surface.

10.6.4 CONSIDERATIONS IN MIXING CHAMBER DESIGN

The above discussion has described the basic mechanisms by which material is removed by abrasive impact. In developing the role of waterjets in carrying the abrasive, one should recognize that, particularly in the removal of metals, the role of the waterjet in displacing material from the surface and in improving the cutting performance of the abrasive is, in fact, limited. Studies have shown in cutting steel that there is no extension of cracks in the surface by plain waterjet impact until water pressures on the order of 1,400 bar have been reached. Under such circumstances virtually all the material removal will occur through the action of the abrasive particles. Optimization of system performance must be directed at optimizing the performance of the particle itself. This means that in the approach to the process, the maximum amount of energy should be transferred from the waterjets to the abrasive particles and from these particles to the target surface.

The most considerable body of work on high pressure, low volume abrasive jet cutting has been carried out by Hashish [10.94]. While he has produced a number of papers describing his findings these have been summarized recently in published reports [10.24]. Regrettably, since this limits the generality of the conclusions he has drawn, almost all of this work, until very recently, has been carried out with the use of the single, side injection of abrasive into a mixing chamber after the water has been accelerated to final velocity (Fig. 10.67).

There are a number of problems with this method of mixing, some of which have significant effects on the results of the cutting process, and which therefore must be considered in judging the analysis of those results which Hashish has drawn. Since these problems are more related to the method of mixing, one must therefore begin by appreciating the problems of this mixing process.

The water enters the mixing chamber through an acceleration nozzle, which directs it out at a pressure which typically ranges from 2,000 to 3,500 bar. At this pressure it is possible for water to cut solid quartzite and basalt without the use of abrasive. The unconfined particles have two problems. Firstly, the impact with the water is likely to shatter them into smaller sizes, and secondly it is difficult for the particle to penetrate through the outer layers of the waterjet stream to obtain maximum acceleration from the faster moving core section.

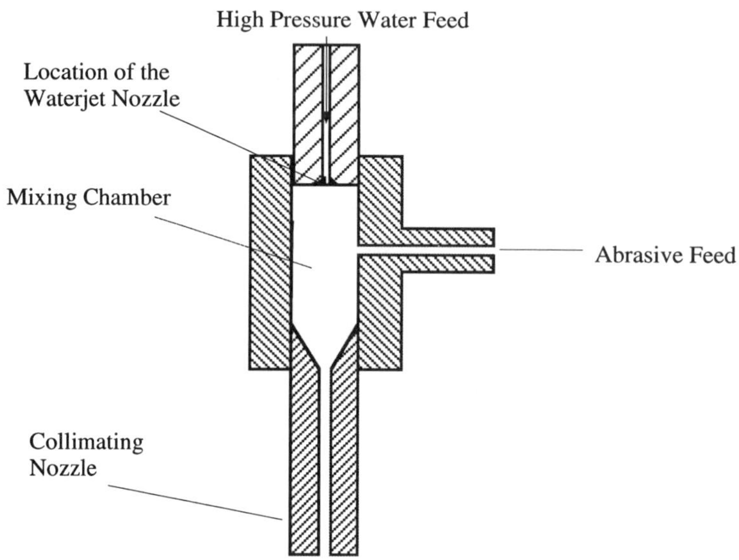

Figure 10.67 Section through the mixing chamber of a conventional abrasive injection system.

The effect of particle size has been partly discussed above, with single particles. The data can be confirmed for abrasive laden waterjets. Because of the problems of abrasive disintegration the exact size of the impacting particles is questionable. However, based on data from Hashish (ibid.) it is possible to derive a curve for the effect of particle size where a garnet abrasive is used to cut cast iron, at a pressure of 2,200 bar, at a traverse speed of 200 mm/min. (Fig. 10.68).

The curve shown through the data points is a little deceptive, since there is a consistent optimum in the data which Hashish has provided over a wide range of abrasive feed rates. This optimum, for a waterjet diameter of 0.45 mm, occurs with an abrasive size of around 180 μm or roughly one third the jet diameter. However it must be remembered that the overall performance is also related to the diameter of the collimating nozzle. Given the significance of this effect it is disturbing to see how many of the particles can be crushed during the mixing process. To find out the scale of this comminution Galecki et al. [10.105] ejected the abrasive stream from an abrasive nozzle along a long, large diameter plastic pipe. In this way the particles were collected without further damage and could be sieved to determine the comminution. The results (Fig. 10.69) showed that only 20 - 30% of the particles survived the mixing process in their original size. The higher the jet pressure, and the smaller the diameter of the collimating nozzle, then the more severe the comminution that occurred. For the sake of illustrating the trend only the two extremes of the data presented in the original paper are presented here. The paper itself gives the result from a 2-factor, 3-level factorial experiment.

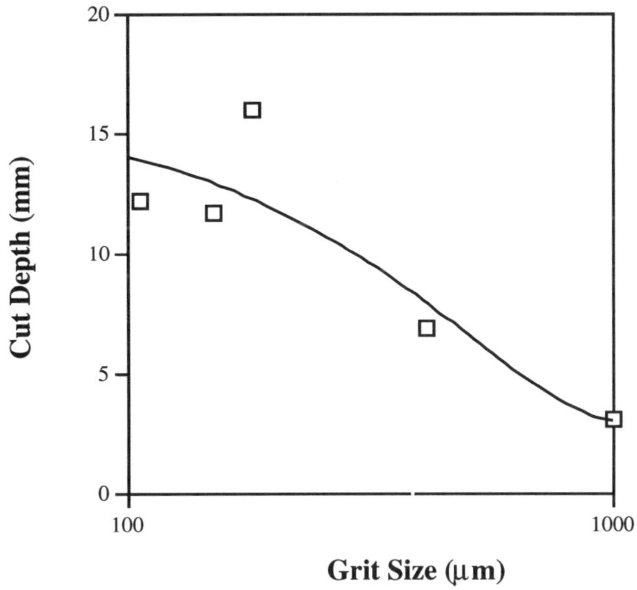

Figure 10.68 Effect of particle size on the depth of abrasive jet cut into cast iron, at an abrasive feed rate of 0.4 kg/min (after [10.24]).

The significance of this can perhaps be examined in terms of the economics of the cutting process. Labus [10.106] was quoted in Chapter 4 as pointing out that, with a single pass through abrasive, a major cost of the cutting process becomes the abrasive itself. This is likely to become increasingly true as environmental laws will make it as costly to dispose of the abrasive as it was to purchase in the first place. Yet with the crushing of the abrasive which occurs with conventional mixing there is little choice, since the reduced size of the particles make them less effective in subsequent cutting processes. Alternatively if the abrasive could be mixed into the water another way, which avoided the high velocity impact between the two then this undesirable crushing can be avoided. In the most dramatic example the water and abrasive are accelerated together in systems like the DIAjet process, without the violent impact of mixing. In this case it has been reported (Fig. 10.70) that the abrasive can be recycled over nine times without significant deterioration in cutting ability. Tests at UMR with garnet have shown that the abrasive can be recycled in cutting dolomite it is important to screen the cuttings and crushed material from the recycled feed, since otherwise this will occupy the system, and cause a significant loss in performance.

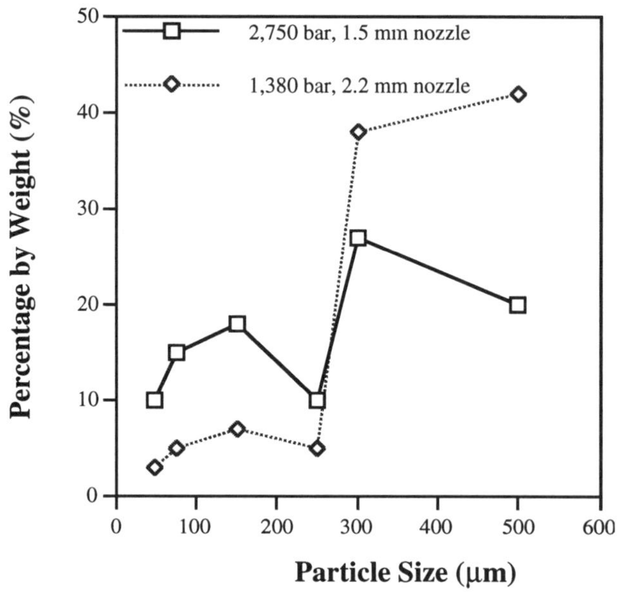

Figure 10.69 Size of particles collected from an abrasive jet after leaving the collimating nozzle - the original particles were approximately 100% at 500 mm size, two mixing conditions are shown [10.105].

The reduction in particle size will also make it more difficult for the abrasive to enter the jet. This point can be seen by reference to data from Hashish. It can be seen (Fig. 10.71) that as the waterjet pressure increases, then over the range reported, that the larger nozzle diameter will give the greater cutting depth. In this size range the abrasive is able to penetrate into the jet and the greater nozzle diameter is transferring more power to the abrasive. However, when the initial waterjet nozzle diameter is further increased (Fig. 10.64) it becomes more difficult, within the geometry of the mixing chamber, and the collimating nozzle for the abrasive to reach the center of the jet stream and less abrasive is accelerated to full velocity, and with increasing diameter, performance of the abrasive jet decreases.

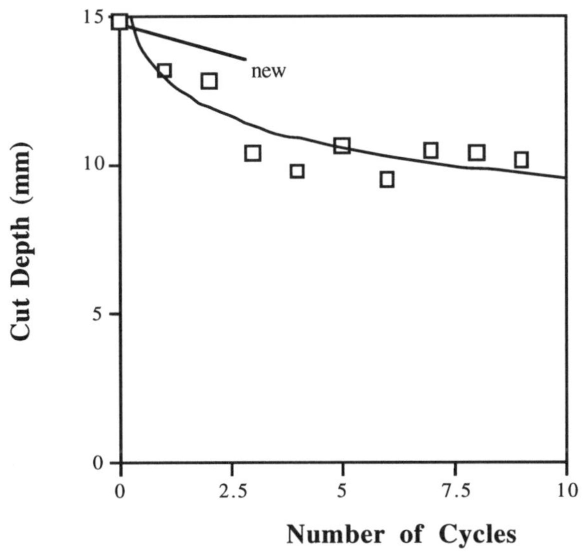

Figure 10.70 Effect of number of cycles through the system on the depth of cut achieved with a slurry abrasive feed [10.107].

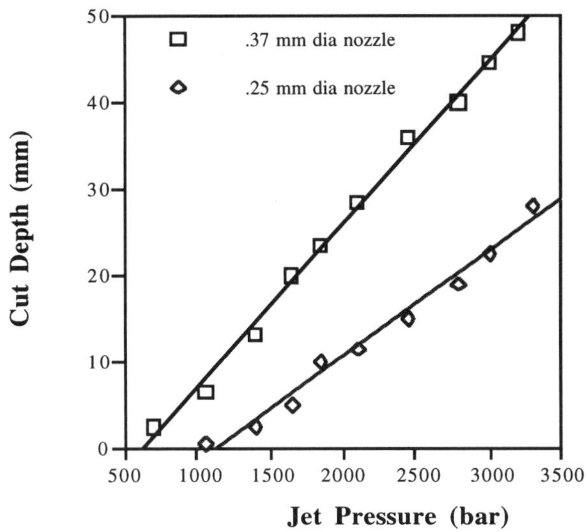

Figure 10.71 Effect of jet pressure in cutting aluminum at 15 cm/min with a 60 mesh garnet at a feed rate of 0.8 kg/min [10.24].

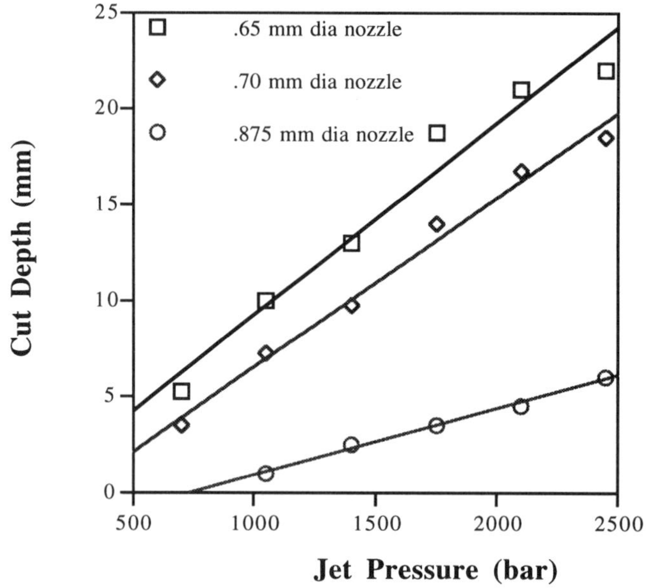

Figure 10.72 Effect of jet pressure in cutting mild steel with larger diameter jets and an abrasive feed of 60 mesh garnet of 0.31 kg/min [10.24].

One can backtrack from this result to examine the cause of the problem. This is that the particles are not properly being accelerated where the jet nozzle diameter, the particle feed rate, and the exit nozzle diameter are not correctly selected. This can be illustrated by a study of the variation in the velocity distribution of the particles leaving the nozzle [10.108]. It can be seen (Fig. 10.73) that there is a broad distribution of particle velocities in the stream leaving the nozzle. The greatest velocity for most particles is achieved at lower abrasive flow rate, larger nozzle diameter and larger collimating nozzle orifice. It should be noted that the particles in this case are steel shot.

The abrasive mixing chamber is acting as a small jet pump in this case. For those to work most efficiently the energy must be effectively transmitted from the water to the other entering fluid (which in this case is usually the air carrier and the abrasive particles). The work discussed above has shown that the abrasives will be most destructive when they impact the surface at the highest velocity and the more destructive jets will need to balance the number of particles striking a surface to create damage with the amount of damage that each makes on impact.

To transfer the energy from the waterjet to the abrasives it must break up and allow mixing with the entraining particles of abrasive. But when one adds polymer to the water this coheres the jet and reduces the ability of the particles to enter the stream and acquire its velocity. It is for this reason that Hashish reports [10.24] that the use of long chain polymers is ineffective when used with conventional abrasive injection systems (Fig. 10.74). In his work the polymer was added to the waterjet stream before it was fed into the mixing chamber. Under such conditions the polymer, which acts to cohere the stream, makes it more difficult for the abrasive to enter the stream and for mixing to occur. The abrasive is inefficiently accelerated and the performance of the jet is reduced.

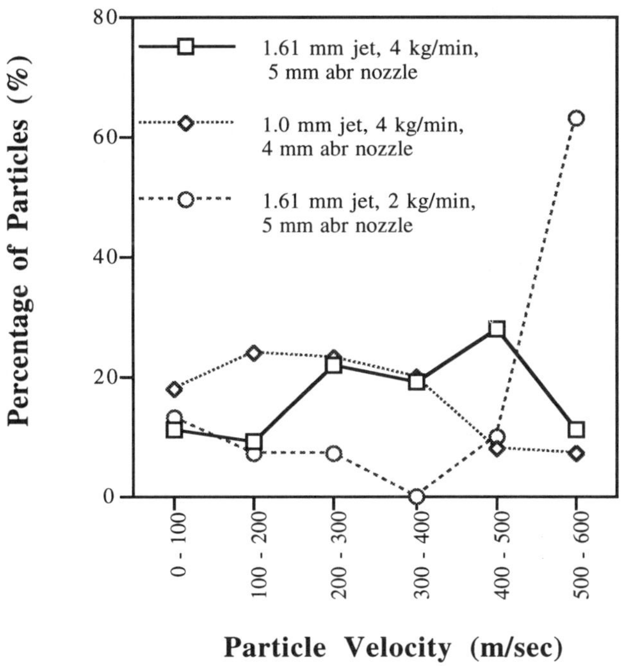

Figure 10.73 Particle velocities on leaving the collimating nozzle (after [10.108]).

In contrast where the polymer and the abrasive are both mixed into the water before it is accelerated, as for example with the DIAjet system [10.109], then a similar long chain polymer was found to enhance performance (Fig. 10.75). In addition, the work by Hollinger [10.25] discussed above has shown the benefits of the combination. It should be noted that in correlating the use of polymer to the amount of abrasive required to cut a given volume of wall this also translates into a reduction in time with the use of the polymer by up to a factor of 3 or more at greater stand-off distances.

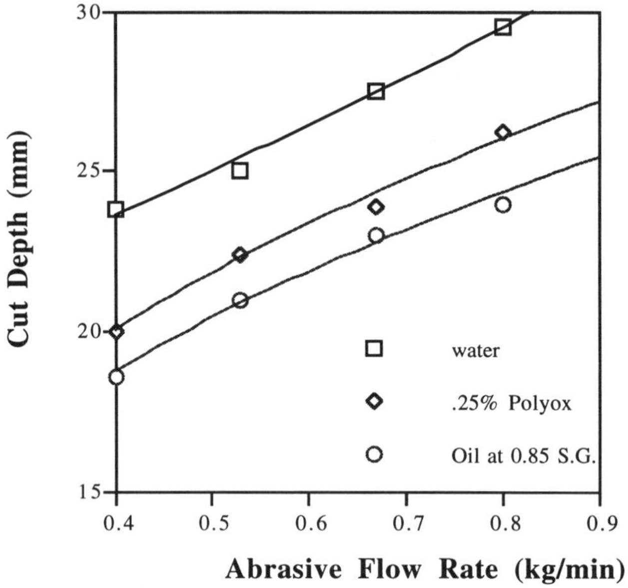

Figure 10.74 Effect of changing the jet fluid on cutting performance with a conventional abrasive waterjet system [10.24].

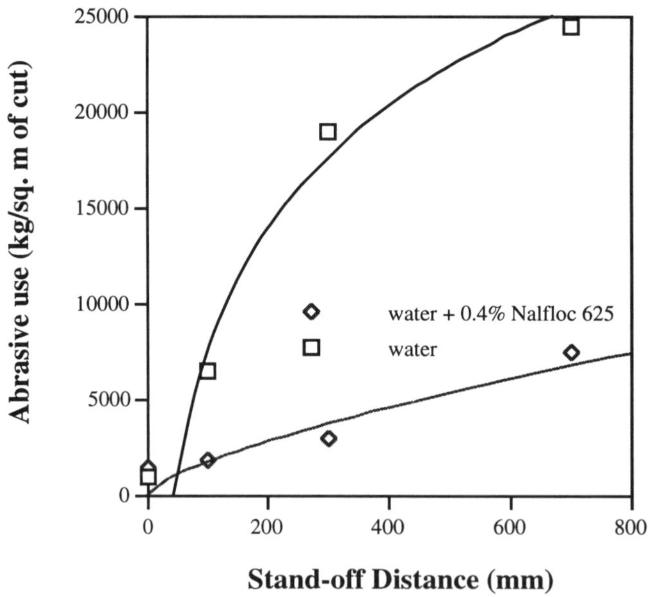

Figure 10.75 The effect of adding polymer in reducing the amount of abrasive required to cut stainless steel, with a 2.0 mm jet at 345 bar with an abrasive concentration of 10.6% [10.109].

These considerations make it of critical importance to design the mixing chamber of a conventional system to provide optimum acceleration, with minimum disruption of the abrasive. In addition it is important to ensure that as great a percentage of the abrasive as possible is distributed in the center of the jet in order to produce the cleanest, deepest cut. Mazurkiewicz et al. have shown [10.110] that in relative terms a smaller diameter waterjet orifice will produce a more concentrated abrasive stream than a larger diameter (Fig. 10.76). Note however that this does not track the relative velocity of the particles as shown above (Fig. 10.73).

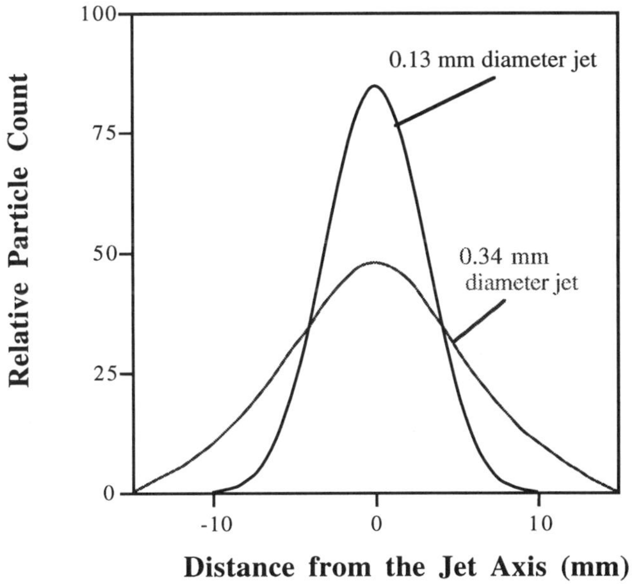

Figure 10.76 Relative grain distribution curves across a 2,800 bar jet, with a collimating nozzle diameter 2.3 mm, abrasive feed of 0.45 kg/min at two jet diameters [10.110].

One should also note that the type of abrasive used does not necessarily appear to materially influence the break-up of the particles through the mixing process, and the subsequent impact on the target surface. For while steel particles show very little degradation through one cycle, Saunders [10.109] has shown that both alumina and sand go through very significant degradation, even with a DIAjet system (Fig. 10.77).

There have been several different designs proposed for changing the nozzle geometry to ensure better mixing with conventional abrasive injection systems, and thereby improving cutting ability. For example as a result of a study of conventional nozzle design at UMR, [10.111] which examined, among other parameters the vacuum pulled in the mixing chamber by jet action, it was clear that the longer the mixing chamber could be, then the more efficient the process. (The ultimate version of this is the drilling device developed by Dr. Savanick [10.112] described in Chapter 8.)

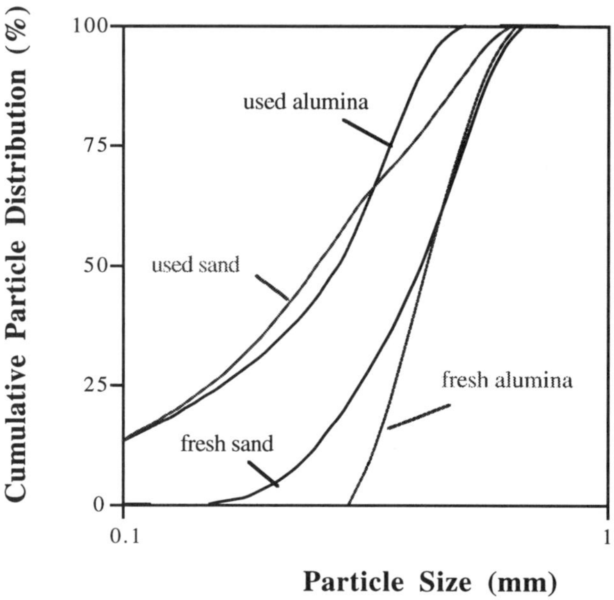

Figure 10.77 Particle size distribution before and after use in a DIAjet system at 345 bar, with a 2 mm diameter nozzle [10.109].

By changing the dimensions of the chamber, improvements in cutting performance of over 75% were reported [10.113]. Two parameters were studied in the program, the relative ratio of the waterjet to the collimating nozzle diameter, and the distance from the abrasive inlet to the collimating nozzle. In order to assess the results the time taken to penetrate a 38.1 mm thick aluminum target was measured. The results of change in length of internal stand-off distance showed (Fig. 10.78) the importance of increasing this length. The work suggested that increasing the collimating nozzle length would improve the power of the abrasive stream. Recent work by Galecki [10.114] has confirmed the latter hypothesis (Fig. 10.79).

The study also considered effect of particle size on optimum nozzle dimensions, and noted that the optimum ratio increased with increase in particle size and particle feed rate. In particular it found that where optimum nozzle ratio was not used for a given particle size, then the gain from increasing the particle diameter would be lost. As with the comments made above on going to too large an inlet diameter for the waterjet to the mixing chamber, this study also found that where the inlet nozzle became too large that less efficient mixing occurred and performance was reduced.

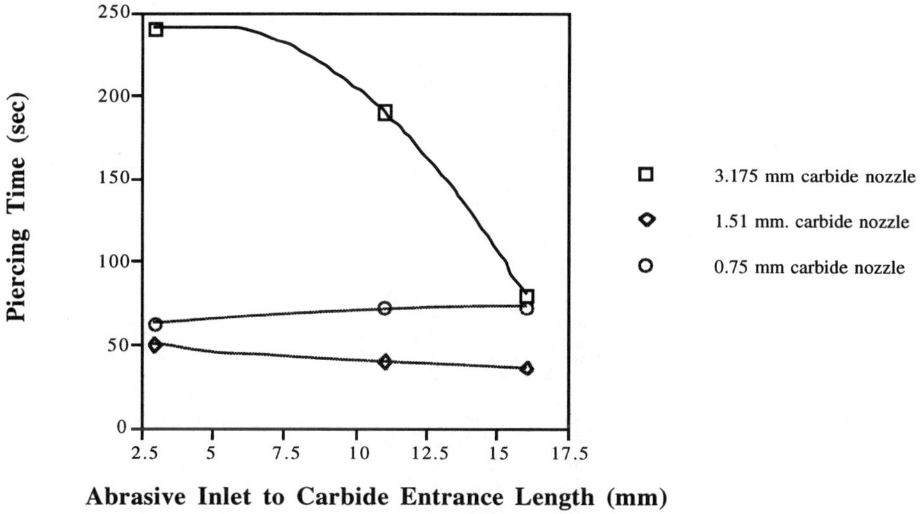

Figure 10.78 Time to pierce an aluminum sample as a function of internal mixing chamber internal stand-off distance (after [10.113]).

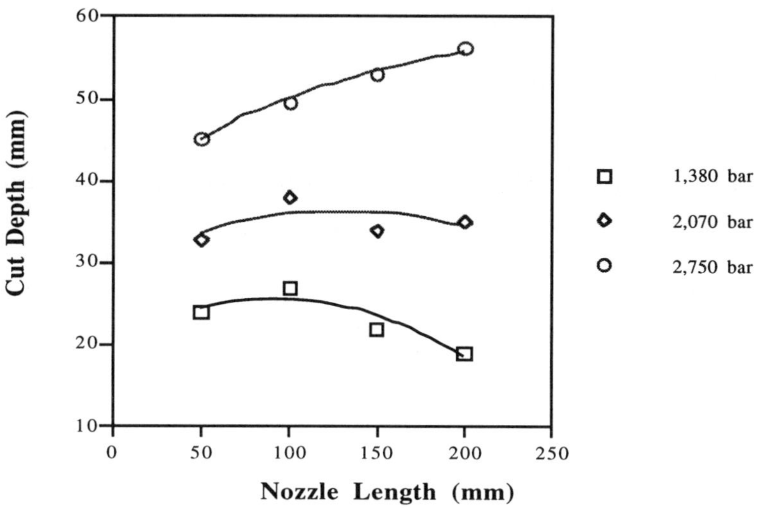

Figure 10.79 Depth cut by steel shot in dolomite, as a function of collimating nozzle length (after [10.114]).

In order to overcome some of these disadvantages investigators have sought to change the configuration of nozzle to improve abrasive mixing. One such [10.115] brought abrasive flow into the center of the jet by focusing converging smaller jet streams within the venturi chamber and drawing abrasive material through the central common focus (Fig. 10.80).

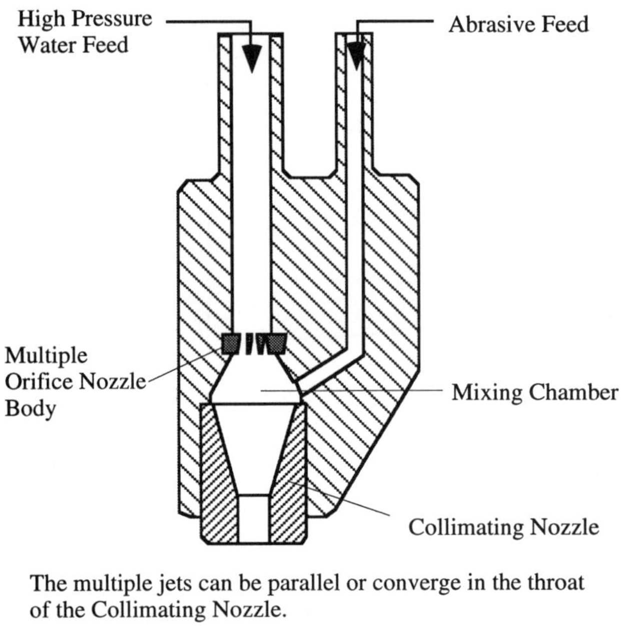

Figure 10.80 Nozzle design for abrasive jet cutting (after [10.115]).

The relative gain that this gives to the performance can be illustrated by contrasting the ability of this jet to cut stainless steel. A study by the Welding Institute [10.116] had used a jet at a pressure of 690 bar with a feed of 2 kg/min of a flint abrasive ranging in size from 0.4 - 1.4 mm to cut through a 10 mm plate at a speed of 35 mm/min. Yie reported [10.117] being able to cut the same material to a depth of 6 mm at a traverse speed of 125 mm using 0.4 kg of 50 mesh garnet at a pressure of 1,000 bar. For a more conventional nozzle system, with an abrasive feed rate of 0.9 kg/min of 60 mesh garnet, a 1,600 bar jet should be able to cut through 10 mm of steel at a speed of 150 mm/min [10.24]. Unfortunately these numbers cannot be absolutely correlated since there is a variation in edge quality with speed, and certain limitations must be placed on acceptable limits, which may be controlling the values reported.

Other designs have also been suggested from time to time in order to improve the performance of the system. Horii, et al. [10.118] have suggested a more even entry through a swirling section (Fig. 10.81). This nozzle has not yet been demonstrated in cutting at higher fluid pressures.

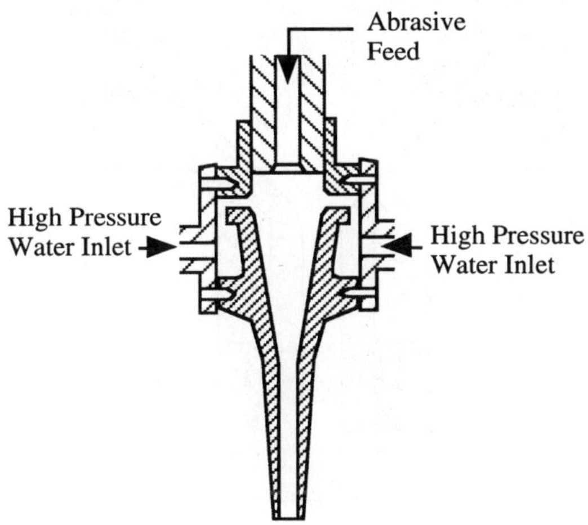

Figure 10.81 Novel abrasive nozzle design proposed by Horii [10.118].

Faber and Oweinah [10.119] have proposed an alternative design in which (Fig. 10.82) the abrasive feed is from opposing sides of the waterjet feed. Interestingly, in that study there was no change in the optimum effective length of the collimating nozzle as a function of abrasive feed rate (Fig. 10.83). The majority of this study was at significantly higher pressures - up to 5,000 bar - than those of most other studies. It was one of the conclusions of the study that trying to use abrasive above a jet pressure of 4,000 bar had no significant improvement in performance, and, in addition, introduced some bad side effects in the process.

One conclusion that can be drawn from the above is that there is still some gain to be achieved by additional improvements in the mixing of the abrasive with the water. The studies have not discussed the problems of properly aligning the jet nozzle and the collimating nozzle, and the results cited above are assumed to have been carried out under perfect alignment. This is, unfortunately, not always the case in an operational environment. There have been several improvements by the different equipment manufacturers in recent years to overcome this problem, however [10.120], 10.121].

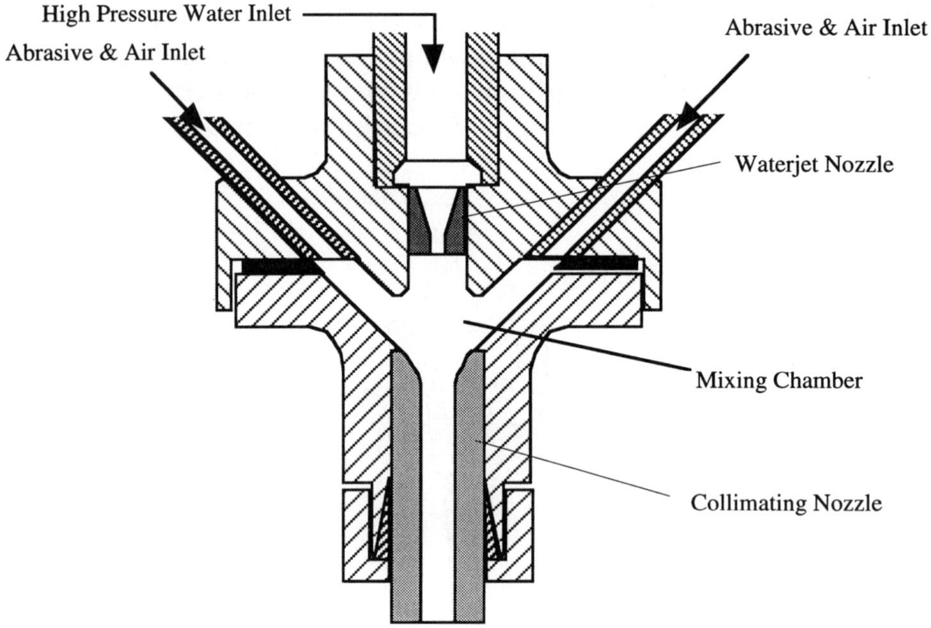

Figure 10.82 Nozzle design studied by Faber and Oweinah [10.119].

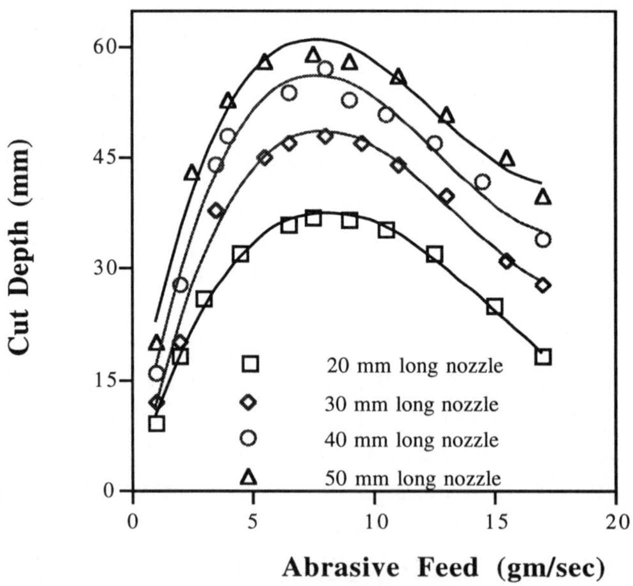

Figure 10.83 Depth of cut by a 5,000 bar 1.2 mm diameter abrasive-laden jet in an Al Mg Si composite as a function of collimating nozzle length [10.119].

There is one additional point which might be mentioned in regard to the mixing process. The normal carrier to bring the abrasive into the mixing chamber is air. This has a disruptive effect on the jet issuing from the nozzle, and thus it has been suggested that the transport fluid should be changed to water. Griffiths [10.122] has shown in a study at lower pressures, and relative to cleaning effectiveness (Fig. 10.84) the addition of water with abrasive adds a greater quantity of material which is accelerated by the high pressure jet as it enters the mixing chamber. As a result there is a significant reduction in the performance of the jet, since the action of the abrasive waterjet is completely due to the impact of abrasive particles, and the presence of additional water, even at high speed, has little positive effect on the process. The reduction in energy applied to the abrasive reduces their velocity, and the performance on impact.

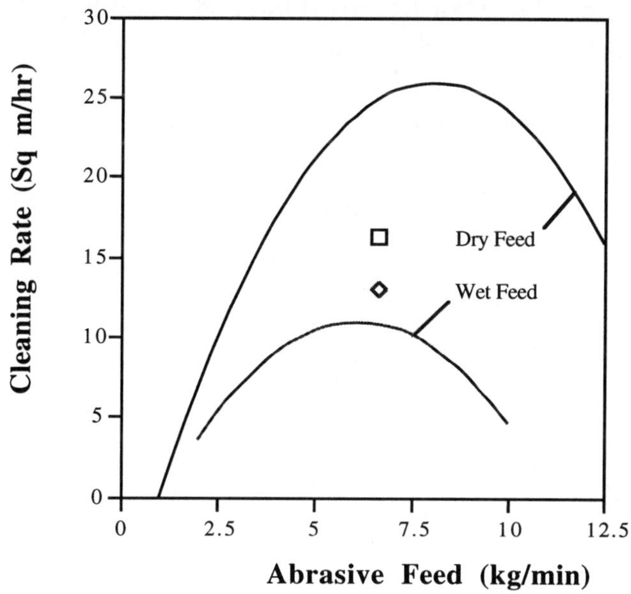

Figure 10.84 Cleaning rate as a function of abrasive feed rate for wet and dry abrasive feeds [10.122].

10.6.5 ABRASIVE JET PARAMETER EFFECTS

There has been a wide range of cutting data now reported on the performance of abrasive laden waterjets in cutting through a variety of materials. In the interests of simplicity only significant aspects of some of this material will be included in this section. As discussed above there remains some wide variations in the results which can be achieved, as a function of abrasive used, nozzle design conditions etc. Results cited below should be treated as those which indicate a trend, rather than considered absolute predictors of abrasive jet performance for any given condition.

The work described above on single abrasive particle impacts stressed the effects of abrasive size and impact velocity on the amount of damage inflicted on the surface. Where the particle is only one in a stream, then it is initially important to first determine the factors of the stream which control material removal. In this regard the work by Hashish is most significant in its comprehensive treatment of the problem [10.24].

Consider first that the controlling parameters of erosion will be the quantity of abrasive falling on a point, and the velocity of its impact. These are normally referred to as the abrasive feed rate (AFR) while the velocity is more frequently identified in terms of the feed pressure of the high pressure water. Hashish has correlated the depth of cut through mild steel with both pressure and abrasive flow rates (Fig. 10.85).

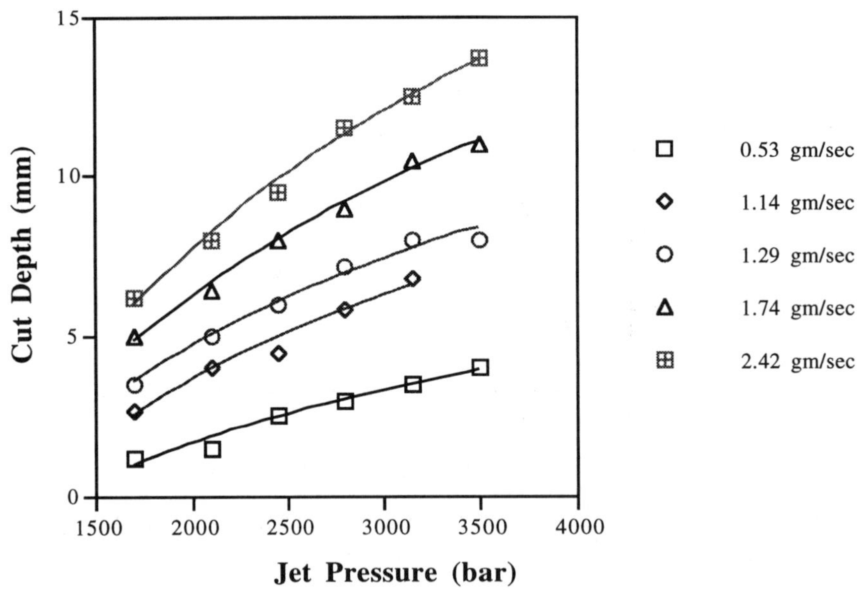

Figure 10.85 Effect of jet pressure and abrasive feed rate on the depth of cut in mild steel (using #60 mesh garnet a jet diameter of 0.25 mm and a feed rate of 150 mm/minute) [10.24].

It is interesting to note that the initially linear improvement in cutting performance begin to diminish as the jet pressure rises to 3,500 bar which tends to correlate with the conclusion of Faber and Oweinah that abrasive jets become less effective above pressures of 4,000 bar [10.119].

Yie has reported the performance of abrasive waterjets in cutting basalt, quartzite and concrete, which respond as more brittle materials. It is interesting to note from the curves generated (Figs. 10.86, 10.87 and 10.88) that while there is a clear optimum to the cutting performance of the jets in basalt within the range tested, such is not the case with concrete.

This may well be due to the differences in material response given that the structure of concrete may well cause it to act more as a ductile material, due to the localized crushing and fracturing which is confined by adjacent free surfaces, rather than the brittle behavior of basalt where the continuous nature of rock means that cracks may extend to their full range.

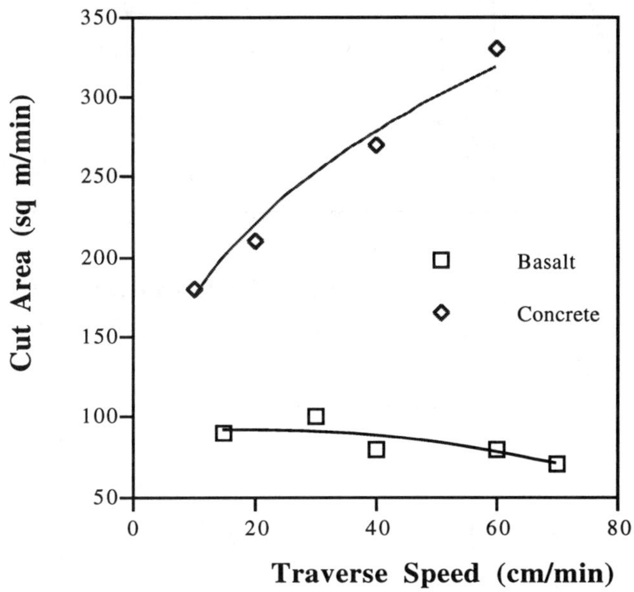

Figure 10.86 Area of cut as a function of traverse speed (after [10.117]).

In developing a predictive equation for the cutting ability of a plain waterjet it is possible to obtain a simple predictive equation for the depth of cut which relates the depth of cut to the jet pressure, the nozzle diameter and the traverse speed of the nozzle over the surface in the form:

$$\text{Depth of cut} = \text{constant} \cdot \text{pressure} \cdot (\text{nozzle diameter})^{1.5} \cdot (\text{traverse velocity})^{-0.33}$$

This curve has been developed over a long range of time with a wide range of target materials. It provides a simple way of evaluating the likely performance of a waterjet in cutting a new target material from the results of a relatively small number of tests.

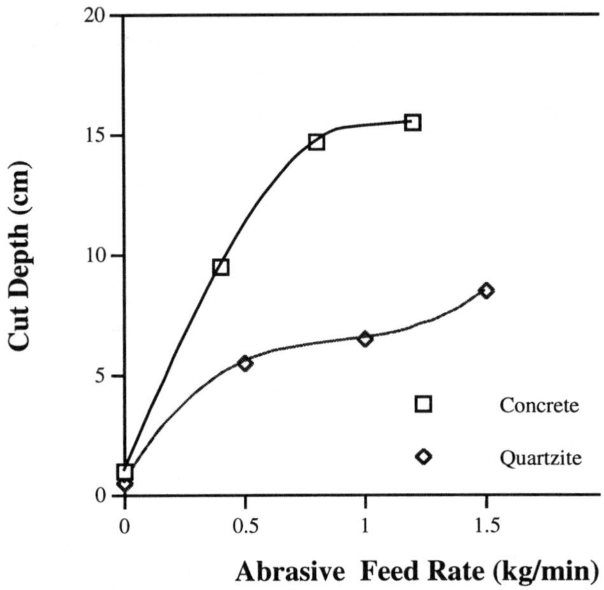

Figure 10.87 Depth of cut as a function of abrasive flow (after [10.117]).

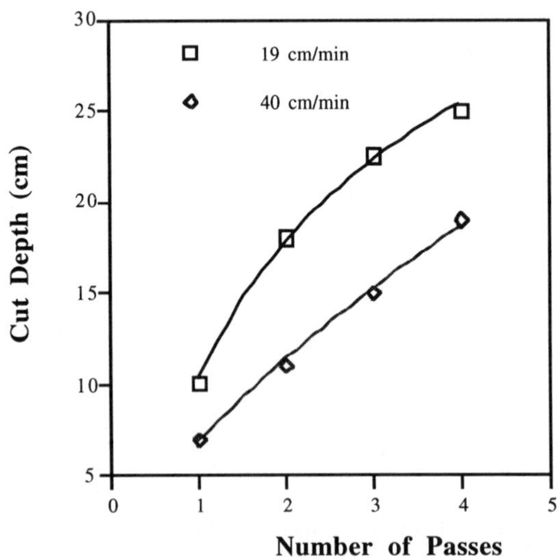

Figure 10.88 Depth of cut as a function of number of passes (after [10.117]).

When one examines the performance of an abrasive jet in cutting material, even recognizing problems of two different failure mechanisms, there are other changes even from an empirical point of view, relative to plain waterjet performance. The curves above indicate that the linear relationship between depth of cut and pressure, over the majority of the commonly used pressure range still remains. However the relationship with nozzle diameter has changed. In terms of the waterjet nozzle larger nozzles can be less efficient (Fig. 10.89) and with the diameter of the collimating nozzle. It should be borne in mind, in evaluating the curve shown, that this is restricted to the particular conditions for the other parameters of test and cannot be expanded to a more general relationship. The difference with the curve prediction for plain waterjet cutting is clear.

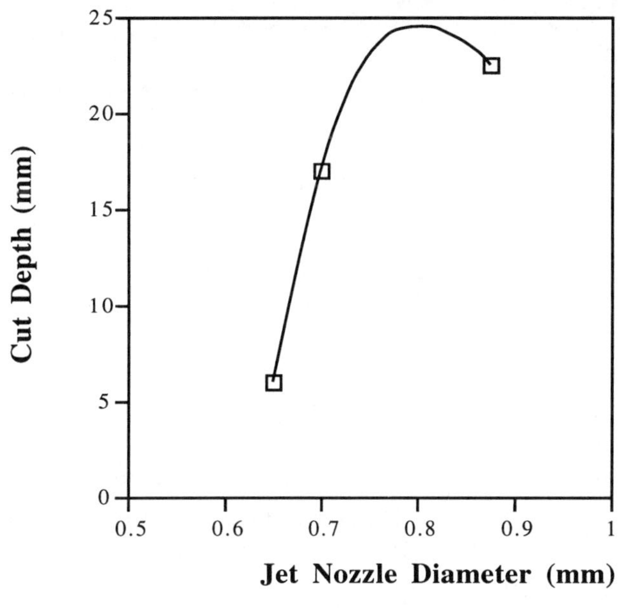

Figure 10.89 Depth of cut in mild steel at 2,500 bar as a function of jet nozzle diameter (after [10.24]).

As has been discussed in the earlier section, abrasive jet cutting speeds are considerably slower than those of conventional jets, and optimal speeds, in terms of the area of the cut created and the volume of material removed, are found are relatively low velocities (Fig. 10.90).

In terms of the simple equation the rules for most effective abrasive cutting have changed from those of plain waterjet cutting. Given further that the damage is largely caused by the abrasive particle impact, rather than that of the water, it is the parameters of that stream which become more important. Yazici [10.123] has defined a term, **erosion**, to determine the relative amount of material removed in cubic cm for each gram of impacting abrasive. This can be considered with the normal calculations of specific energy to determine the most effective method of using the abrasive waterjet system.

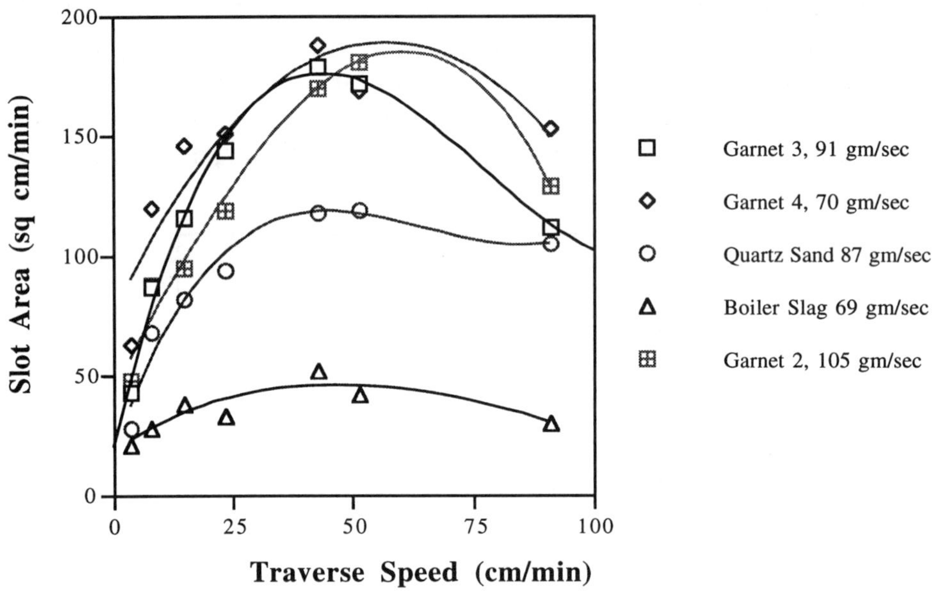

Figure 10.90 Area of slotting rate as a function of traverse velocity [10.123].

The abrasive feed into the mixing chamber will increase jet performance until the point is reached that too much abrasive is entering the chamber, and the feed begins to choke. At this point the depth of cut will begin to fall off (Fig. 10.91).

However, there is no fixed optimum value to the abrasive feed rate, since this is a function of the carrying capacity of the waterjet feeding into the mixing chamber and the efficiency with which the energy in that water can be transferred to the abrasive, either in the mixing chamber or between it and the target. Thus, as the jet pressure increases, so the volume of the water entering the mixing chamber also increases, and the optimum carrying capacity, and thus the power of the jet also increases (Fig. 10.92).

Figure 10.91 Depths of cut in granite at constant water pressure and flow rate, with increasing feed of abrasive [10.123].

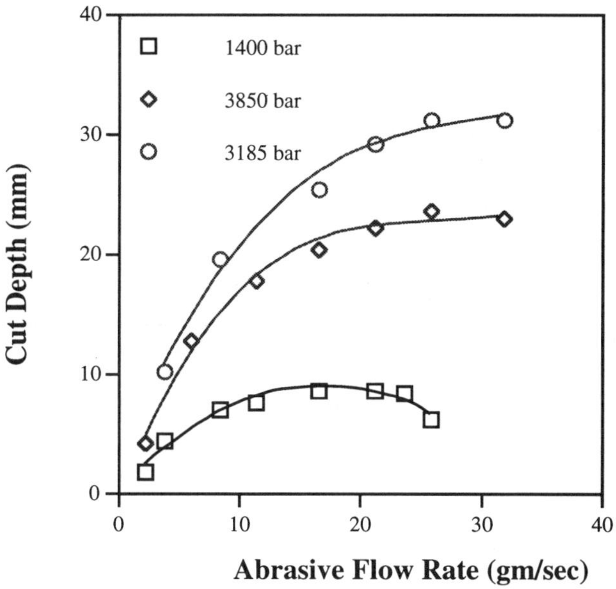

Figure 10.92 Depth of cut in mild steel as a function of abrasive feed rate at different jet pressures (after [10.24]).

As the abrasive feed is increased the specific energy of the process will reflect the improved cutting ability, up to the point that system efficiency falls off with the onset of choking (Fig. 10.93). This point can be seen in the earlier figure (Fig. 10.91) where the final cut made is at significantly less depth than the earlier cuts. It is of interest to note that, while the cuts are at the lower abrasive feed rate did not generate any crack growth out from the slot, at the highest abrasive feed rate there was a significant amount of lateral crack growth from the slot cut out into the surrounding material. To date, there is not enough known about interaction between effects of the jet pressure and abrasive during the cutting process to accurately explain why this occurs. In other materials it is thought that this might be due to an increased role for the waterjet in the cutting process, over that of abrasive. The abrasive is not removing material faster than jet pressure can grow the cracks beyond the zone of immediate jet influence.

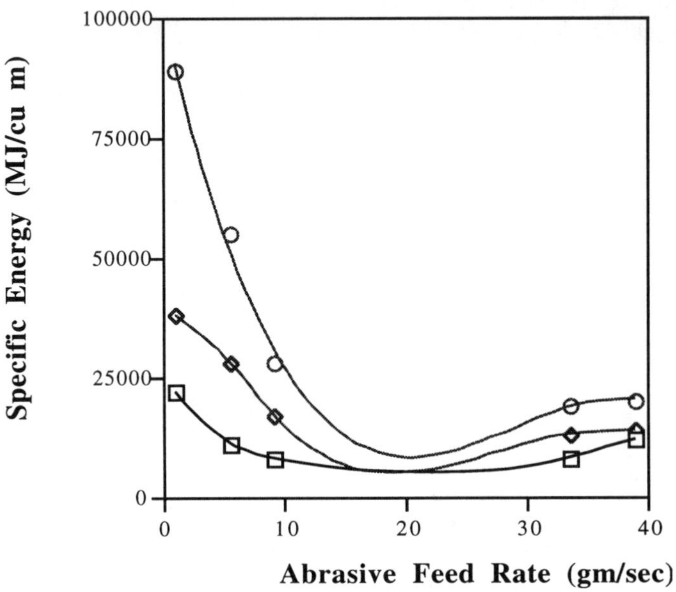

Figure 10.93 Specific energy of cutting granite as a function of abrasive feed rate, conventional abrasive cutting Georgia granite at 15 cm/min [10.123].

If one considers the most efficient use of abrasive an optimal value appears at a much lower level (Fig. 10.94). At the levels of abrasive feed normally considered, there is a significant energy loss from the waterjet stream in accelerating abrasive, the more abrasive there is, the slower it will ultimately move. Since the damage caused by individual particles is related to impact velocity, the slower the particles are going, on a unit volume basis, the less material removed.

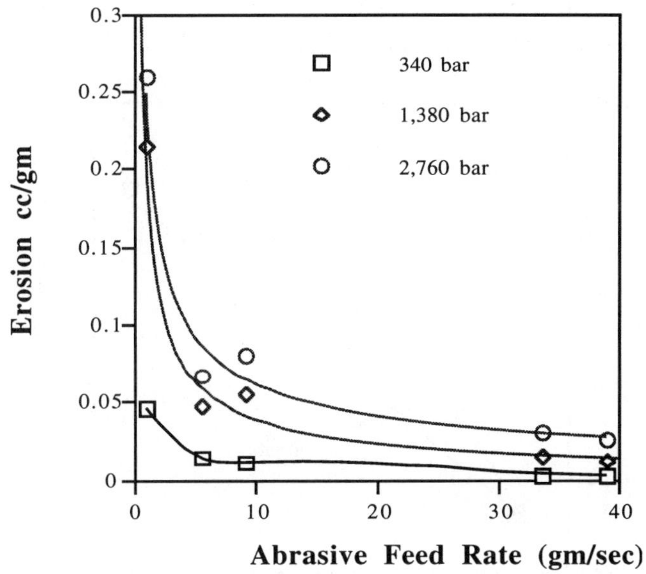

Figure 10.94 Erosion as a function of abrasive feed rate, conventional abrasive cutting Georgia granite at 15 cm/min [10.123].

By the same token increasing the supply jet pressure increases energy available to accelerate the particles, and the erosion therefore increases (Fig. 10.95). One finds that, concurrently, with an increase in jet pressure the amount of material removed by each grain of abrasive increases (a good thing) at the same time as the overall amount of energy required to remove unit volume of material also increases (a bad thing). This means that each case where a decision on the option to choose must be made will need to be decided on its own unique merits. Are equipment, power and maintenance costs cheaper than spent abrasive disposal costs, for example.

The concept of individual particle energy, discussed in the beginning of this part of the chapter also plays a part in determining the best parameters for the cutting jet. Different target materials will require different levels of particle energy in order that material be removed from the surface. Thus in the cutting of the mild steel above (Fig. 10.92) as the abrasive feed reached greater levels the energy which the jet could impart to the individual particles began to fall below that required to remove the steel, and the curve began to show a decline in jet performance. On the other hand, where the material being cut is aluminum, the energy required to remove material is significantly less, and thus, over the same abrasive feed range the reduced individual particle energy was still above that required to remove metal, and the increase in the number of particles caused the depth of cut to continue to increase with abrasive feed rate (Fig. 10.96).

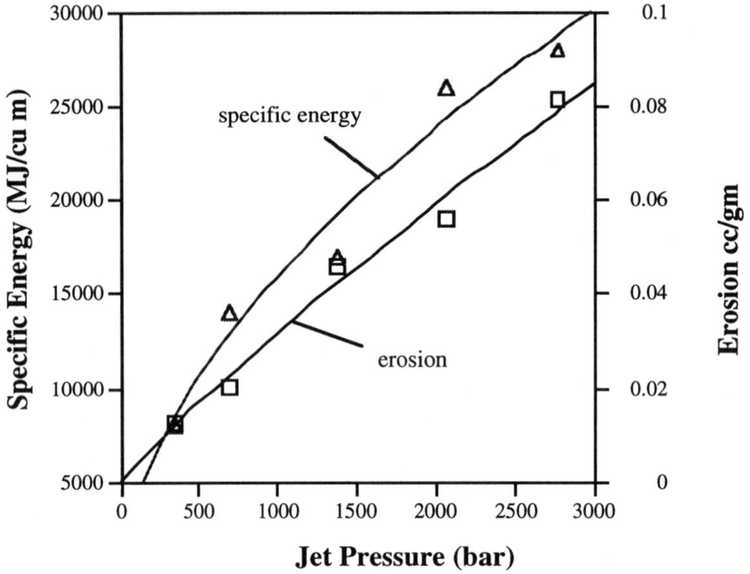

Figure 10.95 Erosion and specific energy of cutting as a function of supply jet pressure (after [10.123]).

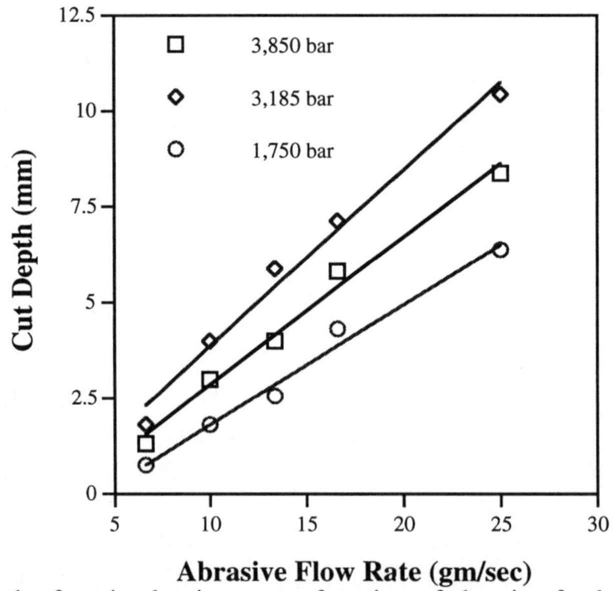

Figure 10.96 Depth of cut in aluminum as a function of abrasive feed rate at different jet pressures (after [10.24]).

In a typical abrasive waterjet cutting operation the abrasive is likely to be a fine grade of garnet (80 - 100 mesh is reported as the most common - [10.24]) and the jet cuts a slot typically 0.75 - 2.5 mm in width with a surface roughness of 2 - 6 μm.

10.6.6 CHOICE OF ABRASIVES

It is the reported practice at the British Welding Institute, when cutting at 1,000 bar and at a flow rate of 25 lpm, to use a flint abrasive up to 50 gm/sec [10.116]. When cutting softer materials smaller particles are used measuring 0.1 - 0.4 mm in diameter. When harder materials and metals must be cut, the system is changed to use 0.4 - 1.4 mm diameter particles.

Faber and Oweinah [10.119] have reported that while copper slag will cut metal three times as rapidly as equivalent feeds of garnet, when the target material is changed to concrete then this situation is reversed.

Griffiths [10.122] has addressed the relationship between particle size and cleaning efficiency, showing that, up to a certain particle size, performance was improved. Beyond that size, he hypothesized that, as a result of an inability to properly accelerate the particle over the distance available, cleaning rate declined (Fig. 10.97).

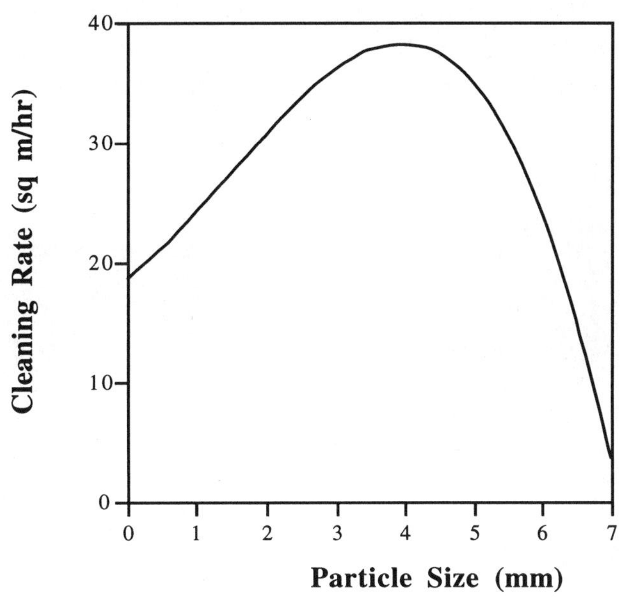

Figure 10.97 Effect of particle size on cleaning rate [10.122].

The role of the abrasive particles is thus a critical one in ensuring that the abrasive waterjet system performs at maximum benefit. This relates not only to particle size, and particle density but also to particle shape. For example, relatively flat shaped particles, but with a leading edge, may be most effective in cutting ductile materials where they would act as the sharp edged planing blade, but would be less effective in inducing the ring fractures on a brittle material. To generate brittle fractures, one needs to concentrate the impact energy on a single point on the surface since in this manner the most intensive fracture can be induced under the contact point.

To illustrate the importance of the proper choice of abrasive consider the use of steel shot as an abrasive for the cutting of either metals, glass or rock [10.125]. In the case of glass cutting the impact of the ball concentrates the initial particle energy at a very small contact point where the ball touches the surface. This intensifies the energy applied to the surface at a small localized point and induces cracking under the point and a surrounding ring fracture mechanism giving the crack system which would best be exploited by subsequent impact. If the same steel ball impacts on a ductile material such as aluminum, the ball may sink into the surface, displacing aluminum around the crater and forming a lip, but without liberating any material from the surface. The ball may impact on a strong elastic material (tool steel) and rebound from the surface without any penetration. If the performance of steel is compared with garnet, in the brittle material the steel is better, but in the ductile and elastic material the garnet cuts a much deeper and more effective path (Fig. 10.98).

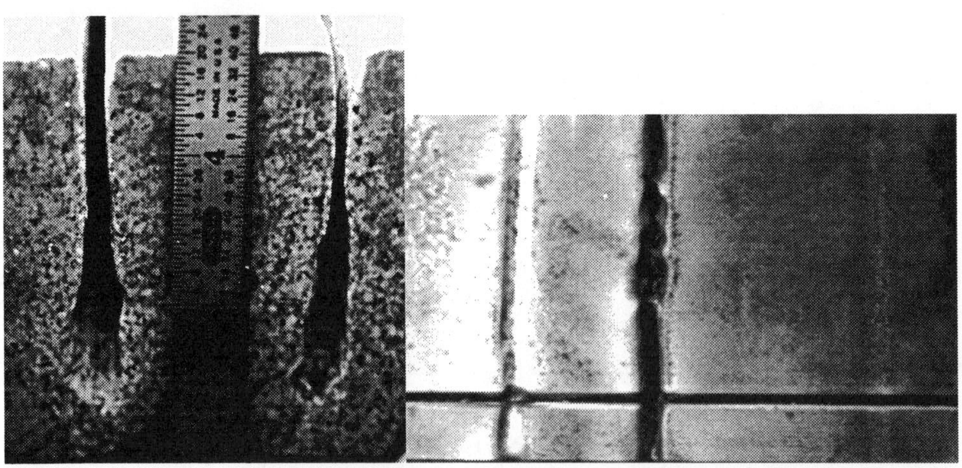

Figure 10.98 Relative garnet and steel shot paths across granite (a brittle material) and tool steel (a ductile material).

Faber and Oweinah [10.119] have studied various of the parameters of different abrasives and shown how these can significantly change the ability of the cutting stream. They considered the difference between broken glass fragments having sharp edges, with cutting ability of glass beads the same size, 158 μm and the same density. The tests were carried out at a pressure of 4,000 bar, a jet nozzle diameter of 0.18 mm, a collimating nozzle diameter of 1 mm, a stand-off distance of 5 mm, and an attack angle of 75° to an Al Mg Si 1 composite, moving under the nozzle at a feed rate of 50 mm/min. The results (Fig. 10.99) clearly show in this material the sharp edged cutting glass is more effective in cutting than point impact attack of glass spheres. It should be noted that the attack angle also favors the ductile cutting of the broken glass over the crack induction from the beads.

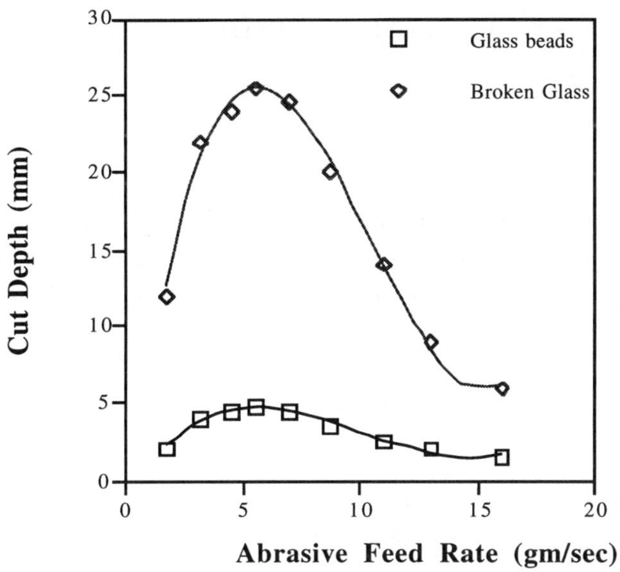

Figure 10.99 Depth of cut into a composite material, as a function of the shape of the abrasive [10.119].

In similar manner they examined effect of grain size, using corundum as abrasive (Fig. 10.100), the effect of grain hardness when cutting both metal composites and synthetic fiber materials (Fig. 10.101), and density of particles used in cutting aluminum composite, contrasting performance of iron and quartz particles (Fig. 10.102). These curves serve to indicate the benefits which can be achieved by properly considering the parameters of the abrasive before starting to cut.

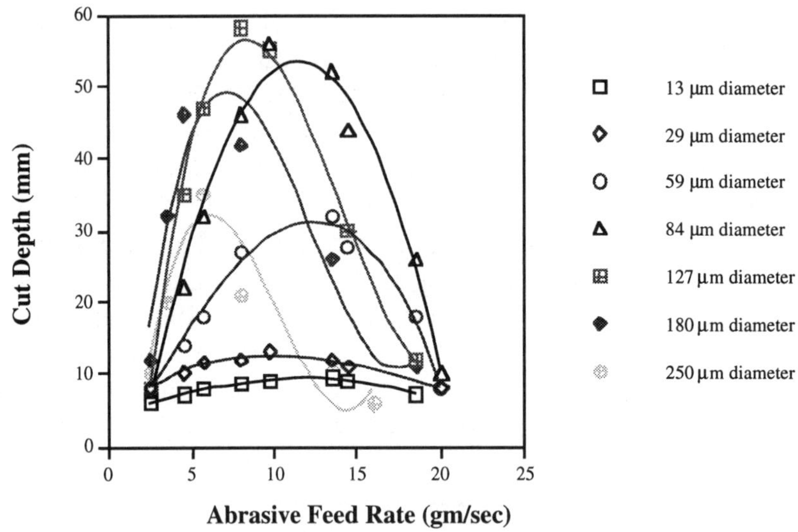

Figure 10.100 The depth of cut as a function of the grain size of the abrasive, when using corundum particles to cut aluminum at 4,000 bar with a traverse speed of 50 mm/min [10.119].

The above curve is somewhat complex to read. What is interesting is as the particle size gets larger, so the optimum cutting performance increases, to a point but so does the feed rate reduce to get that optimum (Fig. 10.101).

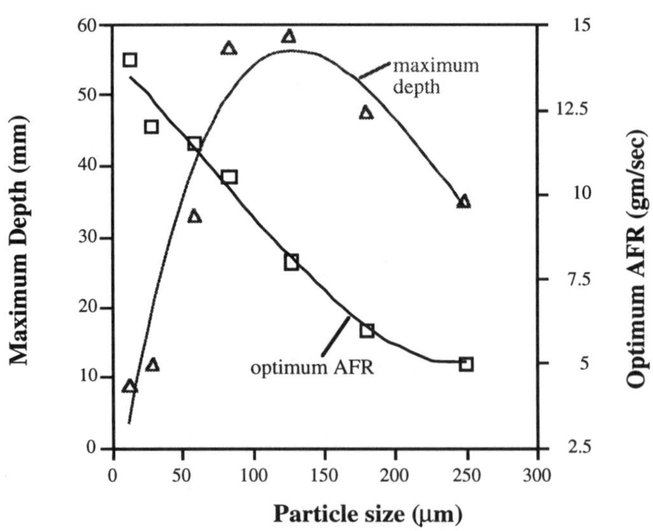

Figure 10.101 Optimum AFR and maximum depth of cut as a function of particle size (after [10.119]).

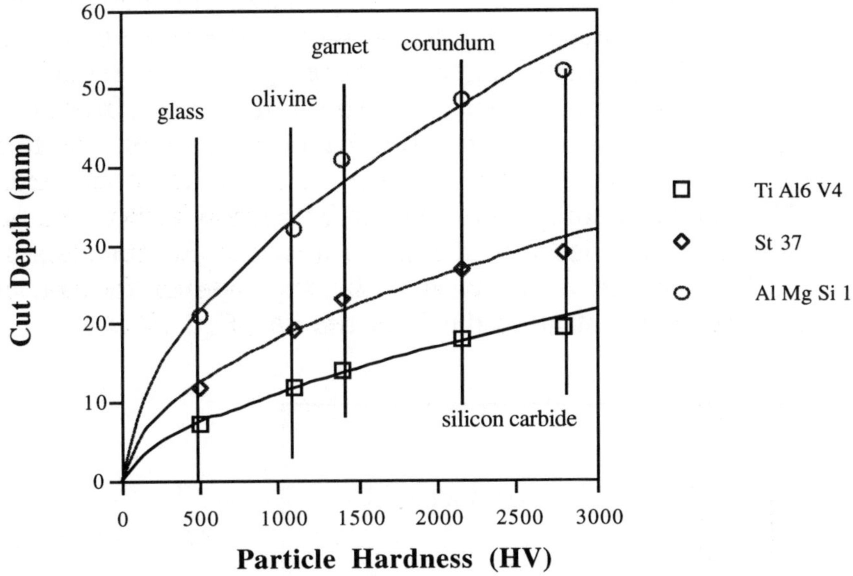

Figure 10.102 Depth of cut as a function of the hardness of the abrasive [10.119].

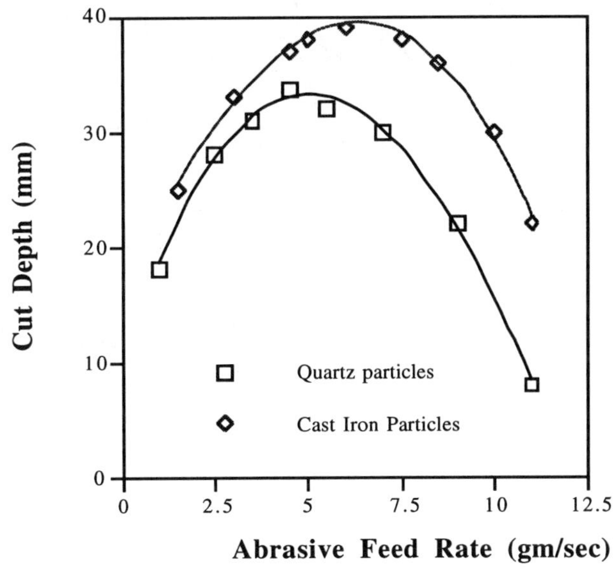

Figure 10.103 Depth of cut as a function of the density of the abrasive cutting Al Mg Si 1 with a 4,000 bar jet at a speed of 50 mm/min [10.119].

Additional comments on the role of density can be obtained by the discussion around Figure 10.60, and that on Figure 10.111.

There is one additional point which perhaps should be made in regard to the use of conventional abrasive injection systems. Although there has been much concern about the performance of these systems, in regard to how this can be maximized, in many cases the benefit that comes from the use of this tool is that it cuts a smooth controlled curve edge. For that to occur, however, the speed at which the traverse is made must be accurately controlled. Hashish has addressed this point in relation both to the amount of abrasive which is used, and also the speed at which the traverse is made [10.126]. The faster the traverse that is made, the rougher the cut, and the greater the abrasive concentration the finer the cut (Fig. 10.104).

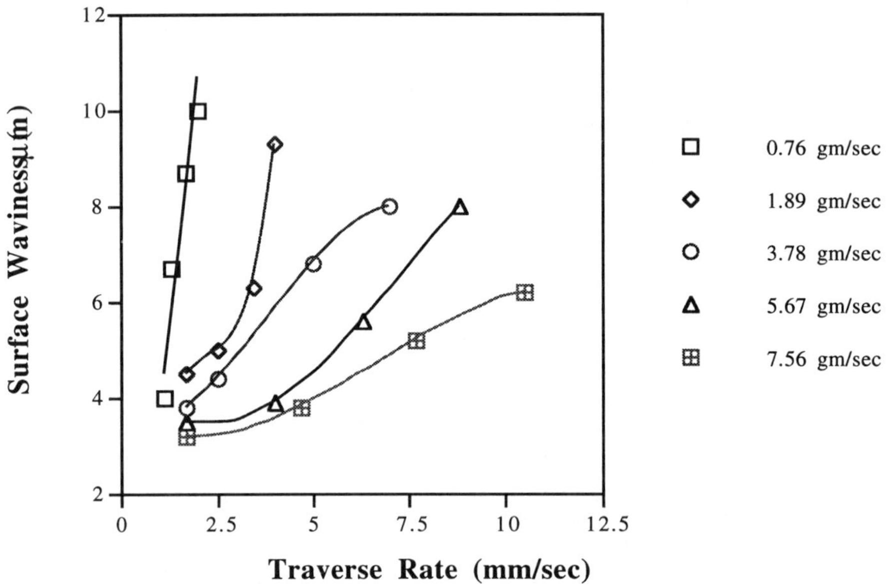

Figure 10.104 Change in traverse speed and AFR on the waviness of a 2,400 bar cut on Inconel using 80 mesh garnet [10.126].

10.6.7 DIAJET ABRASIVE WATERJET CUTTING

This idea of adding the abrasive to the waterjet before it exits the primary nozzle has been introduced in Chapter 4 of this text. Beginning with a Masters thesis at Cranfield [10.127], investigators at BHRA developed a system which has since become known as the DIAjet system [10.128]. This design overcomes many of the disadvantages of the conventional method of entraining abrasive into the waterjet stream. Instead of mixing the abrasive in a chamber, following the acceleration of the water, the abrasive is metered into the flow of water from the pressurizing pump to the nozzle.

Initial experiments were reported in 1986 which showed that the system was capable of cutting through a 13 mm thick mild steel plate at a rate of 51 mm/min when copper slag was used as the abrasive. Subsequently this product has been marketed, with the primary product a unit which operates at 350 bar, with a flow rate of some 70 lpm. Units have become available which work at higher pressures.

A significant study of the parameters controlling the abrasive cutting of rock, using the DIAjet system was undertaken by Yazici as part of his Ph.D. thesis [10.123]. His work concentrated on the evaluation of the DIAjet system, with particular reference to the cutting of rock.

In the above section (10.6.5) the results which Yazici obtained when using a conventional abrasive injection system to cut granite were partially reviewed. These can be contrasted with the results obtained from the use of the DIAjet.

In terms of the depth of cut achieved by the cutting jet the results obtained as a function of jet pressure (Fig. 10.105); nozzle diameter (Fig. 10.106); abrasive feed rate (Fig. 10.107); and traverse velocity (Fig. 10.109) are similar to those obtained with conventional abrasive jet systems.

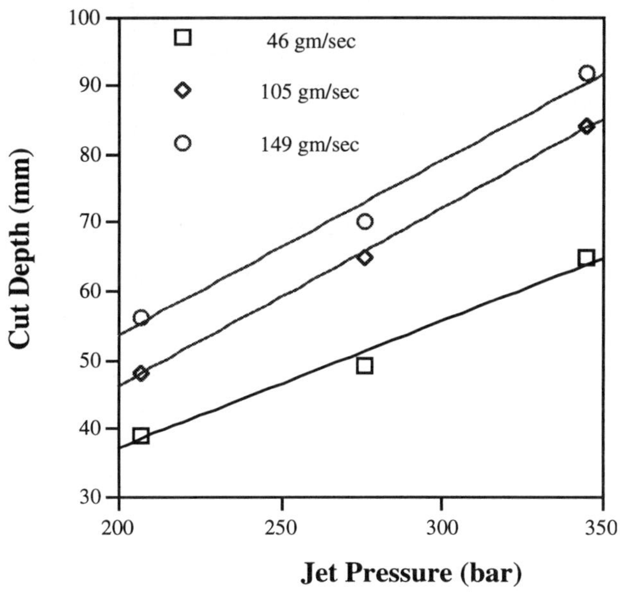

Figure 10.105 Jet pressure as a control on depth of cut in granite using a DIAjet system [10.123].

The only simplifying factor is in the use of only one nozzle to produce the jet so that there is little confusion as to the relative efficiency of the mixing process. Nevertheless, there is little effect from increasing the size of the nozzle, (Fig. 10.106) and this suggests that for many purposes the smaller the nozzle size the more effective the result. It should be borne in mind that these results were obtained with a single size of abrasive. The larger nozzle diameters allow use of a larger abrasive, and this, as has been discussed earlier, has a significant effect on improving performance of the cutting stream. Normally it is recommended that orifice diameter be about three times the diameter of the abrasive particles which are being used.

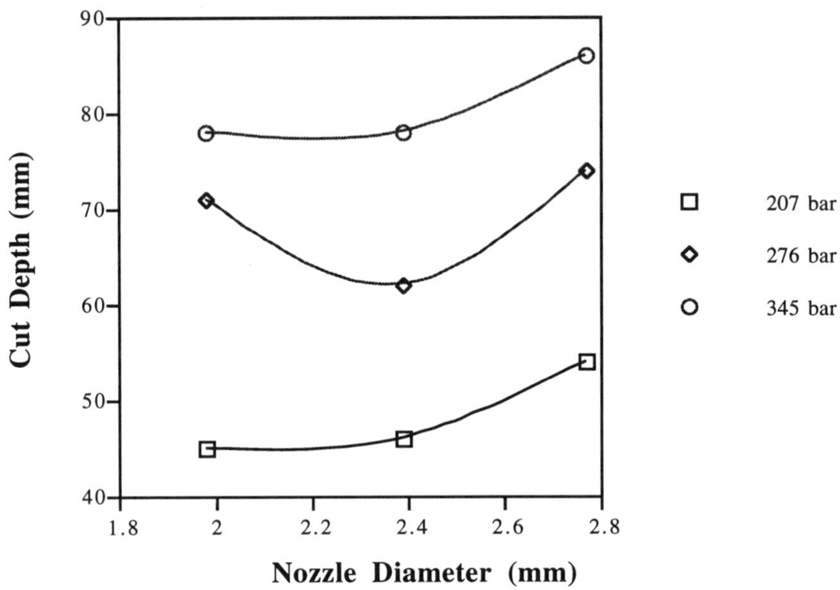

Figure 10.106 Depth of cut as a function of nozzle diameter, using the DIAjet system [10.123].

As with conventional abrasive injection, the depth of cut increases with abrasive feed rate (Fig. 10.107). However, there is, again, a point at which the depth of cut no longer increases due to the reduction in energy transfer to the individual particles of abrasive. A different calculation was made to show that, if the amount of abrasive in a given jet length is calculated, then the maximum concentration of abrasive to achieve the best cutting performance can be found to be a relatively low one (Fig. 10.108).

It is interesting to note that, in terms of traverse velocity, and in the same way as with conventional abrasive jet cutting the relationship is significantly different from that found with plain waterjet cutting. While in waterjet cutting the faster the jet is moved, the more effective, in abrasive cutting there is a clear maximum to the area of cut created per minute (Fig. 10.90), at a relatively low value.

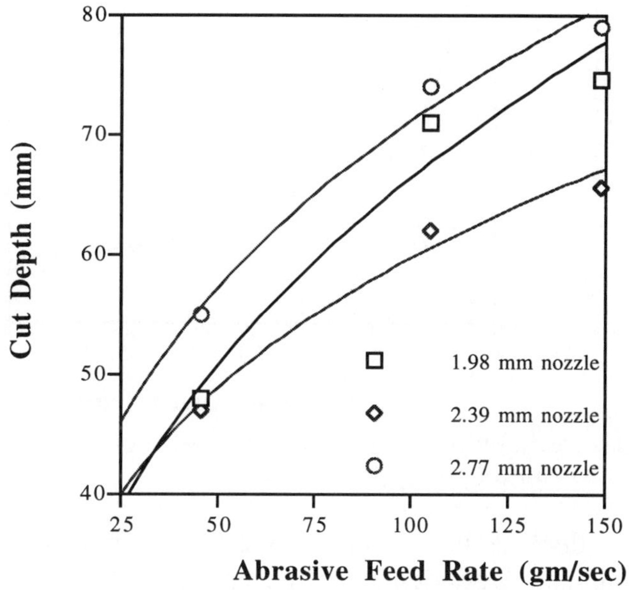

Figure 10.107 The effect of abrasive feed rate on the depth of cut made by a DIAjet into granite [10.123].

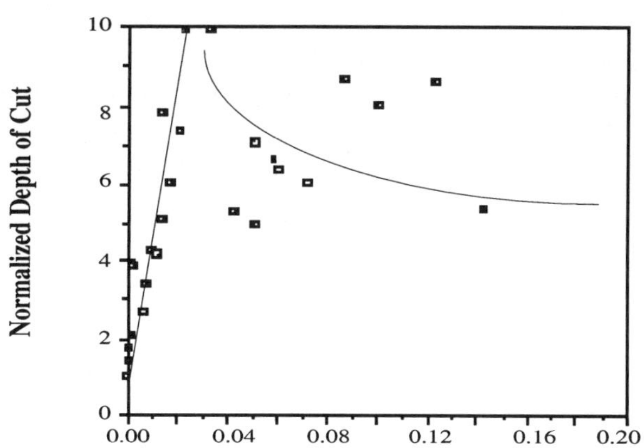

Figure 10.108 The effect of abrasive stream concentration on depth of cut [10.129].

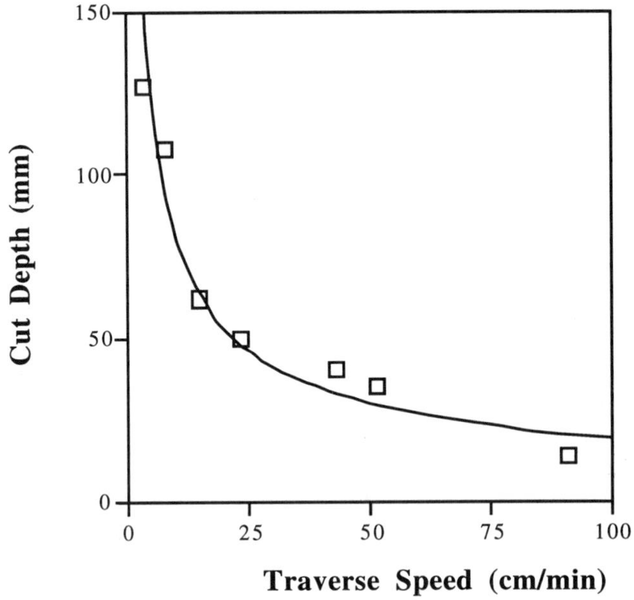

Figure 10.109 The effect of traverse velocity on the depth of cut made by a 276 bar DIAjet on granite at an AFR of 105 gm/sec [10.123].

Although the decline is not great over the range tested beyond that point, this emphasizes dependence on the cracks initiated in the target, and the need for these to reach a significant size for material removal to occur.

It is significant to note also that Yazici examined the size reduction of particles through DIAjet cutting process and found this was significantly less than found in conventional abrasive cutting where up to 90% of the particles are comminuted, almost all before hitting the target. In DIAjet work the amount of reduction is related both to the type of material from which the particle is formed (Fig. 10.110) and the size of the particles before impact (Fig. 10.111). The greater particle size the more likely it is to contain flaws which will fracture under impact with the target surface.

This reduction is a function of both abrasive and target. During the drilling of dolomite and chert it has been found that quartz sand is not greatly degraded during the cutting process, and were a suitable mechanism available to collect and recycle it, this would be quite an effective way of improving process economics. As with conventional abrasive systems there are also better abrasives which could be used (Fig. 10.112). Not only is steel more dense than sand, giving greater depth of cut, it is magnetic and readily stripped from slurry produced in cutting. This will then allow water to be filtered and abrasive and fluid recombined in a second cycle.

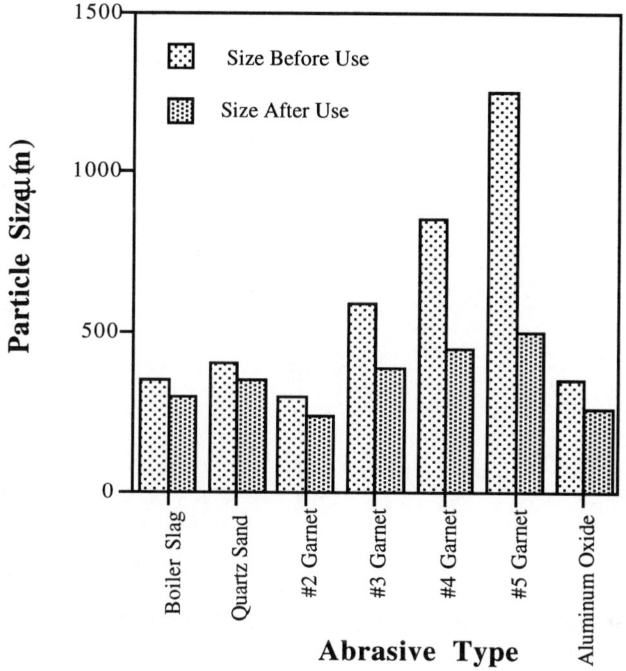

Figure 10.110 Particle size reduction as a function of abrasive type [10.123].

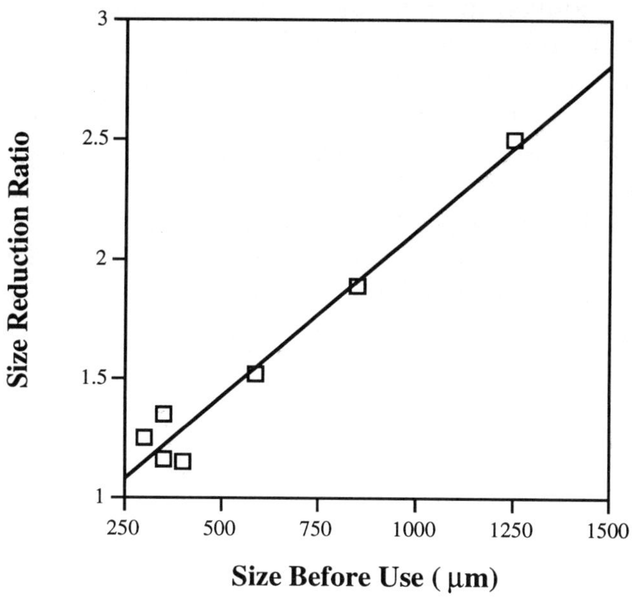

Figure 10.111 Particle size reduction as a function of original size [10.123].

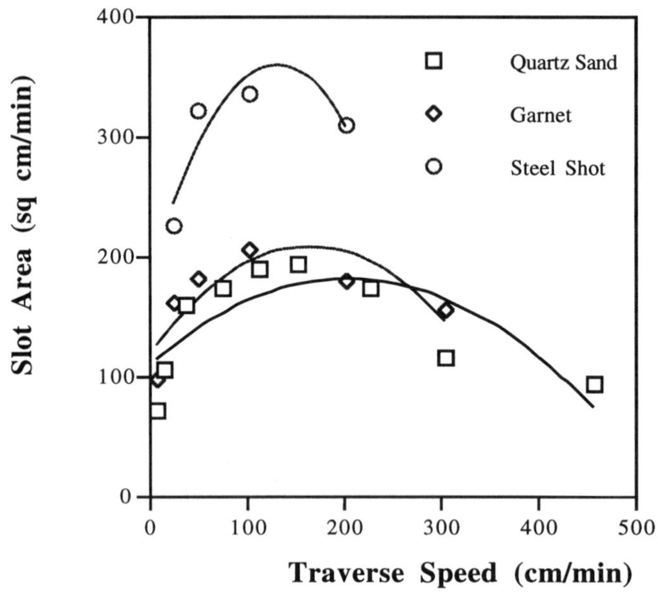

Figure 10.112 Comparison of abrasive performance in cutting dolomite at 350 bar with a 2 mm diameter nozzle [10.130].

However, given that the DIAjet system used significantly more abrasives in a minute than a conventional abrasive injection system it is more critical to evaluate the potential choice of abrasive from an economic as well as a performance point of view. Unless and until effective means are found to recirculate abrasives, then the choice will usually be the slightly less effective but significantly cheaper sand abrasive.

It is worth noting, from that aspect, that the use of steel shot as the abrasive has an additional advantage, not only does it give a deeper cut, and can be recycled, but it does not damage the nozzle and holder as severely as does the quartz or garnet (Fig. 10.113). As shown above (Fig. 10.98) the steel does not damage high strength metals such as tool steel.

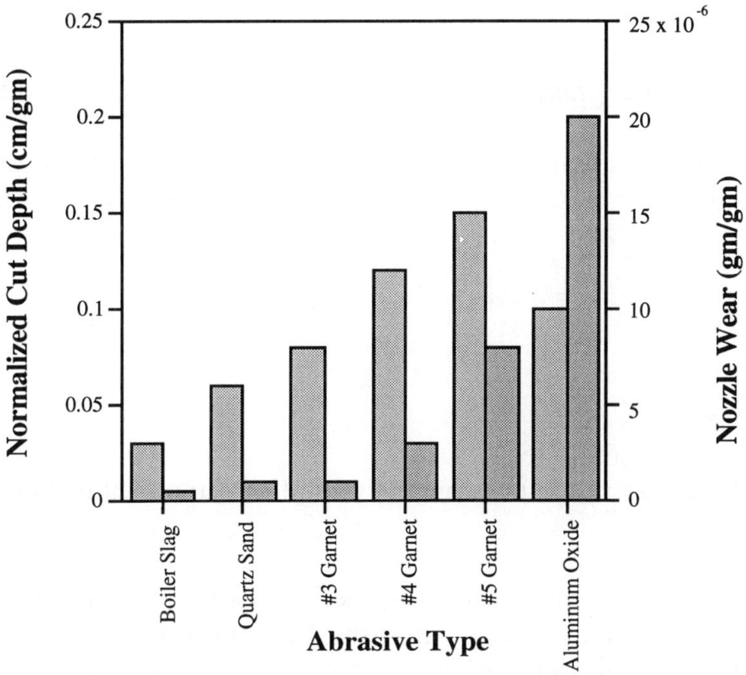

Figure 10.113 Relative performance of different abrasives in cutting and nozzle wear (after [10.120]).

10.6.8 MODELING OF ABRASIVE JET CUTTING

At the beginning of this section dealing with the introduction of abrasives into a jet stream models were presented for the impact of a single particle on a jet surface. Steel shot, at a diameter of 0.3 mm, will provide some 350,000 such impacts for every kilogram of shot added to the water.

Data from the calculations made by Yazici on the erosion of materials [10.123] suggests that the abrasive in a DIAjet system will erode, at 350 bar pressure, a volume of around 0.008 cc for each gram of garnet abrasive used to cut it. Given that the particle is moving at a speed of around 200 m/sec the figures from the single particle impacts suggest that the volume removed should be higher. Many factors influence this relative loss in efficiency. These include significant turbulence within the hole, interparticulate collisions, and flow paths which do not bring the particle into contact with the target at an efficient angle, among others. In the above discussion this will hopefully have been made clear. Some changes can be made to improve performance. Some of these, such as the combination of polymer with the abrasive, have only just begun to be investigated. Other methods of improving performance must still be developed.

At present the mathematical modeling of an abrasive jet system is still a relatively new idea. Hashish [10.89] has developed a model of the form:

$$\frac{h}{d} = 0.282c\frac{N_4}{N_2 N_3} + \frac{(1-N_1)^2}{\frac{N_2 N_3}{1-c} + C_f(1-N_1)}$$

where h is the depth of cut

d is the particle diameter

and the N values relate to subsidiary relationships of the fluid and target properties.

This divides the cutting of the surface into two parts, that which he calls the cutting wear mode, and the deformation wear mode. In the cutting wear mode the material removal is from the impact of particles at relatively shallow angles. This occurs as particles slide down the edge of the cut being made through a sample. The deformation wear mode is the result of impacts at relatively high angles. Such impact occurs as the jet starts to make a new path down through the material, and frequently a "blip" can be seen working its way down the cut edge, in which this deformation wear is occurring. It should be noted that this seems to describe the failure of a ductile material rather than that of the brittle materials such as glass which fail under crack development.

A more empirical predictive equation has been developed by Blickwedel et al. [10.131]. This uses a process coefficient and the threshold pressure of the material as well as the traverse velocity to derive an equation:

$$h = C_s \frac{P - P_o}{v^{\left(0.86 + 2.09/v\right)}}$$

where C_S is a process coefficient

P is the jet pressure, and P_O is the jet threshold pressure (in bar)

v is the nozzle traverse rate in mm/min

Although the equation is simple it is claimed to be used to predict both cutting of brittle materials (Fig. 10.114) and metals (Fig. 10.115).

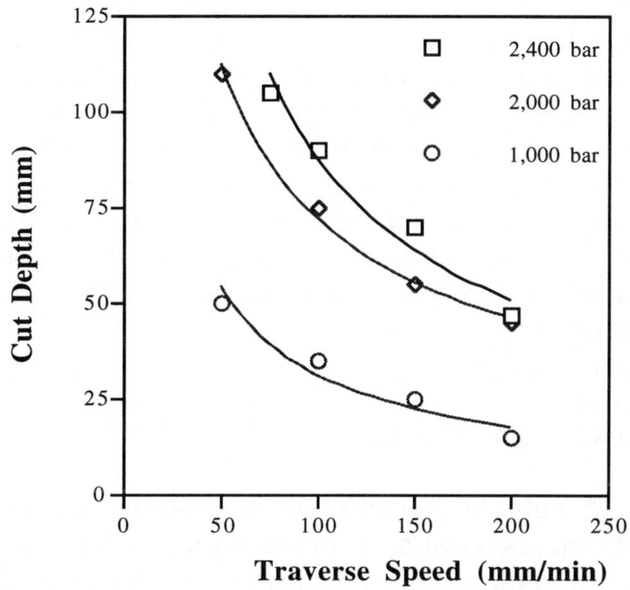

Figure 10.114 Predicted and experimental data on the abrasive cutting of glass - the solid line shows the prediction relative to the data which are the experimental data [10.131].

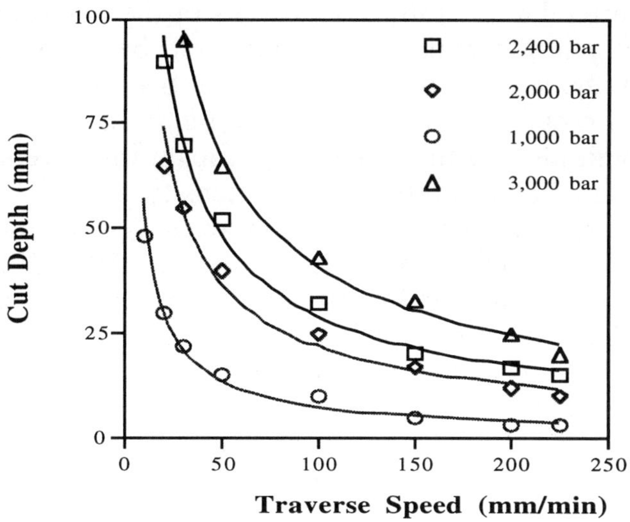

Figure 10.115 Predicted and experimental data on the abrasive cutting of aluminum, the solid line shows the prediction relative to the data which are the experimental data [10.131].

These efforts have been followed by a number of other investigators who have brought different aspects of the process into the evaluation [for example, 10.132, 10.133, 10.134, 10.135].

As yet, it is the current authors opinion that there are too many factors involved in optimizing the performance of these jets to allow a relatively simple model to be accurate over a wide range of conditions. These factors have yet to be properly combined to provide the most efficient cutting system. Further there are at least two different mechanisms of failure involved. For such reasons, in the near future it is likely that only empirically derived relations will be of much use in realistically predicting jet performance. Nevertheless the efforts in deriving these equations will continue to provide valuable insights into the cutting process and they should be encouraged.

10.7 THERMAL PRE-WEAKENING WITH WATERJETS

The discussion of water jet penetration of material has shown that plain waterjets cut into the material by exploiting the fractures already present in the surface. As an attempt to induce further fractures in the surface, and also to exploit thermal shock and its weakening effect, methods of cutting with both thermal weakening and subsequent jet cutting have been tested.

The initial work was carried out at the Bureau of Mines under the direction of Thirumalai [10.136]. The results of the investigation suggested that where the depth of thermal weakening could be maintained beyond the depth at which the jet cut that it would be beneficial. The combined technique would reduce the specific energy required for the jet cutting by an order of magnitude and there were significant improvements in process performance with increase in the heating intensity.

A further study was carried out by Systems Science and Software for the National Science Foundation in the same year [10.137] but beyond that study the process has not been investigated further to this time. Significant problems have been identified with the use of high intensity heating tools, such as the thermal lances discussed in Chapter 7, and thus the use of heat has fallen into disfavor. In addition the techniques described in this chapter have proven to give a sufficient promise of future gain that they have taken over the development of the technology.

10.8 COMMENT

The need to improve the ability of waterjets to cut through more and more difficult materials at faster rates has led to several investigations of processes to enhance performance. At the present time the use of abrasives seems to be the most effective of these techniques. However there are others, such as the introduction of long chain polymers to the water, which have also found a significant niche in industrial use. The technology is still relatively young in its development and the benefits to the achieved from the development of highly intense cavitating jets have yet to be developed.

As many of these process pass from the laboratory into field use the potential benefits to the industry will likely be significant.

10.9 REFERENCES

10.1 Yahiro, T., and Yoshida, H., "On the Characteristics of High Speed Water Jet in the Liquid and its Utilization of Induction Grouting Method," paper G4, 2nd International Symposium on Jet Cutting Technology, Cambridge, UK, April, 1974, pp. G4.41 - G4.63.

10.2 Baumann, V., "On The Execution Of Soil Improvement Method Utilizing Water Jet," paper III-3, International Symposium on Water Jet Technology, Tokyo, Japan, December, 1984, Water Jet Technology Society of Japan.

10.3 Eddingfield, D.L., and Albrecht, M., "Effect of an Air-Injected Shroud on the Breakup Length of a High Velocity Waterjet," Erosion: Prevention and Useful Applications, ASTM STP 664, ed W.F. Adler 1979, pp. 461 - 472.

10.4 Vickers, G.W., Harrison, F.W., and Houlston, R., "Extending the range of Cavitation cleaning jets," paper J1, 5th International Symposium on Jet Cutting Technology, Hanover, Germany, June, 1980 pp. 403 - 412.

10.5 Brunton, J.H., and Rochester, M.C., "Erosion of Solid Surfaces by the Impact of Liquid Drops," In Erosion-Treatise on Materials Science and Technology, C.M. Preece ed, pp. 185 - 248.

10.6 Poulter, T.C., J. Appl. Mech. Vol. 9, p. 31, 1942.

10.7 Brandenburger, E., and de Haller, P., Schweiz. Arch. Vol. 10, p. 331.

10.8 Hancox, N.L., and Brunton, J.H., "The Erosion Of Solids By The Repeated Impact Of Liquid Drops," Proceedings of the Royal Society London, Vol. 260A, pp. 121 - 143.

10.9 Rochester, M.C., and Brunton, J.H., The Influence of the Physical Properties of the Fluid on the Erosion of Solids, CVED/C-MAT/TR10, University of Cambridge, UK, 1973.

10.10 Franz, N.C., "Fluid additives for improving high velocity jet cutting," paper A7, 1st International Symposium on Jet Cutting Technology, Coventry, UK, April, 1972, pp. A7-93 - A7-104.

10.11 Staroselsky, A.V., and Kim, K., "The Effect Of Surface-Active Agents on Rock Cutting With Shear/Drag Bits," 1st North American Rock Mechanics Symposium, Austin, TX, June, 1994, pp. 351 - 358.

10.12 Merle, C., Bouix, M., Sionneau, M., and Vassew, J., "Performances of HP Fluid Jet to cut Food Products," 7th American Water Jet Conference, Seattle, WA., August, 1993, pp. 103 - 118.

10.13 White, A., "Turbulence and Drag Reduction with Polymer Additives," Hendon College of Technology, Research Bulletin, No. 4, 1967.

10.14 Fire Engineering, September, 1969.

10.15 Howells, W.G., "Polymerblasting with Super-Water from 1974-89: A Review," International Journal of Water Jet Technology, Vol. 1, No. 1, March, 1990, pp. 1 - 16.

10.16 Summers, D.A., Disintegration of Rock by High Pressure Jets, Ph.D. thesis, Department of Applied Mineral Sciences, University of Leeds, UK, May, 1968.

10.17 Zakin, J.L., and Summers, D.A., "The Effect of Visco-Elastic Additives on Jet Structure," paper A4, 3rd International Symposium on Jet Cutting Technology, Chicago, IL, May, 1976, pp. A4-47 - A4-66.

10.18 Szymani, R., A study of corrugated board cutting by High Velocity Liquid Jet, MS thesis, University of British Columbia, Sept., 1970.

10.19 Private information from Mr. J. Frank, to Dr. David Summers.

10.20 Howells, W.G., "Polymer Blasting: A Chemists point of view," 2nd U.S. Water Jet Conference, Rolla, MO, April, 1983, pp. 443 - 447.

10.21 Information from HZOIL Corporation, 2509 Technology Drive, Hayward, CA., 94545.

10.22 Cobb, C.C., and Zublin, C.W., Petroleum Engineer International, October, 1985, pp. 56 - 66.

10.23 Zublin, C.W., "Water Jet Cleaning Speeds - Theoretical Determinations," 2nd U.S. Water Jet Conference, Rolla, MO, May, 1983, pp. 159 - 166.

10.23a Allred, R.B., and McCaleb, S.B., "Rx for Gumbo Shale Drilling," SPE paper 4233, 6th Conference on Drilling and Rock Mechanics, Austin, Tx., January, 1973, pp. 35 - 42.

10.24 Hashish, M., "Abrasive Jets," Section 4, in Fluid Jet Technology, Fundamentals and Applications, Waterjet Technology Association, St. Louis, MO, 1991.

10.25 Hollinger, R.H., and Mannheiemer, R.J., "Theological Investigation of the Abrasive Suspension Jet," 6th American Water Jet Conference, Houston, TX, August, 1991, pp. 515 - 528.

10.26 Proceedings of the Conferences on Rain Erosion and Erosion by Liquid & Solid Impact, from the Cavendish Laboratory, Madingley Road, Cambridge CB3 OHE, UK.

10.27 Cook, S.S., Proceedings of the Royal Society, London, series A Vol. 119, 1928, p. 481.

10.28 Fyall, A.A., 2nd Rain Erosion Conference, Lake Constance, Germany, 1967, p. 428.

10.29 Lesser, M.B., and Field, J.E., "The Geometric Wave Theory Of Liquid Impact," Sixth Conference on Erosion by Liquid & Solid Impact, Cavendish Laboratory, Madingley Road, Cambridge CB3 OHE, UK.

10.30 Nebeker, E. B., Presentation, Workshop on Water Jet Cutting Technology, Rolla, MO, 1975, pp. 189 - 198.

10.31 Nebeker, E.B., and Rodriguez, W.E., "Percussive Water Jets for Rock Cutting," paper B1, 3rd International Symposium on Jet Cutting Technology, Chicago, IL, May, 1976, B1-1 - B1-9.

10.32 Mazurkiewicz, M., "The Analysis of High Pressure Water Jet Interruption through Ultrasonic Nozzle Vibration," paper P5, 7th International Symposium on Jet Cutting Technology, Ottawa, Canada, June 1984, pp. 531 - 536.

10.33 Matthewson, M.J., "Theoretical aspects of thin protective coatings," paper 73, Fifth Conference on Erosion by Liquid & Solid Impact, Cavendish Laboratory, Madingley Road, Cambridge CB3 OHE, UK.

10.34 Brook, N., and Summers, D.A., "The Penetration of Rock by High Pressure Water Jets," International Journal of Rock Mechanics and Mining Science, Vol. 6, 1969, pp. 249 - 258.

10.35 Leach, S.J., and Walker, G.I., "The Application of High Speed Liquid Jets to Cutting," Proceedings of the Royal Society, London, 260A, pp. 295 - 308.

10.36 Chahine, G.L., Conn, A.F., Johnson, V.E. Jr., and Frederick, G.S., "Cleaning and Cutting with Self-Resonating Pulsed Water Jets," 2nd U.S. Conference on Water Jet Technology, Rolla, MO, May, 1983, pp. 167 - 175.

10.37 Conn, A.F., "Asbestos Removal with Self-Resonating Water Jets," paper 13, 5th American Water Jet Conference, Toronto, Canada, August, 1989, pp. 133 - 139.

10.38 Chahine, G.L., Conn, A.F., Johnson, V.E. Jr., and Frederick, G.S., "Passively Interrupted Impulsive Waterjets," 6th International Conference on Erosion by Solid and Liquid Impact, Cambridge, UK, September, 1983, pp. 34-1 - 34-9.

10.39 Clark, G.B., Rollins, R.R., Brown, J.W., and Kalia, H., Investigation of the Use of Shaped Charges for Rapid Drilling and Blasting, Final Report from UMR to E.I. duPont, March 1970, RMERC-TR-70-10, 100 pp.

10.40 Birkhoff, G. J., "Explosives with Lined Cavities," Journal of Applied Physics, Vol. 19, June, 1948.

10.41 Watson, A.J., and Moxon, C.J., The Characteristics of High Velocity Water Jets and their Impact with Materials, Dept. of Civil Engineering, University of Sheffield, March, 1986, 9 pp.

10.42 Goyette, R.E., and Worsey, P.N., "A Study of the Penetration of Steel Targets by Explosive Driven Hyper-velocity Water," 1988 Annual Meeting, ADPA, Shreveport, Louisianna, October 1988, 13 pp.

10.43 Bowles, R.E., and Stouffer, R.D., "Jet Velocity Augmentation External to the Nozzle," 1968, ASME Fluids Engineering Conference for Hypervelocity Rock Cutting.

10.44 Mazurkiewicz, M., Barker, C.R., and Summers, D.A., "Adaptation of Jet Accumulation Techniques for Enhanced Rock Cutting," Erosion: Prevention and Useful Application, ASTM STP 664, W.F. Adler, ed, ASTM, 1979, pp. 473 - 492.

10.45 Summers, D.A., Barker, C.R., and Mazurkiewicz, M., "Experimentation in Hydraulic Coal Mining," 1977 AIME Annual Meeting, Atlanta, GA, March, 1977.

10.46 Lin, B., Hagan, P.C., and Roxborough, F.F., "Massive Breakage of Rock by High Pressure Water Jets," 10th International Symposium on Jet Cutting Technology, Amsterdam, Holland, October, 1990, pp. 399 - 412.

10.47 Selberg, B.P., and Barker, C.R., "Dual-Orifice Waterjet Predictions and Experiments," Erosion: Prevention and Useful Application, ASTM STP 664, W.F. Adler, ed, ASTM, 1979, pp. 493 - 511.

10.48 Johnson, V.E., Kohl, R.E., Thiruvengadam, A., and Conn, A.F., "Tunneling, fracturing drilling and mining with high speed water jets utilizing cavitation damage," paper A3, 1st International Symposium on Jet Cutting Technology, Coventry, UK, April, 1972, pp. A3-37 - A3-56.

10.49 Hansson, I., and Morch, K.A., "Comparison of the initial stages of vibratory and flow cavitation erosion," paper 60, 5th Conference on Erosion by Liquid & Solid Impact, Cambridge, UK, September, 1979.

10.50 Benjamin, T.B., and Ellis, A.T., "The Collapse Of Cavitation Bubbles And The Pressures Thereby Produced Against Solid Boundaries," Proceedings of the Royal Society, London, A262, pp. 221 - 240.

10.51 Lauterborn, W., "Liquid Jets from cavitation bubble collapse," paper 58, Fifth Conference on Erosion by Liquid & Solid Impact, Cambridge, UK, September, 1979.

10.52 Discussion on Preece, C.M., Vaidya, S., and Dakshinamoorthy, S., "Influence of Crystal Structure on the Failure Mode of Metals by Cavitation Erosion," Erosion: Prevention and Useful Application, ASTM STP 664, W.F. Adler, ed, ASTM, 1979, page 432.

10.53 Morch, K.A., "The Dynamics of Cavitation Bubbles and Cavitating Liquids," in Erosion - A Treatise on Materials Science and Technology, C.M. Preece ed, Academic Press, 1979, pp. 309 - 355.

10.54 Ellis, A.T., and Starrett, J.E., "A Study of Cavitation Bubble Dynamics and Resultant Pressures on Adjacent Solid Boundaries," paper C190/83, 2nd International Conference on Cavitation, Herriot-Watt University, Edinburgh, Scotland, September, 1983, pp. 1 - 6.

10.55 Summers, D.A., Tyler, L.J., Blaine, J., Fossey, R.D., Short, J., and Craig, L., "Considerations in the Design of a Waterjet Device for Reclamation of Missile Casings," 4th U.S. Water Jet Conference, Berkeley, CA, August, 1987, pp. 51 - 56.

10.56 Brunton, J.H., "Some Mechanisms of Cavitation Damage in Vibratory Systems," paper 59, 5th Conference on Erosion by Liquid & Solid Impact, Cambridge, UK, September, 1979.

10.57 Discussion by various authors on cavitation at the 2nd International Symposium on Jet Cutting Technology, Cambridge, UK, April, 1974, pp. X16, X39, X75.

10.58 Angona, F.A., "Cavitation, a novel drilling concept," International Journal of Rock Mechanics and Mining Science, Vol. 11, No. 4, pp. 115 - 119.

10.59 Beutin, E.F., Erdmann-Jeznitzer, F., and Louis, H., "Influence of Cavitation Bubbles in Cutting Jets," 2nd International Symposium on Jet Cutting Technology, Cambridge, UK, April, 1974, pp. D3-21 - D3-33.

10.60 Vijay, M.M., and Brierley, W.H., "Cutting Rocks and Other Materials by Cavitating and Non-Cavitating Jets," paper 5, 4th International Symposium on Jet Cutting Technology, Canterbury, UK, April, 1978, pp. C5-51 - C5-66.

10.61 ASTM, "Standard Method of Vibratory Cavitation Erosion Test," Designation G32-72, American Society for Testing and Materials, Vol. 10, 1975, Annual Book of ASTM Standards, p. 724 - 727.

10.62 Scott, P.R., Cavitation Resistance of Rock, MS thesis, University of Missouri-Rolla, 1979, 91 pages.

10.63 Barker, B.P. and Putterman, S.J., "Observation of Synchronous Picosecond Sonoluminescence," Nature, Vol. 352, 1991, pp. 318 - 320.

10.64 Suslick, K.S., Ganlenovski, J.J., Schubert, P.F. and Want, H.H., "Alkane Sonochemistry," The Journal of Physical Chemistry, Vol. 87, No. 13, 1983, pp. 2299 - 2301.

10.65 Bankard, M.G., and Mayhall, R.H. Jr., Redesign and Field Operation of a Self Propelled Cavitating Concrete Removal System, Final Report on FHWA/TS 84/207, Daedalean Associates, MD, PB85-105633/XPS, 52 pp.

10.66 Daedalean Associates Inc., Dynamic Response and Explosive Characteristics of Solid Propellants due to Water and Cavitation Impact during Cleaning at NOS, DAI Technical Report No. RSW-8245-002, October, 1983.

10.67 Howard, S.C., Graham, F.C., Hochrein, A.A. Jr., Thiruvengadam, A.P., Research and Development of a Cavitating Water Jet Cleaning System for Removing Marine Growth and Fouling from U.S. Navy Ship Hulls, DAI Technical Report DAI-SCH-7759-601-TR, June, 1978, AD A065463, 55 pp.

10.68 Chahine, G.L., Genoux, Ph.F., and Liu, H., "Flow Visualization and Numerical Simulation of Cavitating Self Oscillating Jets," paper A2, 7th International Symposium on Jet Cutting Technology, Ottawa, Canada, June, 1984, pp. 13 - 32.

10.69 Vyas, B., and Preece, C.M., Journal of Applied Physics, Vol. 47, 1976, , p. 5133.

10.70 Hobbs, J.M., in Erosion by Cavitation or Impingement, ASTM STP 408, 1967, p. 159.

10.71 Summers, D.A., and Sebastian, Z., The Reaming out of Geothermal Excavations, Final Report to SNL on Contract 13-3246, University of Missouri-Rolla, April, 1980, 176 pages.

10.72 Margis, R., Ragan, T., Reisinger, L., Sonoluminescence, Advanced Labratory, Department of Physics, UMR, December, 1991, 43 pp.

10.73 Erdmann-Jesnitzer, F., and Louis, H., in Erosion, Near and Interfaces with Corrosion, ASTM STP 567, A Thiruvengadem ed, 1974, p. 171.

10.74 Lichtarowicz, A., "Cavitating Jet Apparatus for Cavitation Erosion Testing," in Erosion: Prevention and Useful Applications, W.F. Adler, ed., ASTM STP 664, Vail, October, 1977, pp. 530 - 549.

10.75 Summers, D.A., and Peters, J.F., "The Effect of Rock Anisotropy on the Excavation Rate in Barre Granite," paper H5, 2nd International Symposium on Jet Cutting Technology, Cambridge, UK, April, 1974, pp. H5-49 - H5-62.

10.76 Preece, C.M., Vaidya, S., and Dakshinamoorthy, S., "Influence of Crystal Structure on the Failure Mode of Metals by Cavitation Erosion," Erosion:Prevention and Useful Application, ASTM STP 664, W.F. Adler, ed, ASTM, 1979, pp. 409 - 431.

10.77 El Saie, A.A., Investigation of Rock Slotting by High Pressure Water Jet for Use in Tunneling, Ph.D. Dissertation, University of Missouri-Rolla, 1977.

10.78 El Saie, A.A., and Summers, D.A., "A Contrast between Sand, Water, and Cavitation Erosion of Rock," International Conference on Wear of Materials, ASME, Dearborn, MI, April, 1979, pp. 528 - 535.

10.79 Knapp, R.T., Daily, J. W., and Hammitt, F.G., Cavitation, McGraw Hill, 1970, 578 pp.

10.80 Howard, S.C., Kroogle, D.R., and Hochrein, A.A., The Research And Development Of A Cavitating Jet Cleaning System For Removing Marine Growth And Fouling From Offshore Platform Structures: Feasibility Evaluation, Daedalean Associates, December, 1978, DAI-SH-7759-002-TR (AD AO65 464), 60 pp.

10.81 Conn, A.F., Johnson, V.E. Jr., Lindenmuth, W.T., and Frederick, G.S., "Some Industrial Applications of Cavijet Cavitating Fluid Jets," 1st U.S. Water Jet Symposium, Golden, CO., April, 1981 pp. V-2.1 - V-2.11.

10.82 Conn, A.F., "Field Trials Of A Cavitating Jet Fouling Removal Device," 5th Congress On Marine Corrosion And Fouling, Barcelona Spain, May 1980, pp. 39 - 53.

10.83 Houlston, R., and Vickers, G.W., "Surface Cleaning Using Waterjet Cavitation And Droplet Erosion," 4th International Symposium on Jet Cutting Technology, Canterbury, UK, April, 1978, pp. H1,1 - H1,18.

10.84 Conn, A.F., and Gracey, M.T., "Using Water Jets to Clean Explosives and Propellants," paper F4, 9th International Symposium on Jet Cutting Technology, Sendai, Japan, pp. 307 - 322.

10.85 Dickinson, W., Wilkes, R.D., and Dickinson, R.W., "Conical Water Jet Drilling," 4th U.S. Water Jet Conference, Berkeley, CA, August, 1987, pp. 89 - 96.

10.86 Gustafsson, G., Propagation of Weak Shock Waves in an Elliptical Cavity, Doctoral thesis, Lulea University of Technology, Division of Fluid Mechanics, 1985.

10.87 Hawrylewicz, B.M., Puchala, R.J., and Vijay, M.M., "Generation of Pulsed or Cavitating Jets by Electric Discharges in High Speed Continuous Water Jets," 8th International Symposium on Jet Cutting Technology, Durham, UK, September, 1986, pp. 345 - 352.

10.88 Maurer, W.C., Advanced Drilling Techniques, Petroleum Publishing Co., 698 pp.

10.89 Hashish, M., "On the Modeling of Abrasive-Waterjet Cutting," 7th International Symposium on Jet Cutting Technology, Ottawa, Canada, June, 1984, pp. 249 - 266.

10.90 Preece, C.M., Erosion - A Treatise on Materials Science and Technology, Academic Press, 1979, 450 pages.

10.91 Ives, L.K., and Ruff, A.W., Wear, Vol. 46, 1978, pp. 149 - 162.

10.92 Hutchings, I.M., Winter, R.E., and Field, J.E., "Solid Particle Erosion of Metals; The Removal of Surface Material by Spherical Projectiles," Proceedings of the Royal Society, London, Vol. A348, 1976, pp. 379 - 392.

10.93 Adler, W.F., Assessment of the State of Knowledge Pertaining to Solid Particle Erosion, Final Report on USARO Contract DAAG29-77-C-0039, June, 1979, 150 pages.

10.94 Hashish, M., "Data Trends in Abrasive Waterjet Machining", SME Automated Waterjet Cutting Processes, Southfield, MI, May, 1989.

10.95 Evans, A.G., and Wilshaw, T.R., "Quasi-Static Solid Particle Damage in Brittle Solids - 1. Observations, Analysis and Implications," Acta Metallurgica, Vol. 24, pp. 939 - 956.

10.96 Hertz, H., "On the Contact of Elastic Solids" Miscellaneous Papers, McMillan & Co., London, 1896, pp. 146 - 183.

10.97 Evans, A.G., "Impact Damage Mechanics: Solid Projectiles," in Erosion, Treatise on Materials Science and Technology, Vol. 16, C.M. Preece ed, Academic Press, 1979.

10.98 Sheldon, G.L., and Finnie, I., "The Mechanism of Material Removal in the Erosive Cutting of Brittle Material," J. Eng. Ind., November, 1966, pp. 393 - 400.

10.99 Weibull, W., 1939, Ingenior Veterskaps Akademiens Handlings, Stockholm, No. 151, p. 1.

10.100 Knight, C.G., Swain, M.V., and Chaudhri, M.M., "Impact of Small Steel Spheres on Glass Surfaces," Journal of Material Science, Vol. 12, 1977, pp. 1573 - 1586.

10.101 Ripkin, J.F., and Wetzel, J.M., A Study of the Fragmentation of Rock by Impingement with Water and Solid Impactors, Final Report on U.S. Bureau of Mines, Contract HO 210021, February, 1972.

10.102 Chatterton, N.E., Development of a Safer, More Efficient Hydraulic Based Technique for Rapid Excavation of Coal, Rock and Other Minerals, Final Report on U.S. Bureau of Mines, Contract HO232062, PB 245 343, April, 1975.

10.103 Adler, W.F., "Analytical Modeling of Multiple Particle Impacts on Brittle Materials," Wear, Vol. 37, 1976, pp. 353 - 364.

10.104 Bortolussi, A., Yazici, S., and Summers, D.A., "The Use of Waterjets in Cutting Granite," paper E3, 9th International Symposium on Jet Cutting Technology, Sendai, Japan, October, 1988, pp. 239 - 254.

10.105 Galecki, G., Mazurkiewicz, M., and Jordan, R., "Abrasive Grain Disintegration Effect During Jet Injection," International Water Jet Symposium, Beijing, China, September, 1987, pp. 4-71 - 4-77.

10.106 Labus, T.J., "Section 8," Fluid Jet Technology Fundamentals and Applications - A Short Course, 5th American Waterjet Conference, Toronto, Canada, August 28, 1989, pp. 145 - 168.

10.107 Kiyoshige, M., Matsamura, H., Ikemoto, Y., and Okada, T., "A Study of Abrasive Waterjet Cutting using Slurried Abrasives," paper B2, 9th International Symposium on Jet Cutting Technology, Sendai, Japan, October, 1988, pp. 61 - 73.

10.108 Isobe, T., Yoshida, H., and Nishi, K., "Distribution of Abrasive Particles in Abrasive Water Jet and Acceleration Mechanism," paper E2, 9th International Symposium on Jet Cutting Technology, Sendai, Japan, October, 1988, pp. 217 - 238.

10.109 Walters, C.L., and Saunders, D.H., "DIAJET Cutting for Nuclear Decommissioning," paper J2, 10th International Symposium on Jet Cutting Technology, Amsterdam, Netherlands, October, 1990, pp. 427 - 440.

10.110 Mazurkiewicz, M., Olko, P., and Jordan, R., "Abrasive Particle Distribution in a High Pressure Hydroabrasive Jet," International Water Jet Symposium, Beijing, China, September, 1987, pp. 4-1 - 4-10.

10.111 Finucane, L.J., Investigation and Optimization of Hydroabrasive Cutting Head Design, MS thesis, Mechanical Engineering Department, UMR, 1987, 67 pp.

10.112 Savanick, G.A., and Krawza, W.G., "An Abrasive Water Jet Rock Drill," 4th U.S. Water Jet Conference, Berkeley, CA, August, 1987, pp. 129 - 132.

10.113 Mazurkiewicz, M., Finucane, L., and Ferguson, R., "Investigation of Abrasive Cutting Head Internal Parameters," paper B3, 9th International Symposium on Jet Cutting Technology, Sendai, Japan, October, 1988, pp. 75 - 84.

10.114 Galecki, G., and Summers, D.A., "Steel Shot entrained ultra high pressure waterjet for cutting and drilling in hard rocks," Proc. 11th International Symposium on Jet Cutting Technology, St. Andrews, Scotland, September, 1992, pp. 371 - 388.

10.115 Yie, G.G., "Cutting Hard Rock with Abrasive-Entrained Waterjet at Moderate Pressures," 2nd U.S. Waterjet Conference, Rolla, MO, May, 1983, pp. 407 - 423.

10.116 Harris, I.D., "Abrasive Water Jet Cutting and its applications at the Welding Institute," Welding Institute Research Bulletin, Vol. 19, February, 1988, pp. 42 - 49.

10.117 Yie, G.G., "Cutting Hard Materials With Abrasive-Entrained Waterjet - A Progress Report," 7th International Symposium on Jet Cutting Technology, Ottawa, Canada, June, 1984, pp. 481 - 493.

10.118 Horii, K., Matsumae, Y., Cheng, X.M., Takei, M., Hashimoto, B, and Kim, T.J., "Development Of A New Mixing Nozzle Assembly For High Pressure Abrasive Waterjet Applications," paper 12, 10th International Symposium on Jet Cutting Technology, Amsterdam, Netherlands, October, 1990, pp. 193 - 206.

10.119 Faber, K., and Oweinah, H., "Influence of Process Parameters on Blasting Performance with the Abrasive Jet," paper 25, 10th International Symposium on Jet Cutting Technology, Amsterdam, October, 1990, pp. 365 - 384.

10.120 Zaring, K., Erichsen, G., and Burnham, C., "Procedure Optimization and Hardware Improvements in Abrasive Waterjet Cutting Systems," paper 18, 6th American Water Jet Conference, Houston, TX, August, 1991, pp. 237 - 248.

10.121 Singh, P.W., and Munoz, J., "The Alignability of Jet Cutting Orifice and Nozzle Assemblies," Chapter 13, 10th International Symposium on Jet Cutting Technology, Amsterdam, Netherlands, October, 1990, pp. 207 - 219.

10.122 Griffiths, J.J., "Abrasive Injection Usage in the United Kingdom," 2nd U.S. Waterjet Conference, May, 1983, Rolla, MO, pp. 423 - 432.

10.123 Yazici, Sina, Abrasive Jet Cutting and Drilling of Rock, Ph.D. Dissertation in Mining Engineering, University of Missouri- Rolla, Rolla, Missouri, 1989, 203 pp.

10.124 Mason, F., "Water and Sand Cut It," American Machinist, October, 1989, pp. 84-95.

10.125 Summers, D.A., and Yao, Jianchi, "First Steps in Developing an Abrasive Jet Drill," 8th Annual Workshop, Generic Mineral Technology Center, Mine Systems Design and Ground Control, Reno, Nevada, November 4-6, 1990.

10.126 Hashish, M., "On the Modelling of Surface Waviness Produced by Abrasive Waterjets," 11th International Symposium on Jet Cutting Technology, St. Andrews, Scotland, September, 1992, pp. 17 - 34.

10.127 Fairhurst, R.M., Abrasive Water Jet Cutting" MSc thesis, Cranfield Institute of Technology, January, 1982.

10.128 Fairhurst, R.M., Heron, R.A., and Saunders, D.H., "Diajet" - a new abrasive water jet cutting technique," 8th International Symposium on Jet Cutting Technology, Durham, UK, September, 1986, pp. 395 - 402.

10.129 Bortolussi, A., Summers, D.A., and Yazici, S., "The use of waterjets in cutting granite," 9th International Symposium on Jet Cutting Technology, Sendai, Japan, October, 1988, pp. 239 - 254.

10.130 Yao, J., High Pressure Water Jets in Rock Excavation, Ph.D. Dissertation, University of Missouri-Rolla, 1991, 197 pages.

10.131 Blickwedel, H., Guo, N.S., Haferkamp, H., and Louis, H., "Prediction of Abrasive Jet Cutting Efficiency and Quality," 10th International Symposium on Jet Cutting Technology, Amsterdam, October, 1990, pp. 163 - 179.

10.132 Matsui, S., "Prediction Equations for Depth of Cut made by Abrasive Water Jet," 6th American Water Jet Conference, Houston, TX, August, 1991, pp. 31 - 41.

10.133 Hu, F., Yang, Y., Geskin, E.S., and Chung, Y., "Characterization of Material Removal in the Course of Abrasive Waterjet Machining," 6th American Water Jet Conference, Houston, TX, August, 1991, pp. 17 - 29.

10.134 Zeng, J., and Kim, T.J., "A Study of Brittle Erosion Mechanism Applied to Abrasive Waterjet Processes," Chapter 7, 10th International Symposium on Jet Cutting Technology, Amsterdam, Netherlands, October, 1990, pp. 115 - 133.

10.135 Raju, S.P., and Ramalu, M., "A Transient Model for Material Removal in the Abrasive Waterjet Machining Process," paper 9, 7th American Water Jet Conference, Seattle, WA, August, 1993, pp. 141 -155.

10.136 Thirumalai, K., and McNary, O., "Development and Testing of a Thermohydraulic Process for Hard Rock Cutting," AIME Preprint no 73-AM-45, 1973, AIME annual meeting, Chicago, February 25, 1973.

10.137 Pritchett, J.W., and Riney, T.D., "Analysis of Dynamic Stresses Imposed on Rock by Water Jet Impact," paper B2, 2nd International Symposium on Jet Cutting Technology, Cambridge, UK, April, 1974, pp. B2-15 - B2-36.

11 THEORETICAL CONSIDERATIONS

11.1 INTRODUCTION

Throughout the previous chapters and discussions there has been an ongoing thread of thought regarding the way in which waterjets cut through material. This has been tied in with a significant amount of evidence which has been presented to validate the various concepts which have been proposed to explain how waterjets, and the abrasives and other additives which they might include, work. However, many of the rules which have been suggested in earlier chapters for improving the performance of high pressure waterjets have come about as a result of experimental studies and the rules given are thus empirical.

Empirical rules do not, however, ultimately allow one to develop a complete optimization of the process since experiments are, by their nature, usually restricted in the ranges of the parameters and the aspects of the technology which they study. Thus it is useful to develop a theoretical understanding of the basis of waterjet operation since one can, in this way determine whether, in fact, the results one gets from an experiment are the best that can be hoped for, or if, by a change in parameters one might be able to do substantially better.

Unfortunately, there are so many variables and conditions involved in the operation of a waterjet system that it is difficult to develop one all-encompassing theory which describes all the conditions which might occur from the time that a stream of water is first formed into a waterjet which then leaves a nozzle, until the time that material is removed from the target surface. Even in the development of flow from an orifice, the condition of the flow passage upstream, and the shape of the orifice itself will have a significant effect on the jet condition as it leaves the nozzle and the energy content and distribution which the jet retains at surface impact.

This chapter is designed to make the process somewhat easier to understand and to give some idea as to the current state of theoretical development of the processes involved in waterjet application. Early in the history of this new tool the sentiment favored using the highest pressure possible to fragment a target. To provide a basis for further discussion and to shown the initial problems with that approach consider the results of the following laboratory experiment.

As part of a demonstration of some of the parameters controlling the effectiveness of waterjets in cutting material an experiment was designed to observe the effects of change in nozzle diameter and jet pressure on cutting Berea Sandstone. This rock is a "standard" material and has a compressive strength of around 350 bar. It is quite easy for a waterjet to cut.

In order to compare the performance of Intensifier and Triplex pumps the tests were carried out in two suites. For the intensifier jet pressures of 1,000 bar, 2,000 bar and 3,000 bar were used. For the triplex pump jet pressures of 330 bar, 670 bar and 1,000 bar were used. Nozzle diameters for the former were 0.15; 0.25 and 0.35 mm and for the latter 0.76, 1.02, and 1.19 mm. Each nozzle was traversed over the surface at 30 cm/min.

The depths of cut, once the jet had entered the rock was measured and is tabulated as follows:

Table 11.1 Depths of cut into Berea sandstone by jets from an intensifier and a triplex pump at a traverse speed of 30 cm/min

Cutting depth (mm) *Nozzle diameter (mm)*	*Jet pressure (bar)*			
intensifier	1,000	2,000	3,000	
0.15	3.2	7.14	10.3	**6.88**
0.25	5.56	12.7	23	**13.75**
0.35	11.9	26.6	32.5	**23.7**
average	**6.88**	**15.48**	**21.9**	**14.55**
Triplex	330	670	1,000	
0.76	15.08	31.75	41.3	**29.4**
1.02	19.05	42.95	55.6	**39.2**
1.19	28.65	61.1	79.3	**56.4**
	20.9	**45.2**	**58.7**	**41.6**

The volume of rock removed in each cut was assumed to be given by the product of the slot depth, the length of traverse and the width, which was assumed to be three times the nozzle diameter:

Table 11.2 Volume of cut into Berea sandstone by jets from an intensifier and a triplex pump at a traverse speed of 30 cm/min

Intensifier		*Jet pressure*	*(bar)*	
	1,000	2,000	3,000	
nozzle dia (mm)				
0.15	36.8	82.9	120	**80**
0.25	107.5	245.8	445	**266**
0.35	322.6	666.7	882	**624**
average	**155.7**	**331.8**	**482**	**323**
Triplex				
	330	670	1,000	
0.76	876	1,844	2,397	**1,705**
1.02	1,475	3,318	4,302	**3,032**
1.19	2,600	5,560	7,220	**5,126**
average	**1,650**	**3,574**	**4,640**	**3,288**

Once this had been completed the amount of energy in the jet was calculated as a horsepower, based on the pressure and the volume flow rates (based on an assumed coefficient discharge of 1).

Table 11.3 Amount of power in the jets cutting into Berea sandstone from an intensifier and a triplex pump at a traverse speed of 30 cm/min (values in watts)

Intensifier		*Jet pressure*	*(bar)*	
	1,000	2,000	3,000	
nozzle dia (mm)				
0.15	885	2,500	4,600	**2,660**
0.25	2,500	6,960	12,790	**7,400**
0.35	4,820	13,630	25,050	**14,500**
average	**2,720**	**7,700**	**14,170**	**8,190**
Triplex				
	330	670	1,000	
0.76	4,260	12,050	22,140	**12,820**
1.02	8,230	21,420	39,350	**22,780**
1.19	10,450	29,570	54,330	**31,450**
average	**7,430**	**21,010**	**38,610**	**22,350**

From the last two calculations it is easily possible to determine the specific energy of cutting, since this is the amount of energy used (the last table) divided by the volume produced (the immediately preceding table).

Table 11.4 Specific energy of cutting into Berea sandstone by jets from an intensifier and a triplex pump at a traverse speed of 30 cm/min

Intensifier		*Jet pressure*	*(bar)*	
	1,000	2,000	3,000	
nozzle dia (mm)				
0.15	24.1	30.2	38.3	**30.9**
0.25	23.3	28.3	28.7	**26.8**
0.35	14.9	20.4	28.4	**21.2**
average	**20.8**	**26.3**	**31.8**	**26.3**
Triplex				
	330	670	1,000	
0.76	4.9	6.5	9.2	**6.9**
1.02	5.6	6.5	9.1	**7.1**
1.19	4.0	5.3	7.5	**5.6**
average	**4,8**	**6.1**	**8.6**	**6.5**

The numbers shown are for watts/cu mm of rock removed. (This can be multiplied by 1,000 to get the more conventional levels of numbers normally used - see earlier discussions.)

It is interesting to note that the power required for the use of the intensifier is, on average, roughly four times that required for the triplex pump set of tests. The question is thus raised as to why this might be.

To begin the explanation, the different sets of data are first plotted (Figs. 11.1, 11.2). The plots are of averaged values for depth vs pressure and nozzle diameter. The curves are plotted using the Graph III program and the curve fit feature is used under "power" mode to generate the best fit curve.

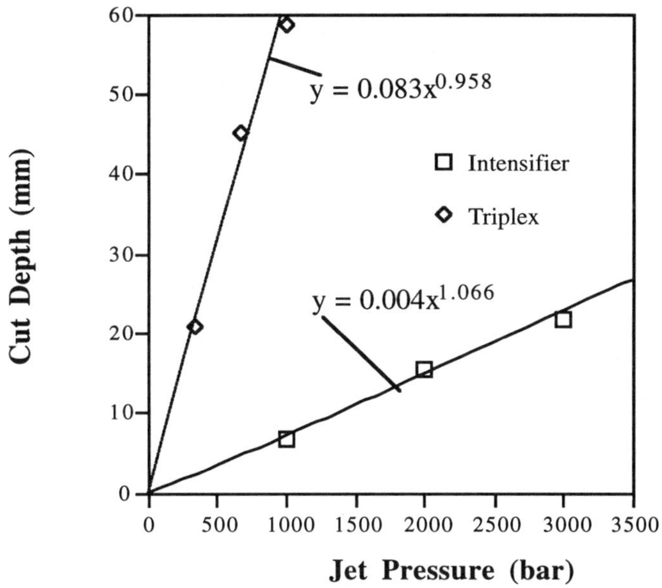

Figure 11.1 Depth of cut as a function of jet pressure for intensifier and triplex pump tests, at 30 cm/min into Berea sandstone.

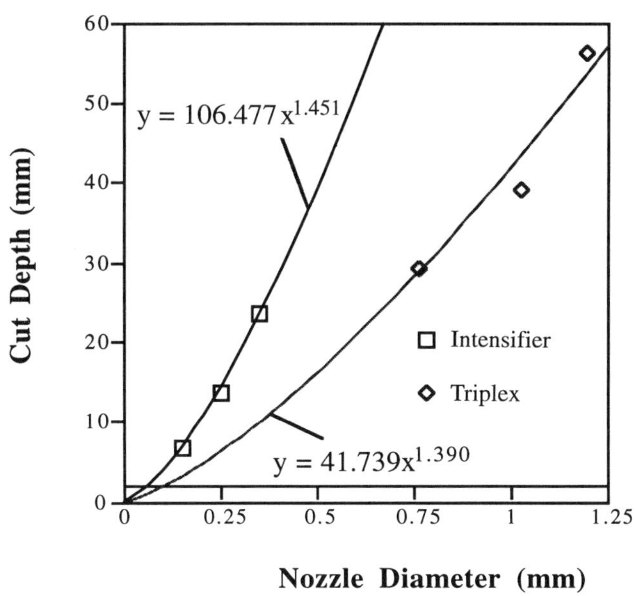

Figure 11.2 Depth of cut as a function of nozzle diameter for intensifier and triplex pump tests, at 30 cm/min into Berea sandstone.

A quick study of the curves suggests that the two cases are not in fact dissimilar, but rather that the data from the two regimes follow the same relationships to these primary parameters. How does this relate to the original question as to the relative efficiency of low and high pressure systems?

Well the depth of cut varies approximately linearly with pressure and to the 1.5 power with nozzle diameter.

$$\text{Volume removed/sec} = \text{depth x width x distance cut}$$

which can be expressed as:

$$= \text{const. pressure}^{1}\text{.diameter}^{1.5}\text{.diameter.speed}$$

$$\text{Energy} = \text{constant x pressure x flow rate}$$
$$= \text{constant x pressure x pressure}^{0.5}\text{ x diameter}^{2}$$

The square root of pressure gives jet velocity and the diameter squared gives the area to combine and give the volume of water flowing.

Thus,

$$\text{specific energy} = \text{energy/volume}$$

$$= \frac{\text{constant x pressure x pressure}^{0.5}\text{ x diameter}^{2}}{\text{const. pressure}^{1}\text{.diameter}^{1.5}\text{.diameter.speed}}$$

which can be simplified to:

$$\text{specific energy} = \text{const . pressure}^{0.5}/(\text{diameter}^{0.5}\text{.speed})$$

which can be simplified to say that the specific energy of waterjet cutting increases with jet pressure and decreases with jet diameter.

As mentioned at the beginning this is an experimentally determined set of relationships, but the data it provides should be borne in mind as the following discussion of early theories of waterjet behavior are discussed, and as a theoretical explanation of waterjetting behavior is developed. One being by considering the amount of energy which arrives at the surface, and how it is distributed.

11.2 JET ENERGY CONSIDERATIONS

It has been an unwritten "rule of thumb" developed by some users of the technology that a waterjet will cut out about 150 to 200 nozzle diameters from the outer end of the jet orifice. This was based, in part, on the work by Leach and Walker [11.1] who showed a rapid decrease in jet pressure beyond that range (Fig. 2.29). The question then becomes, if one spends a great deal of effort in improving the flow into a nozzle and improving the nozzle design, how much improvement can one reasonably expect to achieve. And if, for example, one increases the effective jet cutting range out to 1,000 nozzle diameters is this the best that can be achieved?

Selberg and Barker were faced with this problem in trying to develop better cutting nozzles for use in the second generation of the Hydrominer which was built at UMR [11.2]. However they were able to use earlier theoretical calculations in developing an answer to this problem. The initial theoretical prediction which they based their work on was one developed by Tollmein [11.3]. This had predicted that the jet gets wider as it moves further away from the nozzle, and that the velocity of the jet decreased with distance in an inverse relationship. This in itself is indicative in a change in state of the jet as it moves away from the nozzle (Fig. 11.3).

This type of relationship had been found by a number of investigators, and this result is shown, for example, in studies reported on in an earlier chapter (Fig. 10.6). It can also be seen by a more detailed investigation carried out by Yanaida [11.4] in which the jet spreading was measured by three different methods and found to be steadily increasing as a function of distance from the nozzle (Fig. 11.4).

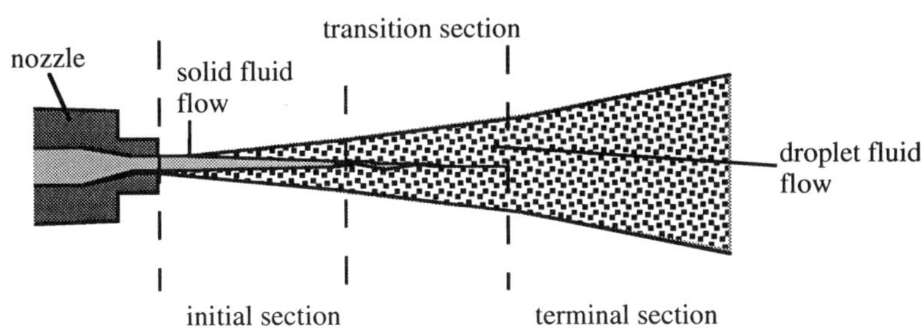

Figure 11.3 Change in jet structure with distance from the nozzle [11.4].

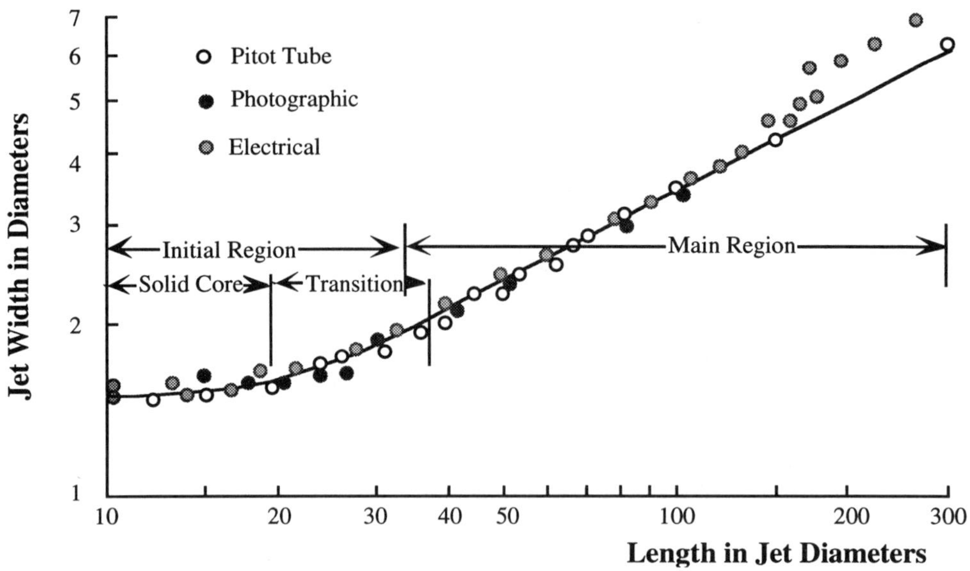

Figure 11.4 Jet width as a function of distance [11.4].

By assuming that the overall kinematic momentum was retained in the jet, even as it grew wider, it was possible for Schlichtling [11.5] to predict, based on these theoretical considerations, the velocity distribution across a waterjet as a function of the distance from the nozzle.

His solution provided predictions for the velocity both along and perpendicular directions to the jet travel in the form:

$$v = \frac{3}{8\pi} \cdot \frac{K}{\varepsilon_o \cdot x} \cdot \frac{1}{\left(1+0.25\eta^2\right)^2}$$

$$u = \frac{1}{4} \cdot \sqrt{\frac{3}{\pi}} \cdot \sqrt{\frac{K}{x}} \cdot \frac{\eta - 0.25\eta^2}{\left(1+0.25\eta^2\right)^2}$$

where:

$$\eta = \frac{1}{4} \cdot \sqrt{\frac{3}{\pi}} \cdot \sqrt{\frac{K}{\varepsilon_0}} \cdot \frac{r}{x}$$

and r is the radius out from the axis

x is the distance along the axis from the nozzle

Selberg and Barker then used a prediction by Hinze that related the centerline jet velocity as a function of the maximum jet velocity at the nozzle exit, in the form:

$$\frac{v_{cl}}{v_{\max}} = A \cdot \left(\frac{d}{x+B} \right)$$

The results from the experiments which Selberg and Barker carried out established the values of the constants, A and B, using an assumed maximum jet velocity at the nozzle derived from the Bernoulli equation. This could then, in turn, allow calculation of the virtual kinematic viscosity (ε_0) and derivation of predicted values for the jet profiles as a function of distance from the nozzle.

They showed that when considerable care was taken in the design of the flow channel into the nozzle and in the nozzle design itself, that the experimental results would lie in a close agreement with the predicted theoretical values (Fig. 11.5).

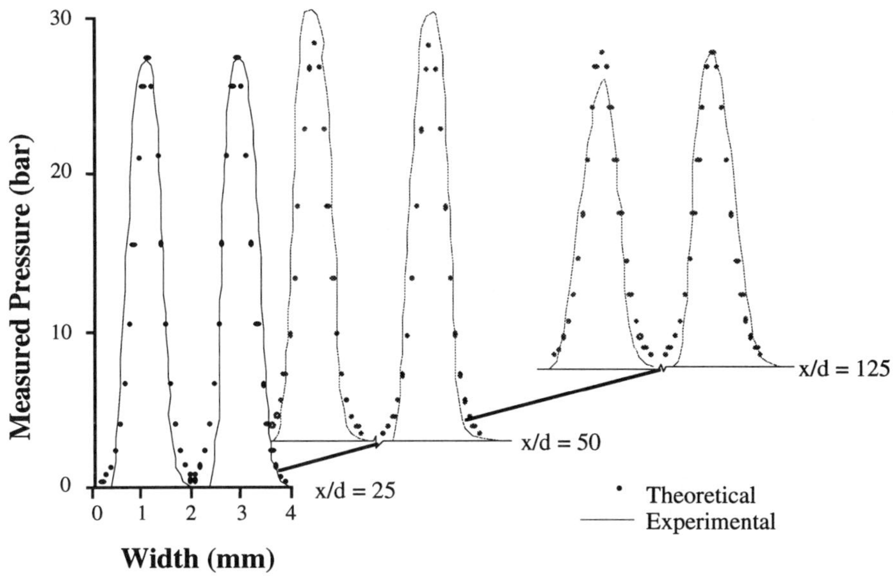

Figure 11.5 A comparison of predicted and actual jet pressure profile results for an 8° diverging, dual orifice nozzle with diameters of 1.016 mm, at 280 bar [11.2].

It should be noted that a study, in the same laboratory only a short time previous to this work [11.6] had found a much wider variation in results. In that study velocity decay was found to vary in an exponential relationship to distance but with the exponent varying from -0.030 - -0.5. However in that experiment the inlet port to the nozzle was not quite as accurately aligned with the outlet diameter of the supply pipe, and the surface of the nozzle was not as highly polished. (To ensure the required accuracy the nozzle in the Selberg study was carefully polished and plated, and the agreement was only evident where a smooth wall to the nozzle was achieved.)

This experiment had two outcomes. Firstly, it provided some additional validation that the theoretical predictions were correct. Secondly, as an outgrowth of this conclusion it also, in this case, indicated that once one had carried out the improvements which the two authors suggest, that there would be very little extra performance gained out of a waterjet system by trying to find additional ways to improve jet flow out of the nozzle.

Of course, had there not been the sought for agreement, one could have drawn one of two other conclusions. One could have deduced that the theory was incorrect, or one could have surmised that there was still a considerable way to go in improving the performance of the nozzle in order to get the optimum jet performance. Even in reviewing this choice the experimental results might provide some indication of the correct choice. For example if the results showed that the jet remained more powerful with distance than the theory predicted, then it is more likely that the theory was not yet fully correct. Or alternatively if the results do not distribute in the way which theory had predicted, this might also indicate that there was some factor in the analysis which had not yet been fully considered in the theoretical development. Barker and Selberg were able (Fig. 2.33) to carry the cutting power of the jet out to over 1,000 nozzle diameters.

It is, therefore, appropriate to consider some of the theoretical approaches which have been taken to predicting the performance of high pressure waterjets in removing material from target surfaces. In line with the discussion above, the basis for the theoretical predictions will be explained, together with the results which they suggest. These will then be compared with some experimental results and the lessons learned from the comparison will then be discussed. During the course of these discussions the role of the energy distribution within the impacting jet will be discussed, when it becomes useful.

11.3 EARLY THEORETICAL APPROACHES

A considerable effort has been made to establish empirical rules for jet cutting and cleaning, but there has also been a strong effort to create a theoretical explanation as to how the process works. The theoretical work can be divided roughly into two camps. The first, and major effort, has been made based primarily on the compressive stress distribution created under an impacting jet; the second looks at the material more as a solid which contains a series of cracks which can be extended to the point that they join together so that it becomes possible for the jet to remove larger pieces of material.

Photographs taken by Daniel, Rowlands and Labus [11.7] during the initial microseconds of impact of a waterjet on a photoelastic target had shown that the jet generated a series of compressive waves into the target. While the dilatational waves rapidly attenuated - with an attenuation coefficient (k in the following relationship).

$$\frac{a.e^{-k.r}}{\sqrt{r}}$$

reported as 0.157 cm^{-1}, in comparison with a reported value for explosive loading of 0.067 cm^{-1}. The shear waves were much more pronounced and did not appear to attenuate as rapidly (Fig. 11.6).

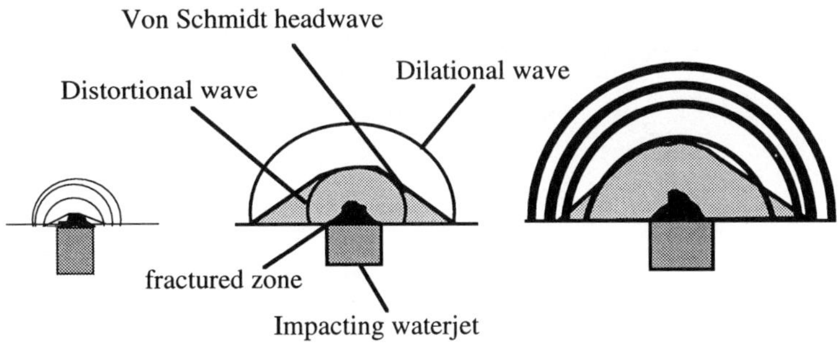

Figure 11.6 Representation of wave propagation shown as isochromatic fringe patterns generated by a waterjet moving at a speed of 2,800 m/sec on impact with CR-39 (after [11.7]) with the nomenclature applied by Singh, et al, [11.8].

To a first approximation is to give the waterjet the same characteristics as a solid impactor. Singh and Hartman [11.8] identified three waves which develop around such an impact point, and suggested that failure can occur, in shear, along the line of the flank or "von Schmidt" shear wave. Given that craters have been observed which roughly follow this shape (Fig. 11.7), it is perhaps, an appropriate initial stepping off point for theoretical development.

Figure 11.7 Dish shaped fragments developed around a waterjet impact point on fine-grained sandstone.

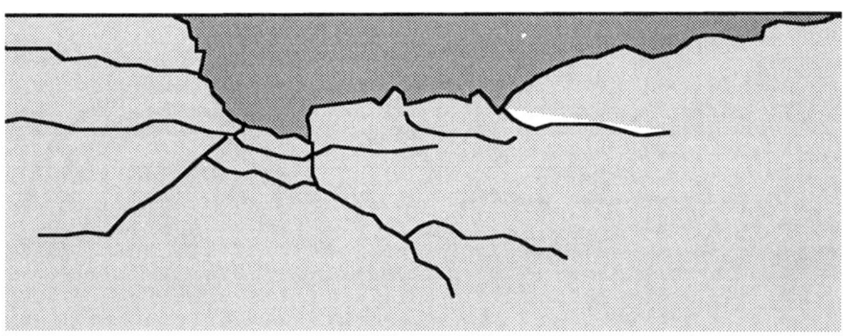

Figure 11.8 Section through a crater generated by single impact simulating rain on a radome showing the developed fracture pattern (after [11.9]).

In their pioneering study of jet penetration of rock Farmer and Attewell [11.10] used the concept of momentum transfer to predict the penetration of a jet into a rock target. Their work drew on earlier studies of projectile penetration into low density targets [11.11] which suggested that the depth of the crater from a single shot of water would be given by the expression:

$$\text{depth of cut} = \frac{\text{K (a constant)}}{\text{a (projectile area)}} \text{. m (projectile mass). (impact velocity - critical velocity)}$$

Where the critical velocity is that at which penetration first occurs in the target material.

Because of turbulence which would occur inside the hole, as water leaving the impact zone would interfere with the sequential length of jet entering the hole, the theory was modified to consider the incoming jet as a series of segments, and this led to the prediction that the depth of penetration (D) could be derived from the relationship that:

$$D = k \cdot d_c \left(\frac{v_0}{c}\right)^{\frac{2}{3}}$$

where: d_c is the diameter of the crater generated

c is the wave speed in water.

Although this gave some indication of the relative performance of a jet on a target, it did not include any consideration of the target response as a means of predicting the actual range of depth to be reached. Additional theoretical development was therefore required.

11.4 POWELL AND SIMPSON'S THEORY

Waterjet structure does not, however, have the simple form of a solid impactor. The velocity, and as a result the impact pressure, varies across the jet, as has been shown in the previous section. Powell and Simpson were the first to consider how this change would be reflected in the target response to the waterjet impact [11.12]. In this analysis the stress distribution was first calculated (Fig. 11.9) for a homogeneous, linearly elastic half-space impacted by a pressurized waterjet, whose normal pressure profile on the surface was approximated by the equation :

$$F(r) = \frac{1}{2} \cdot \rho \cdot v^2 \left[1 - 3\left(\frac{r}{b}\right)^2 + 2\left(\frac{r}{b}\right)^3\right]$$

In the region, radius b, over which the jet loads the surface the jet has radius a, velocity v, density ρ, and b = a 20/3 and where r is the radial distance from the center of the impact zone.

This distribution was based on the experimental measurements made by Leach and Walker [11.1]. Interestingly, in light of the subsequent photoelastic studies by Daniel [11.7], Powell and Simpson presumed that the jet imposed no shear stress on the surface.

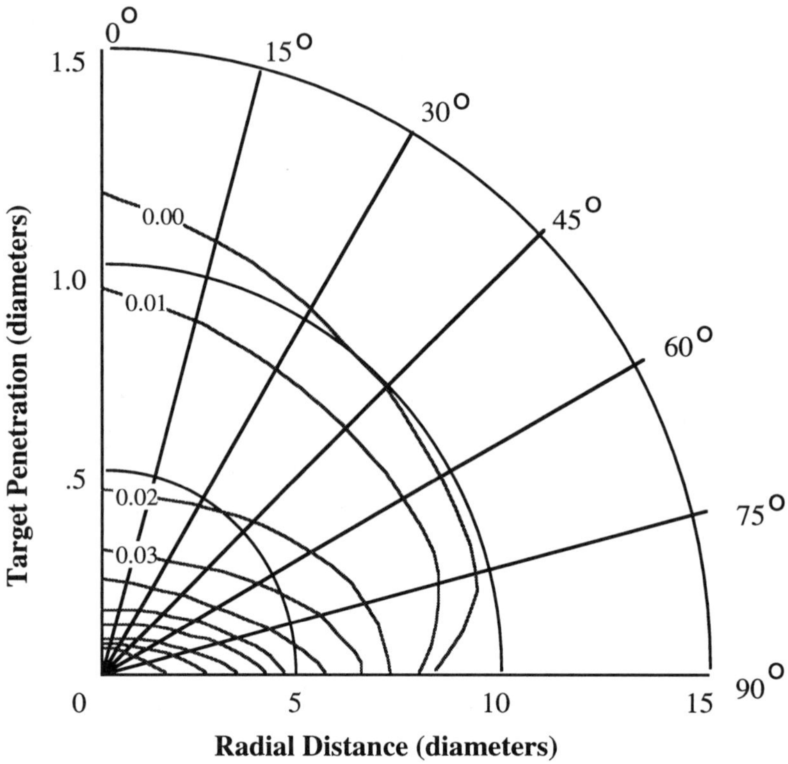

Figure 11.9 Contours of the maximum principal stresses generated under waterjet impact, showing the relative values of stress to the maximum pressure applied by the jet (after [11.12]).

Once the stress distribution was calculated, then the half-space was presumed filled with randomly oriented and distributed incipient cracks, whose presence otherwise had no effect on the stress distribution. Based on a prior calculation of the maximum principal stress (σ_1) and minimum principal stress (σ_3), then the criterion for fracture growth was that:

$$-\frac{(\sigma_1 - \sigma_3)^2}{8(\sigma_1 + \sigma_3)} = \sigma_t$$

where σ_t is the tensile strength of the material.

Based on this evaluation, Powell and Simpson calculated the crack patterns that would be developed in the target under differing levels of applied jet pressure (Fig. 11.10). This indicated that no fracture should occur in the material until the incident jet pressure exceeded 20 times the rock tensile strength, and plotted a projected relationship between the likely depth of cut as a function of the jet pressure expressed as a multiple of this tensile strength (Fig. 11.11).

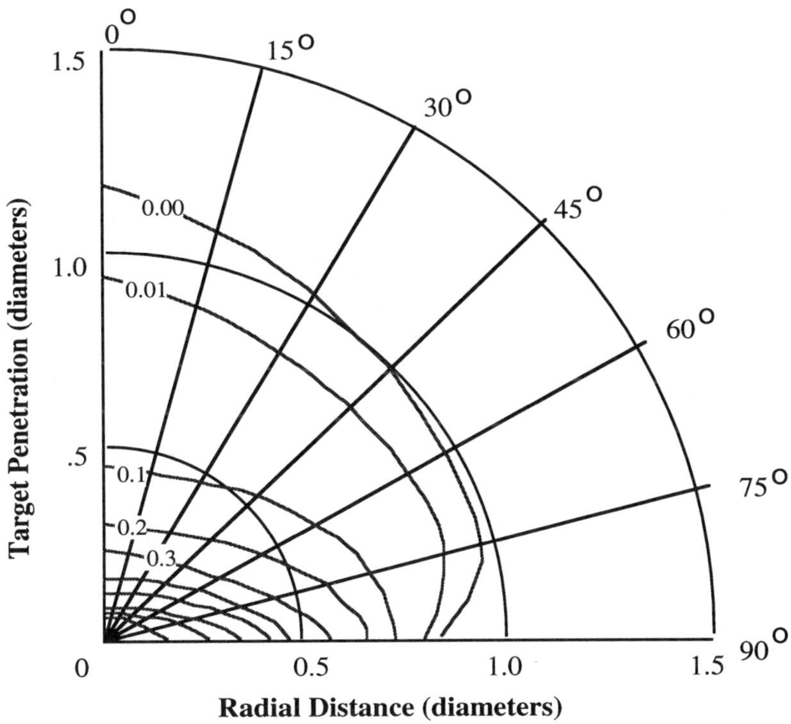

Figure 11.10 Contours for cracks generated in the target below a waterjet impact as a function of position and relative impact pressure (after [11.12]).

While, by the time that this paper was presented, it had been demonstrated (Fig. 7.13) that a jet at a pressure below 700 bar could drill through 20 cm of granite (with a tensile strength of perhaps 175 bar) [11.12] unfortunately this was not well known. It is unfortunate since this initial theoretical review was followed by other research which seemed to corroborate the fact that very high jet pressures were required for effective cutting. Cooley, for example, [11.14] found, using a single pulse water cannon, that the specific energy of excavation reduced as jet pressure was increased from 1.43 - 3.95 times the rock compressive strength, with suggested optimal values at above 10 times the rock compressive strength (see, for example Fig. 7.5). As a result there arose a school of thought which considered that predictive equations for waterjet action should include a dimensionless parameter achieved by dividing jet pressure by rock compressive strength [11.15] and [11.16]. From this it has become common to accept that the uniaxial compressive strength of the material is a meaningful parameter in assessing its cuttability by a waterjet.

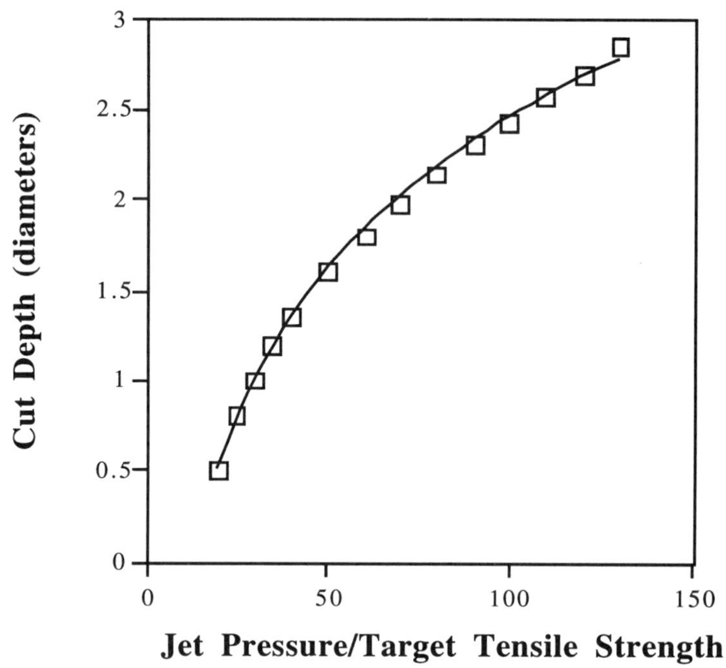

Figure 11.11 Variation in theoretical depth of cut as a function of jet pressure, expressed in terms of the target tensile strength [11.12].

This is an erroneous assumption, as can be shown by drilling two rocks of approximately equivalent compressive strength at the same speed and other jet conditions (Fig. 11.12). Whereas the hole drilled in the smaller grained limestone is small, that in the larger grained sandstone is considerably greater.

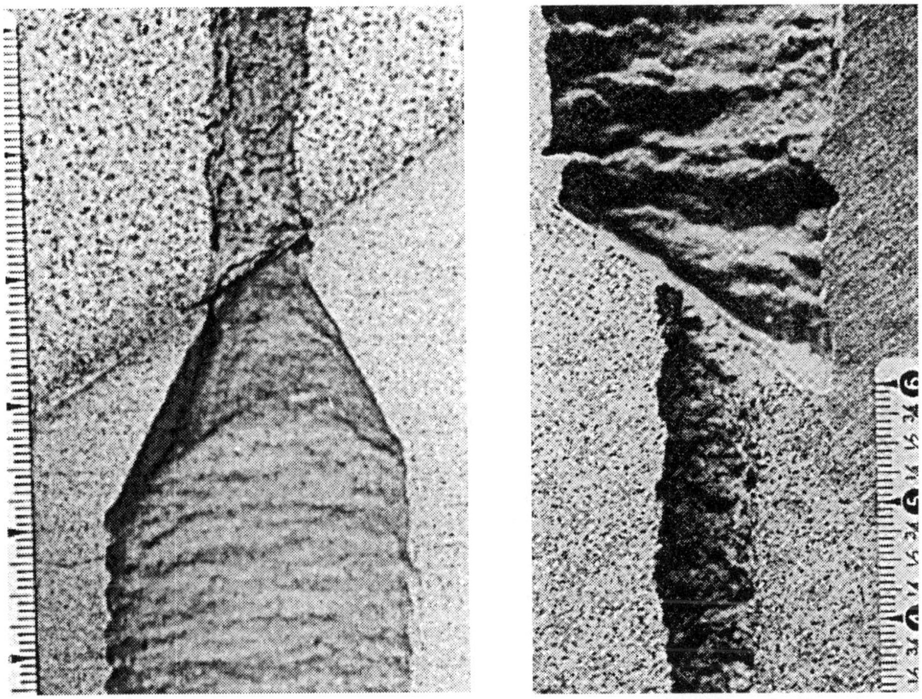

Figure 11.12 Change in hole diameter as a waterjet drills, at constant advance rate and operating parameters, through two rocks of similar compressive strength but different structures.

This result suggests that the critical parameters to consider in the analysis are more related to the structural properties of the target material, rather than its overall mechanical properties. This concept had been originally proposed by Summers [11.13] who noted that Brace [11.17] had found that cracks existed in rocks which could be approximately correlated to the constituent grain size. If the water penetrates these cracks then it will induce an internal pressure. Bieniawski [11.18] had used the criteria first proposed by Griffith [11.19] and [11.20] to determine that these cracks will grow where the stress at the tip exceeds a given value obtained from the expression:

$$\sigma^2 > \frac{2 .\gamma. E}{\pi.c}$$

where γ is the surface energy of the material
E is the material Young's modulus
c is the half the length of the crack

The cracks will grow until they intersect around the individual grains, liberating them and exposing the underlying cracks to jet penetration. With some rocks, such as a coarse sandstone, this penetration of the target by individual grain erosion is common, in others, such as granite, it occurs only in some cases.

11.5 FOREMAN AND SECOR'S THEORY

This approach was further developed, although rather based upon the stress distribution model proposed by Powell and Simpson, by Foreman and Secor [11.21]. While this development of Powell and Simpson's theory added the effect of water presence in the cracks to the stress analysis, it did not consider the cracks as individual entities. Rather it is presumed that once the water hit the rock that it would penetrate into the surface and flow through the rock, under a steady state condition, creating a pressurized fluid within the pores of the rock according to Darcy's Law. Two different components for the loading on rock surface could then be anticipated.

The first component of load was as a result of the conventional effective stress generated under the impacting waterjet due strictly to the total force of the water impinging on the surface. The second stress is located within the rock due to the pore pressure generated by the water migration into the rock surface. The initial pressure distribution under the jet at impact is assumed to be the same as that given above by Powell and Simpson, and initially evaluated by Leach and Walker [11.1]. In order to extend the work and refine it, Foreman and Secor used the finite element code PALOS to gain more accurate approximations for the values of the stresses, strains, and displacements in the rock under this loading condition.

Once the stress condition had been defined, using the above assumptions, the finite element program was instructed to use the Griffith failure criterion to identify the points at which failure might occur, due to tensile stresses generated in the rock. It was assumed, for this model that the rock mass was permeated by flat, elliptical cracks with fracture initiating on some point on the elliptical surface. The point of initiation is a function of the maximum and minimum principal stresses.

Based on these assumptions and calculations, the model indicated that failure would first initiate when the peak pressure reached 18.8 times the tensile strength of the rock and that this would occur at a distance of approximately .75 nozzle diameters beneath the surface (Fig. 11.10). The exact depth was found to depend on the value of Poisson's ratio. The threshold pressure required to initiate rock failure was predicted to be 13.7 times the tensile strength with a Poisson's ratio (υ) of 0.1 and increases to a value of 24.6 times the tensile strength at a υ value of 0.4.

Foreman and Secor recognized that these predicted values were too high, based on the experimental results of Bresee, Cristy, and Mclain [11.22] who had found that the minimum jet pressure required to cut Berea Sandstone, Indiana Limestone, and Georgia Granite were at 140, 245 and 420 bar respectively. Predicted values for this threshold pressure using Foreman and Secor's model ranged from 1.25 - 4 times higher than this.

In order to explain this Foreman and Secor considered that the static theory, while predicting values of the threshold pressure of the right order of magnitude, was based on the assumption that the failure would begin directly beneath the jet while in many instances this does not occur. Accordingly, Foreman and Secor modified the basic equations of Powell and Simpson to consider how pore pressure in the rock would change the failure criteria. Two presumptions were made; firstly that before the failure of any volume of material could occur, all the pores in the affected volume must be filled with fluid. Secondly, that the impacting fluid would flow through the permeable paths and fill the pores in the rock at a rate which could be predicted based upon an application of Darcy's Law.

$$u = \frac{\kappa \cdot P_0}{\mu \cdot l_0}$$

where u is the speed of fluid penetration,
κ is the permeability,
μ is the dynamic viscosity
l is the grain size of the rock.

Under these conditions, the stress distribution within the rock must be modified to take into account the pore pressure. However, in this model the pore pressure was considered to be act purely as a pressure within the pores of the rock (as opposed to a force pressurizing the crack tips). It was considered, therefore, only to act as an additional internal pressure which would contribute to the overall stress condition, with pressure decreasing away from the surface due to the frictional loss of pressure in the fluid as it flowed past the walls of the permeable paths.

In this calculation, the compressibility of the rock mass was a significant factor, since this would considerably effect the extent and rate of the fluid penetration and the response of the rock to pressure. This can be illustrated again by reference to earlier work where it was demonstrated (see Fig. 8.48) that an increased rock stress closed the pre-existing cracks in the rock, reducing the ability of a jet drill to cut into the surface.

The model assumed that compressibility α of the rock could be accurately defined in terms of the bulk modulus of the dry rock mass, K, and the bulk modulus of the individual grains, Kg by the relationship:

$$\alpha = 1 - \frac{k}{k_g}.$$

For the particular case of a rock mass with spherical pores and narrow elliptical cracks, Walsh [11.23] has shown that:

$$\frac{K}{K_g} = \frac{1}{1 + \frac{3}{2}\left(\frac{1-\upsilon}{1-2\upsilon}\right)\eta}$$

where υ is Poisson's ratio
η is the porosity of the rock

The above two relations may be combined so that α is defined by the two measurable quantities, υ and η.

This new set of equations were solved using a finite element code developed for a porous elastic media, examining again the principal stress distribution and the resulting failure contours; except that now the principal stresses were modified by subtracting the pore pressure from each normal stress. Griffiths' theory was again used to predict fracture initiation, which was still considered to be a direct function of the pressure distribution within the rock. The stress distribution, in turn, was related to four mechanical rock properties, the Poisson's ratio, Youngs' modulus, porosity and permeability and two fluid properties, viscosity and permeability.

In order to verify this theory, Foreman and Secor placed thin layers of copper between the waterjet and the target surface. In this manner they determined that there was no damage to a block of Indiana limestone at jet impact pressures of up to 1,400 bar. Where the copper was not present a 280 bar jet issuing through a 1 mm diameter nozzle from the rock cut a cavity 2.5 mm in diameter after 0.5 sec. When the copper was placed between the jet and the surface and with the same firing conditions it was not possible to discern any damage to the surface, although there was a circular impression left on the copper. This was taken to indicate that the failure of the surface is not due to the impact pressure on the surface alone, but that the penetration of the water into the surface is also an important part of the failure process.

A more accurate prediction of the threshold pressure of 2.5 times tensile strength was reached for the Indiana limestone, and a more accurate projection of the crater shape was thus established (Fig. 11.13). This work has since been developed by Crow [11.24] and Rehbinder [11.25].

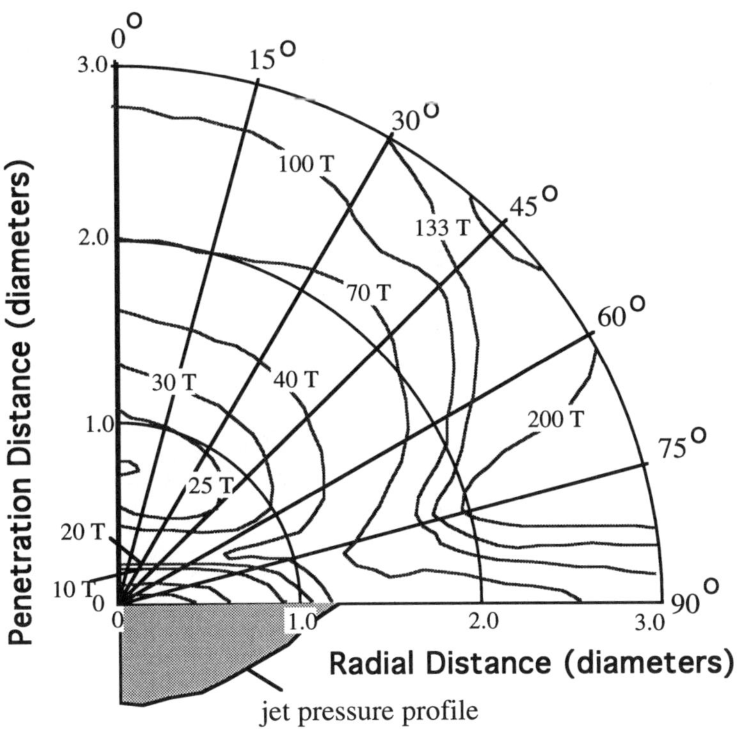

Figure 11.13 Surfaces of fracture initiation predicted by Foreman and Secor, when pore pressure is included in the analysis (after [11.21]).

11.6 CROW'S THEORY

One problem with the development of the theories to cover rock penetration is that the models dealt only with the impact of the jet on the surface of the material. Under many circumstances the penetration of the jet into the surface is quite rapid. In less than one second it is possible for a waterjet to penetrate more than 200 mm into a target, and thus the initial conditions governing jet penetration at the start of the cut are only of transient interest. While an evaluation of initial impact might be of value in identifying the overall mechanisms of material failure, it is an extremely transient aspect of the overall jet penetration. A theory proposed by Crow [11.24] which examined the continuous cutting process as a waterjet traverses over the rock surface, has gained considerable attention.

Crow's premise was that, as the waterjet cut into the rock, that it would establish a steady state condition in which very little shear stress was generated upon the rock surface. The jet was presumed to cut a slot down through the rock, at the same width as the incoming jet, and with the cut face slowly turning from the vertical to the horizontal over the cut depth (Fig. 11.14).

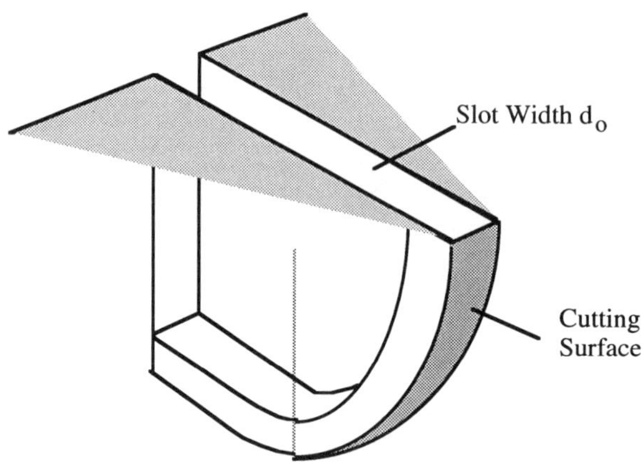

Figure 11.14 Basic cut geometry considered by Crow in developing his theory [11.24].

Crow's conclusion that there would be little shear on the surface was based on the premise that any tension in the water flow across the surface would generate cavitation (Fig. 11.15). Concurrently the high surface pressure exerted by the jet would hold the cavitation bubbles against the surface and simultaneously expose the grains to the direct impact of the water. The model considers the situation that, for constant slot cutting, the grains along the cutting surface on the rock face in contact with the leading edge of the jet are always in a state of incipient failure. Thus, by equating the developed stresses on these grains to the shear strength of the material, it is possible to define the conditions for rock failure and from this the traverse speed at which the jet will cut to a given depth in the target material.

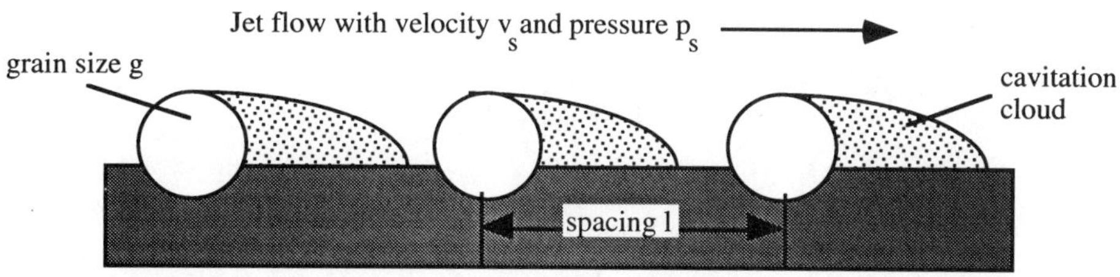

Figure 11.15 Crow's concept of cavitation bubble generation behind the grains of a jet cut [11.25].

If one looks at the segment of an idealized cut under Crow's concept then the general shape of the cutting curve is presumed to be at a positive rake. It is proper to note at this time that Crow's theory does confirm the experimental result that optimal cutting with a waterjet generally requires that the jet be inclined at a positive rake angle forward into the cut, of approximately 11 degrees. One should also note, in the development of Crow's equations, that he assumes that the grain diameter of the rock is reasonably small compared with the size of the waterjet nozzle. This held true for the first experiments he carried out to verify his model, but is not always the case.

The radius of curvature of the cut would control the extent to which the cavitation is created behind the grains. The cavitation number of the jet (defined in Chapter 10) is given by the expression:

$$\sigma = \frac{p_s - p_v}{0.5 \cdot \rho \cdot v^2}$$

Crow shows that this can be equated to twice the relative ratio of the jet thickness at any point to the radius of curvature R. The major feature of this theory is that, as the water flows across the grains, cavitation clouds are generated behind the individual grains which are standing proud of the surface. The greater the radius of curvature the greater the extent of the cavitation and the more of the surface is covered by cavitation. This increases the stress on the grains exposed at the face of the cavitation clouds.

At smaller radii of curvature the pressure of the fluid on the surface is greater, inducing a more rapid collapse of the cavitation cloud. This reduces its size and thus exposes more of the grains to the pressure of the overflowing jet.

Crow is able to simplify the forces on the individual grains to derive an equation of the form:

$$\tau = \frac{\mu_w (p_s - p_v)}{1 + \left(\dfrac{p_s - p_v}{B/2 \cdot \rho \cdot v^2} \right)}$$

where μ_W is a dimensionless coefficient involving the constant B and the coefficient of discharge.

This is the same sort of relationship as that proposed by Coulomb to describe rock failure. In that relationship failure occurs where the shear force exceeds the cohesion of the material plus the product of the angle of internal friction and the normal pressure on the grain. However in many rocks the internal angle of friction is, by Crow's accounting, close to 1 in value, while the value for μ_W in the above expression was calculated to be on the order of 0.6 or less. This would suggest that the shear forces generated by the water would never reach the levels required to fail the rock, and waterjets should not cut material.

In order to overcome this problem Crow considers that the water must be able to penetrate into the rock surface ahead of the traversing jet. Crow rewrites the form of Darcy's law from that used above in describing this process to give:

$$v - u = k \,.\Delta p$$

where k is the permeability of the rock.

Within the pores the water is considered to have zero pressure at the point of furthest penetration, and the driving pressure of the jet at the surface of the rock. Because the cutting condition is considered to be a steady state then the water:air interface within the rock lies at a constant distance from the rock surface, and u is set at zero. By the same token the value for v can be given as v.sin ϕ where ϕ is the angle between the rock surface and the direction of nozzle travel. Crow then uses this to define a new criterion for the failure of the rock:

$$\tau = \tau_0 + \mu_r \cdot \frac{g.v}{k} \sin \phi$$

in this equation g is the size of the average grain in the rock, and k is the rock permeability. From this relationship it is possible to derive what Crow defines as the **intrinsic speed for rock cutting:**

$$c = \frac{k.\ \tau_0}{\mu_r.\ g}$$

Crow was then able to show that with a grain size of 0.125 mm an internal angle of friction of 1 and an intrinsic speed of 310 mm/sec a cut of 12.5 mm could be predicted for Wilkeson sandstone. Predictions were made for jet behavior in this rock (Fig. 11.16) and relatively close agreement found between predicted and actual jet cutting ability.

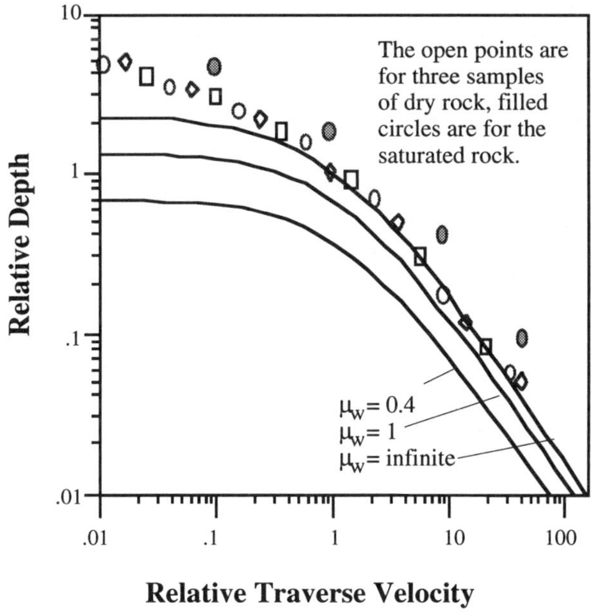

Figure 11.16 Comparison between experiment and Crow's theory in the cutting of Wilkeson sandstone [11.24].

The original calculation was recognized by Crow to contain a small error in the consideration of rock porosity [11.26]. When this was taken into account Crow was able to postulate a maximum rate at which the area of the slot could be created in terms of the equation:

$$(h.v)_{max} = \frac{2.k.\ d_0.\ P_0}{\eta.f.\mu_r.\ g} . (1 - e^{-\mu_w.\phi_o})$$

The phenomena of cavitation have been discussed in an earlier chapter and more exhaustively elsewhere. However, conflicting evidence to this theory is suggested by a study of cavitation damage of a dolomite sample. A test chamber has been built and is described in Chapter 10, in which a 5 cm x 2.5 cm sample of rock was slowly traversed under a waterjet at pressure. The test was carried out under water in the chamber which was pressurized, by valving the drain pipe. By adjustment of the cell pressure, relative to the jet pressure (a ratio which simply approximates to the cavitation number) cavitation could be induced around the flowing jet and caused to collapse on the rock surface. By control of the feed rate of the rock, the residence time of the rock under the jet was changed. As the rock speed was slowed from 5 cm to 1 cm/min a significant change in the cut surface was discerned (see Figs. 10. 43 - 10. 45).

At the highest traverse speed the entire surface was pitted by the collapse of cavitation bubbles. As the residence time increased, however. this even attack did not persist. Rather, the cavitation attack started to concentrate at weakness points and, from these, larger cracks developed. The point was suddenly reached where these cracks attained a critical length and a substantial change in the depth of cut occurred. This was because the main jet pressure was now sufficient to cause the developed cracks to grow, coalesce and remove fragments of material. Cavitation bubble erosion thus takes a certain finite amount of time, over and beyond that available in conventional high pressure waterjet cutting, to give the high cutting rate which it is capable of achieving. The time is not felt by the present author to be present therefore for cavitation to significantly affect normal jet cutting operations significantly.

Crow uses Darcy's Law to predict flow into the target which creates a saturated region, in which the rock is in a continuous state of incipient fracture, one grain diameter beneath the cutting surface. From this he deduces the intrinsic speed of cutting mentioned above. While many granular rocks fail along grain boundaries, in crystalline rock the chips are frequently formed over several adjacent crystals and may measure several inches.

The contrasting arguments against application of this theory are stressed since it has gained considerable recognition and even today is cited by investigators who enter the field. In this regard it is pertinent to draw attention to the discussion held at the Water Jet Workshop in Rolla in 1975

[11.26]. In this presentation Crow modified his failure assumption, so the cavitation bubbles altered the drag forces on the grains, rather than being responsible for the failure. This modified the equation for intrinsic cutting speed and first recognized effect of jet diameter as a parameter.

$$h = \frac{2.\ \mu_w.d_0.\ P_0}{\tau_0} \int^{\phi_0} \frac{e^{-\mu_w.(\phi - \phi_0)}\ .\ \sin\phi)}{1 + (\frac{v}{c})\ .\sin\phi}\, d\phi$$

This expressions states that the depth of the cut is proportional to the diameter of the jet, and also proportional to the pressure of the jet. However, when Crow compared his predicted cutting rates for other rocks there was considerable difference between them and the actual results. The good fit that was obtained to Wilkinson sandstone data of very small grain size became poorer where the rock was switched to a granite and became in his own words "A disaster", when Berea sandstone was used as a cutting target (Fig. 11.17). While the change in grain size may have had influence, Crow himself believed that one of the major problems resulted from the high correlation with permeability that his theory called for. He recognized, at that meeting, a flaw in the theory and was not able to explain it. Crow went on to state [11.28] that the mechanism proposed is "inadequate to describe the actual process....". This point is stressed, since a number of investigators have used this theoretical development as a basis for developing on from it to their own work, not recognizing the inherent flaws in the original work.

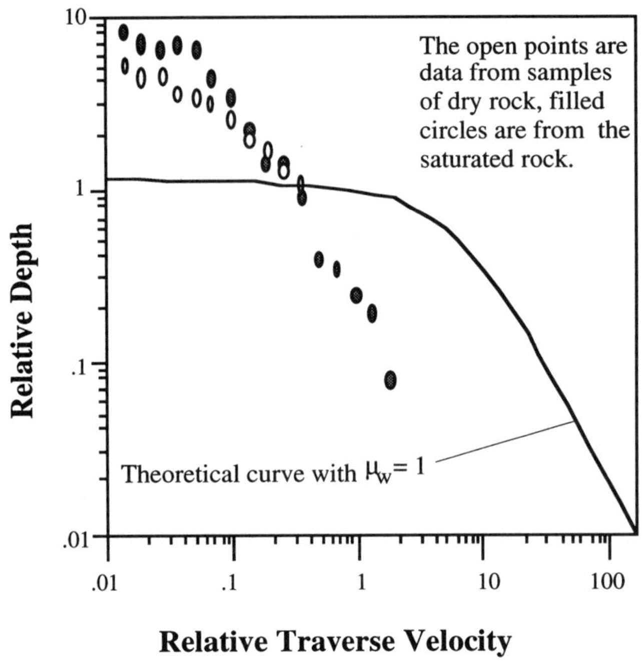

Figure 11.17 Comparison of predicted and actual cutting performance of jets on Berea sandstone using Crow's theory (note the scale is logarithmic) [11.27].

11.7 REHBINDER'S THEORY

The research group attached to Atlas Copco have been involved in jet cutting work for many years. Much of this work was directed at research into the cutting of harder rocks in mining, and this was carried out at the Institute CERAC in Switzerland. However a significant contribution to the theoretical understanding has been provided by Rehbinder who worked for the company in Stockholm [11.25]. In examining the results of jet impact it has been noted that the surface erosion will often occur initially outside the area directly covered by the incoming jet (Fig. 11.18). Dr. Rehbinder considered this in his approach to the problem, which he otherwise addressed basically along the same lines as in the approach described above, simplifying slightly the impact pressure distribution, and then further modifying the assumptions on the effects of the water which is driven into the rock surface ahead of the erosion plane.

Figure 11.18 Distribution of erosion around the jet impact point on aluminum held close to the nozzle.

Rehbinder went back to the original work by Leach and Walker [11.1] to obtain the pressure distribution exerted by the jet on the surface. The basic function which he used is given:

$$\frac{P}{P_o} = e^{-a^2 r^2}$$

And if the momentum of the jet is considered to be still conserved (i.e., before the edges of the jet undergo significant breakup) then this defines the constant a as:

$$a = \frac{1}{\sqrt{2} \cdot r_o}$$

where P_o is the original jet pressure at the nozzle, and r_o is the original jet radius.

From this Rehbinder calculated the fluid pressure distribution which would be exerted within the rock. Based on this pressure distribution, he assumed that the jet acted only on a layer of rock of uniform grain size (δ), porosity and permeability (κ) which lay adjacent to the cutting surface.

Where the surface was presumed to be saturated at the time that the jet arrived then Rehbinder derived a predicted pressure gradient for the fluid within the rock which indicated that, at a radius of 1.81 times that of the jet, that the radial pressure gradient in the fluid would become negative at the surface (Fig. 11.19).

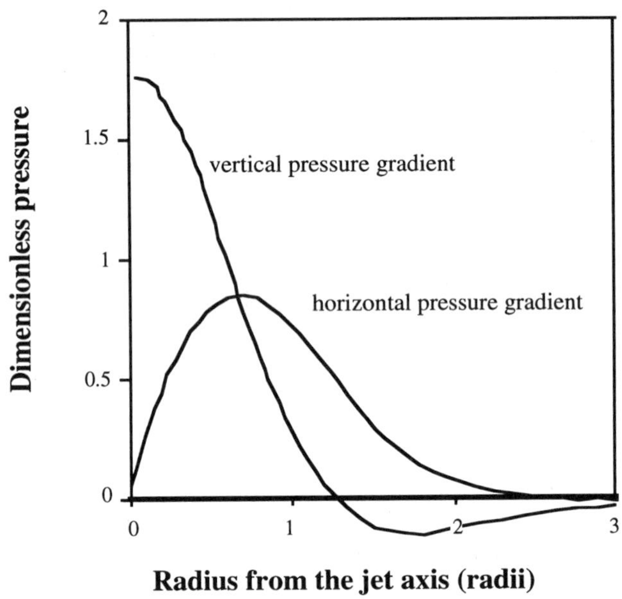

Figure 11.19 Initial pressure gradients in a saturated rock under jet impact, as predicted by Rehbinder [11.25].

Rehbinder then went on to consider what the effect of this pressure distribution would be on the individual grains within the rock mass. He assumed that the pressure drop was linear along the passage ways by which the water was entering the rock. The friction along the walls of the passage is then assumed, over the defined length of the event, to absorb the driving energy which is pushing the water into the rock. This power adsorption is distributed evenly to each grain in the matrix, with the force lost by the fluid being equated to an increased drag force on each grain.

While the number of grains involved is defined by their average size within the zone of influence, it can also be derived from the porosity of the rock under attack. From this Rehbinder derived the force acting on a single grain to be defined by the term:

$$F = \frac{V}{1 - \zeta} \cdot \text{(pressure gradient)}$$

where V is the grain volume and ζ is the porosity.

By substituting the pressure gradient derived from the earlier equations and as shown in the above plot, Rehbinder could derive the pressure exerted by the fluid on each grain around the impact point. As with the pressure gradient this indicated that, beyond the radius of 1.81 times that of the jet, the hydraulic forces would be tensile on the grains, inducing erosion of the grains, under their weaker tensile component, but beyond the original impact area.

The defined boundaries of this problem are, however, only valid for the condition that the jet is impacting a saturated surface, before any significant erosion has occurred. (Which incidentally is an interesting condition given that this also agrees with the erosion pattern shown in Fig. 11.18, but the target material there is aluminum which does not saturate). Once the surface geometry changes, then this model is no longer valid and Rehbinder went on the consider the case of a continuous generation of a slot under a traversing jet. In this case he considered the rock to be initially dry.

The rate at which the slot would be created, provided that the jet was given sufficient time to penetrate the surface, was given by the equation:

$$\frac{dh}{dt} = \frac{\kappa \cdot P(h)}{\mu \cdot l_0}$$

This equation requires the pressure at the depth of the slot to be input. Rehbinder has proposed that this be given by the relationship:

$$P = P_0 \cdot e^{-\beta \cdot h/D}$$

where D is the width of the slot, depth h, and β is a rock defined constant which can be determined experimentally.

This can be used to solve the differential equation and yields the relationship that, as long as the pressure generated at the bottom of the slot exceeds that required for rock removal, the threshold pressure (P_{th}), then the slot geometry will be defined by:

$$\frac{h}{D} = \frac{1}{\beta} \cdot \log_n\left(1 + \frac{\beta . \kappa . P_0}{\mu \cdot l_0 \cdot D} \cdot T\right)$$

This, in turn defines the maximum depth to which the slot can be cut as:

$$\left(\frac{h}{D}\right)_{max} = \frac{1}{\beta} \cdot \log_n\left(\frac{P_0}{P_{th}}\right)$$

From this Rehbinder predicted that slot cutting efficiency would increase with increased traverse speed up to some inherent limit, which was a function of the rock permeability. For permeable rock this limit would be around 100 m/sec, while for relatively impermeable ones it would lie around 1 m/sec. It is interesting to note, in passing that this seems to be (based on some results reported in earlier chapters) about two orders of magnitude high.

Rehbinder initially verified the correctness of his approach, by firing very small slugs of water at the rock surface and examining the craters which resulted. The shape of the craters closely approximated those which were predicted by the theory, both in the location of the failure and also in terms of the size.

Rehbinder then carried out traversing tests using a granite sample moved at relatively high speed under the jet and again found relatively close agreement between the basic theory and the actual results (Fig. 11.20).

Rehbinder also constructed an artificial slot and measured the pressure at the bottom of this in order to define the value for β. The equipment used was relatively straightforward (Fig. 11.21) and resulted in an approximate value of 0.33.

The results were, however, somewhat scattered and indicated that the relative ratio of the nozzle diameter and the slot width was quite critical (Fig. 11.22). This is, to this author's knowledge, the only data available on measurements of this type which would allow independent verification of the theory.

Rehbinder expanded this theoretical study of slot cutting in subsequent work [11.29]. He refined his model by considering that grains are removed at the same rate as the water flows over the surface. In the rock used in this second experiment the value of β was found to be 0.025.

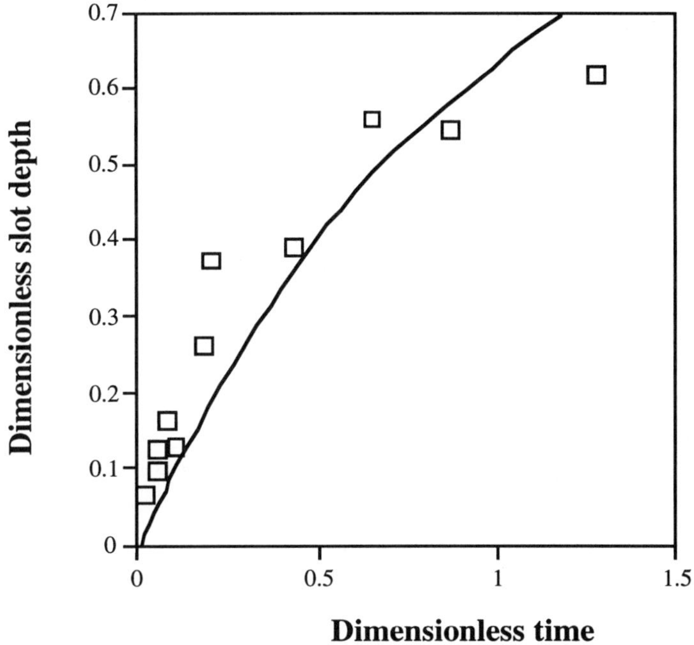

Figure 11.20 Correlation between measured (points) and theoretical results in the cutting of Bohus granite at 2,400 bar with a 1.3 mm jet using Rehbinder's theory [11.25].

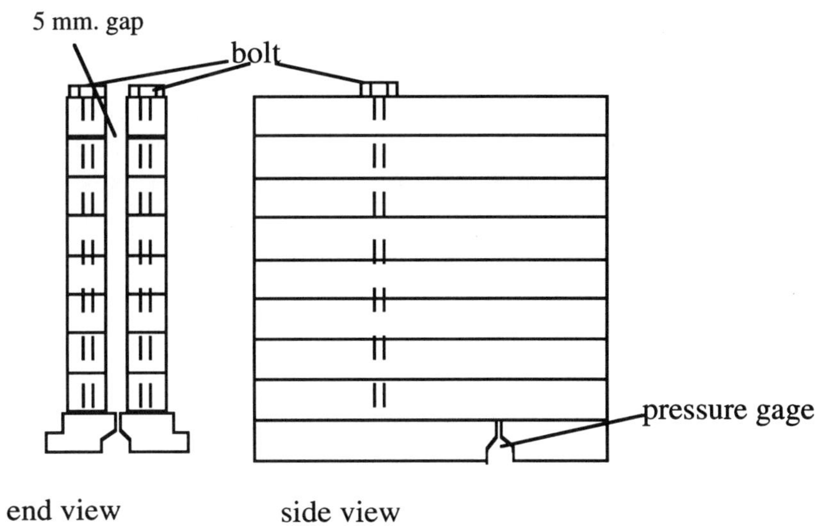

Figure 11.21 Apparatus used by Rehbinder to evaluate the pressure drop as a function of slot geometry [11.25].

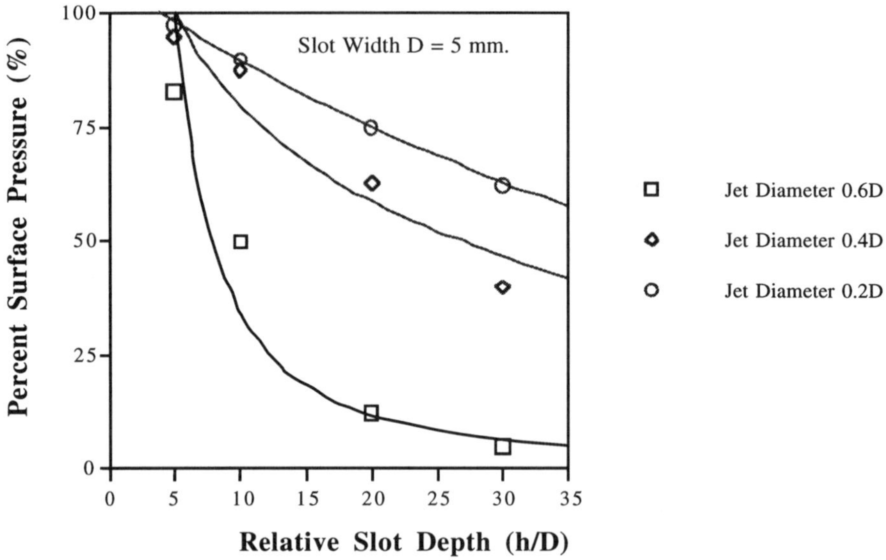

Figure 11.22 Jet pressure at the bottom of a simulated slot as a function of slot geometry [11.25].

Experimental verification of the model was then developed by defining the exposure time of the rock as being the jet diameter divided by the nozzle traverse rate. (This could be increased by also considering the multiple pass condition in which the overall exposure time would be defined as the above value multiplied by the number of passes).

Initial agreement between the theory and experiments were not good for Stockholm granite (Fig. 11.23), it must be remembered that the original theory was developed for rocks where the jet diameter was considerably larger than the individual grains in the rock. In many granites the grain size is significantly larger than the jet, and in this case (as was discussed in Chapter 8) it is only when the jet traverses some distance that the water will reach, and be able to penetrate, cracks in the rock surface.

Rehbinder proposed that a simple experiment be carried out for each rock to be tested in order to evaluate the relationship between permeability, fluid viscosity and grain size. This correction can then be applied to the predictive equation to generate a more accurate estimation of the effective cutting conditions for that rock. Rehbinder suggests that the rock be evaluated in terms of two defined variables its erosion resistance (defined as the grain size divided by the permeability) and the threshold pressure at which material starts to be removed.

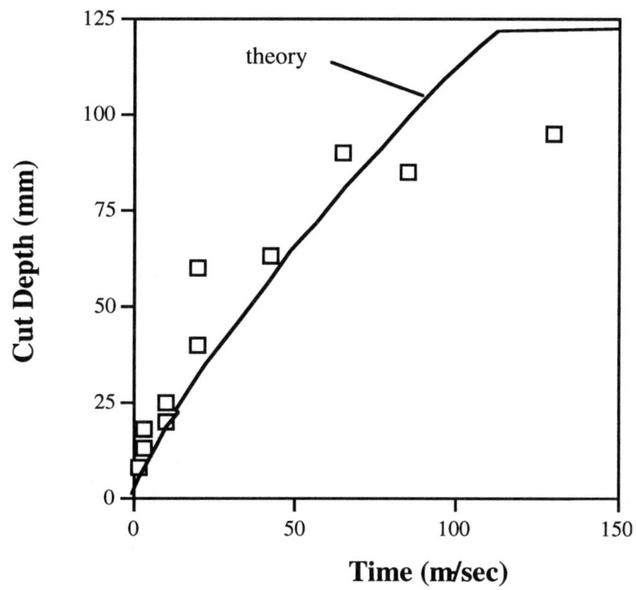

Figure 11.23 Initial correlation between predicted and actual results in the cutting of Stockholm granite with a 2,220 bar jet of 1.3 mm diameter [11.29].

In his third paper on this topic [11.29] Rehbinder then went on to derive the necessary properties to allow a prediction of cutting ability for several rocks. He pointed out that with a typical slot width of three times the jet diameter, and the jet pressure used in cutting is often 2 - 4 times the rock threshold pressure, that the maximum slot depth which can be obtained is on the order of 100 times the jet diameter. The derivation requires a modification to the permeability value, from which the specific erodability of the rock was determined. This was calculated in two forms and these averaged to give a tabulated set of data (Table 11.1)

$$\text{Specific erodibility} = \frac{\kappa \, . \, P_0}{\mu \, . \, l_0}$$

It is pertinent to make a couple of observations. Rehbinder, using fluid flow into the rock, predicts the erosion resistance of the rock as a function of velocity of penetration of fluid into the rock surface. A natural result of this, from his equation for traverse velocity, is a correlation which says that rate and depth of cutting are an inverse function of grain size. It has been an observation of results obtained at UMR that for many rocks, a larger grain rock is easier to cut to a greater depth than finer grained rocks.

Table 11.1 Average specific erodability of selected rocks as derived by Rehbinder [11.30]

Rock	*Average specific erodability cu m/Ns)*
Berea sandstone	8.10^{-8}
Wilkeson sandstone	4.10^{-8}
Darley Dale sandstone	11.10^{-8}
Pennant sandstone	6.10^{-8}
St. Bees sandstone	4.10^{-8}
Horseforth sandstone	10.10^{-8}
Lemuda sandstone	8.10^{-8}
Albrighthausen sandstone	4.10^{-8}
red sandstone	10.10^{-8}
Bohus granite	2.10^{-8}
Stockholm granite	1.10^{-8}
red granite	3.10^{-8}
titania diabase	1.10^{-8}
Kiruna quartz porphyry	2.10^{-8}

Rehbinder's equation also assumes that the flow of water through the rock body reaches a steady state. This is an overly simplistic view of the situation and it is likely that the fluid distribution, and its pressure profile around individual grains has more significance. The forces which are most likely to be critical to erosion of the grains are not the drag forces, which Rehbinder assumes constant over each side of the rock grain, but rather the differential fluid pressures in the cracks on either side of the grain.

11.8 ENERGY INPUT & CRACK GROWTH CONSIDERATIONS

Many of the theories proposed above have concentrated purely on the initial stages of the jet impact upon the surface. Others have considered the conditions which exist during a steady cutting condition. In their development they have looked at the stress conditions which prevail within the target material as it is impacted by the waterjet. In the development of the theories to explain erosion, theoretical development began with an evaluation of the stress field around the impact point. It was subsequently modified by a consideration of the change in stress which is generated by the permeation of water into the target material. Subsequent development examined the role of the permeating fluid on individual grains within the target matrix. From these considerations one approach has developed which has given results with some consistency in predicting cutting behavior. Yet if one examines the experimental results which were inserted at the beginning of this chapter, and considers data derived elsewhere, then perhaps a different explanation can be proposed.

In the initial impact of a waterjet on a smooth surface erosion will occur, close to the nozzle, around the perimeter of the impact zone. This point was recognized by Rehbinder, as discussed above, yet one needs to study the data from an alternative perspective. This is because there are possible alternate factors which might explain the erosion of material. Investigators at Cambridge have shown that a hemispherical-ended waterjet will create a stress, as it impacts a surface, which can exceed 2.5 times that generated in water hammer [11.31]. It is considered that this is generated because, until this ring zone is reached, the surface of the water drop is collapsing on the surface faster than the water already on the surface can escape. In consequence no shear is measured within the zone, and only relatively low values for shear will be generated outside it, until such time as the central segment of water can escape from the cutting zone later in the cutting process. It should be noted, incidentally, that the times given in this discussion are extremely short.

This phenomenon of high pressure impact stress is extremely transient and requires a controlled shape to the jet leading edge. Attempts to develop rock cutting devices based on this have so far failed because of the difficulty in controlling this shape.

As impact pressure is increased, however, circumferential cracks are generated around the impact zone. Field [11.32] has related these to the Hertzian cracks which occur when a solid sphere impacts on a flat surface. Because of the stepped nature of the surface at these cracks the subsequent flow of fluid over the surface will initiate erosion at this location. Hwang and Hammitt [11.33] have shown both the magnitude and transience of these forces by examining the impact of a 2 mm diameter droplet on an aluminum surface. They showed that the maximum depression of the surface occurred 0.5 μsec after impact. The size of the impacting droplet also limited the range of the zone of damage to within 1 cm of the impact surface.

In their analysis they showed that, when impacting aluminum, a droplet moving at 300 m/sec, would generate an impact pressure of 6,300 bar which would compress the material under the droplet by a maximum 0.0112 R_o, where R_o is the initial drop diameter.

This compression would concurrently generate tensile stresses around the compressed zone which would reach a maximum value beyond the initial droplet radius (Fig. 11.24).

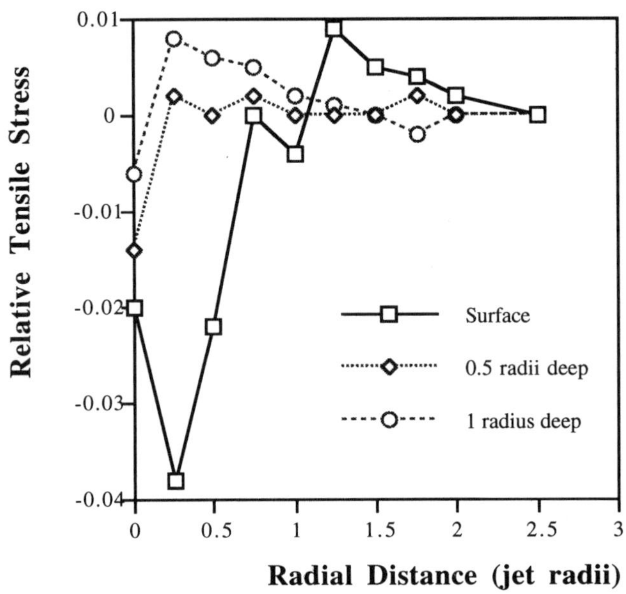

Figure 11.24 Plot of tensile stresses around an impacting water droplet on aluminum [11.33].

The relative position of the surface at several instants after initial impact were also plotted and it can be seen (Fig. 11.25) that the material in the center of the impact zone is initially uniformly compressed with the greatest tensile extension being required on the walls of the depression which is created.

The third factor which will affect the severity of the erosion is that, close to the nozzle, the central core of the jet is still solid and moving at relatively constant velocity. It is only on the edges of such a jet that there is the rapid decay in pressure discussed by Powell and Simpson [11.12]. This pressure decay will incrementally reduce the loading on the surface such that there is a differential stress across the particles concurrent with the curvature of the surface from the unloaded area into the zone of impact. This will further change the stress loading in that surface and increase the differential tensile stresses which will, in consequence, be generated in it.

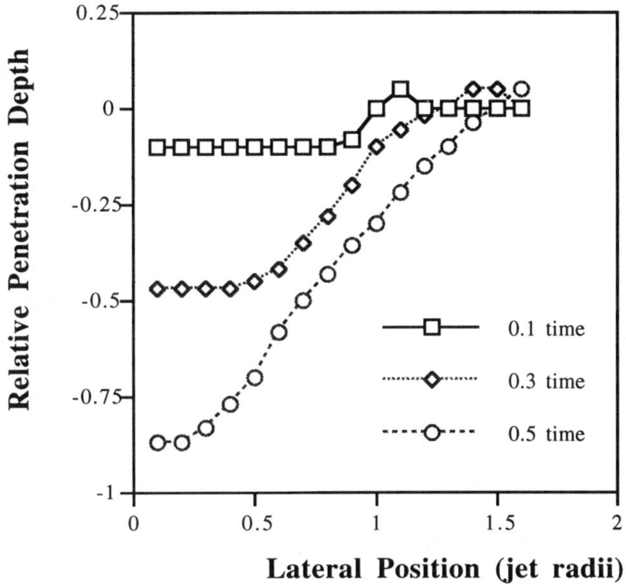

Figure 11.25 Profile of an aluminum surface at several instances after impact by a water droplet (after [11.33]).

However, as the target is moved away from the nozzle the action of the surrounding air will slow the outer regions of the jet, increasing the velocity differential within it, and thus creating a differential loading across the target surface. This, in turn, will induce erosion of the central core of material. However, in the evaluation of this impact event, the point which must be recognized is that significant surface tensile stresses are generated in the surface. Hwang and Hammitt for example found that at 300 m/sec the droplet would generate a compressive stress of 5,200 bar, and a tensile stress with a maximum value of 1,600 bar. This stress is sufficient to open cracks around the perimeter of the impact point, as Field has demonstrated experimentally.

Failure of a smooth, homogeneous, surface is thus initiated by the opening of surface fractures. These fractures are initiated from existing flaws which can be very small. For this reason the presence of a thin surface layer of soft material may inhibit the creation of the very high impact pressures needed to initiate growth of these cracks [11.34]. This could, in part, explain why, in the experiments carried out by Foreman and Secor [11.21], no damage was seen on the surface if a thin copper plate was placed between the impacting jet and the underlying rock surface.

More to the point, however, it should be noted that Field has shown that it is the growth of these flaws that failure is initiated. This point is critical to a true understanding of failure under jet impact, and it must be placed in context within the discussion of the prevailing theories which took place above.

In those theories the major emphasis was in examination of the stress fields created within the body of the target. Failure was presumed to occur at existing flaws in the material, but as a result of a generalized stress generated within the body of the target by the impacting pressure of the waterjet. Water was presumed to flow into the target and, because of its presence, influence the overall pressure distribution, and thus the stress concentrations which would lead to the growth of cracks already existing within the body of the target. The criteria for fracture growth was correctly assumed to be that of crack growth, with the favored theoretical condition being that due to Griffith [11.19]. The validity of this correlation is challenged and it is suggested that the concept of internal failure within the rock mass by growth of "concealed" Griffith cracks is not valid.

As noted above, during the present author's doctoral research program [11.13] a block of granite was placed under a waterjet operating at 700 bar. No discernible cutting of the rock ensued until the rock was rotated around an axis eccentric to that of the jet, when a hole was drilled through the 20 cm thick sample. It was necessary that the jet be exposed to the surface fractures which exist around the grains of the granite, before it could penetrate, at that pressure, into the crack, extend it, and induce material removal.

As with Rehbinder's theory this requires that fluid penetrate into the target surface to induce failure. In contrast with that theory, however, it is suggested that it is not in the flow through the existing continuous paths, that one should be most concerned. As was discussed in the data from Hwang and Hammitt, damage at a surface can occur within less than a microsecond. Within that time frame water has little chance to penetrate far into a material. At the same time where a fluid is penetrating into an open tube it does not generate the same pressures on the walls of the opening as is the case where it is shooting into a closed passage. As an example one could site the case of the Flow hydraulic water cannon used for boulder breaking and tunneling [11.35] or other devices where a waterjet is shot into a drilled hole.

The situation must be distinguished from the steady state condition examined by Leach and Walker [11.1] in which the pressure at the bottom of a hole is shown to drop off rapidly with depth, as the incoming jet is negatively influenced by the equal flow of water back out of the hole. Rather consider the condition where the water in a hole is instantly impacted by a fluid slug. The water will, in an instant, distribute the pressure around the walls of the slot inducing a much greater stress at the root of the hole, and causing crack growth (Fig. 11.26).

Figure 11.26 Crack growth from the bottom of a hole drilled in plexiglass, filled with water and impacted by an airgun pellet.

Waterjet cutting of rock is therefore considered to occur by a process of crack growth initiating from surface cracks which are filled and then pressurized by the impact of the waterjet. This viewpoint is enhanced by Field's result that etching the surface of a glass plate prior to impact, a process which removes surface flaws, dramatically reduces impact damage. This, however, is not, in itself, sufficient to explain why, for example, it is easier to cut granite than marble. It is possible to accept this explains the increase in difficulty found in cutting finer grained materials and the restriction of the cutting path to a line directly under the jet in granular materials. This is in contrast to spallation around the traverse line which can occur in crystalline materials. It is considered this difference is due to relative crack lengths, established by Brace as being equivalent to grain size, and greater density of crack arrest locations in the granular material [11.17]. These, in turn, will control the rate and extent of crack growth, according to Bieniawski [11.18] and thus the cutting rate of the rock.

The controlling physical parameters of the rock therefore are likely to be flaw density (inversely correlated to grain size) and the surface energy of the rock. It is however incorrect to use an average value as being of value. Waterjets attack a material by inducing failure at the weakest point in the surface to which it has access. Much more so than with mechanical tools it is possible for a waterjet to enter and exploit these weakness planes, and thereby to remove, virtually undamaged, stronger phases of the target material, at significantly lower jet pressures than would be required to remove the material if it had to be directly attacked.

This is because, without support from the surrounding softer materials, the harder material will fail through lack of contact with the solid. Such a process was developed, demonstrated and is now commercially available for the cutting of concrete. In this use of waterjets, the pressure used (800 bar) is sufficient to cut the cement paste, but does not cut the aggregate which is however removed by a lack of surrounding support.

An examination of the section through a waterjet cut in granite does not however show the same phenomenon occurring in that rock. Rather the hole shows that there is jet penetration and material removal occurring along the crystal boundaries. The clear evidence is that the waterjets do work by preferentially attacking existing crystal boundaries, first. It might be that the lower surface energy of the mica components (0.0132 cm kg/cm^2) relative to that of the quartz and feldspar components (0.0306 cm kg/cm^2) leads to greater crack growth in those regions.

In addition to these considerations, however, one must also consider the dynamics of many jet cutting operations. Considerable discussion has arisen, over the years, as to the relative efficiency of the process. This evaluation has, for example, played a role in the theoretical development which Rehbinder has made of the optimum performance of a cutting jet.

The problem, however, with many of those analyses is that they must consider, in the steady state condition, the interference between the incoming and outgoing jet. Where the jet is broken into discrete elements, such as, for example at the end of its coherent range, this lack of interference can significantly enhance efficiency and cutting ability. For example Pratt [11.36] has shown that, the specific energy of material removal can be reduced by at least two orders of magnitude (Fig. 11.27).

These results and the foregoing discussion suggest, to the current author, that much greater emphasis must be placed on the developed theories which explain rain erosion of target surfaces and the transient loading that they induce (see for example Engel [11.37], [11.38], and [11.39] and the ongoing work in enhancing material resistance to such loading than in seeking to write a theoretical approach which considers the target material as a continuous solid. Engel has considered stresses which can be induced in the surface by lateral flow of the impacting jet when it crosses the step created by a surface fracture (Fig. 11.28). These, and similar other dynamic conditions are considered to provide a much more realistic model of the failure under a jet.

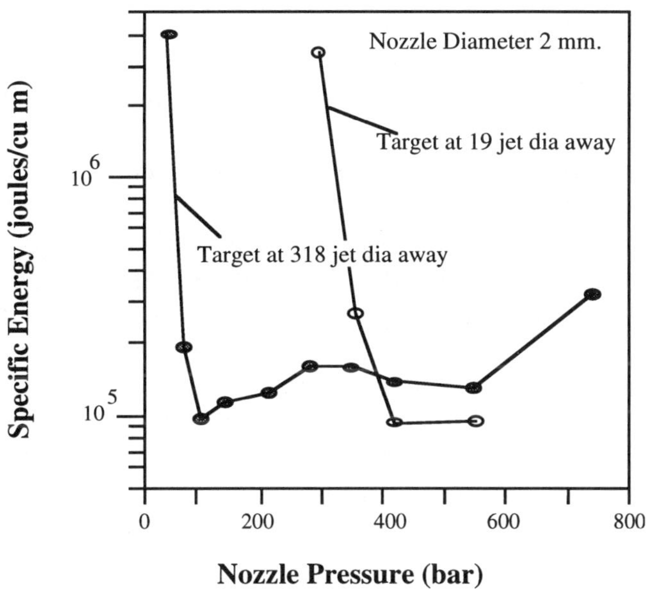

Figure 11.27 Specific energy of rock removal close to and far from the cutting nozzle as a function of jet nozzle pressure [11.36].

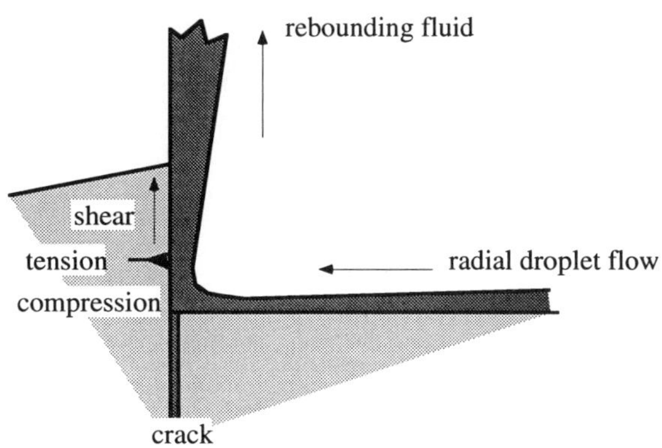

Figure 11.28 Stresses suggested by Engel as developing under radial flow from a droplet over a crack [11.40].

Similarly she has developed a prediction for the interference within the fluid flow [11.38] which indicates that this significantly reduces cutting depth within 15 m/sec, a value consistent with data found experimentally by Summers [11.13] and Page [11.41]. It is Engel's contention [11.39] that it is in the short period after the no-flow condition of initial impact that is most critical to surface erosion. Further she stresses that it is the steepness of the pressure front which is more important than the overall magnitude. From this study she has developed a statistical model which considers impact velocity, drop mass, target strength, impact angle, surface condition (presence of existing flaws), and work-hardening capacity (for metal targets) [11.37].

It is suggested that this model might fruitfully be examined for its application to the current condition. These models do not only describe the way in which waterjets attack granular materials such as rock. It has recently been found [11.42] that the thermal sprayed coatings which provide wear and corrosion protection for engines can be effectively stripped using high pressure water. As these liners are removed by the waterjets it has been noted that the process is one of crack development and growth. As the cracks permeate the coating they cause it to fail and disintegrate into grain-sized fragments which can be swept away and collected. The process of target failure, when either cleaning or cutting, appears to be the same.

11.9 REFERENCES

11.1 Leach, S.J., and Walker, G.L., "The Application of High Speed Liquid Jets to Cutting," Philosophical Transactions, Royal Society of London, 260-A, July, 1966, pp. 295 - 308.

11.2 Selberg, B., and Barker, C.R., "Dual-Orifice Waterjet Predictions and Experiments," Erosion: Prevention and Useful Application, ASTM STP 664, W.F. Adler, ed., American Society for Testing and Materials, 1979, pp. 493 - 511.

11.3 Tollmein, W., Berechnung Turbulenter Ausbreitungsvorgange, NACA TM 1085, National Advisory Council for Aeronautics, 1945.

11.4 Yanaida, K., "Flow Characteristics of Water Jets," paper A2, 2nd International Symposium on Jet Cutting Technology, Cambridge, UK, April, 1974, pp. A2-19 - A2-32.

11.5 Schlichtling, H., Boundary Layer Theory, McGraw Hill, New York, 1968.

11.6 Zakin, J.L., and Summers, D.A., "The Effect of Viscoelastic Additives on Jet Structure," paper A4, 3rd International Symposium on Jet Cutting Technology, Chicago, IL, May, 1976, pp. A4-47 - A4-65.

11.7 Daniel, I.M., Rowlands, R.E., and Labus, T.J., "Photoelastic Study of Water Jet Impact," paper A1, Proc. 2nd Int. Symp. Jet Cutting Tech, Cambridge, UK, April, 1974, pp. A1-1 - A1-18.

11.8 Singh, M.M., and Hartman, H.L., "Hypothesis for the Mechanism of Rock Failure under Impact," 4th Symposium on Rock Mechanics, Pennsylvania State University, PA, 1961, pp. 221 - 228.

11.9 Kinslow, R., "Rain Damage to Supersonic Vehicles," Workshop on the Application of High Pressure Water Jet Cutting Technology, Rolla, MO, November, 1975, pp. 242 - 270.

11.10 Farmer, I.W., and Attewell, P.B., "Rock Penetration by High Velocity Water Jets," International Journal of Rock Mechanics and Mining Science, Vol. 2, No. 2, July, 1965, pp. 135 - 153.

11.11 Wessman, H.E., and Rose, A.W., Aerial Bombardment Protection, Wiley, New York, 1942.

11.12 Powell, J.H., and Simpson, S.P., "Theoretical Study of the Mechanical Effects of Water Jets Impinging on a Semi-Infinite Elastic Solid," International Journal of Rock Mechanics and Mining Science, Vol. 6, July, 1969, pp. 353 - 364.

11.13 Summers, D.A., Disintegration of Rock by High Pressure Jets, Ph.D. thesis, Department of Applied Mineral Sciences, University of Leeds, May, 1968.

11.14 Cooley, W.C., Workshop on the Application of High Pressure Water Jet Cutting Technology, University of Missouri-Rolla, November, 1975, pp. 18 - 35.

11.15 Cooley, W.C., "Correlation of Data on Erosion and breakage of Rock by High Pressure Water Jets," Chapter 33, Dynamic Rock Mechanics, ed. G.B. Clark, Society of Mining Engineers, AIME, New York, 1971.

11.16 Singh, M.M., Finlayson, L.A., and Huck, P.J., "Rock Breakage by High pressure Water Jets," paper B8, 1st International Symposium on Jet Cutting Technology, Coventry, UK, 1972, pp. B8-113 - B8-124.

11.17 Brace, W.F., "Dependence of Fracture Strength of Rocks on Grain Size," 4th Symposium on Rock Mechanics, Pennsylvania State University, 1961, pp. 99 - 103.

11.18 Bieniawski, Z.T., "Mechanism of Brittle Failure of Rock," International Journal of Rock Mechanics and Mining Science, Vol. 4, No. 4, pp. 395 - 430.

11.19 Griffith, A.A., "The phenomena of rupture and flow in solids," Philosophical Transactions, Royal Society London, Vol. A221, 1921, pp. 163 - 198.

11.20 Griffith, A.A., "The theory of rupture," 1st International Congress on Applied Mathematics, Delft, Holland, 1924, pp. 55 - 63.

11.21 Foreman, S.E., and Secor, G.A., "The Mechanics of Rock Failure due to Water jet Impact," Sixth Conference on Drilling and Rock Mechanics, Society of Petroleum Engineers, Austin, TX, 1973, SPE 4247.

11.22 Bresee, J.C., Cristy, G.A., and Mclain, W.C., "Some Comparisons of Continuous and Pulsed Jets for Excavation," paper B9, 1st International Symposium on Jet Cutting Technology, Coventry, UK, April, 1972, pp. B9-125 - B9-132.

11.23 Walsh, J.B, "The Effect of Cracks on the Compressibility of Rock," Journal of Geophysical Research, Vol. 70, January, 1970, pp. 381 - 389.

11.24 Crow, S.C., "A Theory of Hydraulic Rock Cutting," International Journal of Rock Mechanics and Mining Science, Vol. 10, 1973, pp. 567 - 584.

11.25 Rehbinder, G., "Some Aspects of the Mechanism of Erosion of Rock with a High Speed Water Jet," paper E1, 3rd International Symposium on Jet Cutting Technology, May, 1976, Chicago, IL, pp. E1-1 - E1-20.

11.26 Crow, S.C., "The Effect of Porosity on Hydraulic Rock Cutting," International Journal of Rock Mechanics and Mining Science, Vol. 11, 1974, pp. 103 - 105.

11.27 Crow, S.C., Workshop On The Application Of High Pressure Water Jet Cutting Technology, Rolla, MO, November, 1975, pp. 271 - 294.

11.28 Hurlburt, G.H., Crow, S.C., and Lade, P.V., "Experiments in Hydraulic Rock cutting," International Journal of Rock Mechanics and Mining Science, Vol. 12, 1975, pp. 203 - 212.

11.29 Rehbinder, G., "Slot Cutting in Rock with a High Speed Water Jet," International Journal of Rock Mechanics and Mining Sciences, Vol. 14, November, 1977, pp. 229 - 234.

11.30 Rehbinder, G., "Erosion Resistance of Rock," paper E1, 4th International Symposium on Jet Cutting Technology, Canterbury, UK, April, 1978, pp. E1-1 - E1-10.

11.31 Rochester, M.C., and Brunton, J.H., "High Speed Impact of Liquid Jets on Solids," paper A1, 1st International Symposium on Jet Cutting Technology, Coventry, UK, April, 1972, pp. A1-1 to A1-24.

11.32 Field, J.E., "Stress Waves, Deformation and Fracture Caused by Liquid Impact," Phil. Trans. Royal Society, 260A, July, 1966, pp. 86 - 93.

11.33 Hwang, J-B., and Hammitt, F.G., "Transient Distribution of the Stress Produced by the Impact between a Liquid Drop and an Aluminum Body," paper A1, 3rd International Symposium on Jet Cutting Technology, Chicago, IL, May, 1976, pp. A1-1 - A1-15.

11.34 Matthewson, M.J., Ph.D. thesis, University of Cambridge, UK, 1978.

11.35 Kolle, J.J., "Development and Applications of a Hydraulic Probe Generator," 7th American Water Jet Conference, Seattle, WA., August, 1993, pp. 459 - 472.

11.36 Platt, E., Workshop on Waterjet Technology, Rolla, MO, November, 1975, pp. 139 - 148.

11.37 Engel, O.C., A Model for Multiple Droplet Impact Erosion of Brittle Solids, November, 1971, NASA CR-1943, 95 pages.

11.38 Engel, O.G., "Initial Pressure, Initial Flow Velocity, and the Time Dependence of Crater Depth in Fluid Impacts," Journal of Applied Physics, Vol. 38, No. 10, September, 1967, pp. 3935 - 3940.

11.39 Engel, O.G., "Damage produced by high speed liquid drop impacts," Journal of Applied Physics, Vol. 44, No. 2, February, 1973, pp. 692 - 703.

11.40 Engel, O.C., "Impact of Liquid Drops," Symposium on Erosion and Cavitation, ASTM STP 307, 1962, pp. 3 - 16.

11.41 Page, C., "Penetration of Rocks with High Pressure Water Jets," Ph.D. thesis, University of Leeds, September, 1971, 220 pp.

11.42 Zanchuk, W.A., "Ultra-High Pressure Waterjet Stripping of Thermal Sprayed Coatings," Cleaner Times, Vol. 6, No. 6, June, 1994, pp. 15-17.

12 SAFETY, HEALTH AND MEDICAL APPLICATIONS

12.1 INTRODUCTION

The introduction of waterjets into industrial practice carried with it two perceptual problems which initially had to be addressed by those who would both use it, and teach others to do so. The first problem is one of credibility. It became an early practice at the University of Missouri-Rolla, for example, to film experiments so that doubting visitors could see that waterjets would cut through rock and other materials. This had the additional benefit of demonstrating to future operators of the equipment the power which exists within the waterjet stream.

However, once people had grasped that waterjets could cut material, it became important to stress both the beneficial, and potentially risky aspects of this new tool. For once initial acceptance had been achieved, those who began to use the tool on a regular basis sometimes had some problem in grasping the danger that could be created if the tool was improperly used.

This is not necessarily the fault of the industry as a general whole. At the first International Symposium the papers included a discussion by Ward [12.1] on the risks associated with high pressure waterjet use and the peculiar problems which this new device would introduce into industrial practice. However, and much more recently, the current author was invited to give a talk on some of the safety aspects of waterjet use in industry to a meeting in Detroit. Following the presentation of the papers a tour had been arranged to a local industrial site where, among other exhibits, a waterjet robot had been set up to cut a "self-portrait" in plastic.

After running the demonstration one of the participants asked the operator how safe the procedure was. As part of his answer the unit operator racked the nozzle back until it was about 15 cm above the table, turned the equipment on and ran it at a pressure of around 3,000 bar. He then slid his hand along the table, under the jet, at a distance of about 20 cm, to prove that the jet energy dissipated over a short distance and there was little risk of an injury.

The operator could, quite easily, have lost his hand. Cases have been discussed, both formally and informally at jetting user meetings where quite serious accidents have occurred due to the operator underestimating the power contained within a jet stream, and the range which this might have. Several fatalities in which waterjet use has been a contributing factor have also taken place over the years. It is a little strange in that visits to industrial sites show, on the one hand, that operators may use a cleaning lance at a stand-off distance greater than the effective range of the jets, and on the other they may believe that the jet power cannot be extended to pose an operational risk to the user.

It must be a matter of concern to all users of the technique, and to their supervisory management that those involved with waterjet operations fully understand the power of such a jet, the precautions which must be taken in its use, and that they receive the proper supervised training to appreciate the full potential of the tool with which they will be working.

As an illustration of the possible unanticipated risks which a poorly trained operator might experience consider the case discussed above. Under the normal plant working procedure the operator might have run many hundreds of cutting operations in which, with a jet operating at a pressure of, for example, 3,000 bar with a waterjet diameter of 0.15 mm the jet would cut plastic at a 10 mm stand-off, but where there was no cutting power out at a range of 150 mm. This can be illustrated with reference to a back-lit photograph of such a jet, (Fig. 12.1) in which the jet is seen to dissipate over a range of 100 mm.

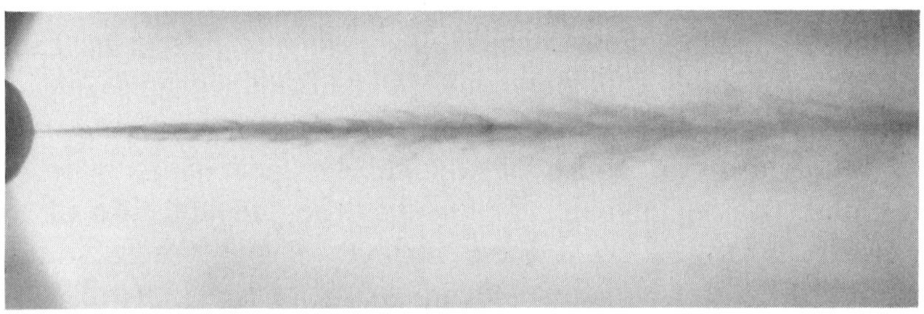

Figure 12.1 Back-lit photograph of a waterjet operating at 3,000 bar pressure, 0.15 mm diameter, the range of the photograph is some 150 mm.

Jet range is controlled, in part by jet pressure and by nozzle diameter. Under normal conditions a jet is considered to retain the majority of its energy for a distance of 150 to 200 diameters from the orifice [12.2]. In this case that would be a maximum distance of 30 mm. However, as has been discussed previously, by improving jet inlet conditions and using some relatively straight forward fluid mechanics principles, Barker and Selberg [12.3] have been able to extend the range of a jet out to over 2000 nozzle diameters (Fig. 12.2). Such a range extension would extend this particular jet out to 300 mm. The management of the factory could well have made that improvement (which would possibly not be visible to an unknowing operator), and in normal cutting the close range cutting ability of the jet would not be necessarily that discernible. However, should the operator again try the "safety" demonstration, the jet would have, with the change, sufficient energy to strip the flesh from his fingers as he ran them across the table.

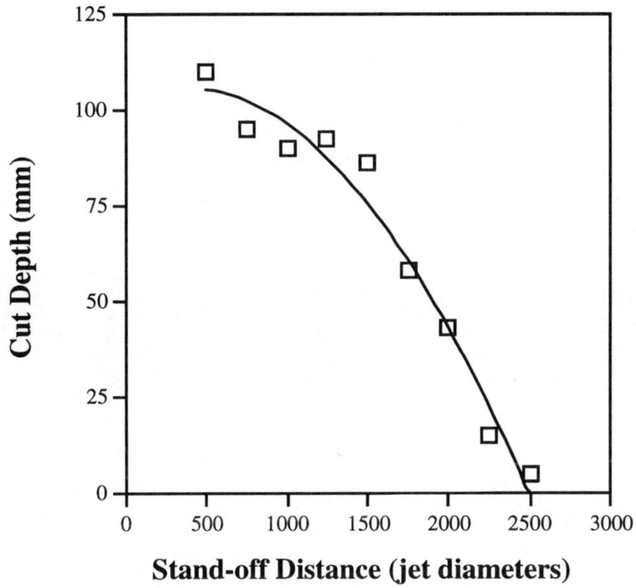

Figure 12.2 Cutting range of an improved waterjet nozzle, diameter 1.0 mm jet pressure 607 bar, penetrating into Berea sandstone (after [12.3]).

Other process changes which might be introduced by the factory could have a similar improved cutting effect. A significant improvement in operating conditions can be achieved if a long chain polymer is introduced into the jet stream (see Chapter 11). This polymer was not in use at the time of the demonstration and the operator was not aware of its potential power. In the early use of such a fluid it is possible to use a higher concentration than that which is most cost effective for the purpose. In one trial of such a system, with the greatest concentration of the polymer "Superwater" which the pump could deliver, a jet of this fluid took the bark from a tree at a distance of over 12 m from the nozzle. That nozzle was larger than that of the current unit, but the pressure at the pump was also less.

A similar order of range extension can be found if the polymer is used with the higher pressure, lower volume system similar to that in use at the above plant. For example, Zakin and Summers [12.4] have reported on the improvement which can be made in jet throw with a variety of polymers, and a simple comparison of the photographic evidence will show the effect (Fig. 12.3). In this case the jet has remained coherent over the full length of the photograph, and continued to be coherent for an additional 50 mm. Again a change in the equipment performance will have been achieved without any clear change in the jet appearance to the uninformed operator. The photographs were taken with an exposure time on the order of a microsecond and with the jet back-lit to minimize the effects of the spray surrounding the jet. The observer does not see this rapidly and would see the jet surrounded by the mist which would obscure its structural change. Again the result would have been that, during the demonstration, the operator could have lost the flesh from his fingers.

Figure 12.3 Back-lit photograph of a polymer laden waterjet, the range of the photograph is 150 mm over which the central core of the jet has remained coherent.

An increase in the size of the nozzle which the operator had used on the previous shift would also have extended the range of jet power to the point that the demonstrating operator's fingers would have been stripped to the bone. Additional means of enhancing jet cutting range have been discussed in earlier chapters and thus will not be further discussed here.

This case is not unique, several similar instances have been informally reported, some of which have resulted in injury to the individual involved. These incidents can only stress the critical need to ensure that all personnel be familiar with the power of the waterjet stream, and the additional risks which can be incurred with the associated fluid streams used for waterjet enhancement such as those where abrasives, polymers or cavitation bubbles are included in the cutting fluid.

The question that this has raised is one which can be summarized in the question "what is the Safe Working Pressure where a waterjet does not cause injury to an individual?" This is considered to be the level of jet pressure at which one can work at which no damage may be anticipated to the operator if hit by a direct stream of fluid.

12.2 DAMAGING JET PRESSURES

This question of the pressure sufficient to create personal damage has been addressed in only a few papers [12.5] [12.6] in regard to the impact of waterjet streams. There are, however, significant data available on the impact of similar fluid streams from which to draw comparative information. For example, many paint spraying systems have used pressurized fluid at the same operating pressure as jetting equipment, through nozzles of the same level of size. Thus, to an initial degree, the types of injury can be considered equivalent, although the quantities of fluid involved may be different.

12.2.1 POSSIBLE DAMAGE MECHANISMS

Jetting injuries from the impact of high-pressure fluid have occurred in sufficient number that it has been possible to determine three separate ways in which the jet may injure the individual. There is the initial mechanical injury created by the passage of the high pressure fluid through the skin and flesh. The second source of damage comes from the additional volume of fluid which has been injected under the skin, and which cannot then escape, but remains trapped within the flesh and under the skin. This fluid will exert a continued pressure on the surrounding flesh, and, since the skin is relatively inelastic and unable to dilate and relieve this pressure, it will continue to exert this pressure until it is relieved externally. Prolonged pressure of this type can lead to the death of the affected tissue, a process known as **necrosis**. The third threat to the continued health of the injured party comes from the foreign material which has been carried into the wound by the waterjet stream. High pressure waterjets are now commonly used to clean a variety of contaminants, and other potentially dangerous substances from tanks, sewers, and other vessels (see Chapter 3). The material being removed may itself be dangerous or it may also carry significant bacteria and other threats to the tissue. This injection of foreign material will also, over time, lead to tissue death.

There are some perceptual problems with the nature of typical waterjet injuries. Because of the very small size of the cutting jet it is possible for the surface appearance of the wound to be very small (Fig. 12.4) and therefore to be treated only very locally.

The treatment may well be only to make a short cut over the immediate injection site, to allow drainage and to clean any fluid and material in the vicinity of the immediate wound that may have been injected from the flesh. Unfortunately flesh provides relatively little resistance to jet impact (see section 12.2.2) and the jet may well have traveled significantly further along the limb or body than may be easily seen. An X-ray of such an injury where the injected material was radio-opaque [12.8] reveals that for a jet impact at the tip of the finger the fluid had penetrated down to the palm of the hand (Fig. 12.5). In other cases the fluid had traveled through the palm and on up into the arm.

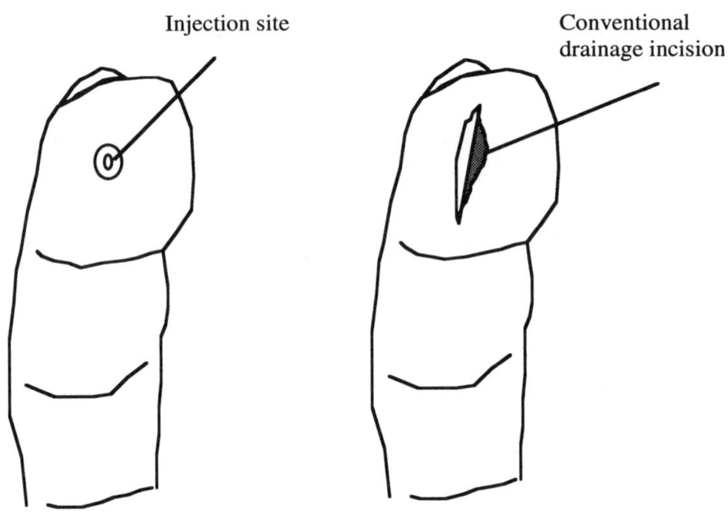

Figure 12.4 Representation of the relative size of a waterjet injury site on a finger and the type of incision often made to drain it (after [2.7]).

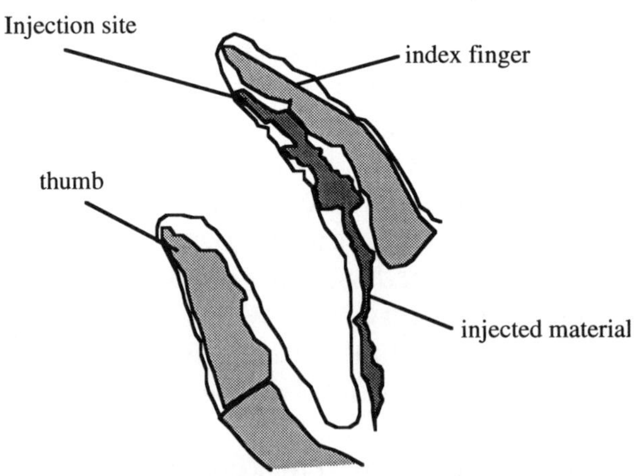

Figure 12.5 Schematic representation of an X-ray showing the range of jet passage from a small injection site (after [12.8]).

The lack of significant skin damage should not be taken as indicative of the situation. In their verbal presentation Neill and George [12.9] discussed the case of a waterjet operator who was struck by a jet at some distance from the nozzle The operator was not wearing safety clothing. The jet had hit the abdomen, and a superficial examination showed some bruising over an diameter of 5 cm. Although there was no major penetration of the skin a semicircle of tiny puncture wounds could be seen on close examination. Because of the concern of the medical officer the individual was taken to a local hospital and held for examination. After 48 hours severe pain developed and an exploratory operation was carried out to see if there was underlying damage. It was found that the jet had penetrated through the pores in the skin, created a hole 4 cm in diameter in the abdominal wall, and perforated underlying intestines in a number of places. Fortunately this could be remediated by removing segments of intestine and the worker was released from the hospital after an additional three weeks.

The jets which are used in most waterjetting applications are small, and can appear to cause only small wounds on the surface of the patient, but to properly deal with the underlying damage it is often necessary to fully explore the path which the jet has cut through the underlying flesh. This frequently requires that an exploratory surgery be undertaken with the full passage of the jet exposed so that it can be properly cleaned.

This may seem to be a relatively drastic procedure for such a wound, but given the contamination to the flesh, and the other threats it is important that the wound be promptly and thoroughly treated. For example Blue and Dirstine [12.10] reported that by the second evening after injury a worker's finger had turned gangrenous and had to be amputated, following a grease gun injection injury. They reported on three such cases where amputations were necessary, in each case because of the additional pressure put on the flesh by the injected material.

A study has shown that, given prompt treatment there is a good chance the wound can be treated and the patient suffers no lasting damage should the injury not be properly treated immediately, the chance of the injured limb having to be amputated increases significantly with time (Fig. 12.6). This work corroborates the finding of Stark and associates [12.12] that time interval between the injury and the time of proper treatment is the major factor in governing the final result. This finding has been challenged by Gelberman and his colleagues, who were unable to confirm this conclusion [12.13]. The recommendation stands and has been corroborated by others [12.14], [12.15] that early extensive surgery to remove the material which has been injected should be the proper treatment.

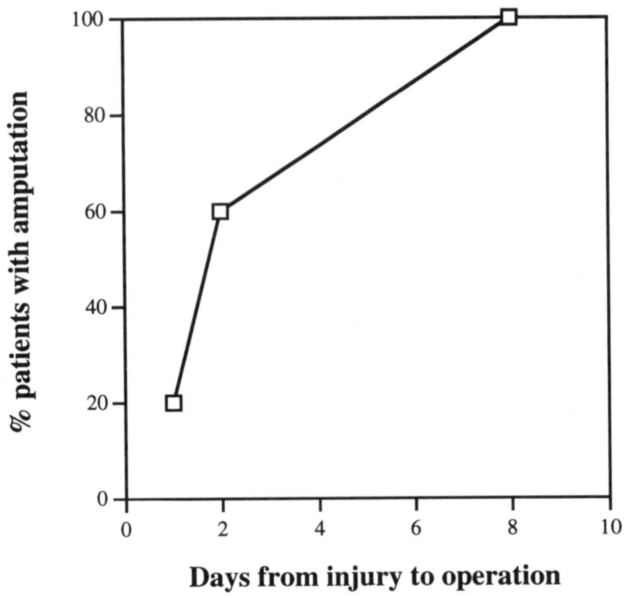

Figure 12.6 Time to treatment and the relationship to the need for amputation [12.11].

12.2.2 NEEDLE-LESS INJECTION SYSTEMS

There has been an interest in injecting fluids into flesh ever since the first snake got hungry. The use of needles is a process which carried the risk of cross-contamination, unless they are used only once, an expensive luxury not always available in many countries. Hingson and Figge [12.16] have reported that the first development of a needle-less injection system was by the French physician Galante, and that Sutermeister in New York, after being struck by a 280 bar jet of oil conceived the idea of using small high pressure systems for medical injections. The article notes that he subsequently put his hand into a "fine" waterjet at 630 bar and felt "pressure but no pain." There was no apparent surface injury but a small blister had developed on the back of his hand. In 1936 Lockhart further developed this technique in which the fluid is driven through the skin under pressure, and from an external orifice rather than using a hypodermic needle [12.17]. This led to the development, through initially F.R. Squibb and Sons, and then R.P. Scherer of the Hypospray injection system.

This technique has since proved invaluable as a means of vaccinating large numbers in a short period of time. However, it was initially necessary to develop a method of providing the jet pressure to drive the jet through the skin, and to determine what pressure was most effective for this use. In the initial studies a small charge of nitroglycerine was used as the pressure source, but this has since been replaced by spring loaded devices such as the Hypospray [12.18].

This tool has a considerable advantage over conventional hypodermic needles. It was reported that over 6,759 patients were inoculated with just two injectors during the Dacca cholera inoculation program in 1959, as opposed to a rate of 100 patients a day which could be served with a single hypodermic needle [12.19]. One jet injector has been reported as being able to do the work of 20 - 30 medical personnel working with hypodermic needles, with rates of treatment reaching up to 500 persons per hour per injector. In addition, because the jet produced is much smaller than a hypodermic (0.075 mm diameter it is 1/16th the size of the needle) trauma associated with the inoculation is reported to be significantly less [12.18].

The smaller jet size, and faster inoculation speed mean that the cost of the jet inoculation can also be significantly cheaper, 14 cents in 1961, in contrast with the price of a hypodermic shot which at that time cost 28 cents. The difference is greater when it is recognized that only 2 cents of the jet shot goes for equipment while 9 cents is required to pay for disposable needles and 13 cents for reusable ones [12.20].

Early work on the necessary pressures to inject fluid through the skin was carried out on cadavers by Figge and Barnett [12.21] who found that a pressure of 270 bar was sufficient to drive jets of fluid, including mercury, distances of up to 2.5 cm through the skin and into the underlying muscle. They also reported that it was easier to drive fluid into the inner surfaces of the limbs, rather than the outer and that cadavers provided greater penetration to jet passage than living flesh. It was found easier to drive a jet through stretched rather than slack skin, and while jets would penetrate through collapsed blood vessels but would bounce around such vessels when filled with saline solution and maintained at typical body pressures.

On a word of caution in this regard, it should be noted that two cases have recently been reported in 1989 and 1990 [12.22], [12.23]. In the first case the patient was hit in the thigh by a jet at a pressure of 560 bar, flowing at around 100 lpm. The jet cut an 18 cm slot in the upper thigh and exited through a 4 cm slot on the other side. Some 20 - 25 cm of the femoral artery had to be replaced. In the latter case a patient was also injured by a high-pressure waterjet, in which the jet hit the thigh and caused "complete disruption of the superficial femoral artery." The patient was successfully treated but the case points out that blood vessels are vulnerable to waterjet attack.

The work of Figge and Barnett (ibid.), and that earlier reported by Higson and Hughes [12.18] showed that the pressure to penetrate the skin will vary from individual to individual, and with age. for example while a pressure of 160 bar is sufficient pressure to inject fluid through the skin into an infant, it may be necessary to use a pressure of 250 bar to ensure adequate injection of fluid through the skin of an adult. It should be noted that this pressure is for a jet diameter of 0.075 mm, with a desired target penetration of up to 25 mm. At larger jet diameters and lower pressures significant surficial damage may still be caused.

From personal observation it is critical to the use of the tool that it is pressed firmly into the skin, for although the jet injection time is very short, (around 0.03 seconds) any movement will cause a cut on the skin. Further Vijay has reported [12.6] that this type of tool is not recommended for patients with a fragile skin, or where the patient has a problem with coagulation. The need to ensure the tool is in contact with the skin has also been illustrated by Calder and Boustred [12.24] who fired 160 ml of fluid at 200 bar through a 0.5 mm nozzle at cadavers. They reported that where the nozzle was not in contact with the surface, decreasing quantities of fluid were injected with increasing stand-off distance (Fig. 12.7). It is interesting to note the sudden drop in penetration at a distance of around 9 cm, (180 nozzle diameters) suggesting this is the point at which the jet is disrupted.

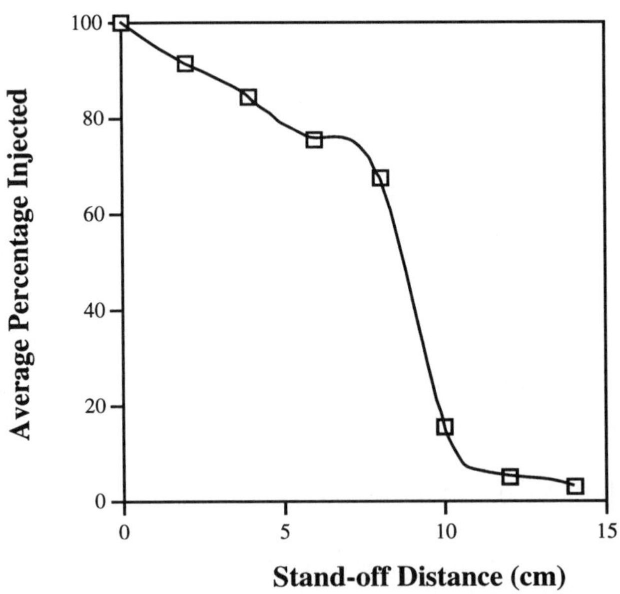

Figure 12.7 Quantity of fluid injected as a function of stand-off distance (after [12.24]).

The fragility of the flesh under the skin can be illustrated with other examples, which are discussed in more detail in Section 12.5. However, one example will illustrate the difficulty in describing a limiting lower pressure at which a jet is incapable of inflicting an injury. Japanese surgeons [12.25] have developed a surgical procedure for removing tissue from the liver. This is an important consideration in liver operations, where conventional surgery will frequently cut through the blood vessels in the organ, subjecting the patient to the trauma of significant blood loss and replacement. Such shock can induce mortality in 5 - 20 percent of patients [12.6]. However, by using a waterjet at a diameter of 0.1 - 0.15 mm and a pressure of 15 - 18 bar it is possible to remove the tissue around the blood vessels without damage to the vessels themselves. In this way the vessels can be exposed, coagulated and tied off before they are cut and the diseased section of the liver excised. By September of 1988 this procedure had been used on over 60 patients, 41 of whom had not required blood transfusions. This procedure did not damage the small blood vessels at pressures below 10 bar, but at 15 bar even the larger vessels in the liver were damaged. This technology has also been used in other countries [12.26], with equipment being now made commercially available from Sweden [12.27] and Japan [12.28] among others.

12.2.3 PULSATING JET ATTACK

At even lower pressures it is possible to induce the gums to bleed when using the pressures generated by a pulsating waterjet cleaning tool, of the type which is available at many drug stores in the United States. This technique was originally proposed in 1911, but did not become popular until a reliable pumping system was developed in the 1960s. This device typically operates at pressures below 6 bar, but even at this pressure it can induce bleeding. Bhaskar [12.29] and Krajewski [12.30] both report on bleeding, in one case in the floor of the mouth, and in the other in the gingival crevice, at these pressures. Lobene [12.31] has found that the damage is enhanced where the pulsation rate for the jet is increased from 800 to 2,400 cycles/min. Similarly he reported that where the jets exerted more than 8 gm of force on inflamed tissue that the patient felt pain. At pressures and pulsation rates below that at which the patient reported pain or bleeding blanching of the tissue was reported to occur.

The use of pulsating waterjets can, however, have other benefits in medical applications. Bhaskar and his colleagues [12.32] found that the alternate compression and relaxation phases of the jet impact on tissue loosens embedded material. This in turn helps to clean wounds, not only of the foreign objects but the 3.5 - 5 bar jet pressure was also found to be more than seven times more effective in removing bacteria than an irrigating syringe. This speeds up the healing process [12.33], as evidenced by the relative presence of contaminated cultures as a function of time (Fig. 12.8). In this case the waterjets contained 250 units/ml of penicillin and 1.42 mg/ml of streptomycin. The treatment has been adopted for use in emergency rooms and for treating victims of combat [12.32].

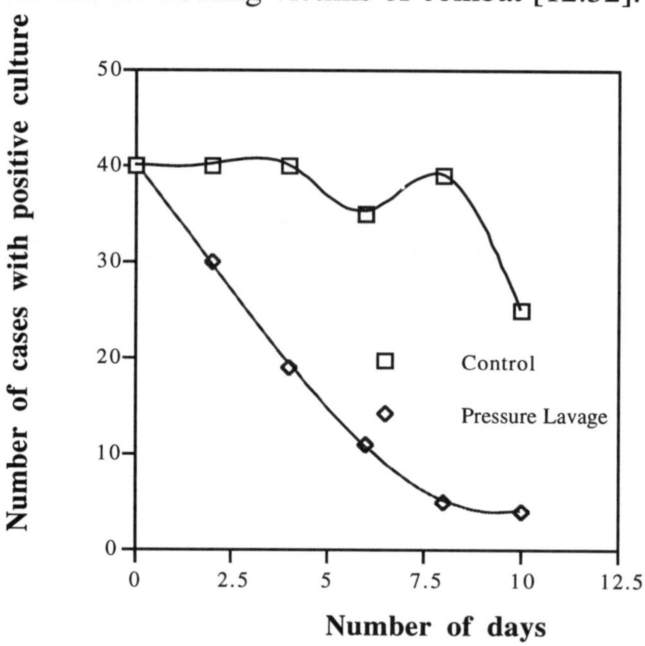

Figure 12.8 Relative presence of positive cultures (*pseudomonas aeruginosa)* after pulsed waterjet lavage of wounds, against a control [12.33].

12.2.4 ABRASIVE JET RISKS

The data provided above have shown that even at pressures as low as 5 bar, that it is possible to induce bleeding in the mouth. There is no significant jet pressure at which the jet can be cited as being safe to use. This is even more true where the jet is used with an abrasive additive. Experiments at UMR have shown (Fig. 12.9) that at a jet pressure of 350 bar, and a flow rate of 40 lpm that it will take an abrasive laden jet less than 1.5 seconds to completely sever the leg of a calf cutting though both hide, flesh and bone.

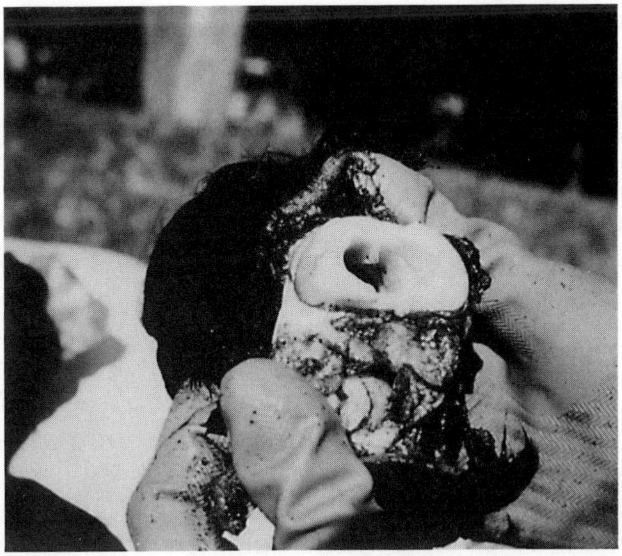

Figure 12.9 Calf's leg cut through by an abrasive-laden waterjet at 350 bar.

It should be noted that the above cut was made on a segment of the leg which had been removed after the animal had been butchered. However, the condition of the cut shows that while the edges of the bone were cleanly cut, the marrow within the bone, being softer, had been removed to a considerable distance on either side of the cut. This is because of rebounding of the abrasive within the cavity while the jet was penetrating from the inside of the bone outward on the distal side of the leg.

Pressures at which such jets become damaging have been cited by Vijay [12.6] who quoted a study by Beasley [12.34] in which 5 gm of 24 micron polymer beads were added to 500 gms of water. This suspension was used in studies on cutting teeth. As the pressure was increased from 3.4 - 4.1 bar the damage pattern changed to one in which tearing, fragmentation, and damage to the tissue could be seen. The procedure also left embedded particles within the tissue. With such low pressures being capable of inducing significant damage the need for precautions in working with this tool is even more significantly underlined.

12.3 PROTECTIVE CLOTHING AND OTHER SAFETY CONSIDERATIONS

The dangers which the presence of waterjets pose to human tissue, even at very low pressures, require that all personnel be provided with comprehensive protection against waterjet impact. The low pressures at which many of the tissues in the head are damaged indicate that this should be an area of special concern. In addition to protecting the individual from the jet action itself, one must also guard against rebounding spray and debris and significant other hazards which may, on occasion, be encountered in the work place. This group of hazards is not always clearly understood. For example, it has been reported in the BHRA Newsletter [12.35] that when there is a build-up of material on the walls of oil rig derricks in the North Sea that the material concentrates naturally occurring radioactive material (NORM) to the point that it might pose a hazard to operators of high pressure cleaning equipment who are not fully protected. The problem not only poses a breathing and ingestion risk to the operator, but there is also the risk of contamination of the environment requiring that special precautions be taken [12.36]. Because NORM can be found as a hard scale which builds up on oilfield equipment even small accumulations will, over time, become significant, particularly if all the equipment is cleaned to a common storage waste site. Although high pressure waterjet washing of these components has been found to be the most effective method of cleaning it still calls for the use of proper safeguards and high levels of protection both in clothing and procedures.

Other hazardous materials may also be encountered in cleaning, Pardey for example [12.37] has cited a case where the material to be removed contained arsenic, a fact not initially known to the operator. Under such circumstances it is therefore prudent to ensure that waterjet equipment operators are properly protected from foreseeable hazards whenever they are operating equipment. This inclusion of the possible respirable components of rebounding spray, together with the increased problem of visibility when the spray contaminates goggles has led to a study to improve such designs for waterjet use [12.38].

These considerations have also led the communities in several countries to develop procedures which they recommend be followed in the operation of high pressure waterjet equipment ([12.39]; [12.40]; [12.41]; [12.42] for example). These recommendations are most generally used with the operation of hand-held equipment, but are also useful guidelines for other applications. The suggestions contained in these recommendations cover a wide variety of different attributes of jet operation. As the technology has developed so it has begun to develop terms which have a special meaning to those in the industry. For this reason it is useful to include such terms in a document of this type. Further items deal with the steps which should be taken in preparing the equipment, the roles of the different people involved in a waterjetting operation, the types of training that might be required, and the procedures to be carried out when using the equipment. It is strongly recommended that any individual planning to use waterjetting equipment obtain copies of the relevant recommendations from one of these groups.

The recommendations of many of these groups deal with the use of protective clothing. It has been suggested that such clothing could fully protect the operator from injury [12.43], there have unfortunately been a number of injuries where jets, at pressures of 420 bar and above, have completely penetrated the protective clothing [12.44], [12.45]. In one case it was established that a 460 bar jet had to be within 25 mm of the p.v.c clothing that the operator was wearing before injury could occur, and that the jet had to be pointed at right angles to the fabric.

12.4 THE ROLE OF TRAINING

At the beginning of this chapter the problems of getting lay people to understand the power of waterjets were briefly reviewed. This difficulty is frequently encountered in people wishing to use high pressure waterjetting equipment. For this reason, particularly since they may not fully appreciate the risks associated with the improper use of the equipment, a proper series of training lectures should be prepared for these new employees or other users of the equipment.

Given that the jobs which they may be asked to undertake with the equipment may vary markedly, and the types of waterjet equipment continue to expand, it is not possible to exhaustively describe the content of this training. Some suggestions as to its content can, however, be made.

At the beginning of the training one should describe the type of equipment which will be used. The terms that are used, in that area, to describe the equipment and its parts should also be covered in this introduction. Even though the trainee may have some experience in the industry it is important to cover this material. Around the United States, for example, the "tip" of the waterjet lance can be either the entire nozzle assembly including nozzle insert and holder, or either one of these two components. Thus the terms which are being used at that location should be described and understood.

Once the different parts of the equipment have been described, the way in which each operates, and the particularly important features of each component should be covered. The way in which the components should be connected should be illustrated, and, where practical, this demonstration should include considerable practical hands-on training for the employee. In this regard when new equipment is obtained from a manufacturer, then the manufacturer should be contacted, and information obtained to ensure that this aspect of the training be comprehensive and cover all the salient equipment performance and features. Frequently manufacturers have material which they can make available to assist in this training.

The training should include physical demonstrations of the power of the waterjets. Frequently this is done by cutting a board, or a concrete block, but more graphic demonstrations where the jet cuts into a piece of meat can often leave a stronger message with the trainee. In the course of the training the force which the jet exerts back on the cutting equipment should also be demonstrated to the trainee, with the more effective method often being by allowing the individual to hold a high pressure lance and to experience, under the proper controlled conditions, the gradual increase in force as the pressure is increased. In this regard a demonstration of the differences in thrust and range of the jet when the nozzle diameter is changed can also be an effective learning experience.

Once the individual has an understanding of the components of the equipment, and their assembly and safe operational use, then the specific features of the work to be undertaken should be described. Again, the wide variety of work for which waterjets have found a use make it difficult to give more than broad guidelines for this part of the training. However, for this reason also the new trainee should work under the supervision of a skilled operator for some time in learning how to properly operate the equipment, before he is given much responsibility for its use. The material which describes the procedures for any task should be prepared in advance in writing, where possible, and discussed with the employee to ensure understanding of the material before beginning to work.

Working sites increasingly require that these procedures be developed and discussed with the site owners, and then committed to a written form before the work can begin and such industrial sites may well have their won guidelines. For those who do not, those issued by the National Associations may be a good place to begin to prepare such a document (see [12.40] to [12.42] for example). The peculiarities of the job site should be given particular consideration. Some precautions are not always obvious. For example Pardey [12.37] has pointed out that when waterjets are used in very cold locations it may be necessary to hire additional personnel to cut away the ice build up on the scaffolding to prevent its collapse during the waterjetting operations.

Operation of waterjetting equipment has other potential hazards which should also be taught to employees. One of these is the noise which can be generated by a waterjetting operation. As Barker, Cummings and Anderson have shown [12.46] noise from an operating jet system can exceed recommended levels, and will increase with a greater distance between the gun and the target (Fig. 12.10). Fan jets were reported to be somewhat quieter than round jets closer to the nozzle but became noisier as the target moved more than 10 cm from the nozzle (Fig. 12.11).

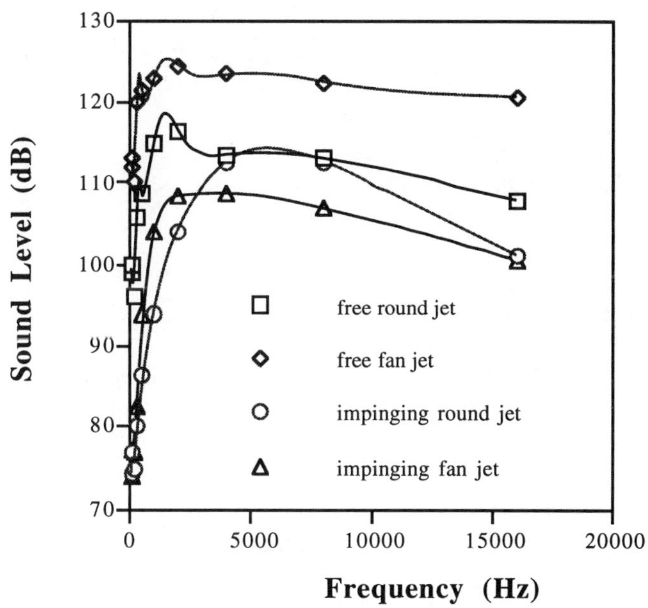

Figure 12.10 Noise frequency distribution 25 cm along both free air and impinging round and fan jets at 700 bar, 1.42 mm diameter jets with impingement plates at 25 cm [12.46].

The problem can be significantly worse in underwater applications where cavitation can be induced in the water, producing high noise levels over a range of frequencies [12.47], [12.48]. Under these conditions special equipment is required to provide protection to the equipment operators, particularly in the head area. This is because of different transmission efficiencies which can be found in these underwater situations. The noise associated with the use of abrasive waterjets is often reduced since the cutting stream is captured on the open side of the part after cutting. This is usually done with a small enclosure, although it can also be accomplished by laying the part on the surface of a water tank, and having the underlying water absorb the particle energy. Studies of the two systems [12.49] have shown that the small catcher containing steel balls to absorb the particle energy produced a noise level of 106.7 dB when the jet was idling and 106.5 dB when the jet was cutting through 1.3 cm thick aluminum. In contrast the water catching tank had a noise level of 115.7 dB while idling and 113.2 dB when the jet was cutting through the metal. Maximum noise was reported at a frequency of 12,500 Hz. These values stress the need for the employee to wear suitable protective equipment.

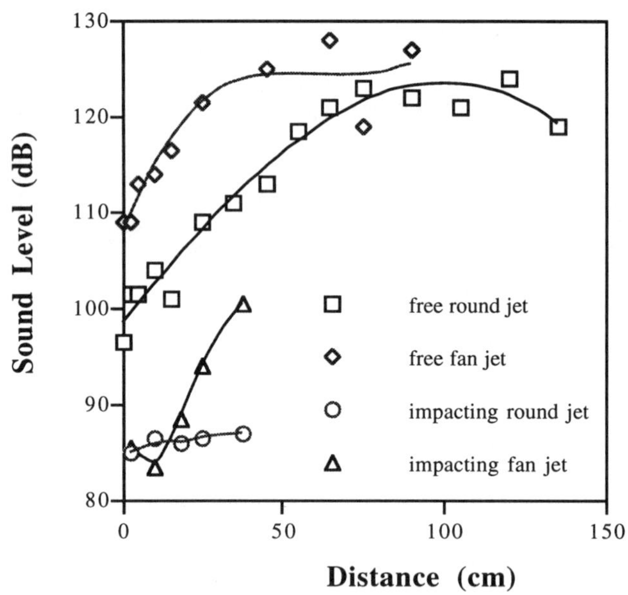

Figure 12.11 Sound levels at 500 hz along both free air and impinging round and fan jets at 700 bar, 1.42 mm diameter with the plates being located at the distance shown for the impinging cases (after [12.46]).

In addition to these points the necessary maintenance of equipment and the need to establish a schedule for inspection of equipment and its regular preventative maintenance should be transmitted to the employee. In this regard it is usually better if the employee is shown how to carry out those simple repairs of equipment which he might be expected to undertake, and when to seek additional help. In this regard it should be stressed that equipment not be repaired while it is still in operation.

12.5 MEDICAL APPLICATIONS

12.5.1 HANDWASHING APPLICATION

Earlier in this text the role of high pressure waterjets in cleaning surfaces has been covered. One of the surfaces not much addressed is that of cleaning the person, and yet it is a market which has, a very significant impact on a nation's economy. There are two major areas where this can be a problem, in the handling of food some twenty-five percent of all food borne illnesses are reported to be due to poor hand washing [12.50]. A typical cost for such an illness to the establishment is quoted as being between $1,400 and $74,000. In 1985 there were nearly 2,000,000 nosocomial infections in the United States and a campaign was set in place to reduce this level [12.51]. Given that these infections can lead to 30,000 deaths directly and 70,000 indirectly each year the one-third of these infections that could be eliminated by proper surveillance and control suggests that improved cleanliness in hospitals and in particular in handwashing practices between patients needs greater effort.

In this regard it is interesting to note the development of an automated handwashing tool which in now being introduced into hospitals. Surveys have shown that medical personnel may only wash their hands as infrequently as 20% of the time they move between patients. This leads to the transfer of infections and several attempts to find ways of improving handwashing have not met with sustained success. An automated device in which 102° F water containing either a surfactant or an antimicrobial agent has recently been marketed [12.52]. Users first insert their arms into two cylindrical receivers, where they are exposed to sprays from a rotating set of jet nozzles. The cleaning solution is sprayed for five seconds and then the hands are rinsed with water for ten seconds, ending the cycle.

It has been reported [12.51] that not only did the introduction of this machine provide a sustained increase in cleanliness for staff at the site (Fig. 12.12) but that staff from outside the area were coming to use the equipment before lunch in order to get "a really good handwashing." This technique has so far proven to give a more sustained improvement in cleanliness at the sites where it was tested than earlier devices.

It is further of interest to note that when the machine was used to remove sodium chloride from the hands, a 10 second cycle was most effective, that use of distilled water improved cleaning by 7 - 8% over tap water, and the use of soaps and other additives did not have any significant effect on the jet cleaning ability at jet pressures of 2.8 and 4.2 bar [12.53].

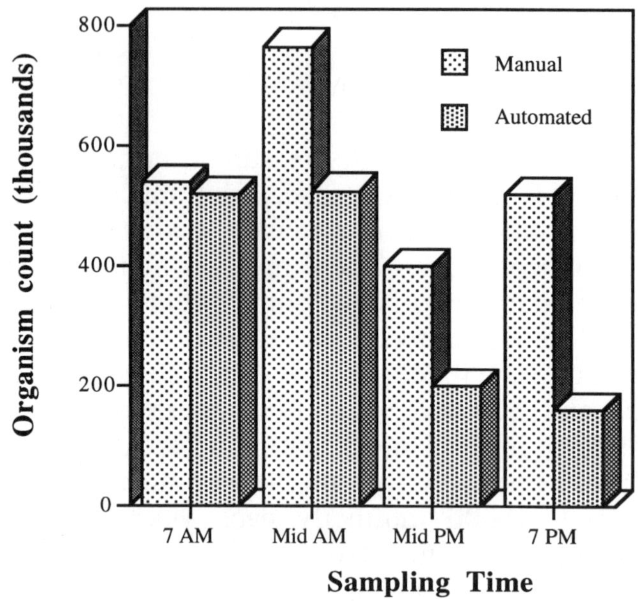

Figure 12.12 Comparison of the number of organisms on the hands of nurses using manual as opposed to automated hand cleaning [12.51].

A more widely used tool in hospitals, especially for cleaning badly contaminated and crushed tissue is the use of the pulsating waterjet in a lavage that has been described above. While these do not operate at pressures much above 10 bar, they have been found quite effective in cleaning wounds of gravel and other dirt and contamination in hospital emergency rooms and have also found military applications [12.54].

12.5.2 DISINTEGRATING BODY STONES

Investigators in Germany [12.55], [12.56] have examined a different application of waterjets to a medical problem. They reported that there are over a million new cases of gallstone disease in the United States each year. Of these some 15% conventional stone extraction is not possible, usually because the stone is too large for it to be passed out of the body. This occurs at a diameter of 2 cm or greater. It was to address this problem that a device was built and tested. In an initial test series [12.55] a jet diameter of 0.2 mm was found to be required to break the stones. It turned out that there were two different types of stones and while dark pigment, calcium bilirubinate, stones could be cut at 150 bar, it required 200 bar to break the lighter cholesterol stones. The dominant factor was found to be the stronger skin in the latter case, and where this was broken the stone broke under internal pressure generated in the stone by the jet.

Details of the development of the device were presented in a subsequent paper [12.56] in which the need to hold the stone up against the nozzle (to prevent injury to the surrounding tissue) was explained and the design to achieve this discussed. Costs were estimated to be significantly less than those required for the previous method of choice, surgery.

In 1986 a third paper was presented [12.57] in which waterjets were tested for use in disintegrating stones, or calculi, developed in the kidneys and urinary tract. While most of these stones can be broken using the concentrated shock waves generated outside the body using a spark discharge in a renal lithotripter this does not work where the stones are near the lower end of the ureter. It was found possible to break such stones using a 0.8 second pulse of water at 385 bar when directed through a 0.18 mm diameter nozzle. The design required the jet be carried to the nozzle through a flexible tube, and the stone be held in place by a Dormia basket of fine wires (Fig. 12.13).

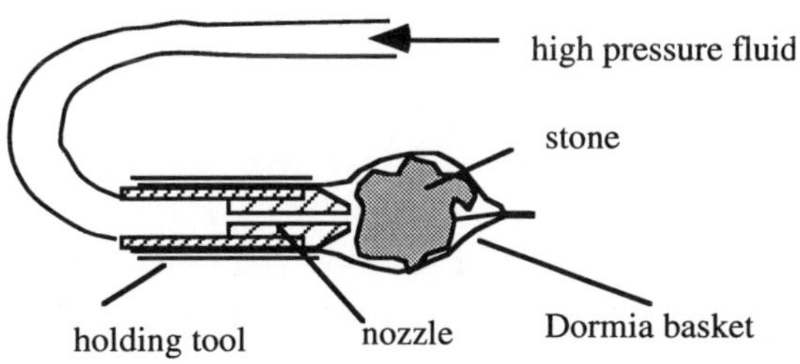

Figure 12.13 Detail of a surgical tool schematic for catching and holding renal calculi in front of a pulsating waterjet (after [12.57]).

12.5.3 SOFT TISSUE APPLICATIONS

One of the treatments which has worked in reducing the size of stones within the bile and urinary tracts has been the use of ultrasound. Recently [12.58] the technique has also been found effective in preliminary tests, for clearing blockages from arteries. In this particular case the relatively high vibrations and narrow collapse range of the cavitation bubbles generated, reduce the range of damage from the tip as it moves through the artery. Care is required since it is possible for the probe to penetrate the wall under certain conditions.

A more interesting application has previously been discussed in terms of the dangers of waterjet use, and this relates to its application in liver and kidney surgery [12.25]. In using a waterjet to remove the tissue around the blood vessels in the liver, it has proved possible to isolate and seal these - usually by cauterizing them - before cutting them. In this way the significant loss in blood can be reduced, with a similar reduction in the trauma to the patient. The jets which have been used are operated at a pressure of 15 - 18 bar, through an orifice of 0.1 - 0.15 mm diameter. This produced a jet 0.2 mm wide downstream of the nozzle, and 0.8 mm wide at a distance of 20 cm. The procedure cannot be universally applied since, in some diseases, such as cirrhosis of the liver, the tissue becomes fibrotic and requires a higher pressure of 18 - 20 bar. This is high enough to damage the finer blood vessels in the liver and can lead to some bleeding. Nevertheless, the procedure has been found effective and is now being adopted as a surfical procedure on human patients.

Vijay [12.6] has reviewed other applications of this tool. One, for example, is in the separation of tumors from the surrounding tissue. These and other interesting applications, such as, for example, it use in treating skin cancer [12.59], are still in the development stage and have yet to find a widespread application.

12.6 COMMENT

Waterjet use has grown across a wide spectrum of applications. While this relatively novel tool has been found to be very helpful in many of these uses, it carries with it potential risks to the untrained user. At the same time the particular applications themselves may carry risks not obvious to one not familiar with that site. Examples of these can include the creation of potentially toxic dust which can be carried into the air where it is breathed into the operator's lungs, and the generation of sparks while cutting or cleaning. Use of the tool can also generate unanticipated problems when it is used in confined spaces.

The productivity and apparent simplicity of this tool has made it popular for use, particularly in cleaning situations. In many of these operations it is therefore important to see that precautions are taken to properly educate and protect the user of the high pressure waterjet equipment from any such risks as they might otherwise be exposed to.

The advantages of waterjet use in medical applications are only just becoming obvious. It is an area which will likely become more developed in future years, as the tools become better developed and their advantages become more obvious. Like many other aspects of waterjet use, this will, however, likely be a slow process to develop, given the limited number of individuals currently studying the problem. It will only be, as with other applications, that as the benefits of this tool become more demonstrated, and the equipment becomes more widely available, that these uses will become popular.

The small size at which tools can be developed, the discriminating nature of jet cutting at different pressures, and the relatively innocuous impact of a jet of saline solution at a pressure of 10 - 20 bar suggest that this role will, before too long, become an important one. At that time medical application papers will likely outnumber the relatively small number of injury reports available in the literature.

12.7 REFERENCES

12.1 Ward, G., "Safety Considerations Arising from Operational Experience with High Pressure Jet Cleaning," paper F1, 1st International Symposium on Jet Cutting Technology, Coventry, UK, April, 1972, pp. F1-1 - F1-24.

12.2 Leach, S.J., and Walker, G.L., "The Application of High Speed Liquid Jets to Cutting," Philosophical Transactions , Royal Society of London, 260-A, July, 1966, pp. 295 - 308.

12.3 Barker, C.R., and Selberg, B.P., "Water Jet Nozzle Performance Tests," paper A1, 4th International Symposium on Jet Cutting Technology, Canterbury, UK, April, 1978, pp. A1-1 - A1-20.

12.4 Zakin, J. L., and Summers, D.A., "The Effect of Visco-elastic Additives on Jet Structure," paper A4, 3rd International Symposium on Jet Cutting Technology, Chicago, IL, April, 1976, pp. A4-47 - A4-66.

12.5 Summers, D.A., and Viebrock, J., "The Impact of Waterjets on Human Flesh," paper H4, 9th International Symposium on Jet Cutting Technology, Sendai, Japan, October, 1988, pp. 423 - 434.

12.6 Vijay, M.M., "A Critical Examination of the Use of Water Jets for Medical Applications," 5th American Water Jet Conference, Toronto, Canada, August, 1989, pp. 425 - 448.

12.7 Craig, E.V., "A New High-Pressure Injection Injury of the Hand," The Journal of Hand Surgery, Vol. 9A, 1984, pp. 240 - 242.

12.8 Nahigian, S.H., "Airless Spray Gun - A New Hand Hazard," Journal of the American Medical Association, Vol. 195, No. 8, February, 1966, p. 176.

12.9 Neill, R.W.K., and George, B., "Penetrating Intra-Abdominal Injury caused by High Pressure Waterjet," British Medical Journal, Vol. 2, 1969, pp. 357 - 358.

12.10 Blue, A.I., and Dirstine, M.J., "Grease Gun Damage," Northwest Medicine, Vol. 64, May, 1965, pp. 342 - 344.

12.11 Parks, B.J., Horner, R.L., and Trimble, C., "Emergency Treatment of High Pressure Injection Injuries to the Hand," Journal of the American College of Emergency Physicians, Vol. 4, 1975, pp. 216 - 217.

12.12 Stark, H.H., Wilson, J.N., and Boyes, J.H., "Grease Gun Injuries of the Hand," Journal of Bone and Joint Surgery, Vol. 43A, 1961, pp. 485 - 491.

12.13 Gelberman, R.H., Madison, J.L., Posch, J.L., and Jurist, J. M., "High Pressure Injection Injuries of the Hand," Journal of Bone and Joint Surgery, Vol. 57A, October, 1975, pp. 935 - 937.

12.14 Karlbauer, A., and Gasperschitz, F., "High-Pressure Injection Injury - A Hand-Threatening Emergency," Journal of Emergency Medicine, Vol. 5, No. 5, 1987, pp. 375 - 379.

12.15 Schneider, L.H., "High Pressure Injection Injuries in the Hand," Jefferson Orthopedic Journal, Vol. 17, 1988, pp. 17 - 19.

12.16 Hingson, R.A., and Figge, F.H.J., "A Survey of the Development of Jet Injection in Parenteral Therapy," Current Researches in Anesthesia and Analgesia, Vol. 31, November, 1952, pp. 361 - 366.

12.17 Lockhart, M.L., U.S. Patent Application no. 69,119, March, 1936.

12.18 Higson, R.A., and Hughes, J.G., "Clinical Studies With Jet Injection - A New Method Of Drug Administration," Current Researches in Anesthesia and Analgesia, Vol. 26, No. 6, November, 1947, pp. 221 - 230.

12.19 Towle, R.L., "New Horizon in Mass Inoculation," Public Health Reports, Vol. 75, 1960, pp. 471 - 476.

12.20 Barrett, C.D., and Molner, J.G., "Experiences with the Hypospray as the instrument of injection," Journal of School Health, 1961, p. 49.

12.21 Figge, F.H.J., and Barnett, D.J., "Anatomical Evaluation of a Jet Injection Instrument Designed to Minimize Pain and Inconvenience of Parenteral Therapy," American Practitioner, Vol. 3, 1948, pp. 197 - 207.

12.22 Walker, W.A., Burns, R.P., and Adams, J., "High Pressure Water Injury: Case Report," Journal of Trauma, Vol. 29, 1989, pp. 258 - 260.

12.23 Bolgiano, E.B., Vachon, D.A., Barish, R.A, and Browne, B.J., "Arterial Injury from a High Pressure Water Jet: Case Report," Journal of Emergency Medicine, Vol. 8, 1990, pp. 35 - 40.

12.24 Calder, I.M., and Boustred, D., "Experiments using High Pressure Fluid Jets on Human Tissues," Forensic Science International, Vol. 26, 1984, pp. 123 - 129.

12.25 Uchino, J., Une, Y., Horie, T., Yokekawa, M., Kakita, A., and Sano, F., "Surgical Cutting of the Liver by Water Jet," Poster paper 1, 9th International Symposium on Jet Cutting Technology, Sendai, Japan, October, 1988, pp. 629 - 639.

12.26 Papachristou, D.N., and Barters, R., "Resection of the Liver with a Water Jet," British Journal of Surgery, Vol. 69, 1982, pp. 93 - 94.

12.27 Anon, "A Water Knife for Liver Operations," information from Hellbergs Dental AB, Helgo, S-355 90 Vaxjo, Sweden.

12.28 Anon, "Aqua Jet MES," information from Sugino Corp., 1700 N. Penny Lane, Schaumberg, IL, 60173.

12.29 Bhaskar, S.N., Cutright, D., and Frisch, J., "Effect of High Pressure Water Jet on Oral Mucosa of Varying Density," Journal of Periodontics, Vol. 40, October, 1969, pp. 593 - 598.

12.30 Krajewski, J., Rubach, W., and Pope, F., "The Effect of Water Pressure Cleaning on the clinically normal gingival crevice," Journal of the California Dental Association, Vol. 43, October, 1967, pp. 452 - 454.

12.31 Lobene, R.R., "A Study Of The Force Of Waterjets In Relation To Pain And Damage To Gingival Tissue," Journal of Periodontology, Vol. 42, No. 3, March, 1971, pp. 166 - 169.

12.32 Bhaskar, S.N., Cutright, D., Hunsuck, E.E., and Gross, A., "Pulsating Water Jet Devices in Debridement of Combat Wounds," Military Medicine, Vol. 136, 1971, pp. 264 - 266.

12.33 Gross, A., Bhaskhar, S.N., Cutright, D.E., Beasley, J.D., and Perez, B., "The Effect of Pulsating Water Lavage on Experimental Contaminated Wounds," Journal of Oral Surgery, Vol. 29, March, 1971, pp. 187 - 190.

12.34 Beasley, J.D., "The Effect of Spherical Polymers and Water Jet Lavage on Oral Mucosa," Oral Surgery, Vol. 32, No. 6, 1971, pp. 998 - 1007.

12.35 Anon, "Will new safety code suffice?" Industrial Jetting Report, July/August, 1990.

13.36 McArthur, A., "Personnel and Environmental Risk Reduction through High Pressure Jet Cleaning of NORM," paper 53, 7th American Water Jet Conference, Seattle, WA, August, 1993, pp. 701 - 728.

12.37 Pardey, P.H., "Field Use of High Velocity Water Jets and the Contribution to Safety and Training," paper F2, 2nd International Symposium on Jet Cutting Technology, Cambridge, UK, April, 1974, pp. F2-11 - F2-18.

12.38 Swan, S.P.D., and Johnson, N.A., "Eye and Respiratory Protection Devices for use in Water Jetting Applications," paper 44, 5th American Water Jet Conference, Toronto, Canada, August, 1989, pp. 455 - 463.

12.39 Lacore, J.P., "Manually Operated High-Pressure Water-jet Cleaning with Water Guns, Lances, Scrapers and Go-Devils," Cahiers de Notes Documentaires - Securite et Hygiene du Travaile, No. 80, Note No. 967-80-75, 1975, pp. 303 - 316.

12.40 Anon, High-Pressure Water Blasting, Construction Safety Council of Ontario, 74 Victoria Place, Toronto, Canada, M5C2A5.

12.41 Anon, Code of Practice for the Use of High Pressure Water Jetting Equipment, Association of High Pressure Water Jetting Contractors, 33, Catherine Place, London, SW1E 6DY.

12.42 Anon, Recommended Practices for the Use of Manually Operated High Pressure Water Jetting Equipment, Water Jet Technology Association, 818, Olive Street, Suite 918, St. Louis, MO, 63101-1598.

12.43 Calder, I.M., and Boustred, D., "High Pressure Water Jet Injury," letter to the British Medical Journal, 28 June 1980, p. 1620.

12.44 Ward, Gardner A., "High Pressure Water Injury," Transactions of the Society of Occupational Medicine, Vol. 16, 1966, p. 30.

12.45 Beaux, J.L.M., "High Pressure Water Jet Injury," British Medical Journal, Vol. 280, No. 6229, pp. 1417 - 1418.

12.46 Barker, C.R., Cummings, A., and Anderson, M., "Jet Noise Measurements on Hand Held Cleaning Equipment," paper D1, 6th International Symposium on Jet Cutting Technology, Guildford, UK, April, 1982, pp. 161 - 178.

12.47 Franklin, R.E., and McMillan, J., AMTE/OUEL Underwater Jet Noise Rig, Oxford University, 1983, PB84-139997, 89 pp.

12.48 Gavrilov, L.R., The Physical Mechanism of Destruction of Biological Tissues with the Use of Focused Ultrasound, Joint Publications Research Service, Arlington, VA, July, 1974, JPRS-62463, 8 pages.

12.49 Merchant, H.C., and Chalupnik, J.D., "Sound Power Measurement of an Abrasive Water Jet Cutting System," 1986 International Conference on Noise Control Engineering, Cambridge, Mass, pp. 241 - 244.

12.50 Anon, "M Tucker to Distribute Automated Handwashing Device," TFS, August 13, 1991.

12.51 Civetta, J.M, Hudson-Civetta, J.A., and Larson, E.L., "Handwashing prevents infection - it really does!" Critical Care Updates, Vol. 1, No. 2, June, 1990, J.B. Lippencott Company.

12.52 Promotional Literature on the CleanTech™ 2000, Meritceh, Englewood, CO, 80112.

12.53 Anon, "Evaluation of Sodium Chloride Removal from the Hands using the CleantechTM 2000 system," report provided by Meritech of Englewood, CO, Summary, April, 1991.

12.54 Gross, A., Cutright, D.E., and Bhaskar, S.N., "Effectiveness of Pulsating Water Jet Lavage in the Case of Contaminated Crushed Wounds," American Journal of Surgery, Vol. 124, 1972, pp. 373 - 377.

12.55 Jessen, K., Phillipp, J., Classen, M., Schikorr, W., and Louis, H., "Endescopic Jet-Cutting A New Method for Stone Destruction in the Common Bile Duct," paper B1, 6th International Symposium on Jet Cutting Technology, Guildford, UK, April, 1982, pp. 39 - 52.

12.56 Classen, M., Leuschner, U., Jessen, K., Louis, H., Haferkamp, H., and Schikorr, W., "Endescopic Jet Cutting of Human Gallstones," paper D4, 7th International Symposium on Jet Cutting Technology, Ottawa, Canada, June, 1984, pp. 211 - 220.

12.57 Aeikens, B., Decker, B., Haferkamp, H., and Louis, H., "Cracking of Ureter Calculi by High Speed Water Jet Pulses," paper 15, 8th International Symposium on Jet Cutting Technology, Durham, UK, September, 1986, pp. 157 - 166.

12.58 Siegel, R.J., Fishbein, M.C., Forrester, J., Moore, K., Decastro, E., Daykovsky, L., and DonMichael, T.A., "Ultrasonic Plaque Ablation - A new method for Recanalization of Partially or Totally Occluded Passages," Circulation, 78, 1988, pp. 1443 - 1448.

12.59 Stoecker, WV., Summers, D.A., Blaine, J.G., "Surgical Liquid Lance Apparatus," U.S. Patent No. 5,037,431, August 6, 1991.

INDEX

D

E

Y

Z